Brock/Springer Series in Contemporary Bioscience

# Large Lakes

# Brock/Springer Series in Contemporary Bioscience

*Series Editor:* *Thomas D. Brock*
*University of Wisconsin–Madison*

*Tom Fenchel*
ECOLOGY OF PROTOZOA: The Biology of Free-living Phagotrophic Protists

*Johanna Döbereiner and Fábio O. Pedrosa*
NITROGEN-FIXING BACTERIA IN NONLEGUMINOUS CROP PLANTS

*Tsutomu Hattori*
THE VIABLE COUNT: Quantitative and Environmental Aspects

*Roman Saliwanchik*
PROTECTING BIOTECHNOLOGY INVENTIONS: A Guide for Scientists

*Hans G. Schlegel and Botho Bowien (Editors)*
AUTOTROPHIC BACTERIA

*Barbara Javor*
HYPERSALINE ENVIRONMENTS: Microbiology and Biogeochemistry

*Ulrich Sommer (Editor)*
PLANKTON ECOLOGY: Succession in Plankton Communities

*Stephen R. Rayburn*
THE FOUNDATIONS OF LABORATORY SAFETY: A Guide for the Biomedical Laboratory

*Gordon A. McFeters (Editor)*
DRINKING WATER MICROBIOLOGY: Progress and Recent Developments

*Mary Helen Briscoe*
A RESEARCHER'S GUIDE TO SCIENTIFIC AND MEDICAL ILLUSTRATIONS

*Max M. Tilzer and Colette Serruya (Editors)*
LARGE LAKES: Ecological Structure and Function

Max M. Tilzer
Colette Serruya
Editors

# Large Lakes

## Ecological Structure and Function

With 224 Figures and 57 Tables

Springer-Verlag
Berlin Heidelberg New York
London Paris Tokyo Hong Kong

Max M. Tilzer
Limnological Institute
University of Constance
D-7750 Constance
Federal Republic of Germany

Colette Serruya
Oceanographic Institute
Tel Shikmona, P.O. Box 8030
Haifa 31080
Israel

ISBN-13:978-3-642-84079-1 e-ISBN-13:978-3-642-84077-7
DOI: 10.1007/978-3-642-84077-7

Library of Congress Cataloging-in-Publication Data
Large lakes : ecological structures and functions / Max M. Tilzer and Colette Serruya, editors.
p. cm. — (Brock/Springer series in contemporary bioscience)
Papers presented at a symposium entitled Functional and structural properties of large lakes, held Constance, West Germany, Sept. 1987.
Includes bibliographical references.
ISBN-13:978-3-642-84079-1
1. Lake ecology—Congresses. I. Tilzer, Max M., 1939–
II. Serruya, C. III. Series.
QH541.5.L3L37 1990
547.5′26332—dc20 90-9646



Cover art: View of Lake Constance.

Production and editorial supervision: Science Tech Publishers.

2131/NY-543210–Printed on acid-free paper.

*This volume is dedicated to the memory of our friend and colleague*

*Peter Kilham*

*who contributed so much to the understanding of large lake ecosystems and finally fell victim to his devotion to African limnology.*

*Big Fish Eat Little Fish*, pen drawing by Pieter Bruegel the Elder (1556). Unlike their terrestrial counterparts, aquatic food-webs are shaped to a great extent by the size relationships of the organisms involved. In addition to determining to a large degree who eats whom, body size has a significant influence on the overall metabolic activity of the organisms involved. In Bruegel's vision we can also recognize our own role in the aquatic trophic cycle and the scientist's role in unraveling the food spectrum of an important terminal consumer. From the Graphische Sammlung Albertina, Vienna.

# Contents

# Contributors

**Farooq Azam** Scripps Institution of Oceanography, University of California, La Jolla, CA 92093-0218, USA

**Gérard Balvay** Institut de Limnologie, 75 av. de Corzent, F-74203 Thonon-les-Bains, France

**Karl Banse** School of Oceanography, WB 10, University of Washington, Seattle, WA 98195, USA

**Martin Barth** Institut für Umweltphysik, Im Neuenheimer Feld 366, D-6900 Heidelberg, FRG

**Michael Becker** Limnologisches Institut, Universität Konstanz, D-7750 Konstanz, FRG

**Gerald L. Bell** Great Lakes Environmental Research Laboratory, 2205 Commonwealth Blvd., Ann Arbor, MI 48105-1593, USA

**Tom Berman** The Yigal Allon Kinneret Limnological Laboratory, Israel Oceanographic & Limnological Research, P.O. Box 345, Tiberias, Israel 14102

**Martin Bohle-Carbonell** Institut für Meereskunde, Troplowitzstr. 7, D-2000 Hamburg 54, FRG

**D. K. Branstrator** Department of Zoology, University of Michigan, Ann Arbor, MI 48109, USA

**Jacques Buffle** Department of Inorganic, Analytical and Applied Chemistry Sciences II, 30 quai Ernest Ansermet, CH-1211 Geneva 4, Switzerland

**Heath J. Carney** Institute of Ecology and Division of Environmental Studies, University of California, Davis, CA 95616, USA

**Byong C. Cho** Department of Microbiology, University of Georgia, Athens, GA 30602, USA

**Uwe Christian** Institut für Umweltphysik, Im Neuenheimer Feld 366, D-6900 Heidelberg, FRG

**Ralf Conrad** Fakultät für Biologie, Universität Konstanz, D-7750 Konstanz, FRG

**Richard De Vitre** Department of Inorganic, Analytical and Applied Chemistry Sciences II, 30 quai Ernest Ansermet, CH-1211 Geneva 4, Switzerland

**Brian J. Eadie** Great Lakes Environmental Research Laboratory, 2205 Commonwealth Blvd., Ann Arbor, MI 48105-1593, USA

**Reiner Eckmann** Limnologisches Institut, Universität Konstanz, D-7750 Konstanz, FRG

**David N. Edgington** Center for Great Lakes Studies and Dept. of Geological & Geophysical Sciences, University of Wisconsin-Milwaukee, Milwaukee, WI 53204, USA

**James J. Elser** Institute of Ecology and Division of Environmental Studies, University of California, Davis, CA 95616, USA

**Anneliese Ernst** Lehrstuhl für Physiologie und Biochemie der Pflanzen, Universität Konstanz, D-7750 Konstanz, FRG

**Gary L. Fahnenstiel** Great Lakes Environmental Research Laboratory, 2205 Commonwealth Blvd., Ann Arbor, MI 48105-1593, USA

**Peter Frenzel** Fakultät für Biologie, Universität Konstanz, D-7750 Konstanz, FRG

**William A. Frez** Great Lakes Environmental Research Laboratory, 2205 Commonwealth Blvd., Ann Arbor, MI 48105-1593, USA

**Sarig Gafny** Institute for Nature Conservation Research, George S. Wise Faculty of Life Sciences, Tel Aviv University, Ramat Aviv, Israel 69978

**Wayne S. Gardner** Great Lakes Environmental Research Laboratory, 2205 Commonwealth Blvd., Ann Arbor, MI 48105-1593, USA

**Avital Gasith** Institute for Nature Conservation Research, George S. Wise Faculty of Life Sciences, Tel Aviv University, Ramat Aviv, Israel 69978

**Meg Gawler** Institut de Limnologie, 75 av. de Corzent, F-74203 Thonon-les-Bains, France

**Federico Giovanoli** Limnologisches Institut, Universität Konstanz, D-7750 Konstanz, FRG

**Charles R. Goldman** Institute of Ecology and Division of Environmental Studies, University of California, Davis, CA 95616, USA

**Moshe Gophen** Kinneret Limnological Laboratory, P.O. Box 345, Tiberias, Israel 14102

**Hans Güde** Institut für Seenforschung und Fischereiwesen, Untere Seestr. 81, D-7994 Langenargen, FRG

**Paul F. Hamblin** Lake Research Branch, National Water Research Institute, Burlington, Ontario, Canada L7R 4A6

**Gerhard Heinz** Institut für Umweltphysik, Im Neuenheimer Feld 366, D-6900 Heidelberg, FRG

**Charles E. Herdendorf** 1507 Cleveland Rd. East, Suite 421, Huron, Ohio 43456, USA

**Eckard Hollan** Institut für Seenforschung und Fischereiwesen, Untere Seestr. 81, D-7994 Langenargen, FRG

**M. Huber-Villegas** Department of Fisheries, Aquaculture & Pathology, University of Rhode Island, Kingston, RI 02852, USA

**Johann Ilmberger** Institut für Umweltphysik, Im Neuenheimer Feld 366, D-6990 Heidelberg, FRG

**Dieter M. Imboden** EAWAG, Überlandstrasse, CH-8600 Dübendorf, Switzerland

**Paul Jäger** Amt der Salzberger Landesregierung, Hydrologischer Dienst, A-5010 Salzburg, Austria

**Alan Jassby** Institute of Ecology and Division of Environmental Studies, University of California, Davis, CA 95616, USA

**Kerry Kelts** EAWAG, Überlandstrasse, CH-8600 Dübendorf, Switzerland

**Peter Kilham†**

**Susan S. Kilham** Department of Biology, The University of Michigan, Ann Arbor, MI 48109-1048, USA

**Joachim Kleiner** Zweckverband Bodenseewasserversorgung, D-7770 Überlingen, FRG

**Hubert Lehn** Institut für Seenforschung und Fischereiwesen, Untere Seestr. 81, D-7994 Langenargen, FRG

**G. G. Leppard** National Water Research Institute, Environment Canada, CCIW, Burlington, Ontario L7R 4A5, Canada

**Gerhard Lindner** Fachhochschule Ravensberg, Postfach 1261, D-7987 Weingarten, FRG

**Augusto Mangini** Heidelberger Akademie der Wissenschaften, Im Neuenheimer Feld 366, D-6900 Heidelberg, FRG

**Helga Müller** Limnologisches Institut, Universität Konstanz, D-7750 Konstanz, FRG

**Karl Otto Münnich** Institut für Unweltphysik, Im Meienheimer Feld 366, D-6900 Heidelberg, FRG

**Toshi Nagata** Otsu Hydrobiological Station, Kyoto University, Shimosakamoto Otsu, 520-01, Japan

**Kazimierz Patalas** Freshwater Institute, 501 University Crescent, Winnipeg, Manitoba, Canada, R3T 2N6

**Hans-Rainer Pauli** Limnologisches Institut, Universität Konstanz, D-7750 Konstanz, FRG

**Jean Pierre Pelletier** Institut de Limnologie, 75 av. de Corzent, F-74203 Thonon-les-Bains, France

---

† Deceased

**Didier Perret** Department of Inorganic, Analytical and Applied Chemistry Sciences II, 30 quai Ernest Ansermet, CH-1211 Geneva 4, Switzerland

**Dagmar Petermann-Seyboldt** Fakultät für Physik, Universität Konstanz, D-7750 Konstanz, FRG

**Walter Pfeiffer** Fakultät für Physik, Universität Konstanz, D-7750 Konstanz, FRG

**Utsa Pollingher** Israel Oceanographic & Limnological Research Ltd., Tel Shikmona, P.O. Box 8030, Haifa, Israel 31080

**Michael A. Quigley** Great Lakes Environmental Research Laboratory, 2205 Commonwealth Blvd., Ann Arbor, MI 48105-1593, USA

**Ekkehard Recknagel** Fakultät für Physik, Universität Konstanz, D-7750 Konstanz, FRG

**Sumner Richman** Lawrence University, Appleton, WI 54912, USA

**John A. Robbins** Great Lakes Environmental Research Laboratory, 2205 Commonwealth Blvd., Ann Arbor, MI 48105-1593, USA

**Jürgen Röhrs** Labor für Geoanalytik GmbH, Richthofenstr. 29, D-3200 Hildesheim, FRG

**Donald Scavia** Great Lakes Environmental Research Laboratory, 2205 Commonwealth Blvd., Ann Arbor, MI 48105-1593, USA

**Michael Schimmele** Institut für Umweltphysik, Im Neuenheimer Feld 366, D-6900 Heidelberg, FRG

**Peter Schlosser** Institut für Umweltphysik, Im Neuenheimer Feld 366, D-6900 Heidelberg, FRG

**Wilfried Schmitz** Institut für Sedimentforschung, Im Neuenheimer Feld 236, D-6900 Heidelberg, FRG

**Jürgen Schneider** Institut für Geologie und Dynamik der Litosphäre, Goldschmidtstr. 3, D-3400 Göttingen, FRG

**Sibylle Schuler** Fakultät für Biologie, Universität Konstanz, D-7750 Konstanz, FRG

**David van Senden** Hydrobiologisches Laboratorium der ETH Zürich, CH-6047 Kastanienbaum, Switzerland

**Colette Serruya** Oceanographic Institute, P.O. Box 8030, Haifa, Israel 31080

**Salvador Serruya** Kinneret Limnological Laboratory, P.O. Box 345, Tiberias, Israel 14102

**Meinhard Simon** Limnologisches Institut, Universität Konstanz, D-7750 Konstanz, FRG

**David C. Smith** Scripps Institution of Oceanography, University of California, La Jolla, CA 92093-0202, USA

**Hans-Henning Stabel** Zweckverband Bodenseewasserversorgung, D-7770 Überlingen, FRG

**Michael R. Talbot** Geological Institute, University of Bergen, N-5007 Bergen, Norway

**Bernwarth Thebrath** Fakultät für Biologie, Universität Konstanz, D-7750 Konstanz, FRG

**Max M. Tilzer** Limnologisches Institut, Universität Konstanz, D-7750 Konstanz, FRG

**Henry A. Vanderploeg** Great Lakes Environmental Research Laboratory, 2205 Commonwealth Blvd., Ann Arbor, MI 48105-1593, USA

**Ulrich Wahl** Fakultät für Physik, Universität Konstanz, D-7760 Konstanz, FRG

**Thomas Weisse** Limnologisches Institut, Universität Konstanz, D-7750 Konstanz, FRG

**Axel Zenger** Institut für Umweltphysik, Im Neuenheimer Feld 366, D-6900 Heidelberg, FRG

# Foreword

The vast majority of the world's lakes are small in size and short lived in geological terms. Only 253 of the thousands of lakes on this planet have surface areas larger than 500 square kilometers. At first sight, this statistic would seem to indicate that large lakes are relatively unimportant on a global scale; in fact, however, large lakes contain the bulk of the liquid surface freshwater of the earth. Just Lake Baikal and the Laurentian Great Lakes alone contain more than 38% of the world's total liquid freshwater. Thus, the large lakes of the world accentuate an important feature of the earth's freshwater reserves—its extremely irregular distribution.

The energy crisis of the 1970s and 1980s made us aware of the fact that we live on a spaceship with finite, that is, exhaustible resources. On the other hand, the energy crisis led to an overemphasis on all the issues concerning energy supply and all the problems connected with producing new energy. The energy crisis also led us to ignore strong evidence suggesting that *water* of appropriate quality to be used as a resouce will be used up more quickly than energy will. Although in principle water is a "renewable resource," the world's water reserves are diminishing in two fashions, the effects of which are multiplicative: enhanced consumption and accelerated degradation of quality.

Large lakes are at the center of this dilemma; they are, in most cases, heavily utilized for urban and rural water supply, as waterways, as fishing grounds, as recreational sites, and, in general and in the majority of cases, are focal points of intense economic activities. Therefore, it is no exaggeration to state that the problems of the world's inland water reserves and their endangered state, are to a significant extent a problem of large lakes.

This volume, however, does not directly address these problems. Rather, it considers large lakes as objects of scientific curiosity, as study sites for physicists, geochemists, and a wide array of biologists. These scientists have a common vision of large lakes as huge reactors, in which physical, chemical, and biological processes interact closely and in an extremely complex fashion. The relevance of the following chapters for the issues of water preservation on a worldwide scale is an indirect one: only if the processes in a "healthy" ecosystem are properly understood, will we be able to successfully attack the formidable challege of maintaining or, if necessary, restoring their quality.

The contributions to this volume were the product of a symposium entitled *Functional and Structural Properties of Large Lakes*, which was held in Konstanz, FRG, in September 1987. This symposium was attended by 138 participants from 15 nations. It was our intention as editors not just to publish a proceedings volume, but to integrate the various chapters as much as possible. Obviously, this was not an easy task because each chapter is an individual contribution, sometimes containing personal biases and flavors that we felt should not be lost during the editorial process. All chapters were peer-reviewed, usually by two colleagues and sometime by even more, and then carefully revised by the respective authors. We wish to thank the many reviewers for their constructive advice. We also want to thank the authors for undergoing the frequently painful, if not annoying, process of having to revise their manuscript, in some instances several times.

We sincerely hope that this volume adequately covers our current understanding of the structural and functional peculiarities of large lakes, and that it will also stimulate further research. If it will help us, moreover, in our concerted efforts to conserve or, if necessary, to restore the quality of these most important natural resources for coming generations, our objectives will have been met.

## Acknowledgments

This volume was edited with the very substantial assistance of Dieter M. Imboden, Walter Geller, Hans-Henning Stabel, and Norbert Pfennig, whose significant contributions are gratefully acknowledged. Moreover, a large number of anonymous reviewers helped with the revision of individual manuscripts. It was a pleasure to work with Thomas D. Brock of Science Tech Publishers, Madison, Wisconsin, USA, and Ruth Siegel, who did a superb and extremely careful job in all matters of technical editing. Last, but not least, Springer-Verlag ensured a volume of sound appearance and excellent quality.

*Max M. Tilzer and Colette Serruya*

*Konstanz and Haifa, January 1990*

# Part 1

# Introduction

This book is concerned with the large lakes of the world. Although it is easy to state in a particular case that a lake is "large" (e.g., Lake Michigan), it is difficult to give a general boundary condition that defines what minimum surface area a lake must have in order to be considered a large lake. It is not the aim of this introductory part to offer such a definition, which necessarily would be arbitrary. Rather, it is the intention here to address this question from two different perspectives: first, in Chapter 1, an inventory of the large lakes of the world will be given, stressing their geographical distribution and morphometry, and the quantities of water stored in them. In Chapter 2, we will look at the specific characteristics of large lakes that are the consequences of the peculiarities of their morphometric and hydrographic features. The essence of Part 1 is that large lakes represent one end of a wide spectrum of lakes, and that no single property beyond its mere size is capable of qualifying a lake to be considered "large." It is nevertheless possible to define an array of both features and properties that characterize large lakes.

# 1

# Distribution of the World's Large Lakes

Charles E. Herdendorf

ABSTRACT An analysis of the world distribution of large lakes has been undertaken. The data base included geographical, geological, morphometric, climatological and limnological information. Only natural lakes (fresh and salt) with a surface area greater than 500 $km^2$ were considered; 253 lakes (including coastal lagoons) were identified. Large lakes occur on all continents except Antarctica. Nearly half of them (48%) are found in North America, and most of these lie above the 40th parallel, attesting to the scouring action of continental glaciers. Tectonic belts, such as the rift valley of East Africa, the Lake Baikal region of Siberia, and the Lake Titicaca area of South America, are the second most common loci of large lakes. Morphometric data were obtained for surface and catchment area, elevation, mean and maximum depth, volume, length and breadth, shoreline length, and orientation of axis for these 253 lakes. These data show that large lakes occupy a surface area of slightly over 1,400,000 $km^2$ and have an estimated volume of 179,000 $km^3$. Furthermore large lakes account for approximately 90% of the total surface area and volume of water held in all lakes of the world. Climatological and limnological data included precipitation, evaporation, basin runoff, water quality, and biological productivity. Information on the latter two parameters, however, is unavailable for many large lakes. About 75% of all large lakes are fresh; the remaining 25% range from brackish to hypersaline. The tropical, freshwater lakes of Africa are among the most biologically productive of the world's lakes.

## 1.1 Introduction

Although relatively few in total number, the 253 large lakes of the world contain nearly 90% of the world's inland surface water supply (Herdendorf, 1982). Thus, large lakes play an important role in the economy and overall prosperity of mankind, being used for many purposes–including domestic and industrial water supplies, irrigation, transportation, water-borne commerce, mineral extraction, waterfowl habitat, fishing, aquaculture, recreation, and waste assimilation. Many of the world's large lakes form international boundaries, and, moreover, most of them have special aesthetic appeal and spectacular beauty.

Given the important attributes of these bodies of water, the present study focuses on the geographic distribution of the world's largest lakes and the interrelationships of their location, geological origin, morphometry, and limnological characteristics. The data base for this analysis is an inventory of large lakes prepared for the U.S. Geological Survey, Office of International Hydrology (Herdendorf, 1984). Key elements of the data base are listed in the appendix at the end of this chapter. This inventory defines large lakes as natural lakes (freshwater and salt water) which have a surface area of at least 500 km². Natural lakes are further defined as essentially static bodies of water, including both inland basins and those separated from the ocean by spits or barrier bars (coastal lagoons). Reservoirs were not included in the inventory, but many exist which have a surface area in excess of 500 km². In total, 253 natural lakes have been identified which satisfy these criteria. Their geographic distribution is shown in Figures 1.1*a* to *h*.

## 1.2 Geographic Distribution

Sixty-four nations contain or share with another country one or more of the large lakes. Table 1.1 presents a breakdown by continent and by nation for these bodies of water.

North America possesses nearly half of all the large lakes but only one-third of the world's total surface area for large lakes. Asia, with less than one-fourth of the lakes, has over 42% of the lake area. This large lake area is mainly due to the presence of the Caspian Sea, which alone accounts for over one-fourth of the area of all large lakes. Africa, the only other major locus for large lakes, accounts for less than one-seventh of the world total by area.

The Northern Hemisphere contains over 87% (220 lakes) of all the large lakes, with most of these occurring north of the 40th parallel (64%, 162 lakes). The Southern Hemisphere contains relatively few lakes and most of them are found in Africa. The African large lakes are distinguished by their large size and the fact that they lie at relatively low latitudes. Table 1.2 gives the number of large lakes that occur within 10° -latitude intervals for both hemispheres, as well as a comparison of lake areas and available land masses for each of these intervals. The dominance of large lakes in the middle latitudes is shown in Figure 1.2.

## 1.3 Morphometry

The published area of a specific large lake, as well as other morphometric data, can vary considerably from author to author. For example, no less than 12 different values have been published for the area of the Caspian Sea (the world's largest lake), ranging from 370,999 to 440,300 km², a difference of nearly 20%. For the inventory, the "most reliable" number was determined by considering the data source, date of publication, consensus of the literature, direct

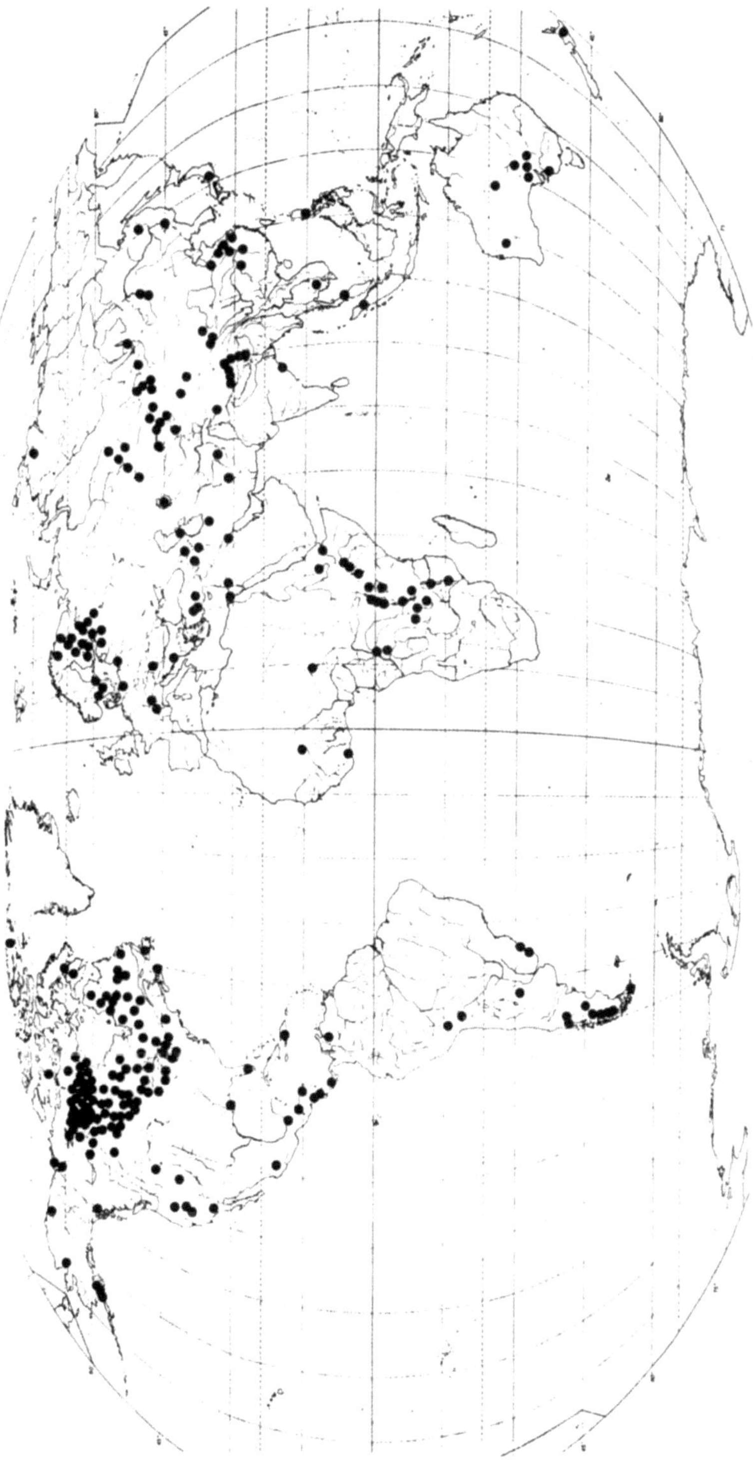

**Figure 1.1** World distribution of large lakes (*a* to *h*). Base maps from Desk Activity Series. © 1990 by Rand McNally & Company, R.L. 90-S-6. (*a*) Large lakes of the world.

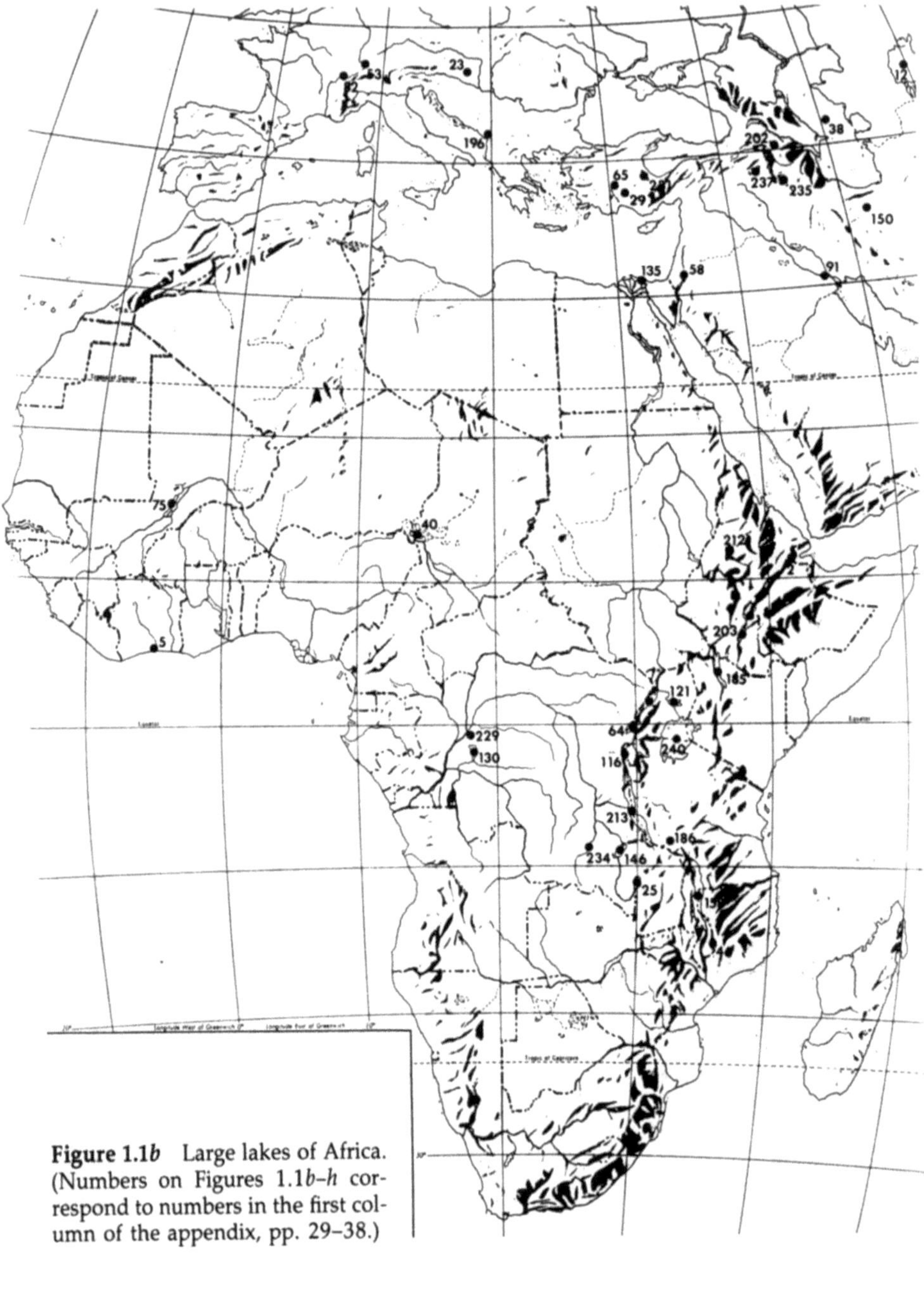


**Figure 1.1*b*** Large lakes of Africa. (Numbers on Figures 1.1*b*–*h* correspond to numbers in the first column of the appendix, pp. 29–38.)

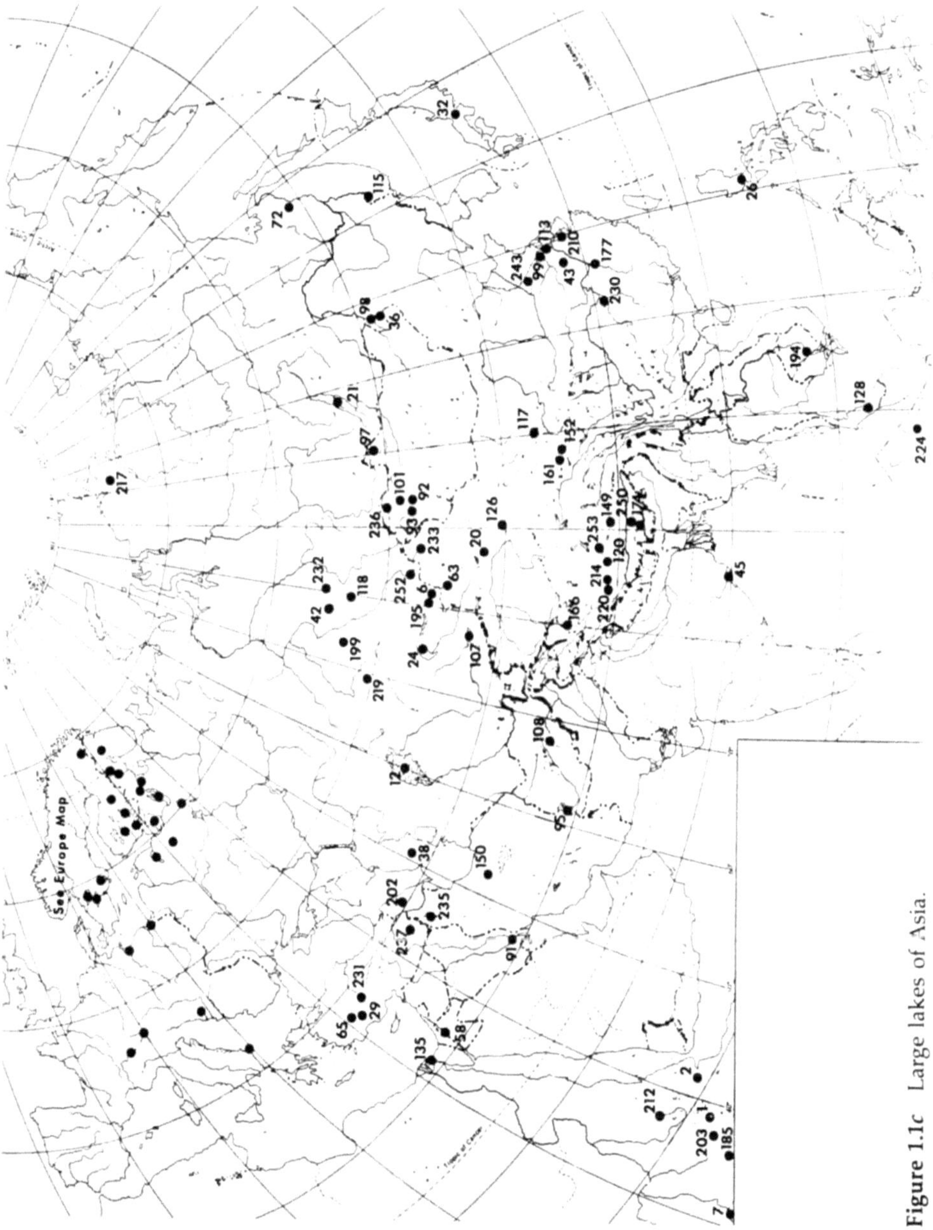


Figure 1.1c Large lakes of Asia.

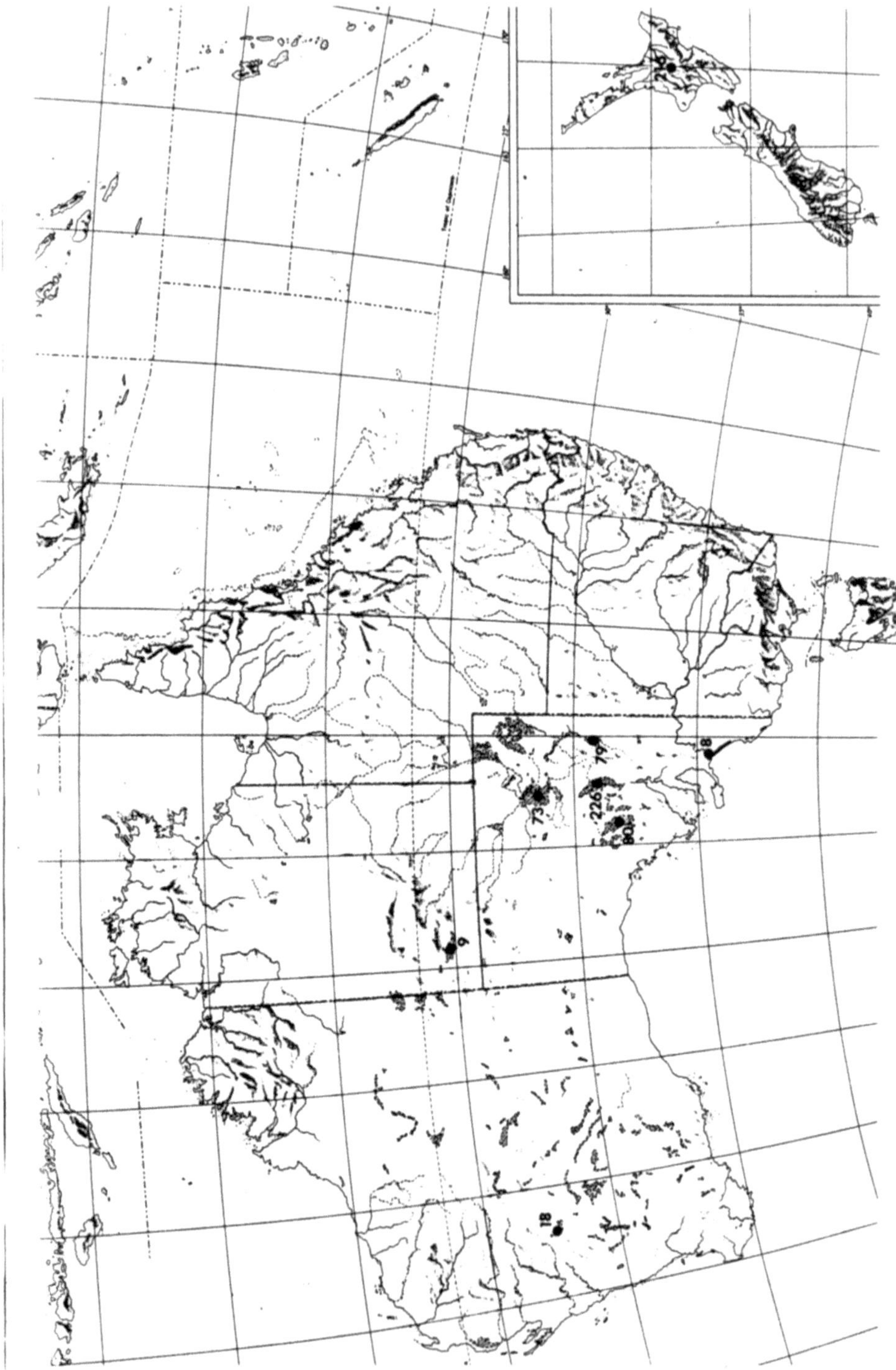


**Figure 1.1*d*** Large lakes of Australia.

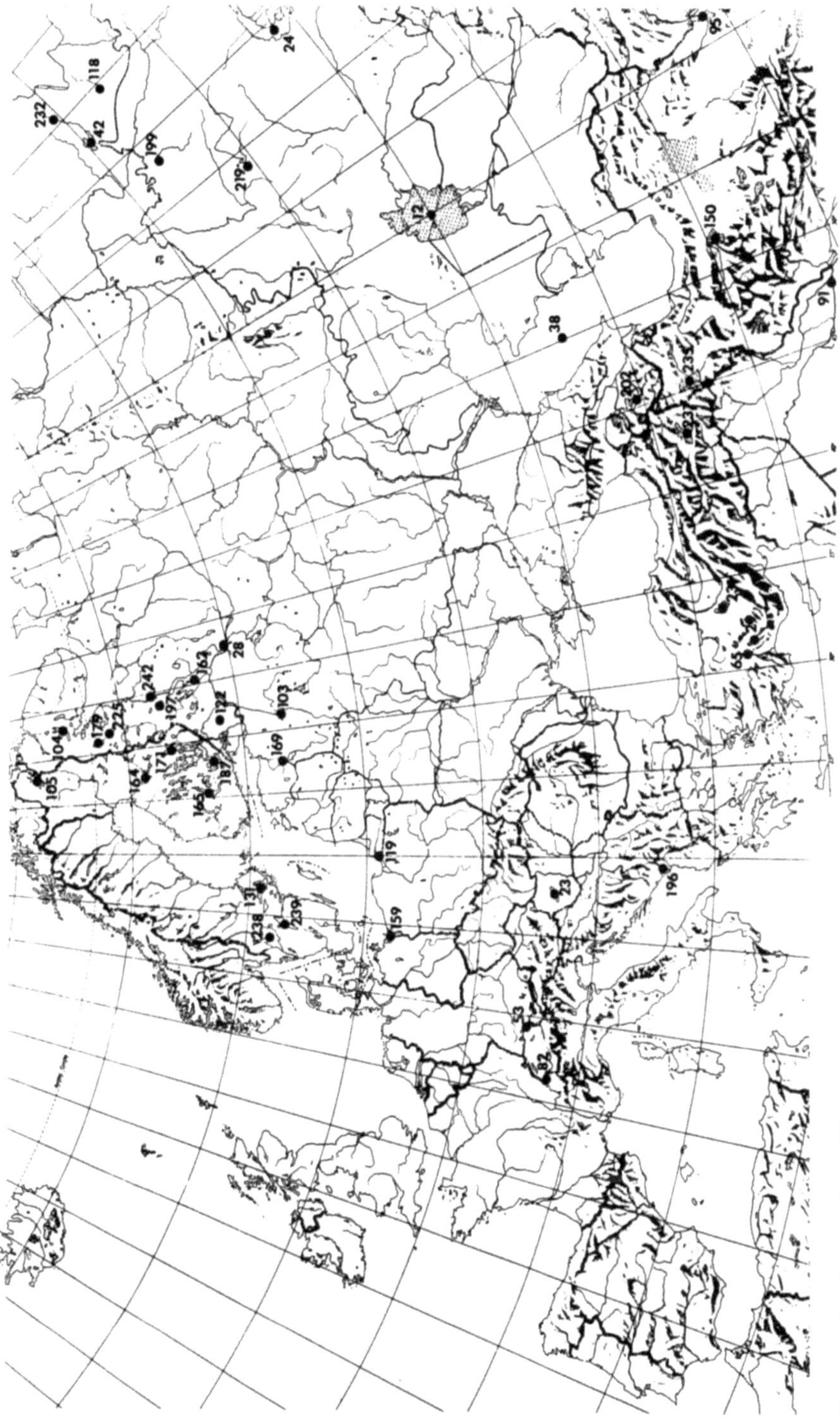


**Figure 1.1e** Large lakes of Europe.

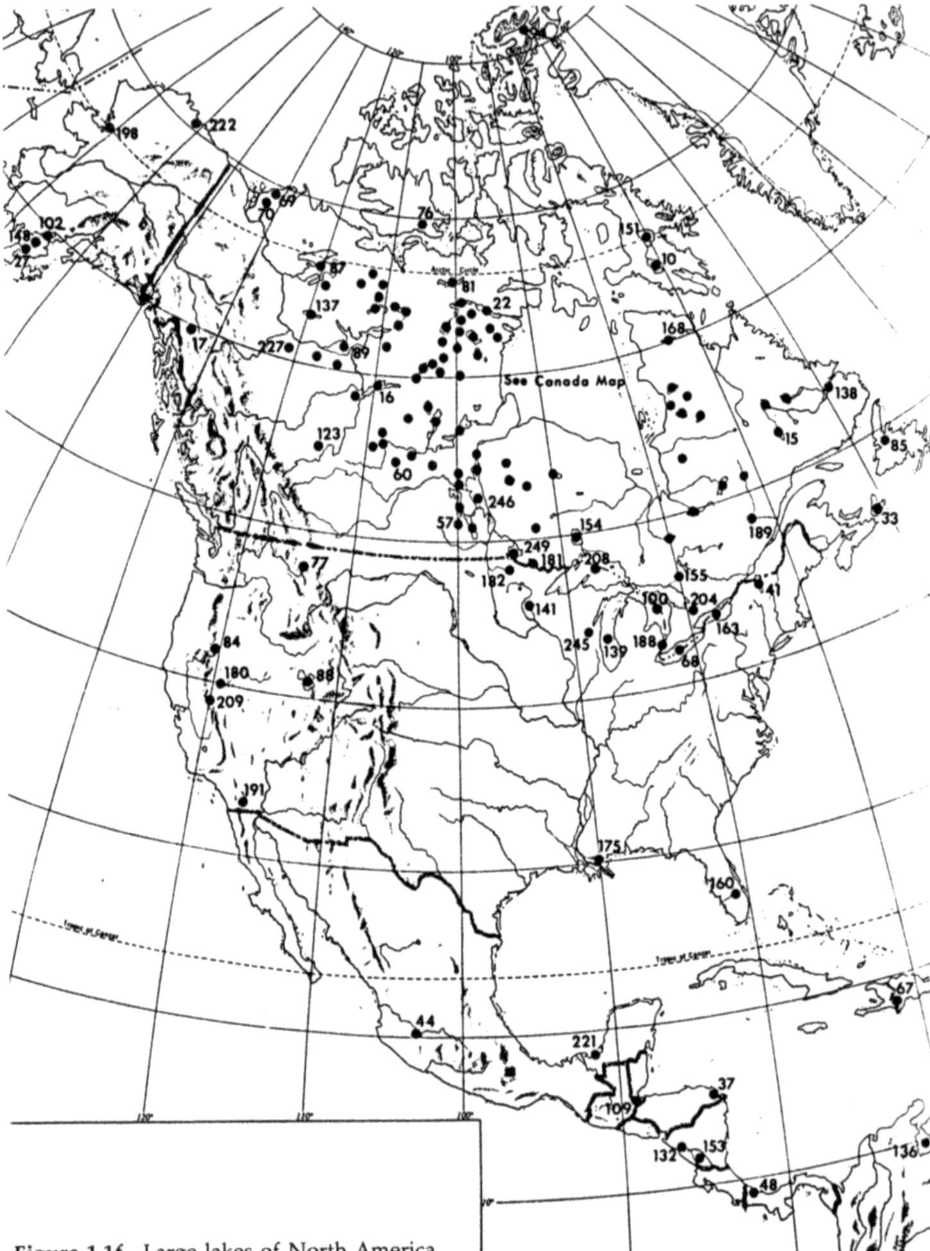


Figure 1.1*f* Large lakes of North America.

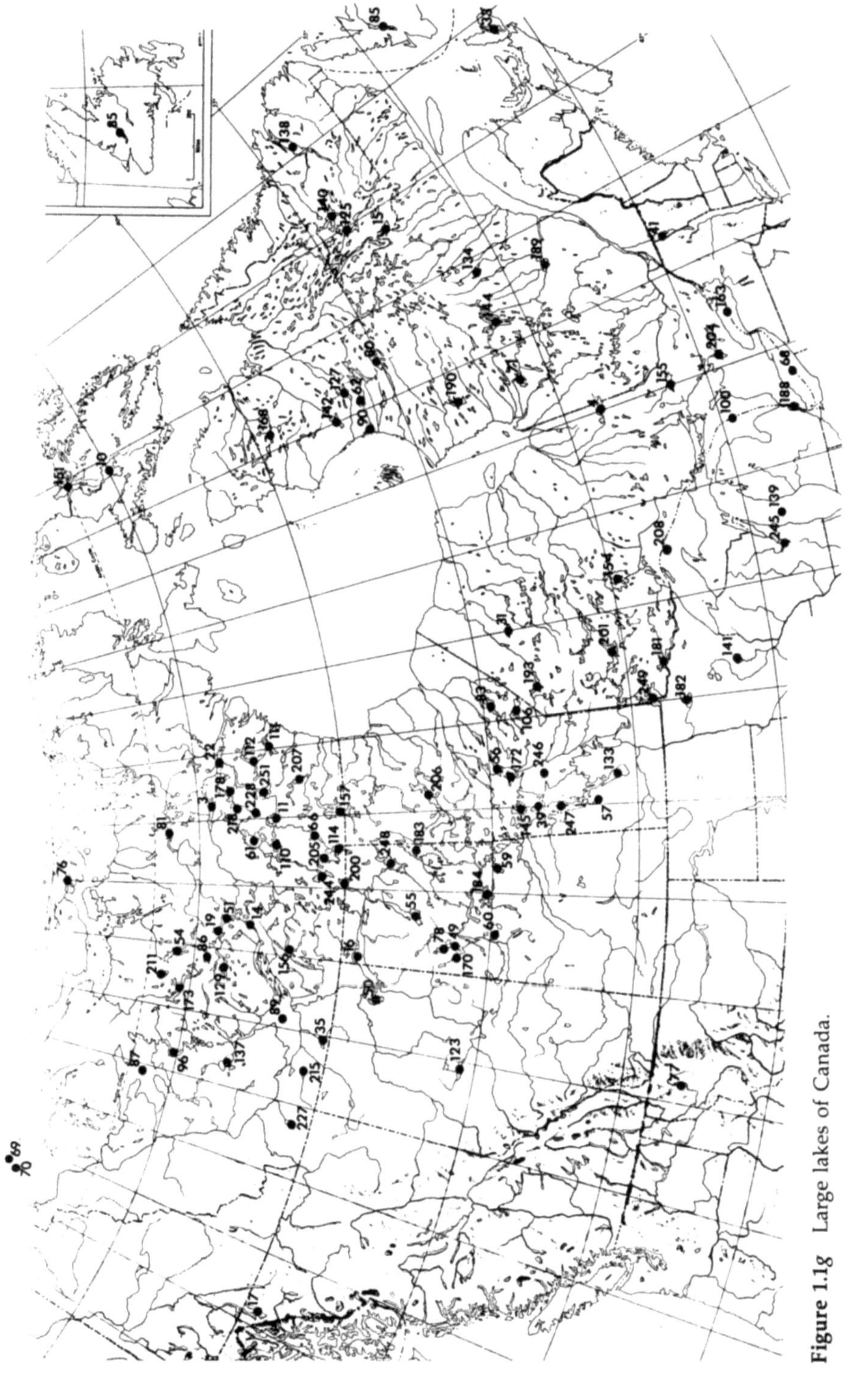

**Figure 1.1g** Large lakes of Canada.

**Figure 1.1*h*** Large lakes of South America.

**Table 1.1** Distribution and area of large lakes by continent

| Location by Nation | Total number | Total area (km²) | Mean area (km²) | % of area of* | |
|---|---|---|---|---|---|
| | | | | Continent | World |
| **Africa** | | | | | |
| Burundi, Tanzania, Zaire, Zambia | 1 | 32,900 | 32,900 | 17.4 | 2.3 |
| Cameroon, Chad, Niger, Nigeria | 1 | 16,600 | 16,600 | 8.8 | 1.2 |
| Djibouti, Ethiopia | 1 | 780 | 780 | 0.4 | <0.1 |
| Egypt | 1 | 1,360 | 1,360 | 0.7 | 0.1 |
| Ethiopia | 3 | 5,310 | 1,770 | 2.8 | 0.4 |
| Ethiopia, Kenya | 1 | 8,660 | 8,660 | 4.6 | 0.6 |
| Ivory Coast | 1 | 780 | 780 | 0.4 | <0.1 |
| Kenya, Tanzania, Uganda | 1 | 68,460 | 68,460 | 36.4 | 4.9 |
| Malawi, Mozambique | 1 | 1,040 | 1,040 | 0.5 | 0.1 |
| Malawi, Mozambique, Tanzania | 1 | 22,490 | 22,490 | 11.9 | 1.6 |
| Mali | 1 | 590 | 590 | 0.3 | <0.1 |
| Rwanda, Zaire | 1 | 2,370 | 2,370 | 1.3 | 0.2 |
| Tanzania | 1 | 2,716 | 2,716 | 1.4 | 0.2 |
| Uganda | 1 | 4,430 | 4,430 | 2.3 | 0.3 |
| Uganda, Zaire | 2 | 7,740 | 3,870 | 4.1 | 0.6 |
| Zaire | 3 | 3,355 | 1,118 | 1.8 | 0.2 |
| Zaire, Zambia | 1 | 4,350 | 4,350 | 2.3 | 0.3 |
| Zambia | 1 | 4,920 | 4,920 | 2.6 | 0.4 |
| *Total* | 23 | 188,851 | 8,211 | 100.0 | 13.5 |
| **Asia** | | | | | |
| Afghanistan | 1 | 520 | 520 | 0.1 | <0.1 |
| Afghanistan, Iran | 1 | 2,080 | 2,080 | 0.4 | 0.1 |
| China | 22 | 39,430 | 1,792 | 6.7 | 2.8 |
| China, India | 1 | 600 | 600 | 0.1 | <0.1 |
| China, Mongolia | 1 | 610 | 610 | 0.1 | <0.1 |
| China, USSR | 1 | 4,190 | 4,190 | 0.7 | 0.3 |
| India | 1 | 1,170 | 1,170 | 0.2 | 0.1 |
| Indonesia | 1 | 1,150 | 1,150 | 0.2 | 0.1 |
| Iran | 2 | 6,550 | 3,275 | 1.1 | 0.5 |
| Iran, USSR | 1 | 374,000 | 374,000 | 63.2 | 26.7 |
| Iraq | 1 | 1,940 | 1,940 | 0.3 | 0.1 |
| Israel, Jordan | 1 | 1,020 | 1,020 | 0.2 | 0.1 |
| Japan | 1 | 688 | 688 | 0.1 | <0.1 |
| Kampuchea | 1 | 2,450 | 2,450 | 0.4 | 0.2 |
| Mongolia | 5 | 9,620 | 1,924 | 1.6 | 0.7 |
| Philippines | 1 | 890 | 890 | 0.2 | 0.1 |
| Thailand | 1 | 1,073 | 1,073 | 0.2 | 0.1 |
| Turkey | 4 | 6,550 | 1,638 | 1.1 | 0.5 |
| USSR | 14 | 136,537 | 9,753 | 23.1 | 9.7 |
| *Total* | 61 | 591,068 | 9,690 | 100.0 | 42.2 |

**Table 1.1** Continued

| Location by Nation | Total number | Total area (km²) | Mean area (km²) | % of area of* Continent | % of area of* World |
|---|---|---|---|---|---|
| **Europe** | | | | | |
| Albania, Yugoslavia | 1 | 600 | 600 | 1.0 | <0.1 |
| Austria, Switzerland, West Germany | 1 | 540 | 540 | 0.9 | <0.1 |
| East Germany, Poland | 1 | 900 | 900 | 1.4 | 0.1 |
| Finland | 5 | 8,408 | 1,682 | 13.3 | 0.6 |
| France, Switzerland | 1 | 580 | 580 | 0.9 | <0.1 |
| Hungary | 1 | 590 | 590 | 0.9 | 0.1 |
| Sweden | 3 | 8,600 | 2,867 | 13.7 | 0.6 |
| USSR | 12 | 42,770 | 3,564 | 67.9 | 3.1 |
| *Total* | 25 | 62,988 | 2,520 | 100.0 | 4.4 |
| **North America** | | | | | |
| Canada | 89 | 201,449 | 2,264 | 41.3 | 14.4 |
| Canada, USA | 8 | 193,857 | 24,232 | 39.7 | 13.8 |
| Dominican Republic | 1 | 500 | 500 | 0.1 | <0.1 |
| Guatemala | 1 | 590 | 590 | 0.1 | <0.1 |
| Honduras | 1 | 1,110 | 1,110 | 0.2 | 0.1 |
| Mexico | 2 | 2,690 | 1,345 | 0.6 | 0.2 |
| Nicaragua | 2 | 9,190 | 4,595 | 1.9 | 0.7 |
| Panama | 1 | 900 | 900 | 0.2 | 0.1 |
| USA | 17 | 77,403 | 4,553 | 15.9 | 5.5 |
| *Total* | 122 | 487,699 | 3,998 | 100.0 | 34.9 |
| **Oceania** | | | | | |
| Australia | 7 | 22,939 | 3,277 | 97.4 | 1.6 |
| New Zealand | 1 | 616 | 616 | 2.6 | <0.1 |
| *Total* | 8 | 23,555 | 2,944 | 100.0 | 1.6 |
| **South America** | | | | | |
| Argentina | 5 | 5,700 | 1,140 | 12.2 | 0.4 |
| Argentina, Chile | 3 | 3,840 | 1,280 | 8.2 | 0.3 |
| Bolivia | 1 | 1,340 | 1,340 | 2.9 | 0.1 |
| Bolivia, Peru | 1 | 8,030 | 8,030 | 17.2 | 0.6 |
| Brazil | 1 | 10,140 | 10,140 | 21.8 | 0.7 |
| Brazil, Uruguay | 1 | 3,750 | 3,750 | 8.1 | 0.3 |
| Chile | 1 | 800 | 800 | 1.7 | 0.1 |
| Venezuela | 1 | 13,010 | 13,010 | 27.9 | 0.9 |
| *Total* | 14 | 46,610 | 3,329 | 100.0 | 3.4 |
| *Grand total* | 253 | 1,400,771 | | | |

* Percent area of large lakes in the respective country relative to total area of all large lakes on the continent and of the world as a whole.

**Table 1.2** Distribution and size of large lakes by latitude

| | | Lake area | | | Land mass[a] | | |
|---|---|---|---|---|---|---|---|
| Latitude | Total number | Total ($km^2$) | Mean ($km^2$) | Percent of total for large lakes | Area ($km^2$) | Percent of total land | Percent of land area occupied by large lakes |
| N90°–80° | 1 | 542 | 542 | 0.1 | 384,000 | 0.3 | 0.1 |
| 80°–70° | 2 | 5,380 | 2,690 | 0.4 | 3,438,000 | 2.3 | 0.2 |
| 70°–60° | 54 | 145,626 | 2,697 | 10.4 | 13,326,000 | 9.0 | 1.1 |
| 60°–50° | 60 | 154,641 | 2,577 | 11.0 | 14,679,000 | 9.9 | 1.1 |
| 50°–40° | 45 | 757,565 | 16,835 | 54.1 | 16,474,000 | 11.1 | 4.6 |
| 40°–30° | 29 | 44,808 | 1,545 | 3.2 | 15,570,000 | 10.5 | 0.3 |
| 30°–20° | 6 | 13,980 | 2,330 | 1.0 | 15,097,000 | 10.1 | 0.1 |
| 20°–10° | 13 | 39,200 | 3,002 | 2.8 | 11,244,000 | 7.6 | 0.3 |
| 10°–0° | 10 | 37,303 | 3,730 | 2.7 | 10,069,000 | 6.8 | 0.4 |
| S 0°–10° | 9 | 116,301 | 12,922 | 8.3 | 10,394,000 | 7.0 | 1.1 |
| 10°–20° | 5 | 37,820 | 7,564 | 2.7 | 9,420,000 | 6.3 | 0.4 |
| 20°–30° | 3 | 9,399 | 3,133 | 0.7 | 9,310,000 | 6.3 | 0.1 |
| 30°–40° | 8 | 29,896 | 3,737 | 2.1 | 4,140,000 | 2.8 | 0.7 |
| 40°–50° | 6 | 6,490 | 1,082 | 0.5 | 971,000 | 0.7 | 0.7 |
| 50°–60° | 2 | 2,000 | 1,000 | 0.1 | 213,000 | 0.1 | 0.9 |
| 60°–70° | 0 | 0 | 0 | 0.0 | 1,788,000 | 1.2 | 0.0 |
| 70°–80° | 0 | 0 | 0 | 0.0 | 8,468,000 | 5.7 | 0.0 |
| 80°–90° | 0 | 0 | 0 | 0.0 | 3,907,000 | 2.6 | 0.0 |
| *Total* | 253 | 1,400,771 | 5,537 | 100.0 | 148,892,000 | 100.0 | 0.9 |

[a] Source: Sverdrup, Johnson, and Fleming (1942)

planimetry of U.S. Defense Mapping Agency maps, and unpublished data supplied by government agencies.

For the purpose of the analysis, the area of a lake is considered as the total area enclosed within the outline of the lake, including any islands. The number and size of the islands, if present, can have varying degrees of influence on the amount of water surface. For example, Great Bear Lake in Northwest Territory, Canada, has 114 major islands but their presence only diminishes the water surface by 1.8%. By contrast, Lake Toba on the Island of Sumatra, Indonesia, contains a large volcanic island (Samosir) which comprises about 70% of the enclosed area of the lake.

Many lakes, particularly those characterized by a closed basin with no outlet, are subject to wide seasonal and year-to-year water level fluctuations. Therefore, their area and depth vary depending on climatic conditions. Most of these lakes are saline lakes and/or playas which originate from deeper lakes (often freshwater) that date back to the Pleistocene epoch when the glacial climate produced periods of increased rainfall and reduced evaporation. Concentrated in the western United States, Middle East, central Asia, central-west-

**Table 1.3** Lakes subject to large fluctuations in surface area

| Lake | Fluctuation (km$^2$) |
|---|---|
| Amadeus | 0– 880 |
| Austin | 0– 829 |
| Balkhash | 17,000–22,000 |
| Bangweulu | 4,000–15,100 |
| Chad | 10,360–25,900 |
| Chany | 2,500– 5,000 |
| Chilka | 910– 1,170 |
| Evoron | 300– 590 |
| Eyre | 0– 7,690 |
| Frome | 0– 2,410 |
| Gairdner | 0– 4,770 |
| Goose | 160– 503 |
| Helmand | 1,165– 3,000 |
| Ilmen | 600– 2,100 |
| Khanka | 4,000– 4,400 |
| Mai-Ndombe | 2,070– 8,210 |
| Pyramid | 450– 570 |
| Rukwa | 750– 3,000 |
| Sap | 2,400–30,000 |
| Scutari | 360– 600 |
| Taymyr | 4,000– 5,000 |
| Torrens | 0– 5,780 |
| Tungting | 3,100–12,000 |
| Urmia | 3,880– 5,960 |

ern South America, northern and eastern Africa, and southern and western Australia, these lakes are remnants of the much larger pluvial lakes (Fairbridge, 1968). As they were reduced in size, outlets were often lost, resulting in dramatic fluctuations in response to variations in rainfall (Lake Eyre, Australia, has gone from a dry bed to a lake of over 7,000 $km^2$ several times during the past century). Fluviatile lakes, such as those located in the Mekong River and Yangtze River valleys of Asia, are also subject to enormous seasonal variations. Lakes with large fluctuations in surface area are listed in Table 1.3.

The large lakes of the world range in area from 500 $km^2$ (by definition) to 374,000 $km^2$. Most of the large lakes are clustered at the small end of this range, with 45% of them being less than 1,000 $km^2$ in area. Less than 8% are greater than 10,000 $km^2$ in surface area. A summary of the area range of the world's largest lakes is presented in Table 1.4.

Shoreline lengths of all lakes were determined using an electronic digitzer and U.S. Defense Mapping Agency topographic maps (1:1,000,000 scale) or were obtained from data supplied by government agencies. Shoreline development (a measure of the irregularity with a value of 1.0 for a circle; see Håkanson, 1981) was calculated for each lake. Volcanically formed Lake Taupo, in New Zealand, is nearly circular and has a shoreline development ratio of 1.7, whereas Lake Paijanne in Finland, formed by tectonic faulting and glacial scour, has an extremely irregular shore with a value of 20.9. In general, however, for large lakes there seems to be a diminished significance of the littoral zone as a whole, regardless of shoreline development.

Large lakes range in elevation from 393 m below sea level (Dead Sea, a graben depression on the Israel-Jordan border) to 4,708 m above sea level (Lake Kyaring, located along the north slope of the Himalaya Mountains in Tibet). Only five of the 253 large lakes lie at elevations below sea level. At the other extreme, only 21 lakes (8%) are found at elevations above 1500 m. Table 1.5 provides a more detailed breakdown of the number of lakes and their

**Table 1.4** Frequency of occurrence of large lakes by area

| Range in area ($km^2$) | Total number | Total area ($km^2$) | Mean area ($km^2$) |
|---|---|---|---|
| 500– 1,000 | 113 | 76,330 | 675 |
| 1,000– 2,000 | 63 | 84,643 | 1,343 |
| 2,000– 3,000 | 21 | 50,192 | 2,390 |
| 3,000– 4,000 | 9 | 30,907 | 3,434 |
| 4,000– 5,000 | 13 | 58,543 | 4,503 |
| 5,000– 10,000 | 15 | 102,768 | 6,851 |
| 10,000– 50,000 | 5 | 291,478 | 22,421 |
| 50,000–100,000 | 1 | 331,910 | 66,382 |
| <100,000 | 1 | 374,000 | 374,000 |
| Total | 253 | 1,400,771 | |

combined areas at 100- or 500-m elevation intervals. These data show that most of the lakes are found between sea level and 500 m (188 lakes, 74%). A second pulse occurs between 500 and 1,500 m (39 lakes, 15%). The influence of continental glaciers and coastal processes is evident in the high concentration of lakes at elevations ranging from sea level to 300 m. In contrast, tectonic lakes are found at both very low and very high elevations.

Detailed bathymetric information is perhaps the most serious deficiency in our morphometric knowledge of large lakes. Without such information, estimates of lake volume are imprecise. Reconnaissance surveys, at least, have been conducted for most large lakes, so that the maximum depths are known with a reasonable degree of certainty. Mean depth (z), however, is essential for calculation of volume (V) based on the relationship with surface area (A), where $V = zA$. Figure 1.3 shows the results of a simple regression of maximum depth versus mean depth for large lakes with complete bathymetric surveys. The relatively high correlations ($r^2 = 0.91$) can permit a reasonable estimate of lake volume if only surface area and maximum sounding data are available. The estimated mean depth of all large lakes is 128 m, yielding an estimated total volume of 179,300 $km^3$.

The maximum depths of 85 of the large lakes are unknown. Of those for which data exist, 50 (30%) have soundings less than 10 m, while 45 (27%) have depths greater than 100 m. At least 52 of the large lakes occupy a cryptodepression (the portion of a lake basin which lies below sea level). Lake Baikal, lying at an elevation of 456 m, has a maximum depth of 1,741 m which yields a cryptodepression of 1,285 m. The Dead Sea's deepest sounding is only 426 m, but it has a cryptodepression of 826 m because the lake surface is

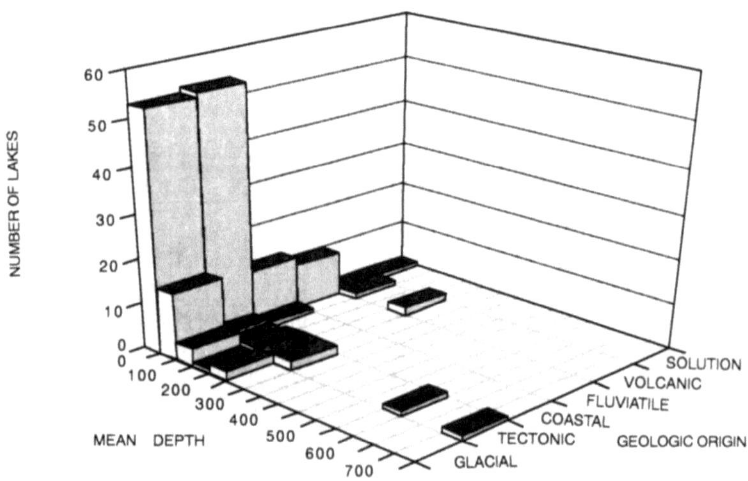


**Figure 1.2** Relationship of area ($km^2$) to latitude of large lakes.

**Table 1.5** Distribution and size of large lakes by elevation

| Elevation (m) | Total number | Total area ($km^2$) | Mean area ($km^2$) |
|---|---|---|---|
| −500 to −400 | 0 | — | — |
| −400 to −300 | 1 | 1,020 | 1,020 |
| −300 to −200 | 0 | — | — |
| −200 to −100 | 0 | — | — |
| −100 to 0 | 4 | 383,140 | 95,785 |
| 0 to 100 | 71 | 259,478 | 3,654 |
| 100 to 200 | 32 | 310,219 | 9,694 |
| 200 to 300 | 36 | 93,507 | 2,597 |
| 300 to 400 | 35 | 67,951 | 1,941 |
| 400 to 500 | 14 | 72,724 | 5,195 |
| 500 to 600 | 9 | 8,072 | 897 |
| 600 to 700 | 2 | 6,364 | 3,182 |
| 700 to 800 | 6 | 43,366 | 7,228 |
| 800 to 900 | 1 | 500 | 500 |
| 900 to 1,000 | 5 | 9,810 | 1,962 |
| 1,000 to 1,100 | 3 | 7,170 | 2,390 |
| 1,100 to 1,200 | 6 | 76,830 | 12,805 |
| 1,200 to 1,300 | 5 | 13,460 | 2,692 |
| 1,300 to 1,400 | 0 | — | — |
| 1,400 to 1,500 | 2 | 2,873 | 1,437 |
| 1,500 to 1,600 | 1 | 1,140 | 1,140 |
| 1,600 to 1,700 | 3 | 12,600 | 4,200 |
| 1,700 to 1,800 | 0 | — | — |
| 1,800 to 1,900 | 2 | 4,100 | 2,050 |
| 1,900 to 2,000 | 1 | 13,600 | 13,600 |
| 2,000 to 2,500 | 1 | 520 | 520 |
| 2,500 to 3,000 | 0 | — | — |
| 3,000 to 3,500 | 1 | 4,460 | 4,460 |
| 3,500 to 4,000 | 2 | 9,370 | 4,685 |
| 4,000 to 4,500 | 5 | 4,480 | 896 |
| 4,500 to 5,000 | 5 | 6,260 | 1,252 |
| *Total* | 253 | 1,400,771 | |

currently 400 m below sea level. Most of the coastal lagoons lie in minor cryptodepressions because their surface elevation approximates that of sea level.

Length and breadth measurements of lakes, in a strict limnological sense, have rather precise definitions (Welch, 1948; Hutchinson, 1957). Because data come from a variety of sources, the inventory defined maximum length as the distance between the two most-distant points on the lake outline. For irregularly shaped lakes this line often passed over extensive areas of dry land, but when compared with the lake area, this "maximum dimension" does provide

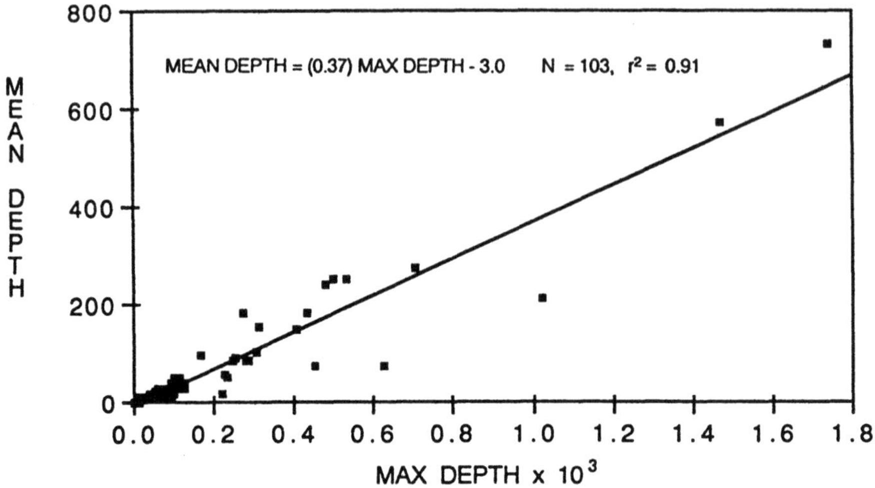


**Figure 1.3** Regression of analysis of maximum depth (m) and mean depth (m) for large lakes.

a rough idea of the shape of the lake. Breadth or width is normally considered to be the maximum length of a straight line connecting points on the lake shoreline at approximate right angles to the line of maximum length. The Caspian Sea, due to its immense size, has the greatest length and breadth measurements, 1,207 km and 483 km respectively. Lakes Tanganyika at 676 km and Baikal at 635 km range next with respect to length, while Lake Huron at 290 km (largely due to Georgian Bay), Aral Sea at 280 km, and Lake Superior at 259 km have the next largest breadths. Most of the large lakes of the world have lengths and breadths well under 100 km.

**Table 1.6** Geological origin of large lakes

| Origin | Total number | Total area ($km^2$) | Mean area ($km^2$) |
|---|---|---|---|
| Glacial | 122 | 410,528 | 3,365 |
| Tectonic | 85 | 786,483 | 9,253 |
| Coastal | 15 | 28,260 | 1,884 |
| Fluviatile | 10 | 22,520 | 2,252 |
| Glacial/tectonic | 8 | 114,470 | 14,309 |
| Volcanic/tectonic | 4 | 12,286 | 3,072 |
| Tectonic/fluviatile | 4 | 7,965 | 1,991 |
| Solution/tectonic | 2 | 1,480 | 740 |
| Glacial/coastal | 2 | 3,769 | 1,885 |
| Tectonic/coastal | 1 | 13,010 | 13,010 |
| *Total* | 253 | 1,400,771 | |

## 1.4 Geological Origin

Hutchinson (1957) offers a detailed classification system for the origin of lakes which includes eleven major categories. Six of these (and their combinations) have been identified as the major causative factors in forming the actual basins which hold the world's large lakes (Table 1.6).

Pleistocene glaciation, particularly in the middle and high latitudes of North America and Europe, is probably the most important process in pro-

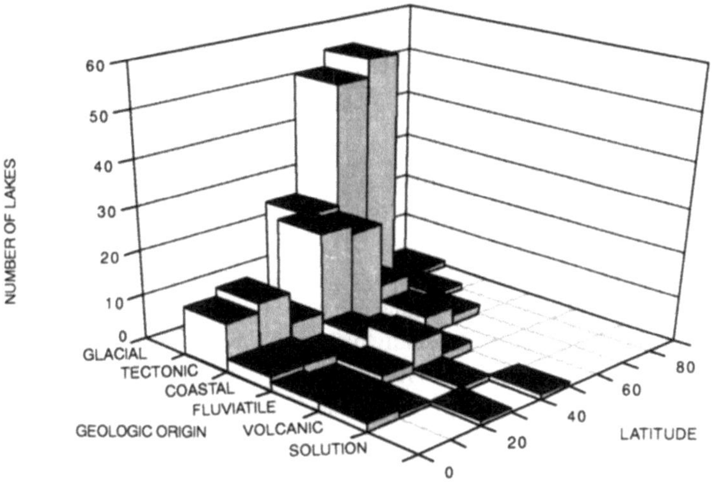


**Figure 1.4** Relationship of geological origin to latitude of large lakes.

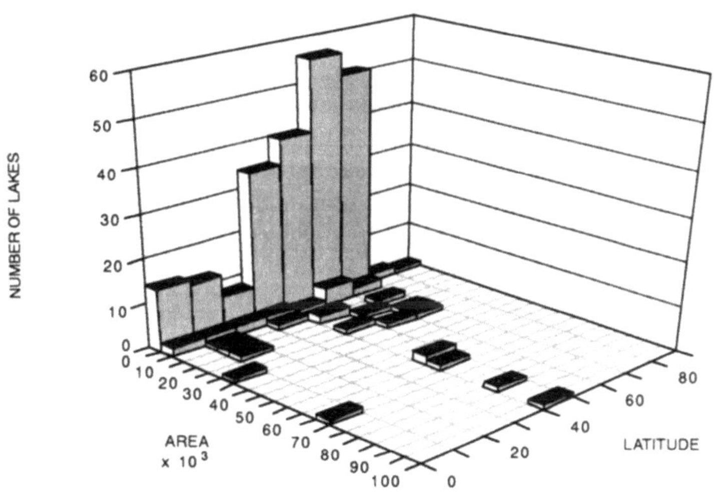


**Figure 1.5** Relationship of geological origin to mean depth (m) of large lakes.

ducing large lakes (Figure 1.4). This process, along with tectonic movement (such as produced the rift valleys of east Africa and Siberia) and volcanic activity, can produce extremely deep lakes (Figure 1.5). Coastal, fluviatile (river), and solution actions produce comparatively shallow lake basins. Of the known and suspected origins, glacial and tectonic basins account for about 85% of the total. On the average, tectonic large lakes are nearly three times larger than glacial large lakes and have a considerably deeper maximum depth.

## 1.5 Climate

Climatological information for the approximate drainage basin for each large lake was extrapolated from Korzoun (1974). Precipitation ranges from 200–300 cm/yr for the tropical lakes of Africa, Central America, and Southeast Asia (however, Lake Llanquihue in the Chilean Andes has the highest precipitation at 320 cm/yr) to 5-20 cm/yr for the lakes of Australia, Central Asia and the Middle East. The maximum actual evaporation is about 125 cm/yr at Lake Toba, whereas the maximum potential evaporation, over 200 cm/yr, occurs in the basins of Lake Chad and Lake Faguibine in north Africa and Lake Eyre in Australia. The lakes with high potential values are therefore subject to desiccation and, at times, particularly for the Australian lakes, are dry. Annual runoff, expressed as a layer of equally distributed water over the entire drainage basin, ranges from 0 to 200 cm/yr. The highest runoff is often associated with the lake basins having the highest rainfall. Climatic conditions can also influence shore erosion rates, water levels, and current patterns in large lakes.

## 1.6 Limnological Relationships

Limnological data in terms of water quality and biological productivity were inventoried for many of the large lakes, including: water transparency, total dissolved solids, alkalinity, pH, circulation, primary productivity, and annual fish yield. Water transparency is typically measured with a Secchi disc (Welch, 1948) and is an indicator of the amount of suspended plankton, organic matter, and inorganic sediment. The deep oligotrophic shield lakes of Canada and Scandinavia, the high plateau lake of Tibet, and deep graben lakes (such as Lake Tahoe with a Secchi depth reaching almost 40 m) are characterized by high transparency. Shallow, eutrophic lakes, such as Lake Beloye in the Soviet Union and Lake Erie, have low transparencies ($<5$ m) coupled with high biological productivity.

Of the 253 large lakes in the world, 75% are freshwater. The remainder grade from brackish water through saline to hypersaline (Table 1.7). Collectively, the world's large lakes store approximately 110,000 $km^3$ of freshwater and 90,000 $km^3$ of brackish and saline water. The Laurentian Great Lakes of North America comprise about 20% of the world's supply of surface freshwater. This concentration is rivaled only by Lake Baikal in Siberia and in the rift-valley great lakes of east Africa. Taken together, these three loci account for over 60% of the water in the earth's freshwater lakes.

Alkalinity of lake water is an index to the nature of the rocks, glacial till, and soils within a drainage basin and the degree to which they are weathered. Although the alkalinity of natural waters is generally the result of bicarbonates, it is usually expressed in terms of $CaCO_3$. Productive large lakes normally have alkalinity values of 100 mg/l or above. The pH of large lakes range from 4.0 to 9.6. Low-pH lakes appear to have low mineralization (dissolved solids and alkalinity) and low biological productivity. The question of eutrophication of large lakes is an interesting one, in that such lakes do not age in the same way as small lakes that undergo succession (Beeton and Edmondson, 1972).

The mixis, or annual circulation pattern, of 91% of the large lakes is known or has been estimated from the latitude/elevation relationship given by Hutchinson and Löffler (1956). Accordingly, 53% of the lakes are classified as dimictic, 23% as polymictic, 16% as monomictic, 5% as meromictic, and 3% as amictic. For those lakes with no bathymetric data, mixing patterns could not be estimated with any degree of certainty.

Biological and water quality information is lacking for many large lakes. However, water chemistry, primary productivity, and fish yield data are available for about 25% of the lakes. Primary productivity is commonly expressed in grams of carbon assimilated per square meter per day (g C $m^{-2}day^{-1}$) and fish productivity is expressed as the annual yield (normally the commercial harvest) in kilograms per hectare per year (kg $ha^{-1}yr^{-1}$). The tropical lakes of Africa (such as Lake Tanganyika with values of 1.75 g C $m^{-2}day^{-1}$ and 125 kg $ha^{-1}yr^{-1}$ are among the most productive of the large lakes. Nixon (1988) attributes the large fish yield of tropical lakes to their lower thermal stability and more efficient wind energy transfer because of a small Coriolis effect at low latitudes. Northern Canadian shield lakes (such as Big Trout Lake, with a fish yield of only 0.75 kg $ha^{-1}yr^{-1}$) are among the lowest in productivity (Figure 1.6).

Large lakes have a great diversity of habitats resulting in high species diversity (Beeton, 1984). Many of the deep tectonic lakes are very old, yielding both ancient and endemic species. Lake Baikal, which dates back to the Tertiary

**Table 1.7** Salinity distribution of large lakes

| Salinity type | TDS[a] (mg/1) | Total number | Total area ($km^2$) | Mean area ($km^2$) |
|---|---|---|---|---|
| Fresh | <2,000 | 189 | 800,144 | 4,234 |
| Brackish | 2,000–20,000 | 17 | 502,234 | 29,543 |
| Saline | 20,000–40,000 | 39 | 64,923 | 1,665 |
| Hypersaline | >40,000 | 8 | 33,470 | 4,184 |
| | *Total* | 253 | 1,400,771 | |

[a] TDS, total dissolved solids.

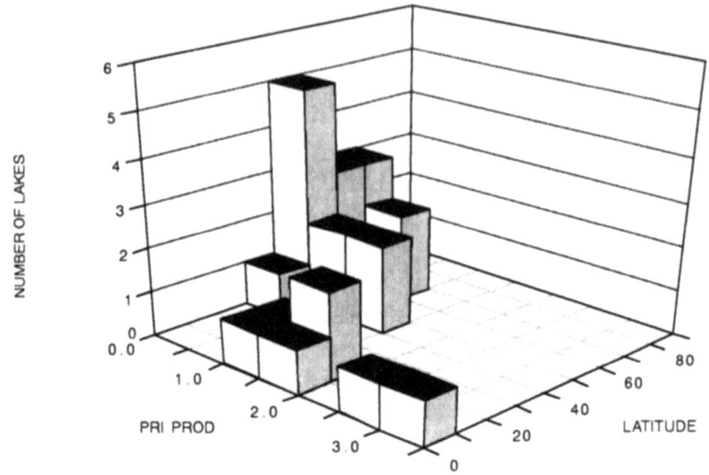


**Figure 1.6** Primary productivity (g C $m^{-2}day^{-1}$) in large lakes in relation to latitude.

Period (age estimates range from 20 to 70 million years), is well known for its endemic fish population, which accounts for nearly 50% of the fishery.

## 1.7 Global Perspective

The global distribution of water, from studies by Nace (1964), is summarized in Table 1.8. From these estimates, it can be seen that although the lakes and inland seas of the world comprise approximately 1,560,000 $km^2$ of surface area and possess a volume of 229,000 $km^3$, they only represent 0.017% of the global water resources. The total number of world lakes is well over one million; however, most of the water is concentrated in a relatively small number of large lakes.

The 253 lakes with surface areas greater than 500 $km^2$ have a total surface area of 1,400,771 $km^2$ and an estimated volume of 179,300 $km^3$. This accounts for 90% and 78%, respectively, of the total surface area and volume of water held in all lakes of the world as estimated by Nace. The fact that the large lakes volume percent is considerably lower than the area percent leads to the speculation that Nace's volume estimate may be too high, and that a total world lake volume of about 200,000 $km^3$ is more realistic.

The 50 greatest lakes of the world, based on area, volume and depth, are listed in Table 1.9. The Caspian Sea ranks highest in area and volume, but only appears third on the depth list. Excluding this giant, Lake Superior heads the area list while Lake Baikal is first in volume and depth. Interestingly, 30 of the 50 deepest lakes of the world have a surface area of less than 500 $km^2$, and, therefore, are not classified here as large lakes. Many of these small, deep lakes were formed by glacial scour, particularly the fjord lakes of Norway, British

**Table 1.8** Global distribution of water

| Location | Area ($km^2$) | Volume ($km^3$) | Percent volume |
|---|---|---|---|
| **Surface water** | | | |
| Freshwater lakes | 860,000 | 125,000 | 0.009 |
| Saline lakes | 700,000 | 104,000 | 0.008 |
| Stream channels | – | 1,250 | 0.0001 |
| **Subsurface water** | | | |
| Soil moisture | | 67,000 | 0.005 |
| | 130,000,000 | | |
| Groundwater | | 8,340,000 | 0.62 |
| Ice caps and glaciers | 18,000,000 | 29,200,000 | 2.15 |
| Atmosphere | 510,000,000 | 13,000 | 0.001 |
| Oceans | 360,000,000 | 1,322,000,000 | 97.2 |
| | *Total* | 1,360,000,000 | 100.0 |

Source: Nace (1964)

Columbia, and New Zealand (Viner, 1987), or by volcanic action (Crater Lake, Oregon, Hutchinson, 1957).

## 1.8 Conclusion

There is nothing magical about using the area value of 500 $km^2$ as a basis for separating large lakes from small lakes. In fact, there are numerous examples of lakes in the 100- to 500-$km^2$ range and smaller which exhibit "large-lake" characteristics. This fact is particularly evident in Table 1.9, which shows that 60% of the world's 50 deepest lakes are less than 500 $km^2$ in area. Likewise, certain large lakes by area (i.e., many fluviatile and coastal-lagoon large lakes) are typified by "small-lake" characteristics. The 500-$km^2$ threshold was merely selected as a convenient statistical criterion, which roughly corresponded to apparent natural and functional differences between large and small lakes. Perhaps depth and/or water volume would provide a more reasonable criterion to discriminate large from small lakes. Unfortunately, of the 253 lakes examined in this analysis, over one-third were lacking sufficient bathymetric data to ascertain these values. Thus, given the current limitations of our limnological knowledge of the world's largest lakes, we must temporarily work with an imperfect classification system.

*Acknowledgements* This research was sponsored by the Ohio Sea Grant Program, project no. R/ES-1 under grant no. NA81AA-D-00095, National Sea Grant College Program, NOAA, U.S. Dept. of Commerce. Sirinimit Bonyuen, Sandra Herdendorf, and Theresa Heitzman assisted in compiling data for the inventory. Base maps for the inventory were graciously supplied by Rand McNally and Co., Chicago, Illinois.

**Table 1.9** Rankings of the world's 50 largest lakes by surface area, volume, and depth

| Area rank | Lake | Area ($km^2$) | Volume rank | Lake | Volume ($km^3$) | Depth rank | Lake | Maximum depth (m) | Area ($km^2$) |
|---|---|---|---|---|---|---|---|---|---|
| 1 | Caspian | 374,000 | 1 | Caspian | 78,200 | 1 | Baikal | 1,741 | 31,500 |
| 2 | Superior | 82,100 | 2 | Baikal | 22,995 | 2 | Tanganyika | 1,471 | 32,900 |
| 3 | Victoria | 68,460 | 3 | Tanganyika | 18,900 | 3 | Caspian | 1,025 | 374,000 |
| 4 | Aral | 64,100 | 4 | Superior | 12,230 | 4 | Nyasa | 706 | 22,490 |
| 5 | Huron | 59,500 | 5 | Nyasa | 6,140 | 5 | Issykkul | 702 | 6,240 |
| 6 | Michigan | 57,750 | 6 | Michigan | 4,920 | 6 | Great Slave | 625 | 28,568 |
| 7 | Tanganyika | 32,900 | 7 | Huron | 3,537 | 7 | Matana | 590 | 190 |
| 8 | Baikal | 31,500 | 8 | Victoria | 2,700 | 8 | Crater | 589 | 54 |
| 9 | Great Bear | 31,326 | 9 | Great Bear | 2,381 | 9 | Toba | 529 | 1,150 |
| 10 | Great Slave | 28,568 | 10 | Great Slave | 2,088 | 10 | Hornindalsvatn | 514 | 51 |
| 11 | Erie | 25,657 | 11 | Issykkul | 1,730 | 11 | Sarez | 505 | 86 |
| 12 | Winnipeg | 24,387 | 12 | Ontario | 1,637 | 12 | Tahoe | 501 | 500 |
| 13 | Nyasa | 22,490 | 13 | Aral | 1,020 | 13 | Chelan | 489 | 140 |
| 14 | Ontario | 19,000 | 14 | Ladoga | 908 | 14 | Kivu | 480 | 2,370 |
| 15 | Balkhash | 18,200 | 15 | Titicaca | 827 | 15 | Quesnel | 475 | 260 |
| 16 | Ladoga | 17,700 | 16 | Kivu | 569 | 16 | Hauroko | 462 | 68 |
| 17 | Chad | 16,600 | 17 | Erie | 483 | 17 | Adams | 457 | 130 |
| 18 | Maracaibo | 13,010 | 18 | Hovsgol | 480 | 18 | Great Bear | 452 | 31,326 |
| 19 | Patos | 10,140 | 19 | Winnipeg | 371 | 19 | Fagnano | 449 | 590 |
| 20 | Onega | 9,700 | 20 | Iliamna | (319)[a] | 20 | Mjosa | 449 | 370 |
| 21 | Rudolf | 8,660 | 21 | Melville | (298) | 21 | Salsvatn | 445 | 45 |
| 22 | Nicaragua | 8,150 | 22 | Onega | 295 | 22 | Manapouri | 444 | 153 |
| 23 | Titicaca | 8,030 | 23 | Maracaibo | 280 | 23 | Nahuel Huapi | 438 | 550 |
| 24 | Athabasca | 7,935 | 24 | Rudolph | 251 | 24 | Dead | 433 | 1,020 |
| 25 | Reindeer | 6,640 | 25 | Toba | 249 | 25 | Tazawa | 425 | 26 |

| Area rank | Lake | Area ($km^2$) | Volume rank | Lake | Volume ($km^3$) | Depth rank | Lake | Maximum depth (m) | Area ($km^2$) |
|---|---|---|---|---|---|---|---|---|---|
| 26 | Issykkul | 6,240 | 26 | Van | (206) | 26 | Teanau | 417 | 352 |
| 27 | Tungting | 6,000 | 27 | Athabasca | 204 | 27 | Como | 413 | 150 |
| 28 | Urmia | 5,800 | 28 | Dead | 188 | 28 | Superior | 407 | 82,100 |
| 29 | Vanern | 5,648 | 29 | Baker | 176 | 29 | Hawea | 384 | 141 |
| 30 | Albert | 5,590 | 30 | Mistassini | (175) | 30 | Wakatipu | 380 | 293 |
| 31 | Netilling | 5,530 | 31 | Argentino | (169) | 31 | Maggiore | 372 | 210 |
| 32 | Winnipegosis | 5,375 | 32 | Vanern | 153 | 32 | Chilko | 366 | 160 |
| 33 | Bangweulu | 4,920 | 33 | Albert | 151 | 33 | Pend Oreille | 366 | 380 |
| 34 | Nipigon | 4,848 | 34 | Fagnano | (125) | 34 | Shikotsu | 363 | 77 |
| 35 | Manitoba | 4,625 | 35 | Tahoe | 124 | 35 | Powell | 358 | 190 |
| 36 | Taymyr | 4,560 | 36 | Nahuel Huapi | (113) | 36 | Llanquihue | 350 | 800 |
| 37 | Koko | 4,460 | 37 | Balkhash | 112 | 37 | Garda | 346 | 370 |
| 38 | Kyoga | 4,430 | 38 | Nicaragua | 108 | 38 | Towada | 334 | 59 |
| 39 | Saimaa | 4,380 | 39 | Reindeer | 96 | 39 | Teletskoye | 325 | 230 |
| 40 | Great Salt | 4,360 | 40 | Geneva | 89 | 40 | Eutsuk | 323 | 270 |
| 41 | Woods | 4,350 | 41 | Edward | 78 | 41 | Wanaka | 311 | 193 |
| 42 | Mweru | 4,350 | 42 | Vattern | 74 | 42 | Orar | 310 | 26 |
| 43 | Peipus | 4,300 | 43 | San Martin | (68) | 43 | Geneva | 310 | 580 |
| 44 | Khanka | 4,190 | 44 | Michikamua | (67) | 44 | Tagish | 307 | 353 |
| 45 | Dubawnt | 3,833 | 45 | Atlin | 67 | 45 | Kurile | 306 | 77 |
| 46 | Mirim | 3,750 | 46 | Koko | (62) | 46 | Walker | 305 | 280 |
| 47 | Van | 3,740 | 47 | Saimaa | 61 | 47 | Titicaca | 304 | 8,030 |
| 48 | Tana | 3,600 | 48 | Taupo | 59 | 48 | Argentino | 300 | 1,410 |
| 49 | Uvs | 3,350 | 49 | Alakol | 57 | 49 | Iliamna | 299 | 2,590 |
| 50 | Poyang | 3,350 | 50 | Champlain | (54) | 50 | Tyrifjorden | 295 | 130 |

[a] Parentheses indicate that the value is an estimated one.

## References

Beeton, A.M. 1984.The world's great lakes. *J. Great Lakes Research* 10(2):106–113.
Beeton, A.M. and W.T. Edmondson. 1972. The eutrophication problem. *J. Fish Res. Board Canada* 29:673–692.
Fairbridge, R.W. (ed.). 1968. *The Encyclopedia of Geomorphology.* Reinhold Book Corp., New York, 1295 pp.
Håkanson, L. 1981. *A Manual of Lake Morphometry.* Springer-Verlag, Berlin, 73 pp.
Herdendorf, C.E. 1982. Large lakes of the world. *J. Great Lakes Res.* 8(2):379–412.
Herdendorf, C.E. 1984. *Inventory of the Morphometric and Limnologic Characteristics of Large Lakes of the World.* Ohio State Univ., Sea Grant Tech. Bull. 17, 78 pp.
Hutchinson, G.E. 1957. *A Treatise on Limnology, Vol. 1: Geography, Physics and Chemistry.* John Wiley & Sons, Inc., New York, 1015 pp.
Hutchinson, G.E. and H. Löffler. 1956. The thermal classification of lakes. *Proc. Nat. Acad. Sci.* 42:84–86.
Korzoun, V.I. (ed.). 1974. *Atlas of World Water Balance.* U.S.S.R. National Committee International Hydrological Decade. English transl.: UNESCO, 1977, 36 pp., 65 maps.
Nace, A.B. 1964. *Global Water Balance.* U.S. Geological Survey Prof. Paper No. 632, 33 pp.
Nixon, S.W. 1988. Physical energy inputs and the comparative ecology of lake and marine ecosystems. *Limnol. Oceanogr.* 33:1005–1025.
Sverdup, H.U., M.W. Johnson, and R.H. Fleming. 1942. *The Oceans—Their Physics, Chemistry and General Biology.* Prentice-Hall, Englewood Cliffs, New Jersey, 1087 pp.
Welch, P.S. 1948. *Limnological Methods.* McGraw-Hill Book Co., New York, 381 pp.

**Appendix to Chapter 1** Large lakes of the world

| Map no. (Fig. 1.1) | Lake | Area ($km^2$) | Max. depth (m) | Volume ($km^3$) | Elevation (m) | Origin | Latitude, ° | Salinity type | Mixis type |
|---|---|---|---|---|---|---|---|---|---|
| 1 | Abaya | 1160 | 13 | 8 | 1268 | Tectonic | 6N | Fresh | Polymictic |
| 2 | Abe | 780 | | | 310 | Tectonic | 11 | Saline | |
| 3 | Aberdeen | 1100 | | | 80 | Glacial | 64 | Fresh | Dimictic |
| 4 | Abitibi | 931 | | | 263 | Glacial | 49 | Fresh | Dimictic |
| 5 | Aby | 780 | | | 1 | Coastal | 5 | Saline | Polymictic |
| 6 | Alakol | 2650 | 54 | 57 | 347 | Tectonic | 46 | Saline | Monomictic |
| 7 | Albert | 5590 | 58 | 151 | 617 | Tectonic | 2 | Fresh | Polymictic |
| 8 | Alexandrina | 580 | 5 | 2 | 1 | Coastal | −35S | Fresh | Polymictic |
| 9 | Amadeus | 880[a] | | | 460 | Tectonic | −24 | Saline | |
| 10 | Amadjuak | 3115 | | | 113 | Solution | 65 | Fresh | Monomictic |
| 11 | Angikuni | 510 | | | 257 | Glacial | 62 | Fresh | Monomictic |
| 12 | Aral | 64100 | 68 | 1020 | 53 | Glacial | 45 | Brackish | Meromictic |
| 13 | Argentino | 1410 | 300 | 169 | 500 | Tectonic | −50 | Fresh | Meromictic |
| 14 | Artillery | 551 | | | 364 | Glacial | 63 | Fresh | Dimictic |
| 15 | Ashuanipi | 597 | | | 259 | Glacial | 53 | Fresh | Dimictic |
| 16 | Athabasca | 7935 | 120 | 204 | 213 | Glacial | 59 | Fresh | Dimictic |
| 17 | Atlin | 774 | 283 | 67 | 668 | Glacial | 59 | Fresh | Dimictic |
| 18 | Austin | 829[a] | | | 460 | Tectonic | −28 | Saline | |
| 19 | Aylmer | 847 | | | 375 | Glacial | 64 | Fresh | Dimictic |
| 20 | Baghrash | 1380 | | | 1038 | Tectonic | 42 | Saline | |
| 21 | Baikal | 31500 | 1741 | 22995 | 456 | Tectonic | 54 | Fresh | Meromictic |
| 22 | Baker | 1887 | 230 | 176 | 2 | Glacial | 64 | Fresh | Dimictic |
| 23 | Balaton | 590 | 11 | 2 | 104 | Tectonic | 47 | Fresh | Monomictic |
| 24 | Balkhash | 18200[a] | 26 | 112 | 343 | Tectonic | 46 | Brackish | Monomictic |

*Continued*

**Appendix to Chapter 1** Continued

| Map no. (Fig. 1.1) | Lake | Area (km²) | Max. depth (m) | Volume (km³) | Elevation (m) | Origin | Latitude, ° | Salinity type | Mixis type |
|---|---|---|---|---|---|---|---|---|---|
| 25 | Bangweulu | 4920[a] | 5 | 5 | 1140 | Tectonic | −11 | Fresh | Polymictic |
| 26 | Bay | 890 | 6 | 3 | 2 | Tectonic | 14 | Fresh | Polymictic |
| 27 | Becharof | 1190 | 92 | 44 | 4 | Glacial | 58 | Fresh | Dimictic |
| 28 | Beloye | 1290 | 20 | 5 | 113 | Glacial | 60 | Fresh | Amictic |
| 29 | Beysehir | 650 | 9 | 2 | 1116 | Tectonic | 38 | Fresh | Dimictic |
| 30 | Bienville | 1249 | | | 391 | Glacial | 55 | Fresh | Dimictic |
| 31 | Big Trout | 661 | 40 | 11 | 213 | Glacial | 54 | Fresh | Dimictic |
| 32 | Biwa | 688 | 103 | 28 | 85 | Tectonic | 35 | Fresh | Monomictic |
| 33 | Bras d'Or | 1099 | 70 | 31 | 1 | Coastal | 46 | Saline | Dimictic |
| 34 | Buenos Aires | 2240 | | | 217 | Glacial | −47 | Fresh | Dimictic |
| 35 | Buffalo | 612 | | | 265 | Glacial | 60 | Fresh | Dimictic |
| 36 | Buyr | 610 | 11 | 2 | 583 | Tectonic | 48 | Fresh | Dimictic |
| 37 | Caratasca | 1110 | 5 | 2 | 1 | Coastal | 15 | Saline | Polymictic |
| 38 | Caspian | 374000 | 1025 | 78200 | −28 | Tectonic | 42 | Brackish | Meromictic |
| 39 | Cedar | 1353 | | | 253 | Glacial | 53 | Fresh | Dimictic |
| 40 | Chad | 16600[a] | 12 | 44 | 240 | Tectonic | 13 | Fresh[b] | Polymictic |
| 41 | Champlain | 1100 | 122 | 54 | 30 | Tectonic | 45 | Fresh | Dimictic |
| 42 | Chany | 2500[a] | 10 | 5 | 105 | Tectonic | 55 | Saline | Dimictic |
| 43 | Chao | 900 | | | 15 | Fluviatile | 32 | Fresh | Dimictic |
| 44 | Chapala | 1140 | 13 | 10 | 1525 | Tectonic | 20 | Fresh | Polymictic |
| 45 | Chilka | 1170[a] | 5 | 4 | 1 | Coastal | 20 | Brackish | Polymictic |
| 46 | Chilwa | 1040 | 5 | 2 | 550 | Tectonic | −15 | Brackish[b] | Polymictic |
| 47 | Chiquita | 1850 | | | 70 | Tectonic | −31 | Saline | Polymictic |
| 48 | Chiriqui | 900 | | | 1 | Coastal | 9 | Saline | Polymictic |
| 49 | Churchill | 559 | 21 | 5 | 421 | Glacial | 56 | Fresh | Dimictic |
| 50 | Claire | 1436 | 2 | 2 | 213 | Glacial | 59 | Fresh | Dimictic |
| 51 | Clinton | 737 | | | 375 | Glacial | 64 | Fresh | Dimictic |

| Map no. (Fig. 1.1) | Lake | Area (km²) | Max. depth (m) | Volume (km³) | Elevation (m) | Origin | Latitude, ° | Salinity type | Mixis type |
|---|---|---|---|---|---|---|---|---|---|
| 52 | Colhue Huapi | 800 | 4 | 2 | 265 | Tectonic | −46 | Fresh | Polymictic |
| 53 | Constance | 540 | 252 | 48 | 396 | Glacial | 47 | Fresh | Dimictic |
| 54 | Contwoyto | 958 | | | 564 | Glacial | 66 | Fresh | Dimictic |
| 55 | Cree | 1440 | 60 | 18 | 487 | Glacial | 57 | Fresh | Dimictic |
| 56 | Cross | 755 | 27 | 6 | 207 | Glacial | 54 | Fresh | Dimictic |
| 57 | Dauphin | 519 | | | 260 | Glacial | 51 | Fresh | Dimictic |
| 58 | Dead | 1020 | 426 | 188 | −400 | Tectonic | 32 | Hypersaline | Meromictic |
| 59 | Deschambaul | 542 | 22 | 3 | 324 | Glacial | 55 | Fresh | Dimictic |
| 60 | Dore | 640 | 20 | 7 | 459 | Glacial | 55 | Fresh | Dimictic |
| 61 | Dubawnt | 3833 | | | 236 | Glacial | 63 | Fresh | Dimictic |
| 62 | Eau Claire | 1383 | | | 283 | Glacial | 56 | Fresh | Dimictic |
| 63 | Ebi | 1070 | | | 213 | Tectonic | 45 | Saline | Dimictic |
| 64 | Edward | 2150 | 117 | 78 | 912 | Tectonic | −1 | Fresh | Monomictic |
| 65 | Egridir | 520 | 13 | 3 | 924 | Tectonic | 38 | Fresh | Polymictic |
| 66 | Ennadai | 681 | | | 311 | Glacial | 61 | Fresh | Dimictic |
| 67 | Enriquillo | 500 | 2 | 1 | −44 | Tectonic | 18 | Saline | Polymictic |
| 68 | Erie | 25657 | 64 | 483 | 174 | Glacial | 42 | Fresh | Monomictic |
| 69 | Eskimo North | 838 | | | 1 | Coastal | 69 | Brackish | Dimictic |
| 70 | Eskimo South | 628 | | | 2 | Glacial | 69 | Fresh | Dimictic |
| 71 | Evans | 547 | 13 | 3 | 241 | Glacial | 51 | Fresh | Dimictic |
| 72 | Evoron | 590[a] | | | 100 | Tectonic | 51 | Fresh | |
| 73 | Eyre | 7690[a] | 4 | 23 | −12 | Tectonic | −29 | Hypersaline[b] | Polymictic |
| 74 | Fagnano | 590 | 449 | 125 | 140 | Glacial | 155 | Fresh | Meromictic |
| 75 | Faguibine | 590 | 10 | 4 | 280 | Tectonic | 17 | Fresh | Polymictic |
| 76 | Ferguson | 588 | | | 11 | Glacial | 69 | Fresh | Dimictic |

*Continued*

**Appendix to Chapter 1** Continued

| Map no. (Fig. 1.1) | Lake | Area ($km^2$) | Max. depth (m) | Volume ($km^3$) | Elevation (m) | Origin | Latitude, ° | Salinity type | Mixis type |
|---|---|---|---|---|---|---|---|---|---|
| 77 | Flathead | 500 | 113 | 24 | 881 | Glacial | 48 | Fresh | Monomictic |
| 78 | Frobisher | 516 | 19 | 2 | 421 | Glacial | 56 | Fresh | Dimictic |
| 79 | Frome | 2410[a] | 1 | 2 | 49 | Tectonic | −31 | Hypersaline | Polymictic |
| 80 | Gairdner | 4770[a] | 1 | 4 | 34 | Tectonic | −32 | Hypersaline | Polymictic |
| 81 | Garry | 976 | | | 148 | Glacial | 66 | Fresh | Dimictic |
| 82 | Geneva | 580 | 310 | 89 | 372 | Glacial | 46 | Fresh | Monomictic |
| 83 | Gods | 1151 | | | 178 | Glacial | 55 | Fresh | Dimictic |
| 84 | Goose | 503[a] | 7 | 2 | 1437 | Tectonic | 42 | Saline | Dimictic |
| 85 | Grand | 537 | 110 | 28 | 85 | Glacial | 49 | Fresh | Dimictic |
| 86 | Gras | 633 | | | 396 | Glacial | 65 | Fresh | Dimictic |
| 87 | Great Bear | 31326 | 452 | 2381 | 156 | Glacial | 66 | Fresh | Dimictic |
| 88 | Great Salt | 4360 | 15 | 19 | 1280 | Tectonic | 41 | Hypersaline[b] | Dimictic |
| 89 | Great Slave | 28568 | 625 | 2088 | 156 | Glacial | 62 | Fresh | Dimictic |
| 90 | Guillaume | 700 | | | 1 | Coastal | 56 | Saline | Dimictic |
| 91 | Hammar | 1940 | 2 | 2 | 10 | Tectonic | 31 | Fresh | Polymictic |
| 92 | Har | 500 | | | 1104 | Tectonic | 48 | Fresh | Monomictic |
| 93 | Har Us | 1760 | | | 1153 | Tectonic | 48 | Fresh | Monomictic |
| 94 | Hazen | 542 | | | 158 | Glacial | 82 | Fresh | Amictic |
| 95 | Helmand | 2080[a] | 11 | 8 | 510 | Tectonic | 31 | Fresh | Polymictic |
| 96 | Hottah | 984 | 70 | 17 | 180 | Glacial | 65 | Fresh | Dimictic |
| 97 | Hovsgol | 2620 | 270 | 480 | 1624 | Tectonic | 51 | Fresh | Dimictic |
| 98 | Hulun | 1590 | 2 | 2 | 1275 | Fluviatile | 49 | Fresh | Polymictic |
| 99 | Hungtze | 2700 | | | 15 | Fluviatile | 33 | Fresh | Polymictic |
| 100 | Huron | 59500 | 229 | 3537 | 177 | Glacial | 45 | Fresh | Monomictic |
| 101 | Hyargas | 1360 | | | 1028 | Tectonic | 49 | Saline | |
| 102 | Iliamna | 2590 | 299 | 319 | 15 | Glacial | 60 | Fresh | Monomictic |
| 103 | Ilmen | 2100[a] | 11 | 12 | 18 | Glacial | 58 | Fresh | Monomictic |

| Map no. (Fig. 1.1) | Lake | Area ($km^2$) | Max. depth (m) | Volume ($km^3$) | Elevation (m) | Origin | Latitude, ° | Salinity type | Mixis type |
|---|---|---|---|---|---|---|---|---|---|
| 104 | Imandra | 900 | 67 | 11 | 127 | Glacial | 68 | Fresh | Amictic |
| 105 | Inari | 1050 | 92 | 15 | 118 | Glacial | 69 | Fresh | Amictic |
| 106 | Island | 1223 | | | 227 | Glacial | 54 | Fresh | Dimictic |
| 107 | Issykkul | 6240 | 702 | 1730 | 1608 | Tectonic | 42 | Brackish | Meromictic |
| 108 | Istada | 520 | | | 2135 | Tectonic | 33 | Saline | Monomictic |
| 109 | Izabal | 590 | 20 | 8 | 8 | Tectonic | 16 | Fresh | Polymictic |
| 110 | Kamilukuak | 638 | | | 266 | Glacial | 62 | Fresh | Dimictic |
| 111 | Kaminak | 600 | | | 53 | Glacial | 62 | Fresh | Dimictic |
| 112 | Kaminuriak | 550 | | | 92 | Glacial | 63 | Fresh | Dimictic |
| 113 | Kaoyu | 700 | | | 15 | Fluviatile | 33 | Fresh | Monomictic |
| 114 | Kasba | 1341 | | | 336 | Glacial | 60 | Fresh | Dimictic |
| 115 | Khanka | 4190[a] | 11 | 19 | 69 | Tectonic | 45 | Fresh | Dimictic |
| 116 | Kivu | 2370 | 480 | 569 | 1460 | Tectonic | −2 | Fresh | Meromictic |
| 117 | Koko | 4460 | 38 | 62 | 3197 | Tectonic | 37 | Saline | Monomictic |
| 118 | Kulundinskoe | 728 | 5 | 1 | 96 | Tectonic | 53 | Saline | Dimictic |
| 119 | Kurisches | 1610 | 7 | 6 | 1 | Coastal | 55 | Saline | Dimictic |
| 120 | Kyaring | 670 | | | 4708 | Tectonic | 31 | Fresh | Dimictic |
| 121 | Kyoga | 4430 | 8 | 27 | 1036 | Tectonic | 2 | Fresh | Polymictic |
| 122 | Ladoga | 17700 | 230 | 908 | 4 | Glacial | 61 | Fresh | Monomictic |
| 123 | Lesser Slave | 1169 | 21 | 14 | 577 | Glacial | 55 | Fresh | Dimictic |
| 124 | Llanquihue | 800 | 350 | 106 | 52 | Glacial | −41 | Fresh | Dimictic |
| 125 | Lobstick | 511 | | | 457 | Glacial | 54 | Fresh | Dimictic |
| 126 | Lop | 3100 | 5 | 5 | 768 | Tectonic | 41 | Saline | Dimictic |
| 127 | Lower Seal | 576 | | | 290 | Glacial | 57 | Fresh | Dimictic |
| 128 | Luang | 1073 | 6 | 2 | 1 | Coastal | 8 | Brackish[b] | Polymictic |

*Continued*

**Appendix to Chapter 1** Continued

| Map no. (Fig. 1.1) | Lake | Area ($km^2$) | Max. depth (m) | Volume ($km^3$) | Elevation (m) | Origin | Latitude, ° | Salinity type | Mixis type |
|---|---|---|---|---|---|---|---|---|---|
| 129 | Mackay | 1061 | | | 431 | Glacial | 64 | Fresh | Dimictic |
| 130 | Mai-Ndombe | 2325[a] | 10 | 12 | 340 | Tectonic | −2 | Fresh | Polymictic |
| 131 | Malaren | 1096 | 64 | 14 | 1 | Glacial | 60 | Fresh | Monomictic |
| 132 | Managua | 1040 | 80 | 33 | 37 | Tectonic | 12 | Fresh | Monomictic |
| 133 | Manitoba | 4625 | 4 | 17 | 248 | Glacial | 51 | Fresh | Dimictic |
| 134 | Manouane | 584 | | | 500 | Glacial | 51 | Fresh | Dimictic |
| 135 | Manzala | 1360 | 1 | 1 | 1 | Coastal | 31 | Saline | Polymictic |
| 136 | Maracaibo | 13010 | 60 | 280 | 1 | Tectonic | 10 | Brackish[b] | Monomictic |
| 137 | Martre | 1776 | | | 265 | Glacial | 63 | Fresh | Dimictic |
| 138 | Melville | 3069 | 256 | 298 | 1 | Glacial | 54 | Saline | Dimictic |
| 139 | Michigan | 57750 | 282 | 4920 | 177 | Glacial | 44 | Fresh | Monomictic |
| 140 | Michikamua | 2030 | 80 | 67 | 460 | Glacial | 54 | Fresh | Dimictic |
| 141 | Mille Lacs | 540 | 11 | 3 | 381 | Glacial | 46 | Fresh | Dimictic |
| 142 | Minto | 761 | | | 183 | Glacial | 57 | Fresh | Dimictic |
| 143 | Mirim | 3750 | 18 | 19 | 1 | Coastal | −33 | Brackish | Polymictic |
| 144 | Mistassini | 2335 | 183 | 175 | 372 | Glacial | 51 | Fresh | Dimictic |
| 145 | Moose | 1367 | | | 255 | Glacial | 54 | Fresh | Dimictic |
| 146 | Mweru | 4350 | 37 | 32 | 922 | Tectonic | −9 | Fresh | Polymictic |
| 147 | Nahuel Huapi | 550 | 438 | 113 | 767 | Glacial | 41 | Fresh | Monomictic |
| 148 | Naknek | 630 | | | 10 | Glacial | 59 | Fresh | Dimictic |
| 149 | Nam | 2500 | | | 4627 | Tectonic | 31 | Saline | Dimictic |
| 150 | Namak | 750 | 1 | 1 | 790 | Tectonic | 35 | Saline | Polymictic |
| 151 | Netilling | 5530 | | | 30 | Glacial | 67 | Fresh | Dimictic |
| 152 | Ngoring | 650 | | | 4270 | Tectonic | 35 | Fresh | Monomictic |
| 153 | Nicaragua | 8150 | 70 | 108 | 32 | Tectonic | 12 | Fresh | Dimictic |
| 154 | Nipigon | 4848 | 165 | 31 | 320 | Glacial | 50 | Fresh | Dimictic |
| 155 | Nipissing | 833 | 22 | 7 | 196 | Glacial | 46 | Fresh | Dimictic |

| Map no. (Fig. 1.1) | Lake | Area ($km^2$) | Max. depth (m) | Volume ($km^3$) | Elevation (m) | Origin | Latitude, ° | Salinity type | Mixis type |
|---|---|---|---|---|---|---|---|---|---|
| 156 | Nonacho | 784 | | | 354 | Glacial | 62 | Fresh | Dimictic |
| 157 | Nueltin | 2279 | | | 278 | Glacial | 60 | Fresh | Dimictic |
| 158 | Nyasa | 22490 | 706 | 6140 | 475 | Tectonic | −12 | Fresh | Meromictic |
| 159 | Oder | 900 | 9 | 3 | 1 | Coastal | 54 | Brackish | Dimictic |
| 160 | Okeechobee | 1810 | 6 | 4 | 6 | Tectonic | 27 | Fresh | Polymictic |
| 161 | Oling | 570 | | | 4270 | Tectonic | 35 | Fresh | Monomictic |
| 162 | Onega | 9700 | 127 | 295 | 33 | Glacial | 62 | Fresh | Dimictic |
| 163 | Ontario | 19000 | 245 | 1637 | 75 | Glacial | 44 | Fresh | Monomictic |
| 164 | Oulu | 928 | 35 | 7 | 122 | Glacial | 64 | Fresh | Monomictic |
| 165 | Paijanne | 1090 | 95 | 18 | 78 | Tectonic | 62 | Fresh | Monomictic |
| 166 | Pangong | 600 | 43 | 10 | 4248 | Tectonic | 34 | Brackish | Dimictic |
| 167 | Patos | 10140 | 5 | 20 | 1 | Coastal | −31 | Brackish | Polymictic |
| 168 | Payne | 533 | | | 134 | Glacial | 59 | Fresh | Dimictic |
| 169 | Peipus | 4300 | 15 | 25 | 30 | Glacial | 57 | Fresh | Dimictic |
| 170 | Peter Pond | 778 | 24 | 11 | 421 | Glacial | 56 | Fresh | Dimictic |
| 171 | Pielinen | 960 | 60 | 9 | 94 | Glacial | 63 | Fresh | Monomictic |
| 172 | Playgreen | 657 | | | 217 | Glacial | 54 | Fresh | Dimictic |
| 173 | Point | 701 | | | 375 | Glacial | 65 | Fresh | Dimictic |
| 174 | Pomo | 880 | | | 4936 | Tectonic | 29 | Fresh | Monomictic |
| 175 | Pontchartrain | 1620 | 5 | 3 | 1 | Fluviatile | 30 | Brackish | Dimictic |
| 176 | Poopo | 1340 | 3 | 2 | 3686 | Tectonic | −19 | Saline | Polymictic |
| 177 | Poyang | 3350 | 20 | 27 | 10 | Fluviatile | 29 | Fresh | Polymictic |
| 178 | Princess | 524 | | | 116 | Glacial | 64 | Fresh | Dimictic |
| 179 | Pya | 660 | 49 | 13 | 101 | Glacial | 66 | Fresh | Amictic |
| 180 | Pyramid | 510[a] | 101 | 28 | 1159 | Tectonic | 40 | Brackish | Meromictic |

*Continued*

**Appendix to Chapter 1** Continued

| Map no. (Fig. 1.1) | Lake | Area (km²) | Max. depth (m) | Volume (km³) | Elevation (m) | Origin | Latitude, ° | Salinity type | Mixis type |
|---|---|---|---|---|---|---|---|---|---|
| 181 | Rainy | 940 | 49 | 9 | 338 | Glacial | 49 | Fresh | Dimictic |
| 182 | Red | 1170 | 9 | 7 | 358 | Glacial | 48 | Fresh | Dimictic |
| 183 | Reindeer | 6640 | 219 | 96 | 337 | Glacial | 57 | Fresh | Dimictic |
| 184 | Ronge | 1413 | 38 | 18 | 364 | Glacial | 55 | Fresh | Dimictic |
| 185 | Rudolph | 8660 | 73 | 251 | 427 | Tectonic | 4 | Brackish | Polymictic |
| 186 | Rukwa | 2716[a] | 1 | 2 | 793 | Tectonic | −8 | Saline | Polymictic |
| 187 | Saimaa | 4380 | 82 | 61 | 76 | Glacial | 61 | Fresh | |
| 188 | Saint Clair | 1210 | 7 | 5 | 175 | Glacial | 42 | Fresh | Dimictic |
| 189 | Saint-Jean | 1003 | 63 | 25 | 98 | Glacial | 49 | Fresh | Dimictic |
| 190 | Sakami | 592 | 110 | 31 | 183 | Glacial | 53 | Fresh | Dimictic |
| 191 | Salton | 950 | 12 | 6 | −70 | Fluviatile | 33 | Saline | |
| 192 | San Martin | 1010 | 170 | 68 | 200 | Glacial | −49 | Fresh | |
| 193 | Sandy | 527 | | | 276 | Glacial | 53 | Fresh | Dimictic |
| 194 | Sap | 2450[a] | 12 | 10 | 5 | Fluviatile | 13 | Fresh | Polymictic |
| 195 | Sasykkol | 740 | 5 | 2 | 350 | Tectonic | 47 | Fresh | Dimictic |
| 196 | Scutari | 600[a] | 60 | 2 | 5 | Solution | 42 | Fresh | Dimictic |
| 197 | Seg | 910 | 97 | 34 | 114 | Glacial | 63 | Fresh | |
| 198 | Selawik | 1400 | | | 1 | Coastal | 67 | Brackish | Amictic |
| 199 | Seletyteniz | 780 | 3 | 2 | 64 | Glacial | 53 | Saline | |
| 200 | Selwyn | 717 | | | 398 | Glacial | 60 | Fresh | Dimictic |
| 201 | Seul | 1658 | 34 | 13 | 357 | Glacial | 50 | Fresh | Dimictic |
| 202 | Sevan | 1360 | 83 | 38 | 1900 | Tectonic | 40 | Fresh | |
| 203 | Shamo | 550 | 13 | 3 | 1235 | Tectonic | 6 | Fresh | Polymictic |
| 204 | Simcoe | 744 | 41 | 12 | 219 | Glacial | 44 | Fresh | Dimictic |
| 205 | Snowbird | 505 | | | 359 | Glacial | 61 | Fresh | Dimictic |
| 206 | South Indian | 2247 | 18 | 16 | 254 | Glacial | 57 | Fresh | Dimictic |
| 207 | South Henik | 513 | | | 184 | Glacial | 61 | Fresh | Dimictic |

| Map no. (Fig. 1.1) | Lake | Area (km²) | Max. depth (m) | Volume (km³) | Elevation (m) | Origin | Latitude, ° | Salinity type | Mixis type |
|---|---|---|---|---|---|---|---|---|---|
| 208 | Superior | 82100 | 407 | 12230 | 183 | Glacial | 48 | Fresh | Monomictic |
| 209 | Tahoe | 500 | 501 | 124 | 1899 | Tectonic | 39 | Fresh | |
| 210 | Tai | 2210 | 5 | 4 | 3 | Fluviatile | 31 | Fresh | Polymictic |
| 211 | Takiyuak | 1080 | | | 381 | Glacial | 66 | Fresh | Dimictic |
| 212 | Tana | 3600 | 20 | 28 | 1840 | Tectonic | 12 | Fresh | Polymictic |
| 213 | Tanganyika | 32900 | 1471 | 18900 | 774 | Tectonic | −6 | Fresh | Meromictic |
| 214 | Tangra | 1400 | | | 4724 | Tectonic | 31 | Saline | |
| 215 | Tathlina | 573 | | | 280 | Glacial | 61 | Fresh | Monomictic |
| 216 | Taupo | 616 | 165 | 59 | 357 | Volcanic | −39 | Fresh | Monomictic |
| 217 | Taymyr | 4560[a] | 26 | 13 | 3 | Glacial | 75 | Fresh | Amictic |
| 218 | Tebesjuak | 575 | | | 146 | Glacial | 64 | Fresh | Dimictic |
| 219 | Tengiz | 1590 | 8 | 5 | 304 | Tectonic | 50 | Saline | Monomictic |
| 220 | Terinam | 810 | | | 4684 | Tectonic | 31 | Saline | |
| 221 | Terminos | 1550 | 1 | 1 | 1 | Coastal | 19 | Saline | Polymictic |
| 222 | Teshekpuk | 820 | | | 2 | Tectonic | 71 | Fresh | Amictic |
| 223 | Titicaca | 8030 | 304 | 827 | 3809 | Tectonic | −16 | Fresh | Monomictic |
| 224 | Toba | 1150 | 529 | 249 | 906 | Volcanic | 3 | Fresh | Monomictic |
| 225 | Top | 990 | 56 | 21 | 109 | Glacial | 66 | Fresh | |
| 226 | Torrens | 5780[a] | 1 | 5 | 30 | Tectonic | −31 | Hypersaline | Polymictic |
| 227 | Trout | 504 | | | 503 | Glacial | 61 | Fresh | Dimictic |
| 228 | Tulemalu | 668 | | | 279 | Glacial | 63 | Fresh | Dimictic |
| 229 | Tumba | 500 | 5 | 1 | 340 | Tectonic | −1 | Fresh | Polymictic |
| 230 | Tungting | 6000[a] | 10 | 18 | 11 | Fluviatile | 29 | Fresh | Polymictic |
| 231 | Tuz | 1640 | 1 | 1 | 925 | Tectonic | 39 | Hypersaline | Polymictic |
| 232 | Ubinskoe | 559 | 3 | 1 | 120 | Tectonic | 56 | Saline | |

*Continued*

**Appendix to Chapter 1** Continued

| Map no. (Fig. 1.1) | Lake | Area (km²) | Max. depth (m) | Volume (km³) | Elevation (m) | Origin | Latitude, ° | Salinity type | Mixis type |
|---|---|---|---|---|---|---|---|---|---|
| 233 | Ulungur | 830 | | | 468 | Tectonic | 47 | Saline | |
| 234 | Upemba | 530 | 4 | 1 | 580 | Tectonic | −9 | Fresh | Polymictic |
| 235 | Urmia | 5800[a] | 16 | 45 | 1275 | Tectonic | 38 | Hypersaline | Polymictic |
| 236 | Uvs | 3350 | 1 | 3 | 759 | Tectonic | 50 | Saline | Polymictic |
| 237 | Van | 3740 | 145 | 206 | 1646 | Tectonic | 39 | Saline | |
| 238 | Vanern | 5648 | 106 | 153 | 44 | Glacial | 59 | Fresh | Monomictic |
| 239 | Vattern | 1856 | 128 | 74 | 88 | Glacial | 58 | Fresh | Dimictic |
| 240 | Victoria | 68460 | 92 | 2700 | 1134 | Tectonic | −1 | Fresh | Polymictic |
| 241 | Viedma | 1090 | | | 250 | Glacial | −50 | Fresh | |
| 242 | Vyg | 1250 | 24 | 7 | 89 | Glacial | 64 | Fresh | |
| 243 | Weishan | 1000 | | | 15 | Fluviatile | 35 | Fresh | |
| 244 | Wholdaia | 678 | | | 364 | Glacial | 61 | Fresh | Dimictic |
| 245 | Winnebago | 560 | 7 | 3 | 228 | Glacial | 44 | Fresh | Dimictic |
| 246 | Winnipeg | 24387 | 18 | 371 | 217 | Glacial | 53 | Fresh | Dimictic |
| 247 | Winnipegosis | 5375 | 12 | 16 | 254 | Glacial | 53 | Fresh | Dimictic |
| 248 | Wollaston | 2690 | 97 | 40 | 398 | Glacial | 58 | Fresh | Dimictic |
| 249 | Woods | 4350 | 21 | 35 | 323 | Glacial | 49 | Fresh | Dimictic |
| 250 | Yamdrok | 800 | | | 4374 | Tectonic | 29 | Fresh | |
| 251 | Yathkyed | 1449 | | | 140 | Glacial | 63 | Fresh | Dimictic |
| 252 | Zaysan | 1800 | 9 | 7 | 386 | Tectonic | 48 | Fresh | Dimictic |
| 253 | Ziling | 1860 | 8 | 6 | 4495 | Tectonic | 32 | Saline | Polymictic |

[a] Lake is subject to large fluctuations (often seasonal) in water level and surface area.
[b] Lake is subject to large fluctuations in salinity.

# 2

# Specific Properties of Large Lakes

Max M. Tilzer

ABSTRACT Working hypotheses, sometimes contradictory, are proposed, which distinguish large lakes from small ones in limnological terms. Large lakes tend to be deep and to have long water retention times. Physical forcing and horizontal heterogeneities both in the physical structure and the communities are greater than in small lakes. Biogeochemical cycling is dominated by internal regeneration to a greater extent than by external inputs. Pelagic communities are more important than both littoral and bottom communities in the overall production process. Diminished nutrient loading causes the generally lower productivity of large as compared with shallow and/or small lakes. Massive algal blooms are extremely rare, also due to more intense mixing of the upper water layers. Only large lakes contain truly pelagic fishes. Whereas the seasonality in the pelagic zone frequently is pronounced, it is virtually absent in the deep benthos due to a steady influx of organic matter from the pelagial.

The specific properties of large lakes *per definitionem* are those which directly or indirectly are connected with the size of the water body concerned. It has already been emphasized that no single isolated property is common to all large lakes of the world. In the following chapters, many specific properties are discussed which may apply to most but not to all large lakes. They have been derived from case studies on one or a few lakes, and attempts are made to examine their potential for generalization.

This introductory chapter summarizes the views of certain limnologists* who expressed their opinions prior to the symposium entitled "Functional and Structural Properties of Large Lakes," as well as the views of the editors of this volume about some useful working hypotheses. Moreover, some of these working hypotheses were clarified during a meeting on *Basic Issues in Great Lakes Research* which was organized by the University of Michigan in Fall, 1986. Thus, this chapter is not a summary of the major conclusions which emerged from the chapters of this book. Rather, it describes a set of working

*T. Berman, E.J. Fee, W. Geller, C.R. Goldman, J.T. Lehman, S.S. Kilham, C.S. Schelske, D.W. Schindler, J.F. Talling, and W.F. Vincent.

hypotheses. Several of these are based on conjecture only, but some appear to be counter-intuitive at first sight. Furthermore, several of the hypotheses put forward below may even contradict each other.

This, however, does not necessarily invalidate any of the hypotheses concerned. It merely means that each of the following theses is applicable only under certain boundary conditions, which do not necessarily apply to *all* lakes that nevertheless have to be termed "large," for other reasons. Strictly speaking, each of the following statements should begin with a clause such as, "In most large lakes . . ."–if not, "In many large lakes . . . ."

Physical features set the stage for the geochemical and biogeochemical cycles as well as the biological processes within the water body. The following sections elaborate our working hypotheses.

## 2.1 Morphology

Large lakes tend to be deeper than small lakes and their shoreline development is reduced. Moreover, the ratio of the volume of the water body to its surface area, that is, its boundaries to the lake bottom and the littoral zone, is considerably greater than in small lakes. The following statements mainly apply to deep, large lakes whereas shallow, large lakes have more properties in common with small lakes.

## 2.2 Hydrography Characteristics

**Water Retention Time** In general, both the area and the volume of the lake are greater not only in absolute terms but also in relation to the area of the watershed. In addition to being heavily influenced by the amount of annual precipitation, the ratio of watershed to lake surface area has a great bearing on the water retention time which tends to be longer in large than in small lakes.

**Internal Water Movements** The large surface area allows increased wind fetch and hence deeper mixing of near-surface water layers even during lake stratification. The large surface area, moreover, allows the build-up of horizontal gradients in air pressure above the lake. Temporal changes in wind and air pressure patterns lead to oscillations of the water masses, both external and internal. Such oscillations are a very common phenomenon in large lakes but are virtually absent or insignificant in small lakes.

**Horizontal Heterogeneities** Spatially variable inputs of energy from the overlying atmosphere, as irradiance and/or as kinetic energy, as well as differences in atmospheric pressure, can cause significant horizontal heterogeneities in the physical properties of the water masses, especially in near-surface layers. As a special case, differential heating of nearshore *versus* offshore areas allows the temporary build-up of zones with markedly altered thermal regimes. Most notable are the "thermal bars" which can occur in large lakes that cool below 4 °C during the winter.

**Heat Budget** Because large lakes tend to be deeper and have smaller surface-to-volume ratios than small lakes, they have greater heat storage capacities and lose smaller proportions of their total heat content during winter. As a consequence, large lakes have a much smaller tendency to freeze over than do small lakes that are located in comparable climatic regimes.

## 2.3 Biogeochemical Cycles

**Internal Cycling** Internal cycling of matter—both nutrient salts and organic material—plays a much greater role, as compared to interactions with the surroundings, in large than in small lakes. As a consequence, processes in large lakes as a whole tend to be less susceptible to short-term perturbations than in small lakes.

**Nutrient Retention** In large lakes, once nutrients have entered the system, they remain within the system considerably longer than in small lakes, owing to prolonged water retention times and more efficient internal regeneration, both by purely chemical processes (e.g., sorption—desorption of nutrient salts on particles, especially phosphate), and biological processes (e.g., lysis and bacterial remineralization of dead organic matter). This causes long-term perturbations within the catchment area to have significantly longer lasting effects in large as compared to small lakes.

**Nutrient Cycling by Settling Particles** On the other hand, in large lakes, nutrients tend to be removed efficiently from the euphotic zone by settling particles, which may either contain nutrients themselves or may scavenge them from the surrounding water by adsorption. This is because sediment resuspension is of lesser significance in large and deep than in shallow and/or small lakes.

## 2.4 Food-web structure

**Pelagic Versus Benthic Communities** Because of the smaller surface-to-volume ratio in large lakes, pelagic communities are usually of considerably greater importance than benthic communities, unless the lake is so shallow that major portions of the water masses are within the euphotic zone.

**Physical Forcing of Biota** Greater wind-fetch and the prevalence of internal water movements cause physical processes to have a greater bearing on biological processes in large than in small lakes.

**Patchiness** Horizontal heterogeneities in the physical structure of the water body can cause significant differences in the growth rate of planktonic communities. These effects can be enhanced by purely physical processes, such as dilution and accretion in areas of water-mass divergence or convergence, respectively. Both can lead to considerable spacial heterogeneities of biomass, as well as of community composition.

**Trophic State** Both external and internal nutrient loading should be smaller in large than in small lakes because of the relatively smaller watershed area and surface-to-volume ratio of the water body, respectively. Hence, large lakes in general should be more oligotrophic than small lakes.

**Primary Productivity** Massive blooms of cyanobacteria are extremely rare in large lakes, not only because large lakes tend to be less eutrophic but also because their upper water layers are mixed more frequently than those of small lakes. The mixed layer in large lakes, moreover, causes shallow phytoplankton to be exposed to lower average irradiance levels than in small lakes. The phytoplankton respond by adaptating their photosynthetic apparatus.

**Seasonality within the Pelagic Community** At temperate and subpolar latitudes, winter mixing in deep, large lakes should lead to a stronger dilution of the planktonic community as the mixing deepens, than is the case in small and/or shallow lakes. This effect should be amplified by the greater seasonal amplitude of mean water-column irradiances to which the phytoplankton community as a whole is exposed. We therefore can expect that the seasonal amplitude of plankton biomass and productivity at any given geographical latitude is greater in large deep lakes than it is in small and/or shallow lakes.

**Seasonality within the Profundal Community** The profundal benthos, by contrast, exhibit less seasonality in large than in small lakes. The main reason probably is the virtual lack of seasonal variations in the environmental conditions at the bottom of deep lakes. Possibly, the seasonality in sedimentary influx from the overlying pelagial layer is compensated to a great extent by enhanced sediment degradation within the water column.

**Physical Forcing of Littoral Communities** Internal seiches cause highly variable environmental conditions in the slope areas over the range of the thermocline depth fluctuations. As a consequence, these areas are frequently virtually devoid of biota.

**Pelagic Fish Predators** Only in large lakes can fish populations foraging for food and sometimes even reproducing in the pelagial be found, whereas in small lakes, fish populations in general live on littoral and/or profundal food.

## 2.5 Summary

While it is true that large lakes represent one end of a wide spectrum of lake sizes, they do possess specific system characteristics that distinguish them from the bulk of small lakes.

## References

Lehman, J.T. (ed.). 1987. Basic Issues of Great Lakes Research. Proceedings of a workshop sponsored by the U.S. National Science Foundation and The University of Michigan. Special Report No. 123 of the Great Lakes Research Division Great Lakes and Marine Waters Center, University of Michigan, Ann Arbor, 174 pp.

# Part 2

# Physical Features of Large Lakes and Their Significance for Biological Processes

The survival, growth, and dominance of small, euplanktonic organisms such as algae, bacteria, and zooplankton depend highly on the motions of their liquid environment which enhance or prevent nutrient supply, access to light, exposure to grazers, and sedimentation. The unique hydrodynamic properties of large lakes therefore are likely to affect the composition of their planktonic communities. Although larger organisms swimming actively such as fish are less dependent on the physical environment than plankton, they may still be greatly affected by it during particular periods; for example, fish cannot occupy anoxic hypolimnia of stratified lakes during summer, and their spawning grounds may be affected by variations in the littoral zone.

Part 2 of this book, therefore, is an attempt to identify the structural and dynamic properties which distinguish large lakes from small ones. Moreover, the time and space scales of hydrodynamic events are compared with the physiological rates of plankters. Thereafter, these relations are illustrated by specific examples based on case studies. Among them are the correlation found between primary productivity and the depth of spring mixing, and the link between the amplitude of variation of the water level and the reproductive success of fish species.

# 3

# Mixing and Transport in Lakes: Mechanisms and Ecological Relevance

Dieter M. Imboden

ABSTRACT Total primary production in the hydrosphere is of similar magnitude as on land (i.e., smaller than on land if expressed per area), but the planktonic biomass amounts to less than 1% of the biomass on land. Several factors distinguish life in water from life on land: buoyancy is a remedy for gravity and guarantees a place in the sun even to the small creature, but the price paid is partial loss of control over position because of currents and turbulence. Nutrient-storage capacity per volume in water is much smaller than in soil, thus nutrient cycling becomes essential and brings physics onto the scene again.

Mixing and transport in lakes has to be discussed in terms of the relative size of physical vs. biological scales. Typical length scales of mixing in large lakes extend over more than 8 orders of magnitude, i.e., from the $10^{-3}$ m-range of viscous energy dissipation to the lake-wide circulation of $10^5$ m or more in length. Likewise, the time scales of mixing cover the range between seconds and days (or even years, if vertical stratification is included). Reaction rates describing biological processes—such as nutrient uptake, photosynthesis, respiration, and mineralization—lie between minutes and months (or years, if the sediment is included), and the length scales lie between $10^{-5}$ m (the size of single cells) and the several thousand meters of algal patches or fish swarms. The scales of the "aquatic weather" and its interference with biology will be discussed in terms of the spatial structure of biomass distribution, the primary production, and the selective properties introduced into the aquatic ecosystem.

## 3.1 Introduction

Life had its origin in the hydrosphere. After the first organisms had appeared in the water it took more than two billion years before the first plants began to conquer the thin interface between atmosphere and pedosphere. The price for the new habitat was high; whereas the hydrosphere provides plants and animals with an environment of remarkable physical and chemical homogeneity, life on land had to adapt to a much wider spectrum of spatial and temporal variation (Table 3.1). The supply of water alters from too much to too little. Though nutrients are more abundant on land than in most aquatic environments, if they become rare on land, they cannot be replenished as easily as in the water where currents provide for rapid distribution. The transparency

**Table 3.1** Conditions for life in pelagic waters and on land

| Factor | Open water | Land |
|---|---|---|
| Control over position | *Limited*. Control by buoyancy or self-propulsion slow relative to water movement | *Normal*. By roots, attachment etc. |
| Gravity | *Small* or absent (buoyancy) | *Large*: supporting structures needed (trees) |
| Light | Low transparency of water | High transparency of air |
| Vertical extinction $\epsilon$ of light | On the order of 0.1 $m^{-1}$ | Negligible |
| Temporal variation of light for individual | Large and unpredictable (depending on position) | Slow and regular |
| Temperature variation | Moderate | Often large or even extreme |
| Water | Abundant | Limits the spreading of life (deserts) |
| Nutrients | Usually growth limiting, but easily distributed within system | Local inhomogeneities co-exist (fertile/barren land) |
| Spreading of species | Easy, by water currents | Usually difficult: easy by "tricks" only (seeds transported by animals, through air, etc.) |
| Ecological niches | Little | Many |
| Number of plant species in total ecosystem[a] | $\approx$ 40,000 | $\approx$400,000 |
| Average standing crop of living biomass[b] | 2 g C $m^{-2}$ | 6,000 g C $m^{-2}$ |
| Total primary production[b] | $50 \times 10^{15}$ g C $yr^{-1}$ | $40 \times 10^{15}$ g C $yr^{-1}$ |
| Primary production ~ per area[b] | 150 g C $m^{-2}$ $yr^{-1}$ | 250 g C $m^{-2}$ $yr^{-1}$ |
| Productivity per biomass | 75 $yr^{-1}$ | 0.04 $yr^{-1}$ |

[a] From Flindt (1986)
[b] Based on values by Bolin and Cook (1983)

for sunlight is much greater in air than in water, but on land space along the vertical axis is difficult to obtain since gravity cannot be compensated for by buoyancy as in the water. In reverse, the open water column in lakes and oceans does not provide the organisms with any support against the ever-moving water.

Weighing all the differences between life on land and life in the open water column, the advantages of terrestrial ecosystems seem to outrun the drawbacks. In fact, productivity per unit area is generally larger on land than in the water. Total primary production on land reaches roughly the same level as in the ocean which has a surface 2½ times larger. But the biomass on land necessary to reach this level of production is more than 300 times bigger than that in the ocean (see Table 3.1). In the competition for light it is mainly the obstacle of gravity which induced land ecosystems to develop a huge "infrastructure," that is the stems of the trees which support the relatively small active biomass, the leaves and needles. The turnover rate of carbon in terrestrial ecosystems is thus much lower than in aquatic systems.

Regarding the predominant role of the physical environment on land, it is well justified to ask what the corresponding restriction is in the water. We do not have to think very hard to find at least a first general answer: The small species in water, the ones which make up for the largest portion of aquatic ecosystems, were named *plankton*. In Greek, πλαζω means *to gad about*, and the word πλαγκτοσ means *to be gadded about*. Thus, the name plankton was chosen to indicate the most apparent condition of their life, the fact that they are driven around by the continual motion of the water in which they live. In short, one could characterize the two kinds of ecosystems by their respective dominant physical force, kinetic for the aquatic system, gravitational for the terrestrial one.

The purpose of this article is to explore in greater detail how the physical environment, the "aquatic weather," interacts with life, that is, how it selects for certain species which are particularly adapted to the specific properties of the different water bodies such as oceans and lakes. Special emphasis is put on water movement and mixing in the open water, the pelagic zone. Though temperature and light are important physical factors, *per se*, their influence on life will only be discussed indirectly in relation to the specific temporal variations caused by vertical transport and mixing.

The general goal of this article provides a natural definition for the term *large lake*. A large lake is a water body in which the pelagic zone makes up the major part of the ecosystem. Quantitatively, the surface area divided by the length of the shore line (the boundary length) can be used as a measure to classify a given lake on a scale covering a range from the large oceans to the smallest lakes and ponds (Table 3.2). This number with dimensions of length is also a measure for the distance from the shore to the center of the lake. [To be exact, for a circular lake it is half the lake radius.] If the number is large compared to the typical width of the littoral zone (the zone in which

**Table 3.2** Morphometric and hydrological differences between oceans and lakes (typical values or order of magnitude)

| | Oceans (total) | Large lakes | Small lakes |
|---|---|---|---|
| Volume ($km^3$) | $1.4 \times 10^9$ | $2 \times 10^2$ | $10^{-2}$ |
| Surface ($km^2$) | $3.6 \times 10^8$ | $10^3$ | 1 |
| Mean depth (m) | $4 \times 10^3$ | $2 \times 10^2$ | 10 |
| Surface area per boundary length (km) | $10^6$ | 10 | $10^{-1}$ |
| Residence time (yr) | | | |
| Surface water | $10^2$ | 1 to 10 | $10^{-1}$ to 1 |
| Total water body | $4 \times 10^4$ | 10 to $10^2$ | $10^{-1}$ to 10 |

the influence from the shore and/or from the lake bottom is felt), the pelagic part of the lake is dominant; the lake can then be characterized as large.

In addition, a large lake must be sufficiently deep in order to develop a thermocline (at least during the summer) which shields most of the surface water—the mixed layer—from direct contact with the sediments. The depth of the seasonal thermocline depends on the lake surface and typical wind speed and roughly varies between 5 and 30 m. Combining the two properties yields the following morphometric definition of a deep lake: *In a deep lake, the cross-sectional area at the depth of the seasonal thermocline is about equal to the surface area of the lake, e.g., not smaller than 90% of the surface area.*

This definition applies to the ocean as well. Indeed, the following considerations strongly rely on similar ideas developed by oceanographers. Yet, lakes exhibit certain mixing characteristics which are not found in the nearly unbounded ocean basins. One typical feature is the standing waves, the so-called seiches; another is linked to the moderate size of most lakes relative to the size of typical meteorological structures (high and low pressure areas) which prevents most lakes from attaining a steady mixing structure but keeps them in constant transition from one weather event to the next. Of course, the fact that most lakes have outlets which are also sinks for chemicals and biomass, distinguishes them from oceans as well, but this point will not be considered here.

## 3.2 Mixing in Large Lakes: Time and Length Scales

**Modes of Mixing Processes** Mixing processes in lakes cover a wide spectrum of different time scales and length scales (Mortimer, 1974). At the small end of these scales, the mixing processes are similar in oceans and lakes. With increasing size and duration of the mixing structure the different dimensions of the water bodies become more apparent.

Since several processes are active at the same time, they cannot be studied individually. Spectral analysis is the method commonly used to isolate the various phenomena. For instance, time series of temperature or current speeds measured at a fixed location in the ocean or in lakes can be decomposed by Fourier transformation into modes of different frequencies. If mixing structures of well-defined time scales exist in the water body (e.g., waves with well-defined periods), they should appear in the spectrum as local maxima at the corresponding frequencies. The same applies to chemical and biological parameters; typical time scales of reactions or biological transformations should manifest themselves as peaks in the spectra of concentration variability.

Time series measured over long-enough periods and containing data points at small-enough intervals are rare, especially for lakes. Horne and Platt (1984) discussed the dominant space and time scales of variability in the physical and biological fields on continental shelves. Their physical information was based on a classic paper by Wunsch (1972) on the spectrum from 2 years to 2 minutes of temperature fluctuations in the main thermocline at Bermuda. A simplified version of this spectrum is shown in Figure 3.1. In zone B, extending over periods between about one day and 30 minutes, the power density drops as $f^{-5/3}$ (f = frequency). This portion of the spectrum represents the so-called "inertial—convective range," reflecting Kolmogorov's (1941) theory of homogeneous, stationary, and isotropic turbulence (see Figure 3.14).

Kolmogorov's theory is based on the spectrum of the spatial variation of a given parameter such as temperature. Suppose that temperature is measured simultaneously all along a horizontal stretch and that—by Fourier transformation—a spectrum is calculated which separates the horizontal variability into different wave numbers, k. The k-values correspond to the frequencies f; their inverse values, $k^{-1}$, measure the length scales of variation. Kolmogorov postulates that at sufficiently large wave numbers, the spectrum of turbulence has a universal shape which depends only on the rate of kinetic energy dissipation $\epsilon$. Behind this concept lies the idea that the kinetic energy enters the system by external forces of scale L, and this scale is much larger than the typical inertial scale of $k^{-1}$. Energy then cascades down through ever-smaller eddies (i.e., through increasing wave numbers) until it is dissipated into heat due to viscous forces. The high frequency end of the inertial range is determined by the rate of viscous energy dissipation (Table 3.3).

Since the spatial spectrum is difficult to measure, it is often calculated from the time spectrum using the so-called "frozen turbulence hypothesis": Assume that a stationary turbulent field is flowing past a recording instrument fixed in space. If the turbulence does not change during its passage, the recorded frequencies f are related to the wave numbers by the relation $f = v_{ad}\, k$, where $v_{ad}$ is the mean advection velocity.

The vertical density stratification found in most oceans and lakes (at least during part of the year) modifies the (k– or f–) spectrum at both ends of the inertial range. The break between zone A and zone B (see Figure 3.1) reflects the attenuation of the modes with $f > N$. N is the Brunt-Väisälä (or stability) frequency which grows with increasing density gradient (see Table 3.3 for

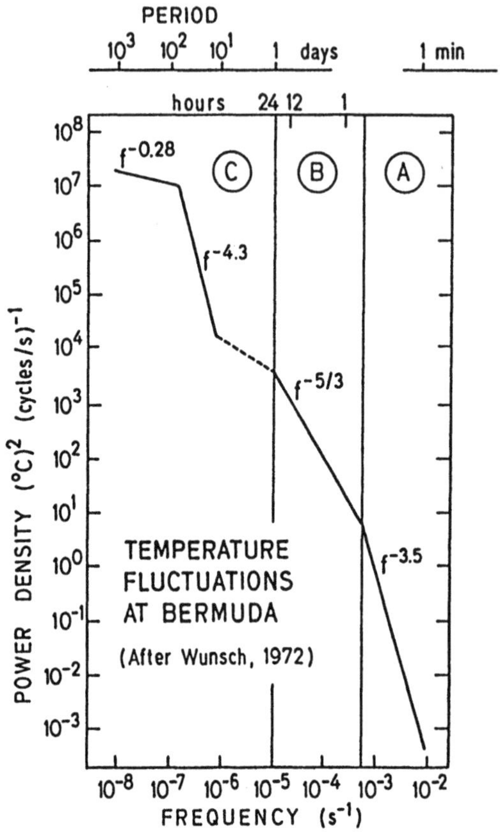


**Figure 3.1** Spectrum from 2 years to 2 minutes of temperature fluctuations in the thermocline of the North Atlantic. The density stratification of the water column shields the high frequency modes (zone A) from the thermistor positioned at the sediment surface on the shelf, i.e., below the mixed layer. Zone B is the inertial range; zone C reflects the major gyre system of the ocean and the so-called mode-type eddies, such as Gulf Stream rings. Simplified after Wunsch (1972).

typical values). In contrast, the low frequency end of the inertial range is determined by the dimension of the mixing zone in which only isotropic turbulence can occur. Due to the smaller mixing depth of lakes the range of isotropic turbulence is commonly more restricted in lakes than in the ocean.

The universal nature of the inertial range appears again in the spectrum of Figure 3.2 calculated from current speed measurements of the mixed layer of Lake Tahoe (Dillon and Powell, 1979). Furthermore, if larger time intervals are used for evaluating the spectra additional structure appears at frequencies below the inertial range. Figure 3.3 shows spectra of fluctuations of hourly mean temperature measured over 1,000 hours (about 40 days) in Lake Zurich (Switzerland) as calculated by Horn et al. (1986). The peaks denoted as mode no. 1 to 5 are internal seiches (standing internal waves) which are typical for

**Table 3.3** Typical sizes of parameters characterizing energy and momentum fluxes in large lakes[a]

| Parameter | Unit | Parameter size: Mixed layer (epilimnion) | Thermocline | Hypolimnion |
|---|---|---|---|---|
| Horizontal space scale, $L_x$ | m | $10^5$ | $10^5$ | $10^5$ |
| Vertical space scale, $L_z$ | m | 20 | 30 | 100 |
| Aspect ratio, $L_z/L_x$ | none | $2 \times 10^{-4}$ | $3 \times 10^{-4}$ | $10^{-3}$ |
| Typical horizontal velocity, $v_x$ | m $s^{-1}$ | 0.1 | 0.02 | 0.02 |
| Vertical eddy diffusivity, $K_z$ | $m^2$ $s^{-1}$ | $10^{-2}$ | $10^{-5}$ | $10^{-4}$ |
| Time scales of mixing— | | | | |
| horizontal, $\tau_x = L_x/v_x$ | s | $10^6$ | $5 \times 10^6$ | $5 \times 10^6$ |
| Vertical, $\tau_z = L_z^2/K_z$ | s | $4 \times 10^4$ | $10^8$ [b] | $10^8$ [b] |
| Brunt-Väisälä frequency, N | $s^{-1}$ | $10^{-3}$ | $10^{-2}$ | $10^{-3}$ |
| Rate of energy dissipation, $\epsilon$ | $m^2$ $s^{-3}$ | $10^{-7}$ | $10^{-9}$ | $10^{-10}$ |
| Viscous energy dissipation scale, $k_v^{-1}$ | m | $2 \times 10^{-3}$ | $5 \times 10^{-3}$ | $10^{-2}$ |
| Coriolis parameter, $f = 1.45 \times 10^{-4} \sin \theta$ | $s^{-1}$ | $1.0 \times 10^{-4}$ | $1.0 \times 10^{-4}$ | $1.0 \times 10^{-4}$ |
| Inertial period, $T = 2\pi/f$ | s | $6.3 \times 10^4$ | $6.3 \times 10^4$ | $6.3 \times 10^4$ |
| Rossby number, $Ro = v_x/(fL_x)$ | none | $10^{-2}$ | $2 \times 10^{-3}$ | $2 \times 10^{-3}$ |

[a] Parameters that depend on the Coriolis acceleration are given for a latitude $\theta = 45°$.
[b] In most lakes, vertical mixing time in the hypolimnion is determined by the annual destruction of the thermocline and not by vertical diffusion.

closed basins but do not appear in oceans (Table 3.4). In oceans, the periods of the seiches depend on lake size, topography, and on the vertical density structure of the water body. The Defant model, a simple method to calculate the period of the seiches, has recently been summarized by Lemmin and Mortimer (1986) and successfully applied to various small lakes. In large lakes, the effects of the earth's rotation are more important; the one-dimensional seiches are increasingly distorted to become rotational waves, but the periods (at least of the low modes having the largest amplitudes) are not significantly altered.

Figure 3.4 gives an overview of seiche periods (first modes) as a function of the length of the lake, $L_x$, and of the relative density difference between hypolimnion and epilimnion, $\Delta\rho/\rho$, calculated by Merian's equation (see e.g., Hutchinson, 1957):

$$T_o = 2L_x/(g'h^*)^{1/2} \qquad g' = g\Delta\rho/\rho\text{: reduced gravity} \tag{3.1}$$

where $h^*$ is the reduced mean depth defined as $h^* = (1/h_e + 1/h_h)^{-1}$. The period of the surface seiches follows from eq. 3.1 by substituting g and h (h = lake mean depth) for g′ and $h^*$.

As discussed before, the modes at low frequency and large wave length are the mixing structures which depend on the shape of the water body and

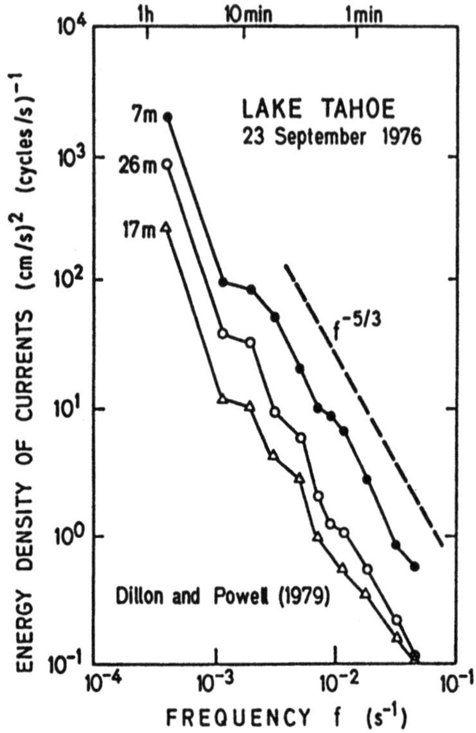


**Figure 3.2** Spectra of current speed measured at different depths in the mixed layer of Lake Tahoe (California/Nevada) calculated for a 5.5-h period. After Dillon and Powell (1979).

are thus the features which—from a physical point of view—are responsible for the differences between oceans, large lakes, and small lakes. In the ocean, the range of the large-scale structures is occupied by the mode-type eddies and gyres (see Figure 3.1). Large lakes may also show the particular finger prints of large-scale motion. The peak found at 100 h in the temperature spectra of Lake Zurich (see Figure 3.3) may originate from a so-called "topographic Rossby wave." Rossby waves are second-class waves which only exist due to the rotation of the earth (see Table 3.4), or more precisely due to potential vorticity gradients. The variable depth of lake basins gives rise to this type of wave (Stocker and Hutter, 1987). One of the first identifications of topographic Rossby waves in lakes was made in Lake Michigan by Saylor et al. (1980), as shown in Figure 3.5.

**Mixing Events** Many mixing processes in lakes are of singular nature; this means that they are triggered by some external or internal force and then evolve along a characteristic pathway determined by the relevant hydrodynamic laws. Of course, no clear-cut distinction between waves (periodic fea-

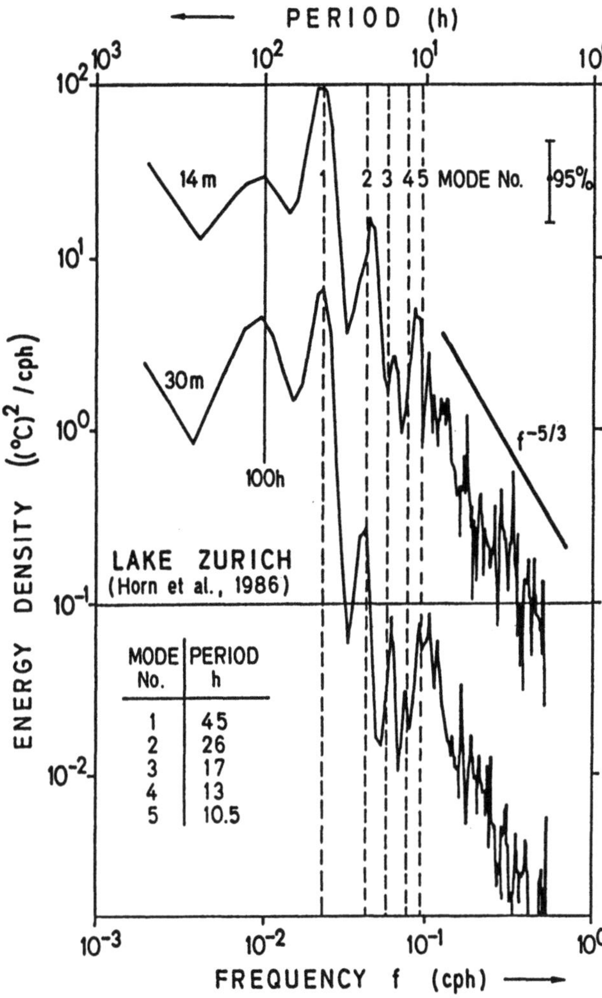


**Figure 3.3** Spectra of fluctuations of hourly mean temperature in Lake Zurich (Switzerland) for the 1,000-hour interval that began on 10 August 1978. The peaks denoted as mode no. 1 to 5 are internal seiches. The inertial period (16.4 h at latitude 47°) lies between these peaks. The maximum at 100 h may be due to topographic Rossby waves. From Horn et al., (1986).

tures) and events (unique features) exists. As an example, Figure 3.6 shows the wind shear acting on the surface of Lake Zug (Switzerland) during a severe storm. The wind force distorted the isotherms in such an extreme way that at the southern end of the lake (from where the wind was blowing) water from 20 m and below moved to the surface, while at the other end the 8°C-isotherm temporarily dropped below 44 m, the depth of the lowest thermistor moored in the lake during the storm. After the wind had stopped, the isotherms began to oscillate about their new equilibrium depth which were not very different

**Table 3.4** Types of waves and their governing factors[a]

| Wave type | Governing factors[b] |
|---|---|
| **First-class waves** | |
| Acoustic waves | Compressibility of water |
| Capillary waves | Surface tension between water and air |
| Surface gravity waves | Gravity, (water depth), (lake size) |
| Internal gravity waves | Gravity, density stratification, (water depth), (lake size) |
| Surface seiches | Gravity, water depth, length or width of lake |
| Internal seiches | Gravity, density stratification, water depth, length or width of lake |
| Kelvin, Poincaré waves (surface or internal) | As above, including earth rotation |
| Nonlinear waves (solitons, bores) | Energy pulse, lake topography |
| **Second-class waves** | |
| Planetary Rossby waves[c] | Variation of Coriolis parameter with latitude |
| Topographic Rossby waves | Variable depth of basin |

[a] First-class waves (or gravity waves) can also exist on a nonrotating earth (slightly modified); second-class waves owe their being to the existence of earth rotation.
[b] Underlined phrases are factors related to lake slope and size and are thus essential for distinguishing waves in small and large lakes, respectively.
[c] Can only occur in large oceans.

from their positions prior to the storm. Thus, the first part of this phenomenon truly had event character. The second part, however, looks more like the internal wave structure usually seen in lakes most of the time.

There exists a variety of other, less spectacular events. Many of them are connected to density-induced mixing as summarized in Figure 3.7. River-induced currents may be significant for the exchange of water from below the thermocline. The river plume, often enlarged in volume by water entrained from the lake surface, pushes hypolimnic water upwards, thus adding to vertical mixing. Figure 3.8 shows such a plunging event measured at the mouth of the major inlet of Lake Lucerne, the River Reuss. The integrated effect of all such events during May and June 1986 amounted to about $100x10^6$ $m^3$ or 8% of the lake volume below 100 m in depth.

The plunging events into the deep hypolimnion of Urnersee are accompanied by a much larger event, the penetration of intermediate water from the neighboring basin (Gersauersee) into the deep hypolimnion of Urnersee (Figure 3.7*d*). Indirect traces of this overflow event were found by Wüest et al. (1988) by careful measurements of the vertical temperature and conductivity structure. Unlike the plunging events (which usually occur during floods), the lateral overflows do not manifest themselves by spectacular and easily mea-

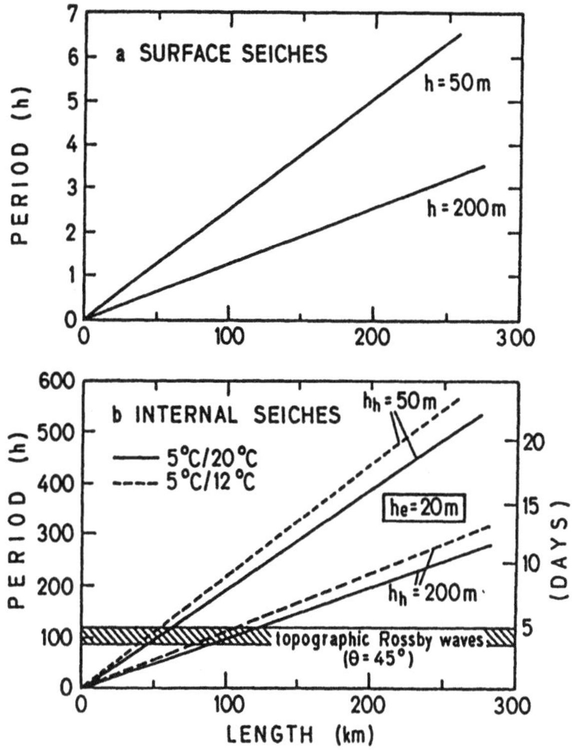


**Figure 3.4** First-mode periods calculated from Merian's equation of (*a*) surface seiches and (*b*) internal seiches, respectively: h= (total) mean depth; $h_e$ and $h_h$=mean depth of epilimnion and hypolimnion, respectively. The internal seiches are calculated for a hypolimnic temperature of 5°C and epilimnic temperatures of 12 and 20°C, respectively. The period of the first-mode topographic Rossby wave at 45° is shown for comparison.

surable changes at the lake surface, but their contribution to the hypolimnic mass flow can be substantial (Figure 3.9).

The daily interaction between meteorological forcing and mixing at the lake surface also belongs to the category of events. The diurnal cycles of heating and cooling cause a complicated temporal and spatial variation of the temperature distribution in the mixed layer (Imberger, 1985). In fact, with increasing spatial and temporal resolution of the measurements it becomes more and more evident that the common expressions *mixed layer* and *thermocline* are only valid in a coarse sense: Mixing always remains finite, structured, and changing. Among all mixing phenomena those in the mixed layer are probably the most relevant ones for ecology. A famous—but by no means the only or most important—example are the Langmuir circulation cells. For a long time, these mixing structures have stimulated scientists to study the impact of mixing on the distribution of biomass (see e.g., Riley, 1946). We will come back to this point in Section 3.4.

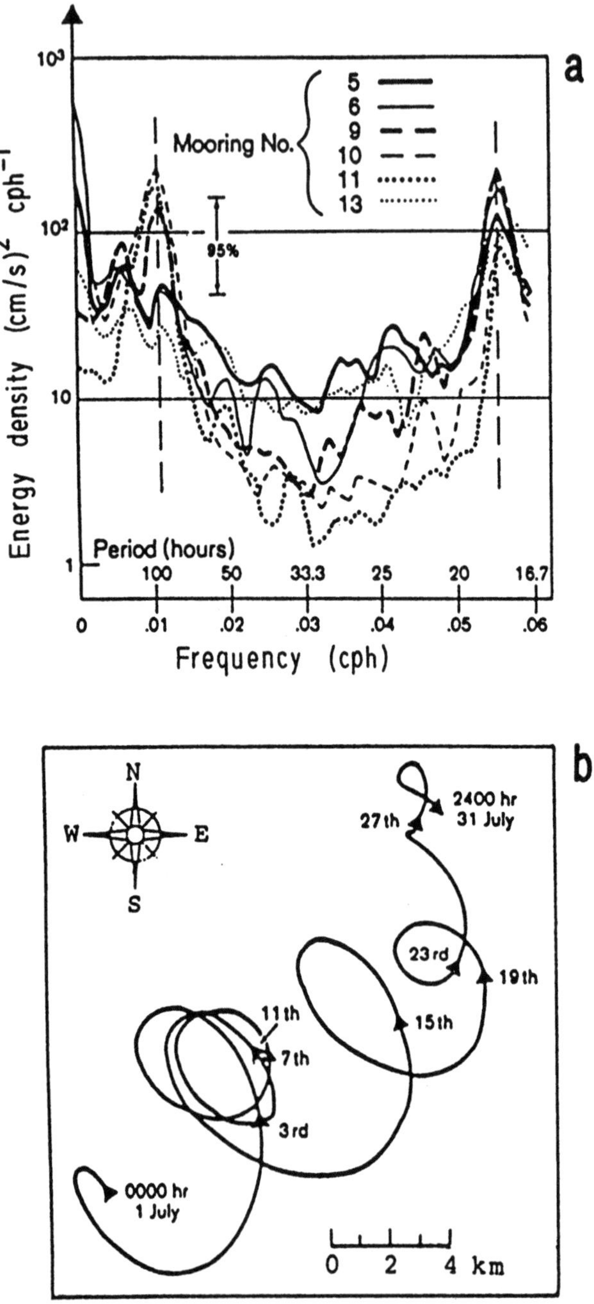


**Figure 3.5** (*a*) Kinetic energy spectra of the eastward velocity component recorded at the 25 m level at six stations in southern Lake Michigan. Vertical lines show the peaks which may represent the inertial mode (16.7 h) and the topographic Rossby wave ($\approx$100 h). (*b*) Hodograph of hourly averaged, low-pass filtered current velocities in Lake Michigan at 32 m depth. From Saylor et al. (1980).

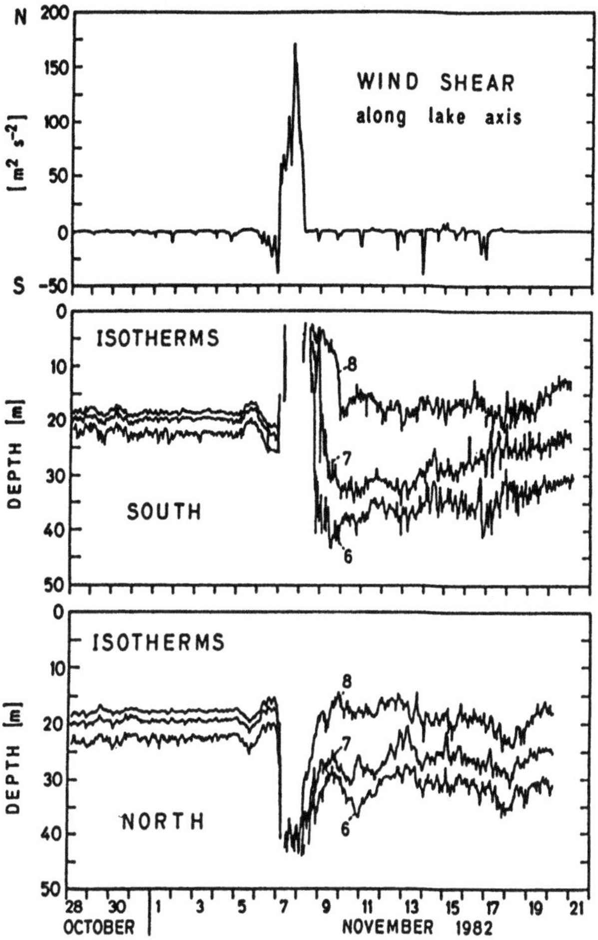


**Figure 3.6** Directed wind shear along the main axis of Lake Zug (Switzerland) and isotherm depths at two stations, South and North, calculated from thermistor chain recordings. A severe storm began on November 7, 1982, after a short period of wind blowing from the north. From Imboden et al., (1988).

**Vertical Versus Diapycnal Mixing** The preceding discussion on isotropic turbulence should not divert our attention from the fact that, regarding its ecological relevance, mixing in lakes is primarily unisotropic. In fact, typical horizontal mixing times are shorter by several orders of magnitude than vertical ones (see Table 3.3). In addition, all parameters which are important for biological processes, such as nutrient concentration and light intensity, are horizontally structured. An algal cell moving 10 m along the horizontal axis hardly feels any stress whereas the same distance in the vertical direction may make all the difference between life and death. The vertical mixing process thus justifies some additional considerations.

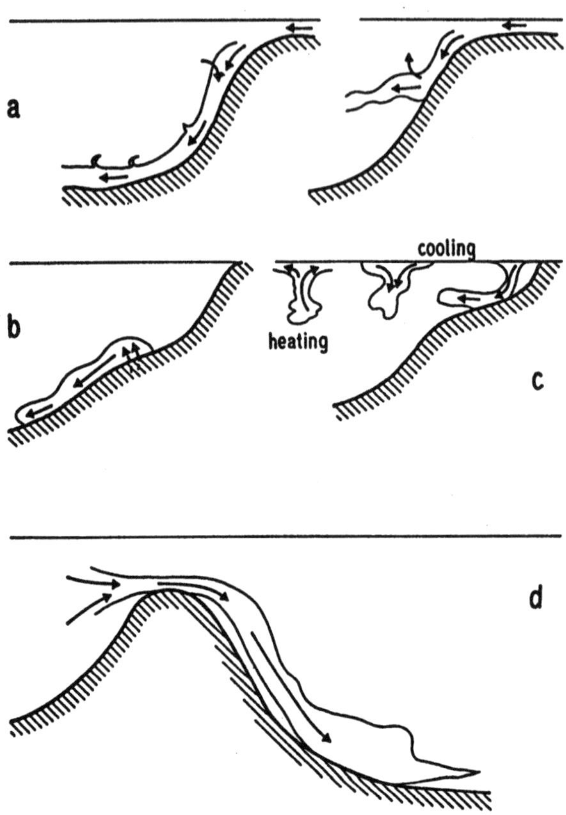


**Figure 3.7** Different types of density-induced mixing processes in lakes: (*a*) density currents (DC) at a river mouth; the river plume either follows the bottom (left) or spreads into an intermediate layer (right), according to the relative densities of river and lake water; (*b*) DC induced by flux of dissolved or particulate chemicals from the sediments into the overlying water; (*c*) DC induced by local cooling or heating; (*d*) DC induced between lake basins with different temperature regimes or different chemical compositions.

There are two kinds of vertical transport; one refers to the velocity component of the (space-fixed) vertical axis, the other to the local isopycnal surface (Figure 3.10). A current meter moored in a lake measures the first kind of movement, to which the term *vertical transport* really belongs. Transport across the isopycnals, "diapycnal mixing," causes water masses of different density to merge irreversibly. Since the isopycnal surfaces may be temporarily distorted due to internal waves, diapycnal mixing is not strictly vertical; it can also result from horizontal water movement.

The effects on plankton of these two kinds of transport are very different. The light intensity varies with the vertical distance from the water surface and thus depends on *vertical* transport alone (if the distortion of the water surface

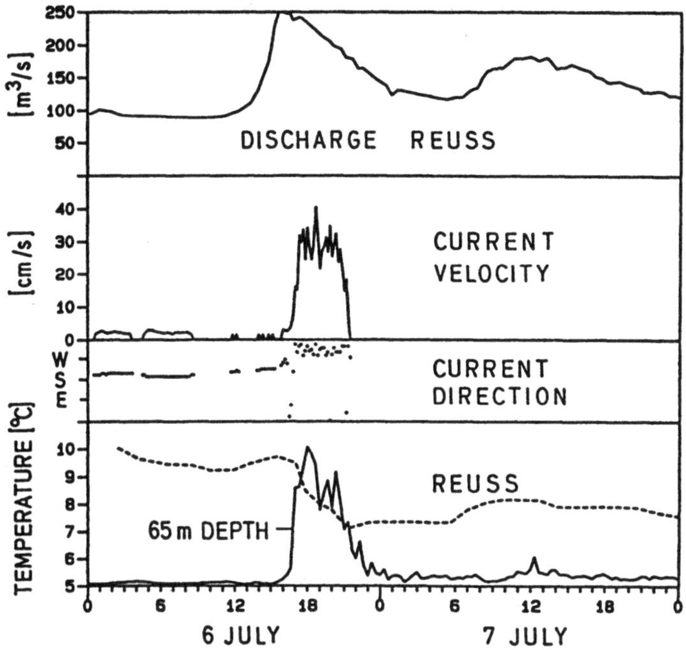


**Figure 3.8** Discharge of the River Reuss into Lake Lucerne; and water current induced by high discharge measured at 65 m depth at 300 m distance from the river mouth. Water temperature at 65 m depth temporarily rises above that of undiluted river water (dashed line), since the plunging river plume entrains warm water from the lake surface. From Wüest et al., (1988).

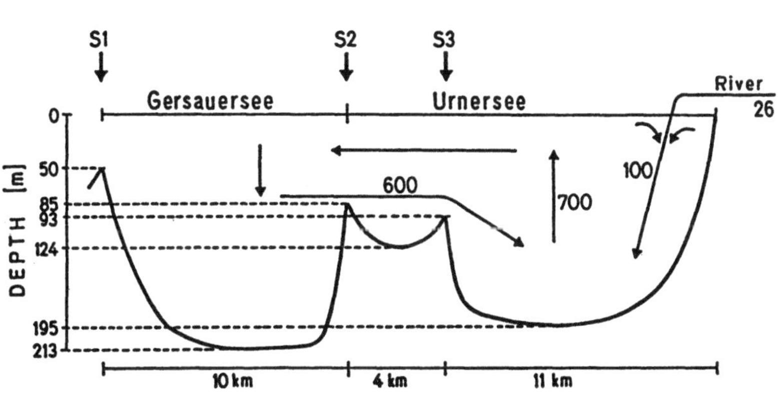


**Figure 3.9** Schematic representation of water-exchange fluxes in two adjacent basins of Lake Lucerne—Urnersee and Gersauersee—during May and June 1986. Since, below about 30 m depth, the water in Gersauersee is colder than in Urnersee and contains more dissolved chemicals, water from the upper hypolimnion of Gersauersee flows over the sills, S1 and S2, into the deep layers of Urnersee. In addition, the plunging events described in Figure 3.8 are also active. During the two months, the combined action of both processes causes a water exchange below 100 m of about 60%. Values are given in $10^6$ $m^3$. From Wüest et al., (1988).

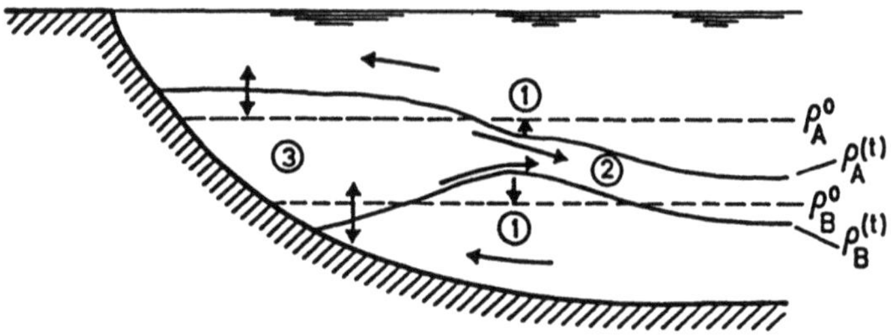


**Figure 3.10** Vertical and diapycnal transport: due to external energy input, the isopycnals $p_i(t)$ (surfaces of constant density) fluctuate around their equilibrium positions $\rho_i^o$, thus causing both vertical (1) and horizontal (2) currents, which have to be distinguished from diapycnal mixing (3), i.e., mixing across isopycnals.

by waves is neglected for simplicity). As we have seen for the case of Lake Zug (Figure 3.6), the vertical amplitudes can be enormous, but most of the movement is reversible—even for such a severe storm as in the example—and thus it is not effective as *diapycnal* transport.

In contrast, water temperature, concentration of dissolved chemicals, and other characteristic properties of the water vary because of *diapycnal* mixing. In the preceding sections several mechanisms were discussed that can lead to diapycnal mixing (e.g., density currents). The long-term effect of all these processes is often integrated into a coefficient of vertical eddy diffusion, $K_z$. Typical values averaged over several weeks were given in Table 3.3. Yet, if calculated for time intervals of hours, i.e., for periods dominated by isolated—e.g., meteorological—events, much larger $K_z$-values can be found (Wüest, 1987).

## 3.3 Physical Characteristics of Phytoplankton

Algae can be characterized by typical rates describing their growth and destruction. Growth rates, the rate at which new biomass is produced per unit time relative to the standing crop of biomass, strongly depend on light, water temperature, and algal species and lie typically between $10^{-1}$ and 1 $d^{-1}$ (Table 3.5). However, uptake of nutrients such as phosphate can occur at much higher rate than growth; the cells are able to store the nutrients until they are used to produce new biomass. Rates of the reverse processes, respiration and mineralization, are generally about one order of magnitude smaller than growth rates. The settling velocity and—for zooplankton and certain kinds of phytoplankton—the velocity of active motion are other important properties for assessing the interplay of physical and biological processes.

Yet, the coarse estimate for these parameters summarized in Table 3.5 conceals the fact that it is the *difference* between various planktonic species that provides the ecological niches which are needed in order to guarantee survival of several species at the same time. According to Reynolds (1984)

**Table 3.5** Parameters describing growth, destruction, and movement of phytoplankton

| Characteristic or parameter | Typical range |
|---|---|
| Growth rate | 0.1 to 1 $day^{-1}$ |
| Nutrient uptake rate | 1 to 100 $day^{-1}$ |
| Respiration/mineralization rate | 0.01 to 0.1 $day^{-1}$ |
| Sinking rate | 0.1 to 1 m/day (extreme values, 0 to 10 m/day) |
| Active movement of zooplankton | Up to 30 m/half-day |
| Time of photoresponse[a] | 0.02 to 0.5 h |
| Time of photoadaptation[b] | |
| From dark to light | 4 to 6 h |
| From light to dark | 2 to 6 wk |

[a] The term *photoresponse* is used for the response of photosynthetic rate P to changing light intensity I along a fixed P(I)-curve. In contrast, *photoadaptation* is the biological adjustment to variable light conditions of changes, such as in photosynthetic pigments and chemical composition of the plankton cell, which eventually lead to a new P(I)-curve (see Figure 3.12). Value from Harris and Piccinin, 1977.
[b] Values from Gallegos et al., 1983.

algae can be characterized by a few main variables describing their selection with respect to environmental conditions. As shown in Figure 3.11, adapted from Reynolds (1987), two variables are of special importance, nutrient concentration—or nutrient stress in Reynolds' terminology—and mixing rate or depth. The second property, mixing rate or depth, is directly linked to the way in which a given algal species adjusts its rate of photosynthesis to absolute level and temporal variation of light intensity, I(t). If mixing depth is small, the algae are not exposed to large and quick changes in I(t)—except for the diurnal and thus predictable changes. Zones of different light level may then co-exist (i.e., the mixed layer and various zones in the thermocline), thus providing niches for several species at the same time. In contrast, if mixing extends to large depths and is intensive, the algae are continuously moving through zones of different light levels; there is not enough time for algal cells to develop different light or growth characteristics. The species which survive in this environment must be able to quickly adapt to the changing light intensity, i.e., they must be able to take immediate advantage of strong light and to survive on little light when transported to greater depth.

Because of its direct impact on photosynthesis, light is truly the key parameter for the development of the aquatic ecosystem. Based on the considerations discussed in Section 3.2, we may expect to find a close connection between vertical (i.e., mode-type) movement, light variation, and photosynthetic rate. From laboratory studies it is known that for constant environmental conditions (water temperature, nutrient concentrations, etc.) a given phytoplankton species exhibits a fixed relationship between productivity P and light

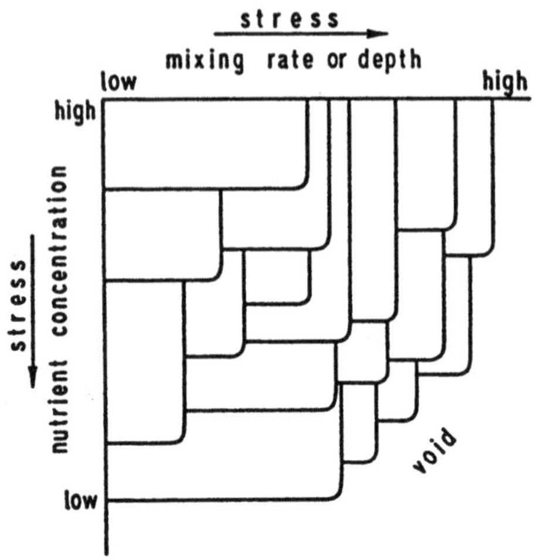


**Figure 3.11** Schematic view of selection of phytoplankton by two environmental stress parameters, mixing rate or depth and nutrient concentration. The extent of tolerance of each species group (represented by a rectangle) is described by the position of the lower right corner of each rectangle. Redrawn from Reynolds (1987).

intensity I, the so-called P(I)-curve (Figure 3.12). When the light varies, the rate changes accordingly. However, the change in P may be delayed with respect to the change in I or even go through some complex temporal variation before reaching a new steady state. We call this process *photoresponse* in order to distinguish it from *photoadaption* discussed below.

Figure 3.13 gives an example of the photoresponse process from an experiment by Harris and Piccinin (1977) in which algae kept in the dark were suddenly exposed to different light intensities and their rate of photosynthesis (measured by oxygen production) determined as a function of time. In these experiments it takes about 5 to 30 minutes until photosynthesis reaches a new

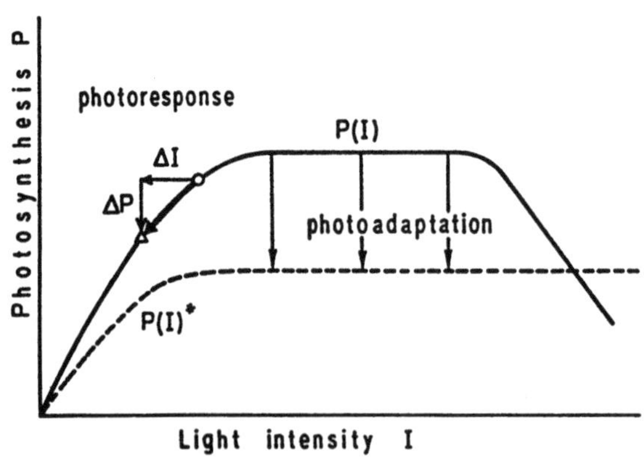


**Figure 3.12** Schematic representation of the relation between rate of photosynthesis P and light intensity I: *Photoresponse* means the change of P with I along a given P (I)-curve; *photoadaptation* is the biological adjustment of the algae to a new light climate, i.e., the development of a new curve, P(I)*.

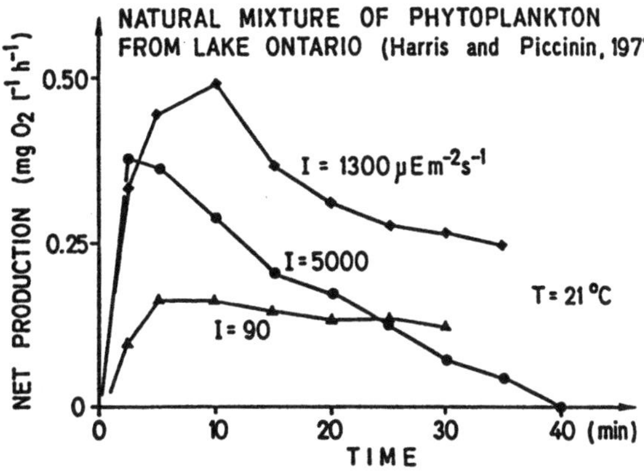


**Figure 3.13** Temporal variation of net photosynthesis for different light intensities. In all experiments, the algae were kept in the dark for 5 minutes and then exposed to differing I-values as indicated in the figure. Photosynthesis rate was measured by the rate of oxygen production. From Harris and Piccinin (1977).

steady value. The response time grows with increasing light intensity. For large intensity ($I=5000\ \mu E\ m^{-2}s^{-1}$) the rate drops to zero after some initial photosynthetic activity. A striking feature of the experiments is the growth optimum that occurs 5 to 15 minutes after exposure to light. In connection with vertical migration of an algal cell due to turbulence this may be important, provided that typical cycling times fall into this range. Löhr (1984) has used these characteristics of photoresponse to model productivity in the impoundments along the River Neckar.

A second process to be distinguished from photoresponse is the biological adjustment of the biomass to new light conditions, the "photoadaptation process." This process becomes important, for instance, if algae grown in the thermocline of a lake at moderate light intensities are suddenly brought to the surface. At the beginning, the photosynthetic apparatus of the cells cannot make use of the high light intensities—to the contrary: The cells may even be inhibited by the light, since their P(I)-curve is still tuned to low light intensities. After some time of adjustment, however, a new P(I)-curve evolves (see Figure 3.12).

Experiments by Gallegos et al. (1983) show that the photoadaptation of algae brought from a high to a low light environment takes about 2 to 6 weeks. For the reverse case, however, for algae brought to a higher light level, the adaptation took only 4 to 6 hours.

## 3.4 Life in Motion

In Section 3.2, a distinction was made between mode-type and event-type mixing processes. The modes were characterized by the time and length scales over which they are effective. Some of these scales are of similar order as the scales describing the dynamics of growth and destruction of plankton. Due to a general principle which is based on theoretical and experimental evidence, the distribution of chemical or biological species is modified by the mixing phenomena of the aquatic environment—provided that the (chemical or biological) reactivity is of similar temporal scale as that of the mixing. The consequences of this observation for the aquatic ecosystem will be discussed in this section.

**Variance of Phytoplankton Distribution** Based on the theory for the behavior of a conservative substance in a stationary, isotropic, turbulent environment developed by Batchelor (1959), Corrsin (1961, 1962) has calculated the spatial spectrum of a tracer undergoing first-order reaction in an isotropic, turbulent field. Denman et al. (1977) have applied this theory to the chlorophyll fluctuation spectrum in the sea. The starting point is the diffusive–advective tracer equation:

$$\partial C/\partial t = D\Delta C - \mathbf{v}\Delta C - RC \tag{3.2}$$

where D is the molecular diffusion coefficient; C, the tracer concentration; **v**, the vector of current velocity; and R, the first-order reaction rate constant. Due to Kolmogorov's turbulence hypothesis (see Section 3.2) the spectrum of the velocity variance in the inertial subrange has the form:

$$E_v(k) = \alpha\, \epsilon^{2/3} k^{-5/3} \tag{3.3}$$

where k is the wave number, $\epsilon$ is the dissipation rate of turbulent kinetic energy (see Table 3.3), and the dimensionless constant $\alpha$ is about 1.5. For large wave numbers—i.e., for fast temporal changes—the tracer follows the motion of the water passively since the reaction is too slow to alter the tracer concentration significantly. In this k-range, the so-called "passive range," the concentration fluctuation spectrum of the tracer, $E_C(k)$, is proportional to $E_v$ (Figure 3.14). For large k-values the passive range is limited by the scale of viscous dissipation: For the case of water movement (turbulence), this scale is given by $k_v = (\epsilon/\nu^3)^{1/4}$ ($\nu \sim 10^{-6}$ $m^2s^{-1}$ is kinematic viscosity); for the concentration of solutes, the scale is given by $k_D = [\epsilon/(\nu D^2)]^{1/4}$ ($D \sim 10^{-9}$ to $10^{-10}$ $m^2s^{-1}$ is molecular diffusivity). At smaller k-values, the turbulent velocity field becomes anisotropic because of the final geometry of the mixed layer. In addition, the reactivity causes the spectra $E_C(k)$ and $E_v(k)$ to diverge for wave numbers larger than $k_R = |R|^{3/2}\epsilon^{-1/2}$. The critical length scale $k_R^{-1}$, above which tracer-specific spatial patterns are to be expected, lies between about $10^3$ m for very reactive tracers at low turbulence and $10^5$ m for slowly reacting tracers at high turbulence (see Figure 3.14).

As shown by Denman et al. (1977), for a tracer with an *in situ* source ($R<0$) as chlorophyll, the spectrum $E_C(k)$ should reach a characteristic maximum at about 1.3 $k_R$ and then sharply fall off at smaller k-values. Powell et al. (1975) have measured both the current and the chlorophyll spectra in Lake

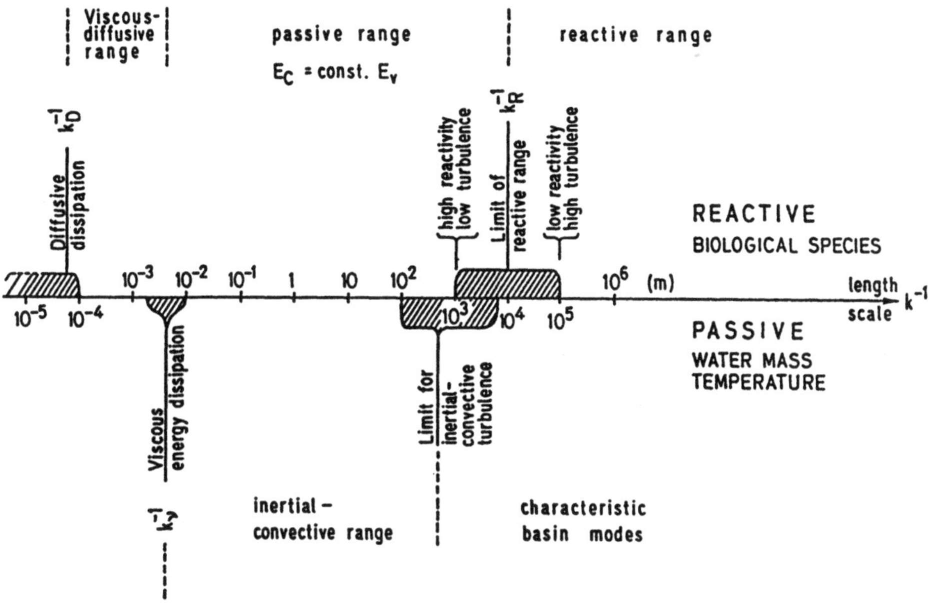


**Figure 3.14** Ranges of the fluctuation spectrum for water movement, or the passive parameter $E_v(k)$, and for a reactive tracer undergoing a first-order reaction $E_C(k)$, respectively, based on the theory of Corrsin (1961, 1962). Note that the scale refers to inverse wave numbers. See text for explanations.

Tahoe (California/Nevada). In both spectra they found a Kolmogorov slope of $-3/5$ for length scales smaller than 100 m (Figure 3.15). The investigators tentatively attribute the peak found in the chlorophyll spectrum at $k \approx 10^{-2}$ $m^{-1}$ to Langmuir circulation and believe that their current meters were not sensitive enough to pick up the Langmuir flow pattern in the velocity spectrum, as well. However, in the light of the theory of Denman et al. (1977), the peak could also result from phytoplankton growth.

The energy dissipation rate $\epsilon$ is estimated from $E_v(k)$ in the inertial range using eq. (3.3) to be $3 \times 10^{-8}$ $m^2s^{-3}$. With $k_{max} = 1.3k_R = 1.3k|R|^{3/2}\epsilon^{-1/2}$, the corresponding growth rate for phytoplankton becomes $|R| = 1.2 \times 10^{-4} s^{-1}$ or about 10 $day^{-1}$. This is certainly an unrealistic production rate. Yet, if $\epsilon$ were reduced to $10^{-9}$ $m^2s^{-3}$—the above value would seem to be quite large for a lake—indeed, the production rate would drop to about 3 $day^{-1}$ which is certainly still high but not completely unrealistic anymore. In addition, the discussion on photoresponse (see Section 3.3) has shown that net growth of biomass is not the only reaction rate characterizing the dynamics of plankton. Other response times, such as the parameters describing photoadaption (see Table 3.5), may also influence the measured patchiness of chlorophyll in Lake Tahoe. As a logical next step, the interaction between turbulence and photosynthesis will thus be discussed in more detail.

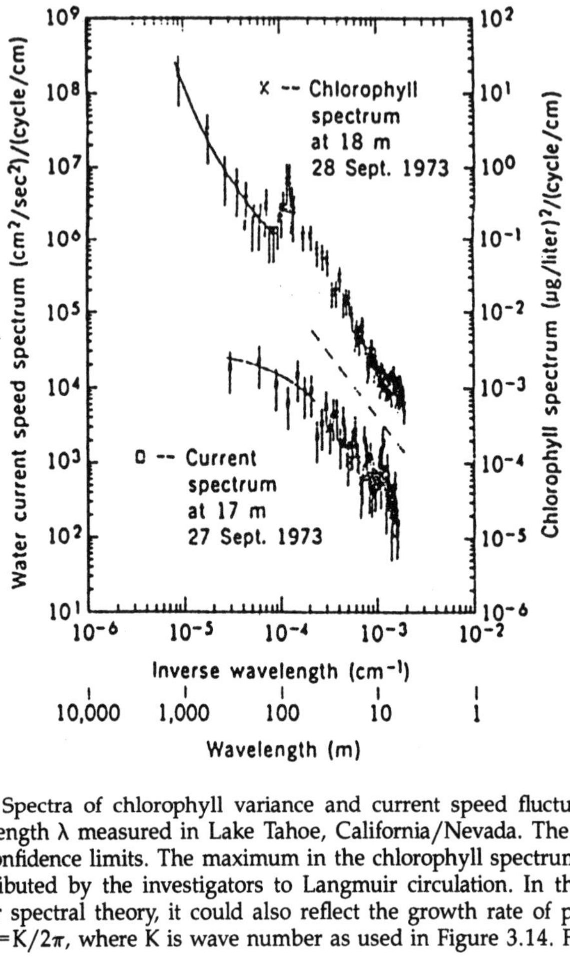


**Figure 3.15** Spectra of chlorophyll variance and current speed fluctuations against inverse wavelength $\lambda$ measured in Lake Tahoe, California/Nevada. The error bars indicate 80% confidence limits. The maximum in the chlorophyll spectrum at about $\lambda$= 100 m is attributed by the investigators to Langmuir circulation. In the light of the reactive tracer spectral theory, it could also reflect the growth rate of phytoplankton. Note that $\lambda^{-1}=K/2\pi$, where K is wave number as used in Figure 3.14. From Powell et al. (1975).

**Turbulence and Photosynthesis** The question of whether turbulence influences productivity in the ocean has puzzled marine scientists for a long time. It is a controversial subject on which one can find quite opposite views in papers published nearly simultaneously and even by scientists working at the same institution. For instance, Lewis et al. (1984) concluded that "dissipation of turbulent kinetic energy . . . was a strong determinant of algal photosynthetic rate in the upper mixed layer." These experiments were carried out in Bedford Basin, a coastal marine inlet near Halifax, Nova Scotia. Two years earlier, Gallegos and Platt (1982) concluded that ". . . for these stations [in the eastern Canadian Arctic], vertical mixing had little quantitative effect on water column primary production." A similar conclusion was drawn by Falkowski and Wirick (1981), based on a simulation model which included the effect of photoadaptation (or rather *photoresponse* in our terminology) and vertical mixing on primary production.

Of course, these results must not necessarily contradict each other since they apply to very different situations and are based on different measuring techniques, and on different "analytical windows." The paper by Lewis et al. (1984) is certainly based on the best experimental evidence. It compared photosynthetic rates for samples taken at the water surface with those at the base of the mixed layer. The authors found that the difference between the two rates disappeared with increasing turbulent kinetic energy dissipation. The result is consistent with the concept of photoadaptation: If vertical mixing exceeds a certain rate, the plankton becomes homogeneous with respect to the P/I-response. However, if mixing is slow, individual plankton communities can develop at different P(I) rates.

Lewis et al. (1984a) have further developed the theories to quantify the relation between vertical mixing and production. In their model, two characteristic, nondimensional numbers are used, the *mixing–photoadaptation ratio:*

$$N_1 = K_z/(h_{mix}^2\gamma) \qquad (3.4)$$

where $K_z$ is the vertical eddy diffusivity, $h_{mix}$ is the mixing depth, and $\gamma$ is the photoadaptation rate constant. This expression measures the ratio of the time of vertical mixing from top to the bottom of the mixed layer to the time needed for photoadaption. If $N_1$ is larger than one, the plankton should be homogeneous with respect to P/I-response.

A second number defined as

$$N_2 = k_s h_{mix} \qquad (3.5)$$

where $k_s$ is the attenuation coefficient for irradiance, measures the degree of homogeneity of light intensity within the mixed layer. In fact, the ratio between irradiance at the bottom and at the top of the mixed layer is $\exp(-N_2)$. Thus, the smaller $N_2$ is, the more homogeneous the irradiance.

The two numbers are used in a one-dimensional, vertical reaction–diffusion model for photosynthesis including photoadaptation. Figure 3.16 summarizes the results in a qualitative way: If $N_1$ exceeds a critical value the phytoplankton is dominated by turbulence. The critical limit itself grows with $N_2$, since the bigger the light gradient across the mixed layer, the more important mixing becomes in order to homogenize the plankton with respect to photoadaptation.

The result of this model analysis confirms the experiments by Lewis et al. (1984) mentioned before. In fact, the vertical diffusivity $K_z$, i.e., the position along the horizontal axis in Figure 3.16, is directly related to the energy dissipation rate $\epsilon$ (Oakey, 1982): $K_z \approx 0.25\ \epsilon/N^2$. This explains why—for a fixed "light gradient number" $N_2$—the effect of photoadaptation disappears with increasing $\epsilon$. It is worth noting that the model is also related to Reynolds' plankton characterization (see Figure 3.11): The position in Figure 3.16 relative to the curve of equal significance also determines the position on Reynolds'

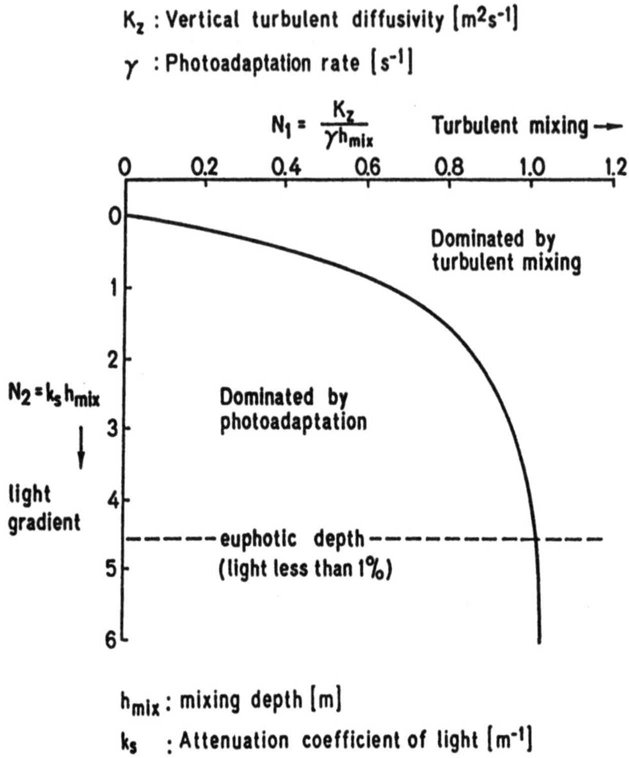


**Figure 3.16** Result of the one-dimensional, vertical diffusion—reaction model for photosynthesis and photoadaptation. Depending on the size of two nondimensional numbers defined in eqs. 3.4 and 3.5, production in the water column is dominated either by turbulence (algal population homogeneous with respect to P/I-response) or by photoadaptation (inhomogeneous population of algae). Note that in the terminology introduced in Figure 3.12 the experiment refers to *photoresponse* rather than to *photoadaptation*. Redrawn from Lewis et al. (1984a).

axis of mixing stress. For algae with large mixing-stress tolerance we expect to find large rates of photoadaptation $\gamma$, thus making the domain of turbulent mixing dominance larger.

Experimental evidence for the connection between biological and physical forcing is sparse. It requires sets of data of the relevant parameters over appropriate time periods (years, seasons, weeks, or days) and with sampling intervals which are correctly tuned to the presumed mechanisms. Whereas for the physical parameters, such data are relatively easy to find for oceans, they are rare for the marine biology parameters and virtually absent for lakes. Horne and Platt (1984) give a few examples for suitable biological time series, among which probably the best is the series of *daily* phytoplankton counts made on samples collected at the Scripps pier (San Diego, California) from 1920 to 1939.

To my knowledge nothing comparable exists for lakes. Though there are long time series of biological parameters for several lakes, these data are usually neither detailed or homogeneous enough to study possible interactions between meteorology, physics, and ecosystem response. As an example, I have analyzed the extensive data collected by the International Commission for Lake Constance (Bodensee). In Figure 3.17 some relevant parameters are plotted for the years 1980 to 1985. One striking feature is the yearly variation in phytoplankton biomass; for no obvious reason, two productive years (1981 and 1984) enclose two less productive ones (1982 and 1983). The mean phosphorus concentration decreased by about 20% during these four years, but one of the more productive years occurred at the end of the period. From a climatic point of view, 1984 was colder than other years. The surface water temperature rose later than usual, indicating the persistence of vertical mixing in the spring. This could explain the larger production in 1984 as a consequence of better vertical nutrient recycling in the first half of the year. However, the warming

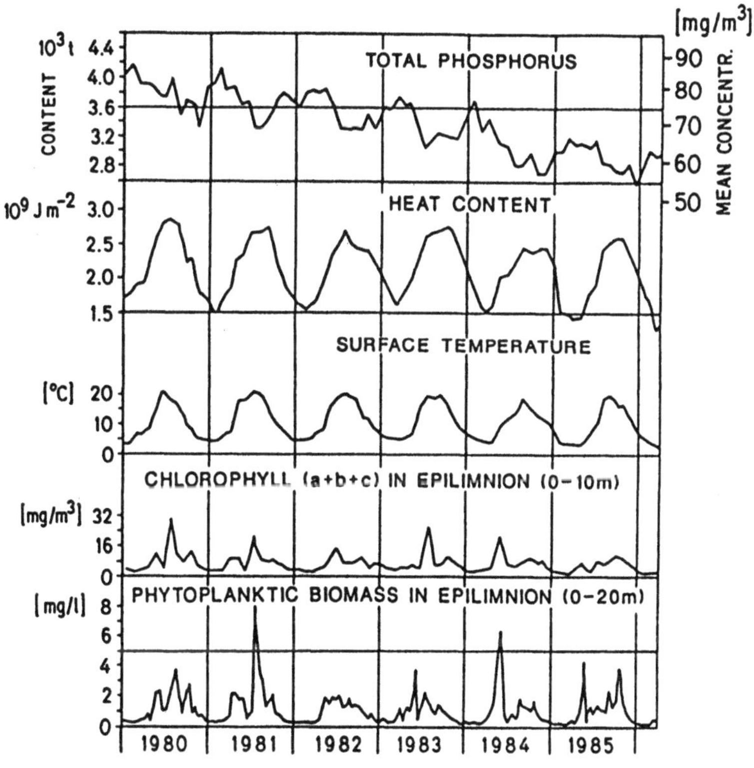


**Figure 3.17** Time series for phosphorus concentration, heat content, surface temperature, chlorophyll (a+b+c) concentration, and phytoplankton biomass for Lake Constance—data from the center of the main basin. From the Annual Report No. 13 (1987) of the International Commission for Lake Constance (Bodensee).

in 1981 was normal, but production still high. The phenomenon remains unexplained though the existence of mysterious biomass variations cannot be denied, either for Lake Constance nor for many other lakes. It is apparent that a wide field of interdisciplinary work awaits attention!

**Biological Patchiness** Besides transport by turbulence, there are other mechanisms which influence the spatial distribution of plankton. One of them is related to the sinking or active movement of plankton. Several authors (Stavn, 1971; George and Edwards, 1973; Ledbetter, 1979) have studied the combined influence of Langmuir circulation and active motion of plankton on the three-dimensional distribution of zooplankton. As mentioned already, Ledbetter used an analytical description for the velocity distribution in the Langmuir cell. In his model as shown in Figure 3.18, he assumes that the zooplankton can actively stay in a certain zone whereas phytoplankton sinks through the cell. In this way patchiness of plankton can be explained.

Another mechanism for phytoplankton patchiness has been proposed by Riley (1976); it is based on the different grazing rates of zooplankton. In Riley's model, the zooplankton migrates vertically in a depth-dependent horizontal advection field which is driven by tidal forces. Since zooplankton migration is controlled by the sun, the continuous phase shift relative to the tidal motion (which has a different period) results in a complex pattern of grazed phytoplankton patches. For lakes, periodic flow patterns such as internal seiches or Kelvin waves, especially if their periods are close to 24 h or some multiple of it, could lead to a similar pattern.

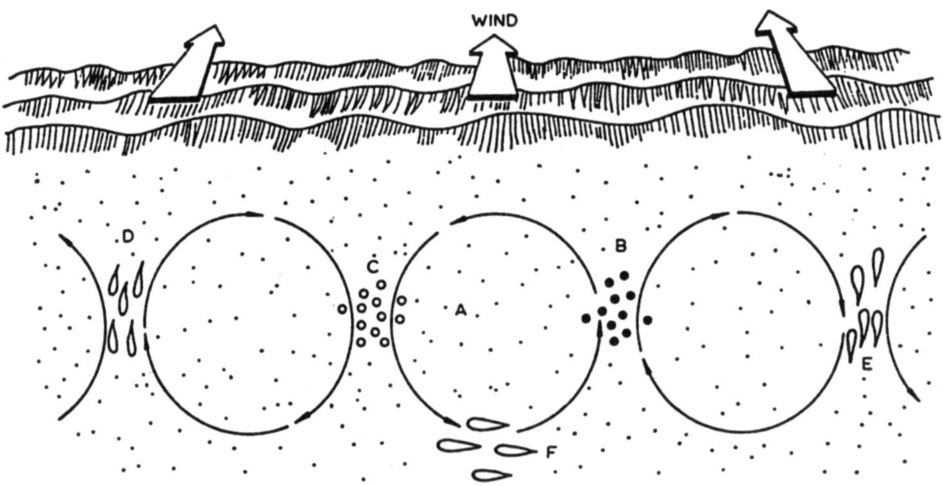


**Figure 3.18** Langmuir circulation and plankton distribution: A, randomly distributed neutrally buoyant particles; B, sinking particles aggregated in upwelling zone; C, floating particles concentrated in downwelling zone; D, E, and F, zooplankton aggregation positions determined by the velocity field in the cell. Redrawn from Stavn (1971) by Ledbetter (1979).

Probably the most important cause of patchiness is related to the non-linear nature of biological processes. In fact, it is easy to show that some initial fluctuation in the concentration field of a parameter is accentuated by a non-linear reaction. Note that such fluctuations are intrinsically related to turbulence. A simple model developed by Steele (1974) serves to illustrate the process: He takes the familiar Lotka-Volterra model for predator—prey interaction (a non-linear model, indeed) and combines it with a spatial diffusion term:

$$\partial P/\partial t = aP - f_1 PH + K_x \, \partial^2 P/\partial x^2 \qquad (3.6a)$$

$$\partial H/\partial t = -bH + f_2 PH + K_x \, \partial^2 H/\partial x^2 \qquad (3.6b)$$

Here, P is the concentration of prey (phytoplankton) with growth rate a; H is the concentration of predator (herbivorous zooplankton) with death rate b; $f_1$ and $f_2$ describe the feeding of H on P; and $K_x$ is the horizontal eddy diffusivity. Assume that at some time t, P and H have an arbitrary spatial variance which we express by the Fourier transformed functions $P^*(k)$ and $H^*(k)$. Their time rates of change are:

$$\partial P^* / \partial t = (a - k^2 K_x) \, P^* - f_1 F^* \qquad (3.7a)$$

$$\partial H^* / \partial t = -\,(b + k^2 K_x) H^* + f_2 F^* \qquad (3.7b)$$

$$F^* = dk' \, P^*(k') \, H^* \, (k-k') \qquad (3.7c)$$

Let us first discuss the corresponding solution without prey—predator interaction ($f_1, f_2 = 0$). For this case the equations are linear. Eq. 3.6*a* describes the autocatalytic growth of phytoplankton under the simultaneous influence of diffusion. From eq. 3.7*a* it follows that only those k-modes of P can grow which fulfill the condition

$$k < (a/K_x)^{1/2} \qquad (3.8)$$

$P^*(k)$ can be interpreted as a phytoplankton patch with size $L = k^{-1}$. Thus, another way to interpret eq. 3.8 is to state that a plankton patch can only grow despite dilution by dispersion if the initial patch size L exceeds a critical size $L_{crit}$:

$$L > L_{crit} = (K_x/a)^{1/2} \qquad (3.9)$$

This is the famous critical-length scale concept first introduced by Kierstead and Slobodkin (1953). Remember that the variable a is the growth rate of the prey.

Since the mathematical structure of more complex ecosystems (i.e., systems with several preys and predators) is similar to that of the simple two-species equations, eqs. 3.6*a* and 3.6*b*, critical length scales can also be developed for higher trophic levels. Typically, growth rates a decrease and thus the characteristic scales increase with the trophic level of the species. This was shown by Steele (1978) in a graph which I have modified by including typical mixing modes of lakes (Figure 3.19). With respect to the comparison between patchiness of biomass of different trophic level and (oceanic) turbulence (Okubo, 1971), we can see that there is a rough equivalence between the scales of physical dispersion and the biological scales associated with the growth rate of organisms. According to Okubo, horizontal eddy diffusivity increases with increasing length scale of the "tracer cloud" being dispersed. As a consequence, time scale t and length scale l of (passive) diffusion are related (see Figure 3.19) by the following equation:

$$l = 1.2 \times 10^{-3} t^{1.15} \qquad (3.10)$$

when l is in meters and time is in seconds. Thus, the patchiness of biological species (phytoplankton, zooplankton, and fish) can also be interpreted as the consequence of the different time scales of growth, $a^{-1}$, of the different trophic levels.

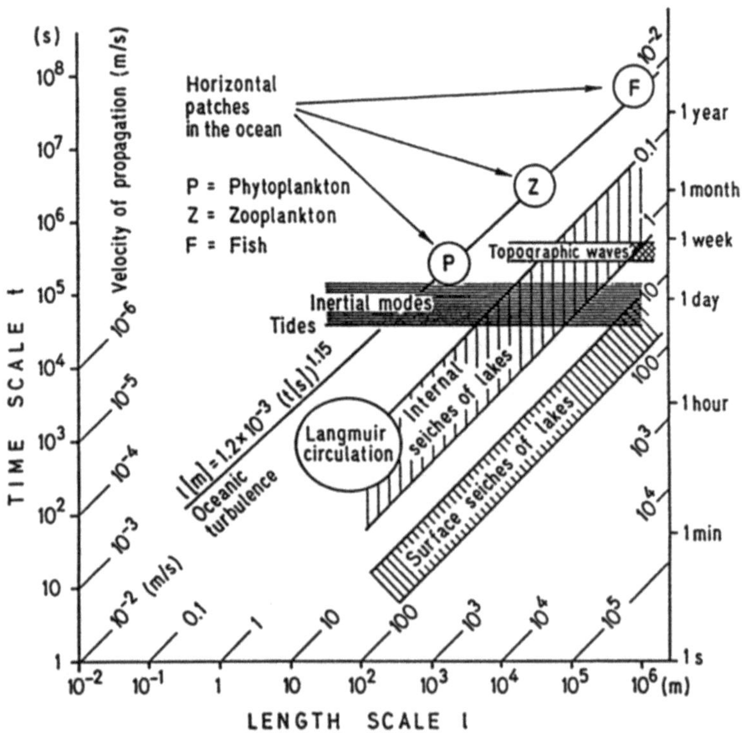


**Figure 3.19** Typical length and time scales of mixing modes in lakes. The diagonals with slope equal to one represent lines of constant velocity of propagation (e.g., surface seiches moving with velocities between about 10 to 100 m $s^{-1}$). Included is Okubo's (1971) relationship between length scale l and time scale t for oceanic turbulence. Typical differences among scales of biological patchiness—for phytoplankton, zooplankton, and fish—taken from Steele (1978) can be interpreted as the direct consequence of the influence of the turbulence scales on species of different growth rates.

The term involving the interaction between P and H, $F^*(k)$, causes an entirely different behavior of the system: $F^*$ changes the stability criterion for growth or decay, mixes the different k-modes, and thus creates new structure in the ecosystem.

No doubt, with these considerations we enter a field of enormous complexity which deals with the interaction of mixing with whole ecosystems, not just with a single species. For many reasons field data are not available—and perhaps may never be! Thus, we are restricted to theoretical considerations and model calculations, but they may be instructive, as well.

As an example, we mention the work by Dubois (1975) in which Steele's model is further developed. Instead of using the Fourier transformation technique, Dubois splits P and H into two parts, mean value and fluctuation. Eqs. 3.6*a* and 3.6*b* are linearized by neglecting the correlation between P and H. In two dimensions, the solution of the model has the form of concentric rings of prey and predator concentration maxima running from the center to the periphery ("prey—predator waves"). Similar results were found by Hilborn (1979) and by Parker (1978) who added the concept of a limiting nutrient to the model.

**Turbulence and Diversity** Probably, the most striking concept of all the theories on the turbulence—ecosystem interaction deals with the question of phytoplankton diversity. Due to the principle of "competitive exclusion" (often called Gause's hypothesis) no two species can occupy the same ecological niche. As stated in the introduction to this chapter, the open water column represents a system with few niches. Hutchinson (1961) concluded that the great diversity observed in aquatic ecosystems contradicts the exclusion principle.

At first sight, mixing of the water column seems to reduce diversity, rather than to increase it, thus making Hutchinson's paradox even more striking. Imagine a well-stratified water body with little vertical mixing; the water layers at different depths represent distinct habitats characterized by different light, temperature, and nutrient conditions. At the onset of a mixing event, these habitats may be destroyed and replaced by a new, deeply mixed water column providing less habitat space than before and reducing species diversity.

Yet, this is only one aspect of mixing, the direct one. As discussed in Section 3.2, an important feature of mixing is its variation in time and space. Mixing is intermittent, random, and turbulent in a general sense. It is hypothesized that the variability of mixing creates niches (although mixing itself may reduce habitats). Here the term *niche* is used in a general way expressing the different tolerance ranges of species with respect to *many* environmental parameters (including nutrients, of course). To express it in simpler terms: Turbulent mixing increases the number of species which can co-exist in a given aquatic environment.

Two examples are given to support the hypothesis, a theoretical and an experimental one. Kemp and Mitsch (1979) have used mathematical ecosystem simulation to demonstrate that turbulence in the water column can alter the interaction pressure between plankton populations in such a way as to allow co-existence of species competing for the same resources. Their model calculations show that plankton species which would not be able to exist simultaneously in the same aquatic environment, were able to coexist as soon as turbulent water movement was added. Laminar flow provided less additional niches than turbulent flow does.

Finally, Ulrich Sommer has drawn my attention to investigations on non-steady state competition; they provide experimental evidence for the mixing hypothesis. Gaedeke and Sommer (1986) have studied the influence of periodic disturbances of different frequency on the maintenance of phytoplankton diversity, as follows: Algal communities were grown in incubators which were diluted at different but regular intervals. The long-term average of all dilution rates was constant. Two nutrients were limiting. Species diversity was determined according to Shannon and Weaver (1949) and plotted against the non-dimensional dilution interval, i.e., the interval divided by the generation time of the plankton (Figure 3.20).

Under these conditions, species diversity shows an interesting behavior: if the dilution interval is smaller than the generation time, the system develops towards the steady state of a continuous culture with diversity in accordance with the competitive exclusion principle. The plankton "feels" the dilution as

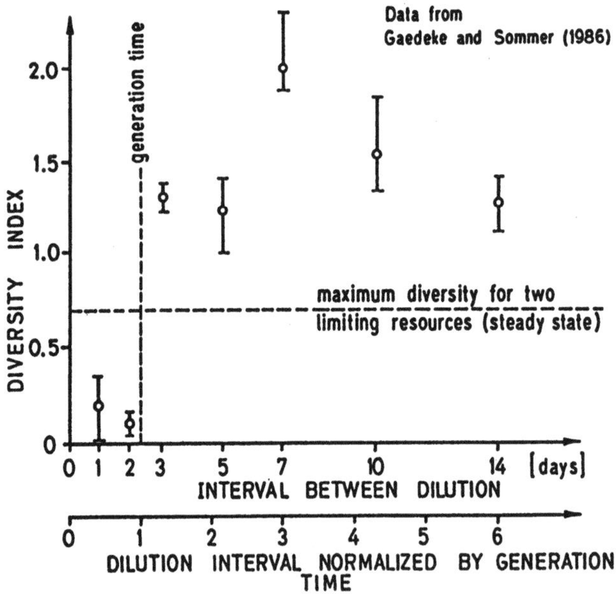


**Figure 3.20** Diversity of phytoplankton in incubator experiments as a function of the dilution interval normalized by the phytoplankton generation time. The vertical bars indicate the total range of diversity at the given dilution interval. After Sommer (1989). Data from Gaedeke and Sommer (1986).

continuous perturbation. If the dilution interval becomes larger than the generation time, diversity increases. The physical perturbation allows co-existence of species which would exclude each other if only the physical environment would be persistent enough in time.

The typical interval between storms or the passage of low-pressure areas over a lake are often in the order of a few days. The storms may have the same effect on the phytoplankton community as the intermittent dilution of the incubators did. As Gaedeke and Sommer concluded, the observed phenomenon seems to be a special form of Conell's (1978) "intermediate disturbance hypothesis" which states that disturbances have a positive influence on the species diversity if they are "intermediate" in frequency and intensity.

## 3.5 Summary and Conclusion

Several factors distinguish life in the water from life on land. The most important among the physical factors is the difference in the leading forces: land ecosystems are dominated by gravity, aquatic (pelagic) ecosystems by kinetic forces. Partial loss of control over position due to mixing and turbulence is one of the crucial environmental factors for the growth of phytoplankton.

Transport and mixing in lakes can be discussed in terms of the time and length scales which are typical for the different processes of water movement occurring in enclosed basins. The scales extend from the basin modes (seiches, Kelvin and Poincaré waves, and topographic waves) which, in large lakes, have typical periods between days and weeks, down to the small-scale turbulence modes, with scales of seconds and millimeters. In between lies the range of inertial turbulence, which has a universal shape in lakes and oceans, but differs in absolute intensity (expressed by the turbulent energy dissipation rate $\epsilon$) by several orders of magnitude.

The dynamics of phytoplankton growth and decay can be discussed in a similar way, that is, by the relevant time scales expressed, for instance, by the reciprocal value of the rates for nutrient uptake, growth, respiration, and mineralization. Spatial structure is often found in the distribution of biomass, i.e., bacteria, phytoplankton, zooplankton, and fish. It results from the interaction of the various time scales of mixing and biomass dynamics, as well as from the nonlinear processes characterizing biological systems.

Experimental evidence for the interaction of biological and physical dynamics in lakes is meager, so far. Yet, comparison to marine investigations lead to the conclusion that mixing is important in lakes, as well. It is postulated as a hypothesis that to a large extent the lacustrine "weather," especially the variance of physical conditions extending from completely calm days to severe storms, as well as the regular variations induced by the seasonal changes in the vertical density stratification, controls the diversity of the planktonic community in the lake. It is possible that intermittent mixing patterns are able to increase the number of co-existing algal species by keeping the system away from steady state, thus explaining at least part of the so-called paradox of plankton. If niches are defined as sets of species properties which are sufficiently different to allow co-existence in spite of the principle of competitive exclusion, one could conclude that the dynamics of mixing patterns create niches in the lake.

I hope that the considerations presented here will help increase mutual understanding between biological and physical limnologist. In the same way that a soil scientist collects meteorological parameters as important background information, an ecological investigation in aquatic systems should, in the future, include physical data as well. In this respect, the physical limnologist meets a great challenge—to make lake physics an integrated part of limnological investigations requires developing a new generation of physical instruments for routine measurements of parameters such as turbulence. Considerable progress has been made in this respect during the last few years, especially in oceanography. If we are ready to meet the requirements, new insights into the nature of aquatic ecosystems certainly await us.

## References

Batchelor, G.K. 1959. Small-scale variation of convected quantities like temperature in turbulent fluid. Part 1. General discussion and the case of small conductivity. *J. Fluid Mech.* 5:113–133.

Bolin, B. and Cook, R.B. (eds.). 1983. *The Major Biogeochemical Cycles and Their Interactions.* SCOPE Report No. 21, Wiley, Chichester.

Conell, J. 1978. Diversity in tropical rain forests and coral reefs. *Science* 199:1304–1310.

Corrsin, S. 1961. The reactant concentration spectrum in turbulent mixing with a first-order reaction. *J. Fluid Mech.* 11:407–416.

Corrsin, S. 1962. Some statistical properties of the product of a turbulent first-order reaction In: J.B. Diaz and S.I. Pai (eds.). *Fluid Dynamics and Applied Mathematics.* Proc. Symp. 1961, Univ. Maryland, Gordon and Breach, p. 105–124.

Denman, K., Okubo, A., and Platt, T. 1977. The chlorophyll fluctuation spectrum in the sea. *Limnol. Oceanogr.* 22:1033–1038.

Dillon, T.M. and Powell, T.M. 1979. Observations of surface mixed layer. *Deep-Sea Res.* 26A:915–932.

Dubois, D.M. 1975. A model of patchiness for prey-predator plankton populations. *Ecol. Modelling* 1:67–80.

Falkowski, P.G. and Wirick, C.D. 1981. A simulation model of the effects of vertical mixing on primary production. *Mar. Biol.* 65:69–75.

Flindt, R. 1986. *Biologie in Zahlen.* 2nd ed., G. Fischer Verlag, Stuttgart.

Gaedeke, A. and Sommer, U. 1986. The influence of the frequency of periodic disturbances on the maintenance of phytoplankton diversity. *Oecologia* 71:25–28.

Gallegos, C.L. and Platt, T. 1982. Phytoplankton production and water motion in surface mixed layers. *Deep-Sea Res.* 29:65–76.

Gallegos, C.L., Platt, T., Harrison, W.G., and Irwin, B. 1983. Photosynthetic parameters of arctic marine phytoplankton: Vertical variations and time scales of adaptation. *Limnol. Oceanogr.* 28:698–708.

George, D.G. and Edwards, R.W. 1973. Daphnia distribution within Langmuir circulations. *Limnol. Oceanogr.* 18:798–800.

Grant, H.L., Williams, III, A.J., and Moilliet, A. 1962. Turbulence spectra from a tidal channel. *T. Fluid Mech.* 12:241–263.

Harris, G.P. and Piccinin, B.B. 1977. Photosynthesis by natural phytoplankton populations. *Arch. Hydrobiol.* 80:405–457.

Hilborn, R. 1979. Some long term dynamics of predator-prey models with diffusion. *Ecol. Modelling* 6:23–30.

Horn, W., Mortimer, C.H., and Schwab, D.J. 1986. Wind-induced internal seiches in Lake Zurich observed and modeled. *Limnol. Oceanogr.* 31:1232–1254.

Horne, E.P.W. and Platt, T. 1984. The dominant space and time scales of variability in the physical and biological fields on continental shelves. *Rapp. P.-v. Réun. Cons. int. Explor. Mer.* 183:8–19.

Hutchinson, G.E. 1957. *A Treatise of Limnology*. Vol. 1, Wiley, New York.

Hutchinson, G.E. 1961. The paradox of the plankton. *Am. Nat.* 95:137–145.

Imberger, J. 1985. The diurnal mixed layer. *Limnol. Oceanogr.* 30:737–770.

Imboden, D.M., Stotz, B., and Wüest, A. 1988. Hypolimnic mixing in a deep alpine lake and the role of a storm event. *Verh. Internat. Verein. Limnol.* 23:67–73.

Kemp, W.M. and Mitsch, W.J. 1979. Turbulence and phytoplankton diversity: A general model of the "paradox of plankton". *Ecol. Modelling* 7:201–222.

Kierstead, H. and Slobodkin, L.B. 1953. The size of water masses containing plankton blooms. *J. Mar. Res.* 12:141–147.

Kolmogorov, A. 1941. The local structure of turbulence in incompressible viscous fluid for very large Reynolds numbers. *Dokl. Akad. Nank SSSR,* 30:299–303.

Ledbetter, M. 1979. Langmuir circulations and plankton patchiness. *Ecol. Modelling* 7:289–310.

Lemmin, U. and Mortimer, C.H. 1986. Tests of an extension to internal seiches of Defant's procedure for determination of surface seiche characteristics in real lakes. *Limnol. Oceanogr.* 31:1207–1231.

Lewis, M.R., Horne, E.P.W., Cullen, J.J., Oakey, N.S., and Platt, T. 1984. Turbulent motions may control phytoplankton photosynthesis in the upper ocean. *Nature* 311:49–50.

Lewis, M.R., Cullen, J.J., and Platt, T. 1984a. Relationships between vertical mixing and photoadaptation of phytoplankton: similarity criteria. *Mar. Ecol. Progr. Ser.* 15:141–149.

Löhr, J. 1984. Einfluss der Tiefendurchmischung auf die Entwicklung von Phytoplankton - dargestellt am Beispiel des Maines. Dissertation, Universität Karlsruhe, Institut für Siedlungswasserwirtschaft.

Mortimer, C.H. 1974. Lake hydrodynamics. *Mitt. Internat. Verein. Limnol.* 20:124–197.

Oakey, N.S. 1982. Determination of the rate of dissipation of turbulent kinetic energy from simultaneous temperature and velocity shear microstructure measurements. *J. Phys. Oceanogr.* 12:256–271.

Okubo, A. 1971. Oceanic diffusion diagrams. *Deep-Sea Res.* 18:789–802.

Parker, R.A. 1978. Spatial patterns in a nutrient-plankton model. *Ecol. Modelling* 4:361–370.

Powell, T.M., Richerson, P.J., Dillon, T.M., Agee, B.A., Dozier, B.J., Godden, D.A., and Myrup, L.O. 1975. Spatial scales of current speed and phytoplankton biomass fluctuations in Lake Tahoe. *Science* 189:1088–1090.

Reynolds, C.S. 1984. *The Ecology of Freshwater Phytoplankton*. Cambridge University Press, Cambridge.

Reynolds, C.S. 1987. The response of phytoplankton communities to changing lake environments. *Schweiz. Z. Hydrol.* 49:220–236.

Riley, G.A. 1946. Factors controlling phytoplankton populations on Georges Bank. *J. Mar. Res.:* 6:54–73.

Riley, G.A. 1976. A model of plankton patchiness. *Limnol. Oceanogr.* 21:873–880.

Saylor, J.H., Huang, J.C.K., and Reid, R.O. 1980. Vortex modes in southern Lake Michigan. *J. Phys. Oceanogr.* 10:1814–1823.
Shannon, C.E. and Weaver, W. 1949. *The Mathematical Theory of Communication*. University of Illinois Press.
Sommer, U. 1989. The role of competition for limiting resources in phytoplankton species replacements. In: U. Sommer (ed.). *Plankton Ecology: Succession in Plankton Communities*. Science Tech Publishers, Madison, WI.
Stavn, R.H. 1971. The horizontal-vertical distribution hypothesis: Langmuir circulation and Daphnia distributions. *Limnol. Oceanogr.* 16:453–466.
Steele, J.H. 1974. Spatial heterogeneity and population stability. *Nature* 248:83.
Steele, J.H. 1978. Some comments on plankton patches. In: J.H. Steele (ed.) *Spatial Pattern in Plankton Communities*. Nato Conf. Ser. IV, Vol. 3, Plenum, New York, pp. 1–20.
Stocker, T. and Hutter K. 1987. *Topographic Waves in Channels and Lakes on the f-Plane*. Springer, Berlin, 176 pp.
Wüest, A. 1987. Ursprung und Grösse von Mischungsprozessen im Hypolimnion natürlicher Seen. Dissertation, ETH Zürich, 144 pp.
Wüest, A., Imboden, D.M., and Schurter, M. 1988. Origin and size of hypolimnic mixing in Urnersee, the southern basin of Vierwaldstättersee (Lake Lucerne). *Schweiz. Z. Hydrol.* 50:40–70.
Wunsch, C. 1972. The spectrum from two years to two minutes of temperature fluctuations in the main thermocline at Bermuda. *Deep-Sea Res.* 19:577–593.

# 4

# On Internal Seiches and Noisy Current Fields—Theoretical Concepts Versus Observations

M. Bohle-Carbonell and David van Senden

ABSTRACT Basin modes or "seiches" are often used to analyze flow fields in lakes. In the following chapter, the reliability of the basin mode concept is investigated for large lakes. Linearity and coherence in space and time are the necessary prerequisites for any flow field to be understood as composed of basin modes. A method for estimating the degree of basin mode character is applied to data from Lake Geneva. The data are analyzed for temporal and spatial coherences. The internal flow field of Lake Geneva is shown to *not* be of basin mode character, but rather it is composed of local sequences of transient current events. It is concluded that these features are not singular to Lake Geneva but that they are typical for large lakes.

## 4.1 Introduction

"Seiches," "basin modes," and "eigen-modes" are terms commonly used to refer to the oscillations of water masses in a closed basin. We prefer the term basin mode as it correctly implies that the characteristics of the wave modes (e.g., frequency and wave length) are in some way determined by the basin topography. Basin modes may be either long surface gravity waves, long internal gravity waves, or topographic waves. Internal waves are influenced by the density stratification.

The basin mode concept has been used extensively to interpret field data, which show fluctuations of water level (seiches) and density interfaces, as inferred from isotherm depths (internal seiches) and as demonstrated for e.g., Lakes Biwa (Kanari, 1975), Ontario (Schwab, 1977), Constance (Bäuerle, 1981), Geneva (Bäuerle, 1985), Zürich (Horn et al., 1986), and Lugano (Stocker et al., 1987). The concept of internal seiches was established in limnology by C. Mortimer (1951) who refers, e.g., to Halbfass (1910). Recently, Mysak (1984, 1985) extended the concept to internal motions driven by topographic waves (topographic waves rely on the conservation of rotation of a water column of changing depth). The interest in the basin mode concept arises from the mathematical possibility of decomposing the temporal and spatial structure of each field of motion into a sum of basin modes. This has been done by Rao (1974) for idealized basins and by Schwab (1975) for Lake Ontario.

Models based on the basin mode concept generally rely on two assumptions: (1) linearity of the flow field and (2) temporal and spatial coherence of the motions over large scales. The first assumption excludes the energy transfer between different modes, while the latter is necessary so that initial perturbations may develop into basin modes. In order to test the second assumption we present results from a stringently conducted spectral analysis of temperature and current time series measurements collected in Lake Geneva. In particular, we use the coherence estimates between different signals to check whether this assumption, and hence the use of the basin mode concept, are justified. The linearity of the flow field is investigated elsewhere (Bohle-Carbonell and Lemmin 1988).

In the first sections of this paper, we review the different types of basin modes and then discuss some previous observations which highlight some of the problems associated with the concept. In Section 4.4, we will describe a method for characterizing the degree of basin mode behavior in a data set. The Lake Geneva data set is described in Section 4.5, and results of the analysis of these data are presented in Sections 4.6, 4.7 and 4.8. A discussion of general findings follows and finally conclusions are drawn.

## 4.2 Basin Mode Structure and Rotation

The spatial structure of seiches in a lake can be influenced by the earth's rotation. In this case the basin modes change from pure gravity waves to rotational gravity waves (quasi-Poincaré and quasi-Kelvin waves). Pressure gradients in these waves partly balance the Coriolis acceleration. (The Coriolis acceleration is present in any coordinate system given relative to a point fixed on the spinning earth, but will not be felt by rapidly changing motions.) Rotational gravity wave seiches are more difficult to model consistently with observations, especially if spatial structures are of interest (Hamblin, 1982).

Rotation does not largely influence the wave period but strongly modifies the wave dynamics (Pedlosky, 1979), particularly the pattern of energy propagation and the dispersion or characteristics of wave reflection. The expected effect of rotation can be measured by the mean of the Rossby radius of deformation, which measures a critical length scale over which rotation is important for the dynamics of the motion. The Rossby radius of deformation, or Rossby radius, can be defined for surface and internal wave motions as:

$$R_d = U / f \tag{4.1}$$

where f is the local Coriolis frequency (1.075 to $10^{-4}$ rad $s^{-1}$) and U is a velocity scale, typically the wave phase speed. For *surface* seiches

$$U = (g\ h)^{1/2} \tag{4.2}$$

where g is the gravitational acceleration and h the depth of the lake. On the other hand, for *internal* seiches in the two-layer approximation

$$U = (g'\ h_e)^{1/2} \tag{4.3}$$

where $g'$ is the reduced gravity of a stratified fluid (gravitational acceleration times density difference between the two layers divided by the mean density), and $h_e = (h_1\ h_2)\ /\ (h_1 + h_2)$ is the equivalent depth for the upper and lower layer depths $h_1$ and $h_2$, respectively. It should be noted that eq. 4.2 follows from eq. 4.3 if one recognizes that a surface seiche is in fact an internal seiche at the air/water interface. More realistic descriptions of the stratification (see, e.g., Pedlosky, 1979) than eq. 4.3 can be used to estimate more accurately the (internal) Rossby radius of a stratified fluid.

A convenient way to distinguish large from small lakes, therefore, is to use the physics of their seiches. This can be done for surface as well as for internal seiches. Seiches in "small lakes" are of the pure gravity wave form, whereas in "large lakes" seiches are rotational gravity waves. For a lake with a length scale $L_b$ this can be crudely stated as follows:

$$\begin{aligned} &\text{For large lakes, } R_d < L_b. \\ &\text{For small lakes, } R_d > L_b. \end{aligned} \tag{4.4}$$

It should be emphasized that the use of this criterion, in combination with eqs. 4.1 and 4.3, is only a "rule of thumb" since the use of the two-layer approximation in the estimate of $R_d$ for internal motions may underestimate the effect of rotation. For example, in most lakes the stratification supports a set of vertical wave modes. In this case, $R_d$ varies with each mode and some modes may show "small" lake behavior while others may fall in the "large" lake category.

The two types of rotational basin modes, quasi-Poincaré waves (PW) and quasi-Kelvin waves (KW), can be distinguished by the following characteristics (Rao, 1966, Pneulli and Pekeris, 1976): with decreasing stratification ($R_d$ decreasing), the periods of PW and KW respectively converge towards the Coriolis period and infinitely long periods. Furthermore, as $R_d$ decreases, the waves are bounded increasingly against the shore line.

The physical form of the internal seiches in any lake is determined by the scales of the lake, the earth's rotation, and the stratification. Small scales, low latitudes, and strong stratification favor the formation of pure gravity waves. Large scales, high latitudes, and weak stratification favor the formation of rotational gravity waves (Csanady, 1975). Consequently, the length scale of a "small" tropical lake may be larger than the length scale of a "large" temperate lake. Furthermore, elongated lakes may show a seasonally varying pattern of both transverse internal seiches of pure gravity wave type and longitudinal internal seiches of rotational gravity wave type.

The period $T_n$ of the *n*th basin mode may be approximated by what is traditionally known as Merian's formula

$$T_n = 2 / n \cdot L_b / (g\ h)^{1/2} \tag{4.5}$$

For internal seiches the same formula applies with $(g\ h)^{1/2}$ being replaced by $(g'\ h_e)^{1/2}$. Another term sometimes referred to is the crossing period, which estimates the time needed by an initial perturbation to cover the distance of one basin scale, e.g., to cross the basin. If the crossing period is short compared to the period for which, e.g., dissipation would have considerable influence on the motion, then an initial perturbation may develop into a basin mode. The numerical value of the crossing period is, making use of eq. 4.5, $T_1/2$.

## 4.3 Some Illustrative Observations

Internal seiche periods are usually estimated from isotherm displacements. It is common that regular oscillations of isotherm depths are observed, along with less regular horizontal currents (Mortimer et al., 1984, Stocker et al., 1987). The potential energy of internal waves is a function of the displacement of isotherms. In pure gravity waves there is equal partitioning between potential and kinetic energy. If the earth's rotation influences the wave dynamics, then, except for pure Kelvin waves, this partitioning is generally broken in favor of the latter (Krauss, 1973). Isotherm depths may indicate mainly periodic fluctuations, which may be only barely detectable in current observations. The periodic nature, particularly the basin mode content, of the observed motions may therefore be overestimated if only isotherm depths are analyzed.

Stocker et al. (1987) illustrated the difficulties encountered when attempting to interpret a data set in terms of the basin mode concept. Two of their observations may be cited: At first sight, isotherm-depth plots showed periodic oscillations. However, the statistics of the spectral analysis (confidence intervals) demonstrated that most oscillations were too small in amplitude to differ significantly from the level of "background noise." In addition, events with strong initial perturbations were found but these did not develop into strong basin modes; where did the initial energy go? A similar observation is reported by Hollan (1984) for Lake Constance.

Data sets from the Lake Geneva comprise a unique "17-station, 14-month set" of continuous water level records (SFDE, 1954), which Mortimer (1974) used to demonstrate the classical and extraordinary event of a revolving internal wave propagating three times in 11 days around the perimeter of the lake. Other events can be found, but none completes a full circuit around the lake (Figure 4.1). One should remember that internal seiches can also be detected in water level fluctuations, as confirmed recently by Sirkes (1987) for the Dead Sea. The phase velocities of revolving water level fluctuations of Lake Geneva suggest a three-day period for its lowest order basin mode. Mortimer et al., (1984) also looked for oscillations with periods shorter than the Coriolis period, using long time series of current and temperature measurements in Lake Ge-

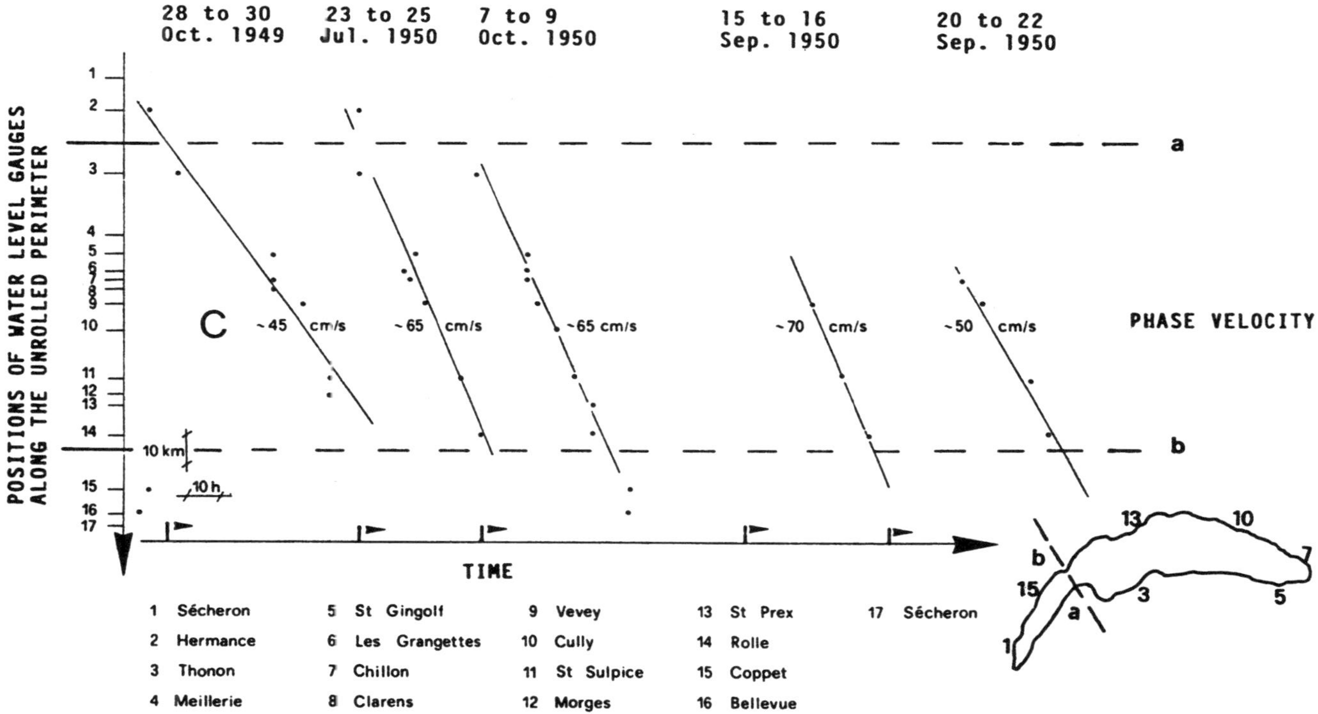


**Figure 4.1** Five events of long-period surface fluctuations propagating anti-clockwise along the perimeter of Lake Geneva. The dots give time and position of the events. The events start either in the southwest corner of the lake's deep basin or at the eastern end of the lake. The events disappear somewhere along the north-western shore line. The data are time series of residuals of six-hourly running means from each of 17 water level gauges (SFDE 1954). The time series are the same as those analyzed by Mortimer (1974) for his classical example of the lake's fundamental internal mode (three cycles from 18 August to 13 September 1950, not shown). The numbers (1 to 17) on the vertical axis indicate the positions (inset) of the water level gauges along the unrolled perimeter. The broken lines refer to the sill between the two basins of the lake. The ticks on the horizontal time axis refer to time at position "a" for each event. The phase speeds "C" were obtained from the slopes of the regression lines (fitted by eye).

neva recorded during the period 1977 to 1978. He concluded that the flow field of Lake Geneva for these periods can be best described as "episodic, short-lived, and characteristically intermittent." In conclusion, the basin mode concept seems to be consistent with shortlived, single events but not with the bulk of all these observations.

## 4.4 Theoretical Considerations for the Analysis of Basin Modes

The essential aim of this investigation is to ascertain whether basin modes form an important part of the observations from Lake Geneva (described in the following section). In this section we develop a technique which provides a measure of the degree of basin mode character in a data set. Since the technique is to be applied to Lake Geneva, it is instructive to first consider the relevant scales for this lake.

The appropriate basin scales $L_b$ of any basin are either the width, length, or perimeter. For the deep basin of Lake Geneva (Figure 4.2), these distances are respectively 10 km, 50 km, or 120 km. Substituting typical values—U = 10 to 60 cm $s^{-1}$, Figure 4.3, or from Mortimer (1984)—into eq. 4.1 gives the internal Rossby radius $R_d$ as 0.9 to 5.5 km. Since $R_c$ is smaller than $L_b$, Lake Geneva can be classified as a "large" lake and hence the appropriate basin scale is either the width (cross modes) or the perimeter (revolving modes) of the lake. The corresponding time scales (crossing period $T_1/2$) cover the ranges 0.2 to 1.2 and 2.5 to 15.0 days, respectively.

By definition, basin modes exhibit spatial coherence over length scales of the order of the basin scales. If a flow field is composed of basin modes at certain frequencies $f_b$ and if the distance $r_l$ between two adjacent observation points (say A and B) is small compared to the basin scales, then we expect the basic structure of the time series recorded at A and B to be similar. In mathematical terms we expect $C_{AB}(f_b)$ of about one, where $C_{AB}(f_b)$ is the coherence estimate between the time series A and B at frequency $f_b$. We now define the "rocking length" or "distance of coherence" $L_r(f_b)$ as

$$L_r(f_b) = r_l/[1 - C_{AB}(f_b)] \tag{4.6}$$

If $L_r(f_b)$ is large compared to $L_b$ the motion is dominated by basin modes; conversely, for short "rocking lengths" basin modes are not important.

In order to label a motion as a basin mode, it is necessary that the wave field last for at least several crossing periods. Otherwise the motion would disintegrate so rapidly that no common basin-wide flow pattern could develop. The interval of time over which the motions remain coherent should be at least two to three periods or about five crossing periods.

This rule (i.e., the factor of 5) seems not to be too restrictive because *aperiodic* random processes exist, which model time series with dominant periods comparable to the length of the time series itself (Turcotte, 1988). Thus,

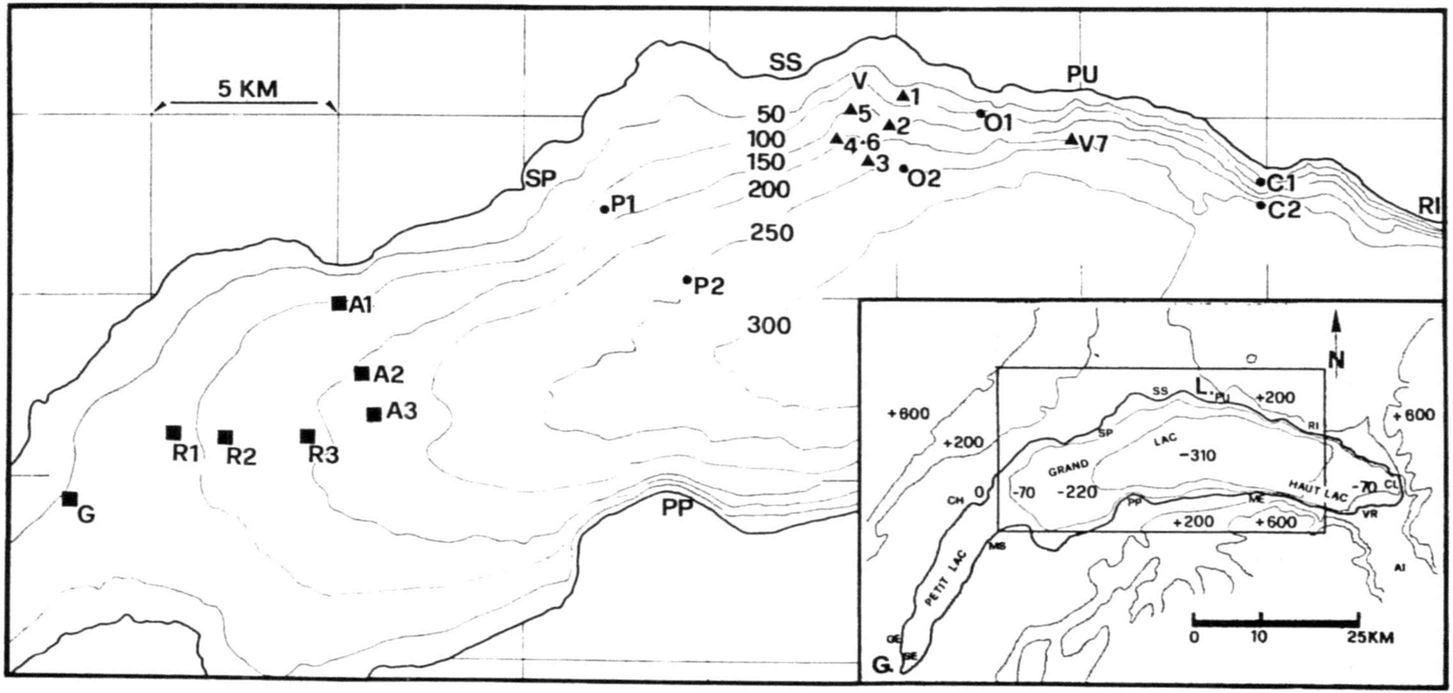


**Figure 4.2** Lake Geneva basin topography and locations of the moorings, water level gauges, and anemometer stations. Water level gauges (SE, SP, CL) and anemometer stations (GE, CH, PU, AI, PP) operated during all campaigns. Moorings Cully (Cl, C2), Ouchy (O1, O2), and St. Prex (P1, P2), and anemometer station (VR) were deployed from October 1981 to March 1982; Allaman (A1, A2, A3), Rolle (R1, R2, R3), and Gland (G) moorings from October 1982 to March 1983; Vidy (V1, . . . , V7) moorings and anemometer stations (SS, RI, ME, MS) from October 1983 to March 1984. The inset gives an overview over the orography of the lake. Contour lines (heights and depths in meters) are given relative to the lake surface.

signals attributed to *periodic* dynamics should have a duration substantially longer than that of the period itself. This question of the relation between the duration of periodic signals, or the length of a wave train, and the underlying physics is linked to technical matters associated with data analysis techniques. There exist techniques for spectral estimation (maximum entropy techniques) which can confidently resolve periods on the order of the total length of the data record. These techniques are of particular use if it is known *a priori* that the analyzed data resulted from periodic processes. If this is not known, then one should use a maximum likelihood technique. (Those not familiar with spectral data analysis should note that there are problems of reliability with all the techniques.) Here we use a Fourier transform technique coupled with statistical techniques (Nuttal 1971) which provide confidence limits for the spectral estimates. In this way we are able to identify, in a noisy background, signals which have a duration longer than their period.

Given these considerations about the temporal character of a basin mode, we now can define the "ringing time," or "coherence drop" $Y(f_b)$ which is essentially a measure of the deviation from basin mode character:

$$Y(f_b) = [1 - C_{AA'}(f_b)] \, / \, (DT\ f_b) \tag{4.7}$$

$C_{AA'}$ is the coherence (or auto-coherence) between two segments, A and A′, of the same time series where A′ lagged behind A by a time DT. The denominator ($DT\ f_b$) is simply the lag expressed as the number of cycles (rings) at frequency $f_b$ (In practice, DT should be chosen in a way that ($DT\ f_b$) is near to one for the analyzed range of $f_b$.). The auto-coherence measures relative changes in the composition of the original time series as a function of the lag. It is expected that $C_{AA'}$ is about one, if the mean composition of the time series is statistically stationary. For decaying oscillations we expect $C_{AA'}$ to decrease monotonically as DT increases. In order to obtain a meaningful threshold value for $Y(f_b)$ we need to set the expected live time of a basin mode. This has to be done in relation to the statistics of the coherence estimate. Here, only coherence estimates above 0.5 are significantly different (confidence level of 95%) from the estimate of 0.0. All coherence estimates below 0.5 therefore are indistinguishable from zero coherence. If we define, as above, a basin mode as having a live time of more than five crossing periods, then this implies that for a motion to be classed as a basin mode $Y(f_b)$ has to be less than 0.2. If $Y(f_b)$ is bigger than 0.2, then the motion is not due to a basin mode.

The data to be analyzed here originated from three experiments conducted during the autumn and winter from 1981 to 1984. These data were collected under conditions of changing mean stratification which effects the periods and spatial structures of gravity waves. Hence, one may doubt whether these data are suitable for the planned analysis. The theory of rotational gravity waves and the rocking length concept suggest that the problem is less serious than might be expected because the moorings were clustered in relatively small subregions of the lake. Thus, we expect reliable estimates of the ringing time and rocking length derived from these data. However, concentration of energy in narrow frequency bands cannot be expected. An analogy to the temporal structure expected here are readings taken from a couple of pendulum clocks which are commonly detuned by the changing ambient temperature.

In summary, basin modes, especially the lower modes of large wave lengths, should be identifiable from the available data by high coherence (in time and space) between time series of similar spectral composition of energy. In particular, the basin modes of Lake Geneva should show rocking lengths of at least 10 km (cross modes) or 120 km (revolving modes) and ringing times of less than 0.2, if we expect live times of the basin modes of 2.5 cycles (five crossing periods).

## 4.5 Data and Lake Geneva: A General Description

Lake Geneva (known locally as Le Léman) is situated at the border of Switzerland and France. The lake's surface area is about 582 km$^2$, and its maximum depth and volume is 310 m and 89 km$^3$, respectively. The climate of the Lake Geneva region is mild, and the lake stays weakly stratified during winter (CIPEL, 1984). The minimum difference between the mean temperatures of near-surface (0 to 50 m) and bottom (200 to 300 m) waters is about 0.5° C and occurs in February. The summer stratification is more than 20 times greater. This corresponds to a seasonal range of the internal Rossby radius from about 10 km in summer to less than 1 km in winter.

A particular interest in lake research is in currents near to the surface. However rotor current meters of the type used in Lake Geneva are sensitive to wind waves, as well as to motions of the mooring caused by the drag of the wind-drift current on the mooring (Halpern and Pillsbury, 1976; Kuhn et al., 1980). Whether or not wind waves and wind-drift bias the measurements depend on the Ekman boundary layer thickness, which is a function of the wind speed. Wind speeds (3-hour mean values) of 10 m s$^{-1}$ or more are less frequent than 1% at Lake Geneva. In contrast, 90% of the wind speeds are less than 5 m s$^{-1}$. Making use of the Ekman layer scaling laws given by Csanady (1982, pp. 11f, 22, and 176) leads to a mean thickness of the Ekmann boundary layer of 7 m or less. The main floats of the moorings were placed deeper than 10 m, i.e., below the Ekman boundary layer.

From 1981 to 1984, three consecutive measuring campaigns were made during autumn and winter of the lake's currents, water temperatures, and winds (not analyzed here) using Aanderaa instruments. Details of the instruments and mooring configurations are described in Bohle-Carbonell (1986a and b). The Aanderaa instruments were recalibrated periodically by the manufacturer, and the sensors showed no significant drift. The moorings sites covered different regions in the north-western part of the main basin (see Figure 4.2). Current meters were deployed at depths of about 15 m, 70 m, and 125 m or 175 m. Data-sampling intervals were 30 min for currents and 60 min for water temperatures, and the resultant time series were of four to five months duration (see caption for Figure 4.2).

The data were preprocessed (error eliminating, smoothing by running mean over 3- and 4-hr averages) to reduce variance at time scales less than 8 hr and to minimize the inconsistency between the current speed (average over sampling interval) and the current direction (instantaneous) measurement. This smoothing time also filters out surface seiches (with periods less than 1.25 hr). However, their current speeds are, by far generally, well below 1 cm $s^{-1}$. To be consistent with the topographic constraints imposed on the flows, the orientations of the current vectors are given in the local shore-tangential and shore-normal (perpendicular) directions. Positive orientations are defined as "clockwise alongshore" and "on shore-normal."

As discussed in the previous section, the main aim of the data analysis is to infer basin mode characteristics from spectra and cross-spectra (presented in the form of coherence and phase spectra). To effectively interpret the results of any analysis, the user has to distinguish between noise and signal. To do this objectively, statistical techniques must be used. This requires that the interpretation of mean values must be stringently handled according to their confidence intervals. For example, means (neighboring spectral estimates) which differ by less than the width of the confidence interval are read as indistinguishable, and the means themselves are not more reliable than any other value within the confidence intervals. There exist rigorous statistical tests which estimate more precisely the difference of two mean values (Stange, 1979), however, these tests would considerably lengthen the necessary computations for a minor net effect.

Spectra of variance, phase, and coherence were computed by means of the technique of short, averaged periodograms (Nuttal, 1971; Kielmann et al., 1979). This technique yields mean spectra and their statistics by averaging over sub-spectra. Statistical quantities calculated were confidence intervals for the estimates of variance, coherence, and phase at each frequency, as well as a test that checks whether the coherence is different from zero. All statistics were done for a confidence level of 95%. The statistical aspects of this technique were used to present only statistically significant features, e.g., phase estimates are shown only if the coherence estimate is significantly different from zero. Preprocessing (linear trend removal, tapering, and prewhitening) of the time series and the averaging over sub-spectra keeps most non-stationarities within the confidence intervals. Instrument resolution gives rise to a minimum resolvable variance level, and for the instruments used here, these levels are 0.5 $cm^2s^{-2}$ and 0.01 $K^2$. Further technical details are described in Bohle-Carbonell (1986a).

Stratification influences the internal wave dynamics. In order to estimate this influence for the observation period, we can examine Figure 4.3 which shows phase spectra of temperature fluctuations. Three different bands of coherent temperature fluctuations were found, at about 1.4 cpd, 0.7 cpd, and 0.2 cpd. The fluctuations were vertically in phase, horizontally out of phase, and propagated anti-clockwise (to the west). Mean phase speeds of about 12 cm $s^{-1}$, 26 cm $s^{-1}$, and 55 cm $s^{-1}$ were estimated from the horizontal phase dif-

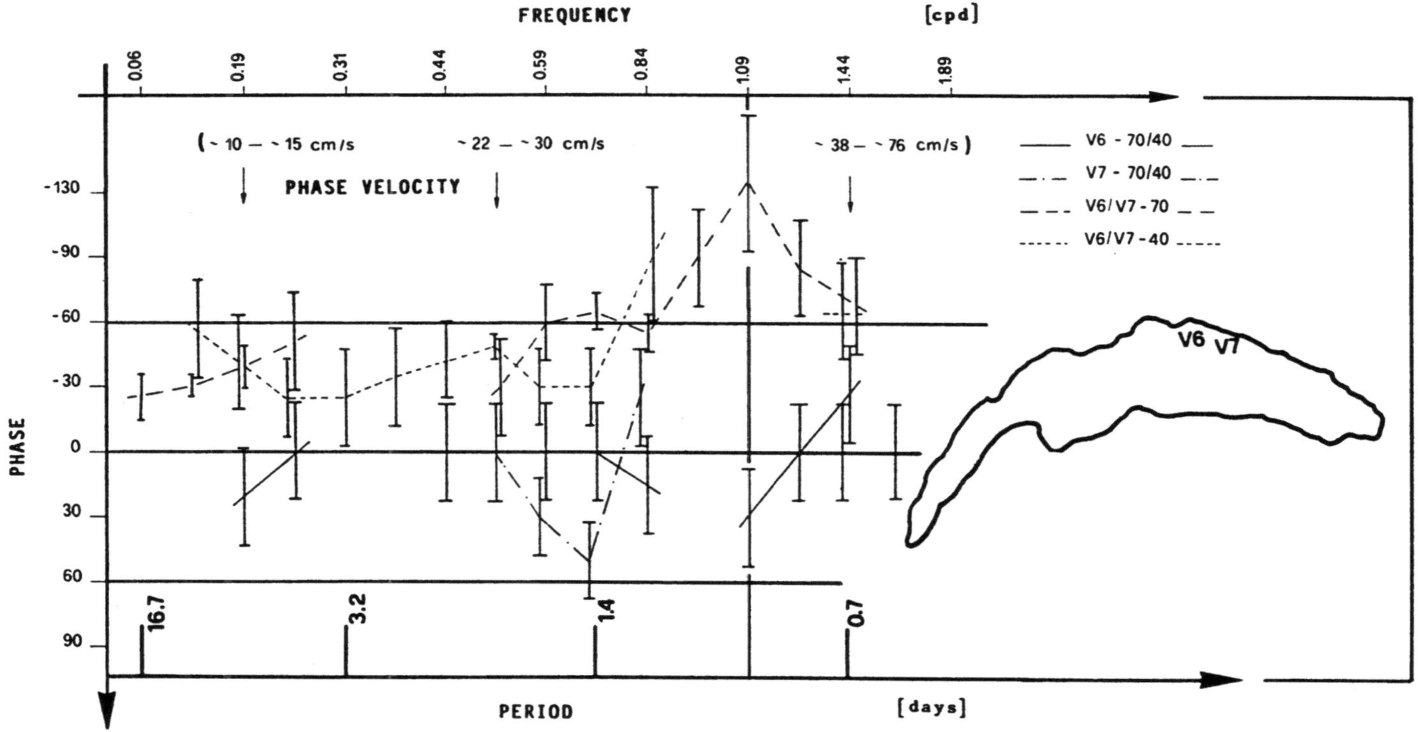


**Figure 4.3** Phase spectra (degrees) of temperature fluctuations recorded at mooring V6 and V7 (see inset) in 40 and 70 m depth (periods between 16.7 days and the Coriolis period 0.7 days). The phase estimates are given only if the corresponding coherence estimate (not shown) at this frequency (cycles/day, cpd) is significantly different from zero (95% confidence level). The phase estimates (mean and 95% confidence interval) describe coherent structures (V6—70/40 and V7—70/40: temperature fluctuations at the same mooring but in different depths. V6/V7—70 and V6/V7—40: temperature fluctuations at the same depth but at different moorings.). Positive phase is to be read as downward or eastward phase propagation. Three estimates of horizontal phase speeds (anti-clockwise propagation) are given for frequencies 0.19, 0.5, and 1.44 cpd (V6 and V7 are 5.8 km apart).

ferences and the distance between the two moorings. Without referring to particularities, if these temperature fluctuations were caused by internal gravity waves, then the internal Rossby radii from eqs. 4.1 and 4.3 were of abut 1.1 km, 2.4 km, and 5.0 km, respectively. It appears, therefore, that for the analyzed frequency range, the internal seiches of Lake Geneva are expected to be of the rotational gravity wave type in both the along- and the cross-shore direction during the observation period.

## 4.6 Spectral Distribution of Energy

Energy spectra of the alongshore current components are shown in Figure 4.4; numerical values are given as amplitudes squared or variances what is the double of the energy. The currents were recorded in shallow depths (15 m) but mainly below the Ekman boundary layer. Significant energy was found mainly in a spectral band centered at periods of about 3 days (frequency 0.3 cpd). All spectra show, independent of the location and year of observation, that energy drops to a minimum level (set as variance of 1 $cm^2s^{-2}$) at periods of about 1.5 days (frequency of 0.7 cpd). Generally, energy decreases with increasing distance from the shore. A secondary maximum was found—for moorings P1, P2, and 02 located well offshore in the central part of the lake—at frequencies greater than the inertial or Coriolis frequency (1.48 cpd; inverse of the Coriolis period). However, the ensemble-mean spectra (lower right figure) shows only one broad band of energy at frequencies below 0.7 cpd. Both energy maxima above are in the frequency ranges of the first-order basin modes of Lake Geneva, as predicted by Bäuerle (1985) and by Mysak (1985).

The absence of narrow bands of high energy in the spectra can be attributed to the effects of changing stratification, which causes a shift in gravity wave period. (Excitation of a broad band of modes is also possible.) These period shifts can be quantified from a model of the rotational internal free oscillation of Lake Geneva (Bäuerle, 1985). The model shows that halving the Rossby radius from 4 km to 2 km corresponds to an effective halving of the frequencies of the lower modes (0.3 to 0.14 cpd for mode 1, 0.5 to 0.25 cpd for mode 2, and 0.75 to 0.38 cpd for mode 3) and a decrease of approximately 1.0 cpd for near-inertial mode periods. Thus, broad bands of high energy in the spectra can be understood within the theory of basin modes formed by rotational gravity waves. It should be noted that Mysak's model (Mysak, 1985) of a particular type of topographic waves does not allow this interpretation, since the wave frequencies are independent of the stratification, in contrast to the ideas of Wang and Mooers (1976).

Variance spectra of the shore-normal current components (now shown) at 15 m and 70 m depth are nearly indistinguishable. Variances scatter generally around 3 and 1 $cm^2s^{-2}$ near the surface (15 m) and at depth (70 m) respectively. Highest values (5 $cm^2s_{-2}$) are found near the surface (moorings P2 and O2 for both near-inertial frequencies and long periods; moorings A2 and A3 for long

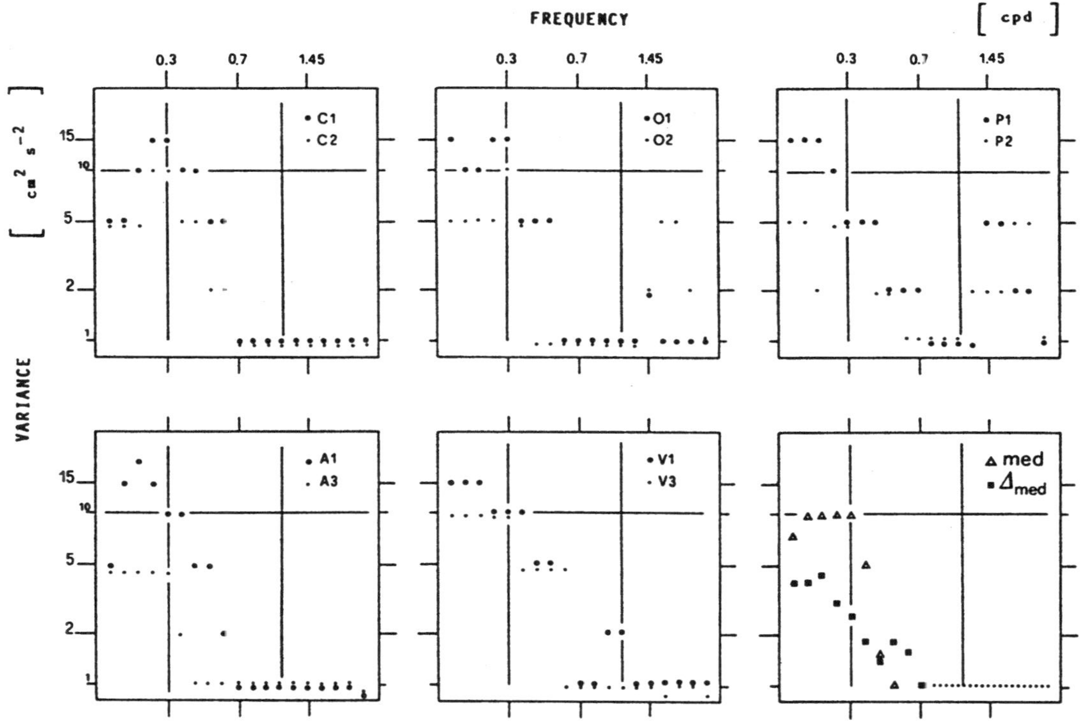


**Figure 4.4** Rectified variance spectra of alongshore current components (amplitude squared, $cm^2s^{-2}$; frequency in cycle/day, cpd). All but the lower right figure show each of the following: spectra of two alongshore current components recorded in shallow depths (15 m) at similar alongshore but different offshore position (further offshore: fine dot). Variance estimates are significantly different from each other (95% confidence level) if the lower value is 0, 1, 2, or 5 $cm^2s^{-2}$ and the higher value is not less than 2, 5, 5, or 15 $cm^2s^{-2}$ respectively. The labels, e.g., C1, refer to the positions of the moorings (see Figure 4.2). The figure at the lower right shows the average spectra (triangle, median; square, median of the absolute residuals). Note the irregular spacing of the axes of each figure, which helps to indicate relevant features, e.g., high variance around 0.3 cpd (left vertical line). The other vertical line is placed at frequency 1.0 cpd to indicate the lower cutoff of the secondary maximum found at some moorings situated further offshore.

periods). Spectra (also not shown) of alongshore current components and temperatures in 70 m depth (and deeper) qualitatively exhibit the same structure as that of the shallower water.

In summary, these energy spectra suggest that an energetically relevant basin mode response, if any, may be limited to frequencies below 0.5 cpd and between 1.25 to 2.1 cpd. Existing models predict basin mode frequencies within both these ranges. The simple absence of distinct spectral peaks does not falsify the basin mode hypothesis because a broad band of high variance might be the consequence of the frequency shift due to the changing stratification.

## 4.7 Spatial Structure of the Flow Field

Mean features of the current field in Lake Geneva will be described by coherence and phase spectra; examples refer mainly to the first measuring campaign. Three topics will be discussed:

1. The phase relationships between the alongshore and shore-normal current components of the current vectors at each location (Figure 4.5), e.g., a phase difference of 90° describes rotating current vectors.
2. The vertical phase relationships (Figure 4.6), e.g., phase differences of 180° across the thermocline are expected.
3. Horizontal phase relationships (Table 4.1), e.g., the sign of the phase difference (in degree) gives the direction of propagation of the fluctuations.

Significant phase differences between the components of local current vectors were found for two frequency bands (see Figure 4.5), one high (1.25 to 2.1 cpd) and one low frequency (below 0.3 cpd). For the high-frequency band, the alongshore and shore-normal current components were generally about −90° out of phase (shore-normal component leads the along-shore component). These clockwise-rotating current vectors are typical for near-inertial gravity waves and were most pronounced at the moorings in the central part of the lake (P1, P2); this was the only place where this feature was found at all depths. For the low-frequency band, significant phase relationships between both current components were found only for moorings O1 and C2 located relatively near (about 1 km) the shore. This indicates that low-frequency, alongshore and shore-normal motions are generally decoupled.

Significant phase differences between currents in different depths (see Figure 4.6) were again found in the same frequency bands. In the high-frequency band phase, relationships were found only at one mooring (P2) in the central part of the lake at near inertial frequencies. The phase difference of 180° across the thermocline shows that the clockwise–turning, near-inertial currents at P2 were of opposite orientation at the depth of 175 m to those above 75 m, as expected for near-inertial internal gravity waves.

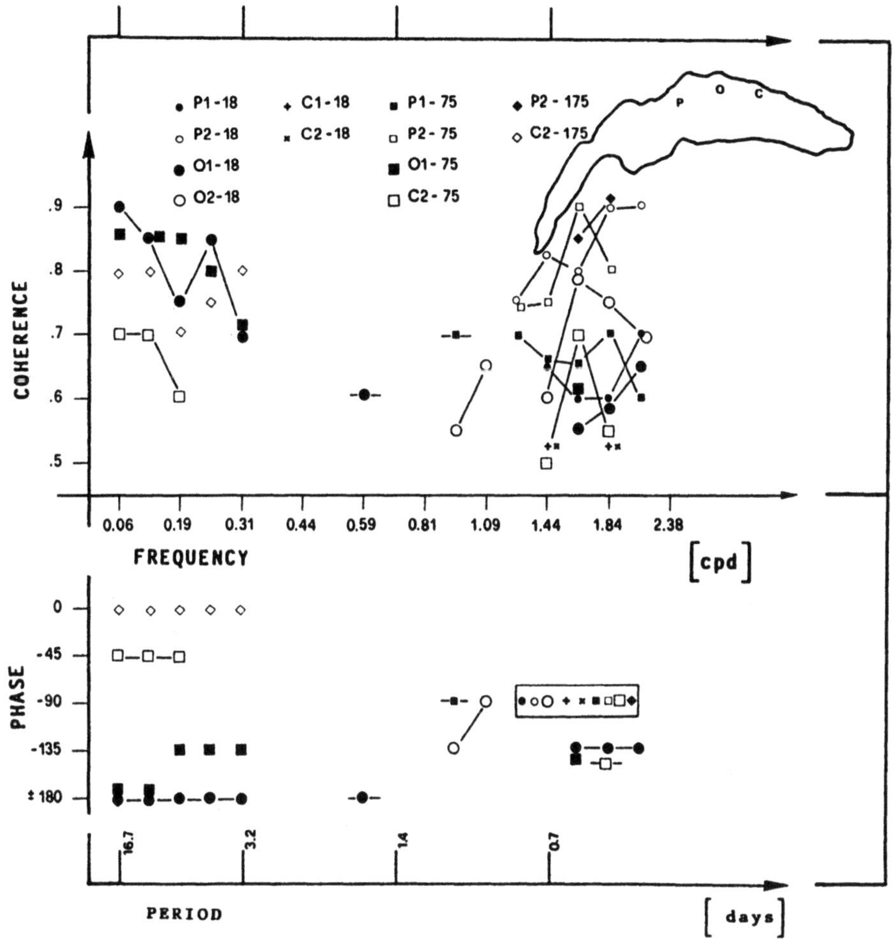


**Figure 4.5** Phase spectra (in degrees) and coherence spectra of alongshore and shore-normal current components recorded at the same mooring and depth during the measuring campaign of 1981 to 1982 [see below]. The figure codes (such as P1–18) refer to the mooring location (see Figure 4.2) and the depth (in meters) of the instrument. Linking lines indicate that the variance spectra have been above 0.5 $cm^2s^{-2}$ for both current components at these frequencies. Negative phase means alongshore component follows shore-normal components. The symbols inside the small rectangle indicate phase is $-90°$ for the enclosed frequency range (1.25 to 2.09 cpd).

Low-frequency currents showed vertical phase relations only at near-shore positions (less than 2 km offshore, see Figure 4.6). Generally, the phase differences scatter between $-45°$ and $+45°$. Currents that are 180° out of phase were found only at C2 for the frequency range 0.25 to 0.31 cpd. The vertical phase relations found at each mooring can differ for the same depths and frequencies, e.g., the alongshore currents (0.06 cpd to 0.19 cpd) are in phase at P1 but 60° out of phase at C2. These observations show that the vertical

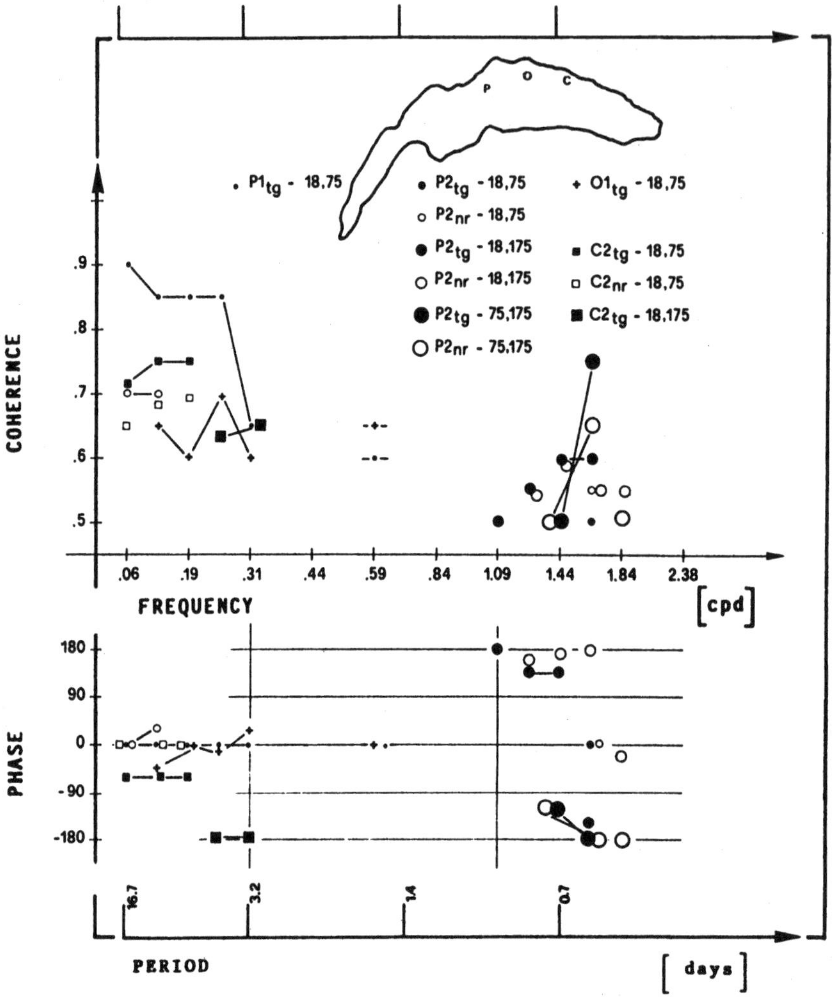


**Figure 4.6** Phase spectra (in degrees) and coherence spectra of current components recorded at the same mooring but at different depths during the measuring campaign of 1981 to 1982 (frequency in cycles/day, cpd). The figure codes refer to the mooring location (see Figure 4.2), the current component (tg, tangential; nr, normal to the shore) and the depth (in meters) of the instruments. Linking lines indicate that the variance spectra of both components at these frequencies have been above 0.5 $cm^2s^{-2}$. Positive phase means the current component in the shallow depth leads the deep current component.

structure of the flow field in Lake Geneva changed horizontally. This feature is inconsistent with models of Lake Geneva's internal, rotational gravity or topographic basin modes. However, it is consistent with the internal topographic wave concept of Wang and Mooers (1976).

The horizontal structure of the flow field is illustrated by the phase dif-

**Table 4.1** Phase spectra of selected current components and temperatures recorded in central parts of Lake Geneva[a,b]

| Frequency | Phase spectrum | | | | | | | | | | | |
|---|---|---|---|---|---|---|---|---|---|---|---|---|
| Time series no. 1 | $P1_{nr}$ | $O1_{tg}$ | $O1_{tg}$ | $C2_{tg}$ | $P1_{Te}$ | $O1_{Te}$ | $O1_{Te}$ | $C2_{Te}$ | $V5_{nr}$ | $V3_{tg}$ | $V2_{tg}$ | $V2_{nr}$ |
| Time series no. 2 | $P2_{nr}$ | $P2_{tg}$ | $P1_{tg}$ | $O1_{tg}$ | $P2_{Te}$ | $P2_{Te}$ | $P1_{Te}$ | $O1_{Te}$ | $V4_{nr}$ | $V4_{tg}$ | $V5_{tg}$ | $V5_{nr}$ |
| **18/15m**[c] | | | | | | | | | | | | |
| 0.06–0.31[d] | /[e] | / | / | +40° (4) | —[f] | — | — | — | 0° (5) | 0° (5) | +10° (5) | / |
| 0.25–0.50 | / | / | / | +40° (4) | — | — | — | — | +10° (5) | −10° (5) | +20° (3) | +10° (1) |
| 1.25–2.09 (cpd) | −10° (4) | 180° (1) | / | / | — | — | — | — | +20° (3) | 0° (5) | / | +20° (1) |
| **75m**[c] | | | | | | | | | | | | |
| 0.06–0.31 | / | / | +60° (2) | +30° (4) | +20° (4) | +100° (3) | +70° (2) | +50° (4) | — | — | — | — |
| 0.25–0.50 | / | / | / | +40° (1) | +20° (1) | / | / | +60° (4) | — | — | — | — |
| 1.25–2.09 (cpd) | / | / | / | / | / | / | / | / | — | — | — | — |

[a] Median phase (in degrees) and number (in parentheses) of significant (95% confidence level) phase estimates per frequency range. Positive phase means time series no. 1 leads.
[b] Labels "P1,"... and subscripts "tg," "nr," and "Te" refer to moorings (Figure 4.2), alongshore and shore-normal current components, and temperatures, respectively.
[c] Instrument depth in m.
[d] Frequency range.
[e] No significant phase estimate in that frequency range.
[f] Combinations of parameters and depths not analyzed due either to strong seasonal changes in temperature or small energies.

**Table 4.2** Phase spectra of selected current components and temperatures recorded in western parts of Lake Geneva[a,b]

| Frequency | Phase spectrum | | | | | | | | | | | |
|---|---|---|---|---|---|---|---|---|---|---|---|---|
| Time series no. 1 | $A1_{nr}$ | $A2_{tg}$ | $A1_{nr}$ | $A1_{tg}$ | $A2_{tg}$ | $A2_{nr}$ | $A3_{tg}$ | $A3_{nr}$ | $A2_{Te}$ | $A1_{Te}$ | $A1_{Te}$ | $A2_{Te}$ |
| Time series no. 2 | $R1_{tg}$ | $R1_{tg}$ | $A2_{nr}$ | $R1_{nr}$ | $A3_{tg}$ | $A3_{nr}$ | $R3_{tg}$ | $R3_{tg}$ | $R1_{Te}$ | $A2_{Te}$ | $R1_{Te}$ | $R3_{Te}$ |
| **15m**[c] | | | | | | | | | | | | |
| 0.06–0.31[d] | /[e] | / | / | / | +20° (5) | +10° (5) | +10° (4) | / | —[f] | — | — | — |
| 0.25–0.50 | / | / | / | / | +20° (2) | +20° (5) | +10° (1) | / | — | — | — | — |
| 1.25–2.09 (cpd) | / | / | +30° (1) | / | +10° (3) | 0° (5) | +20° (2) | +100° (2) | — | — | — | — |
| **70m**[c] | | | | | | | | | | | | |
| 0.06–0.31 | / | / | / | +20° (3) | — | — | — | — | 0° (2) | 0° (3) | +30° (2) | 0° (5) |
| 0.25–0.50 | / | / | / | / | — | — | — | — | +30° (2) | +10° (4) | / | +20° (4) |
| 1.25–2.09 (cpd) | / | / | / | / | — | — | — | — | / | / | / | / |

[a] Median phase (in degrees) and number (in parentheses) of significant (95% confidence level) phase estimates per frequency range. Positive phase means time series no. 1 leads.
[b] Labels "P1,"... and subscripts "tg," "nr," and "Te" refer to moorings (Figure 4.2), alongshore and shore-normal current components, and temperatures, respectively.
[c] Instrument depth in m.
[d] Frequency range.
[e] No significant phase estimate in that frequency range.
[f] Combinations of parameters and depths not analyzed due either to strong seasonal changes in temperature or small energies.

ferences presented in tables 4.1 and 4.2. Phase relationships were found only for frequencies below 0.5 cpd and above 1.25 cpd. Two basic features appear: (1) the phase propagation was anti-clockwise, and (2) horizontally coherent structures were limited mainly to small regions. Several local patterns can be distinguished:

1. Phase differences of strong, low frequency (0.25 to 0.50 cpd) temperature fluctuations at O1 and C2 were significantly different from estimates obtained from currents observed at the same locations. The phase speed of the temperature fluctuations was 50% higher than the phase speed of the current fluctuations. This suggests that the dynamics of current and temperature fluctuations differ.
2. High-energy, low-frequency currents at mooring A1 were decoupled from the currents at neighboring moorings A2, A3, and R3. This suggests that these energetic currents are a local feature.
3. Currents at the closely spaced moorings V1 to V6 were well correlated (moorings V1 to V6 cover an area of about 2 $km^2$).
4. No coherent motions were found for near-inertial temperature fluctuations. All phase differences found at near-inertial frequencies were from currents in shallow depths.

All these observations do not match the features of existing models of basin modes.

## 4.8 Basin Mode Character

The basin mode character of Lake Geneva's flow fields can be described by the "rocking length" $L_r$ ($f_b$) and the "ringing time" $Y(f_b)$, as defined in Section 4.3 (eqs. 4.6 and 4.7 in the following discussion, $L_r(f_b)$ and $Y(f_b)$ will be abbreviated to $L_r$ and Y). Coherence estimates, at a given frequency, were used when both variance spectra at this frequency were above the thresholds of 0.5 $cm^2s^{-2}$or 0.01 $K^2$.

Median values of the ringing time, Y, (Figure 4.7, circles) were estimated from all the time series of significant energy (variance) for each measuring campaign. The confidence limits of the auto-coherence estimates were used to define a range of uncertainty for the medians. Y decreases with a general power law proportional to 0.1 over frequency (in cpd). It follows that the flow field of Lake Geneva changes, on the average, by 10% each day. Motions with periods shorter than 2 days (1 day) may therefore be interpreted as basin modes if the basin mode live time is set to 2.5 (5) cycles, or 5 (10) crossing periods. Long-period motions (5 days and longer, frequency below 0.2 cpd) change so rapidly that their auto-coherences are below 0.5 and hence, indistinguishable from zero after less than one cycle. It results that the basin mode character of

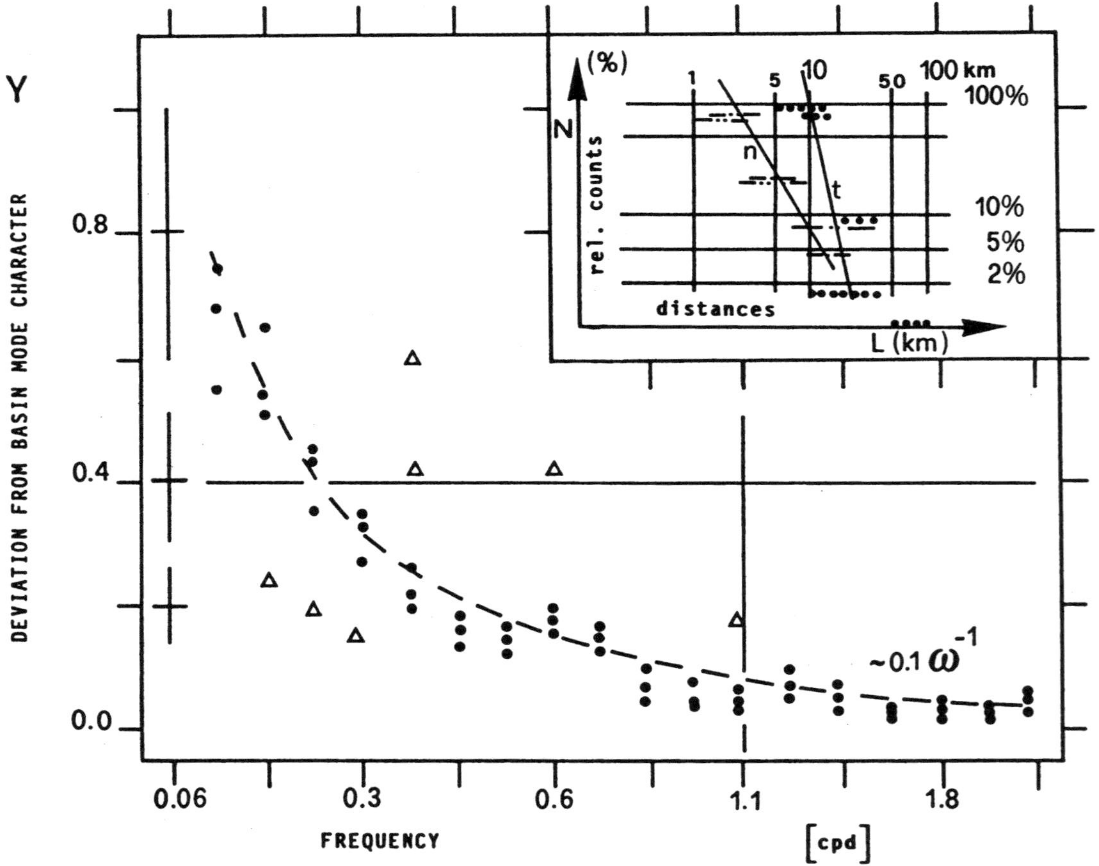


**Figure 4.7** Statistical quantities which measure the degree of non-basin mode character of the motions in time and space (inset). The "ringing time" Y (see eq. 4.7) measures the mean deviation of the motions from basin mode character as a function of frequency (for details see text; circles, median values of each measuring campaign; triangles, some individual estimates which deviate the most). A range of uncertainty (vertical bars) of the median value is estimated from the 95% confidence intervals of the coherence estimates. The broken line gives the general decrease of ringing time Y as function of frequency. N is the percentage occurrence of significantly different (95% confidence level) rocking lengths (eq. 4.6) in alongshore (t) and shore-normal (n) directions. High percentages of small rocking lengths indicate the basin mode character of the motions is not well developed.

motions with periods longer than 1 day or a few days is questionable or lost, respectively.

The rocking lengths $L_r$ were computed from current vectors in shallow depths (15 m) measured during the campaigns 1981 to 1982 and 1982 to 1983. The statistics of the coherence estimate were used to separate the obtained $L_r$'s into different classes. Then, the percentage occurrence N of each class was counted; these data are shown in the inset of Figure 4.7. It appears that less than 2% of the calculated rocking lengths $L_r$ in the alongshore direction (labelled t) were on the order of the basin perimeter (120 km). Less than 10% of rocking lengths in the shore-normal direction (labelled n) were on the order of the basin width (10 km). Rocking lengths on the order of the basin scale ($L_r = L_b$) imply coherence estimates of above 0.8, if, in eq. 4.6, one gives a value of about 0.2 $L_b$ to the mean spacing of the moorings $r_1$. The mean rocking lengths were about 10 km or below 5 km in alongshore and shore-normal directions, respectively. Rocking lengths considerably longer than the basin scales are not observed. Therefore basin modes are not the main contribution to the flow field. This conclusion, however, relies on a "strong" concept of basin mode character because for well-developed basin modes, coherences of about one are expected.

Let us now relax this condition and allow for "weak" basin mode character in which a motion will be accepted as a basin mode if its coherence exceeds 0.6 for a spacing of the observation points of about $L_b$. Using eq. 4.6 and a mooring spacing of 0.2 $L_b$ yields a weak basin mode rocking length of at least 2.5 $L_b$. The value 0.6 of the coherence estimate is chosen because its lower confidence limit is above the threshold, 0.5, of coherences indistinguishable from zero. Considering the composition of the two time series, a coherence of 0.6 essentially implies that 60% of the structure in one series can be attributed to the structure of the other. Or, in terms of the variance, the accordance of both time series drops to 36% (coherence squared).

Again, making use of their confidence limits, rocking length estimates were ordered in classes. Time series of currents (all depths) and temperatures (70–75 m depth) were used to estimate rocking lengths in both alongshore and shore-normal directions for the frequency range 0.06 cpd to 2.38 cpd and for variance spectra significantly different from zero. Average rocking lengths (medians) were calculated from all pairs of time series recorded at neighboring positions at the same depth. The results indicate that the average rocking length was equal to either the basin width or 3/4 of the basin perimeter in shore-normal or alongshore directions, respectively. Thus, the flow field may be interpreted as having weak basin mode character to 100% in the shore-normal and to 75% in the alongshore directions.

To understand the above result, one may assume that the same motion is observed at the northern shore as well as at the southern shore of Lake Geneva. Then observations from each side would have a mean coherence of 0.6 and therefore would differ by 40% in their temporal structure. Similar conclusions hold in the alongshore direction. Also, revolving modes of the quasi-Kelvin wave type will therefore change their temporal structure to more than 50% over one cycle, in agreement with Proudman's (1958) predictions for damped Kelvin waves.

In summary, the main response of the lake is not organized on basin scales. This interpretation holds for both the temporal and spatial structure.

## 4.9 Discussion

Basin mode models state that the motions are organized on length and time scales determined by the lake basin. Furthermore, due to their mathematical formulation, basin mode models suggest a flow field which is highly coherent in time and space. Thus, the lake responds as one—spatially homogeneous and only slowly changing—system. This is the basic fact implicit to any basin mode model. Therefore, in order to test the basin mode character of the response of a lake, one should check the spatial and temporal coherences of the observations.

The main response of Lake Geneva fails this test. This does not exclude the possibility that single events may still be found that exhibit basin mode character. However, Mortimer et al., (1984) looked without major success for channel-type basin modes in Lake Geneva, using nine months of multiple current and temperature records. To date, more than 35 months of data on multiple currents and temperatures have been collected and all observations show noisy fluctuations (U. Lemmin, personal communication).

The set of observations analyzed here shows that the flow field of Lake Geneva consists of fluctuations with energy mainly in two frequency ranges (0.1 to 0.7 cpd and 1.3 to 2.0 cpd, see Figure 4.4). Both ranges are consistent with the frequency intervals proposed for the lowest modes of internal, rotational gravity basin modes—revolving modes and cross modes, respectively. However, the mean rocking lengths and ringing times (see Figure 4.7) of the currents and temperature fluctuations are small compared to values expected for basin modes. The mean spatial structure (Figures 4.5 and 4.6) has features that are inconsistent with the structures theoretically proposed for basin modes (e.g., Mortimer, 1974; Bäuerle, 1985; Mysak, 1985). It appears, therefore, that the flow field of Lake Geneva is mainly not organized on basin scales and therefore is less consistent with basin mode models. In contrast, the nature of the observed flow field seems to be better characterized by the terms "transient in time" and "locally organized in space."

The differences found between the basin mode concept and the observations are less pronounced for near-inertial motions as predicted by Mortimer et al., (1984). Near-inertial motions in the central part of the lake show the spatial structure expected for internal cross-basin modes of Lake Geneva (Bäuerle, 1985); clockwise-turning currents and vertical phase difference of 180° across the thermocline. However, these are quite general features of internal, rotational gravity waves in a stratified fluid, and the special form of a basin mode model is not needed to characterize these flows. They are, e.g., well described by the radiation of near-inertial gravity waves from the near-shore surface layer (Kundu et al., 1983; Kundu, 1984) or by channel models like the ones of Krauss (1979), Renouard (1981) and Mortimer et al., (1984). Some of these models would also explain the observed spatial differences, the propagation of these waves, their dissipation, and their excitation.

It is not our intention to discuss in detail the mechanisms that cause the flow field of Lake Geneva to be transient in time and locally organized in space, but some physical aspects will be cited below.

First, the wind field over the lake is inhomogeneous in space and time. Topographic sheltering of the lake and differential heating cause local winds (LeFranc, 1923; Ganter, 1978). These winds drive local current fields that have to propagate anti-clockwise under the influence of the earth's rotation (Figure 4.3 and Tables 4.1 and 4.2). These local currents fields propagate in patterns that depend on the local forcing, slope, and stratification condition (Carton, 1984). Second, the current field is partly nonlinear (Bohle-Carbonell and Lemmin, 1988), e.g., current events were found with current speeds in the order of the phase velocities (their Figure 5). These nonlinear interactions may transfer energy between motions of different time and length scales. Third, dissipation will be confined, because of the steep slopes of the lake, to narrow near-shore boundary layers. Thus, shore-normal gradients of the currents will be strong and located near shore. Consequently, perturbations caused by the local shoreline, e.g., headland eddies, will be transported horizontally by eddy diffusion far offshore. Wave scattering (Thorpe, 1973) is also likely. Thus, it would be surprising if the flow field of Lake Geneva was found to be composed mainly of basin modes.

The arguments developed in this chapter suggest by their general nature that the features learned from Lake Geneva might be general to lakes in similar conditions. Sirkes' (1987) observations at the Dead Sea give a further insight into necessary conditions for the formation of internal basin modes in large lakes. The width, length, and depth of the deep basins of Lake Geneva and the Dead Sea differ only by about 10%, but the stratification of the Dead Sea is more stable. The phase velocity of internal gravity waves in the dead Sea is increased by a factor of three as compared to waves in Lake Geneva. Thus, the time needed by an initial perturbation to cross the lake's basin is reduced to one-third. Because of its lower latitude, the Coriolis force in the Dead Sea (Coriolis period of 23 hr) is reduced to 70% of its strength in Lake Geneva. Internal waves in the Dead Sea, therefore, are less modified by the earth's rotation and are more similar to pure gravity waves than those in Lake Geneva. This is of importance, as the properties of reflection are simpler for pure gravity waves than for rotational gravity waves. A good reflection of the incident wave is one of the most important prerequisites for the formation of basin modes. Consequently, the Dead Sea—like any well-stratified, low-latitude lake—is a better wave guide for internal gravity waves than Lake Geneva. Regular, basin mode-like waves may develop more easily in the dead Sea. However, even if a given lake is a good wave guide for internal, rotational or pure gravity waves, the external (wind) forcing has to occur at periods and spatial structures comparable to those of the internal basin modes in order to excite them. This seems to be the case at the Dead Sea. In summer, the first and second internal basin mode have periods of about 21 hr and 11 hr, respectively (Sirkes, 1987). These modes are most probably excited by the diurnal and semidiurnal winds that blow persistently along the lake and are linked to the summer land–sea breeze cycle of the Mediterranean.

One may speculate that well-stratified equatorial lakes, such as Lake Kivu (Tietze, 1978), show basin mode-like current patterns in spite of their large size. The small influence of the earth's rotation on gravity waves at these latitudes might at least reduce the complexity of the problem. However, nothing more reliable may be said without a careful analysis of the wind forcing and the stratification. It must be stressed that simple observations of periodic temperature fluctuations at some locations can be misleading, as can be learned from Lake Geneva.

## 4.10 Conclusions

During past decades, the basin mode concept has been widely used to describe internal motions in lakes. We suggest it is questionable whether the quasi-periodic motions observed in many lakes may be interpreted mainly as basin modes. Although temperature fluctuations (isotherm depths) often seem to support the notion "that a considerable portion of the internal motion may be attributed to basin modes," we have shown here that objective correlation analysis of current and temperature data illustrate large deviations from basin mode behavior.

The basin mode concept implies that the whole basin exhibits a coherent flow field. However, fluctuating currents in large lakes, at least those of Lake Geneva, may at best be described as transient in time and only locally organized in space. Thus, in order to accurately describe the apparently noisy continuum of local and transient current events, which seem to be relevant to the dynamics of large lakes, models with more realistic physics are needed.

---

*Acknowledgements* The measurements and a part of the data analysis has been done during MBC's employment at the Laboratoire d'hydraulique (LHYDREP, EPF-L, Switzerland), where C. Perrinjaquet did the programming and a substantial par of the preprocessing of the data. The second author (D.v.S.) was involved with the manuscript preparation and would like to acknowledge that this work was initiated and carried out by M.B.C.

## References

Bäuerle, E. 1981. Die Eigenschwingungen abgeschlossener, zweigeschichteter Wasserbecken bei variabler Bodentopography. *Berichte Institut für Meereskunde*, Kiel, No. 85, 77 pp.

Bäuerle, E. 1985. Internal free oscillations in the Lake Geneva. *Ann. Geophysicae.* 3:199–206.

Bohle-Carbonell, M. 1986a. Hydrodynamik des Genfer Sees—Gleichgewichte und charakterisierende Grössen. *Ecole Polytechnique Fédérale de Lausanne (Switzerland),* Doctoral Dissertation No. 611, Vol. 1, 235 pp.; Vol. 2, 133 pp.

Bohle-Carbonell, M. 1986b. Currents in Lake Geneva. *Limnol. Oceanogr.* 31(6):1255–1266.

Bohle-Carbonell, M., and Lemmin, U. 1988. Observations on non-linear current fields in the Lake Geneva. *Ann. Geophysicae.* 6(1):89–99.

Carton, J.A. 1984. Coastal circulation caused by an isolated storm. *J. Physical Oceanography* 14(1):114–124.

Csanady, G.T. 1975. Hydrodynamics of large lakes. *Ann. Rev. Fluid Mech.* 7:357–386.

Csanady, G.T. 1982. *Circulation in the Coastal Ocean.* Reidel, 279 pp.

CIPEL. 1984. *Le Léman—Synthèse des travaux de la Commission internationale pour la protection des eaux du Lèman contre la pollution 1957-1982.* Secrétariat de la CIPEL, Lausanne (CH), 650 pp.

Ganter, Y. 1978. Contribution a l'étude des brises du Lac Léman. *Rapp. de Travail de l'Institut Suisse de Meteorologie* No. 83, 43 pp.

Halbfass, W. 1910. Gibt es im Madüsee Temperaturseiches? *Int. Rev. Hydrobiology Hydrogr.* 3:1–40.

Halpern, D. and Pillsbury, R.D., 1976. Influence of surface waves on subsurface current measurement in shallow water. *Limnol. Oceanogr.* 21(3):1039–1049.

Hamblin, P.F. 1982. On the free surface oscillation of Lake Ontario. *Limnol. Oceanogr.* 27(6):1039–1049.

Hollan, E. 1984. Strömungen im Bodensee. *Wasserwirtschaft* 74(7/8):366–373.

Horn, W., Mortimer, C.H., and Schwab, D.J. 1986. Wind induced internal seiches in Lake of Zürich observed and modelled. *Limnol. Oceanogr.* 31, 1232–1254.

Kanari, S. 1975. The long-period internal waves in Lake Biwa. *Limnol. Oceanogr.* 20(4), 544–553.

Kielman, J., Willebrandt, J., Holtoroff, J., and Brockmann, C. 1979. POWER1: Spectralprogramm für bivariate reelle Messreihen. *Inst. f. Meereskunde, Kiel.*

Krauss, W. 1973. *Methods and Results of Theoretical Oceanography I: Dynamics of the Homogeneous and Quasihomogeneous Ocean.* Gebrüder Bornträger, Stuttgart, 302 pp.

Krauss, W. 1979. Inertial waves in an infinite channel of rectangular cross section. *Deutsche Hydrographische Zeitschrift* 32:248–266.

Kuhn, H., Quadfasel, D., Scott, F., and Zenk, W. 1980. On simultaneous measurements with rotor, wing and acoustic current meters, moored in shallow water. *Deutsch. Hydrogr. Zeitschrift* 33:1–18.

Kundu, P.K. 1984. Generation of coastal inertial oscillations by time-varying wind. *J. Physical Oceanography* 14(12):1901–1913.

Kundu, P.K., Chao, S.-Y., and McCreary, J.P. 1983. Transient coastal currents and inertio-gravity waves. *Deep Sea Res.* 30(10A):1059–1082.

LeFranc, G. 1923. *Les vents du Leman.* Lib. P. Pelissier, Thonon les Bains, France, 46 pp.

Mortimer, C.H. 1951. The use of models in the study of water movements in lakes. *Proc. int. Ass. Limnology* 11:254–260.

Mortimer, C.H. 1974. Lake hydrodynamics. *Mitt. Intern. Verein. Limnologie* 20:124–197.

Mortimer, C.H., Perrinjaquet, C., and Bohle-Carbonell, M. 1984. Internal oscillatory responses of Lac Léman to wind impulses during 1977/1978 compared with waves in rotating channel models. Ecole polytechnique Fédéral de Lausanne. *Comm. du Lab. d'hydraulique* No. 50, 89 pp.

Mysak, L.A. 1984. Topographic waves in lakes. In: Hutter, K. (Ed.), *Hydrodynamics of Lakes.* Springer-Verlag, WIEH, p. 210–234.

Mysak, L.A. 1985. Elliptical topographic waves. *Geophy. Astrophy. Fluid Dynamics* 31:93–135.

Nuttal, A.H. 1971. Spectral estimation by means of overlapped FFT processing of windowed data. *Naval Underwater Systems Center Report* No. 4169, 39 pp.
Pedlosky, J. 1979. *Geophysical Fluid Dynamics*. Springer-Verlag, New York, 624 pp.
Pnuelli, A., and Pekeris, C.L. 1976. Free tidal oscillations in rotating flat basins of the form of rectangles of sectors of circles. *Proc. Roy. Soc. London Ser. A* 263:19–171.
Proudman, J. 1958. A damped Kelvin-wave of general form. *J. Marine Research* 17:424–428.
Rao, D.B. 1966. Free gravitational oscillations in rotating rectangular basins. *J. Fluid Mechanics* 25:523–555.
Rao, D.B. 1974. Transient response of shallow enclosed basins using the method of normal modes. *Canada Center for Inland Waters, Burlington, Scientific Series* No. 38, 29 pp.
Renouard, D.P. 1981. An experimental study of gravity-inertial waves and wind-induced Kelvin-type upwellings in a rotating system. *J. Physical Oceanography* 11(8):1100–1112.
Schwab, D.J. 1975. A normal mode method for predicting storm surges on a lake. *Center for Great Lake Studies, Milwaukee, Special Report* No. 20, 51 pp.
Schwab, D.J. 1977. Internal free oscillations in Lake Ontario. *Limnol. Oceanography* 25:700–708.
SFDE. 1954. Les denivellations du Lac Léman. Dep. féd. Postes Chemins de Fer, *Communication du service fédéral des eaux* No. 40, 103 pp., 18 appendices.
Sirkes, Z. 1987. Surface manifestations of internal oscillations in a highly saline lake (the Dead Sea). *Limnol. Oceanogr.* 32(1):76–82.
Stange, K. 1979. *Angewandte Statistik*, Vol. 1. Springer-Verlag, 592 pp.
Stocker, K., Hutter, K., Salvade, G., Trösch, J., and Zamboni, F., 1987. Observation and analysis of internal seiches in the southern basin of Lake of Lugano. *Ann.Geophysicae* 5B(6):553–568.
Tietze, K. 1978. Geophysikalische Untersuchungen des Kivusees und seiner ungewöhnlichen Methangaslagerstätte—Schichtung, Dynamik und Gasgehalts des Seewassers. Dissertation, Christian-Albrechts-Universität Kiel, 149 pp.
Thorpe, S.A., 1973. Near-resonant forcing in a shallow two-layer fluid: a model for the internal surge in Loch Ness? *J. Fluid Mech.* 63(3):509–527.
Turcotte, D.L. 1988. Fractals in fluid mechanics. *Ann. Rev. Fluid Mech.* 20:5–16.
Wang, P.D., and Mooers, CH.N.K. 1976. Coastal-trapped waves in a continuously stratified ocean. *J. Physical Oceanography* 6(11):853–863.

# 5

# Long-Term Modelling of Stratification in Large Lakes: Application to Lake Constance

Eckard Hollan, Paul F. Hamblin, and Hubert Lehn

ABSTRACT In the first section of this chapter, the main physical differences between large and small lakes are outlined, and the concepts of thermocline modelling are reviewed. Thereafter, we present a dynamical simulation of the vertical structure of density in Lake Constance as a case study. The application of the mathematical model called DYRESM (Patterson et al., 1984) is justified herein for the simulation of large lakes. As an example, the version of this model developed by Wei and Hamblin (1986), which includes the effect of suspended load, is applied to Lake Constance. The calculations carried out for the period from February 1, 1979 through March 31, 1980 are compared against biweekly routine hydrographic observations, as well as against synoptic transects of temperature and light attenuation taken in November, 1979.

The best agreement between modelled and observed distributions in the hypolimnion is obtained under the following specifications of the model: (1) The vertical eddy diffusivity is directly proportional to the cube of the wind speed over the lake and inversely proportional to the relative potential energy per unit lake volume. (2) The prevailing transverse internal seiching is taken into account for the shear production in the pycnocline. (3) The settling velocity of the suspended load in the river plumes of the main tributaries is assumed to be two orders of magnitude greater than that in the main body of the lake.

## 5.1 Introduction

The large-scale field of motions in lakes is caused by three types of forcing: wind, superficial heat-fluxes, and inflows. The effect of these exciting forces differs significantly between small and large lakes. This makes it generally imperative to modify small-lake models when they are applied to large lakes.

In the following, we analyze the major differences in the dynamics of water movements in small and large lakes. Then, we show how a reservoir thermocline model (DYRESM) can be adapted to describe the stratification of a large lake (Lake Constance).

## 5.2 Physical Aspects of the Difference between Large and Small Lakes

The three forcing mechanisms occur simultaneously in most occasions. However, for analytical clarity, we shall consider them separately.

**Wind-Generated Forcing** The major weather disturbances are very large in comparison with the size of most lakes. As a consequence, the wind field generated by such disturbances becomes a quasi-steady forcing for several hours up to a maximum of a few days in lakes. This causes a current system of similar life-time and a lakewide transport of mass, which indicates that advection is of great importance in lakes.

A major meteorological characteristic of large lakes is that the approximately 1,000-meter-thick planetary boundary layer over these lakes is affected by the water mass underneath; for example, a moderate wind forcing applied to a large lake has its velocity increased by a factor of two and more when compared to the overland value. This does not occur in small lakes and may in fact be less due to shoreline sheltering.

Another remarkable distinction between small and large lakes concerns the effect of the earth rotation on large-scale motions. When a stratified lake, whether small or large, is dynamically disturbed—for example, by wind forcing—the thermocline is tilted. The period of the resulting large-scale motion is size-dependent from a few hours in small lakes to several days in large lakes. Only in large lakes, the time and space scales allow the current to be significantly deflected by the Coriolis force. Similarly, internal seiches, which are standing waves in small lakes, become rotating waves in large lakes. As a consequence, areas of upwelling and downwelling are not situated at corresponding locations in large and small lakes.

The deflection of the wind-induced circulation due to the Coriolis force occurs mainly outside of a zone adjacent to the shore. The width of this zone is characterized by the internal radius of deformation $R^* = c_i/f$, where $c_i$ is the speed of propagation of an internal disturbance in the stratified lake and f is the Coriolis parameter. $R^*$ represents a length scale corresponding to the horizontal range of adjustment in tilting of the thermocline during the excitation, or sudden redistribution, of the current. The hydrodynamical adjustment takes place between the internal pressure forces established by the density stratification and the inertial forces inherent in the current, which is subjected both to the Coriolis force and to the continuity of massflow over the sloping-bottom configuration in the near-shore zone. The parameter $R^*$ typically varies from 2 to about 10 km in mid-latitude lakes.

From the point of view of coastal dynamics, stratified lakes are considered to be large if their widths exceed the distance $2R^*$, defining a deep-water zone lying between the opposite coastal regimes. In the case of Lake Constance, a typical autumnal value of $R^*$ is 3 km. According to its width, which varies between 7 and 12 km, this lake has all characteristics of a large one during the warm season. In the absence of stratification in winter, by contrast, the lake would no longer behave like a large one in this respect, since the corresponding horizontal scale of current adjustment in homogeneous water, the radius of deformation R, is a hundred times larger. However, the absolute size of a large lake implies certain physical properties which distinguish it from small lakes also during the homothermal period. In addition to the lakewide deflection effect of the Coriolis influence on drift currents, there is an increase of the wind speed by a factor of about two for moderate winds over the lake due to the long fetch. By contrast, wind forcing over small lakes is generally weaker and, moreover, strongly controlled by sheltering from the upwind shore.

**Thermally-Induced Circulation** In the absence of large-scale wind fields, the local variations due to meteorological heterogeneity over a large lake surface may cause horizontal differences of superficial cooling or heating and generate thermal circulation. These conditions do not occur in small lakes simply because of the obviously more homogeneous meteorological fields over small size areas. Another difference can be expected from the superposition of the wind-generated currents and free convection in large, deep lakes, particularly during the cold season. This interaction is presumably more efficient than in small lakes, since the wind-driven circulation and the accompanying turbulence are enhanced and, hence, increase the cooling of the surface layer. This heat deficit amplifies the convective processes and may result in a larger vertical range of convection. This, in turn, will influence the maximal depth of penetration of the vertical exchange by convective cooling in winter. This depth of overturn may have considerable bearing on the quantity of available nutrients and, consequently, on primary productivity during the ensuing period of stratification.

**Inflow—Outflow** The circulation caused by the main tributaries is relatively weaker in large than in small lakes, except for those small lakes which are subjected to an extremely low inflow regime. However, we have to consider the long-term effects of inflow-induced advection during the periods of calm weather. This weak transport contributes particularly to the formation and distribution of different water masses in the hypolimnion of deep lakes. The spatial variability of such circulation is greater in large lakes, even during events of high discharge, since the volume of the lake water in the layer of intrusion is far too great to be set rapidly into motion in its entirety by the input of momentum. Recent high-resolution profiling measurements in large lakes have highlighted the pronounced variability due to inflows which is observed in deep layers of large lakes.

This brief outline of the differences between large and small lakes is not exhaustive. Rather, it emphasizes the inbuilt dynamical processes, dependent on size, which give large lakes specific physical properties able to greatly affect both chemical and biological phenomena.

## 5.3 Review of Temperature-Modelling Concepts for Lakes

Since the early empirical approach of Berger (1955) who estimated that the maximal depth of mixing of small and intermediate lakes varies as the fourth root of their surface area, remarkable progress has been made in modelling long-term stratification in lakes.

Basic theoretical considerations concerning the vertical mixing processes have first been applied to oceans. Reviews of this early stage of research in physical oceanography and of the relevant modification of the models for coastal waters are given by Turner (1973) and Bowden (1983). Later, these methods were successfully adapted and modified for lakes (Fischer et al., 1979; Orlob, 1983; and Harleman, 1986).

The turbulent kinetic energy in the water generated by wind-induced shear and breaking waves or by convective cooling is the main cause for the formation of the upper mixed layer. In quantitative terms, a fraction of the vertical turbulent energy flux is transformed into relative potential energy of the water column when deep, dense water is entrained into the mixed layer. Thus, the increase in depth of the mixed layer depends on the event-related stirring energy and the density difference between surface and deep waters. The mathematical treatment of vertical turbulent fluxes in the weakly stratified water columns of lakes has developed according to two different approaches: the eddy diffusion hypothesis and the integral or mixed-layer concept.

**Empirical Thermocline Models** Thermocline models based on eddy mixing coefficients derived from field measurements of thermal structure and introduced tracers, have been advanced by Baba (1974), Walters et al. (1978), Kilworth and Carmack (1979), and, more recently, by Henderson-Sellers (1985). In small and shallow reservoirs the arbitrary specifications of eddy diffusivity were successful. The thermal stratification in reservoirs with surface inflows and bottom outflow is generally dominated by vertical convection and less influenced by vertical diffusion. In contrast, the thermal profile in lakes is very sensitive to this diffusion (Harleman, 1986). The less-than-satisfactory results from these ad hoc thermocline models in deeper and larger water bodies have led to the development of the semi-empirical parameterizations of mixing.

One approach describes the mixing process in terms of a conventional mixing-length hypothesis and a turbulent velocity scale. In turn, these properties are obtained from the solution of higher order turbulent closures. A comprehensive general review of the widely adopted theoretical solutions for this purpose is presented by Blumberg (1986). One member of the family of these solutions, the so-called $k$-$\epsilon$ model, has been applied to the seasonal development of thermal stratification in a lake by Svensson (1978). The epilimnion was well modelled but the $k$-$\epsilon$ model did not simulate the vertical flux of heat across the thermocline to the hypolimnion. A weak point of this approach consists in the choice of the unknown model coefficients, which have been determined for boundary layer flows mainly in laboratory experiments, and thus, may not adequately describe the nature of turbulent mixing away from boundaries.

Most recently, Lam and Schertzer (1987) calculated the first long-term simulation of the thermocline in the central and eastern basin of Lake Erie for a period of as long as 16 years. They applied an areally averaged, empirical model similar to Henderson-Sellers (1985) approach and assumed a modification of the vertical diffusion concept, which took account of the characteristic three-layered stratification (epi-, meta-, and hypolimnion) encountered in both basins. The comparison of the calculations with measurements yielded remarkable agreement. This success resulted particularly from the proper parameterization chosen for Lake Erie. The previously mentioned limitations of this method are highlighted by the review on one-dimensional thermocline models given in the same context of the Lake Erie Study—1979–1980 by Royer et al. (1987).

**Integral Thermocline Models** The surface layer formed by the action of vigorous wind stirring or by convective cooling generally has a homogeneous temperature. The turbulent kinetic energy equation averaged across this layer is the basis for integral models. The details of the energy transfers in the mixed layer have been developed for the oceanic regime (Pollard et al., 1973). Further achievements since then are outlined in the work of Barkmann (1987). The earlier results were applied to lakes by Green (1976) and others. A major improvement to integral models for lakes was the incorporation of the role of the shear due to internal waves at the base of the mixed layer which was done by Spigel and Imberger (1980). They assumed that this shear increases linearly from the onset of wind forcing up to one-quarter-period of the internal wave, whereupon the shear ceases. Ivey and Patterson (1984) further improved this approach by adding a mixed layer at the lake bottom.

The integral approach, however, can only be used for mixed layers, which, in deep lakes, may occupy only a small fraction of the water column. Persistent, but very weak stratification may occur in the deep hypolimnion of fjord-like large lakes, as reported by Vollenweider (1964) for southern Alpine lakes. A first step to modelling the interior layers is a semi-empirical approach based on the increase of potential energy of the water column during the turbulent mixing process and the decay of turbulence. The method of Fischer et al. (1979) assumes that the dissipation of turbulent kinetic energy is constant with depth, whereas Harleman (1986) introduces a dependence on the local stability. This assumption is supported by oceanographic observations (Gregg, 1987) and by theory (Henyey et al., 1986). An earlier form of the semi-empirical model has been reasonably successful in intermediate-sized lakes (Patterson et al., 1984). For large deep lakes, it must be emphasized that the formation of very weak stratification in the lower layers of the hypolimnion still is unresolved. The difficulty arises from the presumably very irregular distribution and the inhomogeneous structure of the weak turbulence occurring in these deep water masses.

## 5.4 A Case Study: Lake Constance

Lake Constance is central Europe's second-largest lake (volume, 47.6 km$^3$; length, 62 km; width, 13 km; and mean depth, 100 m). Over the past 36 years, temperature, concentration of suspended material, and other limnological parameters have generally been measured biweekly at several locations in this lake. The first two variables, along with simultaneous measurements of surface meteorology and hydrology of four of the main tributaries to the lake, form the basis of the present study. Although the response of the lake to external forcing consists mainly of short-term, three-dimensional dynamical processes (Hollan and Simons, 1978), observation shows that the annual stratification cycle is predominantly one-dimensional. From the long-term limnological data set, we have chosen the period from February 1, 1979 through March 31, 1980 for a simulation using a one-dimensional dynamic model. There were also synoptic profiles obtained during two surveys in 1979, using a high resolution CTD-probe, by the Institute of Applied Physics at Kiel. These measurements have enabled us to make a comparison between the one-dimensional model calculations and the three-dimensional distribution of temperature and attenuation of light in the lake.

We use a simulation model called DYRESM, which was developed by Imberger and Patterson (1981) and by Fischer et al. (1979) to hindcast the water quality of small storage reservoirs in Australia. This model has been adapted by Patterson et al. (1984) and applied to a very elongated, intermediate-sized lake in the mountainous region of Canada. Wei and Hamblin (1986) recognized the importance of including suspended solids as a component of water quality, particularly in lakes situated in glacierized basins. Such lakes are subject to sudden and substantial sediment loading, as in the case for Lake Constance.

Neither the standard simulation model nor the modified version have thus far been applied to large lakes, except for the short-term simulation by Ivey and Patterson (1984) in Lake Erie. The purpose of the present study is to evaluate the applicability of the three-component version of DYRESM to large lakes by means of direct comparison of the model results with field observations. The analytical details of the model are published in a separate paper (Hollan et al., 1990).

## 5.5 Field Observations for Lake Constance

The main body of Lake Constance has an elongated basin (Figure 5.1). The Alpenrhein and the Bregenzerach rivers are its main tributaries, and their annual average discharges comprise about 73% and 16% of the total flow,

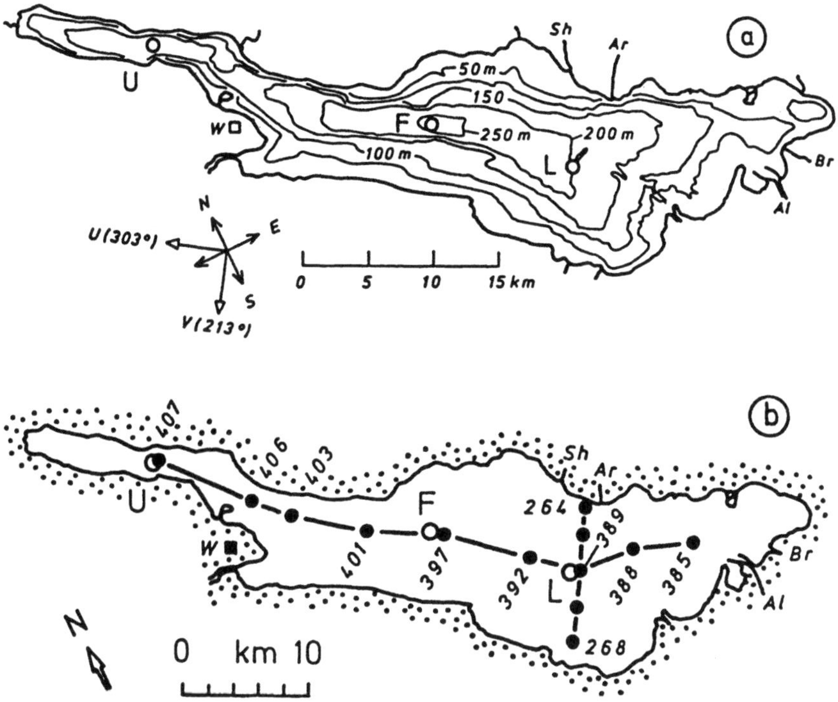


**Figure 5.1** Bathymetry of Lake Constance and measurement sites. (*a*) Routine survey stations (U, F, and L) selected for the comparison with the calculations; weather station (W) at Konstanz; inflow sites of the rivers Alpenrhein (Al), Bregenzerach (Br), Argen (Ar), and Schussen (Sh). A 50-m contour interval is used for depth. The lake-oriented coordinate system for decomposition of the wind is shown at the lower left. (*b*) Transects with the Kiel-Multisonde, November 20 and November 29–30, 1979. The sites of the vertical profiles are given by the filled circles, and the station numbers are in italics. Open circles indicate the stations U, F, and L.

respectively. They enter at the eastern side of the lake and subject the lake to the Alpine hydrological regime of sudden high discharges with considerable load of suspended solids. Furthermore, the two minor inflows on the northern shore, the Argen and the Schussen rivers, have also been included in our calculations. As an example, Figure 5.2 shows the discharge and associated time history of temperature, conductivity, salinity, and suspended solid concentration of the Alpenrhein during the simulation period. The dependence of inflow density on salinity and suspended sediment load is taken from the empirical formulae of Wagner and Wagner (1978), which are based on river discharge. The coefficients used were updated by Bührer and Wagner (1982). Since this new relation is based on the actual constituents of dissolved solids, the less-informative conductivity measurements have not been used for the calculation of density. In the other three inflows, only temperature and discharge have been measured, and the inflow density has been determined in

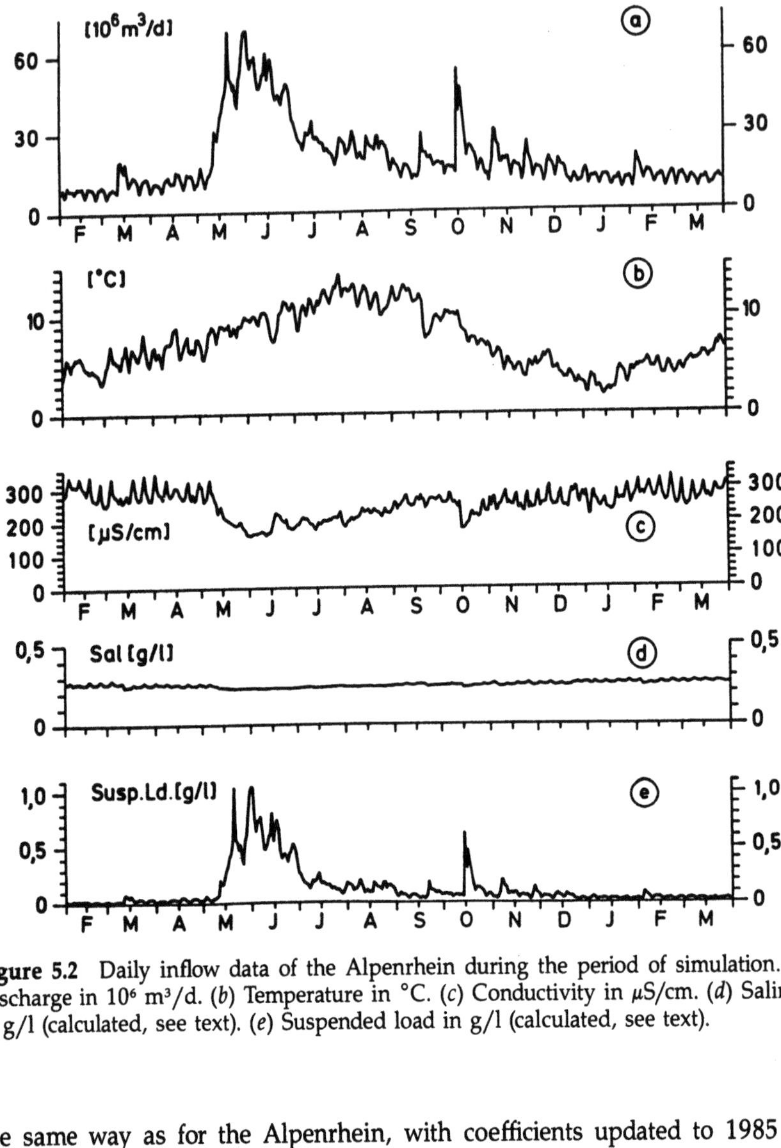


**Figure 5.2** Daily inflow data of the Alpenrhein during the period of simulation. (*a*) Discharge in $10^6$ m$^3$/d. (*b*) Temperature in °C. (*c*) Conductivity in $\mu$S/cm. (*d*) Salinity in g/l (calculated, see text). (*e*) Suspended load in g/l (calculated, see text).

the same way as for the Alpenrhein, with coefficients updated to 1985 by Wagner (personal communication). As shown in Figure 5.2, the inflow of the Alpenrhein varies by a factor of approximately six, and high suspended sediment concentrations coincide with high daily discharges. In the other three rivers the salinity maxima corresponding to low discharges are significantly higher than in the Alpenrhein.

Meteorological observations consisting of daily averaged values of wind speed, air temperature, relative humidity, fraction of cloud cover, and shortwave solar radiation were provided by the German Weather Service from its station at Konstanz (see Figure 5.1*a*). These data are considered to be representative of the main features of the air mass over the lake except for the occasional föhn winds which act mainly on the eastern part of the lake. The incoming long-wave radiation was derived from the mean daily cloud cover and air temperature according to the formula of Swinbank (TVA, 1972). Besides the daily meteorological averages, components of six-hour averaged winds are required for the calculation of shear at the base of the epilimnion due to basin-scale internal waves. These components of wind are resolved along the longitudinal axis of the lake (303°) and the transverse axis (213°), as indicated in Figure 5.1*a*.

Temperature profiles were measured at approximately biweekly intervals; the measurements were made with a standard platinum thermometer at a depth resolution of 1 meter down to 50 m and beyond that, with reversing thermometers at greater depth intervals. The three selected stations are situated in the center of the western basin, at the central deepest point of the main basin, and in the middle of the transect of maximum width as shown in Figure 5.1*a*. In addition, at the two western stations, measurements of light attenuation were carried out on every other cruise, from the surface down to the 60 m depth, using a transmissometer from Philipp Schenk products, Vienna (1965).

In order to compare with the calculated temperature profiles, lakewide-averaged observations were determined from the measurements at the three selected sites by area-weighted averaging in every same-measurement depth. The three open lake stations minimized the effect of smearing of the profiles due to tilting of the isotherms, which is more pronounced in the near-shore region. These data are displayed along with the model results in Figures 5.3 and 5.4*b*. The light transmissivity data were converted to sediment concentration following the relation $S = -\ln(TR/100)/0.14$, where S is the suspended load in mg/l and TR is the transmissivity in % for light of 400 nm on a beam length of 25 cm in the range $TR > 17\%$. This equation is valid for Lake St. Clair and the St. Lawrence River (Canada) and has been applied to Lake Constance since specific data for it is lacking. Finally, the two transects of the short-term survey selected for comparison with the simulation on November 20 and on November 29 through 30 are depicted in a separate map in Figure 5.1*b*.

## 5.6 Model Description and Extensions

The governing physical processes and numerical procedures incorporated in the DYRESM model are outlined by Fischer et al. (1979) and by Imberger and Patterson (1981). Subsequent model extensions relevant to this application are discussed by Patterson et al. (1984) for lakes, by Wei and Hamblin (1986) for suspended sediments, and by Patterson and Hamblin (1988) for ice-covered lakes. The essential aspect of the numerical scheme is that the vertical structure is represented by a variable number of layers of different thicknesses. Layers form or disappear according to the development or reduction of the stratification. This type of Lagrangian grid scheme avoids the artificial smearing of the natural step-like structure at the base of the epilimnion, that occurs when fixed-grid schemes are applied.

The salient feature of the physics of the model is that the vertical turbulent mixing formulation is semi-empirical in character. The mixing formulation for the epilimnion is given in terms of the standard treatment of the turbulent kinetic energy equation. Two semi-empirical mixing laws, based on the underlying principle that a certain fraction of the dissipated turbulent energy is transformed into potential energy of the lake, were examined for both the thermocline and the hypolimnion. The first approach is characterized by direct relation to the sum of the wind energy input per day and the daily release of potential energy of the inflowing rivers, and an inverse relationship to the depth-dependent stability. In the second case suggested by Harleman (1986), the essential difference is that the mixing is inversely proportional to the mean potential energy per unit volume of the lake instead of its local stability.

Another important model extension results from the main wind forcing from south-westerly directions. Due to the geographical orientation of the lake (see Figure 5.1*a*), the transverse wind component is, in general, about twice the longitudinal component. Therefore, it was necessary to extend the production of shear at the base of the epilimnion to include the transverse component of the wind. This shear was allowed to build up until either one quarter-period of the transverse internal seiche had elapsed or the transverse wind component changed sign—whichever occurred first.

A further modification is the incorporation of the effect of particle settling in the river inflows, when these waters descend from the lake surface to their level of horizontal spreading. The bulk sediment settling velocity in these plunging river plumes was assumed to be two orders of magnitude greater than the lake settling velocity of $10^{-2}$ mm $s^{-1}$. The best agreement between observed and calculated density profiles was achieved for a sediment sinking rate of 5 mm $s^{-1}$ in the river plumes. The density of the lake waters has been determined from their temperature, salinity, and suspended load by using the formulae given by Wagner and Wagner (1978) for the main tributaries. The light attenuation coefficients were derived from the field transmissivity observations down to 30 m depth.

## 5.7 Comparison of Model Results with Observations

The initial values, representing the starting point of the calculations on February 1, 1979 were vertically almost uniform field data. Initial salinity data were taken from a seven-year average of monthly salinity measurements (G. Wagner, personal communication). The model was run continuously until March 31, 1980. From calculated temperature profiles, we selected those depicting the strongly stratified season between July 23, 1979 and December 12, 1979 (Figure 5.3). This comparison exhibits a reasonable overall agreement with the observations except for the profiles on September 17 and October 22, 1979, when the epilimnion temperature and thickness are underestimated. On the other hand, the thermal stratification is evidently well simulated on August 7. The

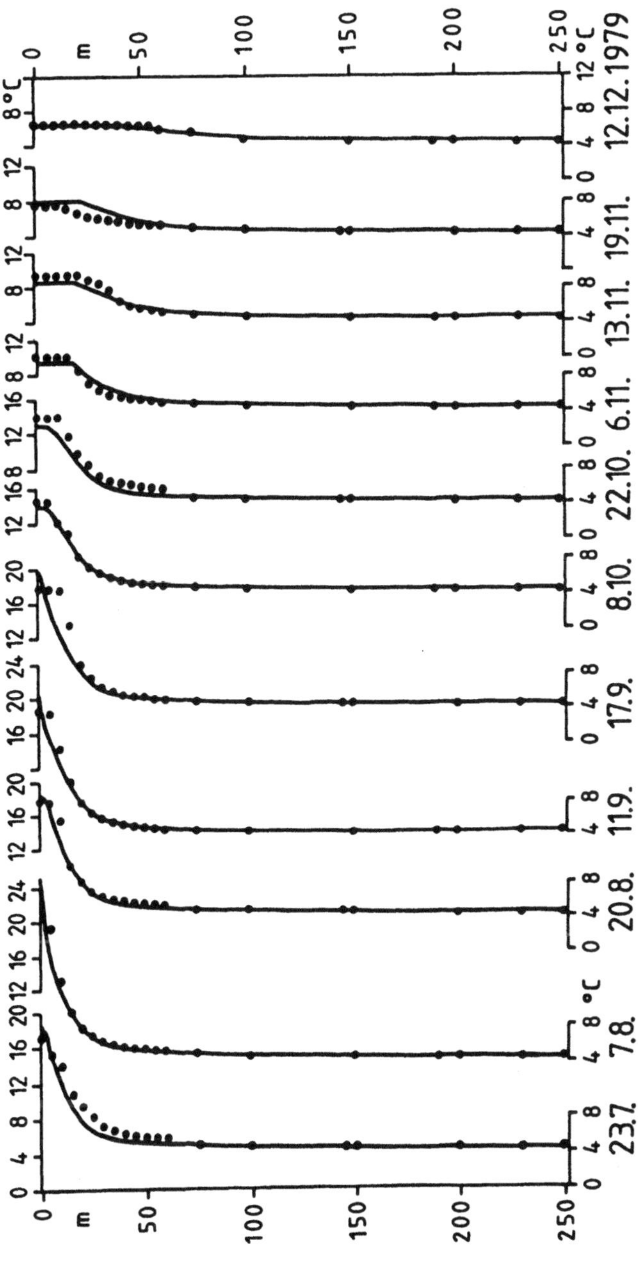


**Figure 5.3** Comparison of the calculated and observed temperature profiles from July 23 (23.7.) through December 12, 1979 (12.12.). Solid line: calculation. Dots: measurements.

model results are complemented by the temperature-depth time histories shown in Figure 5.4 for the whole simulation period. The calculated isotherms clearly demonstrate that the annual cycle of thermal stratification in Lake Constance is reproduced quite well over the entire simulation period. Statistically, the volume-weighted *rms* (root-mean-square) temperature error, averaged over the whole period, equals 0.45°C.

The synoptic temperature transect, obtained by the high-resolution profiler (Kroebel, personal communication), demonstrates the nearly horizontal uniformity of the temperature structure on November 20 (Figure 5.5*a*), in spite of a relatively weak stratification and strong wind events during the first half of this month.

The epilimnion had deepened to 25 m, which compares well with the calculation (Figure 5.5*b*), though its temperature differs by about 0.5°C. A rain storm between November 5 and 10 caused a strong inflow event with increased suspended load. This is clearly reflected in the isopleths of attenuation of blue light on the same transect (Figure 5.5*c*). The increased levels of attenuation associated with this event are found at the depth of 35 m. After a ten-day period of autumnal cooling under calm conditions, there is remarkable agreement between the measurements of the near-synoptic longitudinal transect and the numerical approach displayed in Figure 5.6. The calculated deep temperatures slightly exceed the observations, as in the previous comparisons.

Up to this point, the calculations have employed Harleman's (1986) vertical mixing specification outlined in the preceding model description. A further numerical experiment to determine the sensitivity of the simulation to the vertical mixing below the epilimnion was carried out, assuming the inversely proportional dependence of mixing on the local stability as in the original DYRESM formulation for small lakes and reservoirs. We found that the volume-weighted *rms* temperature error was about twice the previously mentioned error. The Harleman formulation resulted in decreased mixing during the well-stratified summer period and increased mixing during the spring and autumn, relative to the standard approach.

Finally, we want to demonstrate that the numerical model may be employed to generate hypotheses on the nature of time- and depth-dependent distributions of other components of water quality. These hypotheses may be tested by properly designed field experiments. In the present study, two parameters have been taken into account, namely, suspended sediments and salinity. Figure 5.7 shows the calculated slow settling of the increased suspended matter, which enters the lake from May onward. Bearing in mind that the represented quantity is a horizontal average, it is apparent from Figure 5.7 that the loading from the previously mentioned storm in November is too small to be distinguished from the aftermath of the main sediment input in spring. The other example, which is shown in Figure 5.8, is the annual cycle of the calculated dissolved solids. There is a pronounced summer minimum in the thermocline region and near-temporal constancy in the hypolimnion. This overall annual behavior is in qualitative agreement with the seven-year averaged ob-

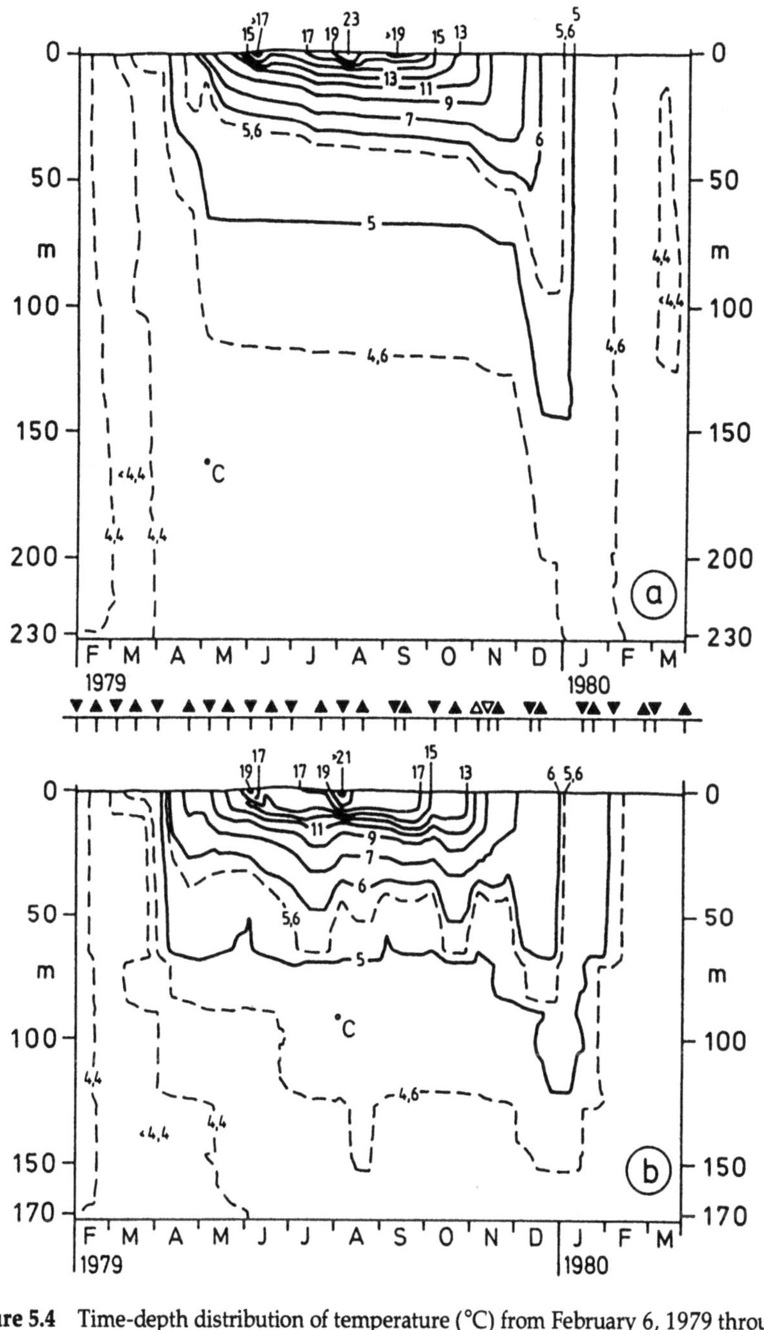

**Figure 5.4** Time-depth distribution of temperature (°C) from February 6, 1979 through March 24, 1980. (*a*) Calculation. (*b*) Observation. The measurement dates are indicated on the separate scale marked by solid and open triangles. Upward-pointing solid triangles indicate volumetrically weighted averages for the stations U and F, while solid-downward triangles are averages of stations F and L. Open triangles represent measurements from only one station (upward, F; downward, L).

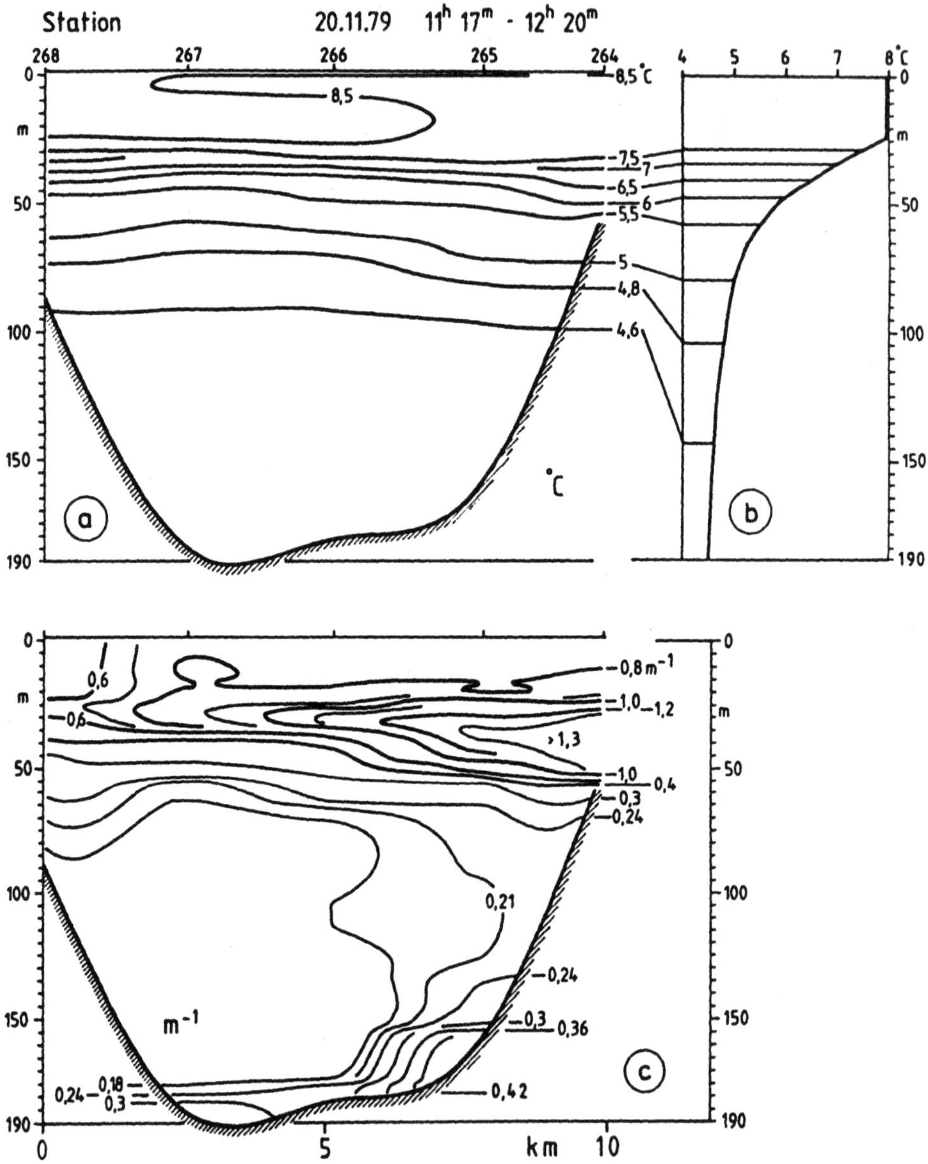


**Figure 5.5** (*a*) Transverse temperature cross sections, stations 264–268 (see Figure 5.1*b*). (*b*) Calculated temperature profile on November 20, 1979. (*c*) Attenuation of blue light (455 nm).

servations made available by Wagner (personal communication). In winter 1979/80, deep convective overturn appears to be restricted to the upper 130 m depth; this is consistent with CTD-probe measurements carried out by Kroebel, Hollan, and Rathlev on March 12, 1980.

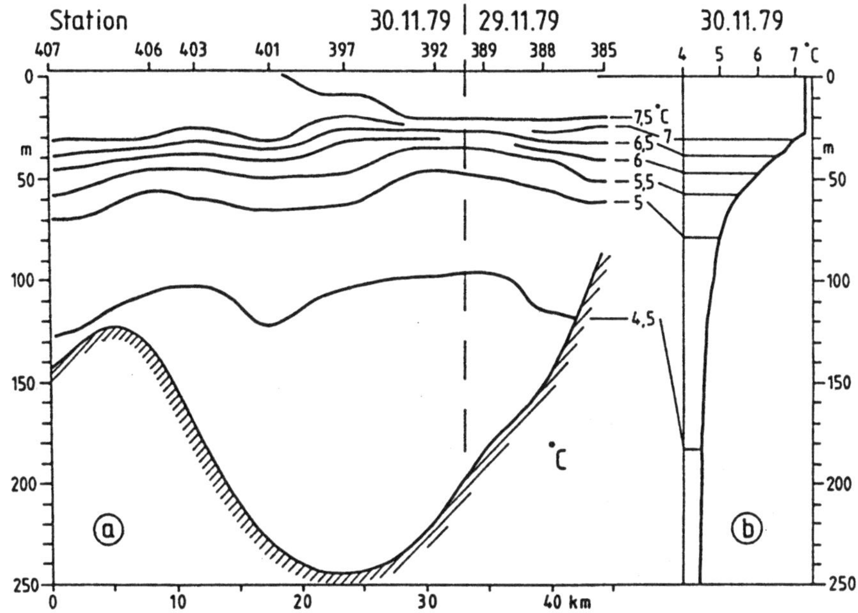


**Figure 5.6** (*a*) Longitudinal transect of temperature (°C). (*b*) Calculated profile on November 30, 1979. The measurement stations on this central transect (see Figure 5.1*b*) are selected from a lake-wide survey on a zig-zag shaped course on November 29 (stations 385–389 from 17 hr through 18.47 hr) and on November 30 (stations 392–407 from 10 hr through 17.11 hr).

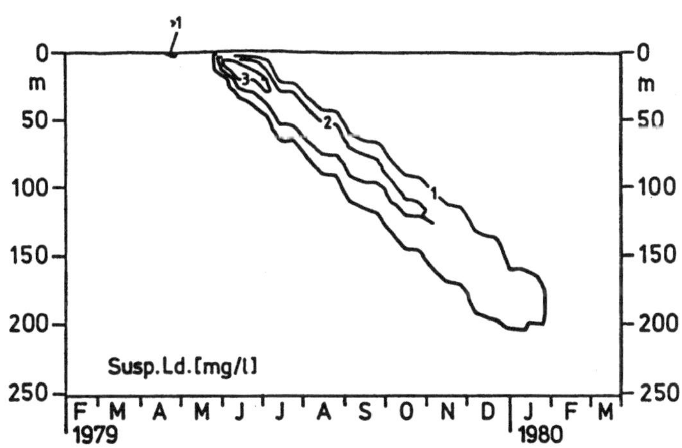


**Figure 5.7** Time-depth distribution of the suspended sediment (mg/l) from February 6, 1979 through March 24, 1980. Same numerical experiment as was shown in Figure 5.4*a*.

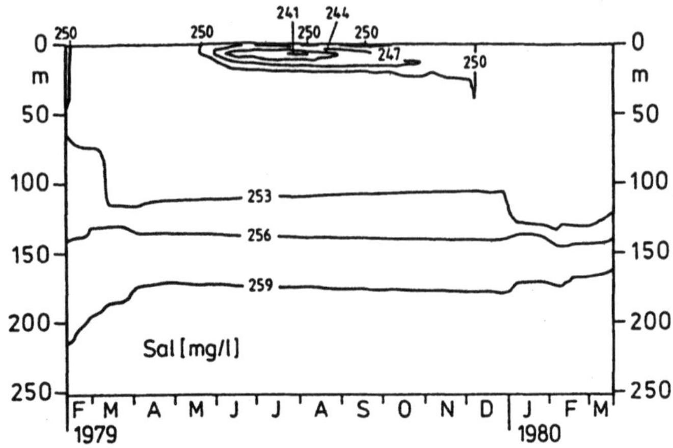


**Figure 5.8** Time-depth distribution of salinity (mg/l) for the same simulation period and calculation as in Figure 5.7.

## 5.8 Conclusions

The good correspondence between the DYRESM model results and the field observations indicates that, with appropriate modification—such as improved mixing in the deep layer and inclusion of transverse wind forcing of internal waves—the one-dimensional modelling approach may yield useful results for large lakes. Furthermore, the dynamical simulation proved the consistency between the observed meteorological and hydrological fields and the physical response. The study also indicates a number of possible improvements for routine data collection for Lake Constance. In particular, dissolved solid concentrations and suspended sediment profiles should be added to the regular sampling program, which should include measurements of the outflow. More information is also necessary on suspended-sediment particle size, settling velocity, mineralogy, and organic fraction in the lake and at the mouths of the major inflows. The knowledge of the spatial distribution of meteorological forcing over the lake should be improved by additional measurements at other sites on the shore and direct measurement of longwave radiation should be made.

This study emphasizes the need for improved specification of vertical mixing in long-term simulations in large lakes. Evidence indicates that semi-empirical solutions for mixing are fruitful. However, we must know more about the parameterization of vertical mixing processes below the mixed layer. Moreover, the parameterization of the entrainment of lake water into the complicated plunging river plumes in large deep lakes should be more detailed than is prescribed in DYRESM. The increase of the sediment sinking rate in these plumes, which has to be assumed in this model, is probably partly due to this inadequate approach (Patterson, personal communication). There is much work to be done in order to adjust the small lake models to the specific conditions prevailing in large lakes.

The ability of the model to simulate the annual cycle of stratification suggests that the long-term climatology of the lake could be studied, based on existing data at least for the past 30 years. It would also be valuable to conduct applications to other large lakes in order to substantiate the various parameterizations of the physical processes involved. A comparison of the model performance on a wide range of forcing conditions and basin geometries would improve the understanding of the stratification in large lakes. For small lakes, a version of the model including dissolved and particulate phosphorus as well as dissolved oxygen has been developed (McCrimmon, 1988).

---

*Acknowledgments* The authors record their gratitude to Prof. Werner Kroebel from the University of Kiel for providing three isopleth diagrams. R.C. McCrimmon of McMaster University assisted in the calculations, and Urszula Zawielak, Karlsruhe, did the drawings. The investigation was sponsored by the GKSS-Forschungszentrum Geesthacht at Geesthacht, FRG, through the bilateral agreement on scientific and technological exchange between Canada and Germany of 1971.

## References

Baba, N. 1974. *A Numerical Investigation of Lake Ontario Dynamics and Thermodynamics.* Ph.D. Thesis. Princeton University, 246 pp.

Barkmann, W. 1987. Der Einfluß der Wärmebilanz auf die Struktur der säsonalen Grenzschicht. *Ber. Inst. f. Meereskunde.* No. 171, Kiel, 161 pp.

Berger, F. 1955. Die Dichte natürlicher Wässer und die Konzentrations-Stabilität in Seen. *Arch. Hydrobiol., Suppl.* 22:286–294.

Blumberg, A.F. 1986. Turbulent mixing processes in lakes, reservoirs and impoundments, pp. 79–104 in: Gray, W.G. (ed.). *Physics-based Modeling of Lakes, Reservoirs and Impoundments.* American Society of Civil Engineers, New York, 308 pp.

Bowden, K.F. 1983. *Physical Oceanography of Coastal Waters.* Ellis Horwood, Chichester, 302 pp.

Bührer, H. und Wagner, G. 1982. Die Belastung des Bodensees mit Phosphor- und Stickstoffverbindungen und organischem Kohlenstoff im Abflußjahr 1978/79. *Ber. Int. Gewässerschutzkomm. Bodensee.* 28, 68 pp.

Fischer, H.B., List, E.J., Koh, R.C.Y., Imberger, J., and Brooks, N.H. 1979. *Mixing in Inland and Coastal Waters.* Academic Press, London, 483 pp.

Green, A.W. 1976. Evolution of the Lake Ontario thermocline during International Field Year in the Great Lakes 1972, p. 67–78 in Watson, I.A.E. (ed.). *The Dynamics of Stratification and of Stratified Flow of Large Lakes.* International Joint Commission, Regional Office, Ontario.

Gregg, M.C. 1987. Diapycnal mixing in the thermocline: a review. *J. Geoph. Res.* 92, C5:5249–5286.

Harleman, D.R.F. 1986. Hydrothermal modeling of reservoirs in cold regions: status and research needs. *Cold Regions Hydrology Symposium, American Water Resources Assoc.,* Fairbanks, p. 39–50.

Henderson-Sellers, B. 1985. A new formulation of eddy diffusivity thermocline models. *Appl. Math. Model* 9:441–446.

Henyey, F.S., Wright, J., and Flatté, S.M. 1986. Energy and action flow through the internal wave field: an eikonal approach. *J. Geophys. Res.* 91:8487–8496.
Hollan, E. and Simons, T.J. 1978. Wind-induced changes of temperature and currents in Lake Constance. *Arch. Met. Geophys. Bioklim.* Ser. A27:333–373.
Hollan, E., Hamblin, P.F., Lehn, H., and McCrimmon, R.C. 1990. Observation and model of the annual stratification cycle in Lake Constance. *Arch. Hydrobiol., Suppl.* (In press.)
Imberger, J. and Patterson, J.C. 1981. A dynamic reservoir simulation model—DYRESM 5, p. 310–360 in Fischer, H.B. (ed.). *Transport Models for Inland and Coastal Waters.* Academic Press, New York.
Ivey, G.N. and Patterson, J.C. 1984. A model of the vertical mixing in Lake Erie in summer. *Limnol. Oceanogr.* 29:553–563.
Kilworth, P.D. and Carmack, E.C. 1979. A filling-box model for river dominated lakes. *Limnol. Oceanogr.* 24:201–217.
Lam, D.C.L. and Schertzer, W.M. 1987. Lake Erie thermocline model results: comparison with 1967–1982 data and relation to anoxic occurrences. *J. Great Lakes Res.* 13:757–769.
McCrimmon, R.C. 1988. An oxygen model for ice-covered and ice-free reservoirs. Thesis, M. Eng., Dept. Civil Engineering. McMaster University, Hamilton, Canada. 225 pp.
Orlob, G.T. (editor) 1983. *Mathematical Modeling of Water Quality: Streams, Lakes and Reservoirs.* John Wiley & Sons, New York, 518 pp.
Patterson, J.C., Hamblin, P.F., and Imberger, J. 1984. Classification and dynamic simulation of the vertical density structure of lakes. *Limnol. Oceanogr.* 29:845–861.
Patterson, J.C. and Hamblin, P.F. 1988. Thermal simulation of lakes with winter ice cover. *Limnol. Oceanogr.* 33:323–338.
Pollard, R.T., Rhines, P.B., and Thompson, R. 1973. The deepening of the wind-mixed layer. *Geophys. Fluid. Dynamics.* 4:381–404.
Royer, L., Chiocchio, F., and Boyce, F.M. 1987. Tracking short-term physical and biological changes in the central basin of Lake Erie. *J. Great Lakes Res.* 13:587–606.
Spigel, R.H. and Imberger, J. 1980. The classification of mixed layer dynamics in lakes of small to medium size. *J. Phys. Oceanogr.* 10: 1104–1121.
Svensson, V. 1978. A mathematical model of the seasonal thermocline. *Institute of Technology*, Report No. 1002, University of Lund, Sweden.
Turner, J.S. 1973. *Buoyancy Effects in Fluids.* Cambridge University Press, Cambridge, U.K., 367 pp.
TVA. 1972. Heat and mass transfer between a water surface and the atmosphere. *Tennessee Valley Authority*, Rep. No. 0-6803, Norris, Tenn., 173 pp.
Vollenweider, R.A. 1964. Über oligomiktische Verhältnisse des Lago Maggiore und einiger insubrischer Seen. *Mem. Ist. Ital. Idrobiol.* 17:191–206.
Wagner, G. and Wagner, B. 1978. Zur Einschichtung von Flußwasser in den Bodensee-Obersee. *Schweiz. Z. Hydrol.* 40:231–248.
Walters, R.A, Carey, G.F., and Winters, D.F. 1978. Temperature computation for temperate lakes. *Applied Mathematical Modelling.* 2:41–48.
Wei, C.Y. and Hamblin, P.F. 1986. Reservoir water quality simulations in cold regions. *Cold Regions Hydrology Symposium, American Water Resources Assoc.*, Fairbanks, p. 167–177.

# 6

# Spring Mixing Depth as a Determinant of Annual Primary Production in Lakes

Charles R. Goldman and Alan Jassby

ABSTRACT Year-to-year fluctuations in spring mixing depth are a major source of interannual variation in primary production at Lake Tahoe, California-Nevada. The morphology of the Tahoe basin predisposes the lake to a variable spring mixing depth. The relation between mixing depth and annual production is mediated by the transport of nitrate and possibly viable biomass from below the euphotic zone.

## 6.1 Introduction

Knowledge of the spring circulation period and its importance for many temperate lakes dates back to the studies of Lake Geneva (Lac Léman) by Forel (1895), who proposed the first lake typology based on thermal conditions. The specific role of major circulation events in fertilizing surface waters and modifying the light regime experienced by primary producers was understood, first in marine waters, more than 50 years ago (Marshall and Orr, 1927; Gran and Braarud, 1935). Mixing processes, occurring at many different frequencies, have since been shown to cause changes of algal community composition or primary productivity in both lake and ocean waters (Viner, 1985; Tett and Edwards, 1984).

Despite the acknowledged importance of spring mixing events, relatively little quantitative evidence has been collected on the existence and ramifications of variability in the spring circulation period from one year to the next. Part of the problem lies in assessing the intensity of this circulation. Interannual fluctuations in mixing intensity are difficult to identify in those lakes that mix deeply every year and become nearly isothermal, but do not homogenize completely (such lakes are included among oligomictic lakes by Lewis, 1973). As a result, much of the evidence for variable spring mixing comes from dimictic lakes experiencing temporary meromixis, where temperature or oxygen measurements clearly indicate complete circulation in some years, and only partial circulation in others (e.g., Wöthersee: Findenegg, 1937; Schleinsee: Einsele, 1941; Linsley Pond: Hutchinson, 1941; Great Bear Lake: Johnson, 1966; Martin Lake: Wetzel, 1973; Hall Lake: Culver, 1977; Lawrence Lake: Wetzel, 1981). Sometimes, a comparison of the effects of mixing intensity between different years can be made even in the absence of temporary meromixis. For example, the extent of sediment resuspension after ice breakup, which reflects mixing intensity, was shown to have profound effects on vernal and summer phytoplankton abundances in Lake Erie (Chandler, 1940, 1942a, 1942b, 1944) and Lake Michigan (Scavia et al., 1986). All of these studies demonstrate at least a potential role for the spring circulation period as a cause of interannual variation in production, but the quantitative significance remains unknown.

In order to assess the quantitative contribution of spring mixing to year-to-year variability in primary production, long-term data series for both mixing intensity and productivity are required. Long series for either variable are quite rare. Goldman (1974) had suggested earlier the potential importance of variable mixing depth in controlling primary productivity from year to year at Lake Tahoe, but insufficient data were available at the time. In the present investigation, we examined the long-term data set for Lake Tahoe, which includes measurements of primary productivity since 1959 (continuously since 1967) and estimates of the maximum mixing depth every year since 1973. On the basis of these data series, we were able to show a close linkage between year-to-year variability in productivity and the extent of spring mixing.

Lake Tahoe is a large, subalpine lake located in a graben near the crest of the Sierra Nevada Range in Northern California (39°N, 120°W). The lake has a surface area of 500 km$^2$, a maximum depth of 505 m, and a mean depth of 313 m. The onset of cultural eutrophication in ultra-oligotrophic Lake Tahoe has been documented over the last three decades (Goldman, 1974, 1981, 1985). On the average, annual primary production has increased at a mean rate of about 6% per year, with a concomitant loss in secchi transparency of 0.44 m per year. Eutrophication at Tahoe reflects the growing human population in the watershed and an accompanying loss of vegetation, increase in soil erosion, and rise in nutrient loading from runoff and atmospheric deposition.

## 6.2 Methods

Primary productivity has been measured continuously at Lake Tahoe for the past 19 years using a standardized $^{14}C$ method (Goldman, 1963). Data are collected weekly in summer and biweekly during the colder months (the lake does not freeze over in winter). At least ten depths are sampled between 0 and 105 m. The water column is also sampled at 50-m intervals (and occasionally 25-m intervals) from 100 m to the bottom.

The time and extent of mixing in Lake Tahoe can be estimated by examining successive depth profiles of dissolved nitrate, which has been measured regularly since 1973 (Paerl et al., 1975; Goldman and de Amezaga, 1984). Depth of mixing is taken to be the maximum point at which an upper layer of approximately uniform nitrate concentration meets a lower layer of rapidly increasing nitrate. Two independent observers examined the data and arrived at the same conclusions regarding the time and extent of mixing, which always reached its maximum depth in March or early April. Previously published estimates of mixing depth may differ slightly because of different criteria, but we believe the values reported here to be the most objective. When available, other deep-water stations and other variables were examined to check the conclusions. Due to the sampling frequency in time and depth, the time of maximum spring mixing has an accuracy of about 2 weeks, and the depth of mixing an accuracy of about 25 m. Tahoe is approximately 450 m deep at the deep-water sampling station for nitrate; the methods used here thus cannot distinguish among mixing events penetrating below 450 m.

## 6.3 Results

As noted above, primary productivity in Tahoe is dominated by an increasing trend (Figure 6.1); what concerns us here are the fluctuations about this long-term trend. We examined the relationships between annual primary production and other variables in Lake Tahoe. Because trend or other sources of autocorrelation can cause spurious correlations or obscure valid associations, we first filtered each series to remove autocorrelation and used the residual "pre-whitened" series for cross-correlation (Box and Jenkins, 1976). The strongest peak occurred in the lag zero cross-correlation between primary productivity

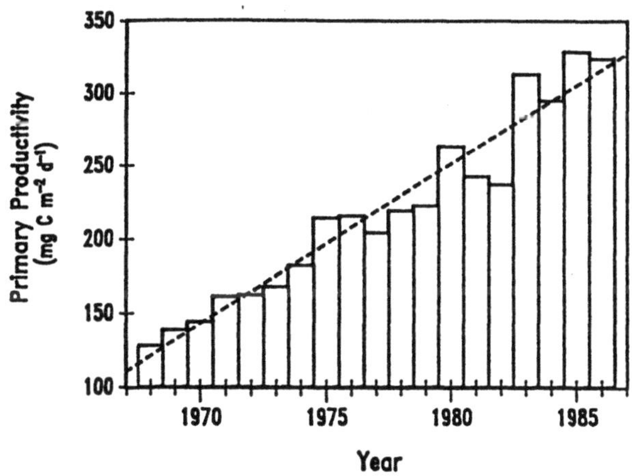


**Figure 6.1** Annual average of primary productivity at Lake Tahoe from 1968 through 1986. The trend line estimated from linear regression is also shown.

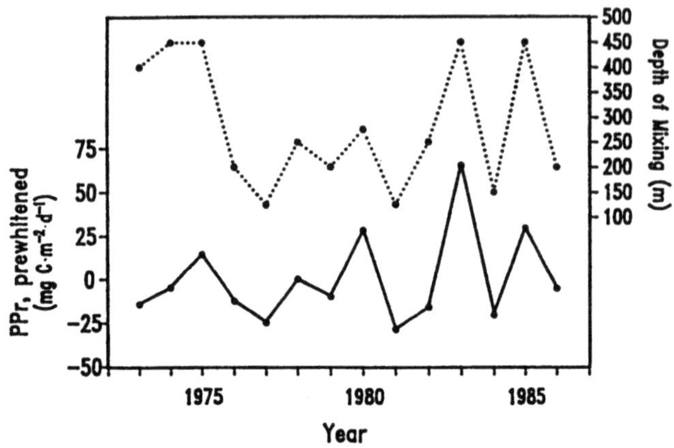


**Figure 6.2** Time series of average annual primary productivity (PPr) and spring mixing depth from 1973 through 1986 at Lake Tahoe. The primary productivity values have been prewhitened by filtering with a Markov process to remove autocorrelation.

(filtered with a Markov process) and spring mixing depth (no filtering was required to remove internal structure in the time series for mixing depth). A plot of the two variables illustrates their striking correspondence (Figure 6.2); spring mixing depth and prewhitened annual primary production have a significant direct association ($r=0.68$, $P<0.01$).

## 6.4 Discussion

Mixing events in lakes are critical processes in maintaining productivity levels. At certain points in the annual cycle of heating and cooling, particularly during the spring season in many temperate lakes, the density of water approaches uniformity throughout the water column. Winds, aided by convection, can then stir the lake, bringing up materials from deep waters to participate again in primary production in the euphotic zone. The data presented here indicate that this mixing event not only maintains a characteristic trophic state, or level of production, but also, through year-to-year variability in mixing depth, is tightly linked to interannual fluctuations about this level. Preliminary analyses of the seasonal pattern of productivity in Lake Tahoe suggest that the effect of mixing depth extends primarily from March through August. These analyses also indicate that the variability of mixing depth is related to weather events in late winter and early spring.

**Basin Morphometry** What characteristics predispose a lake to incomplete and, hence, variable spring mixing? Several types of factors can be recognized, including basin morphology; geographic location, which includes the systematic effect of latitude and altitude on net radiation, air temperature, and winds; and basin exposure, including the presence or absence of an ice cover. Of these factors, basin shape is particularly interesting for the case of Tahoe. In general, incomplete mixing becomes more likely as maximum depth increases and wind fetch decreases. The relative depth $Z_r$, defined as the maximum depth as a percentage of mean surface diameter, is a convenient scalar quantity summarizing this effect of basin morphology on the likelihood of incomplete mixing. Herdendorf (1982) tabulates the data necessary to calculate $Z_r$ for 164 of the 253 largest natural lakes ($\geq$500 km$^2$) in the world. Tahoe has the largest value, $Z_r$=2.0, and so, in comparison to other large lakes, is the most prone to incomplete mixing by virtue of its shape.

When we consider all lakes, Tahoe is no longer unique with regard to $Z_r$. Of the 50 deepest lakes in the world, for example, Tahoe ranks 23rd in terms of $Z_r$ (calculated from the data of Herdendorf, 1982); and many small lakes have a $Z_r$ greater than that of Tahoe. Lake Nyos, Cameroon, where a rare, deep mixing event in 1986 resulted in massive release of stored carbon dioxide from the hypolimnion and the asphyxiation of about 1700 people in the surrounding basin, has an extremely high relative depth: $Z_r$=15 (Kling and Evans, 1987; Kling et al., 1987).

**Internal Loading** The stimulation of productivity by spring mixing and other circulation events is usually attributed to the return of regenerated nutrients from deep waters to the euphotic zone. Bioassays of Lake Tahoe water using the $^{14}C$-uptake technique have implicated nitrogen, phosphorus, trace minerals such as iron and zinc, and perhaps natural chelating agents in the stimulation of primary productivity (Goldman, 1981). During the first two decades of study beginning in 1959, bioassay experiments consistently showed stimulation by nitrogen and iron additions, but little or no response, and often even inhibition, when phosphorus was added. Positive responses to phosphorus were observed only when nitrogen was added simultaneously. Recently, however, the lake phytoplankton has shown an increasing positive response to phosphorus addition.

Despite the increasing bioassay response to phosphorus, several separate pieces of evidence suggest that nitrate-nitrogen ($NO_3$-N) is the primary nutrient mediating the effect of spring mixing on annual production. First, internal loading of $NO_3$-N far exceeds external loading of N. Although the exact amounts vary from year to year, approximately 70 meq m$^{-2}$ $NO_3$ is mixed into the euphotic zone (0–105 m) during the winter and early spring (Byron and Goldman, 1986), whereas annual loading of N from runoff and deposition amounts to only about 20 meq m$^{-2}$. Second, the actual quantity of nitrogen returned to the euphotic zone clearly is affected by the depth of spring mixing (Figure 6.3): even if the two anomalous points in Figure 6.3 (1978 and 1980) are included, the association between the two variables is significant ($r=0.71$, $n=14$, $P<0.01$). Finally, if we exclude the anomalous years identified in Figure 6.3, euphotic-zone $NO_3$ and the interannual change in primary production also have a significant link ($r=0.65$, $n=12$, $P<0.05$). Together with the bioassay

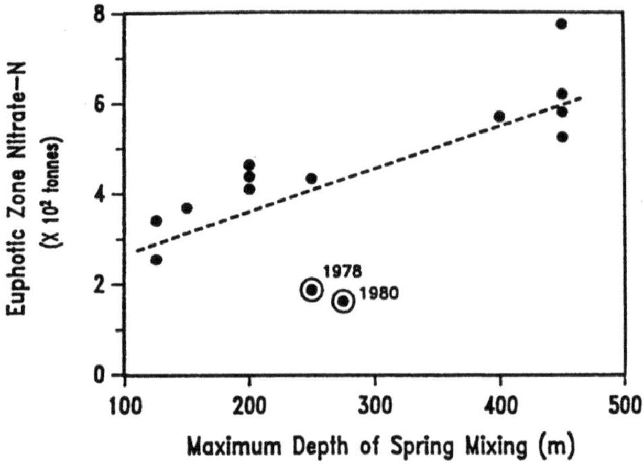


**Figure 6.3** Nitrate-nitrogen in the euphotic zone (0–105 m) of Lake Tahoe versus the depth of spring mixing. Nitrate data are taken from the same profile used to estimate maximum spring mixing depth from 1973 through 1986.

data showing nitrogen sensitivity, these different considerations implicate transport of aphotic $NO_3$ into the euphotic zone as the link between spring mixing and interannual production changes.

Previous studies have shown that algal cells in the deep waters of Tahoe below the euphotic zone are viable and still capable of photosynthesis (Tilzer et al., 1977; Vincent, 1978). This raises the possibility that spring mixing events are also inoculating the surface water with phytoplankton. The precise effect of spring mixing then would depend not only on its depth and the extent of winter nutrient regeneration in deep waters, but also on the survival rate of phytoplankton that have sunk below the euphotic zone during previous years. We are now analyzing the detailed phytoplankton enumerations for more information on these possibilities.

*Acknowledgments* We thank T. Powell, P.J. Richerson, P. Marjanovic, and J.J. Elser for their helpful comments and suggestions; P. Arneson for assistance with the data management and graphics; G. Malyj for manuscript preparation; and the National Science Foundation (NSF grant BSR-8705170) and Environmental Protection Agency (EPA cooperative agreement CR814581) for support.

## References

Box, G.E.P. and Jenkins, G.M. 1976. *Time Series Analysis, Forecasting and Control*. Rev. ed. Holden-Day, San Francisco. 575 pp.

Byron, E.R. and Goldman, C.R. 1986. *Lake Tahoe Interagency Monitoring Program Sixth Annual Report: Water Year 1985*. Tahoe Research Group, Univ. California, Davis. 55 pp.

Chandler, D.C. 1940. Limnological studies of Western Lake Erie. I. Plankton and certain physical-chemical data of the Bass Islands region, from September 1938 to November 1939. *Ohio Journal of Science* 40:291–336.
Chandler, D.C. 1942a. Limnological studies of Western Lake Erie. II. Light penetration and its relation to turbidity. *Ecology* 23:41–52.
Chandler, D.C. 1942b. Limnological studies of Western Lake Erie. III. Phytoplankton and physical-chemical data from November, 1939 to November, 1940. *Ohio Journal of Science* 42:24–44.
Chandler, D.C. 1944. Limnological studies of Western Lake Erie. IV. Relation of limnological and climatic factors to the phytoplankton of 1941. *Transactions of the American Microscopical Society* 63:203–236.
Culver, D.A. 1977. Biogenic meromixis and stability in a soft-water lake. *Limnology and Oceanography* 22:667–687.
Einsele, W. 1941. Die Umsetzung von zugeführtem, anorganischen Phosphat im eutrophen See und ihre Rüchwirkungen auf seinen Gesamthaushalt. *Zeitschrift für Fischerei* 39:407–488.
Findenegg, I. 1937. Holomiktische und meromiktische Seen. *Internationale Revue der Gesamten Hydrobiologie* 35:586–610.
Forel, F.A. 1895. *Le Léman: Monographie Limnologique. Tome 2. Mécanique, Chimie, Thermique, Optique, Acoustique*. F. Rouge, Lausanne. 651 pp.
Goldman, C.R. 1963. The measurement of primary productivity and limiting factors in freshwater with Carbon-14, p. 103–113 in Doty, M.S. (editor), *Proc. Conference on Primary Productivity Measurements, Marine and Freshwater*. Division of Technology Information Report TID-7633, Atomic Energy Commission, Washington, D.C.
Goldman, C.R. 1974. *Eutrophication of Lake Tahoe Emphasizing Water Quality*. EPA-660/ 3-74-034. U.S. Govt. Printing Office, Washington, D.C. 408 pp.
Goldman, C.R. 1981. Lake Tahoe: two decades of change in a nitrogen deficient oligotrophic lake. *Internationale Vereinigung für Theoretische und Angewandte Limnologie Verhandlungen* 21:45–70.
Goldman, C.R. 1985. Lake Tahoe: a microcosm for the study of change. *Illinois Natural History Survey Bulletin* 33:247–260.
Goldman, C.R. and de Amezaga, E. 1984. Primary productivity and precipitation at Castle Lake and Lake Tahoe during twenty-four years, 1959–1982. *Internationale Vereinigung für Theoretische und Angewandte Limnologie Verhandlungen* 22:591–599.
Gran, H.H. and Braarud, T. 1935. A quantitative study of the phytoplankton in the Bay of Fundy and the Gulf of Maine (including observations on hydrography, chemistry and turbidity). *Journal of the Biological Board of Canada* 1:279–467.
Herdendorf, C.E. 1982. Large lakes of the world. *Journal of Great Lakes Research* 8:379–412.
Hutchinson, G.E. 1941. Limnological studies in Connecticut. IV. The mechanism of intermediary metabolism in stratified lakes. *Ecological Monographs* 11:21–60.
Johnson, L. 1966. Temperature of maximum density of fresh water and its effect on circulation in Great Bear Lake. *Journal of the Fisheries Research Board of Canada* 23:963–973.
Kling, G.W. and Evans, W.C. 1987. Followup on the 1986 Lake Nyos gas disaster. Unpublished MS. Yaoundé, Cameroon.
Kling, G.W. et al. 1987. The 1986 Lake Nyos gas disaster in Cameroon, West Africa. *Science* 236:169–175.
Lewis, W.M. 1973. The thermal regime of Lake Lanao (Philippines) and its theoretical implication for tropical lakes. *Limnology and Oceanography* 18:200–217.
Marshall, S.M. and Orr, A.P. 1927. The relation of the plankton to some chemical and physical factors in the Clyde Sea area. *Journal of the Marine Biological Association of the United Kingdom* 14:837–868.
Paerl, H.W., Richards, R.C., Leonard, R.L., and Goldman, C.R. 1975. Seasonal nitrate

cycling as evidence of complete vertical mixing in Lake Tahoe, California-Nevada. *Limnology and Oceanography* 20:1–8.

Scavia, D., Fahnenstiel, G.L., Evans, M.S., Jude, D.J., and Lehman, J.T. 1986. Influence of salmonine predation and weather on long-term water quality trends in Lake Michigan. *Canadian Journal of Fisheries and Aquatic Sciences* 43:435–443.

Tett, P. and Edwards, A. 1984. Mixing and plankton: an interdisciplinary theme in oceanography. *Oceanography and Marine Biology: an Annual Review* 22:99–123.

Tilzer, M.M., Paerl, H.W. and Goldman, C.R. 1977. Sustained viability of aphotic phytoplankton in Lake Tahoe (California-Nevada). *Limnology and Oceanography* 22: 84–91.

Vincent, W.F. 1978. Survival of aphotic phytoplankton in Lake Tahoe throughout prolonged stratification. *Verhandlungen Internationale Vereinigung für Theoretische und Angewandte Limnologie* 20:401–406.

Viner, A.B. 1985. Thermal stability and phytoplankton distribution. *Hydrobiologia* 125: 47–69.

Wetzel, R.G. 1973. Productivity investigations of interconnected lakes. I. The eight lakes of the Oliver and Walters chains, northeastern Indiana. *Hydrobiological Studies* 3: 91–143.

Wetzel, R.G. 1981. Longterm dissolved and particulate alkaline phosphatase activity in a hardwater lake in relation to lake stability and phosphorus enrichments. *Internationale Vereinigung für Theoretische und Angewandte Limnologie Verhandlungen* 21: 369–381.

# 7

# Behavior of a Medium-Sized Basin Connected to a Large Lake

Axel Zenger, Johann Ilmberger, Gerhard Heinz,
Michael Schimmele, Peter Schlosser, Dieter Imboden, and
Karl Otto Münnich

ABSTRACT Regular measurements—extending over about one year—of water temperature, of electric conductivity, and of the concentration of the stable isotope helium-3 are analyzed with regard to the lateral exchange of water between the main part of Lake Constance—the Obersee—and its smaller neighboring basin, the Überlingersee. In the absence of a significant density stratification (as occurs in late winter and early spring), different wind speeds over the two basins cause different vertical mixing leading to horizontal concentration gradients between the basins. Wind-induced or density-driven horizontal currents are not strong enough to completely reduce the gradients. In contrast, during periods of strong vertical stratification (summer and autumn), internal waves extending along the whole lake cause periodical horizontal currents and mixing and thus lead to homogeneous concentration distributions along the horizontal axis. The seasonal variation of horizontal gradients couples the nutrient balance of Überlingersee to the chemistry of the main lake.

## 7.1 Introduction

Lake Constance has been subject to several physical research programs (see e.g., Hollan and Simons, 1978); various models have been developed to describe the physics of the lake. In 1986, a multidisciplinary research program was initiated to study in some detail the Überlingersee, i.e., the smaller of the two principal basins of Lake Constance. In this study, measurements, as well as mathematical models, are aimed at increasing our understanding of the functional relationships of the lacustrine ecosystem.

Since the Überlingersee can be treated as a general example of a relatively small basin coupled to a larger lake (the Obersee), the question arises of how the small basin is influenced by the larger lake with respect to its physical, chemical, and biological behavior. The special question discussed in the present paper deals with the physical interaction of the two coupled lakes, e.g., with the mechanisms of inter-basin mixing.

## 7.2 Topography and Sites of Measurement

Lake Constance, which is one of the largest lakes in Europe, is made up of three individual parts: the Obersee, which is the main body; the Überlingersee; and the Untersee (Figure 7.1). A moraine wall at a depth of about 100 m near the Isle of Mainau is taken to be the boundary between the Überlingersee and the Obersee, whereas the Untersee is connected to the main lake only by the Seerhein River. The moraine wall separates the deep water of the Überlingersee from the main basin. For the waters above the sill, this boundary is rather arbitrary. Sometimes, Überlingersee water is found near Meersburg, and sometimes, tongues of Obersee water cross the Mainau boundary in the direction of Überlingen. This effect is described in greater detail in Section 7.6.

Topographically, the Überlingersee is a narrow channel with steep sides and an approximately parabolic cross section, while the Obersee looks rather like a flat, reversed pyramid. The mean depth of the Überlingersee of 81 m

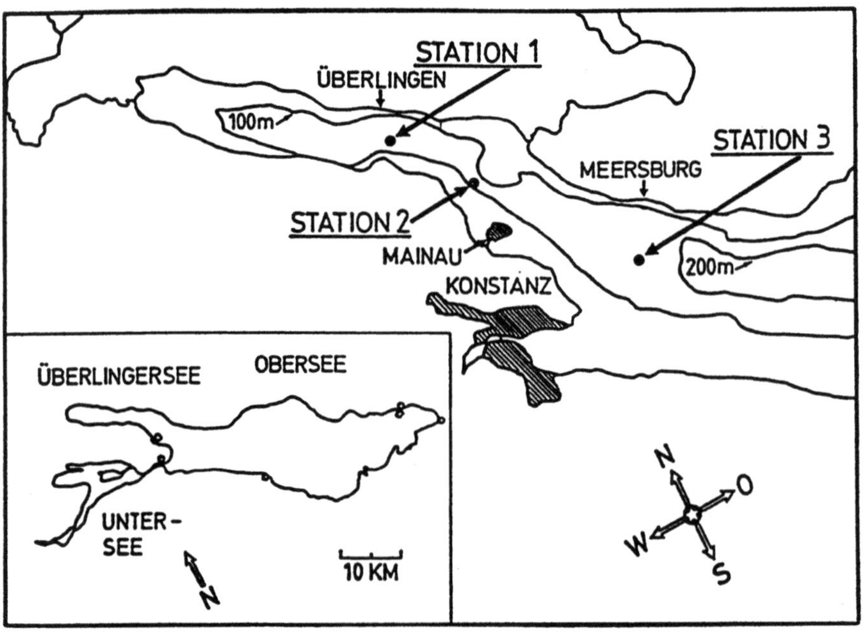


**Figure 7.1.** Map of the Überlingersee indicating locations of the sampling stations. Inset: map of the three parts of Lake Constance.

is more than half of its maximum depth of 146 m, whereas the Obersee has mean and maximum depths of 100 m and 252 m, respectively (Penck 1894). The basin shape is not the only significant difference; whereas the Überlingersee is surrounded by hills and is therefore protected from the prevailing winds, the Obersee has rather flat banks.

In order to record spatial differences between the Überlingersee and the Obersee, temperature, $^3He$, and electric conductivity were measured monthly at the mid-lake station of the Überlingersee (station 1, Figure 7.1) and near Meersburg (station 3), in the western part of the Obersee. The Meersburg station may not really be representative of the whole Obersee but it nevertheless indicates the type of the main lake water that enters the Überlingersee. In order to check lateral homogeneity, temperature and conductivity were measured twice a month at four additional stations in the Überlingersee close to station 1 and station 2, and at two stations in the Obersee in the vicinity of station 3. At the mid-lake station (see Figure 7.1), three buoys were installed for continuous recording of meteorological parameters and water temperature at 22 different depths. Two current meters were installed near the Isle of Mainau (station 2).

## 7.3 Methods Used to Study Mixing Processes

**Tracer Methods** Mixing phenomena in lakes can be assessed by observing suitable substances with known in situ production or consumption rate, called "tracers," which are not homogeneously distributed in space. The temporal development of the concentration of such substances is given by

$$\frac{\partial C}{\partial t} = \vec{\nabla}\,(\mathbf{K}\cdot\vec{\nabla}\,C + D\cdot\vec{\nabla}\,C) - v\cdot\vec{\nabla}\,C + Q \qquad (7.1)$$

where: C(x,y,z,t), is the tracer concentration; **K**(x,y,z,t), the tensor of eddy diffusivity; D, the molecular diffusion coefficient; Q, the production term of C (a negative sign means a sink); and $\vec{v}$(x,y,z,t), the vector of current velocity. Compared to eddy diffusion, molecular diffusion D is negligible except very close to boundaries (e.g., in the surface boundary layer), and possibly in strongly stratified small lakes (Quay et al., 1980). Because of the special topographic and hydrodynamic conditions of Lake Constance, horizontal eddy diffusivities taken from the literature (e.g., Murthy, 1975) cannot be used to assess an exchange rate between the Überlingersee and the Obersee. In order to describe the horizontal mixing phenomena, it is therefore necessary to measure the temporal development of suitable substances.

In the present study, three tracers—temperature, electric conductivity, and $^3He$—are used. As subsequently discussed, these three tracers differ considerably in their production term Q.

*Temperature* The heat content of a lake is primarily changed by the flux of energy across the lake surface. The main processes are long- and short-wave radiation, as well as the flux of latent and sensible heat. Since the latter strongly depends on wind speed and fetch, we expect the development of temperature to be different for the Überlingersee and Obersee, respectively.

Temperature profiles (precision, 0.01 K) were measured twice a month at the mid-lake station (1) of Überlingersee, near the Isle of Mainau (2), and near Meersburg (3). To include cross-lake variations, three profiles across the lake were taken at each station. In addition, temperature was measured quasi-continuously at 22 depths at the mid-lake station of Überlingersee (station 1).

*Conductivity* In Lake Constance the electric conductivity of water is mainly determined by the ion pair $Ca^{2+}$ and $HCO_3^-$. The balance of these ions is controlled by river inflow, by biological activity, by sedimentation, and by ions released from the sediment. Therefore, electric conductivity has various sources and is not a conservative tracer.

The conductivity measurements were carried out twice a month at stations 1, 2, and 3 (see Figure 7.1); the measurements have an accuracy of 1 $\mu$S/cm. To compare the conductivity of water at different temperatures, all data are converted to values at 20° C ($\kappa_{20}$).

Since the ion content of the river water is very different in the two lake basins and since the Obersee is being diluted by water of low conductivity from the Rhine River in summer, different conductivity values are to be expected for the two basins. Therefore, conductivity can serve as a tracer for lateral mixing between the two basins of the lake.

*$^3$Helium* The tritium/$^3$He method is based on the combined measurement of tritium and its decay product, $^3$He. This method was first applied in physical limnology by Torgersen et al. (1977) and has been used by Weiss et al. (1978) in Lake Constance for the study of gas exchange and internal mixing in winter 1977/78.

Tritium (T), the radioactive isotope of hydrogen, decays into $^3$He by beta emission with a half-life of 12.43 years. The natural background in precipitation of about 5 TU* is small compared to the "bomb" tritium released into the atmosphere by nuclear weapon testing, mainly between 1952 and 1963. By 1963, the tritium concentration of the surface water of Lake Constance had increased to values of about 1,000 TU (Weiss et al, 1979). In 1964/65, the tritium concentration in the hypolimnion (below 50 m) was 450 TU (Weiss et al. 1979). From 1965 to 1985, tritium in the lake decreased gradually to about 60 TU, mainly because the lake water had been replaced by river water of lower tritium concentration. Today, tritium in the deep waters is rather uniform since the variations in the input concentration have become small.

In the lake, the tritium is steadily converted into $^3$He by radioactive decay

* TU, "tritium unit." 1 TU indicates a T/H ratio of $10^{-18}$ (Taylor and Roether, 1982).

(this is the usually called "tritiogenic" $^3He$). The $^3He$ produced remains in the water body, provided that it is "closed" (no removal of $^3He$ by gas exchange and mixing occurs). Consequently, given the tritium concentration T(0) at time $t_o$, at time t the concentration of $^3He$ is given by:

$$^3He(t) = T(0)\,(1-\epsilon^{-\lambda t}) + {}^3He(0) \tag{7.2}$$

Where $^3He(0)$ is the initial $^3He$ concentration at time $t_o$ and $\lambda$ is the decay rate of tritium. The solution of eq. 7.2 is based on the assumption that the tritium concentration T(0) remains constant between the time $t_o$ and t. For times t, which are small compared to the radioactive mean life $\tau$ ($\tau=1/\lambda \simeq 18$ years) we can approximate eq. 7.2 by

$$^3He(t) = T(0)\, t/\tau + {}^3He(0) \tag{7.2a}$$

From eq. 7.2, we define the $^3He/T$ age t of a water mass by

$$t = \tau\,[\,^3He(t)/T(0)\,] \tag{7.3}$$

where $^3He(t)$ and T(0) are both given in tritium units (TU). In other words, the water age t is being set to zero whenever the water mass has lost all tritiogenic $^3He$ by gas exchange with the atmosphere. In fact, the $^3He$ concentration defined by eq. 7.3 is actually a (tritiogenic) excess $^3He$ concentration over the equilibrium value of $^3He$ dissolved in water, if in full equilibrium with the atmosphere. This equilibrium concentration is always subtracted here. As soon as the water mass is closed off from the atmosphere, the clock starts running again. The absolute value of a $^3He/T$ close-off age has an uncertainty of 2 to 3 weeks due to the 5% error in measurement of tritium. However, the error for comparing ages is only about 5 days.

In this approximation, the water "age" t is proportional to the tritiogenic $^3He$ concentration (see eq. 7.3). The age approximates the time elapsed since the water was last ventilated completely. If the ventilation is only partial, the water simply starts with a non-zero initial age. The only significant sinks for $^3He$ are gas exchange and the river outlet, both acting on the water layers at the surface of the lake. Whereas, due to heat exchange and solar radiation, the temperature at the water surface can vary significantly in a short time, the surface concentration of $^3He$ usually changes more slowly. Therefore, the change of the $^3He$ concentration at a given depth (i.e., $\partial C/\partial t$ in eq. 7.1) caused by vertical mixing with surface water depends less strongly on the detailed knowledge of the helium surface concentration than in the corresponding case using temperature.

Vertical profiles of $^3He$ consisting of samples from ten depths were usually determined in the Überlingersee (station 1) and Obersee (station 2).

**Additional Measurements** Information on the hydrodynamic processes in the lake and on the acting meteorological forces is needed to understand the driving mechanisms of vertical mixing and horizontal exchange. Such information is gained from two current meters moored at the station Mainau (station 2), from a meteorological buoy at the mid-lake station of the Überlingersee, and from information about the drift of fishing nets in the eastern part of the Überlingersee.

## 7.4 Dynamics of the Spatial Distribution of Temperature, Electric Conductivity, and $^3He$ in Lake Constance

In this section, data for temperature, electric conductivity, and $^3He$ are presented. Since the concentration of any tracer is influenced simultaneously by vertical and horizontal mixing (see eq. 7.1), lateral gradients must be discussed in order to assess the relative importance of lateral versus intra-basin vertical mixing.

**Observations of Temperature in Überlingersee and Obersee** Because of its influence on water density, temperature is one of the most important factors controlling mixing in lakes. Temperature gradients primarily determine the current patterns and the vertical exchange.

The temperature stratification in Lake Constance starts in April or May and lasts until December. During this period, due to internal waves, the temperature at a given depth fluctuates periodically around an average value. The amplitudes of the fluctuations depend on the vertical temperature gradient and can reach up to $\pm 5°$ C. In summer, at a given site, the temperature variations caused by internal waves largely exceed the lateral differences between the two parts of the lake; thus, it is impossible from our data set to trace the processes responsible for horizontal exchange.

Internal waves are less important or completely absent during the winter period when the water is not stratified. Horizontal mixing is then mainly driven by external meteorological forcing. From the beginning of January to the end of May, the vertical temperature difference between the epilimnion and the hypolimnion of the Obersee is only a few tenths of a degree. In contrast, the Überlingersee shows much larger vertical temperature variation in winter (up to 1.3° C between 5 and 40 m, as shown in Figure 7.2). The weak inverse stratification of the Überlingersee in winter 1986/87, associated with low surface temperatures, demonstrates that in this part of the lake, vertical mixing is not sufficient to overcome the stratifying effect of heat loss at the lake surface.

**Seasonal Variation of Conductivity in the Two Basins** In spring and summer, electric conductivity ($\kappa_{20}$) varies between 230 and 270 $\mu$S/cm in the epilimnion and between 285 and 300 $\mu$S/cm in the hypolimnion of Lake Constance. Generally, the electric conductivity increases with depth and reaches its maximum above ground. In winter, the vertical gradient is reduced by vertical mixing resulting in maximum variation over the whole water column of only about 10 $\mu$S/cm.

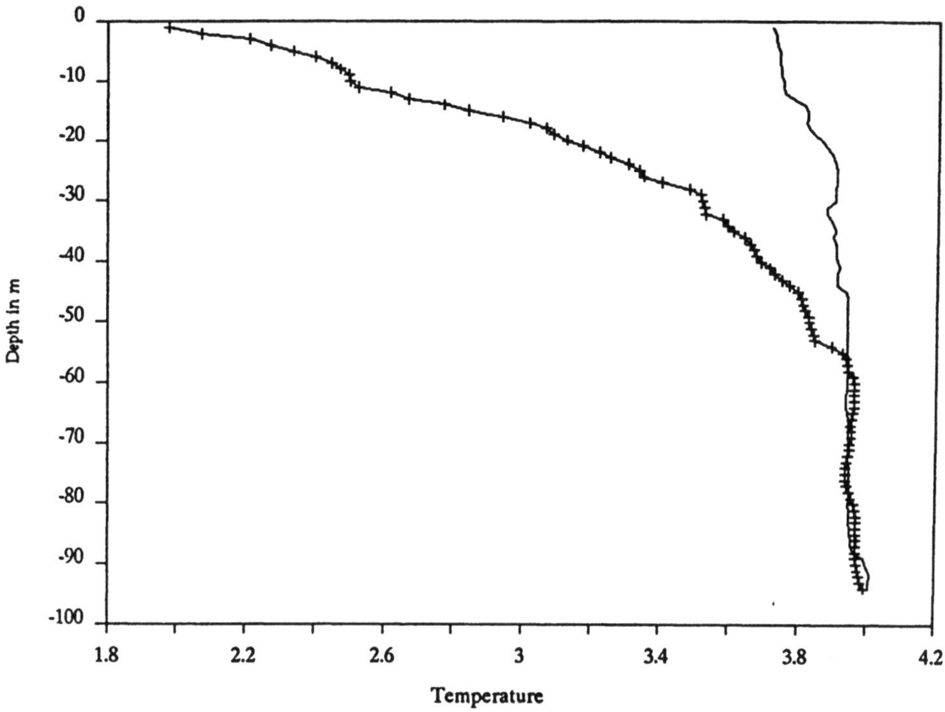


**Figure 7.2** Comparison of the vertical temperature distribution in the Überlingersee and Obersee on February 5, 1987.

For better comparison between the two lake basins, Figure 7.3 shows vertically averaged conductivities for the deep hypolimnion (between 60 and 110 m depth). In this depth range, the influences of biological processes and of dissolution at the lake bottom on electric conductivity are not important.

To visualize the influence of vertical displacement of the water column by internal waves, the boundaries of the 60 to 110 m depth interval were displaced by $\pm 10$ m. In Figure 7.3 the hatched area represents the variation due to vertical water displacement. Yet, even if this uncertainty is taken into account, significantly higher electric conductivities were found in the Überlingersee than in the Obersee during the whole year.

The difference may be caused by the chemical composition of the water entering the lake through the inlets. In the summer, the Rhine River, by far the most important inlet of Lake Constance, usually has a lower $\kappa_{20}$ than the lake does. In winter, when the inflow is small, the conductivity of the Rhine River reaches about 315 $\mu$S/cm (Int. Gewässerschutzkommission, 1982), i.e., larger than $\kappa_{20}$ in the lake. Part of the Rhine water enters the hypolimnion of the Obersee, but the influence of the Rhine on the Überlingersee is small. In contrast the tributaries of the Überlingersee have larger $\kappa_{20}$ values than the lake does.

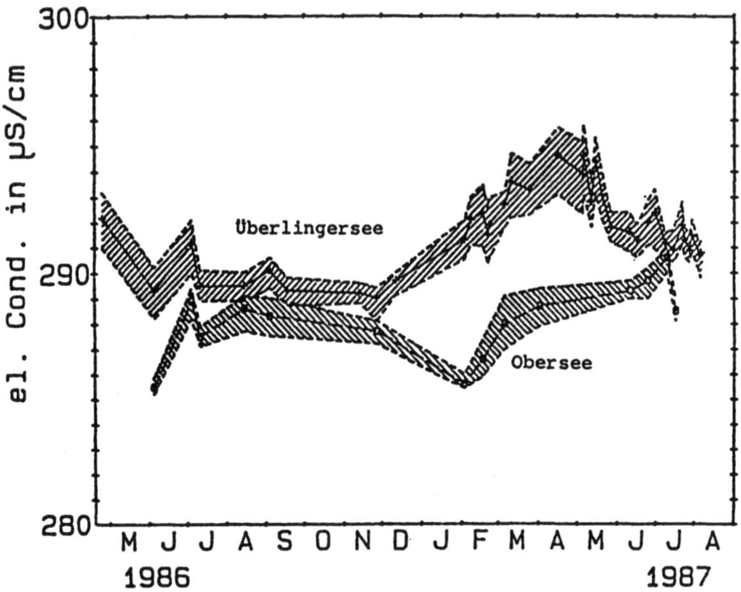


**Figure 7.3** Yearly variation of conductivities between 60 m and 110 m depth in the Überlingersee and Obersee.

The lateral differences between the two basins show a clear seasonal trend; they are smaller in summer, i.e., during the period of strong vertical density stratification. It is hypothesized that in summer the vertical density stratification causes the wind energy to be stored in internal waves. This results in stronger horizontal mixing and reduces horizontal gradients. In contrast, the absence of vertical density gradients in winter causes horizontal mixing to become less efficient. The $^3$He data discussed in the next section seem to confirm this hypothesis.

**Seasonal Change of $^3$He Age in the Überlingersee** In Figures 7.4 and 7.5, the temporal development of the $^3$He in the Überlingersee is presented. Figure 7.4 shows the consequences of the winter mixing between October 1985 and June 1986. In October 1985, the lake was still well stratified. For the $^3$He age this means that a "young" surface layer (indicating a relatively intensive gas exchange with the atmosphere) lies above "old" hypolimnic water. With decreasing thermal stability, vertical mixing becomes stronger, leading to a decrease of the "age content" of the water column until June 1986. From October 1985 to January 1986, the decrease of the $^3$He age below 30 m is accompanied by an increase in the age in the upper 30 m. High $^3$He ages are observed in the surface layer if, due to storm events and lack of stability, upward mixing of old water—being rather rich in $^3$He—is very strong; meanwhile, at

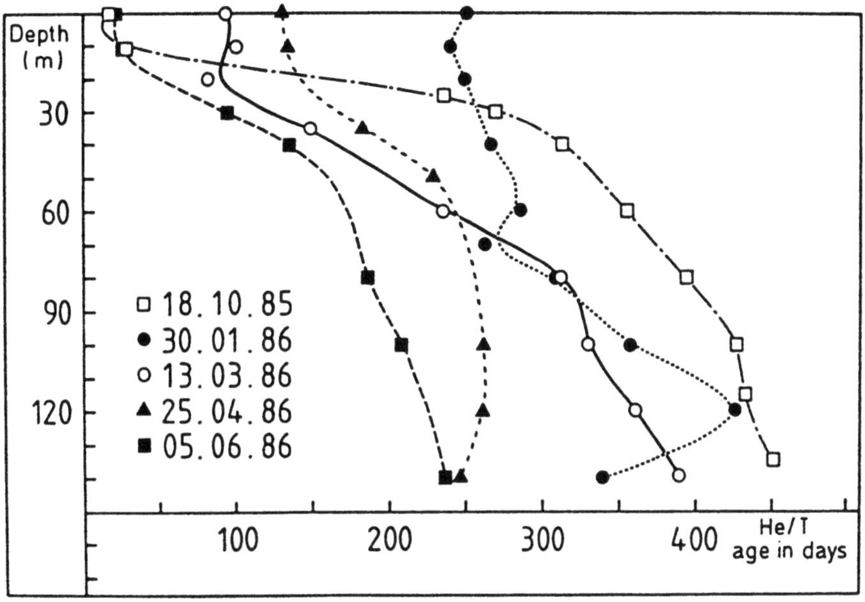


**Figure 7.4** Temporal development of $^3$He age in the winter season 1985/86 in the Überlingersee.

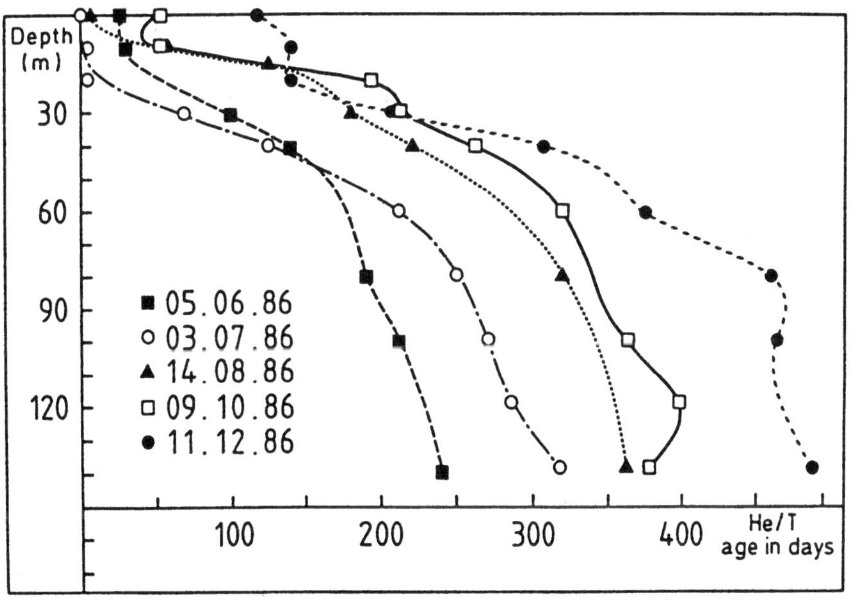


**Figure 7.5** Temporal development of $^3$He age in the summer season 1986 in the Überlingersee. Between May and June 1986 the thermal stratification becomes strong enough to reduce the vertical exchange across the thermocline. Therefore, the $^3$He age increases in the hypolimnion from June onward. At the same time, the $^3$He concentration of the epilimnion decreases by gas exchange and due to reduced mixing across the thermocline.

the surface, gas exchange with the atmosphere is not fast enough to turn the "water age clock" to zero, immediately.

In January 1986, the water in Überlingersee shows an age maximum at 120 m which cannot be explained by radioactive decay or vertical diffusion. At the same time, a minimum was found in the vertical profile of dissolved oxygen at 120 m depth. Since, at 140 m $^3$He and oxygen concentrations are equal in Überlingersee and Obersee, we suppose that water from Obersee has replaced the bottom water in Überlingersee.

After June 1986, the age below 20 m increases again (see Figure 7.5), and the hypolimnion can be thought of as a closed system until November. During this period, the temporal development of $^3$He is mainly determined by the constant volume source, which is the tritium decay. Vertical diffusion has little influence on the vertical $^3$He age distribution. In Figure 7.6 the $^3$He age is plotted against time for three depths (20 m, 40 m, and 100 m). After June, at 100 m and 40 m depth, the age increases approximately as the real time does (for comparison, shown by the uppermost line in Figure 7.6), indicating the aging of a water mass isolated from mixing and gas exchange. In contrast, at 20 m, $^3$He is steadily lost to the surface layer. By June, the latter has nearly reached age zero, that is equilibrium with the atmosphere (see Figure 7.5). In summer, the $^3$He concentration at 20m is modified by the motion of the thermocline and by gas exchange. This can be seen from the increase of the $^3$He age at 20 m, which is significantly smaller than the closed-system aging line (Figure 7.6). The "high" age in October at 20 m was caused by an upwelling of about 5 m together with a strong vertical $^3$He gradient in this depth range.

**Differences In Water Age Between Überlingersee and Obersee** Figure 7.7 compares the age profiles in Überlingersee and Obersee during the period from October 1985 to July 1986. In October 1985, when the lake was fully stratified, the $^3$He profiles in both basins were nearly equal (Figure 7.7*a*). In January 1986, after reduction of the density stratification, vertical mixing—and thus $^3$He loss to the atmosphere—apparently became more efficient in the Obersee, resulting in a smaller $^3$He age than in the Überlingersee at nearly all depths (Figure 7.7*b*). The lateral age difference increased until March 1986 (Figure 7.7*c*). Starting in April (Figure 7.7*d*), the $^3$He ages of the two basins became more alike again and were indistinguishable after June (Figures 7.7*e* and 7.7*f*.)

Presumably, the lateral difference between the two basins may have been even larger than shown by Figure 7.7, since, under certain meteorological conditions, the sampling site in Obersee is influenced by water from Überlingersee. This possibly was the case at the end of April 1986 when, after the passage of a storm, the $^3$He age in Obersee temporarily became greater than can be accounted for by the local $^3$He production from March to April. In June, $^3$He in Obersee was again back to normal.

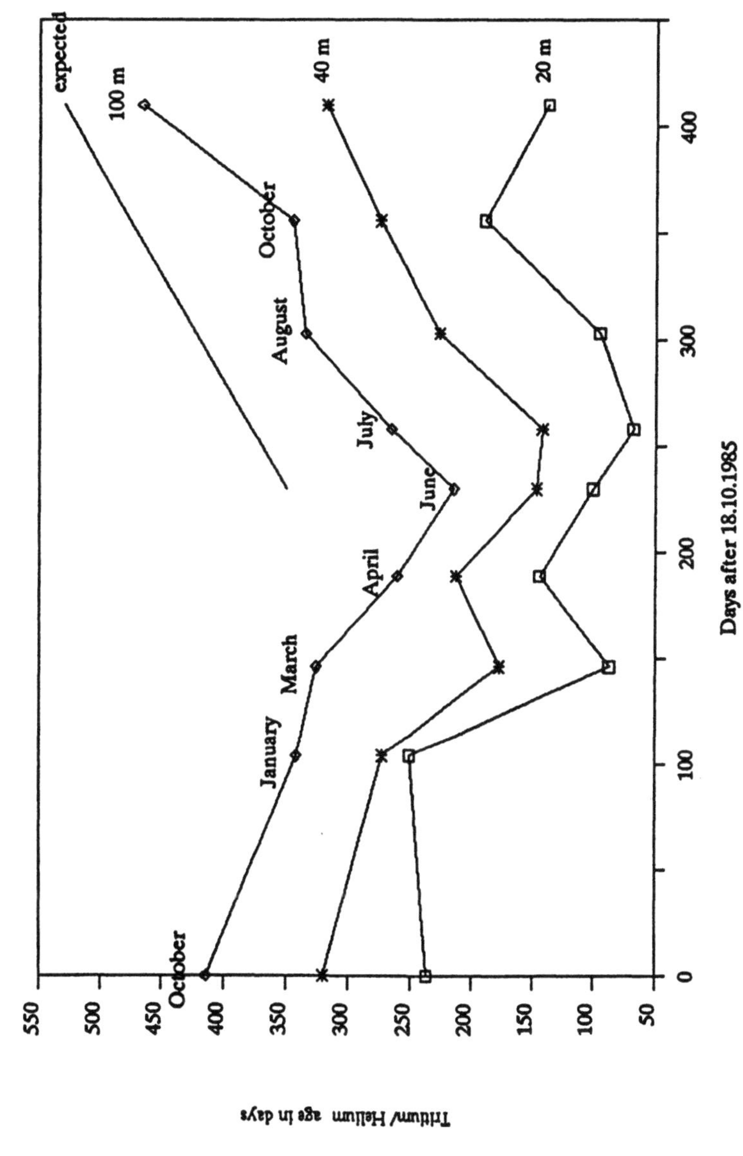


**Figure 7.6** Temporal development of the $^3He$ age in the Überlingersee at 20 m, 40 m, and 100 m. The uppermost line represents the expected increase of $^3He$ age with time for a closed system (i.e., one month per month).

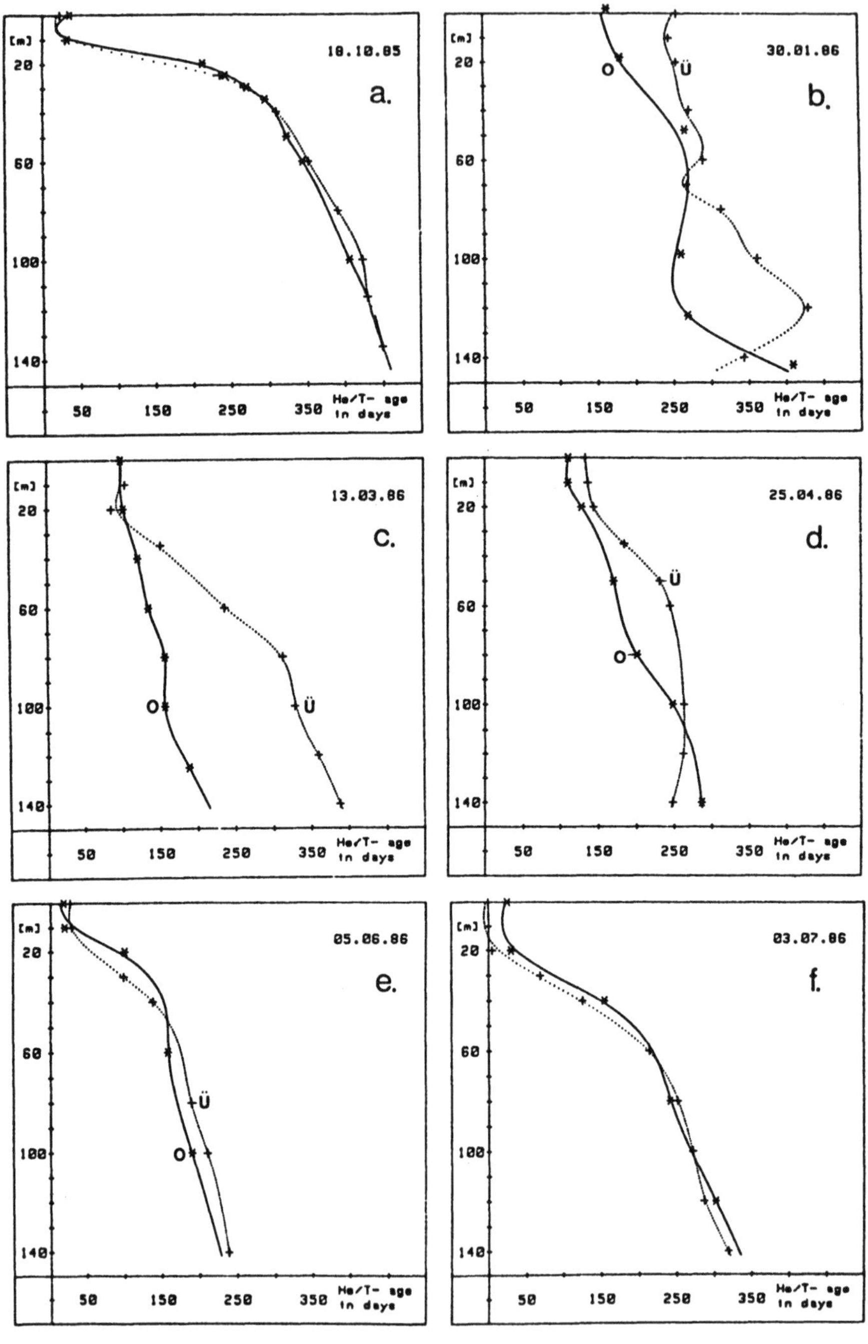


**Figure 7.7** Comparison of vertical ³He-age profiles in the Überlingersee (Ü) and Obersee (O) between October 1985 and July 1986.

**Observed Lateral Inhomogeneities: Summary** With regard to the distribution of the observed tracers in the two basins (Obersee and Überlingersee), two time regimes can be observed: the winter period (without or with weak stratification) and the summer period (with stratification). The horizontal concentration gradients are quickly destroyed during summer. Coinciding with the absence of horizontal $^3$He gradients in summer, we find that the difference in the hypolimnic electric conductivity is reduced between Überlingersee and Obersee to 1 to 2 $\mu$S/cm. In winter, vertical mixing processes are more efficient in Obersee than in Überlingersee, creating lateral differences in $^3$He, electric conductivity, and temperature between the two parts of the lake, especially after strong wind events. The different tracers all confirm that these differences are only equalized rather slowly.

In the next two sections, we will discuss some possible mechanisms that may be responsible for the occurrence and destruction of lateral gradients between the two basins of Lake Constance.

## 7.5 Mechanisms for the Development of Horizontal Gradients

In 1986, horizontal gradients of $^3$He and conductivity were observed until June (see Figures 7.3 and 7.7). As speculated in Section 7.4, these gradients may be caused by different processes of vertical mixing in Obersee and Überlingersee. As will be shown in this section, the local wind fields over the two parts of the lake are different, indeed.

The intensity of vertical mixing in lakes primarily depends on the interplay between the turbulent energy input by wind and the thermal energy input from the atmosphere causing vertical density gradients. In winter, the intensity of vertical mixing is connected to the local wind energy input; horizontal inhomogeneities of energy input by wind cause horizontal differences in vertical mixing. In summer, however, the wind energy is stored in internal waves and redistributed across the lake. Vertical mixing is then related to the horizontal pattern of energy dissipation from internal waves. Therefore, a direct correlation between the wind forces and vertical mixing can only be expected during winter.

Investigations on wind speed at Lake Constance go back to Kopfmüller's (1922) study on the thermal wind systems which are typical for summer conditions. Huss and Stranz (1970) also described the winds during the summer. The occurrence of strong winds at Lake Constance was discussed by Mühleisen (1977). He showed that strong winds preferentially occur from the west and are observed less frequently at Unteruhldingen (in the eastern part of the Überlingersee) than on the northern shore of the Obersee. The effect of another type of strong wind, the Föhn winds blowing from the south, is only felt in the eastern part of the Obersee, near Lindau.

Mühleisen's observations yield the first support for our hypothesis explaining the different vertical mixing patterns in the two basins of Lake Constance. Further evidence is produced by a comparison between the wind data collected at our meteorological buoy (Überlingersee, station 1) and the data from the meteorological station at Constance (from the *Deutscher Wetterdienst*). The latter data are regarded as being representative of the mean wind field at Obersee. However, the absolute size of wind speeds, one measured at 4.4 m height above the free water surface, and the other measured on shore 17 m above ground, cannot be directly compared. Under neutral conditions, the horizontal wind speed U(z) varies with distance z above ground (or above the water surface) according to the following equation:

$$U(z) = \text{const} \cdot \ln(z/z_0) \tag{7.4}$$

where $z_0$ is the "roughness height," a measure of the structure of the surface (land or water). Above urban areas, $z_0$ is typically of the order of 0.7 to 1 m (Pasquill, 1974), while above the water surface it lies between $10^{-5}$ and $10^{-4}$ m (Graf and Prost, 1979). Using eq. 7.4, the wind speeds from both observation sites can be converted to a level above ground which is sufficiently high to suppress the local properties of the ground.

Wind speeds converted to the 60-m level are presented in Figure 7.8*c* for the two sites. Due to the uncertainty of the roughness heights $z_0$ of the land and the lake station, respectively, an error of less than 10% is introduced into this conversion. If, instead of the 60-m level, the reference height is assumed to be 150 m, the converted wind speed of the Obersee station (Constance) increases by about 20%, while the converted data from the Überlingersee remain virtually the same because of the much smaller roughness height, $z_0$. Thus, the converted wind speed from Obersee shown in Figure 7.8*c* represents a lower limit of the free wind speed above Obersee.

As shown in Figure 7.8, during strong winds (daily mean $> 3$m/s) the wind direction at Obersee and Überlingersee are virtually identical. For winds blowing from the northeast (the period from January 10 to 11), the wind speed measured at the meteorological buoy in the middle of the Überlingersee is in accordance with the data derived from the measurements at Constance. This can be taken as a confirmation for the wind speed conversion presented above. For wind directions from the west and southwest (the period from January 1 to 6) the wind speed in Überlingersee is generally smaller than in the Obersee. This might be due to the shadowing effect of the "Bodan Rück," a 200-m-high chain of hills located southwest of the Überlingersee. Since the input of wind energy into water is proportional to the third power of the wind speed (assuming a similar roughness height for the water surface of Überlingersee and Obersee, respectively), and since strong winds are mainly blowing from the west and southwest, vertical mixing presumably was more efficient in Obersee than in Überlingersee in winter 1985/86. Thus, horizontal concentration gradients had a chance to develop. The effect of different wind mixing on $^3He$ is schematically shown in Figure 7.9. For $^3He$, the different wind energy input is the only effect producing horizontal gradients. For other tracers, such

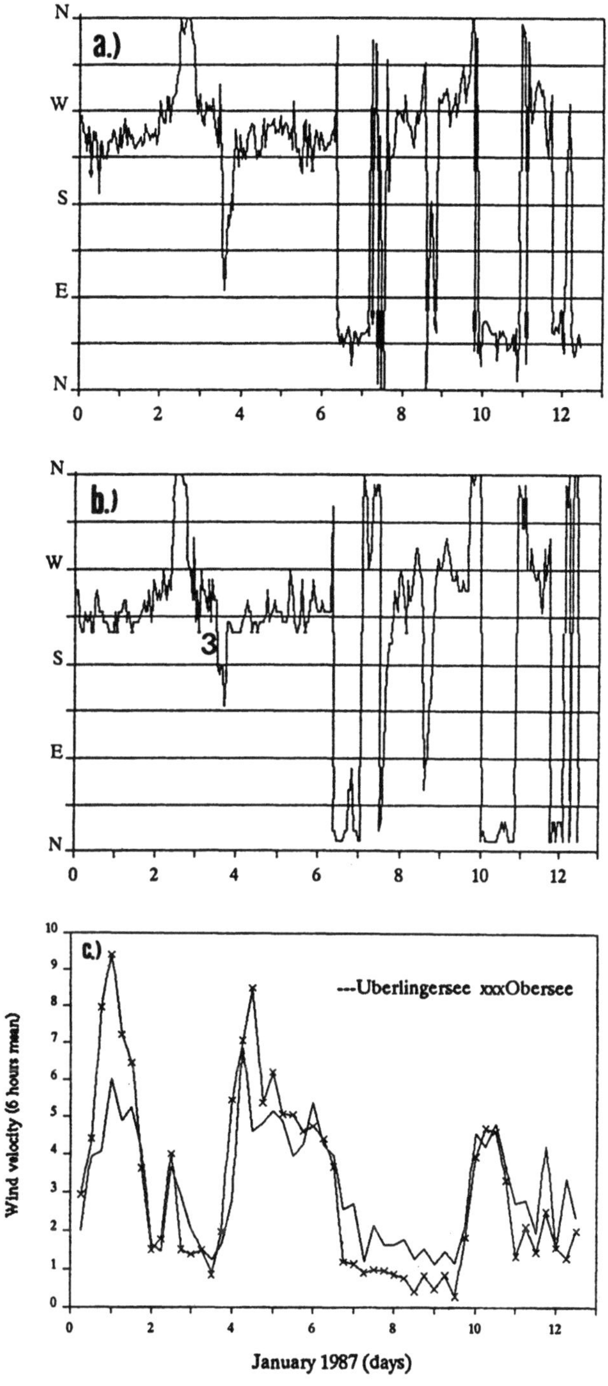


**Figure 7.8** (*a*) Wind direction at the meteorological buoy in the Überlingersee. (*b*) Wind direction at the meteorological station at Constance. (*c*) Wind speeds at the two locations.

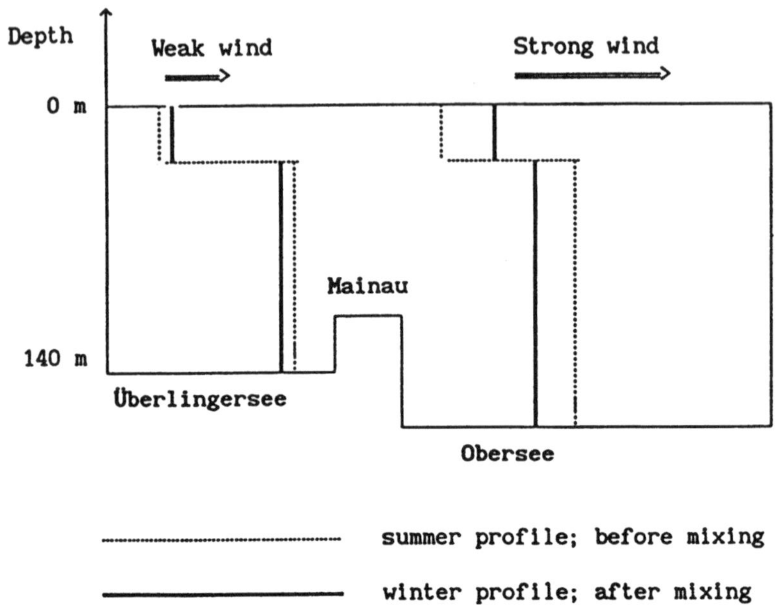


**Figure 7.9** Due to the lower wind speed above the Überlingersee, vertical mixing is less efficient than in the Obersee. A comparison of the vertical $^3$He profiles between the two parts of the lake demonstrates the development of horizontal gradients.

as electric conductivity, additional effects—for example, differences in river input—ought to be considered, as well.

## 7.6 Mechanisms of Horizontal Exchange

In summer, water temperatures recorded at fixed depths are characterized by periodic oscillations (Figure 7.10). The oscillations last from May to November, that is, for the whole stratification period of the lake. The period of the ground mode changes from 110 hr in June to 90 hr in August. The oscillation is damped with a constant of about 0.1 $d^{-1}$ but is re-excited frequently enough to keep it going during the whole period.

It is possible that the measured temperature oscillations are due to the effects of standing internal gravity waves ("internal seiches") extending over the whole length of the lake. The rather simple model of A. Defant (1918), modified for internal waves by Mortimer (1979), was successfully tested by Lemmin and Mortimer (1986) on several lakes. For a lake of given topography, the model yields the frequencies and amplitudes of the interface oscillation for a two-layer lake with given density difference between the layers. Figure 7.11 shows the model calculation for Lake Constance. A vertical amplitude of the thermocline displacement of 10 m at the western end of the Überlingersee

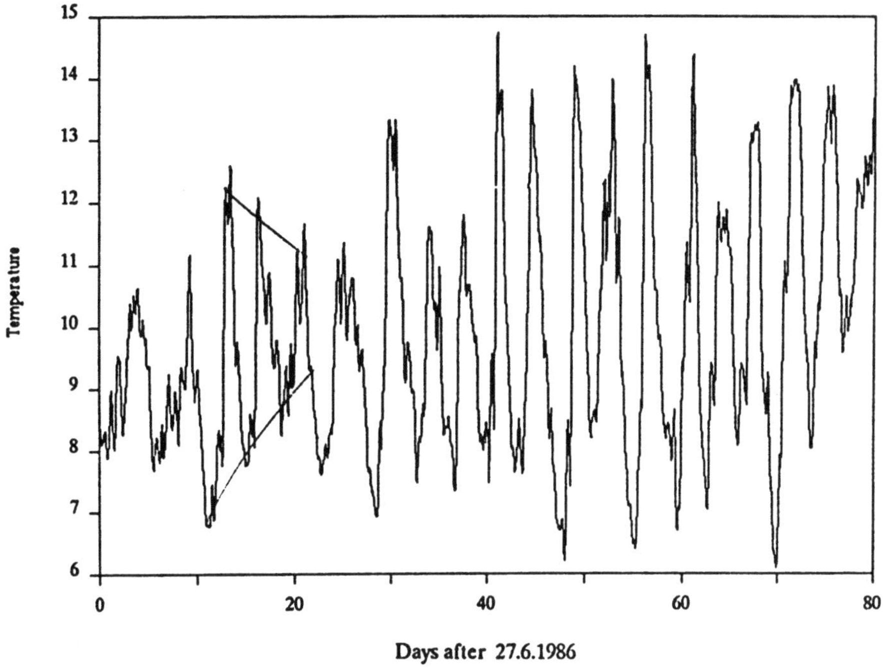


**Figure 7.10** Variation of water temperature at 14 m depth at station 1 in the Überlingersee. The enveloping curves characterize a linearly damped oscillation with damping constant 0.1 $d^{-1}$.

would yield a maximum amplitude of the thermocline of 3 m at the eastern end of the Obersee. Displacements of 10 m are observed rather frequently in the Überlingersee. The large amplitudes in the Überlingersee are a consequence of its small basin cross section. For the same reason, large current velocities are predicted in both the epilimnion and the hypolimnion by the Defant model, particularly at the straits near the Isle of Mainau where the cross section is rather narrow (see Figure 7.1).

Thanks to the friendly cooperation of some fishermen, we were able to trace the horizontal water displacement in the epilimnion resulting from internal seiches. The fishermen noted the positions upon release and recovery of their freely floating nets (up to 960 m long, and 10 m wide), which were usually exposed in the lake between 10 and 20 m depth for about 12 hrs. As shown in Figure 7.12, the fishing-nets data agree with the internal wave motion derived from temperature measurements at station 1. A rise in water temperature corresponds to an epilimnic transport of water *into* the Überlingersee, while a drop in temperature means outward flow. This is a rough and qualitative indication of the action of internal seiches.

The Defant model adequately reproduces the time constant of the observed oscillation as well as the horizontal epilimnic velocities at the straits near the

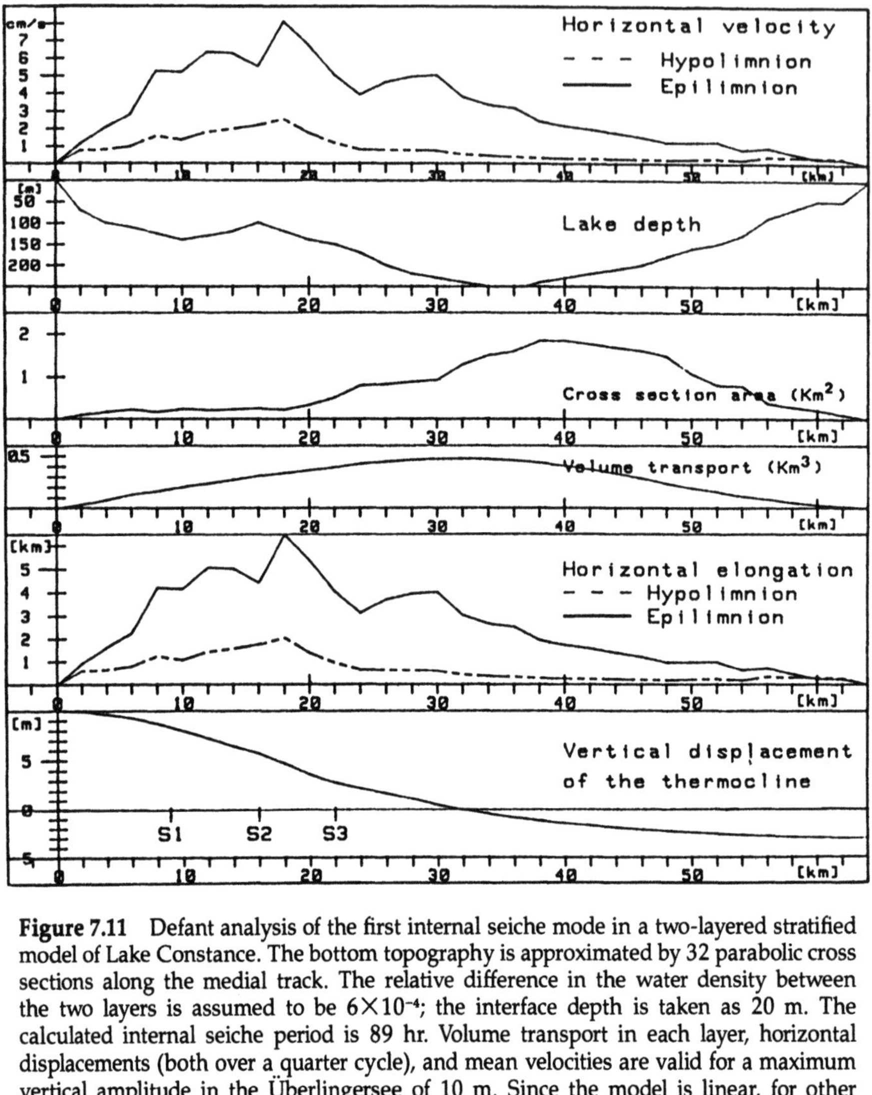


**Figure 7.11** Defant analysis of the first internal seiche mode in a two-layered stratified model of Lake Constance. The bottom topography is approximated by 32 parabolic cross sections along the medial track. The relative difference in the water density between the two layers is assumed to be $6 \times 10^{-4}$; the interface depth is taken as 20 m. The calculated internal seiche period is 89 hr. Volume transport in each layer, horizontal displacements (both over a quarter cycle), and mean velocities are valid for a maximum vertical amplitude in the Überlingersee of 10 m. Since the model is linear, for other amplitudes all model functions change linearly.

Isle of Mainau. Furthermore, the measurements show good correlation between the current direction (station 2) (Figure 7.13*a*) and the temporal variation of the temperature at 26 m (station 1) (Figure 7.13*b*); a rise in water temperature is connected with a flow of hypolimnic water from Überlingersee into Obersee. Yet, the measured hypolimnic water currents demonstrate that the Defant model does not explain all the observed phenomena; the water velocities during inward flow (to the Überlingersee) are different in magnitude than during outward flow (Figure 7.13*c*). The Coriolis force and topographical effects may

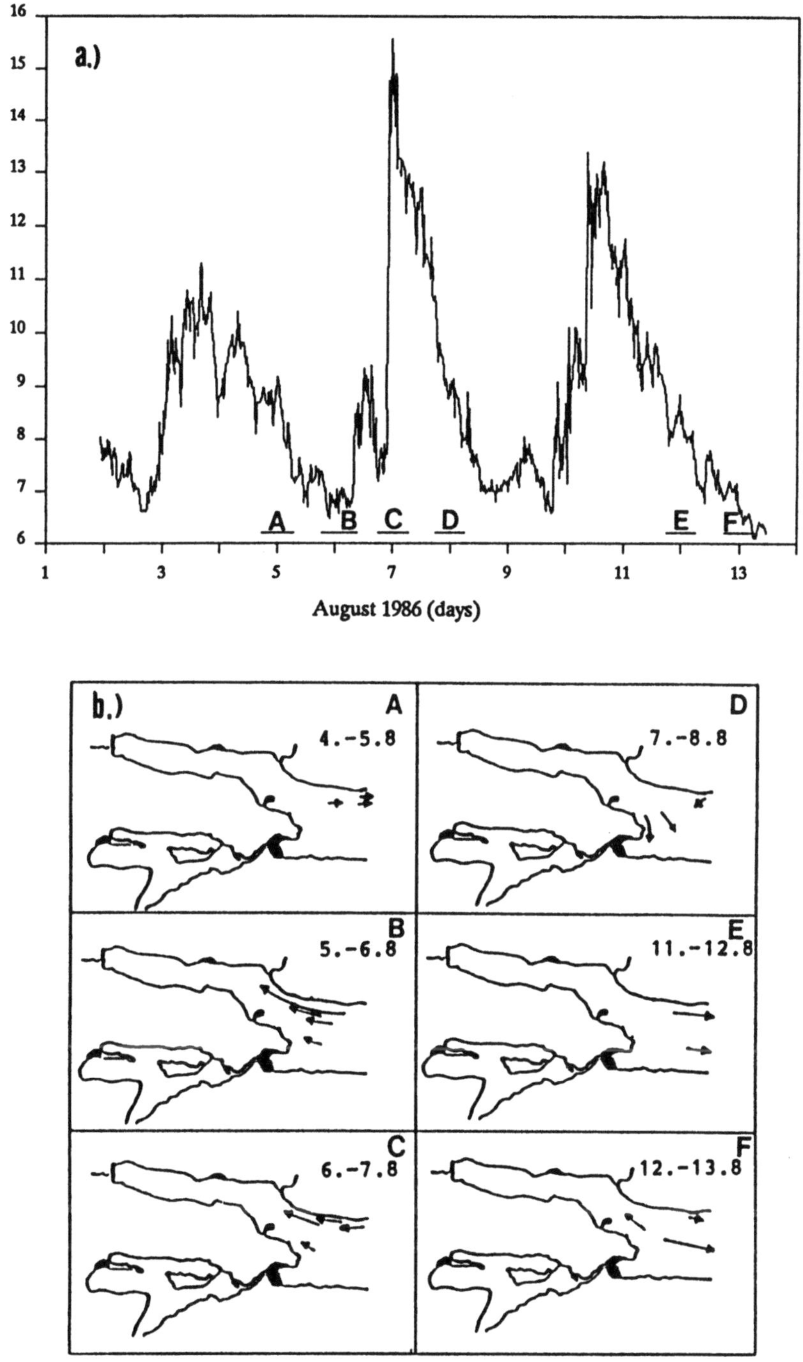


**Figure 7.12** (*a*) Temperature at 17.4 m depth at station 1 in the Überlingersee for August 2 to 13, 1986. (*b*) The horizontal displacement of the fishing nets for each period (A to F) indicated in *a* is shown by small arrows.

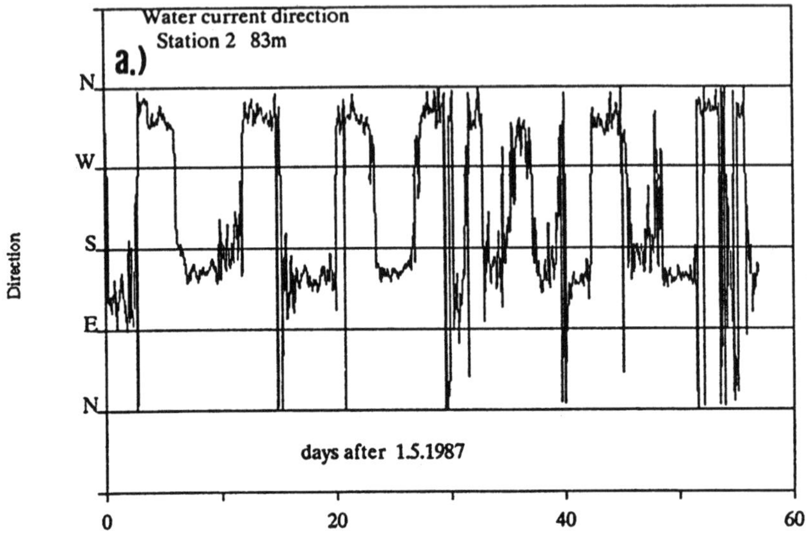


**Figure 7.13** (*a*) Direction of water currents at station 2 (Mainau) at 83 m depth. Current direction indicates where the water is flowing to; northwest means direction Überlingersee, southeast means direction Obersee. (*b*) Water temperature at station 1, 26 m depth, for the same period. (*c*) Water velocity at station 2 at 83 m depth for the same period.

be responsible for this asymmetry. Further measurements are necessary to verify the predictions of a modified Defant model which includes the Coriolis force (Defant, 1954; Mortimer, 1979).

Since the flow of water is not completely reversible, the periodic transport of water across the straits of Mainau over a distance of up to 7 km in the epilimnion and up to 2 km in the hypolimnion provides an efficient exchange between the two parts of the lake. This became apparent in the distribution of $^{3}He$ and electric conductivity after the onset of internal waves (see Figures 7.3 and 7.7). Investigations with artificial tracers are planned to give more detailed information on horizontal mixing between the two basins of Lake Constance.

Due to the absence of a density stratification, no distinct internal waves occur in winter. Other phenomena cause horizontal mixing—such as wind-driven and density currents—the latter being caused by the difference in vertical mixing intensity between the Überlingersee and the Obersee.

To compare the effect of wind-induced versus density-driven currents, we examined the currents at 80 m depth at station 2 between January and March 1987. Currents above 1.1 cm/s (the threshold speed of the current meters) were found to be correlated with wind events. During the remaining time when only current direction was recorded by the instruments, no preferred direction was observed. Therefore, if density currents occurred, they would play a minor role compared to wind-induced currents.

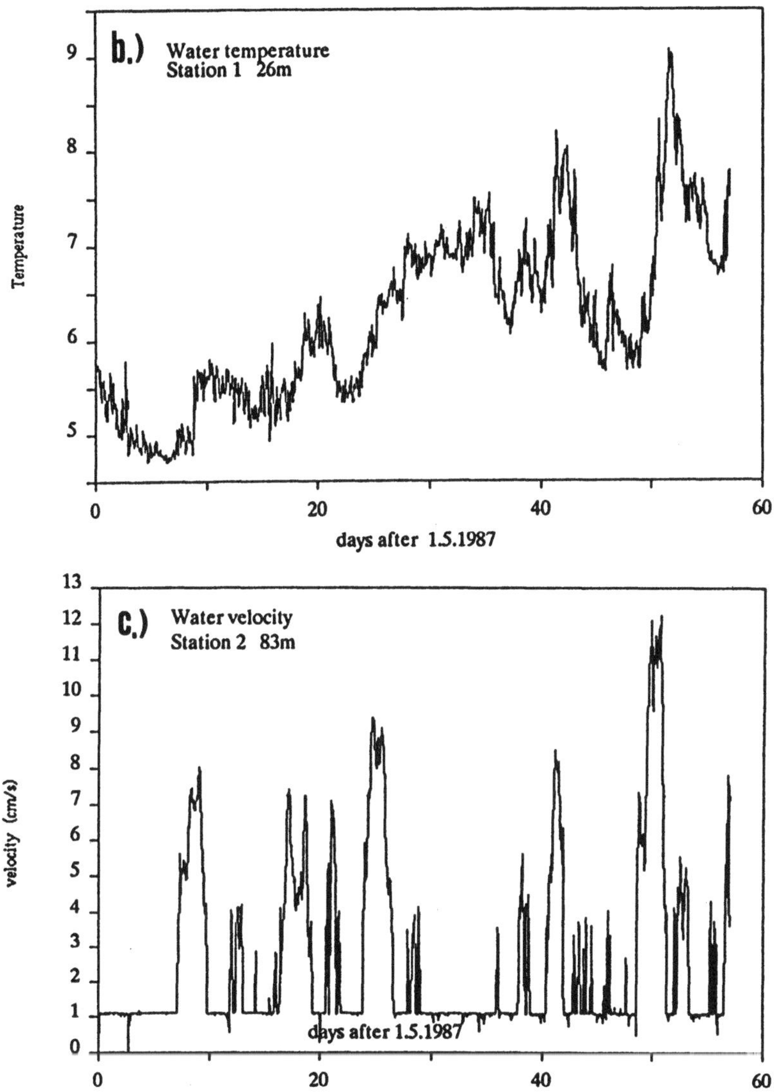


**Figure 7.13** *b* and *c*

## 7.7 Conclusion

With respect to the interaction between the two basins of Lake Constance, the Obersee and the Überlingersee, two regimes can be distinguished: the summer regime, characterized by the dominance of internal wave motion, causes an effective horizontal exchange between the basins; the winter regime is characterized by different vertical mixing in the two basins. Neither wind-induced lateral mixing nor density currents are strong enough to destroy the horizontal gradients during this regime.

The question arises whether for a given substance the lake is horizontally well mixed. The answer depends not only on the horizontal mixing intensity, but also on the substance-specific reaction time scale. For substances with production/sink rates which are small compared to the typical frequency of internal waves, the two parts of the lake can be supposed to be quasi-homogeneous in summer, even if the production rate is spatially not constant. However, substances which react fast will show lateral patchiness; in this case, internal waves may cause large time fluctuations at a fixed station. This is important for the interpretation of biological data; internal waves, due to the considerable displacement of the water masses that they produce, may effect the balance of nutrients and biomass in Überlingersee and cause temporal fluctuations which are of purely physical origin.

---

*Acknowledgments* The meteorological data of the Wetterwarte Konstanz were kindly placed at our disposal by the German Weather Service. This work is funded by the Deutsch Forschungsgemeinschaft in the Sonderforschungsbereich 248 "Stoffhaushalt des Bodensees."

## References

Defant, A. 1918. Neue Methode zur Ermittlung der Eigenschwingungen von abgeschlossenen Wassermassen. *Ann. Hydrogr. Berlin* 46: 78–85.

Defant, F. 1954. Theorie der Seiches des Michigansees und ihre Abwandlung durch die Wirkung der Corioliskraft. *Archiv für Meteorologie u. Geophysik*, Ser. A.:218–24.

Fischer, K.H. 1980. Aufbau eines 3-Helium Messsystems und Anwendung der Helium-3 Methode. Dissertation, Heidelberg.

Fuchs, G., Roether, W., and Schlosser, P. 1987: Excess 3-Helium in the ocean surface layers. *J. Geophys. Research.* 92:6559–6568.

Graf, W.H. and Prost, J.P. 1979. The aerodynamic drag; first experiments on Lake Geneva. In: Graf, W.H. and Mortimer, C.H. (eds.): *Hydrodynamics of lakes.* Elsevier, Amsterdam 303–312.

Hollan, E. and Simons, T.J. 1978. Wind-induced changes of temperature and currents in Lake Constance. *Arch. Met. Geoph. Biokl.* Ser. A, 27:333–373.

Huss, E., and Stranz, D. 1970. Die Windverhältnisse am Bodensee. *Pure and Applied Geophys.* 81:323–356.

Internationale Gewässerschutzkommission. Report No. 28, 1982. Die Belastung des Bodensees mit Phosphor und Stickstoffverbindungen und org. Kohlenstoff im Abflußjahr 1978/79.

Kopfmüller, A. 1922. Der Land- und Seewind am Bodensee. *Das Wetter* 39:97–107; 49:33–41, 65–78, 108–115.

Lemmin, U. and Mortimer, C.H. 1986. Tests of an extension to internal seiches of Defant's procedure for determination of surface seiche characteristics in real lakes. *Limnol. Oceanogr.* 31:1207–1231.

Lynch, M.C.F. and Kay, D.J. 1981. Performance of a mass spectrometer for determining low tritium levels from $^3He/^4He$ measurements. *Methods of Low Level Counting and Spectrometry.* IAEA-SM-252/47, p. 511–523, IAEA, Vienna.

Mortimer, C.H. 1974. Lake Hydrodynamics. *Mitt. Int. Ver. Theor. Angew. Limnol.* 20:124–197.
Mortimer, C.H. 1979. Strategies for collection and analysis of lake data essential for interactive coupling with models. In: Graf, W.H. and Mortimer, C.H. (eds.): *Hydrodynamics of Lakes*, Elsevier, Amsterdam, 1979a.
Mühleisen, R. 1977. Starkwinde an und auf dem Bodensee. *Meteorol. Rdsch.* 30:15–22.
Murthy, C.R. 1975. Horizontal diffusion characteristics in Lake Ontario. *Journal of Physical Oceanography* 6:76–84.
Pasquill, F. 1974. *Atmospheric Diffusion*. Ellis Harwood Limited, John Wiley, New York.
Penck, A. 1984. *Morphometrie des Bodensees*. Geogr. Gesellschaft, München.
Quay, P.D., Broecker, W.S., Hesslein, R.H., and Schindler, W. 1980. Vertical diffusion rates determined by tritium tracer experiments in the thermocline and hypolimnion of two lakes. *Limnol. Oceanogr.* 25:201–218.
Taylor, C.B., and Roether, W. 1982. A uniform scale for reporting low-level tritium measurements in water. *Int. J. Appl. Radiat. Isot.* 33:377–382.
Torgersen, T., Top, Z., Clarke, W.B., Jenkins, W.J., Broecker, W.S. 1977. A new method for physical limnology—tritium/helium-3 ages results for Lake Erie, Huron and Ontario. *Limnol. and Oceanogr.* 22:181–193.
Weiss, W., and Roether, W. 1975. Der Tritiumabfluß des Rheins 1961–1972. *Sonderdruck aus Deutsche Gewässerkundliche Mitteilungen*, 19th Jahrgang, 1, February 1975.
Weiss, W., Roether, W.. and Bader, G. 1976. Determination of blanks in low level tritium measurements. *Int. J. Appl. Radiat. Isotopes* 27:217–225.
Weiss, W., Fischer, K.H., Kromer, B., Roether, W., Lehn, H., Clarke, W.B., and Top, Z. 1978. Gas exchange with the atmosphere and internal mixing in Lake Constance (Obersee). *Verh. der Gesellschaft für Ökologie*, 7th Jahresversammlung Kiel, 1977.
Weiss, W., Lehn, H., Münnich, K.O., and Fischer, K.H. 1979. On the deep water turnover of Lake Constance. *Arch. Hydrobiol.* 86:405–422.
Wetzel, R.G. 1975. *Limnology*. W.B. Saunders Company, London.

# 8

# Effects of Water Level Fluctuation on the Structure and Function of the Littoral Zone

Avital Gasith and Sarig Gafny

ABSTRACT Water level fluctuations (WLFs) are most pronounced in impoundments, but are also observed in natural lakes over a wide size range, both on a seasonal and on an interannual time scale. In small and/or shallow lakes, WLFs can affect the metabolism and structure of the entire lake system. In most cases of large deep lakes, only the littoral region is affected. The effects on the biota, either beneficial or adverse, depend on local conditions.

In Section 8.1, effects of WLFs on the physical and chemical environment and on the community of the littoral are described in general. Thereafter, we examine the effects of long-term WLFs in Lake Kinneret, Israel. The relatively large amplitude of WLFs in this lake results in periodic changes of the littoral slope and of the substrate composition. This latter factor affects the breeding of the dominant fish species of the lake. At lower water levels, semi-aquatic vegetation that grows well in the exposed littoral provides vegetative habitats after it is inundated.

WLFs should be looked at as an environmental disturbance maintaining the temporal and spatial heterogeneity, which, in general, is characteristic of the littoral zone.

## 8.1 Introduction

The littoral zone of lakes is most commonly defined as the lakeward region of macrophyte vegetation (Wetzel, 1983). To include cases where benthic algae replace macrophytes, the littoral zone can be more generally defined as the shallow region extending from the shoreline to the lower limit of the euphotic zone. The latter approach underscores aspects of the metabolic significance of the littoral zone but neglects other functions, such as fish-habitat availability for colonization, foraging, spawning, and cover.

WLF is a natural phenomenon of inland waters resulting from periodic imbalances between water inputs (inflow, precipitation) and outputs (outflow, evaporation). Seasonal WLFs are most pronounced in lakes and reservoirs with relatively short retention times in conjunction with seasonal precipitation patterns. Long-term irregular WLFs mainly occur in semi-arid and dry tropical regions with distinct wet and dry seasons (e.g., John, 1986). Water use for drinking and irrigation, and more often, for power generation and flood control, are additional causes of water level fluctuations in man-made, as well as in natural lakes.

In the following, effects of WLF on structural and functional properties of the littoral environment are examined. As a case study, WLFs in Lake Kinneret, Israel, are described.

## 8.2 Effects of WLF on the Littoral Environment

The effects of WLFs on the structure and function of the littoral zone in lakes have not been extensively studied. Most of the information originates from studies of water level manipulation in ponds and reservoirs (Frazer, 1972).

### Littoral Features Affected by WLFs

*Physiography* Shallow systems with gradually sloping shorelines are more affected than lakes with steep littoral slopes, since larger areas are flooded or exposed, respectively (Frey, 1967; Moss and Moss, 1969; McLachlan, 1970; Howard-Williams and Lenton, 1975; Herdendorf and Fay, 1988). Two physiographic parameters are modified by WLF:

1. Shoreline morphometry, especially in convoluted shorelines where changes of water level can significantly modify the area of lagoons and consequently change habitat availability and affect biomass production (Monro and Larkin, 1950; Pieczynska, 1972; Mitchell and Rogers, 1985; Osborne et al., 1987).
2. Littoral slope and substrate composition (Pieczynska, 1972). In general, gentler slopes and softer substrates are associated with falling water levels and vice versa.

*Water Quality* Water quality can be greatly affected by WLFs, as a result of the introduction of organic matter and nutrients from re-flooded terrestrial areas (Nees, 1964; Cooper, 1966; McLachlan, 1970; Hestand et al., 1973; Osborne et al., 1987), the increased concentration of dissolved materials due to water loss by evaporation (McLachlan et al., 1972), and the increased turbidity due to shoreline erosion and resuspension of bottom sediment (e.g., Grimas, 1962; Swanson, 1967; Geen, 1974; Walker and Tyler, 1984; Mitchell and Rogers, 1985; Osborne et al., 1987), or to rapid re-flooding of shallow lakes (Hestand and Carter, 1974).

*Biotic Characteristics* WLFs affect species composition and abundance and biotic interactions, as shown in the following examples:

1. Macrophytes: reduction of vegetation beds following drawdown of lake level is used as a management tool in fish ponds, small lakes, and hydroelectric reservoirs (e.g., Quennerstedt, 1958; Bennet, 1962; Grimas, 1962; Lantz et al., 1964; Mathis, 1965; Beard, 1969; Hestand et al., 1973, Nichols, 1975; Galdsby and Sanders, 1977; Hejny and Husak, 1978, Cooke, 1980).

   Changes in the composition and reduction in the productivity of vegetation as a result of WLFs have been described by Robel, 1962; Ager and Kerce, 1970; Clayton, 1982; Mitchell and Rogers, 1985. These effects have been explained by reduction in habitat availability, erosion, deterioration of light conditions, and desiccation. Proliferation of semi-aquatic vegetation in exposed areas have also been reported (e.g., Potapov, 1959—abstract in Frazer, 1972; Pieczynska, 1972; Hestand and Carter, 1974).
2. Benthic invertebrates: WLFs generally cause a reduction in species diversity and abundance of littoral invertebrates (e.g., Fisher and LaVoy, 1972). Some species have only a little ability to retreat with the lowering water and, therefore, become trapped in plant mats or stranded in small depressions (Bennet, 1954; Aass, 1960; Il'ina, 1963—abstract in Frazer, 1972; Cooper, 1966). In contrast, species which are able to migrate with the changing water level (e.g., chironomids) or are independent of the substrate (e.g., mobile benthos) may increase in abundance (Dahl, 1932, abstract in Frazer, 1972; Grimas, 1962; Cowell and Hudson, 1968; Geen, 1974; Martin et al., 1981). Littoral invertebrates moving out of the drying beds into the open water are often subject to predation (e.g., Bennet, 1954; Cooper, 1966). In extreme situations, benthic fauna has been completely eliminated following the destruction of the weed bed and direct exposure to wave action (Geen, 1974).
3. Fish: *a. Spawning*: there is evidence for serious delay of spawning following winter drawdown of lake level which resulted in freezing to the bottom, preventing fish, such as walleye, from entering the lake until after ice break-up (e.g., Monro and Larkin, 1950; Geen, 1974). Lake level lowering may also cause egg mortality by stranding egg beds (Monro and Larkin, 1950; Aass, 1964—abstract in Frazer, 1972; Geen 1974) or by forcing the fish to spawn in less favorable areas, thus exposing the eggs to increased predation (e.g., Martin, 1955). Absence of appropriate substrate may inhibit ovulation in some fish species (Stacey, 1984)

   *b. Habitat*: flooding of the shoreline vegetation may have beneficial effects; breeding success is enhanced and more suitable substrates for spawning, as well as protective cover for the early life stages, are provided (e.g., Martin et al., 1981). Limited, brief drawdowns of the water level can increase microhabitat complexity and species diversity (e.g., Godshalk and Barko, 1988).

*c. Food availability*: impaired fish growth and population decline following WLF have been attributed to reduced food availability (Miller and Paetz, 1959; Aass, 1960; Tyurin, 1961—abstract in Frazer, 1972; Nilsson, 1966—abstract in Frazer, 1972; Fuerest, 1970—abstract in Frazer, 1972). There is some evidence that WLFs may force changes in feeding habits of fish and change predator-prey relationships, frequently by increasing piscivory (e.g., Nilsson, 1961; 1964). As is the case for invertebrates, drawdown of the water level may force small fish to move out of the littoral region, exposing them to pelagic predation (e.g., Bennet, 1954; Heman et al., 1969). Changes in food availability and trophic interactions can lead to changes in species composition (e.g., Eschmeyer, 1949; Lantz et al., 1964; Geen, 1974). Increases and decreases in fish yields are claimed to be positively correlated with higher and lower lake levels respectively (Il'ina, 1963—abstract in Frazer, 1972).

There is evidence that fish feeding may be most efficient at "intermediate" levels of vegetation density (Crowder and Cooper, 1982). In this respect, WLF may have a beneficial effect on fish by limiting vegetative growth in the littoral zone. WLFs have actually been used for manipulating fish population densities in reservoir fisheries (e.g., Wood, 1951; Sheild, 1958; Groen and Schroeder, 1978; Martin et al., 1981).

**Characteristics of WLFs Affecting the Littoral Environment** The effects of WLFs on the littoral environment are influenced by the amplitude, frequency, periodicity, direction (flooding or drawdown), and timing of the fluctuations in relation to littoral dynamics. The impact of WLFs may be expected to be directly related to the amplitude of the fluctuations. The frequency of fluctuation determines the duration of the effect (exposure or inundation) and the time available for the responses of the biota. Thus, the effect of lake drawdown, for example may be either beneficial or detrimental to the littoral ecosystem, depending on the duration of the fluctuation (Godshalk and Barko, 1988). The fluctuation cycle is not necessarily symmetrical (unequal rates of flooding and drawdown); furthermore, the frequency of fluctuations may vary from diel (e.g., pump storage systems) to seasonal fluctuations.

The periodicity of fluctuations determines the degree of predictability of observed effects. The high "pulse stability" of the marine analogue, the intertidal ecosystems (Odum, 1969), is accompanied by high diversity and production. The much-lower predictability of WLFs in freshwater ecosystems may explain the much-lower levels of diversity and production observed in the littoral regions of fluctuating freshwater ecosystems.

The direction of water level change, i.e., rise or fall of lake level, frequently determines whether effects on the community are beneficial or adverse. Flooding may increase habitat availability and improve habitat conditions for spawning (e.g., Martin et al., 1981) or plant development, whereas drawdown may achieve just the opposite (Monro and Larkin, 1950; Geen, 1974). However, a given change in water level may have different consequences in different situations. For example, drawdown associated with erosion and resuspension of bottom sediments reduces light penetration and limits plant development (Swanson, 1967; Geen, 1974; Walker and Tyler, 1984; Mitchell and Rogers, 1985; Osborne et al., 1987). The same effect can be caused by rapid flooding of a shallow lake (Hestand and Carter, 1974). On the other hand, continued plant growth throughout the lake basin sometimes follows drawdown in shallow lakes when light conditions remain unaltered (Hestand et al., 1973).

The timing of the fluctuations in relation to the dynamics of the littoral ecosystem influences the magnitude of the impact. For example, if drawdown coincides with the spawning period, spawning beds can be exposed, leading to collapse or decline of fish populations (e.g., Monro and Larkin, 1950; Martin, 1955; Geen, 1974). Similarly, maximum reduction of submerged plant biomass may be expected if drawdown and exposure occur at periods of maximum plant growth (spring, summer) rather than at periods of growth arrest (winter). Flooding or drawdown of a certain intensity, occurring during periods of strong wave action is likely to have a different impact than similar WLFs during a quiescent period.

In conclusion, the impact of WLF in littoral zones is, to a large degree, site-specific. This is due to the large number of interacting factors which define any littoral site.

**Impact of WLF in Relation to Lake Size** The littoral to pelagial ratio is a key factor in the evaluation of the impact of a given WLF disturbance in relation to lake size. Wetzel and Hough (1973) and Howard-Williams and Allanson (1981) have emphasized the metabolic role of the littoral in small lakes. In addition, it has been shown that weed beds provide cover and support to a diverse community of grazers and predators (Keast, 1984; Carpenter and Lodge, 1986; Miller et al., 1988). Thus, drawdowns of water level in small and/or shallow lakes could lead to reduced diversity and overall decline of ecosystem productivity (Bartell and Breck, 1979; Breck and Kitchell, 1979; Carpenter and Lodge, 1986).

In contrast, in large, deep lakes, the littoral zone represents a negligible fraction of the lake area as a whole and, thus, plays a minor role as a source of primary production. However, due to its small proportion in the ecosystem, the littoral zone may be expected to act as a limiting resource of the habitats that provide cover, sites for colonization, spawning, and foraging grounds. Further reduction in habitat availability following lake drawdown, may lead to direct or indirect decreases in the diversity and abundance of littoral biota. Lodge et al. (1988) suggest that, as the lake size increases, the impact of littoral fish on the pelagial environment diminishes. This hypothesis should be restricted to situations where the dominant pelagic populations do not depend on littoral habitats at a certain stage of the life cycle.

## 8.3 A Case Study: Lake Kinneret

Lake Kinneret is a warm, monomictic lake located at ca. 210 meters below sea level in the northern part of the Jordan Valley. Its surface area is 170 km², its maximum depth 43 m. The length of shoreline is 53 km. The shoreline development index of 1.16 indicates a very regular shoreline.

Archaeological data indicate that, at least since 1000 B.C., the maximum water level has oscillated between −209 and −209.5 m and that the natural amplitude of the regular annual WLFs did not exceed 2 m (Serruya, 1978). Since the early 1930's, however, man-made modifications in the lake and its watershed (Serruya, 1978) have resulted in artificial control of the lake level and an increase of the yearly WLF up to 3 meters (−209 to −212 m). The lake is presently used as the major regulatory reservoir of the Israeli water supply system. It has recently been decided to increase the lake storage capacity in order to catch the winter floods of the Jordan River by lowering the lake water level down to −213 m in autumn. As a result, the multiannual amplitude of the WLF can reach 4 m.

**Littoral Zone** Considering the lower boundary of benthic algae (Dor, 1970), the littoral zone in Lake Kinneret extends to a depth of 3 m (between −209 to −212 m) and occupies an area of about 450 ha, that is, less than 3% of the lake area. Therefore, despite its relatively small surface area, Lake Kinneret has properties otherwise characteristic of large lakes—namely, a large pelagic zone as compared to the littoral.

The littoral zone of Lake Kinneret is characterized by the absence of submerged macrophytes, except for sporadic appearances in sheltered areas (Gophen, 1982). This can be attributed to strong wave action, relatively large WLFs and unfavorable light conditions (Waisel, 1967; Gophen, 1982).

The bottom substrate varies around the perimeter of the lake (Serruya, 1978). It includes boulders and stones of various sizes consisting of basalt, limestone, and flint. Finer substrates, mainly composed of sand, silt, and clay are also found.

Substrate size (picked at random and its three-dimensional parameters measured), littoral slope (determined by depth sounding), and plant biomass (determined by harvesting plots of 1 m²) were measured along 70 transects perpendicular to the shore, which subdivided the lakeshore into sections of approximately 750-m-length each (Figure 8.1). The study extended from October 1986 to September 1987.

**Results and Discussion** During summer 1986, the lake level dropped

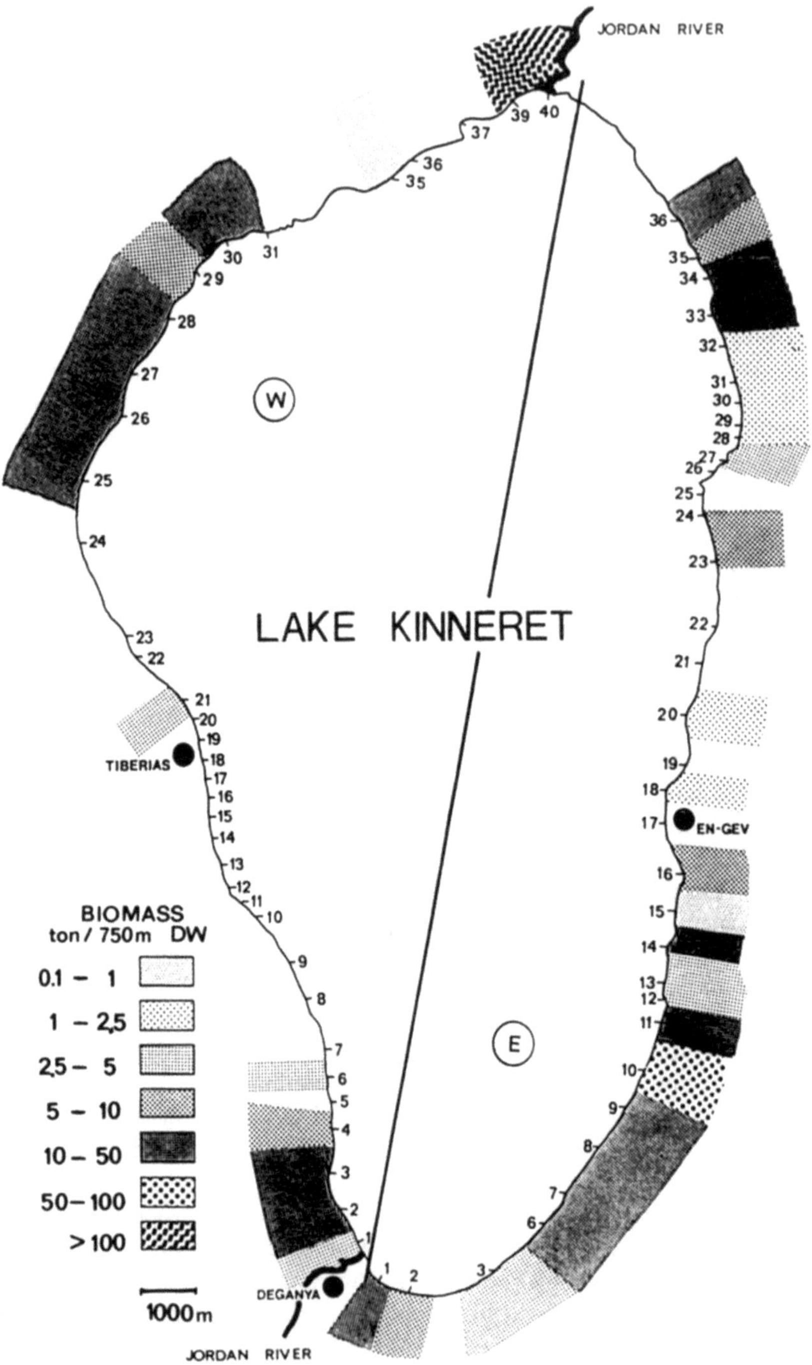


**Figure 8.1** Plant biomass distribution on the exposed shoreline of Lake Kinneret. The numbers indicate the transect locations: W—western shoreline, E—eastern shoreline.

by 1.5 m from −211 m down to a record low (−212.46 m). From October 1986 to April 1987, it rose by 3 m to a normal high (−209.50 m). The water hydrograph for January 1986 to August 1987 is shown in Figure 8.2.

*Area exposed*: the lowest lake level reached at the end of summer 1986 was associated with exposure of an area of 450 ha, which was similar in size to that of the entire littoral zone as previously defined for a water level of −209 m. The offshore distance of the exposed littoral varied from a few meters up to 300 m (Table 8.1). The areas exposed on the eastern side of the lake were generally larger (median value of 66 m) than those exposed on the western side (median value of 37 m).

**Littoral Slope**: the average slope of the littoral zone in selected transects is shown in Table 8.1. The western shores are generally steeper (median values of 6.5%) than the eastern ones (median value of 2.1%). Relatively gentle slopes (1–5%) are dominant (42% of the transects), followed by steep slopes (5–10%), in 29% of the cases. In 16% of the transects, the slop was very steep (>10%) and, in about 13% of the transects, the slope was very gentle (<1%). The difference in littoral slopes around the lake corresponded to the difference of the areas exposed. The littoral slope often varies within the transect, usually sloping more gently at lower lake levels.

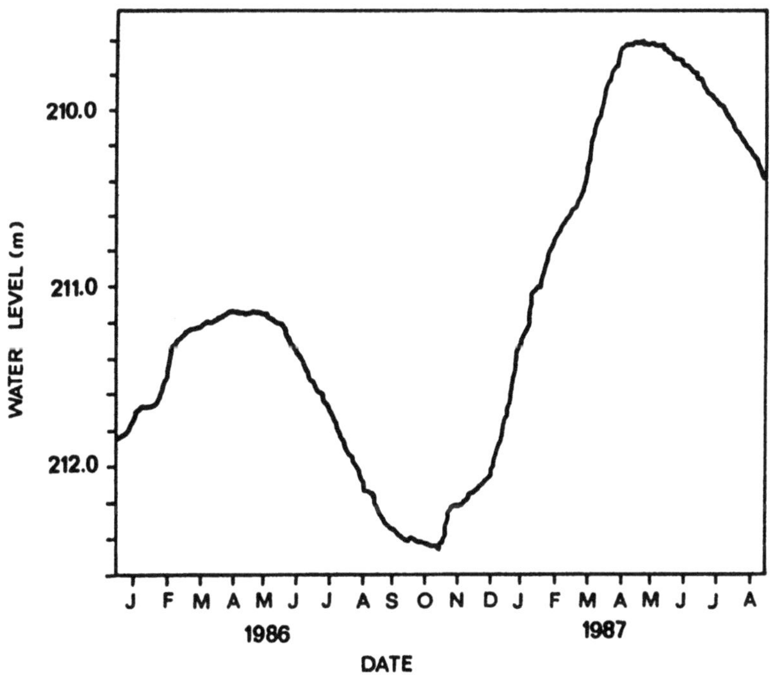


**Figure 8.2** Water level fluctuations (meters below sea level) in Lake Kinneret in 1986/87.

**Table 8.1** Average slopes and maximum distances of exposed areas (between −209.5 and −212.5 m lake levels) in transects[a] around the shoreline of Lake Kinneret

| Eastern transects | | | Western transects | | |
|---|---|---|---|---|---|
| Transect no. | Slope (%) | Distance (m) | Transect no. | Slope (%) | Distance (m) |
| E1 | 1.7 | 161 | W3 | 1.3 | 92[b] |
| E3 | 2.0 | 84[b] | W4 | 3.5 | 45[b] |
| E6 | 1.4 | 106[b] | W5 | 5.2 | 52 |
| E7 | 0.7 | 123[b] | W6 | 3.0 | 40[b] |
| E8 | 0.4 | 200[b] | W8 | 7.8 | 37 |
| E10 | 2.0 | 154 | W10 | 11.2 | 33 |
| E11 | 2.3 | 65[b] | W21 | 8.1 | 38 |
| E13 | 4.2 | 66 | W22 | 6.5 | 26[b] |
| E19 | 12.0 | 22 | W24 | 10.9 | 28 |
| E20 | 5.9 | 57 | W25 | 1.2 | 236 |
| E21 | 8.7 | 28 | W27 | 0.5 | 93[b] |
| E23 | 15.7 | 20 | W28 | 5.7 | 52 |
| E25 | 5.2 | 57 | W29 | 6.0 | 47 |
| E30 | 2.1 | 138 | W30 | 3.2 | 85 |
| E31 | 3.4 | 76 | W31 | 4.0 | 67 |
| E32 | 2.4 | 115 | W32 | 18.8 | 17 |
| E33 | 0.8 | 275 | W33 | 8.8 | 30 |
| E34 | 0.6 | 301 | W34 | 6.5 | 46 |
| | | | W36 | 8.6 | 33 |
| | | | W39 | 0.5 | 312[b] |

[a] The locations of the transects are shown in Figure 8.1.
[b] The littoral zone is bordered by a wall or embankment.

*Substrate* the substrate distribution is given in Tables 8.2 and 8.3. In general, coarser substrate is correlated with steeper slopes. Moreover, the proportion of finer substrates (more sand and less boulders)—and gentler slopes—continuously increases with receding lake levels (Figure 8.3). Within an amplitude of 3 m, the decline from the highest to the lowest lake level results in an increase of the proportion of sandy shorelines from 6 to 43%. In addition, fewer littoral areas are bordered by walls and embankments (from 25% of the cases at high level to 0% at low level).

*Lakeshore vegetation* Massive development of semiæquatic vegetation was observed in certain lakeshore areas during the low-water phase. Plant biomass distribution on exposed shores around the lake was shown in Figure 8.1. It varied from less than 0.1 ton dry weight (DW)/750 m on stony bottom (40% of the shores) to more than 100 ton DW/750 m on sandy or silty substrates. The average biomass per unit area varied from less than 0.01 to 1.3 kg/m$^2$,

**Table 8.2** Substrate composition (dominant elements) at different water levels in selected transects around the shoreline of Lake Kinneret

| Transect no. | Water level (m) | | | | | | |
|---|---|---|---|---|---|---|---|
| | −209.5 | −210 | −210.5 | −211 | −211.5 | −212 | −212.5 |
| W3 | W* | W | S | S+SS | MS+SS+S | S+C | C+S |
| W8 | LS+B | LS | LS | LS | LS | LS | LS |
| W13 | W | W | W | W | W | W | W |
| W22 | W | W | G | G | B+LS+SS | B+LS+SS | B+LS |
| W24 | MS+P | MS+P | MS+P | MS+P | LS+MS | LS+MS | B+LS |
| W25 | W | W | S | S | S | S | S |
| W32 | B+LS | B+LS | B+LS | B+LS | B+LS | B+LS | B+LS |
| W39 | S+SI | S+SI | S+SI | S+SS | S+LS | LS+B | LS+B |
| E1 | S+P+SS | P+SS | S+P+SS | S+P+SS | G+S+SS | C+S+G | C+P+SS |
| E6 | W | W | S+SS | S+SS | S+SS | S | S |
| E10 | W | W | W | W | S+LS | S | S |
| E12 | W | LS+B | LS+MS+B | LS+MS+B | MS | SS | S |
| E16 | W | W | MS+SS+P | MS+SS+P | MS+SS+P | S | S |
| E19 | G | G | P+G | P+G | P+G | P+G | P+G |
| E22 | LS | LS+MS | MS+SS | SS+B | SS+G | P+G | P+G |
| E25 | MS+P+G | MS+P | MS+G+S | S+MS | S+LS+B | S+MS | S |
| E30 | G+P | G+P | G+P | G | S | S | S |
| E32 | MS+S | MS+S | S | S | S | S+SS | S+SS+G |
| E34 | G+S | LS | SS+MS | P+SS | MS+LS | MS+LS | MS |
| E36 | S+P | S+P | S | S | S | S | S |

* B, boulders; C, clay; G, gravel; LS, large stones; MS, medium stones; P, pebbles; S, sand; SI, silt; SS, small stones; W, wall. Sizes are given in the caption for Figure 8.3.

increasing away from the water apparently reflecting the longer duration of exposure.

The estimated total lakeshore biomass was 1,000 tons ash-free dry weight, i.e., equivalent to about 500 tons of carbon. That is less than 1% of the annual phytoplankton production.

The relative contribution of the major plant species of macrophytes in a declining order of importance was as follows:

| | |
|---|---|
| *Cyperus alopecuroides* | 30% |
| *Tamarix jordanensis* | 28% |
| *Cyperus conglomeratus* | 18% |
| *Typha angustata* | 15% |
| *Phragmites australis* | 7% |

The highest biomass was contributed by *C. alopecuroides*, a species that, under normally high water levels, is mainly restricted to the Jordan River delta. During the period of low water levels, it proliferates on exposed sandy shores.

**Table 8.3** Regional distribution of bottom substrate (dominant elements) and range of average slope in the littoral zone of Lake Kinneret

| Region | Transect | Substrate composition | Slope (%) |
|---|---|---|---|
| Northwest | W30–W35 | boulders, large–small stones (basalt) | 3–19 |
| Tabgha–Ginosar | W25–W26 | sand, silt | 0.5–1.2 |
| Midwest | W22–W24 | boulders, large–small stones, (limestone) | 6.5–11 |
| Southwest | W2–W23 | large–small stones (limestone, basalt) | 1.7–11 |
| South | W1–E3 | clay, sand, gravel, gravel, pebbles (basalt, limestone, flint) | 2 |
| Southeast | E6–E14, E16–E18 | sand, silt, pebbles, small stones | 0.7–4.5 |
| Mideast–northeast | E19–E32 | large–small stones, pebbles, gravel, sand | 2.1–12 |
| North | E36–W39 | sand, silt | 0.5–0.6 |

Unprecedented development of young *Eucalyptus* was also recorded in some locations.

For several months following inundation, the vegetation provides cover, sites for colonization, and food for the littoral biota until it decomposes. In the years following consecutive high lake levels, vegetative littoral habitats become very scarce. The periodic existence of emergent vegetation may be of great importance for the biota in lakes such as Lake Kinneret, which are largely devoid of submerged vegetation.

*Fish*: an unexpected relation between WLF and near-shore spawning of an endemic, pelagic fish (*Mirogrex terraesanctae*) in Lake Kinneret could be demonstrated. This species spawns during the winter in very shallow water (5 to 50cm) on stony substrates. The lower depth of spawning is determined by the development of benthic diatoms on stones. During winter, the continuously rising lake level provides new stones clear of algae that are suitable for spawning. WLFs thus increase the potential area of spawning grounds for this fish species and, hence, may have a positive influence on its overall population size. The pelagic *Mirogrex terraesanctae* can be considered to be specially adapted to and dependent on WLFs in the littoral. Increase in population size of this fish, suspected in recent years, may be partly attributed to the increases in the amplitude of the WLFs. This case study exemplifies the possible importance of littoral habitats in a large lake situation and the potential influence of WLFs.

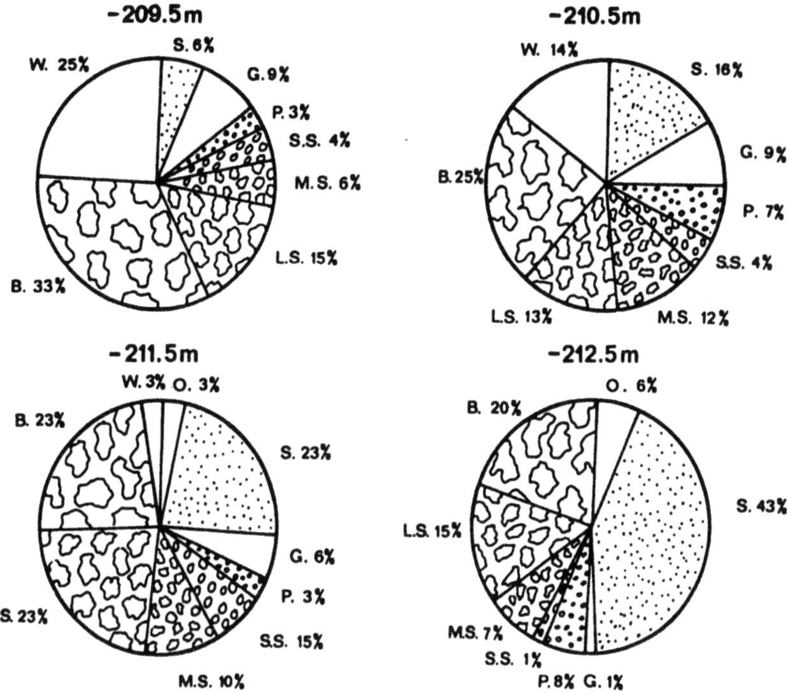


**Figure 8.3** Relative proportion of substrate distribution at the shoreline at four different lake levels. S—sand: up to 0.2 cm; G—gravel: 0.2–1 cm; P—pebbles: 1–5 cm; SS—small stones: 5–10 cm; MS—medium stones: 10–20 cm; LS—large stones: 20–35 cm; B—boulders: >35 cm; O—others: silt, clay, and other fine substrates.

## 8.4 Conclusion

The study of the littoral region of Lake Kinneret has provided us with valuable information on the modifications of the littoral slope and substrates that may occur when the lake level is allowed to fluctuate as much as 4 m. In addition, quantitative information on macrophyte growth suggests that in Lake Kinneret—despite the fact that pelagic production dominates—the vegetative littoral zone plays an important role in providing substrates for colonization, cover, and foraging grounds for fish which is far in excess of its importance as a source of primary production. Finally, the discovery of the strategy of spawning of M. *terraesanctae,* which requires fresh stony substrate provided by the increasing lake level, may have an impact on lake management. this zooplanktivorous, pelagic species maintains the highest biomass of all fish species in the lake. The recent decline of the zooplankton biomass and the subsequent increase of small-size algae, which result in degradation of water quality, are attributed to the increase in the population of this fish (Gophen, 1984). Our findings may offer a new approach for manipulation of the reproductive success of this species through regulation of the lake level.

The study of the littoral zone of Lake Kinneret underscores the heterogenic and dynamic nature of this lake region. It highlights conclusions of other researchers—for example, of the changes of the vegetative cover and of fish habitats at receding levels. Moreover, this study conveys a special message concerning large lakes; the littoral zone of such lakes affects the food-web at high trophic levels by providing qualitative "services" to both benthic invertebrates and fish (i.e., spawning grounds and shelter) and controlling, to a certain extent, their reproduction and survival.

Littoral habitats may play a disproportionate role in large lacustrine ecosystems. Disturbances caused by WLFs may contribute to the temporal and spatial heterogeneity of this region. Further studies are needed to elucidate the littoral—pelagial relationship in large lakes and the possible dependency of higher trophic levels on littoral resources.

---

*Acknowledgements* We wish to thank T. Berman and the staff of the Kinneret Limnological Laboratory for their kind assistance. We thank C. Serruya for helpful remarks and suggestions. This study was supported by a grant from Israel Oceanographic and Limnological Research and Tel Aviv University. This report is a contribution of the Institute for Nature Conservation Research, George S. Wise Faculty of Life Sciences, Tel Aviv University (INCR/5/88).

## References

Aass, P. 1960. The effect of impoundment on inland fisheries. In: *IUCN/FAO International Union for Conservation of Nature and Natural Resources. Seventh Technical meeting, Athens, September 1958.* Vol. IV. Soil and Water Conservation, Natural Aquatic Resources, p. 69–76.

Ager, L.A. and K.E. Kerce. 1970. Vegetational changes associated with water level stabilization in Lake Okeechobee, Florida. In: *Proc. 24th Ann. Conf. Southeast. Assoc. Game and Fish Commissioners,* p. 338–351.

Bartell, S.M. and J.E. Breck. 1979. Simulated impact of macrophyte harvesting on pelagic phosphorus cycling in Lake Wingra. In: J.E. Breck, R.T. Prentki and O.L. Loucks (eds.), *Aquatic Plants, Lake Management and Ecosystem Consequences of Lake Harvesting.* Inst. Environ. Studies, Madison, WI, p. 229–249.

Beard, T.D. 1969. Impact of an overwinter drawdown on aquatic vegetation in Murphy Footage, Wisconsin. *Res. Rep. 43,* Dept. of Natural Resources, Madison, Wis. 16 pp.

Bennett, G.W. 1954. The effect of a late-summer drawdown on the fish population of Ridge Lake, Coles Country, Illinois. *Trans. North Am. Wildl. Conf.* 19:259–270.

Bennett, G.W. 1962. *Management of Artificial Lakes and Ponds.* Reinhold Publ. Co., New York, 283pp.

Breck, J.E. and J.F. Kitchell. 1979. Effect of macrophyte harvesting on simulated predator-prey interactions. In: J.E. Breck, R.T. Prentki and O.L. Loucks (eds.). *Aquatic Plants, Lake Management and Ecosystem Consequences of Lake Harvesting.* Inst. Environ. Studies, Madison, WI, p. 211–228.

Carpenter, S.R. and D.M. Lodge. 1986. Effects of submersed macrophytes on ecosystem processes. *Aquat. Bot.* 26:341–370.

Clayton, J.S. 1982. Effect of fluctuation in water level and growth of *Lagarosiphon major* on aquatic vascular plants in Lake Rotoma, 1973–80. *New Zealand J. Mar. Freshwater Res.* 16:89–94.

Cooke, G.D. 1980. Lake level drawdown as a macrophyte control technique. *Water Resources Bulletin* 16:317–322.

Cooper, G.P. 1966. Fish production in impoundments. *Res. Dev. Rep. Mich. Dep. Conserv.,* 104. 21pp.

Cowell, B.C. and P.L. Hudson. 1968. Some environmental factors influencing benthic invertebrates in two Missouri River reservoirs. In: *Proceedings of the Reservoir Fishery Resources Symposium, Athens, Georgia, April 5–7, 1967,* p. 541–555.

Crowder, L.B. and W.E. Cooper. 1982. Habitat structural complexity and the interaction between bluegills and their prey. *Ecology* 63:1802–1813.

Dor, I. 1970. Production rate of the periphyton in Lake Tiberias as measured by the glass-slide method. *Isr. J. Bot.* 19:1–15.

Eschmeyer, R.W. 1949. The fisheries picture with special reference to the TVA impoundment: Part 2. *Prog. Fish Cult.* 11:267–271.

Fisher, S.G. and A. LaVoy. 1972. Differences in the littoral fauna due to fluctuating water levels below a hydroelectric dam. *J. Fish. Res. Bd. Canada* 29:1472–1476.

Frazer, J.C. 1972. *Water Levels, Fluctuation and Minimum Pools in Reservoirs for Fish and Other Aquatic Resources, an Annotated Bibliography.* FAO Fisheries Technical Paper No. 113. (FIRI/T113). 42pp.

Frey, D.G. 1967. Reservoir research—objectives and practices with an example from the Soviet Union. In: *Reservoir Fishery Resources Symposium presented by the Reservoir Committee of Southern Division.* American Fisheries Society, p. 26–36.

Geen, G.H. 1974. Effect of hydroelectric development in Western Canada on aquatic ecosystems. *J. Fish. Res. Bd. Canada* 31:913–927.

Godshalk, G.L. and J.W. Barko. 1988. Effects of winter drawdown on submersed aquatic plants in Eau Galle Reservoir, Wisconsin. *Proc. 22nd Ann. Meeting Aquat. Plant Cont. Res. Prog.,* p. 100–111.

Goldsby, T.L. and D.R. Sanders. 1977. Effects of consecutive water fluctuations on submerged vegetation of Black Lake, Louisiana. *J. Aquat. Plant Managem.* 15:23–28.

Gophen, M., 1982. Unusually dense watermilfoil (*Myriophyllum spicatum* L.) vegetation in the southern basin of Lake Kinneret (Israel) in 1979. *Aquat. Bot.* 13:307–315.

Gophen, M. 1984. The impact of zooplankton status on the management of Lake Kinneret (Israel). *Hydrobiologia* 113:249–258.

Grimas, U. 1962. The effect of increased water level fluctuation upon the bottom fauna in Lake Blasjon, northern Sweden. *Rep.Inst. Freshwat. Res. Drottningholm.* 44:14–41.

Groen, C.R. and T.A. Schroeder. 1978. Effects of water level management on walleye and other coolwater fishes in Kansas reservoirs. *Amer. Fish. Soc. Special Publication* 11:278–283.

Heman, M.L., R.S. Campbell, and L.C. Redmond. 1969. Manipulation of fish population through reservoir drawdown. *Trans. Am. Fish. Soc.* 98:293–304.

Hejny, S. and S. Husak. 1978. Ecological Effect of Fishpond Amelioration. In: Dykyjova, D. and J. Kvet (eds.). *Pond Littoral Ecosystem,* p. 409–415.

Herdendorf, C.E. and L.A. Fay. 1988. Development of an environmental sensitivity index for costal areas of Lake Erie, North America. *Verh. Internt. Verein. Limnol.* 23:380–385.

Hestand, R.S., B.E. May, D.P. Schultz, and C.R. Walker. 1973. Ecological implications of water levels on plant growth in a shallow water reservoir. *Hyacinth Contr. J.* 11:54–58.

Hestand, R.S. and C.C. Carter. 1974. The effect of a winter drawdown on aquatic vegetation in a shallow water reservoir. *Hyacinth Contr. J.* 12:9–12.

Howard-Williams, C. and G.M. Lenton. 1975. The role of the littoral zone in the function of a shallow tropical lake ecosystem. *Freshwat. Biol.* 5:445–459.

Howard-Williams, C. and B.R. Allanson. 1981. An integrated study on littoral and pelagic primary production in a Southern African costal lake. *Arch. Hydrobiol.* 92:507–534.

John, D.M. 1986. The inland waters of tropical West Africa. An introduction and botanical review. *Arch. Hydrobiol.* 23:1–244.
Keast, A. 1984. The introduced macrophyte, *Myriophyllum spicatum*, as habitat for fish and their invertebrate prey. *Can. J. Zool.* 62:1289–1303.
Lantz, K.E., J.T. David, J.S. Houghes, and H.E. Schafer. 1964. Water level fluctuations, its effect on vegetation control and fish population management. *Proc. 18th Ann. Conf. S.E. Assoc. of Game and Fish Comm.* 18:483–494.
Lodge, D.M., J.W. Barko, D. Strayer, J.M. Melack, G.G. Mittelbach, R.W. Howarth, B. Menge, and J.E. Titus. 1988. Spatial heterogeneity and habitat interactions in lake communities. In: Carpenter S.R. (ed.). *Complex Interactions in Lake Communities.* Springer-Verlag, NY., p. 181–208.
Martin, N.V. 1955. The effect of drawdown on Lake Trout reproduction and the use of artificial spawning beds. *Trans. North Am. Wildl. Conf.* 20:263–271.
Martin, D.B., L.J. Mengel, J.F. Novotny, and C.H. Walburg. 1981. Spring and summer water levels in a Missouri River Reservoir: Effect on age-0 fish and zooplankton. *Trans. Amer. Fish. Soc.* 110:370–381.
Mathis, W.P. 1965. Observation on control of vegetation in Lake Catherine using Israeli carp and a fall and winter drawdown. *Proc. 19th Annu. Conf. S.E. Assoc. of Game and Fish Comm.*, p. 197–205.
McLachlan, A.J., P.R. Morgan, C. Howard-Williams, S.M. McLachlan, and D. Bourn. 1972. Aspects of the recovery of a saline African lake following a dry period. *Arch. Hydrobiol.* 70:324–340.
McLachlan, S.M. 1970. The influence of lake level fluctuation and the thermocline on water chemistry in two gradually shelving areas in Lake Kariba, Central Africa. *Arch. Hydrobiol.* 66:499–510.
Miller, R.B. and M.J. Paetz. 1959. The effect of power, irrigation and stock water developments on the fisheries of the south Saskatchewan River. *Can. Fish. Cult.* 25:13–26.
Miller, A.C., D.C. Beckett, and C.M. May. 1988. The value of submersed plants to macroinvertebrates. *Proc. 22nd Ann. Meeting Aquat. Plant Cont. Res. Prog.*, p. 87–93.
Mitchell, D.S. and K.H. Rogers. 1985. Seasonality/aseasonality of aquatic macrophytes in Southern Hemisphere inland waters. *Hydrobiologia* 125:137–150.
Monro, D.A. and P.A. Larkin. 1950. The effect of changes to the natural water level and water course on wildlife. *Trans. British Columbia Nat. Resour. Conf.* 3:267–272.
Moss, B. and J. Moss. 1969. Aspects of the limnology of an endorheic African lake (Lake Chilwa, Malawi). *Ecology* 50:109–118.
Nees, J.C. 1964. Development and status of pond fertilization in central Europe. *Trans. Amer. Fish. Soc.* 76:335–358.
Nichols, S.A. 1975. The use of overwinter drawdown for aquatic vegetation management. *Wat. Res. Bull.* 11:1137–1148.
Nilsson, N.A. 1961. The effect of water level fluctuations on the feeding habits of trout and char in the Lake Blasjon and Jormsjon, north Sweden. *Rep. Inst. Freshwater Res. Drottningholm* 42:261–283.
Nilsson, N.A. 1964. Effect of impoundment on the feeding habits of brown trout and char in Lake Ransaren (Swedish Lappland). *Verh. Int. Ver. Theor. Angew. Limnol.* 15:444–452.
Odum, E.P. 1969. The strategy of ecosystem development. *Science* 164:262–270.
Osborne, P.L., J.H. Kyle, and M.S. Abramski. 1987. Effect of seasonal water level changes on the chemical and biological limnology of Lake Murray, Papua New Guinea. *Aust. J. Mar. Fresh. Res.* 38:397–408.
Pieczynska, E. 1972. Ecology of the eulittoral zones in lakes. *Ekol. Pol.* 20:637–732.
Quennerstedt, N. 1958. Effect of water level fluctuation on lake vegetation. *Verh. Int. Ver. Theor. Angew. Limnol.* 13:901–906.

Robel, R.J. 1962. Changes in submersed vegetation following a change in water level. *J. Wildl. Managem.* 26:221–224.
Serruya, C. 1978. *Lake Kinneret: Monographiae Biologicae.* W. Junk bv Publishers, The Hague. Vol. 32, 501 pp.
Sheild, J.T. 1958. Experimental control of carp reproduction through water drawdowns in Fort Randall Reservoir, South Dakota. *Trans. Am. Fish. Soc.* 87:23–33.
Stacey, N.E. 1984. Control of the timing of ovulation by exogenous and endogenous factors. In: Potts, G.W. and R.J. Wootton (eds.), *Fish Reproduction Strategies and Tactics.* Academic Press, p. 207–222.
Swanson, G.A. 1967. Factors influencing the distribution and abundance of *Hexagenia* nymphs (Ephemeroptera) in a Missouri River reservoir. *Ecology* 48:216–225.
Waisel, Y. 1967. A contribution to the knowledge of phanerogamus vegetation of Lake Tiberias. *Bull. Sea Res. Stn. Haifa* 44:3–16.
Walker, T. and P. Tyler. 1984. Tropical Australia, a dynamic limnological environment. *Verh. Int. ver. Limnol.* 22:1727–1734.
Wetzel, R.G. 1983. *Limnology,* 2nd ed. Sounders College Publ., Philadelphia, PA. 767 pp.
Wetzel, R.G. and R.A. Hough. 1973. Productivity and the role of aquatic macrophytes in lakes. An assessment. *Pol. Arch. Hydrobiol.* 20:9–19.
Wood, R. 1951. The significance of managed water levels in developing the fisheries of large impoundments. *J. Tenn. Acad. Sci.* 26:214–235.

# Part 3

# Particle Transport and Chemical Fluxes

This section deals with various aspects of sedimentation in large lakes. First, mechanisms which generate sediment facies characteristic of large lakes are described; most notable are interflow sediments, which are deposited in the vicinity of major inlets, and turbidites. Thereafter, sediment resuspension in large lakes at considerable water depth is discussed, as well as examples of subsequent relocation (focusing) using radiotracer techniques.

The consequences of sediment dynamics on the retention of contaminants in the water column and on the transport of nutrients and other chemical constituents are examined.

The radioactive fallout following the nuclear power plant accident at Chernobyl has affected European large lakes. Residence times of settling matter within the water column are determined by following radionuclide pathways. Moreover, it has been possible to trace radionuclides in various components of the food-web.

Finally, aspects—both theoretical and applied—of the past history of large lakes are touched on, with emphasis on preindustrial eutrophication.

# 9

# Horizontal Transport and Sedimentation by Interflows and Turbidity Currents in Lake Geneva

Federico Giovanoli

ABSTRACT Turbidity currents and interflows are the main mechanisms of transport of suspended particles from rivers to lakes. Turbidity currents originate from singular events—for example, a delta deposit destabilized by unusually high flood waters and its subsequent transport through delta channels to the basin plain. The sedimentation rate maximum is at the end of the channel and decreases laterally from the channel axis. On the basin plain,turbidity currents mingle with the autochthonous sedimentation, which has a sedimentation rate that is smaller by several orders of magnitude.

In contrast, interflow sedimentation rates decrease exponentially with distance from the river mouth. The inflow is affected by Coriolis forces and interflow sediments are dominant in the areas extending to the right of the inflow. The composition of the sediments evolve with distance to the inflow due to different settling velocities of the components.

In the case of Lake Geneva, turbidity currents occur in a Rhone delta channel approximately twice a week in summer (Lambert and Giovanoli, 1988). They are absent in winter. Below the interflow, the sedimentation rates of total solids are about 100, 30, and 5 g $m^{-2}$ $d^{-1}$ at 1.85. 4.75, and 9.3 km from the river mouth, respectively. In the interflow, the composition of particulate matter is similar to that of the Rhone, except the organic matter content, which is four times higher. This is explained by the relatively low settling velocity of organic matter. This interpretation is supported by the observed increase of organic matter content in sediment traps and in bottom sediments with distance. With sedimentation rates of about 0.15 g $m^{-2}$ $d^{-1}$, the contribution of the autochthonous organic matter is minor.

## 9.1 Introduction

Many large lakes have only a few major tributaries. Consequently, the input of allochthonous sediment is unevenly distributed over the whole lake basin.

In the absence of upstream lakes as settling ponds, inflowing waters are frequently turbid and the input of solid matter to a large lake is considerable.

Near a river mouth, in addition to the formation of a delta itself (Bates, 1953; Coleman, 1976; Wright, 1977; Brunsden and Prior, 1984), the following two processes control the transport pathways and distribution of particles on the lake bottom. These are: deposition (settling out) of bedload and the coarser fractions of the suspended particles; and entrainment of lake waters (mixing). Depending on the density contrast between river water and lake water, overflows, interflows or underflows may also occur (Wetzel, 1975; Sturm and Matter, 1978; Fischer et al., 1979).

*Overflows* are common for rivers entering the oceans and for "warm" rivers (e.g., Omo River into Lake Turkana, Yuretich, 1969), as well as for rivers draining large lakes (e.g., Niagara River into Lake Ontario, Saylor et al., 1981).

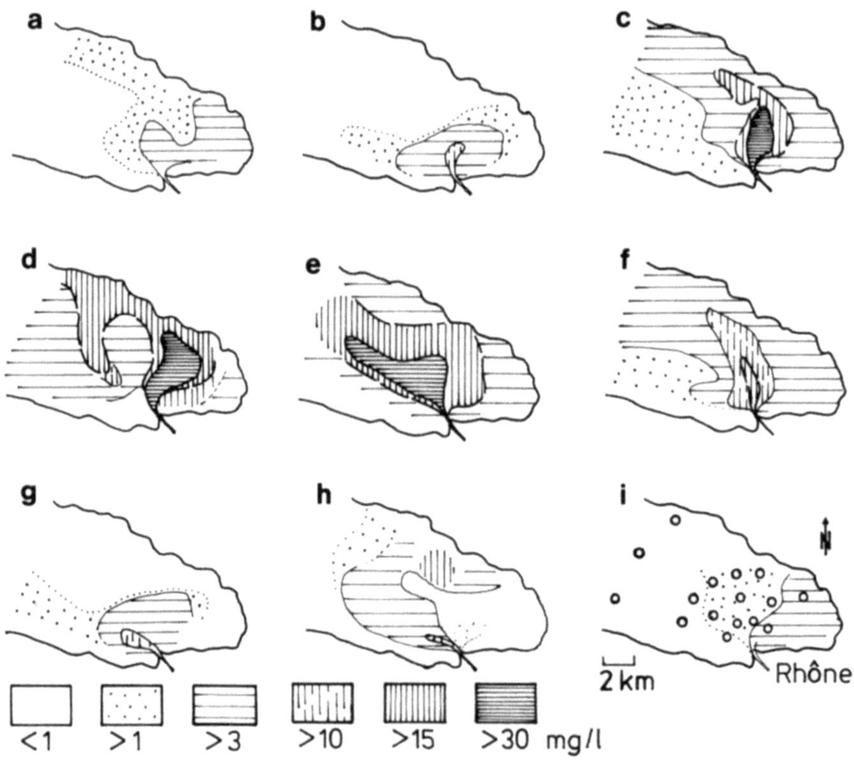


**Figure 9.1** The spreading of the alpine Rhone interflow. The isolines indicate the maximum concentrations of suspended particles (in mg/l) in a vertical profile. In figure 9.1 *i*, the point at which the river enters the lake is shown; the circles indicate the sampling station locations. The sampling period and the depth interval of the maxima for each figure are as follows: *a*, March 22–31, 80–120 m; *b*, April 26–May 11, 0–25 m; *c*, May 24–28, 12–24 m; *d*, June 8–15, 14–18 m; *e*, July 20–23, 10–22 m; *f*, Sept. 20–27, 11–23 m; *g*, Nov. 11–19, 28–39 m; *h*, Dec. 8–13, 30–70 m; *i*, Feb. 23–25, 90–155 m.

Sediment-laden rivers usually intrude into dilute lake waters due to density contrasts increased by sediment suspensions. If lakes are density stratified, these intrusions may form *interflows* which spread along a horizontal plane. Interflows often extend over considerable distances and at depths with the steepest density gradient. Sediments resulting from interflows are generally spread over large areas.

*Underflows* carry water and sediment suspension to the deeper part of lakes. Underflows rarely last more than a few hours. Continuous flow is unlikely because river density does not generally remain greater than that of deep lake water for extended periods. Although rare, underflow events can account for as much solid matter transport to the deeper basins as can the longer-term, continuous interflows.

The transport path of interflows (Figure 9.1) is controlled not only by density stratification and horizontal density differences, but also by Coriolis forces and internal currents. The momentum of the inflow dampens only a few hundred meters from the river mouth (Giovanoli and Lambert, 1985). In contrast, an underflow trajectory is controlled by the topography of the delta front or basin plain (Figure 9.2).

This fundamental difference explains why, in many cases, the two flow types separate in plane-view and produce independent sediment facies (Figure 9.3). In the northern hemisphere, interflows are forced to the right of the inflow, where, consequently, the interflow facies is well represented (Figure 9.4). Turbidity currents are commonly channelized in the delta foreslope with a characteristic "Thalweg."

A second important difference is that the river waters quickly lose their

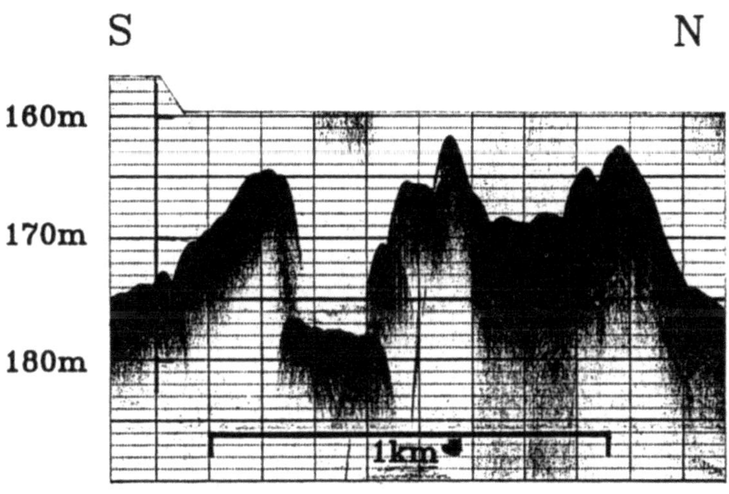


**Figure 9.2** The upper delta channel of the alpine Rhone in Lake Geneva, 2.75 km off the river mouth; channel bottom at 178-m water depth. Secondary channel at the right.

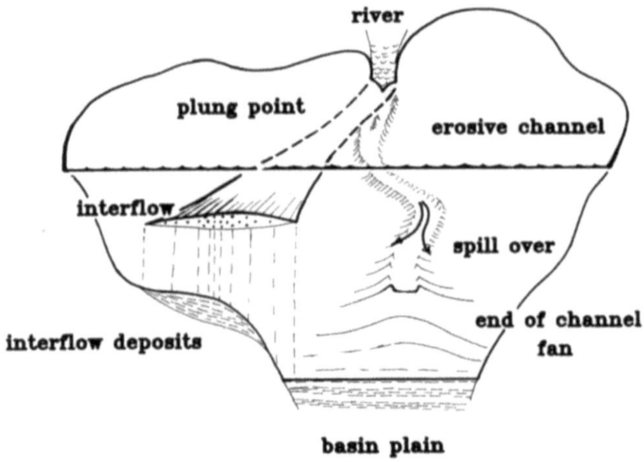


**Figure 9.3** Schematic representation of the two main transport pathways and the related sedimentary facies.

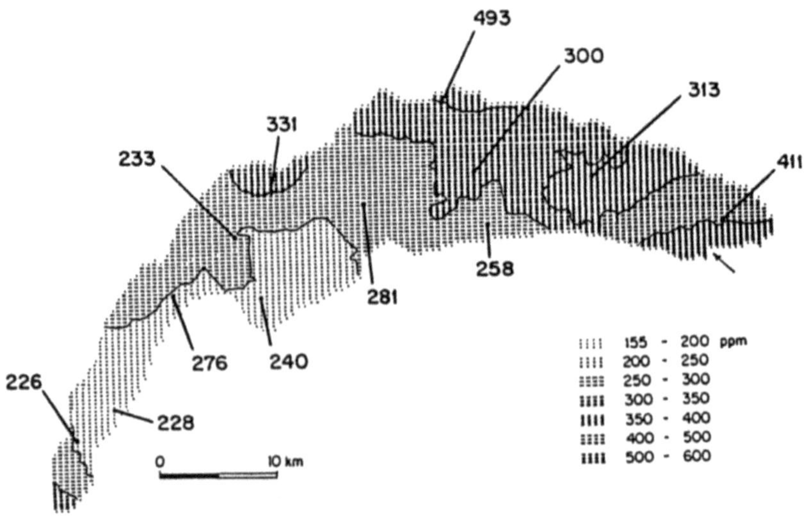


**Figure 9.4** Distribution of apatite-phosphorus in Lake Geneva surface sediments (Jaquet, 1978), as an expression of the interflow facies. Numbers refer to values in cores (Dominik et al., 1981b). High values at the northern shore represent newly formed minerals produced by pollution from a point source.

physical and chemical identity with increasing transport (Giovanoli and Lambert, 1985). Interflow water masses are dissipated by horizontal spreading and entrainment of lake waters. In contrast, channelized turbidity currents maintain their flow velocity and identity until they reach the deepest part of the basin.

Depending on slopes, turbidity currents may even accelerate (Hamblin and Carmack, 1978).

Different mechanisms of sediment transport are recorded by distinctive bottom deposits (sedimentary facies). After the basic contribution of Kuenen and Migliorini (1950), marine *turbidite facies* have been extensively described (Mutti and Ricci Lucchi, 1972). Reviews by Walker and Mutti, 1973; Kelts and Arthur, 1981. Ludlam (1967), Sturm and Matter (1978), Lambert and Hsu (1979), and Kelts and Hsu (1980) treat mass movements, turbidites, and river suspension flows in lakes.

*Interflow facies* deposits are rarely recognized, perhaps due to the absence of oceanic counterparts. Therefore, the role they play in large-lake sedimentation may have been overlooked. For this reason, in the following case study we briefly describe the delta-type deposits of Lake Geneva and focus on the processes leading to interflow deposits.

## 9.2 Delta and Turbidity Current Deposits of the Alpine Rhone River

The sand bars, foresets, and topsets of the Rhone river mouth will not be discussed here. These deposits however, provide unstable masses of sandy material which are important sources of suspended matter for turbidity currents.

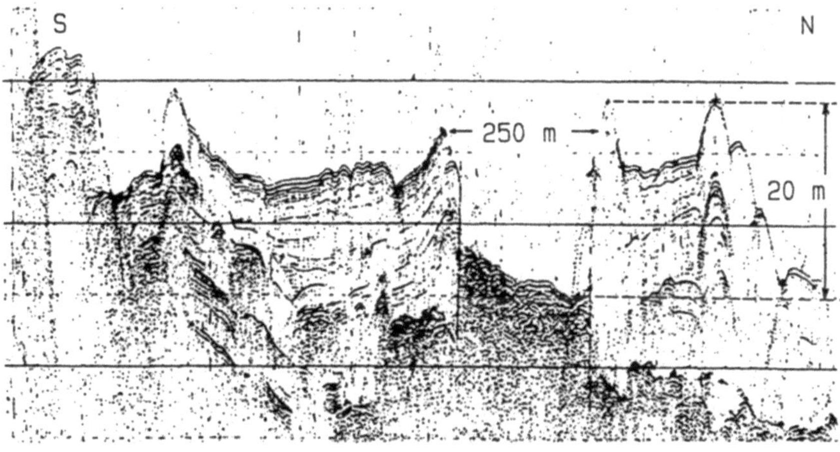


**Figure 9.5** Old erosional channels, filled up with younger sediments in a seismic cross section at about 3 km from the mouth of the Rhone at 170-m water depth (courtesy of J.-P Vernet, Inst. Forel, Univ. Geneva, unpublished data).

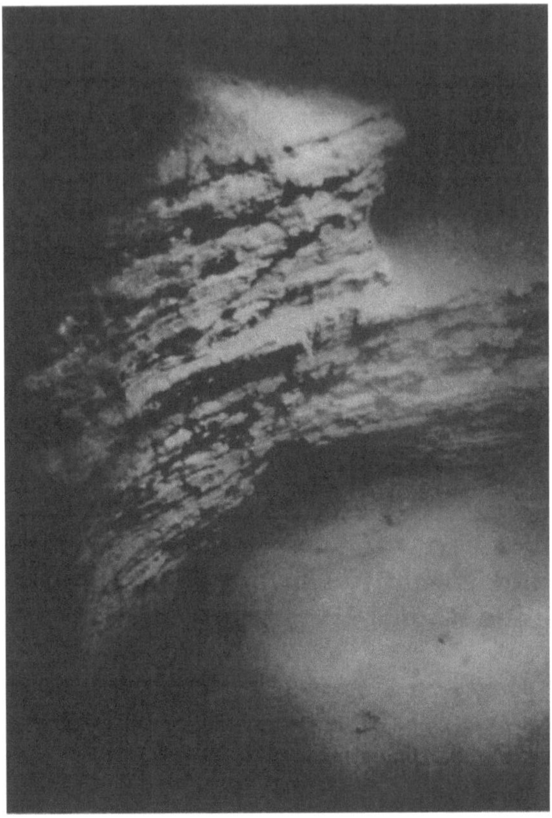

**Figure 9.6** The upper channel—an erosional feature, as indicated by the exposure of older strata along the vertical channel sides. View from the submarine F.A. Forel. Vertical sequence is about 5 m high. (courtesy of J. Piccard)

Five facies are found:

1. River mouth deposits (sand bars, foresets, and topsets)
2. Delta channels as erosional features
3. Channel levees as spillover deposits
4. End of the channel deposits
5. Channel fan and basin plain deposits

In the upper delta, the erosional character of the channels can be deduced from seismic cross sections (Figure 9.5). Several generations of channels have been cut in the delta and subsequently filled. Channel sides are steep, and old strata are visible in submarine dives (Figure 9.6). The age and duration of the channel formation are not clear. For instance, the Rhine flowing into Lake

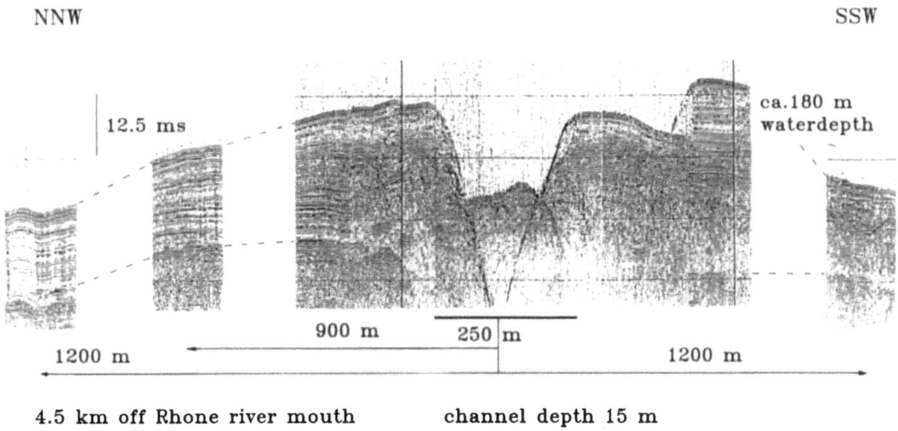


**Figure 9.7** The depositional origin of the channel levees in the middle part of the delta. The fact that the levees are formed by spillover is indicated by the formation of the morphological high and the thinning away of the sediment thickness with distance from the channel axis.

Constance formed a new river mouth in 1900 and has not yet developed a new channel. It is probable that the initial formation depends on some rare high-flow events, which are expected once or twice in a hundred years.

In the middle part of the delta, channels do not erode into older strata and levees are formed by accumulation. Sedimentation rates decrease with the distance from the channel levees (Figure 9.7). This indicates that here the spillover from turbidity currents is the source of sediments.

In the lower part of the delta, the coarse material settles in the channel, thus filling it slowly and forming the end of the channel deposits (Houbolt and Jonker, 1968). From this point on, the current can spread over the central (or basin) plain. The deposits fill all topographical irregularities.

With the exception of river mouth deposits, we note that from the river mouth to the basin plain, sedimentation rates reach a maximum near the channel end and decrease laterally with increasing distance from the axis.

The basin plain facies is characterized by alternating series of turbidities and of "normal" sediments (Figure 9.8). The latter have a very low sedimentation rate. This facies is therefore characterized by two very different types of sediments, which contrast strongly in sedimentation rate. All core analyses that assume a constant and small sedimentation rate—as it is frequently found in studies of pore water, pollution, or paleolimnology—should avoid this facies.

The origin of the turbidites is related either to suddenly increasing river discharges (high flow), or to slope failures of the river mouth deposits and possibly of the channel walls. This is documented by measurements of two different kinds of turbidity currents (Lambert and Giovanoli, 1988): one with

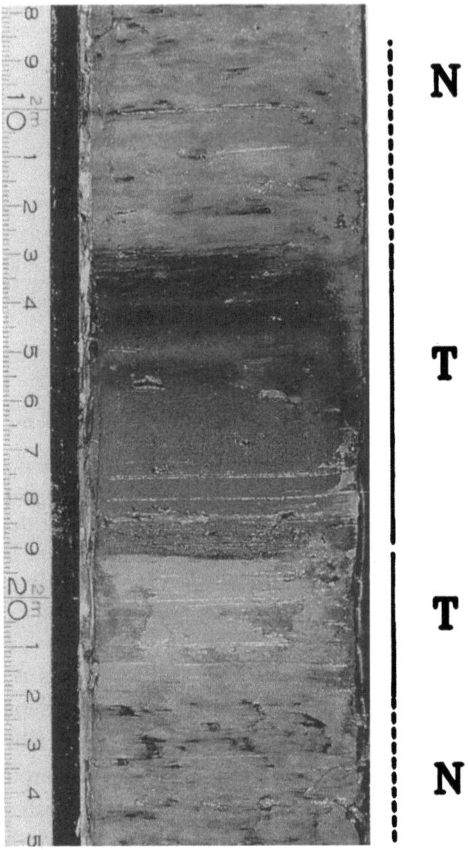


**Figure 9.8** The turbidites of the basin plain. Alternating turbidites (T) with "normal" (N) sediments. Sedimentation rate of 0.06 g $cm^{-2}$ $y^{-1}$, Vernet et al., 1984).

a positive temperature anomaly related to direct input of high flows, and the other one without a temperature anomaly. The latter is caused by slope failures in the hypolimnion. There is no correlation between magnitude of river discharge and maximum velocity in the turbidity current. Turbidity currents in a channel were measured about twice a week in summer; maximum velocities (15-min mean) were about 90 cm $s^{-1}$ (Lambert and Giovanoli, 1988). Slope failures can be triggered both by high flows and by the rapid accumulation of new material, but additional forces such as earthquakes, storm waves and seiches can also cause slope failures.

The magnitude and the frequency of turbidity currents are related in a way commonly found in nature; frequent small events and rare large ones. The frequency distribution of current-generated turbidites therefore should dis-

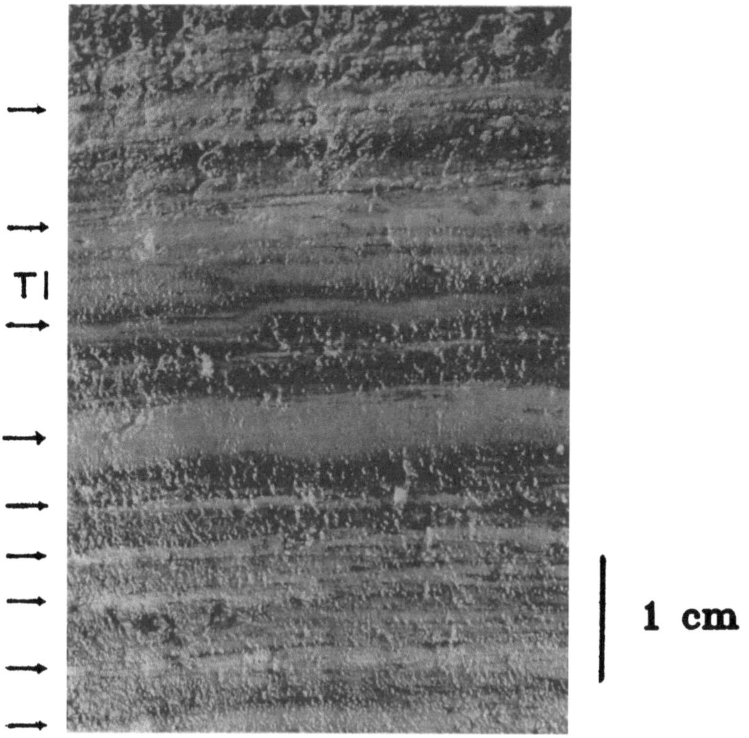


**Figure 9.9** The interflow sediments at station L14 (9.25 km from the Rhone River) are regularly laminated. One year includes several laminae; annual rhythm is indicated by the repetition of the same lamina as indicated by the arrows. The sedimentation rate of about 4 to 5 mm $y^{-1}$ of already compacted sediments is in good agreement with that of a $^{137}Cs$-dated core (Wang et al., 1986). T indicates a turbidite possibly originating from the basin slope. The thick, light-gray laminae are interpreted as deposits from years with exceptionally high flows.

play a Poisson distribution. The frequent small turbidity currents will produce very fine laminations, which are difficult to distinguish from interflow laminae. The horizontal evolution of the thickness of these laminations, rather than composition or even grain size makes it possible to separate them from interflow or other deposits.

It is important to note that, in addition to underflows from delta areas, slope failures are major sediment suppliers to basinal areas. In some lakes, they account for half of the sediment load on the basin plain (Ludlam, 1974). They are also visible in Lake Geneva interflow facies (Figure 9.9). The sediments of the central lake plain of Lake Geneva are clearly dominated by inflow-generated turbidites from the Rhone and the Dranse rivers.

## 9.3 Interflow Deposits: General Features

The interflow facies is characterized by an exponential decrease in sedimentation rate with increasing distance from the river mouth (Figure 9.10). The sediments settle out as a blanket over the old topography. A year may comprise several laminae (Figure 9.11; also see Figure 9.9) which document intermittent river input. River suspensions depend on rainfall and its pre-history in the catchment area or, in alpine areas, on glacial melt water during warm summer days.

Annual sets of laminae occur when summer and winter inputs vary greatly and when there is an overprint by autochthonous sedimentation. Sedimentation rate is high in comparison to the rest of the lake, and laminations are very regular and distinct. The most characteristic facies is expected to be found a few kilometers from the river mouth on topographic highs or on gentle slopes protected from the direct influence of the turbidites. Because an interflow facies has asymmetrical thickness relative to the axis of the lake, higher sedimentation rates along the north coast are expected. This is supported by dated sediment

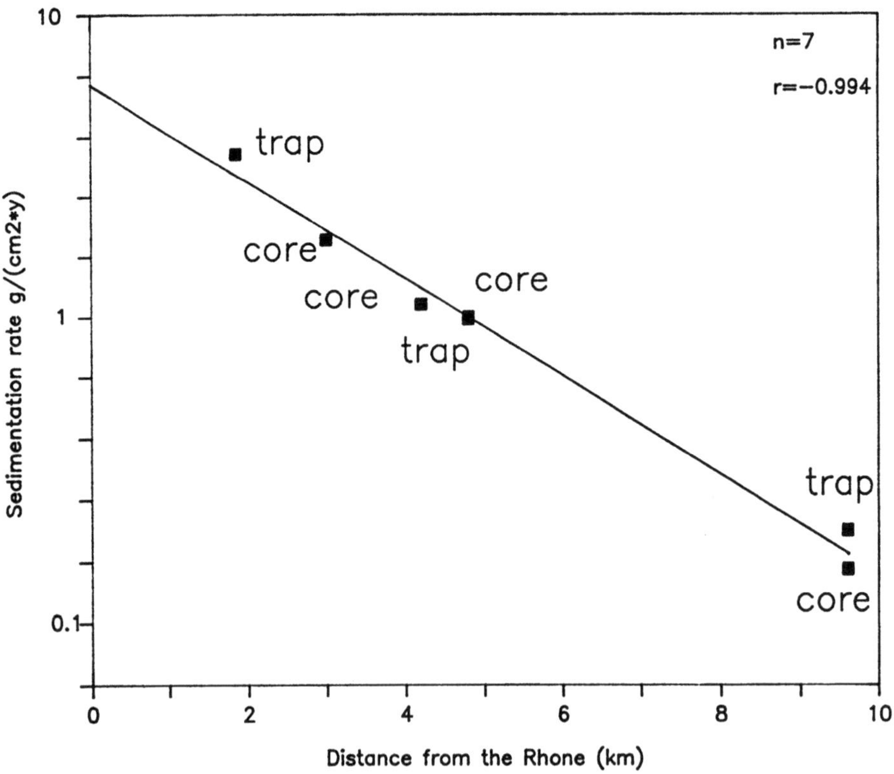


**Figure 9.10** Exponential decrease of rate of sedimentation with distance from the river mouth, in the interflow facies.

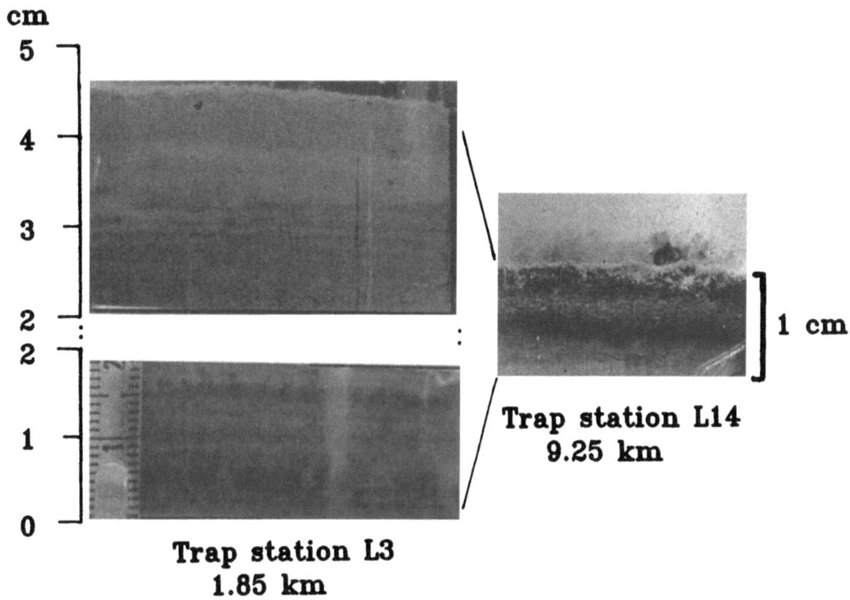


**Figure 9.11** The trap sediments during the summer high inflow of the Rhone. Left side: Station L3 (1.85 km from the Rhone), right side: station L14 (9.25 km). Total exposition at left is 62 days (31.5–5.7 and 5.7–3.8.1983); at right 62 days (3.6.–4.8.1983). Note several laminae; the top layers are aggregates of zooplankton, and a fecal pellet is visible at right.

cores from Lake Constance (Dominik et al., 1981a), which document a minimum sedimentation rate along the Swiss shoreline which is on the left side of the inflow.

## 9.4 Evolution of Particulate Matter during Interflows and Subsequent Sedimentation

**From the Rhone River to an Interflow** The *concentration* of particulate matter in the Rhone river varies between 10 and 5000 mg/l (*Hydrologisches Jahrbuch der Schweiz*). It is controlled mainly by high melt-water flow rates during the summer (from snow and glaciers). During the three summer months, about 90% of the solids are transported; roughly 40% are sandy (>63$\mu$m). Storm events are most common from spring to late autumn; they are superimposed on the long-term melt-water maximum. The natural cycle is modified by regular discharges for electrical power production, which add a component of diurnal, weekly, and seasonal cycles. The total storage capacity of the reservoirs is 1150 $\times$ $10^6$ $m^3$, or 1/5 of the mean annual discharge. Most power production occurs in winter; this doubles the natural mean winter discharge and reduces the maxima during summers (Loizeau, 1987). The latter are sig-

**Table 9.1*a*** Major elements as oxides, carbonates, and organic matter in the Rhone River: purely allochthonous components, Allo = $Al_2O_3 + Fe_2O_3 + K_2O + MgO + TiO_2$

| | Water discharge | Susp. fine | Susp. total | Concentration* (weight percent) | | | | | | | | Ratio (%/%) | |
|---|---|---|---|---|---|---|---|---|---|---|---|---|---|
| | $m^3/s$ | mg/l | mg/l | $SiO_2$ | $Al_2O_3$ | $CaCO_3$ | $Fe_2O_3$ | $K_2O$ | MgO | $TiO_2$ | POM | $SiO_2$/Allo | $CaCO_3$/Allo |
| **Spring (n=17)** | | | | | | | | | | | | | |
| Average | 178 | 58 | 77 | 48.0 | 18.4 | 14.1 | 6.0 | 4.2 | 3.0 | 0.5 | 4.19 | 1.50 | 0.44 |
| Standard deviation | 44 | 38 | 62 | 1.4 | 1.0 | 2.4 | 0.6 | 0.3 | 0.5 | 0.2 | 1.67 | 0.09 | 0.09 |
| In % of the average | 25 | 66 | 80 | 2.9 | 5.3 | 16.9 | 10.4 | 8.1 | 15.2 | 47.4 | 39.8 | 6.1 | 20.3 |
| **Summer (n=12)** | | | | | | | | | | | | | |
| Average | 407 | 278 | 458 | 54.4 | 18.5 | 9.0 | 6.4 | 4.3 | 3.5 | 0.7 | 1.32 | 1.64 | 0.28 |
| Standard deviation | 87 | 198 | 311 | 2.1 | 0.8 | 3.3 | 0.8 | 0.2 | 1.1 | 0.2 | 0.90 | 0.11 | 0.11 |
| In % of the average | 21 | 71 | 68 | 3.8 | 4.2 | 36.4 | 12.2 | 3.8 | 32.1 | 32.1 | 68.1 | 6.6 | 40.6 |
| **Autumn (n=5)** | | | | | | | | | | | | | |
| Average | 216 | 99 | 123 | 52.2 | 20.0 | 8.1 | 7.3 | 4.4 | 3.4 | 0.6 | 2.04 | 1.47 | 0.23 |
| Standard deviation | 42 | 94 | 92 | 3.0 | 1.3 | 2.1 | 0.5 | 0.4 | 0.4 | 0.1 | 1.09 | 0.13 | 0.06 |
| In % of the average | 19 | 95 | 74 | 5.8 | 6.3 | 25.6 | 6.2 | 7.9 | 11.9 | 17.5 | 53.4 | 8.5 | 24.6 |
| **Winter (n=8)** | | | | | | | | | | | | | |
| Average | 133 | 17 | 22 | 48.4 | 18.6 | 11.4 | 6.5 | 4.2 | 3.6 | 0.4 | 5.09 | 1.46 | 0.35 |
| Standard deviation | 38 | 6 | 9 | 1.5 | 0.6 | 2.9 | 0.7 | 0.2 | 0.8 | 0.2 | 1.17 | 0.08 | 0.10 |
| In % of the average | 29 | 36 | 41 | 3.0 | 3.5 | 25.2 | 11.2 | 4.5 | 21.1 | 35.1 | 22.9 | 5.6 | 29.5 |

* Other elements measured do not show seasonal variations: NaO mean 1.4%, with a coefficient of variation of 20% due to analytical limits of the EDAX; S as element, approx. 0.3%; and Mn as oxide, approx. 0.09%.

**Table 9.1*b*** Correlation matrix for the major components in the Rhone River*

| | | | | | |
|---|---|---|---|---|---|
| **$SiO_2$** | | | | | |
| 0.87 | **$Al_2O_3$** | | | | |
| 0.71 | 0.80 | **$Fe_2O_3$** | | | |
| 0.86 | 0.93 | 0.70 | **$K_2O$** | | |
| 0.56 | 0.55 | 0.68 | 0.50 | **MgO** | |
| 0.41 | 0.36 | 0.45 | 0.30 | 0.34 | **$TiO_2$** |

* n=46; a value of r = 0.28 is significant on the 95% level.

nificant for the transport of the coarse solids, as seen in a comparison of transport estimates for the 1960's, 3 to 7 million T $y^{-1}$ (Vernet et al., 1971) versus the present value of 2 to 3 million T $y^{-1}$ (Burrus, 1984).

The *chemical composition* of the suspended matter of the Rhone shows little variation (Table 9.1*a*) with the exception of irregular fluctuations in calcium carbonate ($CaCO_3$) and seasonal variations in organic matter content (POM). The fluctuations in carbonate content are probably due to local erosion processes in the catchment area, which are most differentiated in carbonate terrains. Variations in organic matter content relate to the dilution of lowland erosion material by sediments derived from glacial erosion. Factor analyses for discharge and for concentration of suspended solids and major elements (Mg, Al, Si, P, K, Ca, Ti, and Fe) and OM—in total, 12 variables and 46 measurements—singles out two groups of variables, which together account for 76% of the variance. These two groups are discharge and concentration of suspended solids (39%) and the elements Si, Al, Fe, K, and Mg (37%). Calcium and organic matter define two other factors accounting for 11% and 7% of the variability, respectively. It is most remarkable that elemental composition is not influenced by concentration changes, despite the geological heterogeneity of the watershed (Serruya, 1969). This holds true even considering that analyses were performed only on the fraction smaller than 63 $\mu$m. This can be explained by the homogenizing effects of river transport.

The highest correlation is found between Al and K, followed by Al and Si (Table 9.1*b*), suggesting that they are bound by silicate minerals such as micas, chlorites, and potassium feldspars. The lack of correlation between Ca and Mg is also significant. Wang et al. (1986) estimated that the dolomite content did not exceed 3% dry weight in a core taken in an area affected by the interflow. In Table 9.1 all elements are expressed as oxides except Ca, which is calculated as carbonate, and organic matter (OM). The latter is related to organic C by a constant factor of 0.5 (Bloesch, 1974).

Among the elements of the second group (Si, Al, Fe, K, and Mg) only Si derives from both allochthonous and autochthonous sources. As a consequence, the elements other than Si, are grouped together, including Ti as a minor constituent. These are used as a normalizing factor (Allo) for the allochthonous fraction in the lake (see Tables 9.1 and 9.2). Taking the sum, rather than one of the elements, reduces the effect of random fluctuations in any one of the components.

Silica and organic matter show slight but significant differences between winter and summer. Higher summer values of Si are explained by an increase of quartz in glacial erosion products. Organic matter (OM) input to Lake Geneva displays a distinct annual cycle with a maximum in winter and a minimum in summer (Figure 9.12*a*), in complete contrast with the autochthonous OM cycle in the lake. This may be interpreted as the "dilution" of the soil erosion and the anthropogenic products of the lowlands by highland glacial erosion products.

**Suspended Matter in the Interflow** The *concentrations* of suspended matter in an interflow correlate well with that in the Rhone (Giovanoli and Dominik, 1984). In the part of the lake affected by an interflow, and over a reference area of about 80 $km^2$, the "standing crop" of suspended matter changes from 6,000 T in winter to 27,000 T in summer (Giovanoli and Dominik, in prep.). In an interflow, mean grain size is around 7 $\mu m$ (standard deviation $+13$ and $-4$ $\mu m$), with little variation in time and no significant size decrease with distance from the river mouth (annual mean at 2 km is $6.9 \pm 1.6$ $\mu m$, and at 4.75 km, $6.7 \pm 2.1$ $\mu m$).

The *composition* of the suspended matter in an interflow is very similar to that of the Rhone River suspension. This is demonstrated in the ratios $SiO_2$/Allo and $CaCO_3$/Allo (Table 9.2). However, organic matter concentration in the interflow and in the Rhone are significantly different (Mann-Whitney U test, $p < 0.0001$). The mean value in the interflow is 4.0 times higher than in the Rhone. The difference persists for each season but the enrichment factor varies from 2 to 7. The deviation from the mean enrichment is significantly lower in winter ($p < 0.025$) and significantly higher in autumn ($p < 0.037$). The data show no changes with distance from the Rhone inlet. The interflow has higher amounts of suspended organic matter than the Rhone inflow; a condition discussed together with sediment trap results in section 9.5.

Above and below the interflow, the two ratios $Sio_2$/Allo and $CaCO_3$/Allo, together with organic matter content, fall near initial values of the Rhone, but during summer only (see Table 9.2).

The effect of autochthonous particles is visible in some samples: $SiO_2$/Allo ratio is high above and below the interflow in spring, and also in November, although to a lesser extent. $CaCo_3$/Allo exhibits one unique peak in September and is enriched throughout the summer but only above the interflow. These peaks are caused by diatom blooms in spring and lacustrine chalk precipitation in autumn. The observations were verified by SEM (Dominik et al., 1984).

Above the interflow, suspended matter contains most organic particles during autumn and fewer during the summer. This implies mixing of the interflow with overlying water. Below the interflow, organic matter is also always

**Table 9.2** Seasonal evolution of major elements derived from mixed origin (allochthonous and autochthonous) in the Rhone River; above, within, and below the interflow, and in the trap[a]

| | Spring | Summer | Autumn | Winter |
|---|---|---|---|---|
| **$SiO_2$/Allo ratio of dry weight** | | | | |
| Rhone River | 1.50 | 1.64 | 1.47 | 1.46 |
| Lake: | | | | |
| Above | 4.20 | 1.90 | 1.7, 2.6[b] | 1.65 |
| Interflow | 1.65 | 1.54 | 1.50 | 1.44 |
| Below | 3.92 | 1.82 | 1.57 | 1.79 |
| Trap L3 upper | 1.80 | 1.75 | 1.58 | 1.79 |
| Trap L3 lower | 1.78 | 1.77 | 1.63 | 1.68 |
| Trap L10 upper | 2.26 | 1.67 | 1.58 | 1.89 |
| Trap L10 lower | 2.04 | 1.76 | 1.60 | 1.80 |
| **$CaCO_3$/Allo ratio of dry weight** | | | | |
| Rhone River | 0.44 | 0.28 | 0.23 | 0.35 |
| Lake: | | | | |
| Above | 0.49 | 0.72 | 4.4, 0.5[b] | 0.29 |
| Interflow | 0.38 | 0.38 | 0.18 | 0.30 |
| Below | 0.43 | 0.45 | 0.30 | 0.32 |
| Trap L3 upper | 0.48 | 0.33 | 0.33 | 0.47 |
| Trap L3 lower | 0.46 | 0.35 | 0.32 | 0.51 |
| Trap L10 upper | 0.43 | 0.38 | 0.36 | 0.47 |
| Trap L10 lower | 0.46 | 0.38 | 0.38 | 0.49 |
| **Particulate organic matter (POM) in weight percent** | | | | |
| Rhone River | 4.19 | 1.32 | 2.04 | 5.09 |
| Lake: | | | | |
| Above | 40.20 | 25.50 | 72, 90[b] | 28.50 |
| Interflow | 11.10 | 5.80 | 14.20 | 14.90 |
| Below | 19.20 | 9.80 | 19.90 | 21.80 |
| Enrichment factor[c] | 2.7 | 4.4 | 7.0 | 2.9 |
| Trap L3 upper | 3.04 | 1.19 | 3.48 | 4.43 |
| Trap L3 lower | 3.38 | 1.24 | 3.41 | 3.60 |
| Trap L10 upper | 6.85 | 1.75 | 3.41 | 5.74 |
| Trap L10 middle | 5.54 | 1.73 | 3.32 | 4.25 |
| Trap L10 lower | 5.17 | 1.82 | 3.31 | 4.25 |
| Trap L14 upper | 11.84 | 4.31 | 8.94 | 6.26 |
| Trap L14 middle | 10.11 | 4.36 | 6.11 | 5.94 |
| Trap L14 lower | 9.21 | 4.35 | 6.94 | 5.73 |

[a] $SiO_2$ and $CaCO_3$ are normalized with Allo (see Table 9.1a).
[b] September, November
[c] Interflow/Rhone (yearly mean 4.0)

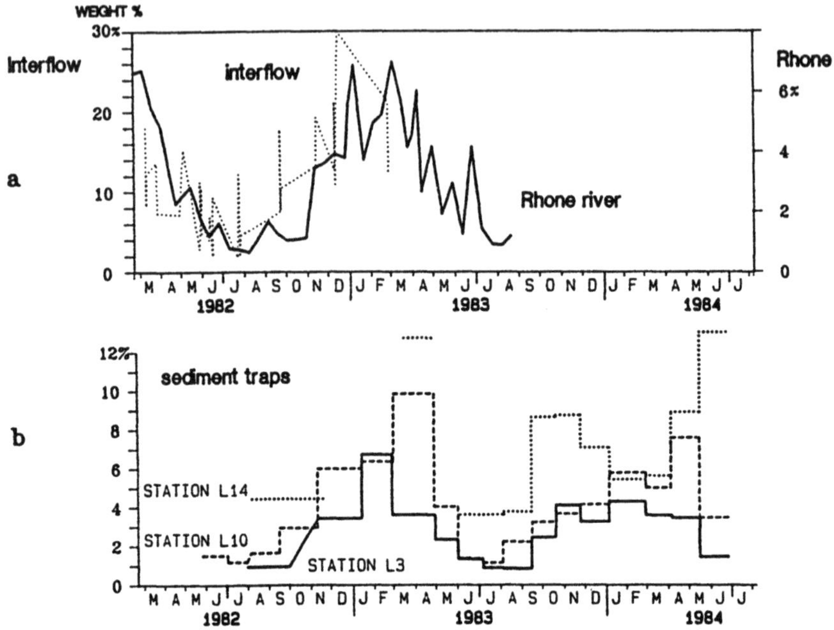


**Figure 9.12** Organic matter content of suspended matter in Lake Geneva. *a*. Seasonal variation in the content of organic matter suspended in the River Rhone–water entering Lake Geneva (solid line), and in the interflow water within the lake (dotted line). *b*. Seasonal variation in the content of organic matter in the settling particles trapped at station L 3; 1.85 km, station L 10; 4.75 km, and station L 14; 9.25 km off the inlet of the River Rhone.

enhanced relative to the interflow suspensions. This is interpreted as the result of the predominance of allochthonous components in the interflow.

**Sedimentation under the Influence of the Interflow** The *composition* of material collected below the interflow at the station closest to the river mouth (L3, 1.85 km) is not significantly different from that of the Rhone (see Table 9.2). In contrast to the interflow measurements, not only the fluctuations in organic matter concentration match those of the Rhone—higher concentrations in winter, lower concentrations in summer (see Figure 9.12*b*), but also the absolute values. At station L10 (4.8 km), the influence of the Rhone inflow remains dominant, especially in summer. Autochthonous material—$SiO_2$ from diatoms, and organic matter—is added in spring. The amount of calcite precipitation is more difficult to estimate from the trap samples because $CaCO_3$/Allo ratios change little (see Table 9.2).

Trap site L14 (9.25 km) is not along the axis of the interflow. It represents a site affected but not dominated by the Rhone inflow. The organic matter content is significantly higher than at station L3 but still follows the Rhone pattern, with low values in summer and high values in winter. The addition of further peaks in spring and autumn from autochthonous particles results in a bimodal distribution. This is important for the identification of laminated sediments below this trap site as annual varves (see Figures 9.9 and 9.11). The result of a one-year deposition consists of several dark and light laminae. The winter and spring laminae are generally darker, while the summer laminae are lighter with less organic matter content. Very fine white laminae probably represent autochthonous calcite precipitation, but could only be analyzed in thin sections. The exceptionally well-preserved structures are explained by the high sedimentation rate, which excludes significant bioturbation. This is a general feature of interflow deposits and occurs in Lake Constance, as well as other lakes.

## 9.5 Comparison of the Composition of Suspended Matter in the Rhone River and the Sediment

The Rhone, the interflow, and the sedimented matter all show the same $SiO_2$/Allo and $CaCO_3$/Allo ratios. Organic matter content (OM) in the suspended particles also show a maximum concentration in winter and a minimum concentration in summer. This covariant pattern demonstrates the dominance of allochthonous over autochthonous contributions of the OM (see Figure 9.12). The absolute concentrations of the OM differ between the Rhone and the interflow and also between the interflow and the sediment trap material. In contrast, the Rhone and the trap samples are similar. This observation could be explained by three different mechanisms that are *not* mutually exclusive:

1. Fractionation of the suspended matter at the river mouth due to different densities of the components (enrichment of the lighter organic matter).
2. Differential settling velocities for allochthonous organic particles as compared with inorganic particles (extension of the fractionation to a longer transport path).
3. Contribution of the autochthonous organic rich particles.

The fractionation of the suspended matter at the river mouth seems to be a less dominant process because sediment traps directly below the interflow and near the river mouth have only about 25% less organic matter content in spring and 10% less in summer than the Rhone suspension (see Table 9.2). In the sediment traps, OM increases with distance from the Rhone. At 9.25 km, the enrichment in OM approximates that between Rhone and interflow suspension. This evolution with distance also occurs in surface sediments (Figure 9.13, Jaquet et al. 1983), shown by a zonal pattern of increasing organic matter concentration outward from the Rhone toward NW.

Differences in settling velocities between organic and inorganic matter (different vertical residence times) are estimated by flux-budget calculations.

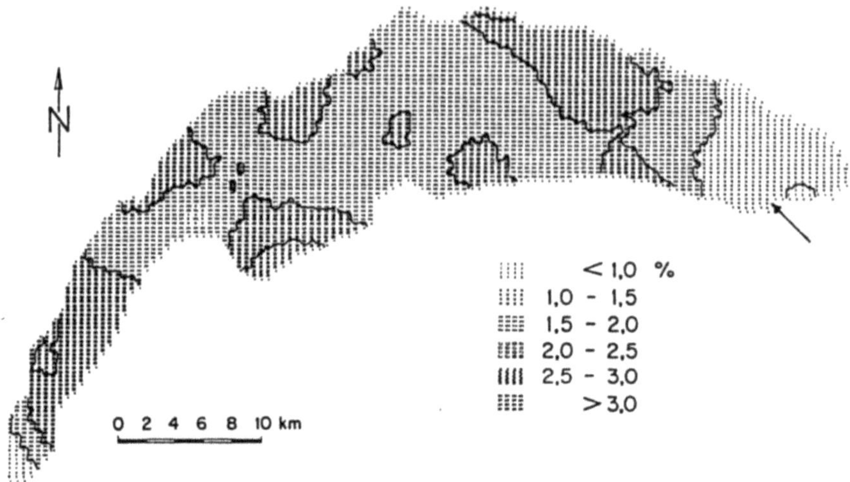


**Figure 9.13** Organic carbon (in weight % of dry sediment) in the superficial sediments of Lake Geneva. Note the evolution from the Rhone inflow (arrow) to WNW, following the interflow (Jaquet et al., 1983).

**Table 9.3** Sedimentation rates of total solids and of particulate organic matter (POM) in sediment traps at different water depths and at increasing distances from the Rhone River[a] and at the lake mid-point

| Station | Water depth (m) | Distance (km) | Sedimentation rate ($g\ m^{-2}\ d^{-1}$) Total solids | POM |
|---|---|---|---|---|
| L3 | 50 | 1.85 | 98.3 | 1.36 |
| | 130[b] | | 129.4 | 2.19 |
| L10 | 50 | 4.75 | 26.6 | 0.56 |
| | 115 | | 26.8 | 0.58 |
| | 180[b] | | 26.0 | 0.60 |
| L14 | 50 | 9.3 | 5.3 | 0.28 |
| | 150 | | 5.4 | 0.28 |
| | 250[b] | | 5.8 | 0.30 |
| SHL 2[c] | 60 | Lake mid-point | 2.4 | 0.13 |
| | 200 | (1986) | 2.7 | 0.13 |
| | 300[b] | | 3.4 | 0.17 |
| SHL 2[c] | 60 | Lake mid-point | 2.2 | 0.11 |
| | 200 | (1987) | 2.9 | 0.15 |
| | 300[b] | | 3.4 | 0.16 |

[a] Data for the traps were taken from 3.6.1983 to 3.6.1984.
[b] This is 3 to 10 m above the lake bottom.
[c] Data from Gandais, 1988.

Results from a budget model for the total suspended matter (Giovanoli and Dominik, in prep.) recalculated for organic matter, indicate that the vertical residence time of the organic matter is 2 to 3 times longer than that of total suspended matter. This agrees with the evolution pattern for the concentration of organic matter with distance to the Rhone. The third, above-mentioned mechanism is independent of inflow. However, comparing fluxes, it is evident that the autochthonous organic matter fraction is almost negligible near the Rhone inlet (Table 9.3). This flux thus only affects the concentration in the interflow if its residence time is extremely long as compared to the Rhone organic matter.

Furthermore, in regions where the influence of the Rhone interflow is small, such as in the central plain of the lake, organic matter content is low (see Figure 9.13).

## 9.6 Conclusions

In large lakes with important tributaries, two sedimentary facies occur which are controlled by the transport pathways of suspended matter:

1. A delta-complex facies with underwater channels, levee deposits, end-of-channel deposits, and turbidites in the central part of the lake (basin plain).
2. An interflow facies, often separate from the delta complex, deposits mainly to the right of the inflow in the northern Hemisphere.

In the distal delta-complex facies, sedimentation rate reaches a maximum at the channel end, whereas, in the interflow facies, sedimentation rate decreases exponentially with distance from the mouth of the river. In the case of Lake Geneva, this decrease amounts to two orders of magnitude within about 10 km.

Inflow composition controls the composition of the sediment within these two facies. In the interflow facies, an evolution occurs with distance due to different settling velocities of organic and inorganic material. In Lake Geneva, this lateral evolution explains an increase of organic matter content for the bottom sediments with distance from the Rhone River.

Seasonal variations in the inflow are reflected in the rate of evolution of $SiO_2$ and of organic matter contents. The latter shows a maximum in winter which matches the interflow pattern and the fluctuations in traps up to about 4.5 km from the river inlet. This explains the nature of the laminations of the sediments—with high OM concentration in the thinner (dark) winter and spring laminae and the thicker summer laminae with lower OM concentration.

In the trap furthest from the inlet, autochthonous particles become more important, as shown by the normalized element concentrations. The sum of the purely allochthonous elements—$Al_2O_3$, $Fe_2O_3$, $K_2O$, MgO, and $TiO_2$—are used as a normalizing factor (Allo) for the components with mixed origin.

---

*Acknowledgements* I am very grateful to Kerry Kelts and to the editors for polishing my English style. This work was supported by the Swiss National Science Foundation, Project "Apports allochtones dans le delta du Rhône." The planning and realization of the field work was done together with Dr. James Z. Dominik at the Institute F.A. Forel, University of Geneva, under director Prof. J.-P. Vernet. I thank Dr. D. Burrus and the technicians of the institute for their considerable assistance.

## References

Bates, C.C. 1953. Rational theory of delta formation. *Bull. Americ. Assoc. Petrol. Geol.* 37/9:2219–2162.

Bloesch, J. 1974. Sedimentation und Phosphorhaushalt im Vierwaldstättersee (Horwer Bucht) und im Rotsee. *Schw. Z. Hydrol.* 36:71–186.

Brunsden, D. and Prior, D.B. (eds.) 1984. *Slope Instability*. John Wiley and Sons Ltd, New York.

Burrus, D. 1984. Contribution à l'étude du transport du phosphore dans le Rhône alpin. Dissertation No. 2135, Univ. Genève, 100 pp.

Coleman, J.M. 1976. Deltas-*Processes of Deposition and Models for Exploration*. Burgess Publishing Co., Minneapolis, 124 pp.

Dominik, B., Dominik, J., Giovanoli, F., and Vernet, J.-P. 1984. Qualitative seasonal variations of the suspended solids in eastern Lake Geneva: allochthonous versus autochthonous components. *3rd. Intern. Sympos. on Interactions between Sediment and Water.* Geneva, p. 310–313.

Dominik, J., Mangini, A., and Müller, G. 1981a. Determination of recent deposition rates in Lake Constance with radioisotope methods. *Sedimentology* 28:653–677.

Dominik, J., Viel, M., and Vernet, J.-P. 1981b. Evolution des Formes du Phosphore dans les Sediments. In: *Rapports sur les Etudes et Recherches Entreprises dans le Basin L'aaemanique*. Commission Internationale pour la Protection des Eaux du Léman—CIPEL. Lausanne, Switzerland, p. 187–202.

Fischer, H.B., List, E.J., Koh, R.C.Y., Imberger, J., and Brooks, N.H. 1979. *Mixing in Inland and Coastal Waters*. Academic Press, New York and London, 485 pp.

Gandais, V., and Vernet, J.-P. 1988. Déterminations des Flux de Nutriments sous Forme Particulaire dans le Léman à l'Aide de Trappes à Sediments. In: *Rapports sur les Etudes et Recherches Entreprises dans le Basin Lémanique*. Commission Internationale pour la Protection des Eaux du Leman—CIPEL. Lausanne, Switzerland, p. 97–118.

Giovanoli, F. and Lambert, A. 1985. Die Einschichtung der Rhone im Genfersee: Ergebnisse von Strömungsmessungen im August 1983. *Schw. Z. Hydrol.* 47/2:159–178.

Hamblin, P.F. and Carmack, E.C. 1978. River induced currents in a fjord lake. *J. Geophys. Res.* 82/C2:885–899.

Houbolt, J.J.H.C. and Jonker, J.B.M. 1968. Recent sediments in the eastern part of the lake of Geneva (Lac Léman). *Geologie en Mijnbouw* 47/2:131–148.

*Hydrologisches Jahrbuch der Schweiz*. Eidgenössische Drucksachen und Materialzentrale. Bern, Switzerland.

Jacquet, J.-M. 1978. Formes du Phosphore dans les Sediments Lémanique en 1978. In: *Rapports sur les Etudes et Recherches Entreprises dans le Basin Lémanique*. Commission

Internationale pour la Protection des Eaux du Léman—CIPEL, Lausanne, Switzerland, p. 310–326.

Jaquet, J.-M., Rapin, F., Davaud, E., and Vernet, J.-P. 1983. Géochimie des sédiments du Léman. *Materiaux pour la géologie de la Suisse—Hydrologie*. Kümmerly and Frei, Geographischer Verlag, Bern, Switzerland, No. 30, 70 pp.

Kelts, K. and Arthur, M.A. 1981. Turbidites after ten years of deep-sea drilling—wringing out the mop? *SEPM Spec. Publ.* No. 32, p. 91–127.

Kelts, K. and Hsü, K.J. 1980. Resedimented facies of 1875 Horgen slumps in Lake Zürich and a process model of longitudinal transport of turbidity currents. *Eclog. Geol. Helv.* 73/1:271–281.

Kuenen, Ph.H. and Migliorini, C.I. 1950. Turbidity currents as a cause of graded bedding. *J. Geol.* 58:91–127.

Lambert, A. and Hsü, K.J. 1979. Non-annual cycles of varve-like sedimentation in Walensee, Switzerland. *Sedimentology* 26:453–461.

Lambert, A. and Giovanoli, F. 1988. Records of riverborne turbidity currents and indications of slope failures in the Rhone delta of Lake Geneva. *Limnol. Oceanogr.* 33/3:458–468.

Loizeau, J.-L. 1987. Analyses et correlations des sediments fins du delta du Rhone dans le lac Léman. Thesis, Univ. Genève, 47 pp.

Ludlam, St. D. 1967. Sedimentation in Cayuga Lake, New York. *Limnol. Oceanogr.* 12:618–632.

Ludlam, St. D. 1974. Fayettville Green Lake, New York 6. The role of turbidity currents in lake sedimentation. *Limnol. Oceanogr.* 19/4:656–664.

Mutti, E. and Ricci Lucchi, R. 1972. Le torbiditi del'Appenino settentrionale: introduzione all' analisi de facies. *Soc. Geol. Ital. Mem.* 11:161–199.

Saylor, J.H., Benett, J.R., Boyce, F.M., Liu, P.C., Murthy, C.R., Picket, R.L., and Simons, T.J. 1981. Water movements. In: Aubert, E.J. and Richards, T.L. (eds.), IFYGL—The International Field Year for the Great Lakes. NOAA, Ann Arbor, Michigan, p. 247–319.

Serruya, C. 1969. Les dépots du lac Léman en relation avec l'évolution du bassin sédimentaire et les caractères du mileu lacustre. Dissertation No. 1439, Université de Genève, 254 pp.

Sturm, M. and Matter, A. 1978. Turbidites and varves in Lake Brienz (Switzerland): deposition of clastic detritus by density currents. *Spec. Publ. Int. Assoc. Sediment.* 2:147–168.

Vernet, J.-P., Meybeck, M., Pachoud, A., and Scolari, G. 1971. Le Léman: Une synthèse bibliographique. *Bull. B.R.G.M.*, Ser. 2, Sect. 4/2:47–84.

Vernet, J.-P., Dominik, J., and Favarger, P.-Y. 1984. Texture and sedimentation rates in Lake Geneva. Environm. Geology 5/3:143–149.

Walker, R.G. and Mutti, E. 1973. Turbidite facies and facies associations. In: Middleton, G.V. and Bouma, A.H. (eds.), *Turbidites and Deep water Sedimentation*. Soc. Econ. Paleon. Mineral. Pacific Sect., p. 119–157.

Wang, Z.J., El Ghobary, H., Giovanoli, F., and Favarger, P.-Y. 1986. Interpretation of metal profiles in a sediment core from Lake Geneva: metal mobility or pollution. *Schw. Z. Hydrol.* 48/1:1–17.

Wetzel, R.G. 1975. *Limnology*. W.B. Saunders Company, Philadelphia, 743 pp.

Wright, L.D. (1977). Sediment transport and deposition at river mouths: A synthesis. *Geolog. Soc. America Bulletin* 88:857–868.

Yuretich, R.F. 1969. Modern sediments in Lake Rudolf, Kenya. *Sedimentology* 26:313–331.

# 10

# Significance of Sediment Resuspension and Particle Settling

Brian J. Eadie, Henry A. Vanderploeg, John A. Robbins, and Gerald L. Bell

ABSTRACT Seasonal particle characteristics and sediment-trap-measured resuspension rates are examined for the Laurentian Great Lakes and compared with other large, deep lakes. Results are used to estimate the influence of particle-related processes on the current chemical composition of these lakes. Even in deep systems, such as the Laurentian Great Lakes, particle settling times are relatively short and compounds with a high affinity for particulate matter are efficiently scavenged and removed to the sediments. After reaching the bottom, the settled materials are mixed by the feeding activities of bottom-dwelling organisms into an homogenized pool representing years-to-decades of recent sedimentation. It is apparent from the relatively slow decline of the concentrations of these particle-associated constituents in water and biota that sediments are a leaky sink; small concentrations persist for decades because of processes that can remobilize materials from the bottom.

## 10.1 Introduction

The annual cycle of particle production and transport plays a major role in the seasonal and long-term behavior of nutrients and contaminants in lakes. Compounds entering the lakes are removed to the sediments at a rate proportional to their affinity for settling particles (Eadie and Robbins, 1987). Particle residence times in the water column are relatively short. Even in deep systems, such as the Laurentian Great Lakes, particle settling times are less than one year (Wahlgren et al., 1980). Constituents such as fallout Sr-90 and chloride have built up over time in the water because they have a low affinity for particulate matter and thus are very inefficiently transferred to sediments. In contrast, particle-associated contaminants—such as plutonium isotopes, Cs-137, and hydrophobic organic compounds—have been efficiently scavenged and removed to the sediments. After reaching the bottom, the settled materials are mixed by the feeding activities of bottom-dwelling organisms into an homogenized pool representing years to decades of recent sedimentation (Robbins, 1982).

Concern with the long removal times of recently controlled trace contaminants and nutrients (e.g., PCB, DDT, and phosphorus) in lakes and an increasing interest in restoration have led to a closer examination of the processes involved in the exchange of these and other materials between the water and the large inventory stored in the lakes' sediments. It is apparent from the relatively slow decline in the concentrations of particle-associated constituents in water and biota that sediments are a leaky sink; small concentrations persist in the water for decades because of processes that can remobilize materials from the bottom.

Four processes responsible for redistributing sediments in small lakes have been identified (Hilton, 1985): slumping, intermittent complete (lake) mixing, peripheral wave action (in the littoral zone), and random (shallow water wave) redistribution. In large lakes, surface wave and littoral zone processes are less important, while intermittent mixing (especially seasonal overturns) and internal waves (Mortimer, 1971; Chambers and Eadie, 1981) are the dominant processes. Current speeds of up to 13 $cm \cdot s^{-1}$ have been reported at 1 m above the bottom for the 100-m-deep Lake Michigan station (Figure 10.1) during the summer (Saylor and Miller, 1988) and winter currents of over 25 $cm \cdot s^{-1}$ were observed during the unstratified period (Saylor, personal communication). These currents have sufficient energy to remobilize the unconsolidated floc from the sediment-water interface.

In this chapter, we will examine the seasonal particle characteristics and resuspension rates (as measured by sediment traps) at a 100-m-deep station in Lake Michigan and compare them with a more limited data set recently

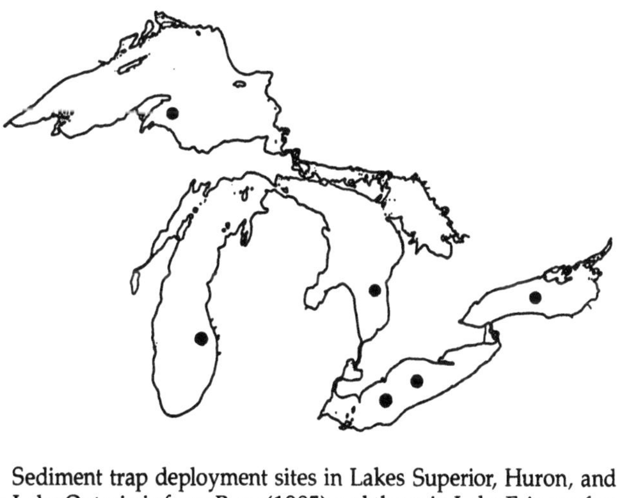

**Figure 10.1** Sediment trap deployment sites in Lakes Superior, Huron, and Michigan. The station in Lake Ontario is from Rosa (1985) and those in Lake Erie are from Charlton and Lean (1987).

collected for stations of similar depth in Lakes Huron and Superior and with other data reported for large, deep lakes. We will then use these results to estimate the influence of particle-related processes on the current chemical composition of these lakes.

Sediment trap and water samples were collected at the stations shown in Figure 10.1. The traps used were 10.2-cm-diameter cylinders with a 5:1 aspect ratio above the top of a funnel. The samples were collected in $CHCl_3$-poisoned 500-ml bottles located below the funnel (see Chambers and Eadie, 1981, and Eadie et al., 1984 for details of collection, preservation, and analysis). Water samples were collected in 5-L Niskin bottles. Total suspended matter (TSM) was measured by filtering 4 L through preweighed glass fiber filters.

The Laurentian Great Lakes are seasonally stratified, carbonate-buffered systems with mean depths ranging from 19 (Erie) to 149 m (Superior) and hydraulic residence times ranging from 3 to 180 years. Contaminants entering the lakes become intimately involved in the seasonal cycle of particle production and transport as illustrated in sediment-trap-measured mass flux isopleths for the 100-m-deep Lake Michigan station (Figure 10.2). During the winter, when the lakes are isothermal or weakly stratified, materials from underlying sediments are nearly uniformly distributed throughout the water column. Stratification begins in the late spring (May–June) and is accompanied by a succession of plankton blooms (from diatoms to greens) that appear to benefit from nutrients resupplied earlier from the bottom by vigorous vertical mixing. With the completion of stratification, the epilimnion is effectively isolated from un-

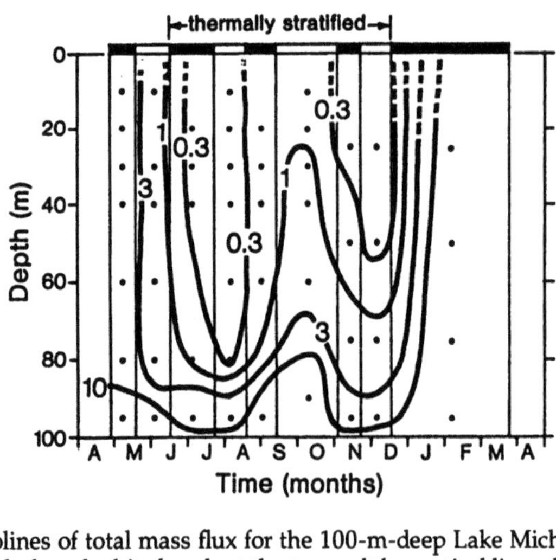


**Figure 10.2** Isolines of total mass flux for the 100-m-deep Lake Michigan trap station. The alternating black and white bands at the top and the vertical lines delimit the periods of collection. Solid black circles indicate locations of traps. The first and last deployment captured extensive resuspended sediment. Soon after stratification, the water column clears as particles rapidly settle out. $CaCO_3$ precipiation dominates the increased mass flux during August and September.

derlying waters. Key nutrients are depleted and the water column is cleared as detritus sinks through the epilimnion. Beginning in August, calcium carbonate begins to precipitate in the warm surface waters of three of these lakes (Ontario, Michigan, and Huron), producing a new population of rapidly settling particles (Strong and Eadie, 1978). This process is usually complete by the end of September, although the signal persists in the hypolimnion as these particles settle out. The aftermath of this annual "whiting" event is well-scavenged epilimnetic water possessing the minimum amount of suspended matter. The low concentration and flux of particulate matter in early fall is partially the consequence of the formation of (high-density) calcite in the size range efficiently grazed by zooplankton (Vanderploeg et al., 1987). Zooplankton biomass is high during late summer and these animals ingest a significant fraction ($\sim 0.1$ $d^{-1}$) of the TSM. The calcite is indiscriminately ingested along with very slowly settling detritus and repackaged into rapidly settling fecal pellets. In the late fall, when the lake overturns, resuspended sediment materials begin to reappear in near-surface waters.

## 10.2 Particle Characteristics and Size Distribution

The distribution of particle sizes and their chemical composition mediate processes such as the settling and transport of particle-associated constituents. Particles enter the lake via long-range atmospheric transport, terrestrial runoff from the basin watershed, and in-situ production processes, such as primary productivity and $CaCO_3$ precipitation. In an examination of 26 Canadian lakes of varying eutrophy, Sprules et al. (1983) reported a bimodal particle-size distribution—with peaks at 8–16 $\mu m$ and approximately 500 $\mu m$—which is similar to nearshore and to temperate-to-polar offshore marine environments.

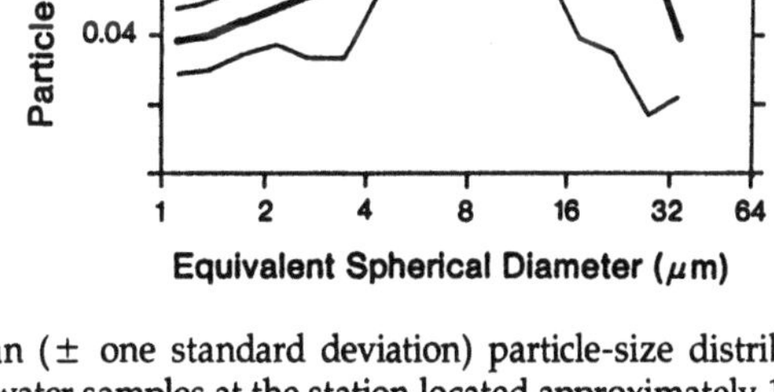


**Figure 10.3** Mean (± one standard deviation) particle-size distribution for 35 Lake Michigan surface water samples at the station located approximately 10 km inshore from the sediment trap station. Samples were analyzed by Coulter counter.

A similar size distribution (Figure 10.3) for particles less than 80 μm has been measured for 35 samples collected from a depth of 4 m near the Lake Michigan trap station over the period from 1977–1985 (see Vanderploeg, 1981a, and b for details of the Coulter counting procedure and individual sample spectra). Since seasonal data were very consistent from year to year, they were combined and 10-day blocks were averaged. Throughout the year, particles with a nominal diameter of less than 16 μm constituted the major fraction of the TSM volume. The peak at 8 μm is a result both of calcite abundance during late summer and of non-calcareous seston in this size category throughout the year (Vanderploeg et al., 1987).

The seasonal total of particle volumes (Figure 10.4*a*) has a large pulse in the 4- to 32-μm-size range in July, when the seston was dominated by small, low-specific-gravity (ca. 1.2 g·cm³) diatoms and flagellates (Vanderploeg, 1981a; Scavia and Fahnenstiel, 1987), resulting in the maximum particle volume for the year and relatively low mass. During late August to September, $CaCO_3$ particles (specific gravity = 2.7 g·cm³) increase in the 2- to 16-μm-size categories and constitute a significant fraction of the total TSM mass. In a series of experiments using a mild-acid-addition technique to distinguish between calcite and other seston, Vanderploeg et al. (1987) examined the size spectra of the $CaCO_3$ during this whiting period. The $CaCO_3$-size-distribution

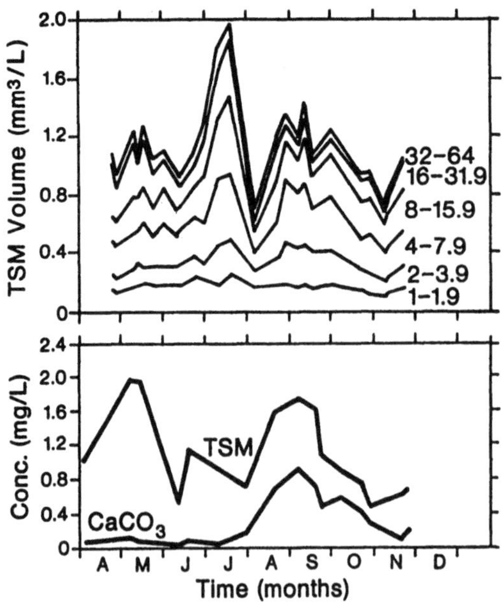


**Figure 10.4** (*a*) Seasonal measurements of particle volumes in the identified nominal size ranges at a station approximately 10 km inshore from the Lake Michigan trap station. The curves represent cumulative particle volumes. (*b*) Seasonal concentration of surface-water total suspended matter (TSM) at the same site. The lower curve represents the fraction of the TSM analyzed as $CaCO_3$.

peaked at 7 $\mu$m. The concentration of near-surface TSM was systematically measured at the Lake Michigan trap station. TSM mass peaked in the thermally unstratified early spring (see Figure 10.4*b*) during the period of intense resuspension and again during the late summer due to precipitation of $CaCO_3$ (Strong and Eadie, 1978).

## 10.3 Particle Setting Velocities

There is an important distinction between the particulate matter in the lake collected as the total suspended matter (TSM) associated with a water sample and that collected as settling matter in a sediment trap. The former represents an instantaneous estimate of the population of particles in the water (if the sample is large enough to capture the rare, large particles), while the latter is a time-integrated sample of settling particles that, because of their size and/or density, have sufficiently high settling velocities to be captured.

Although imperfect, because of the different sampling time scales, a mean settling velocity for the ensemble of particles can be estimated by dividing trap-measured mass flux by the average concentration of TSM collected at the trap location at the times of deployment and retrieval. Attempts have been made to improve this estimate by using a depth-weighted mean TSM for the interval above the trap (Bloesch and Sturm, 1986). In the Laurentian Great Lakes, there is little (less than two-fold) difference in TSM concentration throughout the vertical profile until the benthic nepheloid layer (BNL) is reached. In this near-bottom layer, the sources of TSM and trapped material are primarily from local resuspension and horizontal transport (Eadie et al., 1984); a depth-weighted TSM would therefore result in an erroneously high value for settling velocity.

Using TSM values from trap locations, settling velocities have been calculated for the three open-lake stations (Figure 10.5). During the stratifed period, the settling velocities in the upper 35 m of the water column are similar for the three lakes (mean = $0.52 \pm 0.16$ m·d$^{-1}$), not significantly different from the value of 0.76 m·d$^{-1}$ reported by Rosa (1985) for epliminetic settling of particulate matter in offshore Lake Ontario, nor from the 0.55 m·d$^{-1}$ calculated for two offshore stations (7 and 10) in Lake Erie (Charlton and Lean, 1987). The higher settling velocities (2 to 3 m·d$^{-1}$) in the BNL of our three lakes are again similar to that estimated for Lake Erie (1.6 m·d$^{-1}$). Using the same approach, similar values (epilimnion = 0.66 and BNL = 1.87 m·d$^{-1}$) were reported for the Swiss lake, Lake Zug (Bloesch and Sturm, 1986), and can be estimated (epilimnion = 0.75 and BNL = 4.0 m.d$^{-1}$) for the Greifensee (Lee et al., 1987). These calculated epilimnetic-ensemble particle-settling velocities fall within the range measured with in-situ settling chambers ($-0.32$ to 1.68 m.d$^{-1}$) during stratified conditions in Lake Erie (Burns and Pashley, 1974). This surprising uniformity may be due to the biological repackaging of the wide spectrum of particulate matter into a less-diverse array of fecal material.

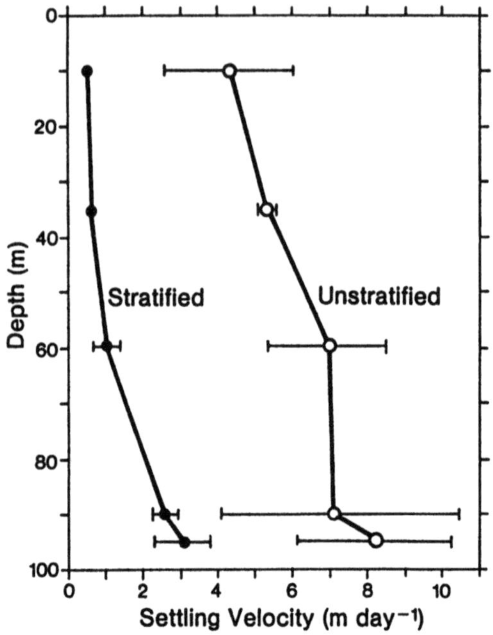


**Figure 10.5** Particle-settling velocity averages for the stations in Lakes Michigan, Huron, and Superior (see Figure 10.1). Those calculated for the stratified period are similar to other reported values. The high rates throughout the water column during the unstratified period and for the BNL during the period of stratification indicate extensive recycling of particulate matter.

During the unstratified period, settling velocities in the upper water column are approximately 5 m·day$^{-1}$, an order of magnitude higher than during stratification. At this rate, the particle residence time in the water column is only 20 days. Within the (ca. 20-m-thick) BNL, the particle residence time is only a few days during the unstratified period and about a week during thermal statification. These rates imply that a large recharging of the particle pool, either via horizontal transport or local sediment resuspension, occurs throughout the year. Contributions from these two sources have not yet been discriminated.

## 10.4 Particle Fluxes

Seasonal particle fluxes reported for Lakes Michigan (Eadie et al., 1984; Eadie and Robbins, 1987), Ontario (Rosa, 1985), and Erie (Bloesch, 1982 and Charlton and Lean, 1987) exhibit pronounced seasonal differences. During the stratified period, the mass fluxes are low ($<$ 1g.m$^{2}$·d$^{-1}$) near the surface and increase exponentially near the bottom. When thermal stratification breaks down, the mass flux increases dramatically.

An example for offshore Lake Michigan (see Figure 10.2) is used to illustrate the seasonal cycle of particle flux in some detail. Fluxes near the surface show a seasonal pattern that is in general agreement with the TSM measurements (see Figure 10.4*b*). The maximum fluxes correspond to the TSM maxima in April to May, and a second, smaller peak is clearly shown in the period of $CaCO_3$ precipitation in August to October. The July period that showed an increase in particle volume but not mass (Figure 10.4) occurs at a time of low mass flux. Small flagellated algae were found to dominate the particle pool at this time (Scavia and Fahnenstiel, 1987). These algae have very slow settling velocities. The collection covering November 1 through December 15 had the minimum flux and TSM; the deepening of the thermocline may dilute the biogenic particles in the surface mixed layer, and the $CaCO_3$ has disappeared via settling. Large fluxes are evident in the last collection, covering the (generally unstratified and ice-free) winter months.

The same general pattern of mass flux has now been measured in Lakes Huron and Superior (Figure 10.6). In this representation, three years of Lake Michigan data have been averaged for ease of comparison. The flux profiles for the Huron station are similar to the Michigan station data, although somewhat smaller in magnitude. Mass fluxes in the BNL are approximately twice as high in Michigan, both in summer and winter. The Lake Huron station is in the area of highest recent sediment accumulation (Robbins, 1980) and is expected to represent a region of high flux for deep water. Recent sediment accumulation at the Michigan site is lower (Robbins and Edgington, 1975) and the station is deeper. Station depth, distance from shore, and slope all play a role in mediating local mass fluxes. Near the surface, the summer flux profiles for all three stations are similar. However the summer BNL flux and the entire winter mass flux profile at the Lake Superior station are much smaller than at either of the other locations. It would appear that there is significantly less material in the Lake Superior BNL, an indication that resuspension and sed-

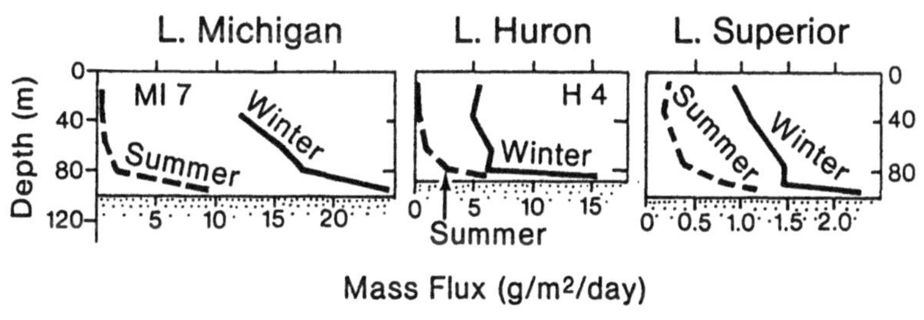


**Figure 10.6** Sediment-trap-measured mass flux profiles for the stratified and unstratified periods for the stations shown in Figure 10.1. Note that the mass flux scale for Lake Superior is smaller than for Michigan and Huron. Results extend earlier conslusions for Lake Michigan of a decoupling of surface waters from the influence of sediment resuspension during the period of thermal stratification and the occurrence of a large amount of sediment resuspension, strongly influencing lake water chemistry, during the unstratified period.

iment-water exchange rates are lower in this lake. Their overall importance have to be related to other internal recycling processes and external loads.

## 10.5 Carbon and Nutrient Fluxes

Trap measurements of carbon and nutrient fluxes are valuable in estimating budgets and for calculating chemical energy transport to the benthos. It has been argued that traps deployed at the base of the thermocline (ca. 35 m) during the stratified period collect a representative sample of the primary flux (Eadie et al., 1984) of (mostly) autohthonous material. Extrapolation of these fluxes to estimate "loads" must be done with care and a consideration for factors that affect the variable to be calculated. For example, the flux of organic carbon calculated by this method would probably be significantly underestimated because the spring bloom generally peaks prior to stratification. With this caveat in mind, fluxes of organic carbon, nitrogen, and total phosphorus for the stratified period are presented in Table 10.1.

The values for Lake Michigan are within the range reported for the four offshore Lake Michigan stations collected in 1980 (Eadie et al.,1984). These values can be considered as the flux of organic matter entering the hypolimnion during this period. Except for the phosphorus in Lake Superior, the chemical flux values are surprisingly consistent among the lakes, but are considerably less than the mean values reported by Rosa (1985) for offshore Lake Ontario. This difference is presumed to be due to higher productivity in Ontario. Carbon fluxes of this magnitude are sufficient to support an abundant benthic biomass. The benthic biomass at a location close to the trap station in Lake Michigan was measured at 3.39 $g \cdot m^{-2}$ (Nalepa et al., 1985).

Gardner et al.(1985) estimated that this abundance of *Pontoporeia hoyi* (the most abundant benthic organism) was equivalent to 13.000 $cal \cdot m^{-2}$ of food for the fish in Lake Michigan. Although not as well studied, the other lakes have sufficient carbon flux to support similarly robust benthic communities.

The carbon and phosphorus values for Lake Michigan are again within the range previously reported, although there were substantial changes in plankton communities (Scavia and Fahnenstiel, 1987) between the trap collections in 1980 and 1984–85. Nitrogen fluxes were almost twice the previously reported mean (Eadie et al., 1984), possibly due to the fact that the earlier

**Table 10.1** Nutrient fluxes at 35 m depth during the period of stratification for the three stations shown in Figure 10.1

| Nutrient (mg $m^{-2} \cdot d^{-1}$) | Lake Michigan | Lake Huron | Lake Superior | Lake Ontario |
|---|---|---|---|---|
| Organic carbon | 74 | 97 | 78 | 176 |
| Nitrogen | 4.0 | 4.1 | 3.2 | 26. |
| Phosphorus (total P) | 0.40 | 0.31 | 0.11 | 2.3 |

nitrogen values were measured by the Kjeldahl procedure, while those reported above were measured by combustion on a CHN analyzer.

## 10.6 Sediment Resuspension

The rate of sediment resuspension can be calculated from the trap-measured flux profiles, corrected for primary flux by subtracting the near-surface (< 35m) value for the stratified period from each flux measurement. The assumption that these upper water values are uncontaminated by resuspension is supported by measurements of fallout Cs-137 (Eadie and Robbins, 1987). Cs-137 entered the lakes during atmospheric testing of nuclear weapons, primarily during the 1960s, and currently virtually all Cs-137 resides in the sediments. During the stratified period, our measurements show no Cs-137 in traps above the BNL, while measurements in samples collected during the unstratified period have a Cs-137 activity that is similar to that of fine-grained sediments and nearly constant throughout the water column. Other investigators use the same conceptual approach but employ other (and sometimes nonconservative) tracers to make this calculation (Bloesch, 1982; Rosa, 1985). To estimate the flux at the sediment-water interface ($J_0$) we use a least-squares fit to the flux profiles to the model, resulting in the following equation:

$$J = J_0 \exp^{-bz} \tag{10.1}$$

where J (units, $g \cdot m^{-2} d^{-1}$) is the trap-measured flux of resuspended matter at z(m), the trap height above bottom, while b($m^{-1}$) can be viewed as the ratio of settling velocity to eddy diffusivity. The depth-weighted, mean value of resuspension, R ($g \cdot m^{-2} \cdot d^{-1}$), is then the integral of this function (evaluated between 35 m and the bottom for the stratified period and surface to bottom for the unstratified period) divided by the depth. Calculated values for the three lakes are presented in Table 10.2.

These calculations show a similarity between Lakes Michigan and Huron,

**Table 10.2** Sediment resuspension parameters for eq. 10.1 calculated by using the least-squares-fitting, trap-measured flux of the three stations in Figure 10.1

| | Value during stratified period | | | Value during unstratified period | | |
|---|---|---|---|---|---|---|
| | Lake Michigan | Lake Huron | Lake Superior | Lake Michigan | Lake Huron | Lake Superior |
| $J_0$ ($g \cdot m^{-2} \cdot d^{-1}$) | 10.5 | 5.7 | 1.1 | 11.6 | 10.8 | 2.2 |
| $b_z$ ($m^{-1}$) | 0.049 | 0.060 | 0.029 | 0.0092 | 0.016 | 0.012 |
| $r^2$ (corr. coef.) | 0.93 | 0.94 | 0.92 | 0.58 | 0.47 | 0.71 |
| $D_z$ ($cm^2 \cdot s^{-1}$) | 7.6 | 9.6 | 9.0 | 126 | 57 | 41 |
| R* ($g \cdot m^{-2} \cdot d^{-1}$) | 1.5 | 0.74 | 0.15 | 7.0 | 5.2 | 1.1 |

* R is the estimate of the depth-weighted upward flux from the sediments.

and nearly an order-of-magnitude smaller value of R for resuspension in Lake Superior. It is unclear how much of this difference is due to Superior's greater mean depth (149 m; 86 m for Michigan and 60 m for Huron), lower productivity (40–100 $gC \cdot m^{-2} \cdot y^{-1}$; 140–225 for Michigan and 80–90 for Huron), lower runoff from the granitic shield basin, or to other factors contributing to a smaller pool of active particles.

Eddy diffusivity ($D_z$) values were calculated using the average settling velocity (3 $m \cdot d^{-1}$ and 8 $m \cdot d^{-1}$ for the stratified and unstratified periods, respectively) for the bottom 10 m (see Figure 10.5). The calculated $D_z$ values are 10 to 100-fold larger than values estimated from radon profiles in the Baldeggersee (Imboden and Joller,1984) and other estimates (Lerman, 1979). One possible explanation for this discrepancy is that steady state is assumed in the calculations of both the exponent b (eq. 10.1) and settling velocities. If the traps are loaded by short-time-scale events (eg., storms) and TSM is measured only during calmer weather (after material has settled out), then calculated settling velocities and diffusion coefficients would be overestimated. In-situ estimates of settling velocities, such as those of Burns and Pashley (1974), and TSM data under turbulent conditions both need to be measured to test this hypothesis.

The magnitude of recycling of nutrients via sediment resuspension can be calculated by fitting the measured profiles of chemical flux to the same exponential model as described above for resuspended mass calculation. Values for the depth-weighted rate of sediment nutrient resuspension (R) for the unstratified period are given in Table 10.3

Earlier (Eadie et al.1984) we reported that resuspension of sediment-bound phosphorus was large compared to new loads for Lake Michigan. Our measurements indicated that 15 to 20% of Lake Michigan winter-trap-collected phosphorus was extractable by 0.1 *N* NaOH, a fraction shown to be readily bioavailable to diatoms (Williams et al., 1980); thus, estimates of this recycling process are important for lake management. Estimates of this component of the phosphorus cycle are shown in Table 10.4 for the three lakes and are compared to both recently measured loads and the target loads agreed upon jointly by the US and Canada through the International Joint Commission (IJC).

Even accounting for the fact that the resuspension term is only for the (6-

**Table 10.3** Calculated values of nutrient resuspension from sediments during the unstratified period for the three lake stations

| Nutrient ($mg\ m^{-2} \cdot d^{-1}$) | Lake Michigan | Lake Huron | Lake Superior |
|---|---|---|---|
| Organic carbon | 250. | 212. | 58. |
| Nitrogen | 60. | 27. | 13. |
| Phosphorus | 7.2 | 5.8 | 1.2 |

**Table 10.4** Comparison of the calculated flux of phosphorus for Lakes Michigan, Huron, and Superior

| | Phosphorus flux (mg $m^{-2} \cdot d^{-1}$) | | |
|---|---|---|---|
| | Lake Michigan | Lake Huron | Lake Superior |
| Resuspension | 7.2 | 5.8 | 1.2 |
| Estimated load (1981)* | 0.24 | 0.23 | 0.11 |
| IJC target load* | 0.26 | 0.20 | 0.12 |

* From DePinto et al., 1986.

to 9-month) unstratified period and that a smaller fraction of the resuspended phosphorus is readily extractable compared to new load (ca. 15% vs. 50%), it is clear that the recycling rate of sediment-bound phosphorus is much larger than that for newly introduced phosphorus.

The results presented illustrate that, although the Laurentian Great Lakes are deep, the process of sediment resuspension is very important in the cycling of compounds with a high affinity for particulate matter. The annual reinjection of relatively high fluxes of particulate matter from the sediments into the water column helps to precipitate out newly introduced contaminants, but detrimentally re-exposes lake water to contaminants stored in the resuspendible pool. Currently, it is estimated that approximately 90% of the PCB in Lake Michigan is in the sediment inventory. The coupling of bioturbation and resuspension keeps this contaminant (and others) in intimate contact with the overlying water. In attempting to assess the overall importance of this process, seasonal timing is also important. The major phytoplankton bloom occurs prior to lake stratification. In effect, resuspension sets the initial nutrient conditions for lake water in the spring. The efficiency of capture of these nutrients—via primary productivity within the euphotic zone—controls the size of the pool of recyclable epilimnetic nutrients after strafication begins.

*Acknowledgements* We would to express our appreciation to the crew of the RV Shenehon for their assistance in trap deployment and retrieval, to Dr. Wayne Gardner for his comments on this manuscript, and to Dr. Jürg Bloesch for his thorough and constructive review. GLERL contribution 603.

## References

Bloesch, J. 1982. Inshore-offshore sedimentation differences resulting from resuspension in the eastern basin of Lake Erie. *Can. J. Fish and Aq. Sci.* 39:748–759.

Bloesch, J. and Sturm, M. 1986. Settling flux and sinking velocities of particulate phosphorus and particulate organic carbon in Lake Zug, Switzerland. In: P. Sly (ed.), *Sediment and Water Interactions*, Springer-Verlag, p. 481–490.

Burns, N.M. and Pashley, A.E. 1974. In Situ Measurement of the Settling Velocity Profile of Particulate Organic Carbon in Lake Ontario. *J. Fish. Res. Board Can.* 31:291–297.
Charlton, M.N. and Lean, D.R.S. 1987. Sedimentation, resuspension and oxygen depletion in Lake Erie (1979). *J.Great Lakes Res.* 13:709–723.
Chambers, R.L. and Eadie, B.J. 1981. Nepheloid and suspended particulate matter in south-eastern Lake Michigan. *Sedimentology* 28:439–447.
DePinto, J.V., Young, T.C., and McIlroy, L.M. 1986. Great Lakes water quality improvement. *Environ. Sci. Technol.* Vol. 20, No. 8, p. 752–759.
Eadie, B.J., Chambers, R.L., Gardner, W.S., and Bell, G.L. 1984. Sediment Trap Studies in Lake Michigan: Resuspension and Chemical Fluxes in the Southern Basin. *J. Great Lakes Res.* 10(3):307–321.
Eadie, B.J. and Robbins, J.A. 1987. The Role of Particulate Matter in the Movement of Contaminants in the Great Lakes. In: R. Hites and Eisenreich, S. (eds.), *Sources and Fates of Aquatic Pollutants,* Advances in Chemistry Series No. 216, American Chemical Society, Washington, D.C., p. 319–364.
Gardner, W.S., Nalepa, T.F., Frez, W.A., Chichocki, E.A., and Landrum, P.F. 1985. Seasonal patterns in lipid content of Lake Michigan macroinvertebrates. *Can. J. Fish and Aq. Sci.* 42:1827–1832.
Hilton, J. 1985. A conceptual framework for predicting the occurence of sediment focusing and sediment redistribution in small lakes. *Limnol. Oceanogr.* 30:1131–1143.
Imboden, D.M. and Joller, Th. 1984. Turbulent mixing in the hypolimnion of Baldeggersee (Switzerland) traced by natural radon-222. *Limnol. Oceanogr.* 29(4):831–844.
Lee, C., McKenzie, J.A., and Sturm, M. 1987. Carbon isotope fractionation and changes in the flux and composition of particulate matter resulting from biological activity during a sediment trap experiment in Lake Greifen, Switzerland. *Limnol. Oceanogr.* 32:83–96.
Lerman, A. 1979., *Geochemical Processes Water and Sediment Environments.* John Wiley and Sons, Inc., New York, 481 pp.
Mortimer, C.H. 1971. *Large Scale Oscillatory Motions and Seasonal Temperature Changes in Lake Michigan and Lake Ontario.* Spec. Rpt. No. 12, Center for Great Lakes Studies, Univ of Wisconsin-Milwaukee, Milwaukee, WI., 106 pp.
Nalepa, T.F., Quigley, M.A., Childs, K.F., Gauvin, J.M., Heatlie, T.S., Parker, M.P., and Vanover, L. 1985. *Macrobenthos of Southern Lake Michigan.* NOAA Data Report ERL GLERL-28, GLERL, Ann Arbor, MI., 273 pp.
Robbins, J.A. and Edgington, D. 1975. Determination of recent sedimentation rates in Lake Michigan using Pb-210 and Cs-137. *Geochim. Cosmochim. Acta* 39:285–304.
Robbins, J.A. 1980. *Sediments of Southern Lake Huron: Elemental Composition and Accumulation Rates.* Ecological Research Series. EPA-600/3-80-080, Washington, D.C., 211 pp.
Robbins, J.A. 1982. Stratigraphic and dynamic effects of sediment reworking by Great Lakes zoobenthos. *Hydrobiologia* 92:611–622.
Robbins, J.A. 1986. *Sediments of Saginaw Bay, Lake Huron: Elementa Composition and Accumulation Rates.* Spec. Rep. 102 of Great Lakes Res. Div., Univ. of Mich., Ann Arbor, MI., 103 pp.
Rosa, F. 1985. Sedimentation and sediment resuspension in Lake Ontario. *J. Great Lakes Res.* 11(1):13–25.
Saylor, J.H. and Miller, G.S. 1988. Observations of Ekman Veering at the bottom of Lake Michigan. *J. Great. Lakes. Res.* 14:94–100.
Scavia, D. and G.L. Fahnenstiel 1987. Dynamics of Lake Michigan phytoplankton: mechanisms controlling epilmnetic communities. *J. Great Lakes Res.* 13:103–120.
Sprules, W.G., Casselman, J.M., and Shuter, B.J. 1983. Size Distribution of Pelagic Particles in Lakes. *Can. J. Fish. Aquat. Sci.* 40:1761–1769.
Strong, A.E. and Eadie, B.J. 1978. Satellite Observations of calcium carbonate in the Great Lakes. *Limnol. Oceanogr.* 23:877–887.

Vanderploeg, H.A. 1981a. Seasonal particle-size selection by *Diaptomus sicilis* in offshore Lake Michigan. *Can. J. Fish. Aquat. Sci.* 38:504–517.
Vanderploeg, H.A. 1981b. Effect of the algal length/aperture length ratio on Coulter analysis of lake seston. *Can. J. Fish. Aquat. Sci.* 38:912–916.
Vanderploeg, H.A., Eadie, B.J., Liebig, J.F., Tarapchak, S.J., and Glover, R.M. 1987. Contribution of calcite to the particle-size spectrum of Lake Michigan seston and its interactions with the plankton. *Can. J. Fish. Aquat. Sci.* 44:1898–1914.
Wahlgren, M.A., Robbins, J.A. and Edgington, D.N. 1980. Plutonium in the Great Lakes. In:Hanson, W.C. (ed.), *Transuranic Elements in the Environment.* Technical Information Center/U.S. Department Energy, Washington, D.C., p. 639–683.
Williams, J.D.H., Shear, H., and Thomas, R.L. 1980. Availability to *Scenedesmus quadricauda* of different forms of phosphorus in sedimentary materials from the Great Lakes. *Limnol. Oceanogr.* 25(1):1–11.

# 11

# Time Scales of Sediment Focusing in Large Lakes as Revealed by Measurement of Fallout Cs-137

D. N. Edgington and J. A. Robbins

ABSTRACT The sediments of large lakes provide a sink for many of the trace contaminants that have been mobilized in the environment. Several models have been developed to describe the resultant concentration profiles in sediment cores—in terms of input fluxes, sedimentation rates, and mixing depths. Implicit in these models are the assumptions that sedimenting particles are disconnected from prior input events and that transport is only by vertical advection, even though the measured integrated fluxes indicated that there had been considerable sediment focusing. Recent measurements of Cs-137 and Pb-210 in sediment cores from Green Bay in Lake Michigan and least-squares best fits for sedimentation rates and mixing depths from these data suggest that these simple models are incorrect and do not adequately reflect the resuspension, horizontal transport, and redeposition of bottom sediments. Furthermore, such processes are not confined to shallow water; a comparison of total Cs-137 content, sedimentation rates, and mixing depths measured in a series of cores taken from Lake Michigan at the same locations ten years apart indicate that while the overall integrated Cs-137 content of the sediments, corrected for radioactive decay, has remained constant, there has been an increase within and a decrease at the periphery and outside of the depositional zones. Over the same period there was, as would be predicted from the simple models, an increase in the thickness of the measured mixing depth.

## 11.1 Introduction

Lake Michigan, the world's fifth largest lake, is an essentially closed system created during the retreat of the glaciers from the region about 10,000 years ago. As shown in Figure 11.1, since that time, large areas of the lake bottom have remained denuded of sediments, while up to 10 m of fine-grained materials have accumulated in discrete depositional basins (Lineback and Gross,

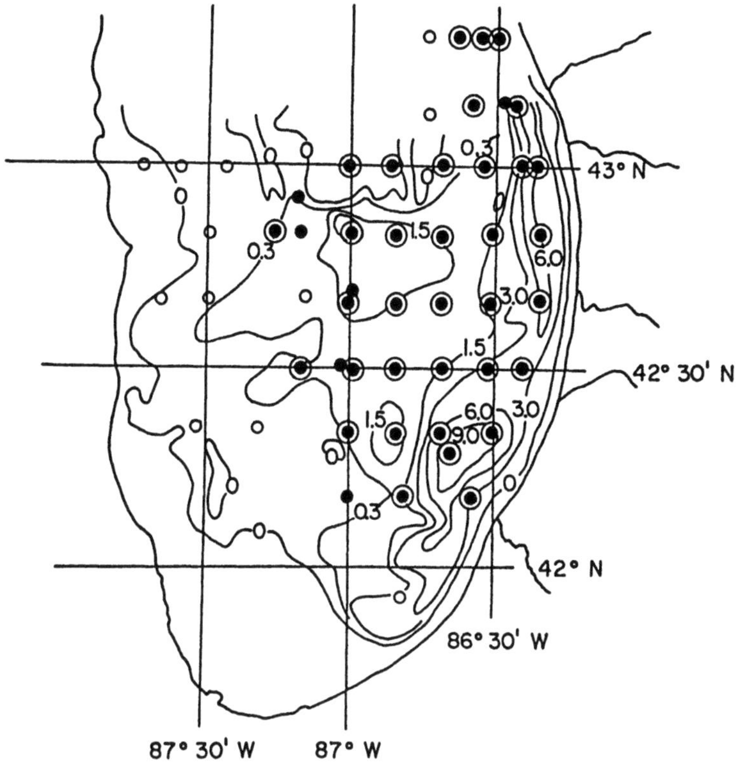


**Figure 11.1** The distribution pattern of recent sediments in southern Lake Michigan showing the thickness of the Waukegan member in meters (redrawn from Lineback and Gross, 1972) and the stations sampled in 1972 (open circles) and 1982 (solid circles).

1972). In general, patterns of sediment accumulation correspond to bathymetric features in the Great Lakes, but a notable exception is the large area of deposition in the southeastern part of Lake Michigan running parallel to the shoreline. This markedly asymmetric feature can be attributed to a combination of preferential loading from the continual erosion of the bottom and bluffs on the western side of the lake and an average pattern of water circulation, which transports particles southwards along this shoreline and produces a stationary gyre-like pattern in the southeastern quadrant of the southern half of the lake. These features are clearly discernable in satellite images of the turbidity of the surface water (Mortimer, 1988). A representative image, Figure 11.2, shows, as evidenced by its high reflectivity, a narrow band of resuspended sediment along the whole of the western shore, the result of a major spring storm blowing from the northeast. This suspended sediment has been driven around the bottom of the lake as a result of the counter-clockwise circulation. From there it has moved northwards and separated from the shore. The contribution of additional suspended sediments from the major tributaries on the eastern side is clearly visible.

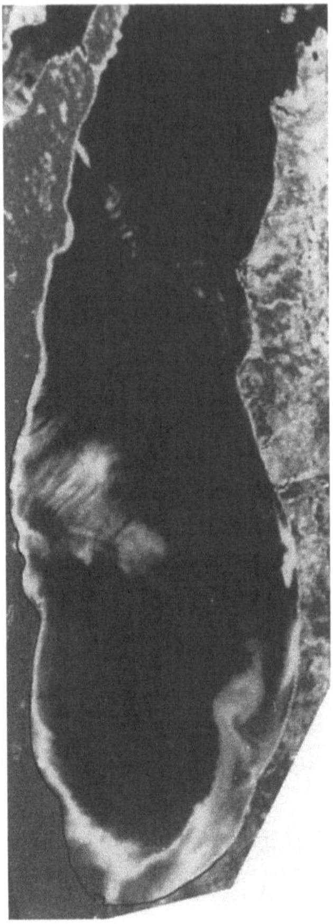

**Figure 11.2** Evidence of a major sediment resuspension event in southern Lake Michigan. Image recorded by the CZCS on May 16, 1981 in the green band.

Another important process affecting the distribution of sediment in the lake is episodic resuspension and redeposition. We have documented the evidence for massive erosion and slumping events in sediment cores taken from the deepest areas of the lakes related to major storm events (Edgington and Robbins, 1976a). An extensive set of studies using sediment traps has shown that, during the winter months when the lake is isothermal, fine-grained material from the sediment/water interface is resuspended and distributed homogeneously within the overlying water (Eadie and Robbins, 1987). With the onset of thermal stratification, this suspended particulate material disappears entirely from the epilimnion and largely from the hypolimnion, as a result of settling and redeposition on the lake bottom. This annual cycle of resuspension and redeposition induces a net migration of fine-grained particulate matter to the less turbulent areas of the lake bottom. Such intermittent sediment redistribution events clearly must play an important role in controlling the ultimate fate of sediments and their associated contaminants.

The horizontal redistribution of source materials and their preferential accumulation in depositional basins is commonly referred to as "sediment focusing." This term was first used by Likens and Davis (1975) to describe the observed distribution of sediments in Mirror Lake, New Hampshire. Lehman (1975) modeled the effects of various scenarios for sediment focusing on the rates of sediment accumulation at the center of a small lake. More recently, Ludlam (1984) showed that uniform varving of sediments in the profundal areas of Fayetteville Green Lake, N.Y., is disturbed by episodic turbidity flows of sediment from the littoral zone. In large lakes, such as the Laurentian Great Lakes, there is little or no sedimentation in the littoral zone and the observed sediment focusing occurs entirely within the profundal sediments.

In general, particle-reactive tracers and contaminants become associated with the finest grained sediments and are codeposited with them. In two previous papers we have shown that the accumulation of Pb-210 and fallout radionuclides (Cs-137 and Pu-239, 240) have patterns of accumulation which are virtually congruent with variations in the thickness of the postglacial sediment shown in Figure 11.1 (Robbins and Edgington, 1975; Edgington and Robbins, 1976b). Focusing was recognized then as an important process affecting the inventory of these tracers in individual sediment cores. As a result of focusing, the inventories of these radionuclides at any particular coring location showed little relation to their known lake-wide average input fluxes. We found that the inventories of these radionuclides varied from a small fraction of to several times more than the expected average areal loading. In addition, cores taken from even the deepest parts of the basin have been shown to preserve, in their Pb-210 profiles, records of major disturbances related to storm events (Edgington and Robbins, 1976a), similar to those observed with varved sediments in shallow lakes (Ludlam, 1984).

Since this initial survey of the distribution of fallout radionuclides was conducted almost 10 years after the injection of the major fraction of the total source-term, it was not possible to distinguish whether the transfer of Cs-137 to the sediments was due to association with, and settling of, particles suspended in the water or due to direct association with the sediments at the sediment/water interface (Santschi et al., 1986). Unlike its behavior in many other aquatic environments, Cs-137 is as strongly adsorbed as plutonium by sediments in the Great Lakes (Alberts, J.J., personal communication, 1988; Edgington and Nelson, 1986), with values of greater than $10^5$ for the distribution coefficient ($K_d$, $l \cdot kg^{-1}$, = conc. in sediment / conc. in water). Since Wahlgren et al. (1980) have shown that the residence time of plutonium in the water is less than two years and that Cs-137 appears to be essentially irreversibly adsorbed by these sediments (Alberts, J.J., personal communication), the residence time of Cs-137 is probably even shorter than that of plutonium. As a result of this strong association with particles, it would be expected that transfer to the bottom sediments on particles would be the dominant mechanism of removal from the water. Model calculations confirm the observation that contaminants such as Sr-90, which have a $K_d$ value $\leq 1000$, remain almost totally in the water.

What is perhaps less obvious is that, unless the focusing is essentially instantaneous, sediment profiles may not accurately reflect the history of loadings. Detailed studies of the Cs-137 and Pu-239 profiles in sediment cores taken from the Great Lakes show that the best theoretical fits to the known input fluxes are obtained with an effective residence time of less than two years. Essentially all the models which have been developed to interpret measured sediment profiles of contaminants with time-dependent input fluxes have assumed a pristine source of sediment and that the flux across the sediment/water interface had no horizontal component. The effects of the reintroduction into the water of Cs-137 as a result of particle resuspension from the sediment/water interface has now been measured by Eadie and Robbins (1987). The effects of the horizontal movement and redeposition of this material have been ignored in previous one-dimensional sedimentation models. This is, of course, understandable since there was a virtual absence of any time-dependent data that could be used to calculate the contribution of inputs from such processes to sediment profiles.

## 11.2 Large-Scale Focusing of Sediments

The measurements of Cs-137 in several series of sediment cores collected over the last decade and a half from Lake Michigan provide an unique opportunity to assess the importance of the long-term redistribution or focusing of particles and their contaminant load, as well as the time scale of this process. The initial survey of radionuclides in the lake was conducted in 1972. The coring sites are shown in Figure 11.1. Details of the field and analytical methods have been published elsewhere (Robbins and Edgington, 1975). The areal variation in the vertically integrated total Cs-137 content, as estimated from individual sediment cores, is shown in Figure 11.3*a*. The areal distribution indicates both a strong focusing of this, then a comparatively recently introduced, contaminant, and a close correspondence between the total Cs-137 content and the post-glacial deposition of sediment as shown in Figure 11.1.

Since a major fraction of the fallout radioactivity was introduced into the lake as a large pulse between 1961 and 1965, the observed distribution pattern indicated that this focusing process operated on time scales of less than about five years. Furthermore, the distribution of Cs-137 within sediment cores, taken from areas of high sedimentation rate where details of the loading history had been preserved, confirmed that this initial focusing process had to have been rapid, and that the time scale could have been as short as a few months to a few years.

While there have been no new significant inputs of fallout radionuclides to the lakes since the early 1970's (H.A.S.L., 1977), continuous meaurements over this time period have shown that the concentrations of both Cs-137 and

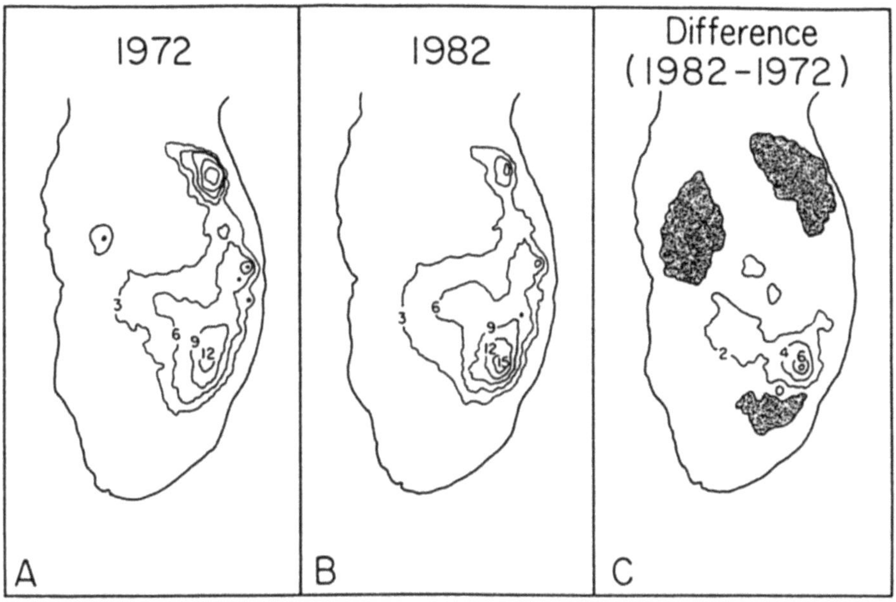


**Figure 11.3** Changes in the distribution of Cs-137 in the sediments of southern Lake Michigan (expressed as pCi.cm$^{-2}$) between 1972 and 1982. The isopleths were drawn using a computer contouring program. (*a*) The distribution in 1972. Note the position of the maximum deposition of Cs-137 and compare it with the sediment distribution shown in Figure 11.1. (*b*) The distribution in 1982. Note in particular that the maximum concentration is now greater at the center of the depositional zone than it was in 1972. (*c*) The difference in distribution between 1982 and 1972. The shaded areas represent negative values.

Pu-239,240 in the water have declined with a residence time of about 20 years (Edgington and Nelson, 1985; Robbins, unpublished data) indicating a continued, but decreasing supply to the water column. Recent studies by Eadie and Robbins (1987) have suggested that the continued source of radionuclides to the water can be attributed to the resuspension of bottom sediments during the unstratified period. Continuing resuspension of sediments containing Cs-137 might be expected to occur preferentially at the margins of depositional basins and thus imply an ongoing process of focusing. In fact, if the rate of resuspension is greater than the rate of accumulation of sediments at the margins, then one might expect that a transient tracer, such as Cs-137, could become more narrowly focused than the host sediments. To further explore these ideas in relation to the role of sediments as a final sink, as well as the efficacy of the simple models we had proposed to predict the fate of contaminants, we resampled the area in 1982 using exactly the same techniques as used in the earlier survey. The coring sites resampled in this later survey as well as the new sites are indicated in Figure 11.1. To correct for possible changes in the calibration of the gamma-ray spectrometers over the ten-year period, selected samples from the 1972 survey, that had been archived, were recounted. After correction for radioactive decay, it was found that small corrections ($<$ 5%) were necessary to account for analytical differences.

Computer-generated contour plots of the 1972 and 1982 total vertically integrated Cs-137 content ($pCi.cm^{-2}$) are shown in Figure 11.3*a* and *b*, respectively. Since Lake Michigan is essentially a closed basin, not unexpectedly, the total Cs-137 contained within the southern part of the lake (south of 43°20′ N), obtained by integration of the areas included within the contours in each figure, after correction of the 1972 data for radioactive decay, did not change significantly ($<$10%) during the decade; this indicates that there had been no further net transfer of Cs-137 from the water to the sediments. However, the total Cs-137 content of the principal depositional area increased significantly. In fact, the total Cs-137 content measured at stations in the center of the depositional area showed that new inputs during the decade (when the atmospheric flux was essentially zero) more than offset losses due to radioactive decay. The net difference between the 1982 content and the 1972 content, corrected for radioactive decay, is shown in the computer-generated contour plot in Figure 11.3*c*. Evidently, net movement of this radionuclide toward the center of the depositional area has occurred at the expense of that stored at the margins. Thus, during the ten-year period between 1972 and 1982, well after the passage of the initial large pulse into the system, focusing of the contaminant has continued to a very significant degree. It may be expected that cores collected toward the margins of depositional areas or basins will continue to show losses of sediment and Cs-137, while those collected from within the basins will continue to gain sediment and Cs-137. These conclusions are consistent with observations that the sedimentation rate increases from the periphery to the center of depositional areas.

## 11.3 Effects of Focusing on Vertical Profiles of Radionuclides

The process of long-term focusing has important implications for the interpretation of contaminant profiles. For those cores that are collected from the centers of basins, the concentration of Cs-137 will not decrease as rapidly from depth in the core toward the sediment/water interface as would be predicted by the sedimentation and diagenetic models currently in use (c.f., Robbins and Edgington, 1975; Robbins, 1978). In particular, elevated near-surface concentrations of a transient contaminant, such as Cs-137, could be falsely ascribed to bioturbation (Robbins and Edgington, 1975; Christiansen and Bhunia, 1986) or to changes in sedimentation rate (Oldfield and Appleby, 1984). In our earliest paper we had inferred that this mixed layer could result from physical or biological mixing processes, but later Robbins and coworkers (Robbins et al., 1977; Fisher et al., 1980; Robbins, 1982) showed that several organisms, widely distributed in sediments, occurred in sufficient numbers to redistribute materials at the sediment/water interface. Moreover, they found that the range of penetration of organisms into the sediments was comparable to the mixing depths inferred from Pb-210 or Cs-137 profiles. The effect of bioturbation is to create a region of constant concentration below the sediment/water interface, the magnitude of which is lower, and higher, than would be expected for Pb-210, and Cs-137, respectively, in the absence of mixing.

Long-term focusing of sediment could be expected to have little or no effect on Pb-210 profiles because of their steady-state nature resulting from the constant new input of this particular radionuclide from the atmosphere to the system and the reversible nature of the adsorption process for this radionuclide. However, for the case of a transient tracer, such as Cs-137, which appears to be irreversibly adsorbed to these sediments, focusing will result in an apparent continuing input of sediment, labeled with Cs-137, that will result in a higher-than-predicted concentration of this radionuclide in the sediments immediately below the sediment/water interface. For cores collected in areas where there has been extensive long-term sediment focusing, the application of steady-state sedimentation rate—sediment mixing models (e.g., Robbins and Edgington, 1975) will result in the calculation of correct values of the sedimentation rate and mixing depth from the Pb-210 profile, while the analysis of the Cs-137 profile will result in an underestimate of the sedimentation rate and an overestimate of the mixing depth.

While we could obtain self-consistent values of sedimentation rate and mixing depth at many coring locations from Pb-210 and Cs-137 profiles, there were many other locations, particularly from areas where sediment redistribution would have been expected to be extensive, where we could not. This lack of self-consistent data has been particularly evident for cores collected from Green Bay, a relatively shallow embayment in the northwestern part of Lake Michigan (Deering and Edgington, 1985). Examples of typical Pb-210 and Cs-137 profiles measured in two cores, collected from the southern end of Green Bay (stations 12 and 25), are shown in Figures 11.4 and 11.5. The values of sedimentation rates and mixing depths calculated from these profiles, using the steady-state model, are compared in Table 11.1. While the Pb-210 profiles exhibited the expected exponential decrease with depth as well as relatively high sedimentation rates, consistent with the known sediment loadings and distribution in the bay, the analysis of the Cs-137 profiles suggested far lower sedimentation rates and very large mixing depths. Self-consistent values were obtained when the sedimentation rates were calculated from the position of the 1962 horizons in the Cs-137 profiles, with a small correction for a mixed depth calculated from the Pb-210 profile, thus assuming that a large fraction of the excess Cs-137 in the surface sediment layers resulted from the long-term focusing process.

This behavior has been seen before. In an earlier paper, Robbins et al. (1978) observed, for a core collected in 1972 from the center of a depositional basin in central Lake Erie, high concentrations of Cs-137 in the surface sediment layers, similar to those found in the cores from Green Bay, which were

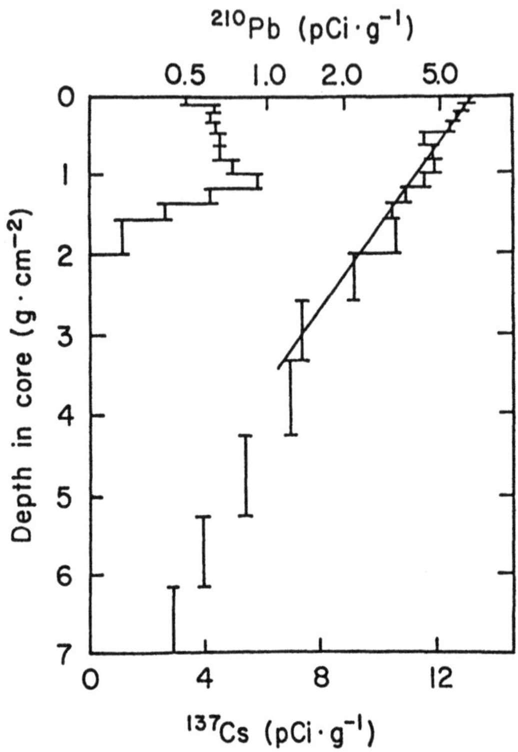


**Figure 11.4** Pb-210 and Cs-137 profiles measured in a sediment core collected at station 12 in Green Bay in 1982. The profiles in this and the succeeding figures are plotted versus cumulative mass of sediment per unit area. The lines drawn through the Pb-210 data points represent the best least-squares-fit using the sedimentation rate model of Robbins and Edgington, 1975.

inconsistent with both the very pronounced Cs-137 maximum found deeper in the core and the almost perfect exponential decrease in Pb-210 activity. To obtain self-consistent results we arbitrarily attributed the apparent mixing-depth anomaly with Cs-137 to horizontal redistribution effects. We suggested then that "... lake-wide redistribution and integration of Cs-137 would account for the high concentrations found in the surficial sediments. This process appears to occur preferentially for Cs-137 because of the time-dependent nature of its inputs. Because the influx of Pb-210 has been constant and the redistribution processes are essentially continuous, the flux of Pb-210 to surface sediments has attained a steady-state value and therefore decreases exponentially" (Robbins et al., 1978).

The contours shown in Figure 11.3*c* clearly indicate that there has been a significant increase in the amount of CS-137 deposited in the center of the major depositional area of the southern basin of Lake Michigan. In the early study of these sediments, station 29 was identified as having Pb-210 and Cs-

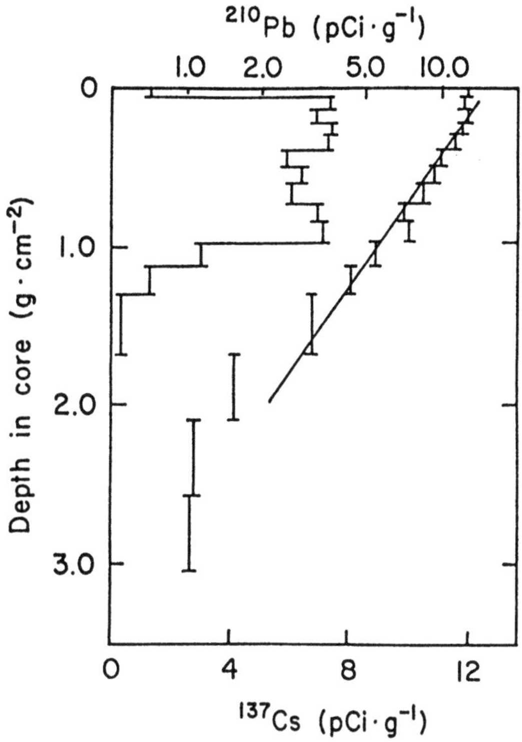


**Figure 11.5** Pb-210 and Cs-137 profiles measured in a sediment core collected at station 25 in Green Bay in 1982.

137 profiles that provided very self-consistent sedimentation rates and mixing depths (Robbins and Edgington, 1975). The data shown in Figure 11.6 are a comparison of the Cs-137 profiles obtained at this station in 1972 and 1982. A summary of the sedimentation parameters calculated from the model is given in Table 11.2. These data indicate that while the mass sedimentation rate has

**Table 11.1** Comparison of sedimentation parameters calculated for two sediment cores from Green Bay

| Station number | 12 | 25 |
|---|---|---|
| Sedimentation rate ($g \cdot cm^{-2} \cdot y^{-1}$) | | |
| Pb-210 | 0.058 | 0.033 |
| Cs-137 (model) | 0.029 | 0.019 |
| Cs-137 (1962 horizon) | 0.061 | 0.036 |
| Mixing depth (yr) | | |
| Pb-210 | 0 | 9 |
| Cs-137 (model) | 32 | 36 |

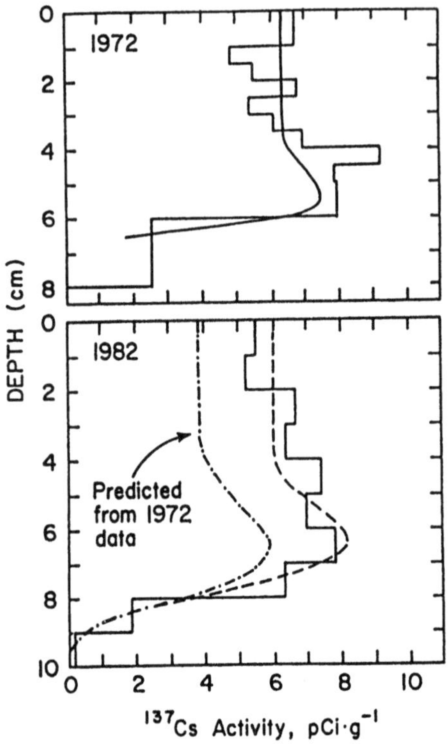


**Figure 11.6** A comparison of the distribution of Cs-137 in two sediment cores collected at station 29, at the center of the major depositional zone, in 1972 and 1982. *a*) The solid line represents the best least squares fit to the 1972 data. (*b*) The dashed line represents the best least squares fit to the 1982 data. The dot-dash line is the predicted vertical profile of Cs-137 in the sediments at station 29 in 1982 using the sedimentation parameters measured in 1972 (see Table 11.2) in the model of Robbins and Edgington (1975).

remained essentially constant, thus validating the use of a steady-state model, the thickness of the mixed layer appears to have increased by a small amount.

The ratio of the activity in the surface-mixed layer to that in the subsurface maximum layer should decrease in a predictable manner that is dependent on the radioactive decay of Cs-137 and the further dilution of activity in the mixed layer due to the input of new sediment. Any difference between this predicted ratio and the observed or model-estimated ratios can be attributed, in the

**Table 11.2** Comparison of sedimentation parameters for Cs-137 profiles in cores collected at station 29

| | Sedimentation rate | | Mixing depth | | | Activity ratio | | |
|---|---|---|---|---|---|---|---|---|
| Date | ($cm \cdot yr^{-1}$) | ($g \cdot cm^{-2} yr^{-1}$) | (cm) | (yr) | Flux[a] | Obs. | Best fit | Pred. |
| 1972 | 0.30 | 0.059 | 4.0 | 16 | 1.35 | 0.77 | 0.84 | 0.91 |
| 1982 | 0.24 | 0.063 | 4.5 | 22 | 2.88 | 0.75 | 0.74 | 0.65 |

a This is the flux-normalizing factor (see Robbins and Edgington, 1975), which is a measure of the amount of focusing that has occurred.

absence of other inputs, to the effect of the focusing or horizontal transport of Cs-137 contaminated sediment from outside of the depositional zone. Predicted activity ratios were calculated for both the 1972 and 1982 Cs-137 profiles using the simple sedimentation rate–mixing depth model of Robbins and Edgington (1975) and having as inputs the known annual fluxes of Cs-137 to the lake (H.A.S.L., 1977) and the sedimentation rates and mixing depths estimated from the 1972 Pb-210 and Cs-137 profiles at station 29.

The observed and predicted values of the activity ratio are also given in Table 11.2. Between 1972 and 1982 the predicted ratio decreased from 0.91 to 0.65, but the observed ratio decreased only from 0.77 to 0.75, indicating a continued influx of Cs-137 to this location. Since the observed ratios in both 1972 and 1982 are less than the ratios predicted from the known source-term, both of the Cs-137 profiles at station 29 provide further evidence for the effect of long-term sediment focusing on the overall sedimentation process in Lake Michigan.

For purposes of illustration we have compared Cs-137 distributions in cores collected at the center of the depositional basin where the sedimentation rate is the highest. At this site we would expect to observe profiles showing the most sensitive response to the changes in apparent mixed depth. Obviously, however, a comparison of two profiles in two cores collected ten years apart does not provide a basis for the desired statistical rigor. However, since sedi-

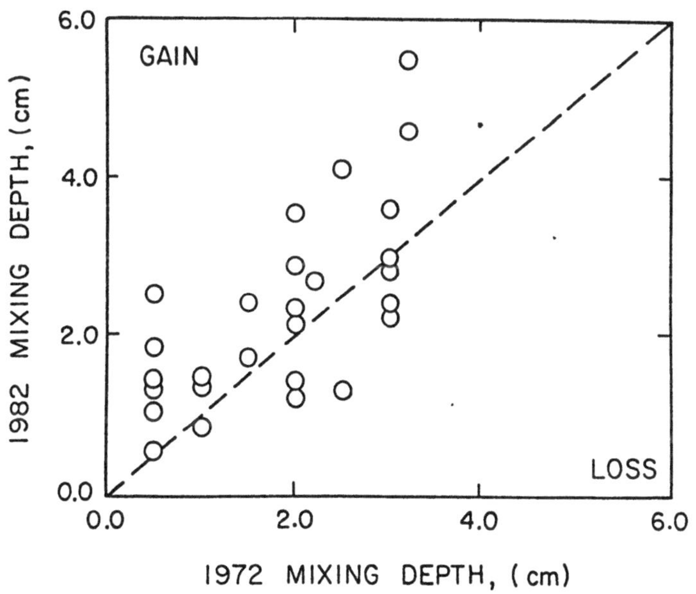


**Figure 11.7** A comparison of the values of the apparent mixing depth calculated from a linear least squares fit of the observed Cs-137 profiles. The dashed line is the expected result for the simple model.

mentation rates and mixing depths were calculated from each Cs-137 profile it is possible to compare the change in mixing depth at all stations. This comparison is shown in Figure 11.7. It should be noted that, not surprisingly as sediment cores were predominantly taken from depositional zones by design, there was an increase in the apparent mixed layer depth between 1972 and 1982 at a large proportion of the stations, with the average depth increasing from 1.7 ± 1.0 cm to 2.3 ± 1.0 cm. This small, but statistically significant, increase is consistent with the postulated effect of long-term focusing of transient tracers.

In addition, one of us (Robbins, unpublished data) has been able to analyze a more extensive time-series of cores from a station in Lake Erie. The results indicate that the effective depth of the mixed layer is increasing with time, thus demonstrating the continuing horizontal movement of bottom sediments with their associated contaminant load.

---

*Acknowledgements* This work was funded by the University of Wisconsin Sea Grant Institute under grants from the National Sea Grant College Program, National Oceanic and Atmospheric Administration, U.S. Department of Commerce, and from the State of Wisconsin. Federal grant NA84AA-D-00065, project R/GB-15. The authors thank D. Schwetz, J. Benante, L. Deering, R. Paddock, and P. Anderson for their valuable contributions to the field work and analysis of the samples.

## References

Christiansen, E.R. and Bhunia, P.K. 1986. Modelling radiotracers in sediments: comparison with observations in Lakes Michigan and Huron. *Journal of Geophysical Research* 91:8559–8571.

Eadie, B.J. and Robbins, J.A. 1987. Role of particulate matter in the movement of contaminants in the Great Lakes. In: Hites, R.A. and Eisenreich, S.J. (eds.), *Sources and Fates of Aquatic Pollutants*. Advances in Chemistry Series No. 216., American Chemical Society, Washington, D.C., p. 319–364.

Edgington, D.N. and Robbins, J.A. 1976a. Patterns of deposition of natural and fallout radionuclides in the sediments of Lake Michigan and their relation to limnological processes. In: Nriagu, J.O. (ed.), *Environmental Biogeochemistry*. Ann Arbor Science, Ann Arbor, Mich., p. 705–729.

Edgington, D.N. and Robbins, J.A. 1976b. The behavior of plutonium and other long-lived radionuclides in Lake Michigan. II. Patterns of deposition in the sediments. In: *Impacts of Nuclear Releases into the Aquatic Environment*. International Atomic Energy Agency, Vienna, p. 245–260.

Edgington, D.N. and Nelson, D.M. 1986. The persistence of pollutants in large lakes: the lessons from the study of radioactivity. In: Sibley, T.H. and Myttenaere, C. (ed.), *Application of Distribution Coefficients to Radiological Assessment Models*. Elsevier Applied Science Publishers, London, p. 250–266.

Fisher, J.B., McCall, P.L., Lick, W.J., and Robbins, J.A. 1980. Vertical mixing of sediments by tubificid oligochaetes. *Journal of Geophysical Research* 85:3997–4006.

H.A.S.L. 1977. Final tabulation of monthly Sr-90 fallout data, 1954–1976. *Environmental Quarterly*. United States Department of Energy Report HASL-329.

Lehman, J.T. 1975. Reconstructing the rate of accumulation of lake sediment: the effect of sediment focusing. *Quaternary Research* 5:541–550.

Likens, G.E. and Davis, M.B. 1975. Post-glacial history of Mirror Lake and its watershed in New Hampshire, U.S.A.: an initial report. *Int. Verhhein. Theor. Angew. Limnologie* 19:982–993.

Lineback, J.A. and Gross D.L. 1972. Depositional patterns, facies and trace element accumulation in the Waukegan Member of the late pleistocene Lake Michigan Formation in southern Lake Michigan. *Environmental Geology Notes, No. 58*. Illinois State Geological Survey. 25 pp.

Ludlam, S.D. 1984. Fayetteville Green Lake N.Y. Varve chronology and sediment focusing. *Chemical Geology* 44:85–100.

Mortimer, C.H. 1988. Discoveries and testable hypotheses arising from Coastal Zone Color Scanner images of southern Lake Michigan. *Limnology and Oceanography* 33:205–228.

Oldfield, F. and Appleby, P.G. 1984. Empirical testing of Pb-210 dating models for lake sediments. In: Haworth, E.Y. and Lund, J.W.G. (ed.), *Lake Sediments and Environmental History*. University of Minnesota Press, Minneapolis, Minn., p. 93–124.

Robbins, J.A. 1978. Geochemical and geophysical applications of radioactive lead. In: J.O. Nriagu (ed.), *Biogeochemistry of Lead in the Environment*. Elsevier-Holland, p. 285–393.

Robbins, J.A. 1982. Stratigraphic and dynamic effects of sediment reworking by Great Lakes zoobenthos. In: Sly, P.G. (ed.), *Developments in Hydrobiology, Vol. 9, Sediment/Water Interaction*. Proceedings of 2nd International Symposium on Sediment Water Interactions, Kingston, Ont. *Hydrobiologia* 1982; 611–622.

Robbins, J.A. and Edgington D.N. 1975. Determination of recent sedimentation rates in Lake Michigan using Pb-210 and Cs-137. *Geochimica Cosmochimica Acta* 39:285–304.

Robbins, J.A., Krezowski, J.R., and Mozley, S.C. 1977. Radioactivity in sediments of the Great Lakes: post-depositional redistribution by deposit-feeding organisms. *Earth and Planetary Science Letters* 36:325–333.

Robbins, J.A., Edgington, D.N. and Kemp, A.L.W. 1978. Comparative Pb-210, Cs-137l and pollen geochronologies of sediments from Lakes Ontario and Erie. *Quaternary Research* 10:256–278.

Santschi, P.H., Nyffeler, U.P., Anderson, R.F., Schiff, S.L. and O'Hara, P. 1986. Response of radioactive trace metals to acid-base titrations in controlled experimental ecosystems: evaluation of transport parameters for application to whole-lake radiotracer experiments. *Canadian Journal of Aquatic Science* 43:60–77.

# 12

# Characterizing Autochthonous Iron Particles and Colloids—the Need for Better Particle Analysis Methods

D. Perret, R.R. De Vitre, G.G. Leppard, and J. Buffle*

ABSTRACT The complexity involved in studying the biogeochemical cycles of particles and colloids in freshwaters is the result of several different factors: the great number of appearance and disappearance fluxes of these species, the difficulties associated with distinguishing and measuring those fluxes, and the difficulty of sampling colloidal particles without denaturation before their characterization. On the other hand, information obtained in laboratory studies of synthetic colloids are not always representative of the mechanisms encountered in heterogeneous natural environments. This article, taking autochthonous iron-rich colloids as a model, summarizes the state of the art concerning current knowledge of these colloids and emphasizes the need to develop new techniques specifically oriented towards analysis of natural colloids and their fluxes while taking care to minimize artifacts introduced by presently employed techniques. Finally, as a pertinent illustration, both the problems and the future uses of well-known filtration techniques are considered.

The transport of trace elements by colloids and particles plays a key role in controlling their biogeochemical cycling in lakes, estuaries, and oceans (Whitfield and Turner, 1987; Sigg, 1985; and refs. therein). The mechanistic details of these processes, however, are still poorly understood, largely because of a lack of appropriate techniques enabling determination of the nature, concentration, chemical reactivity, and physical fluxes of specific particulate species (Buffle et al., 1987). These processes are discussed below with particular emphasis being placed on those parameters which must be measured if we are to elucidate, in detail, the role played by particles (Section 12.2). In order to illustrate the specific types of problems that may be encountered, this discus-

*Corresponding author.

sion employs the case of iron oxyhydroxide particles which are formed at the redox boundary of eutrophic lakes and whose characteristics are described in Section 12.1. The possible artifacts which can be introduced by available techniques are illustrated using the specific but important case of filtration (Section 12.3).

## 12.1 Nature of Autochthonous Iron Oxyhydroxide Produced at the Redox Boundary of Eutrophic Lakes

**Literature Survey** It is well established that iron hydroxide particles have strong adsorption properties (Crosby et al., 1983) for a large number of cations (Swallow et al., 1980; Balistieri and Murray, 1982) and anions, including phosphate (Lijklema, 1980) and humic substances (Tipping, 1981a). It can therefore be expected that the biogeochemical cycle of iron may play an important role in controlling the bioavailability of nutrients and, thereby, the trophic level of lakes (Mayer et al., 1982). This is likely to be particularly important for stratified lakes possessing an anoxic deep layer promoting the rapid cycling of iron at the redox boundary (see Figure 12.4). Assessment of the quantitative importance of this cycle, however, requires that each of the various fluxes in Figure 12.4 be determined. Davison in particular (Davison and Woof, 1984b; Davison, 1985; and refs. therein) has studied and discussed in detail the cycles of Fe (see Figures 12.1 and 12.4), Mn, and S at the redox boundary of eutrophic lakes. Fe particles have been studied by several authors (Tipping et al., 1981b; Crosby et al., 1983; Koenings, 1976), and they are often found to be a hydroxophosphate of iron (Mayer et al., 1982; Buffle et al., 1989) with a P/Fe ratio of up to 1:3. More recently, it has been shown (see below; Buffle et al., 1989) that they contain approximately equal amounts of Fe(II) and Fe(III). Particle sizes ranging from 1 to 10 $\mu$m have been reported (Laxen and Chandler, 1983), but these values were derived from filtration experiments, possibly leading to overestimation of sizes by a factor of 10 or more (see Section 12.3). Pore sizes of 20–200 Å and specific surface areas of 100 $m^2 \cdot g^{-1}$ have been measured by Crosby et al. (1983) for particles synthesized by oxidation of Fe(II) under conditions mimicking natural ones.

Although of great interest, the information reported in the literature is presently limited by the fact that the measurements were either made in the laboratory (where complete reproduction of natural conditions is not always easy) or were subject to possible important artifacts (e.g., using sediment traps for determination of fluxes or filtration for size distribution measurements; see Sections 12.2 and 12.3). The results presented below were obtained (i) from direct field measurements made (ii) on particles naturally formed within the water column while (iii) paying special attention to minimize artifacts from known sources. The study site was Lake Bret in the Canton of Vaud, Switzerland.

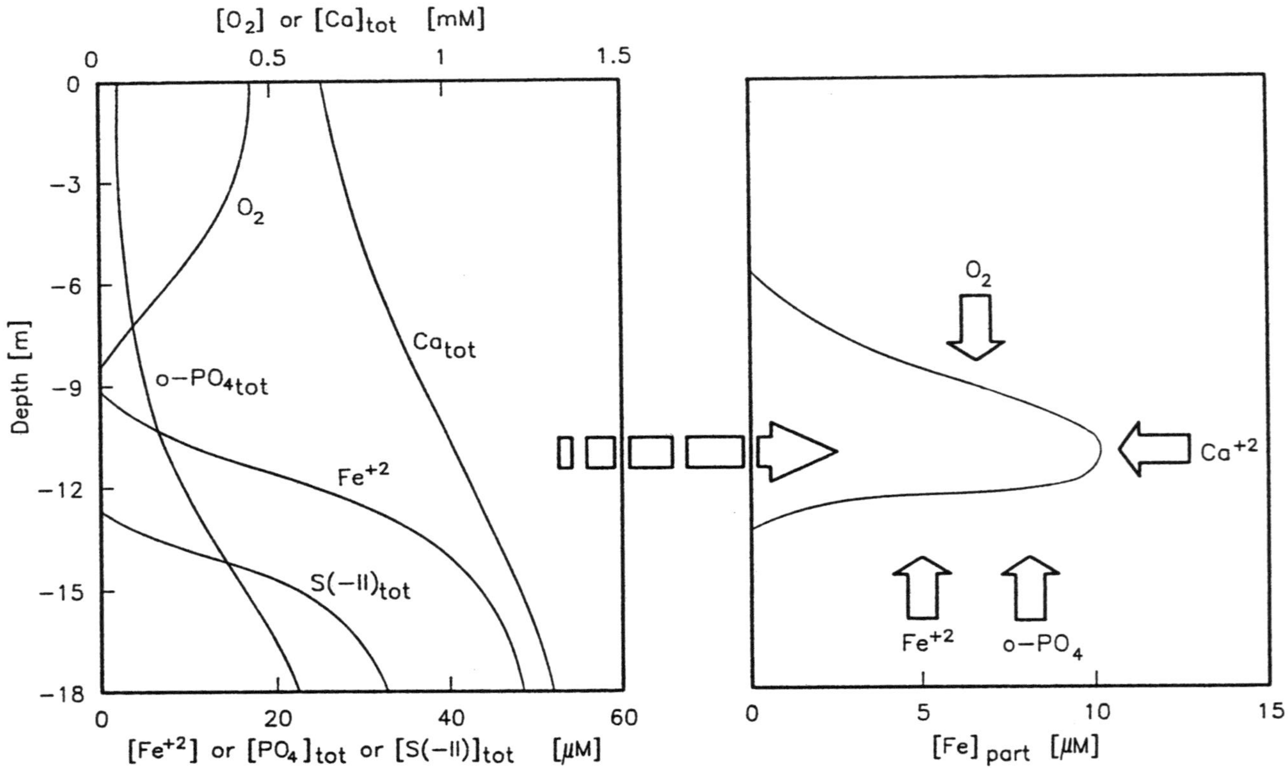


**Figure 12.1** Typical concentration profiles of species related to the iron cycle in Lake Bret during the summer; the transport of ferrous iron, oxygen, phosphate, and calcium into the transition zone leads to the formation of an autochthonous colloidal iron-rich species [Fe(II)/Fe(III)/OH/$PO_4$/Ca].

**Physico-chemical Characteristics of the Lacustrine Iron-rich Colloids**

Morphological and size characterization of colloids, as well as their elemental composition on a particle-specific basis, were accomplished by transmission electron microscopy coupled to an energy dispersive spectrometer (TEM-EDS) (Leppard et al., 1988; Buffle et al., 1989). These results were corroborated and complemented by determination of the particle composition, on a wet chemical analysis basis, using a multi-method speciation approach (De Vitre et al., 1988). The chemical composition of the lake water (Figure 12.1) and reactivity of the iron colloids was assessed, *in situ*, using polarographic methods (Buffle et al., 1988b; De Vitre et al., 1988).

*Iron Particle Classes:* The TEM-EDS results show that most of the iron-rich particles contain three major elements (up to five in some cases) with other minor elements represented in very low proportions. They may be grouped into the following four classes (the proportions being based on a particle number basis): 49% contain Fe/P/Ca, 23% contain Fe/P/Ca/Si/Al, 15% contain Fe/P/Ca/Si, and 13% contain Fe and other elements. These various associations have been discussed elsewhere (Buffle et al., 1989) and it has been suggested that "pure iron particle units" are best represented by the class containing only Fe/P/Ca, the other classes containing inclusions of clay leaflets or diatom frustules. The characteristics of these "pure" particles are discussed below.

*Morphology:* Figure 12.2 shows that the structure of these iron particles is amorphous and porous, i.e., no discrete particle limit is observed, and that they are composed of spherical globules (which sometimes include dense globulets). Under high-resolution conditions it can be seen that the minimum grain size of these globules is 1–2 nm.

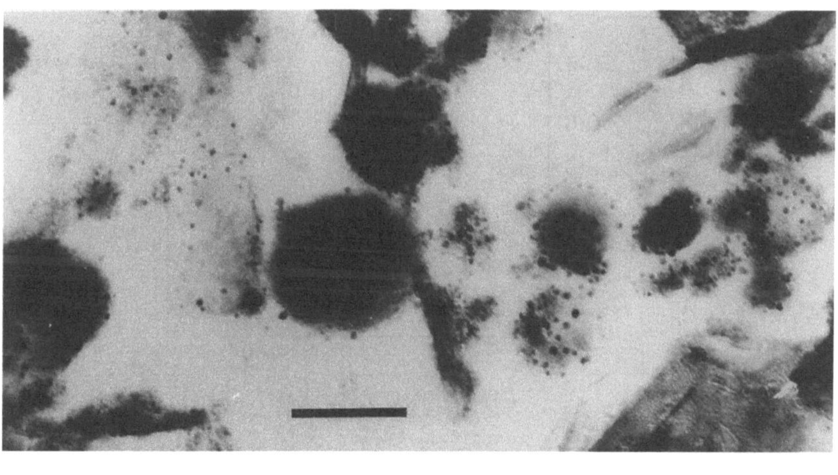

**Figure 12.2** Electron micrograph of iron-rich globules isolated from Lake Bret. Bar is 0.15 $\mu$m long.

*Size Distribution:* The spherical globules fall generally in the size range 45–310 nm (Figure 12.3), while the globulets are found between 7–16 nm. Aggregates of globules are also seen. It is important to note that the size of all globules is below the pore size of the membranes (0.45 $\mu m$) on which particles were retained during the sampling procedure. This particular point is discussed in Section 12.3, but, in any case, it is likely that the absence of particles smaller than 45 nm does not reflect the actual water column situation but rather the fact that these very small globules were not retained on the sampling membranes.

*Particle-Specific Elemental Composition:* EDS measurements have been done on 71 discrete particles previously isolated by means of the coupled TEM to ensure that they belonged to the Fe/P/Ca class. Values of $0.25 \pm 0.06$, $0.19 \pm 0.05$, and $1.34 \pm 0.32$ were obtained for the molar ratios P/Fe, Ca/Fe, and P/Ca, respectively.

*Wet Chemical Field Analysis:* The P/Fe ratio found above was confirmed by analysis of particulate $PO_4$ and Fe concentrations directly on the water sample. This, coupled with LAMMA (laser microprobe mass analyser) determinations, confirmed that the P in the globules observed by TEM is in the form of $PO_4$. Furthermore, wet analysis also showed that the Fe(II)/Fe(III) ratio in the iron particles is close to one. Both ratios were found to be statistically constant with season and depth.

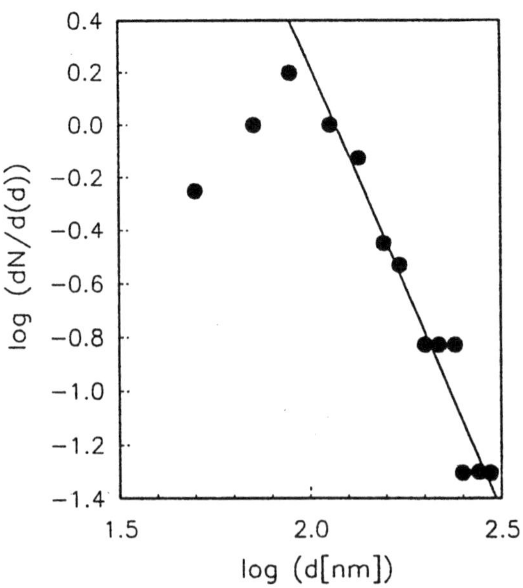


**Figure 12.3** Size distribution of 71 iron-rich globules, obtained by TEM examination and expressed according to the Pareto law, $\log[dN/d(d)] = cte - \beta \log(d)$, where N equals the number of globules with a size within a diameter interval d(d), and $\beta$ expresses the process of elimination of the particles in the water column.

*Chemical Reduction Rate:* Results consistent with zero-order kinetics were observed for the dissolution of at least 10% of Fe present in the colloids when ascorbic acid and S(−II) were used as reductants under lacustrine conditions. With S(−II) at lake water concentrations, dissolution rates of $3.4 \cdot 10^{-9}$ mole of Fe per hour were measured (De Vitre, unpublished results) which corresponds roughly to a reaction time of 8 hours for the reduction of 10% of the particulate Fe.

All these results are consistent with those of laboratory studies on the formation and reactivity of iron hydroxide particles and polymers (Pysik and Sommer, 1981; Schneider and Schwyn, 1987; Schneider, 1988) and are discussed elsewhere (Buffle et al., 1989). It can simply be pointed out that the remarkable constancy of the Fe/P/Ca and Fe(II)/Fe(III) ratios in these particles with time and depth, coupled with the EDS observation that $PO_4$ is not only adsorbed at the surface but belongs to the matrix of the particles, suggests that these particles are composed of a limited number of well-defined components.

## 12.2 The Nature of the Important Factors to be Measured in the Iron Cycle and the Available Methods

**Factors Characterizing the Iron Cycle** In order to assess the exact role of iron particles in the circulation of trace elements and nutrients, two types of information are needed:

1. Those related to the reactivity of trace elements and nutrients with the particles, including, in particular, the electrical charge of the particle, adsorption equilibria and rate constants as a function of pH, the specific surface area, porosity, and size distribution of the particle, and the diffusion coefficients of trace elements inside the particle.
2. Those related to the detailed characterization of the iron cycle and, in particular, the formation and elimination fluxes of the iron particles and their relative importance compared to those of other absorbing particles. These latter factors are, in fact, as important as adsorption strength, since the transport of adsorbable elements depends on the number and lifetime of settling particles.

For most of the *factors of type 1*, information is available in the literature (see Section 12.1), although it is most often based on laboratory studies. Comparison of our results with those of Crosby et al. (1983) suggests, however, that laboratory and field particles are similar if conditions close to natural systems are respected in the former case. Very little is presently known about the diffusion coefficient of elements inside such particles or about the relative scavenging role of the different size classes of particles, in particular the very small ones. Two remarks can, however, be made:

- Because of the high porosity of the particles, it is very likely that adsorption occurs at sites present *within* the particle structure, and not only at its surface. Furthermore, because of the small sizes of the particles, it may also be expected that diffusion times inside the particles will be short and will not appreciably affect adsorption equilibria.
- Determination of the relative scavenging role of the various size classes of particles can be assessed by transforming the particle size into a particle volume distribution (Lerman, 1979).

Very little is known regarding the *factors of type* 2 (iron cycle fluxes), largely because of experimental difficulties (see below). The results obtained with our data (see Section 12.1 and refs. therein) and cited in the literature, however, enable one to estimate the most important steps of the cycle as well as the orders of magnitude of the corresponding reaction times (Figure 12.4).

In Figure 12.4:

- Eddy diffusion times ($\tau_D$) are estimated from typical values of vertical eddy diffusion coefficients (0.03–10 $cm^2 \cdot s^{-1}$) (Lerman, 1979) and the concentration gradients of the species of interest in the lake (e.g., Figure 12.1).
- The oxidation time of Fe(II) ($\tau_{Ox}$) depends heavily on the exact nature of the oxidation reaction by $O_2$, noncatalyzed (Schneider and Schwyn, 1987; Davison and Seed, 1983; Sung and Morgan, 1980), catalyzed by Fe(III) hydroxide (Sung and Morgan, 1980) or influenced by other factors (e.g., $PO_4$, ionic strength and alkalinity (Sung and Morgan, 1980; Sarikaya, 1979)). Fe(II) may also be oxidized by $MnO_x$ formed by oxidation of Mn(II) by $O_2$ in the same water strata. The oxidation time indicated in Figure 12.4 is estimated from the literature values corresponding to aquatic conditions similar to those prevailing in Lake Bret.
- Particle formation times ($\tau_P$) are estimated from Schneider (1988) for globules containing Fe(III), ($\tau_{P,III}$), and from experimental observations for FeS precipitation ($\tau_{P,II}$).
- The time needed for coagulation of Fe/P/Ca globules with other settling particles of the water column is presently unknown. The occurrence of this process is, however, likely, due to the relatively small electrical charge of iron hydroxide particles at the lacustrine pH (7.5) and is strongly suggested by the fact that the size distribution follows a Pareto law with $\beta = 3.2$ (Figure 12.3); such values of $\beta$ are indicative of elimination of particles by perikinetic coagulation (Kavanaugh et al., 1980).
- The time indicated for reduction of globules by S(−II) ($\tau_R$) results from experimental observations (Section 12.1).
- The sedimentation times of Fe/P/Ca globules ($\tau_{S,G}$) are estimated from

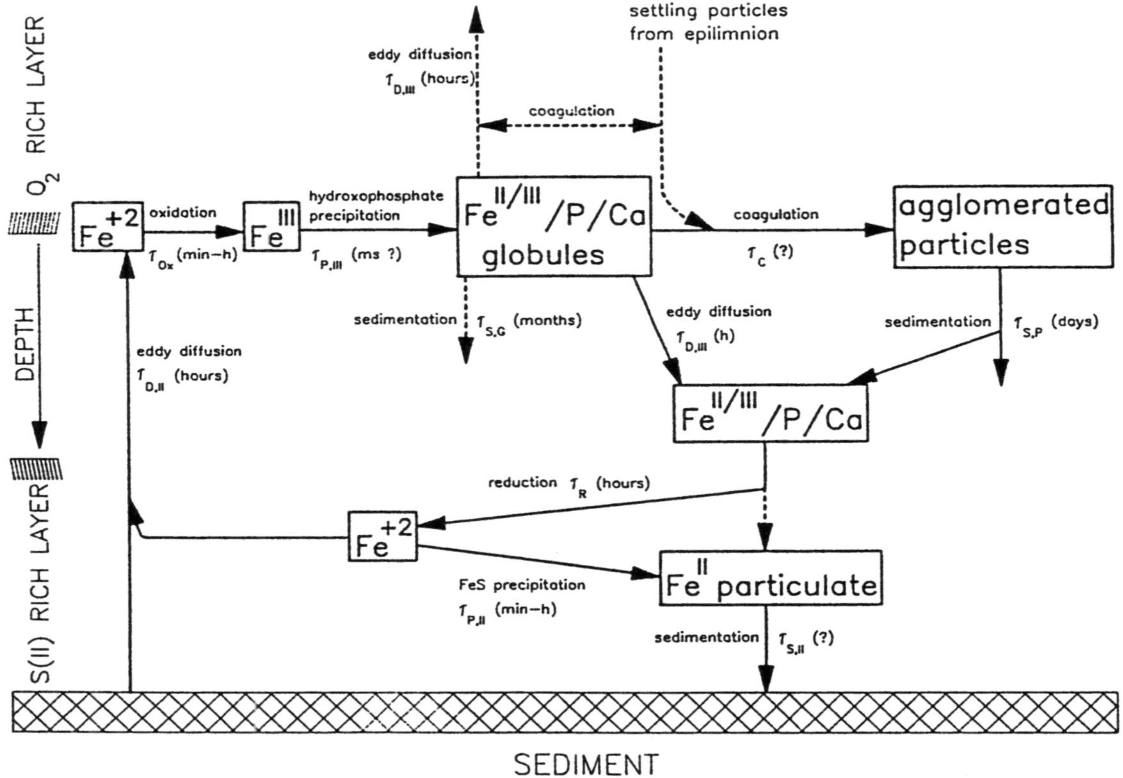


**Figure 12.4** Schematic overview of the different physical and chemical fluxes involving the iron species at the oxic-anoxic boundary. The subscripts of the reaction time $\tau$ refer to: *first subscript*—eddy diffusion (D), coagulation (C), sedimentation (S), precipitation (P), oxidation (Ox), or reduction (R); *second subscript*—$Fe^{II}$ particles (II), $Fe^{III}$-containing particles (III), coagulated particles (P), or Fe/P/Ca globules (G).

typical values of sedimentation coefficients (Lerman, 1979) and the size range observed for globules (Figures 12.3 and 12.8). The sedimentation rate of FeS particles is as yet unknown.

Although the above reaction times are only rough estimates, Figure 12.4 allows an important fact to be pointed out. Experimental evidence suggests that, during the three months of summer, the concentration of iron particles (Figure 12.1) can be considered as the result of a quasi-stationary state in the iron cycle. If this is true, the rate of the elimination process must be of the same order of magnitude as that of the slowest step in the chain of the particle production mechanisms (eddy diffusion and/or oxidation), i.e., hours). Free globule sedimentation ($\tau_{S,G}$, i.e., months) can therefore be ruled out, and even sedimentation of agglomerated particles ($\tau_{S,P}$, i.e., days—weeks) is not very likely since these processes are probably much too slow. The most probable processes, therefore, are either reductive dissolution of setting agglomerates or downward eddy diffusion of globules due to the concentration gradient produced by reductive dissolution at depth. The existence of a high concentration gradient of iron particles at the limit of the sulfidic layer (Figure 12.1) strongly supports the important role of this reductive dissolution.

**Limitations of Available Methods for Particle Studies** Figure 12.4 also emphasizes the intricacy of the iron cycle at the redox boundary, which is sharply at odds with the very poor resolution of presently available methods for particle sampling, characterization, and determination of fluxes.

Of the six types of flux-producing processes (see Fig. 12.4 and its discussion) only *eddy diffusion* (the $\tau_D$'s) can be measured unambiguously *in situ* (Lerman, 1979). As regards the four *"chemical fluxes,"* (i.e., oxidation, $\tau_{Ox}$ (Sung and Morgan, 1980; Davison and Seed, 1983); reduction, $\tau_R$ (Waite and Morel, 1984; Zinder et al., 1986; Stone and Morgan, 1987; Pysik and Sommer, 1981); precipitation, $\tau_P$'s; and coagulation, $\tau_C$ (Weilenmann, 1986)), reported data were largely obtained from laboratory studies (information for the latter two being especially scarce). These four processes involve surface (heterogeneous) reactions and are often highly dependent on particle concentration, the presence of "foreign" particles in solution (precipitation by heterogeneous nucleation, coagulation of particle mixtures, catalytic oxidation), microbial action (oxidation, reduction), and the presence of dissolved adsorbing and complexing agents (catalytic reductive dissolution). All of these effects are difficult to control individually and their exact environmental roles are difficult to estimate by laboratory simulations. *In situ* chemical flux determinations are, therefore, extremely necessary. Finally, *sedimentation fluxes* are most often determined by means of sedimentation traps (Simpson, 1982; Bloesch and Burns, 1980), which, as is discussed in these latter papers, have several limitations including, in particular, an often poorly controlled trapping efficiency (dependent on trap shape, turbulency, sediment resuspension, etc.). Of particular importance for studies of the iron cycle are that:

1. Traps must generally be left *in situ* for long periods (days) implying that they do not give real-time information but, rather, are a time-integrating collection method. They are also a depth-integrating method since they collect particles from all depths above the trap.
2. Traps are also nonselective. This is a very important limitation when studying the iron cycle since autochthonous iron oxyhydroxides are minor particulate components when compared to others such as calcite, clays, and allochthonous iron oxides. Since the latter two contain relatively large amounts of Fe, the sedimentation fluxes specifically due to the iron cycle of Figure 12.4 are determined by the difference of the Fe fluxes above and below the redox boundary layer (Davison et al., 1982). This leads to high uncertainties because of the much larger amount of Fe in clays compared to Fe/P/Ca particles.
3. Because traps must be left *in situ* for days, important biological and chemical transformations may occur during the sampling period. This is particularly important for Fe(III) particles which may easily be reduced to soluble Fe(II), thereby allowing its diffusion back into the water column resulting in severe underestimation of the fluxes.

As a result of these very important limitations, the development of flux determination methods which are selective for particular chemical species is certainly necessary if we are to gain a more detailed insight into the iron redox cycle.

In *particle characterization,* the most important problems are linked to sampling and fractionation. In most cases, filtration or centrifugation is used and the particulate material thus obtained is examined by characterization techniques, such as chemical analysis (Laxen and Chandler, 1983), X-ray fluorescence, electrophoresis (Tipping, 1981a), spectroscopy (IR, EPR, Mössbauer: Davison and Dickson, 1984a), TEM-EDS (see Section 12.1) or light scattering (Rees, 1987; see also Simpson, 1982, for a good review). In most cases, the application of such techniques to the real situation existing in the water column is compromized by particle denaturation during the initial sampling (and possibly the pre-concentration) step. Centrifugation problems are due to the absence of size selectivity for particles larger than 0.1 $\mu$m and the difficulty or impossibility of controlling denaturations due to coagulation during centrifugation (Kavanaugh et al., 1977; Salim and Cooksey, 1981), chemical reactions caused by differences in chemical conditions in the centrifugate with respect to the bulk solution, and, finally, the difficulty in maintaining, during centrifugation, the partial pressures of $O_2$, $CO_2$ and $H_2S$ prevailing *in situ*. Problems associated with filtration are discussed in Section 12.3. In any case, the development of characterization methods applicable *in situ* without, or with a minimum of sample handling (De Vitre et al., 1988; Buffle et al., 1988b), is highly desirable.

## 12.3 Filtration of Lacustrine Colloids: Can It Reflect Particle Size and in What Way?

As was mentioned above, filtration is one of the most ubiquitous operations in any study of aquatic samples, despite its important limitations and possible artifacts. Appropriate fractionation strategies require knowledge of the exact nature and relative importance of these limitations which can be grouped into three categories:

1. Contamination of samples by trace components present in the membrane or losses of trace compounds by adsorption at the membrane surface (Truitt and Weber, 1979).
2. Denaturation of water sample colloids between field collection and their subsequent study if storage and filtration periods are too long (Laxen and Chandler, 1982).
3. Denaturation of colloids at the membrane surface due to possible coagulation, retention by the membrane, and, ultimately, gel formation (clogging) (Suki et al., 1984).

When filtration is used for studying the properties of aquatic colloids, such as those of the iron cycle, the first type of problem is not limiting since it concerns only components present at the sub-micromolar level. The second type of problem can also be made negligible if filtration is done directly in the field or, even better, at depth (De Vitre et al., 1987). Consequently, we shall focus here on the artifacts produced at the membrane surface.

**Problems and Potential Remedies** Studies of membrane filtration processes focusing on their application to water analysis are very scarce, but much information can be gleaned from the extensive studies performed for industrial applications. Numerous theoretical or empirical models have been used to explain the fact that solution flux through a membrane is generally observed to decrease with time. They are based on changes in osmotic pressure (Michaels, 1968; Kozinski and Lightfoot, 1972; Trettin and Doshi, 1979), occurrence of interactions between different solutes or between the solute and the membrane (Fane et al., 1983; Matthiasson, 1983), variation of hydraulic resistances (Suki et al., 1984), or particle collisions which lead to coagulation within the membrane (Green and Belfort, 1980). These industrial-condition-oriented models are generally not directly applicable to limnological studies since particle concentrations are much lower and their nature is much more diverse (large heterogeneity, as shown in Figure 12.5). Coagulation rate (and consequently gel formation) is decreased by the first condition but increased by the second.

Nevertheless, drawing upon these models, it may be schematically stated that two basic processes may occur at the membrane surface (Figure 12.6)

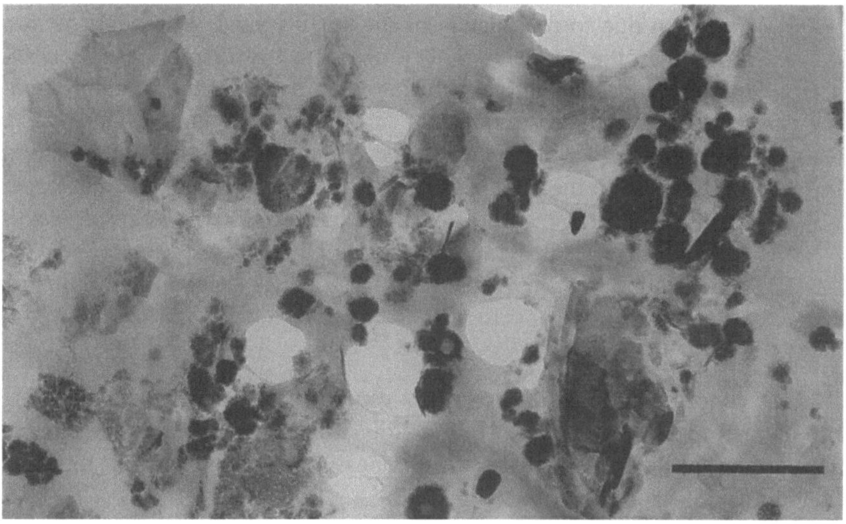

**Figure 12.5** Electron micrograph showing the large heterogeneity of a natural sample containing iron-rich colloids in a matrix of calcite crystals, clays, and organic fragments. Bar is 0.45 $\mu m$ long.

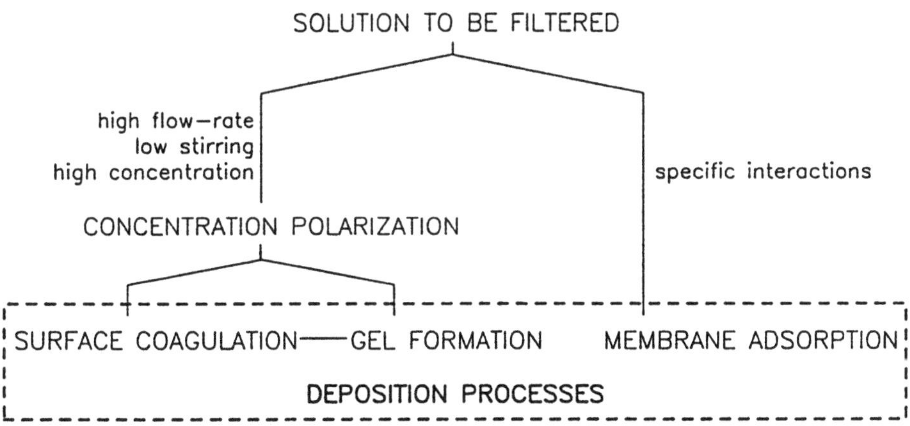


**Figure 12.6** Main causes of decrease in flow rate during filtration. "Deposition processes" is a global term for retention on a membrane.

leading to the retention of colloids, which, if only their size were to be considered, should pass through the membrane pores. These basic processes are:

1. Retention by adsorption due to specific interaction between the colloid and the membrane surface. One may expect this process to depend on the chemical nature of the membrane.

2. Retention due to coagulation of the small colloids into agglomerates larger than the pore size. This process may result from the "*concentration polarization*" effect, which always occurs at the solution-membrane interface on the retentate side of the membrane (Figure 12.7). Those particles which cannot pass through the pores accumulate in a layer close to the membrane and their concentration becomes larger than that of the same particles in the bulk solution. As a result of this concentration gradient, these particles will diffuse back towards the bulk solution but, the larger these particles, the smaller their diffusion rate will be.

The concentration polarization effect can be represented by the following approximate equation (Meares, 1976; Buffle, 1988a), which gives the ratio of the particle concentrations at the membrane surface ($C_0$) to that in the bulk solution (C):

$$C_0/C = \exp(J \cdot \sigma \cdot \delta / D)$$

where J ($cm \cdot s^{-1}$) is the flow rate of solution through the membrane per unit

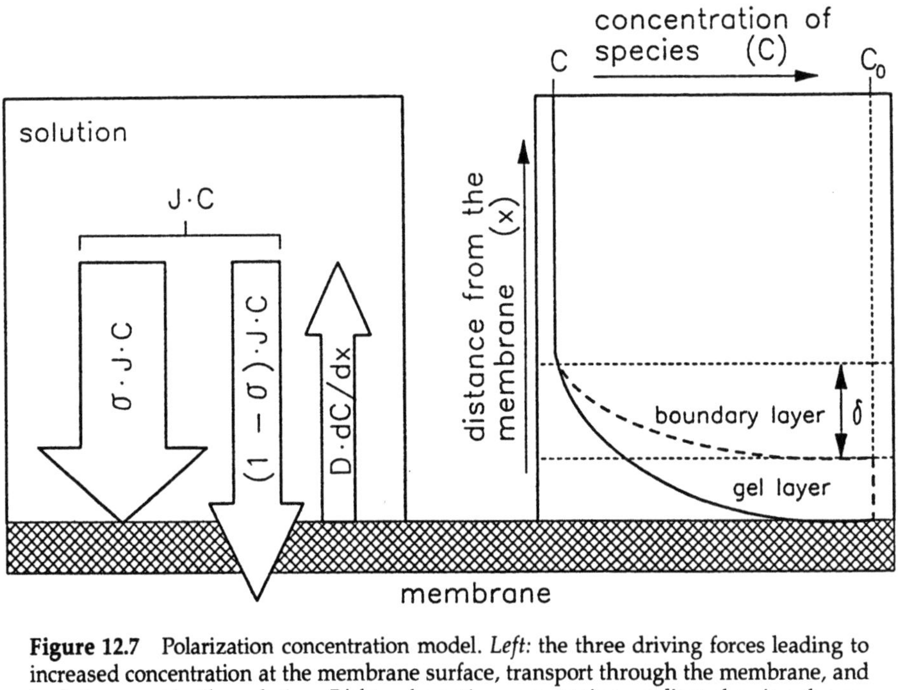


**Figure 12.7** Polarization concentration model. *Left:* the three driving forces leading to increased concentration at the membrane surface, transport through the membrane, and back-transport to the solution. *Right:* schematic concentration gradient showing the existence of a boundary layer, and, in extreme cases, a gel layer (dashed curve) at the membrane surface; C, concentration in the bulk solution; $C_0$, concentration at the membrane surface.

surface area, $\sigma$ is the retention coefficient ($\sigma = 1$ for fully retained particles; $\sigma = 0$ for particles not retained at all), $\delta$ is the diffusion layer thickness (cm), and D ($cm^2 \cdot s^{-1}$) is the diffusion coefficient of the particle. This equation enables one to estimate that $C_0$ may take on values many orders of magnitude larger than C if the calculation is made for particle sizes of 0.04–0.3 $\mu$m (Figure 12.3) and with solution flow rates typical of those used in syringe filtration (100-300 $ml \cdot min^{-1}$) (Buffle et al., 1988b). With such a high particle concentration, coagulation becomes very rapid and produces agglomerates, which are deposited on the membrane and may finally form a clogging gel (Figures 12.6 and 12.7). Note that, in contrast to the specific adsorption effect, the coagulation effect is expected to be rather independent of the chemical nature of the membrane. Specific adsorption effects are more difficult to minimize than coagulation ones, but our results suggest that the latter are often predominant when filtration conditions are not chosen with care.

To minimize coagulation effects, all parameters must be selected so as to decrease the $C_0/C$ ratio. The above equation shows that this can be achieved by:

- Using very low solution flow-rates, J, through the membrane. For filtrations involving large volumes this necessarily implies that the filtering surface area be increased (e.g., membranes with large surfaces, hollow fibers) if the filtration time is to remain acceptable; this, however, also increases the risks of specific surface adsorption.
- Increasing the diffusion flux of particles back into solution by decreasing $\delta$ through induced turbulency (e.g., by efficient stirring at the membrane surface). Note, however, that turbulency has the simultaneous adverse effect of increasing the rate of orthokinetic coagulation of colloids in solution.
- Limiting the concentration factor, $V_0/V_f$, where $V_0$ and $V_f$ are, respectively, the volumes in the filtration cell initially and at the end of filtration. This prevents the bulk concentration from increasing excessively.

It should be noted that, at present, none of these conditions are normally applied in the case of routine use of syringe filtration.

**Tests of Filtration Factors for Iron Particles**

*Results* The above factors have been systematically tested by studying the results of filtration of the iron particles formed at the redox boundary (see Figures 12.1 and 12.4). Figure 12.8 shows the (apparent) size distributions of these colloids obtained by two different filtration modes. In syringe filtration (Figure 12.8*a*), the untreated water sample is filtered on membranes with different pore sizes, without solution stirring and with uncontrolled but "normal" flow rates (200 $ml \cdot min^{-1} = 1\ cm \cdot s^{-1}$ for a filter area of 3.3 $cm^2$; note that these

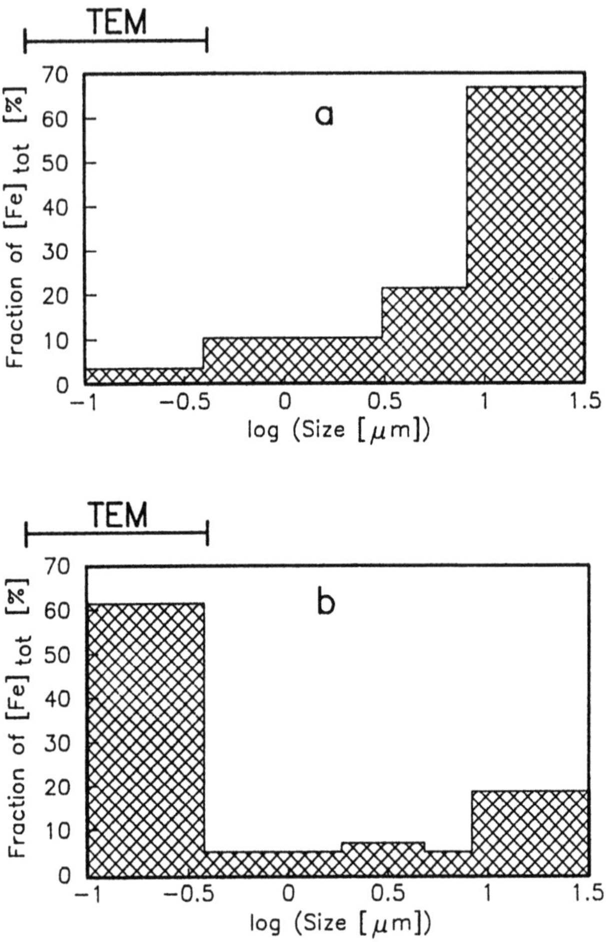


**Figure 12.8** Size distributions of iron-rich colloids in Lake Bret obtained by (*a*) syringe filtration (cellulose nitrate membranes, fast flow rate, no cascade, no stirring, field filtration), and (*b*) cascade filtration (polycarbonate membranes, low flow rate, cascade mode, stirring, in situ filtration).

values are high compared to those discussed below). In this case, size fractions are computed by differences between the various filtrates. At depth cascade filtration (Figure 12.8*b*; De Vitre et al., 1987) was performed with a custom-made submersible apparatus. In this case, the same water sample is passed successively through a series of membranes with decreasing porosities under conditions of slower flow rates for successive membranes, stirring at the membrane surface, and under an $N_2$ atmosphere. Clearly the use of these conditions, which all favor minimization of artifacts, yields particle sizes which are about one order of magnitude smaller than the ones given by syringe filtration.

The fact that the above difference is mainly due to surface coagulation has been shown by TEM observations of membranes used at different flow rates; an example can be seen in Figure 12.9 which shows the nature of the "particles" retained on filters. By using scanning electron microscopy (SEM; low resolution) to get a top view of the surface membrane after filtration, particles of a few μm seem to be retained (Figure 12.9*a*) while, when TEM (high resolution) is used to get profile micrographs of thin sections through the membrane (Figure 12.9*b*), it is clear that these "particles" are, in fact, agglomerates of much smaller ones.

Figure 12.10 shows the separate effects of flow rate (J), concentration factor ($V_0/V_f$), and nature of the membrane, these being discussed below. It should simply be noted that Figure 12.10*b* shows the fraction of iron retained on the membrane, even though for the membrane pore size used (3 μm), none of the iron particles (see Figure 12.3) should be retained based on their size. Any retention is therefore due to secondary effects such as those given above (retention by adsorption, coagulation).

It must also be stressed that, due to the large heterogeneity of water samples (see Figure 12.5), the reproducibility of the results (over more than 150 filtration operations) was found to be about 10%. This implies that any filtration operation must necessarily be repeated several times for statistically

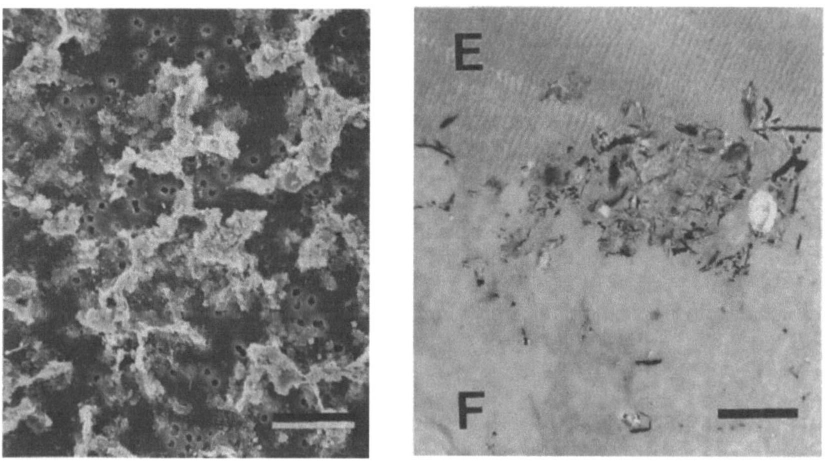


a b

**Figure 12.9** (*a*) Scanning electron micrograph and (*b*) transmission electron micrograph of particles retained on filtration membranes, showing the aggregation and coagulation problems produced during filtration, and the possible overestimation of particle size when using SEM (low resolution) instead of TEM. SEM picture shows a membrane surface, while TEM picture corresponds to a perpendicular view of a membrane surface (F = filter material; E = external solution on the top part of the membrane); clearly, the "pack" retained by the membrane is an aggregate of very small particles which should have been able to pass through the pores. Bar in a) is 30 μm long. Bar in b) is 3 μm long.

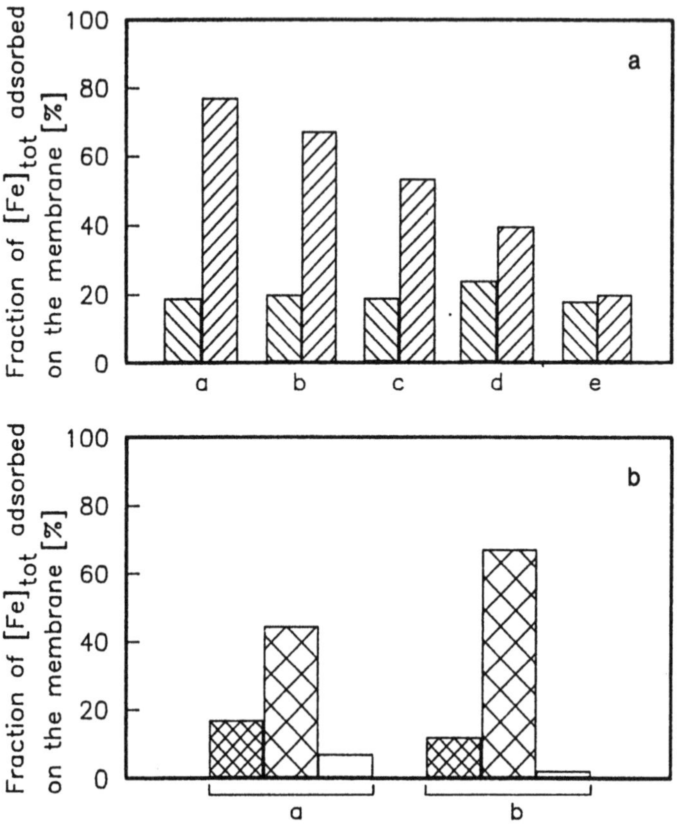


**Figure 12.10** (*a*) Effect of the flow rate on the fraction of iron retained on the membrane. Membrane types: a, polycarbonate; b, cellulose nitrate; c, cellulose acetate; d, polyvinylidene fluoride; e, acrylic copolymer. Nominal pore size is 0.2 $\mu$m; = 1 cm·h$^{-1}$, = 20 cm·h$^{-1}$. (*b*) Effect of the ratio of $V_0/V_f$ (= initial volume/final volume in the cell) on the fraction of iron retained on the membrane. Membrane types: a, polycarbonate; b, cellulose nitrate. Nominal pore size is 3 $\mu$m; = 1.1 (20 ml/18 ml), = 5.0 (20 ml/4 ml), = 1.7 (200 ml/120 ml); in the third case, stirring was more efficient, leading to a lower value of iron retained on the membrane.

valid results to be obtained. All results presented here are averages of triplicates.

*Discussion* The results of cascade filtration are much closer to those of TEM than those of syringe filtration (see Figures 12.3 and 12.8). As suggested before, this is due to the fact that, in the former case, all conditions favor the minimization of artifacts: oxygen contamination—and therefore possible oxidation of Fe(II)—is avoided by at depth operation; rather low solution flow rate is imposed by the less porous membrane; efficient stirring decreases $\delta$; and $V_0/V_f$ never exceeds five. In addition, cascade filtration progressively eliminates the large particles from the successive filtrates, thus decreasing the probability of coagulation.

Figure 12.10*a* shows how the flow rate, J, is a predominant factor in producing artifacts by surface coagulation irrespective of the nature of the membrane. SEM observations suggest that with the smallest flow rate (1 $cm \cdot h^{-1}$) the surface coagulation process is negligible. It must be noted, however, that, by using such very low flow rates for filtering large volumes, the time necessary to perform the filtration becomes so long that coagulation may occur in the filtration cell itself (Perret, unpublished results).

These results have also shown (Buffle et al., 1988b) that stirring somewhat decreases the proportion of Fe retained on the membrane (i.e., the coagulation effect), but only by 10% (with the classical magnetic stirring used). More efficient turbulency might be created at the membrane surface (Matthiasson and Sivik, 1980), but, as mentioned above, this might also favor coagulation in solution. Figure 12.10*b* shows that, as expected from the artifacts interpretation based on coagulation, an increase in the concentration factor, $V_0/V_f$, leads to an increase in iron retention.

Finally, the effect of the chemical nature of the membrane must be considered. The data in Figure 12.10*a* indeed show that retention significantly depends on the membrane nature and seem to suggest that polycarbonate membranes (Nuclepore) are the worst offenders of the membranes tested. Taken by themselves, the results shown in Figure 12.10*a* are, however, misleading in that respect. Indeed other data (Perret, unpublished results) show that an inverse effect of the membrane nature is observed when membranes of higher porosity (3 $\mu m$) are used. Although the exact reason of this inverse behavior is not yet fully understood, the important point to stress here is that the overall retention in cascade filtration is always high at high flow rates, irrespective of the chemical nature of the membrane.

**Conclusions** All of the above results show that important precautions must be taken if iron colloids are to be filtered without artifacts. Specifically, filtration must be performed at low and controlled flow rates ($\leq 10$ $cm \cdot h^{-1}$), preferably in cascade, with small concentration factors, in stirred solutions, and under an $N_2$ atmosphere. Although iron colloids might be particularly sensitive to coagulation due to their low electrical charge at the lake pH (7.5), it is quite likely that these precautions are applicable to most aquatic colloids and, indeed, these results are in agreement with previous observations (Danielson, 1982).

These results have important implications concerning the use of filtration for size distribution measurements, whether using cascade filtration followed by chemical analysis of the filtrates or by microscopic observations of particles retained on filters. In the latter case, it has already been discussed that the retained "particles" observed by SEM (see Figure 12.9) are in fact agglomerates of much smaller colloids. Great caution must therefore be used when determining size distributions from SEM observation of particles on filters, as is common practice. The use of cascade filtration for size distribution measurements must also be accompanied by important precautions. Tests of the influence of filtration conditions, in particular flow rate, is highly recommended. In any case, TEM is certainly by far the best method presently available and a few TEM observations, at least, should complement any other method. On the other hand, filtration remains a useful tool for operational fractionation, provided that the above filtration conditions are well controlled to minimize artifacts and that particle size is not inferred from the pore size of the membrane used. It must be stressed that well controlled filtration conditions imply, in particular, *avoiding* syringe filtration, because of its high and ill-controlled flow rate.

## References

Balistieri, L.S. and Murray, J.W. 1982. The adsorption of Cu, Pb, Zn, and Cd on goethite from major ion seawater. *Geochim. Cosmochim. Acta* 46:1253–1265.

Bloesch, J. and Burns, N.M. 1980. A critical review of sediment trap technique. *Schweiz. Z. Hydrol.* 42:15–55.

Buffle, J., Zali, O., Zumstein, J., and De Vitre, R.R. 1987. Analytical methods for the direct determination of inorganic and organic species: seasonal changes of iron, sulfur, and pedogenic and aquagenic organic constituents in the eutrophic Lake Bret, Switzerland. *Sci. Tot. Environ.* 64:41–59.

Buffle, J. 1988a. *Complexation Reactions in Aquatic Systems: an Analytical Approach*, Ellis Horwood, U.K., 710 pp.

Buffle, J., De Vitre, R.R., Perret, D., and Leppard, G.G. 1988b. Combining field mesurements for speciation in non-perturbable water samples: application to the iron and sulfide cycles in a eutrophic lake. *In* Kramer, J.R. and Allen, H.E. (eds.), *Metal Speciation: Theory, Analysis and Application.* Lewis Publishers, Michigan, Chapter 5.

Buffle, J., De Vitre, R.R., Perret, D., and Leppard, G.G. 1989. Physico-chemical characteristics of a colloidal iron phosphate species formed at the oxic-anoxic interface of an eutrophic lake. *Geochim. Cosmochim. Acta* 53:399–408.

Crosby, S.A., Glasson, D.R., Cuttler, A.H., Butler, I., Turner, D.R., Whitfield, M., and Millward, G.E. 1983. Surface areas and porosities of Fe(III) and Fe(II) derived oxyhydroxides. *Environ. Sci. Technol.* 17:709–713.

Danielson, L.G. 1982. On the use of filters for distinguishing between dissolved and particulate fractions in natural waters. *Water Res.* 16:179–182.

Davison, W., Woof, C., and Rigg, E. 1982. The dynamics of iron and manganese in a seasonally anoxic lake: direct measurement of fluxes using sediment traps. *Limnol. Oceanogr.* 27:987–1003.

Davison, W. and Seed, G. 1983. The kinetic of the oxidation of ferrous iron in synthetic and natural waters. *Geochim. Cosmochim. Acta* 47:67–79.

Davison, W. and Dickson, D.P.E. 1984a. Mössbauer spectroscopic and chemical studies of particulate iron material from a seasonally anoxic lake. *Chem. Geol.* 42:177–187.

Davison, W. and Woof, C. 1984b. Study of the cycling of manganese and other elements in a seasonally anoxic lake, Rostherne Mere, U.K. *Water Res.* 18:727–734.

Davison, W. 1985. Conceptual models for transport at a redox boundary. *In* Stumm, W. (editor), *Chemical Processes in Lakes.* J. Wiley and sons, N.Y., p. 31–49.

De Vitre, R.R., Bujard, F.J., Bernard, C., and Buffle, J. 1987. A novel *in situ* cascade ultrafiltration unit specifically designed for field studies of anoxic waters. *Int. J. Env. Anal. Chem.* 31:145–163.

De Vitre, R.R., Buffle, J., Perret, D., and Baudat, R. 1988. A study of iron and manganese

transformations at the $O_2/S(-II)$ transition layer in a eutrophic lake (Lake Bret, Switzerland): a multimethod approach. *Geochim. Cosmochim. Acta* 52:1601–1613.
Fane, A.G., Fell, C.J.D., and Waters, A.G. 1983. Ultrafiltration of protein solutions through partially permeable membranes: the effect of adsorption and solution environment. *J. Membr. Sci.* 16:211–224.
Green, G. and Belfort, G. 1980. Fouling of ultrafiltration membranes: lateral migration and the particle trajectory model. *Desalination* 35:129–147.
Kavanaugh, M.C., Zimmermann, U., and Vagenknecht, A. 1977. Determination of particle size distributions in natural waters: use of a Zeiss Micro-Videomat image analyser. *Schweiz. Z. Hydrol.* 39:86–98.
Kavanaugh, M.C., Tate, C.H., Trussell, A.R., Trussell, R.R., and Treweek, G. 1980. Use of particle size distribution measurements for selection and control of solid/liquid separation processes. *In* Kavanaugh, M.C. and Leckie, J.O. (editor), *Particulates in Water.* Adv. Chem. Ser. 189, ACS, Washington, DC, Chapter 14.
Koenings, J.P. 1976. *In situ* experiments on the dissolved and colloidal state of iron in an acid bog lake. *Limnol. Oceanogr.* 21:674–683.
Kozinski, A.A. and Lightfoot, E.N. 1972. Protein ultrafiltration: a general example of boundary layer filtration. *A.I.Ch.E.J.* 18:1030–1040.
Laxen, D.P.H. and Chandler, I.M. 1982. Comparison of filtration techniques for size distribution in freshwaters. *Anal. Chem.* 54:1350–1355.
Laxen, D.P.H. and Chandler, I.M. 1983. Size distribution of iron and manganese species in freshwaters. *Geochim. Cosmochim. Acta* 47:731–741.
Leppard, G.G., Buffle, J., De Vitre, R.R., and Perret, D. 1988. The ultrastructure and physical characteristics of a distinctive colloidal iron particulate isolated from a small eutrophic lake. *Arch. Hydrobiol.* 113:405–424.
Lerman, A. 1979. *Geochemical Processes,* J. Wiley and sons, N.Y., 481 pp.
Lijklema, L. 1980. Interaction of orthophosphate with iron (III) and aluminum hydroxides. *Environ. Sci. Technol.* 14:537–540.
Matthiasson, E. and Sivik, B. 1980. Concentration polarisation and fouling. *Desalination* 35:59–103.
Matthiasson, E. 1983. The role of macromolecular adsorption in fouling of ultrafiltration membranes. *J. Membr. Sci.* 16:23–36.
Mayer, L.M., Liotta, F.P., and Norton, S.A. 1982. Hypolimnetic redox and phosphorus cycling in hypereutrophic Lake Sebasticook, Maine. *Water Res.* 16:1189–1196.
Meares, P. 1976. The physical chemistry of transport and separation by membranes. *In* Meares, P. (editor), *Membrane Separation Processes*, Elsevier, Amsterdam, Chapter 1.
Michaels, A.S. 1968. New separation technique for the CPI. *Chem. Eng. Progr.* 64:31–43.
Pysik, A.J. and Sommer, S.E. 1981. Sedimentary iron monosulfides: kinetics and mechanisms of formation. *Geochim. Cosmochim. Acta* 45:687–698.
Rees, T.F. 1987. A review of light scattering techniques for the study of colloids in natural waters. *J. Contamin. Hydrol.* 1:425–439.
Salim, R. and Cooksey, B.G. 1981. The effect of centrifugation on the suspended particles of river waters. *Water Res.* 15:835–839.
Sarikaya, H.Z. 1979. Interactions between ferrous iron oxidation and phosphate. *In* Curi, K. (editor), *Treatment Disposal of Liquid and Solid Industrial Wastes,* 3rd German Environmental Engineering Symposium, Pergamon Press, Oxford, p. 143–160.
Schneider, W. and Schwyn, B. 1987. The hydrolysis of iron in synthetic, biological and aquatic media. *In* Stumm, W. (editor), *Aquatic Surface Chemistry,* J. Wiley and sons, N.Y., p. 167–194.
Schneider, W. 1988. Iron hydrolysis and the biochemistry of iron: the interplay of hydroxide and biogenic ligands. *Chimia* 42:9–20.
Sigg, L. 1985. Metal transfer mechanisms in lakes: the role of settling particles. *In* Stumm, W. (editor), *Chemical Processes in Lakes,* J. Wiley and sons, N.Y., p. 283–310.

Simpson, W.R. 1982. Particulate matter in the oceans: sampling methods, concentration, size distribution and particle dynamics. *Oceanogr. Mar. Biol., Ann. Rev.* 20:119–172.

Stone, A.T. and Morgan, J.J. 1987. Reductive dissolution of metal oxides. *In* Stumm, W. (editor), *Aquatic Surface Chemistry*, J. Wiley and sons, N.Y., p. 221–252.

Suki, A., Fane, A.G., and Fell, C.J.D. 1984. Flux decline in protein ultrafiltration. *J. Membr. Sci.* 21:269–283.

Sung, W. and Morgan, J.J. 1980. Kinetics and products of ferrous iron oxygenation in aqueous systems. *Environ. Sci. Technol.* 14:561–568.

Swallow, K.C., Hume, D.N., and Morel, F.M.M. 1980. Sorption of Cu and Pb by hydrous ferric oxide. *Environ. Sci. Technol.* 14:1326–1331.

Tipping, E. 1981a. The adsorption of aquatic humic substances by iron oxides. *Geochim. Cosmochim. Acta* 45:191–199.

Tipping, E., Woof, C., and Cooke, D. 1981b. Iron oxide from a seasonally anoxic lake. *Geochim. Cosmochim. Acta* 45:1411–1419.

Trettin, D.R. and Doshi, M.R. 1979. Limiting flux in ultrafiltration of macromolecular solutions. *Chem. Eng. Commun.* 4:507–522.

Truitt, R.E. and Weber, J.H. 1979. Trace metal ion filtration losses at pH 5 and 7. *Anal. Chem.* 51:2057–2059.

Waite, T.D. and Morel, F.M.M. 1984. Photoreductive dissolution of colloidal iron oxides in natural waters. *Environ. Sci. Technol.* 18:860–868.

Weilenmann, U. 1986. *The role of coagulation for the removal of particles by sedimentation in lakes.* Ph.D. Thesis 8018, Zürich Polytechnical School (Switzerland).

Whitfield, M. and Turner, D.R. 1987. The role of particles in regulating the composition of sea water. *In* Stumm, W. (editor), *Aquatic Surface Chemistry*, J. Wiley and sons, N.Y., p. 457–488.

Zinder, B., Furrer, G., and Stumm, W. 1986. The coordination chemistry of weathering; 2) dissolution of Fe(III) oxides. *Geochim. Cosmochim. Acta* 50:1861–1869.

# 13

# Pathways and Residence Times of Radiotracers in Lake Constance

A. Mangini, U. Christian, M. Barth, W. Schmitz, and H. H. Stabel

ABSTRACT We obtained information on the behavior of Cs and Ru isotopes in Lake Constance derived from the reactor accident at Chernobyl from measurements of settling fluxes, sediment cores, and water profiles. Sediment trap data suggest that a part of Cs (32%) and a larger fraction of Ru (53%) were scavenged by settling particles shortly after the fallout reached the lake surface. The overall epilimnetic residence times in May were 12–17 days (Cs) and 8–13 days (Ru). For the entire summer, the residence times are estimated to be 33–53 days for Cs and 15–38 days for Ru. The residence times of the Chernobyl radiotracers in the epilimnion ($\approx$20 m) of Lake Constance were close to those of Pb-210 (30–40 days), as obtained from previous investigations. In the hypolimnion ($\approx$80 m), the residence time for dissolved Pb-210 of 1–1.5 years is longer than the residence time of Cs-137 (5.5 months), suggesting that settling particles scavenged Pb-210 less effectively than Cs-137. Sediment cores show a deficiency of (Chernobyl-derived) Cs-137 in comparison to Ru in the deepest part of the lake and enrichment of Cs in sediments at shallower depth, reflecting the longer residence time of Cs in the water column. The inventories of radiotracers Pb-210 and (bomb-derived) Cs-137 in the sediments suggest irregular accumulation of sediment on the bottom of Lake Constance.

## 13.1 Introduction

Radioactive elements, such as Pb-210 (a daughter of atmospheric radon of natural origin) and Cs-137 (from nuclear tests and nuclear plant leakages), are useful tracers for understanding the behavior of anthropogenic pollutants in marine and lake systems. The questions of interest for trace elements entering the aquatic systems are:

1. How are trace elements scavenged from the water column?
2. How long does it take until they are completely removed from the water column?

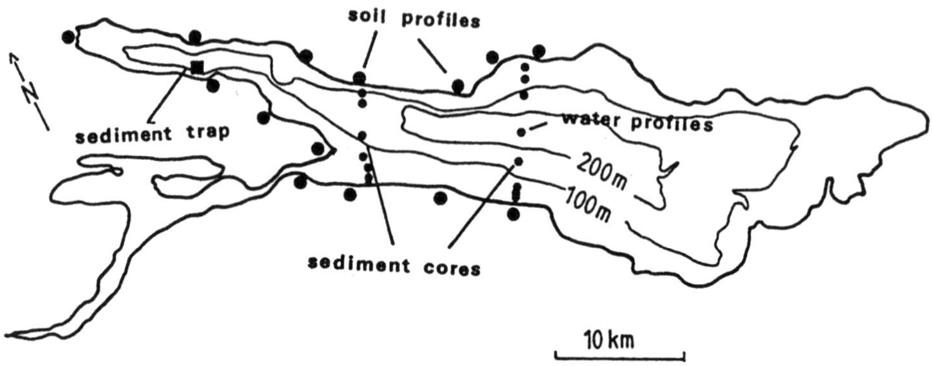


**Figure 13.1** Locations in Lake Constance of the soil and water profiles, the sediment cores, and the sediment trap sampled for the study of the pathways of Cs and Ru isotopes from the Chernobyl reactor accident.

3. Does the depositional flux into sediments vary in different areas of the lake?
4. Does the depositional flux into sediments vary with lake size?

The advantage of performing these studies on radionuclides instead of using stable trace elements is based on the well-known input functions of natural tracers and nuclear products, and on the very sensitive detection of radioactivity.

Following the Chernobyl reactor accident, we obtained information on the behavior of the radioisotopes Cs-137, Cs-134, Ru-103 and Ru-106 in Lake Constance by measuring settling fluxes, sediment cores, and water profiles at the locations shown in Figure 13.1. The residence times derived from these radiotracers are compared with those for Pb-210 obtained from our earlier studies (Dominik et al., 1981; Appel, 1982; Barth, 1984).

## 13.2 Cs-137 from Nuclear Bomb Tests and Cs and Ru Isotopes from the Chernobyl Reactor Accident

The first appearance of Cs-137 in central European lake sediments corresponds to the beginning of nuclear tests in 1953. The Cs-137 fallout rate then continuously increased until 1963, the time marker corresponding to the nuclear test ban treaty. In the period after 1963, the Cs-137 concentration in sediments generally dropped to lower values, reflecting smaller atmospheric concentrations. The deposition of bomb-Cs-137 in the Lake Constance area amounts to 3,600 ± 1300 $Bq/m^2$ (decay-corrected), determined from soil profiles sampled near the cities of Konstanz and Lindau (Hemmerich, 1980; Appel, 1982). This standing crop of Cs-137 is expected in the sediment of Lake Constance. The plot of the observed decay-corrected standing crop of Cs-137 against the sediment accumulation rate (Figure 13.2) shows that the standing crop of Cs-137 is proportional to the sediment accumulation rate. Some cores are deficient in Cs-137; other cores have Cs-137 in excess of the atmospheric deposition. This suggests that the depositional fluxes of Cs-137 and the sedimentation rate differed at different localities in the lake.

In recent years, there have been several accidents involving nuclear power stations, the Chernobyl accident of April 26, 1986 being the most spectacular. Radioactive fallout reached Lake Constance (central Europe) on May 1, 1986. Aerosols loaded with radioactive material were washed out by heavy rainfall. Total areal deposition of Cs-137 on Lake Constance from soil profiles was 17,000 ± 3,000 Bq/m², ranging from about 14,000 Bq/m² at the southern shoreline to 28,000 Bq/m² at the northern shoreline of the lake. Total deposition of Ru-103 was determined to be 22,000 Bq/m², the areal patchiness being similar to that of Cs-137.

The ratios of the various isotopes were calculated to be as follows:

Cs-137/Cs-134 = 1.98 ± 0.05
Ru-103/Ru-106 = 4.4 ± 0.2
Ru-103/Cs-134 = 2.5 ± 0.2

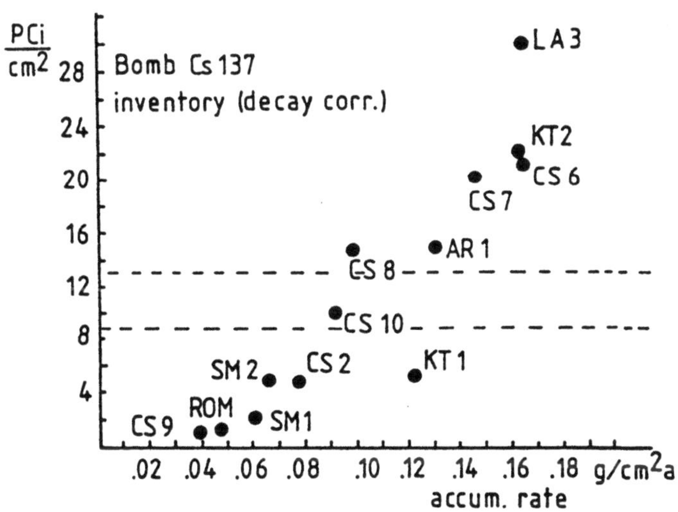


**Figure 13.2** Plot of the standing crop of Cs-137 (decay corrected) in sediment cores from different localities in Lake Constance against the sediment accumulation rate (data from Dominik et al., 1981, and Christian, 1987). The horizontal lines indicate the range of standing crops expected from the fallout of nuclear bomb tests (data from Appel, 1982; and Hemmerich, 1980). The Cs-137 data from Table 3 in Dominik et al. (1981) were multiplied by a factor of 2 to compensate for an error in the original data (H. Livingston, 1982, and personal communication; and Christian, 1987).

These longer-lived radioisotopes are powerful tracers for geochemical studies in the lake. At least two different types of studies can be performed: 1) A short-term experiment to determine how fast these radiotracers are scavenged from the water column and transported into sediments, and 2) a study of the accumulation rate of the radiotracers in different areas of the lake.

**The Flux of Chernobyl Cs and Ru into Sedimentation Traps** Sedimentation traps were deployed at 20-m and 120-m depth in the Überlingersee, a small bay of Lake Constance. During the period of consideration these traps were sampled weekly. At 20-m depth (i.e., the maximum extension of the euphotic zone), the trap was deployed beginning on March 5, whereas the deeper trap was exposed from mid-June onwards. Therefore, only the upper trap provided a complete record of the sedimentation process during the period shortly after the Chernobyl radiotracer input. The trap configuration and sample preparation are described by Kleiner et al. (1984). In spring and summer 1986, there were two peaks of high fluxes of settling material in Lake Constance, the first culminating about mid-May whereas the latter occurred in mid-August (Figure 13.3*a*). At both events, the sinking particles consisted predominantly of calcium carbonate, while organic matter (9 to 13.5% of dry weight) and silicon (diatomaceous frustules) were of minor importance. In mid-June, there was an allochtonous input of minerogenic particles (as indicated by the aluminum content) which contributed only a few percent of the total annual settling flux of dry weight material. These seasonal patterns of sedimentation fluxes were similar to those found earlier (Stabel, 1985).

In 1986, the authigenic precipitation of calcite, diatomaceous silicon, and particulate organic matter was well separated in time from the allochtonous input, which in Lake Constance is triggered by melt water deriving from the alpine snow cover. This separation was favored by the meteorological conditions at the end of April 1986, when a long period of high pressure over central Europe induced temperatures above-average and allowed the fallout cloud to drift westward (Smith and Clark, 1986). Of course, there is considerable variation in the regional decomposition rate of the various materials at the lake bottom due to changes in the contribution of allochtonous material with distance from the mouth of the Alpenrhein, the main tributary. At the locality where the traps were deployed, calcite amounts to 36% of the total bottom sediment matter, whereas in the central basin of Lake Constance, the calcite content amounts to about 20–30% (Dominik et al., 1981).

The fluxes of Cs and Ru into the 20-m trap in the period from the beginning of May to September (Tables 13.1 and 13.2) were approximately 32% and 53% of the depositional fluxes measured in nearby soil profiles on the southern shore. The fluxes of Ruthenium into the trap paralleled that of dry weight matter (Figures 13.3*a* and 13.3*b*), with a maximum during the first two weeks and a steep decrease in the following weeks. The flux of Cs-137 (Figure 13.3*c*) also exhibited a maximum during the first two weeks; however, the decrease in the successive period was not as steep as for Ru. We conclude that

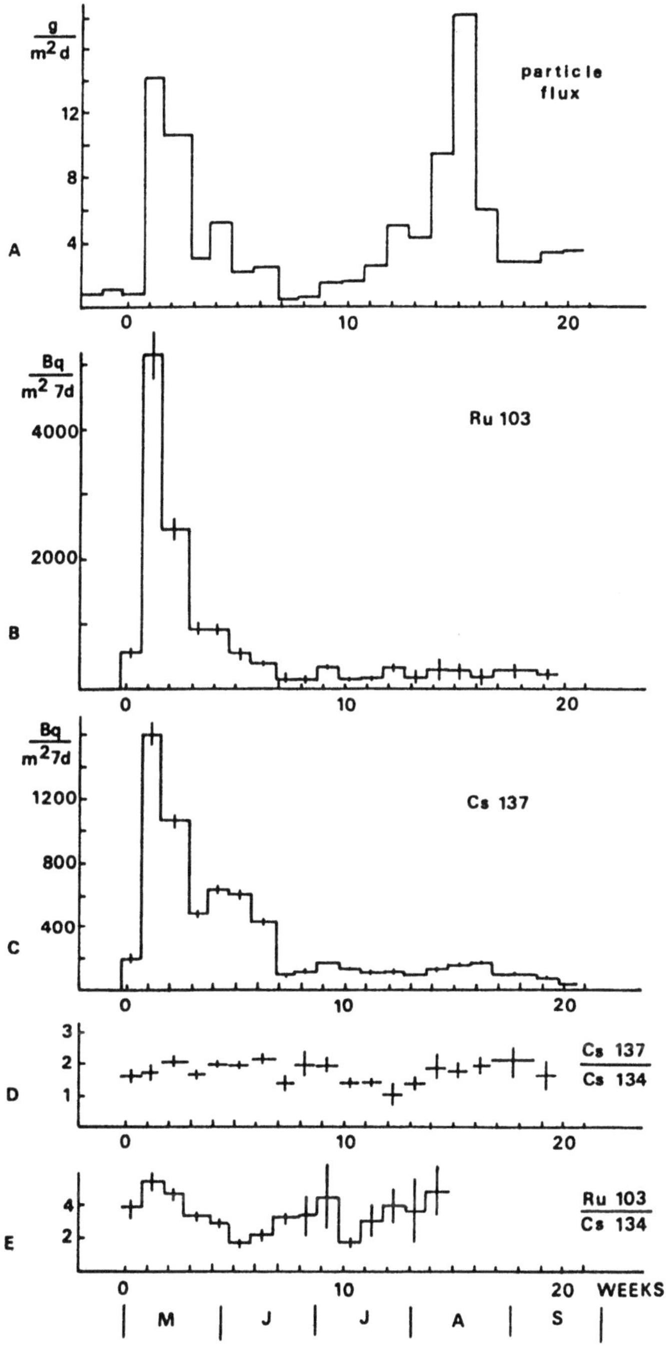


**Figure 13.3** Fluxes of particles and isotopes into the sediment trap deployed at 20 m in the Überlingersee in the weeks after the input of the Chernobyl fallout. Values are decay-corrected to May 1, 1986. (*a*) Total particles. (*b*) Ru-103. (*c*) Cs-137. (*d*) Activity ratio of Cs-137 to Cs-134. (*e*) Activity ratio of Ru-103 to Cs-134

**Table 13.1** Cs-137 and Ru-103 fluxes (decay-corrected to May 1) into the sedimentation traps at 20 m depth in the Überlingersee

| Sampling interval | Particle flux (mg/m² d) | Cs-137 (Bq/m² 7d)* | Cs-134 (Bq/m² 7d)* | Cs-137/Cs-134 | Ru-103 (Bq/m² 7d)* | Ru-103/Cs-134 |
|---|---|---|---|---|---|---|
| 28.4– 6.5 | 949.3 | 190 ± 20 | 120 ± 20 | 1.6 ± 0.3 | 470 ± 90 | 3.9 ± 0.7 |
| 6.5–12.5 | 14296.3 | 1540 ± 90 | 930 ± 100 | 1.7 ± 0.2 | 5120 ± 460 | 5.5 ± 0.5 |
| 12.5–21.5 | 10687.8 | 1070 ± 30 | 535 ± 30 | 2.00 ± 0.13 | 2500 ± 170 | 4.7 ± 0.3 |
| 21.5–27.5 | 3102.8 | 445 ± 20 | 275 ± 16 | 1.65 ± 0.13 | 930 ± 100 | 3.4 ± 0.2 |
| 27.5– 3.6 | 5360.7 | 605 ± 15 | 308 ± 10 | 1.96 ± 0.10 | 920 ± 55 | 3.0 ± 0.3 |
| 3.6–10.6 | 2194.3 | 570 ± 20 | 300 ± 15 | 1.90 ± 0.15 | 525 ± 90 | 1.8 ± 0.1 |
| 10.6–18.6 | 2515.1 | 400 ± 12 | 185 ± 9 | 2.2 ± 0.2 | 430 ± 50 | 2.3 ± 0.4 |
| 18.6–24.6 | 521.6 | 60 ± 6 | 35 ± 8 | 1.7 ± 0.4 | 120 ± 60 | 3.4 ± 1.7 |
| 24.6– 1.7 | 684.7 | 80 ± 8 | 40 ± 9 | 2.0 ± 0.4 | 140 ± 60 | 3.5 ± 1.5 |
| 1.7– 8.7 | 1491.4 | 140 ± 10 | 70 ± 9 | 2.0 ± 0.2 | 320 ± 50 | 4.6 ± 1.8 |
| 8.7–16.7 | 1629.4 | 100 ± 7 | 70 ± 8 | 1.4 ± 0.2 | 130 ± 40 | 1.8 ± 0.4 |
| 16.7–23.7 | 2597.0 | 80 ± 6 | 55 ± 6 | 1.5 ± 0.2 | 180 ± 40 | 3.2 ± 1.0 |
| 23.7–30.7 | 5048.1 | 100 ± 9 | 70 ± 10 | 1.4 ± 0.2 | 290 ± 70 | 4.1 ± 1.0 |
| 30.7– 5.8 | 4346.5 | 65 ± 7 | 40 ± 6 | 1.6 ± 0.3 | 150 ± 70 | 3.7 ± 1.8 |
| 5.8–12.8 | 9517.0 | 100 ± 8 | 50 ± 9 | 2.0 ± 0.4 | 200 ± 100 | 4.0 ± 2.0 |
| 12.8–19.8 | 18135.5 | 125 ± 7 | 55 ± 7 | 2.3 ± 0.3 | 280 ± 100 | 5.0 ± 2.0 |
| 19.8–26.8 | 6088.6 | 145 ± 9 | 70 ± 7 | 2.1 ± 0.2 | 140 ± 70 | 2.0 ± 1.0 |
| 26.8– 9.9 | 2817.7 | 90 ± 9 | 35 ± 8 | 2.6 ± 0.6 | — | — |
| 9.9–16.9 | 3468.9 | 45 ± 7 | 25 ± 7 | 1.8 ± 0.5 | — | — |
| 16.9–23.9 | 3571.3 | 30 ± 4 | 20 ± 7 | 1.5 ± 0.5 | — | — |

* The period of exposure of the traps was 7 days.

**Table 13.2** Cs-137 and Ru-103 fluxes (decay-corrected to May 1) into the sedimentation traps at 120 m depth in the Überlingersee

| Sampling interval | Cs-137 ($Bq/m^2$ 7d) | Cs-134 ($Bq/m^2$ 7d) | Cs-137/Cs-134 | Ru-103 ($Bq/m^2$ 7d) | Ru-103/Cs-134 |
|---|---|---|---|---|---|
| 11.6–24.6 | 11 ± 2 | 5 ± 2 | 2.2 ± 0.9 | 30 ± 5 | 6.0 ± 3.0 |
| 24.6– 8.7 | 34 ± 4 | 15 ± 4 | 2.3 ± 0.6 | 130 ± 60 | 8.7 ± 4.0 |
| 8.7–16.7 | 70 ± 6 | 46 ± 6 | 1.5 ± 0.3 | 165 ± 40 | 3.6 ± 0.9 |
| 16.7–23.7 | 427 ± 11 | 230 ± 9 | 1.86 ± 0.10 | 800 ± 240 | 3.5 ± 0.9 |
| 23.7–30.7 | 47 ± 6 | 32 ± 6 | 1.5 ± 0.3 | 50 ± 25 | 1.5 ± 0.8 |
| 30.7– 5.8 | 32 ± 4 | 25 ± 5 | 1.4 ± 0.3 | — | — |
| 5.8–12.8 | 43 ± 4 | 35 ± 6 | 1.3 ± 0.3 | 170 ± 50 | 4.8 ± 1.5 |

there are different co-precipitation behaviors for these elements—Ru seems to adsorb very quickly to the autochtonous particles as compared to Cs and, thus, exhibits properties very similar to that of the particles. In contrast, Cs had a longer residence time in the epilimnion. The longer residence time of Cs as compared to Ru is suggested also by the trend of the Ru-103/Cs-134 ratio with time. The ratio at the time of deposition was 2.5 ± 0.2. As shown in Figure 13.3*e*, the Ru-103/Cs-134 ratio increased to 5.5 during the first two weeks and then decreased to a value of about 2 after the sixth week. Thereafter, it increased again to a value of about 4. From this we conclude that during the first two weeks, Ru was more effectively removed by particles than Cs was. The increase in the Ru-103/Cs-134 ratio to more than 2.5 after the sixth week could be caused either by Ru enrichment or by Cs depletion. One explanation could be that the sedimenting matter contained some remobilized sediments which were enriched in Ru. The average value of the Ru-103/Cs-134 ratio from May to September was 4.2 ± 0.6.

**Chernobyl Cs and Ru in the Sediment Cores and Water Column**

The water profile sampled in summer 1986 (about five months after the input) shows low values for Cs in the epilimnion and in the deeper part of the hypolimnion (Figure 13.4*a*). We assume that the low concentration in the epilimnion was caused by rapid scavenging with particles that removed a larger amount of radionuclides. A higher Cs concentration was recorded in the metalimnion. A similar vertical distribution was obtained by independent measurements performed by Peter Santschi (personal communication). It is likely that Cs originally—after input on May 1—was more or less evenly distributed within the upper strata of Lake Constance. Due to the high authigenic particle formation rate during May (occurring solely within the uppermost water layers), a significant portion of the Cs content was co-precipitated with the solids, thus diminishing the epilimnetic standing crop.

The second profile sampled in winter after the turnover of the lake (Lake Constance is a warm, monomictic lake, which mixes most deeply during March

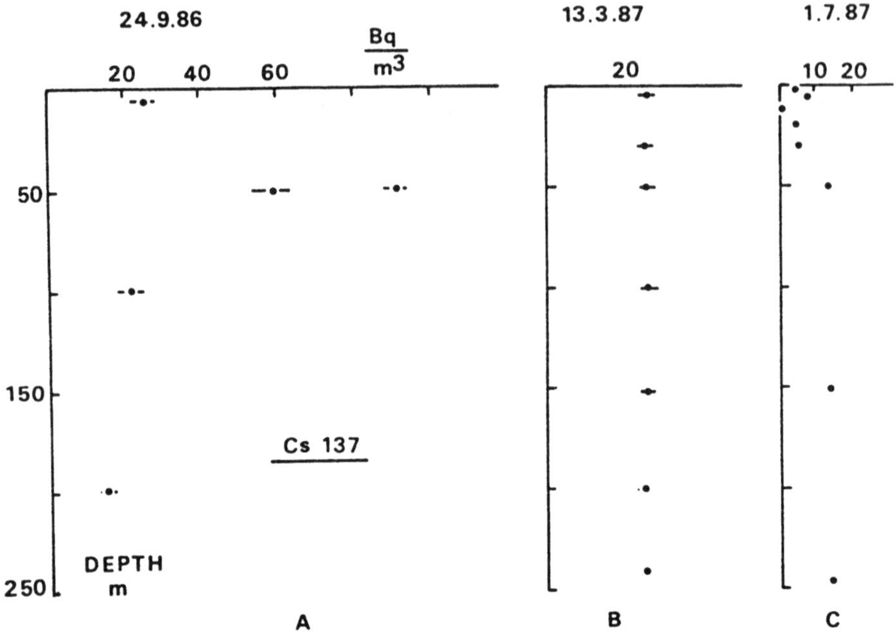


**Figure 13.4** Distribution of Cs-137 in the water column in Lake Constance: (*a*) 6 months, (*b*) 10 months (winter profile), and (*c*) 14 months after the radiotracer input.

and April) shows a nearly constant vertical distribution of Cs-137 (Figure 13.4*b*). The standing crop of Cs in the water column by this time was about 13% of the total deposition input. The decline continued to summer 1987, when only 6% of the initial input remained in the water column. Again, the lowest concentrations were observed in the epilimnion (Figure 13.4*c*).

The elements Cs and Ru differed in their behavior in the water column of Lake Constance, as shown by the inventories of the radionuclides in sediment cores taken on a transect across the deepest part of the lake about 6 months after the radiotracer input. The standing crop of Cs-137 in the sediments in the center of the lake is 7,000 ± 1,000 $Bq/m^2$, which is about 35% of the areal input. In the littoral and in the shallower depths we found 14,200 ± 2,000 $Bq/m^2$, a value close to that in the soil, as shown in Figure 13.5. The activity ratio of Ru-103/Cs-134 was about 4 in the center of the lake (mean of five core samples) and, therefore, was significantly higher than the ratio of the atmospheric precipitation of 2.5 ± 0.2 derived from soil measurements. Instead, there is a similarity with the Ru-103/Cs-134 ratio of 4.4 ± 0.4 measured in the settling material (see above). The activity ratio of Ru-103/Cs-134 of larger than 2.5 indicates that a fractionation between Ru and Cs has occurred in the water column, with Ru being enriched in the particulate phase. The lower Ru-103/Cs-134 ratios measured in the sediments from shallow parts of the lake (i.e., ratios $\leq 2.5$) suggest an unfractionated deposition. The inventory

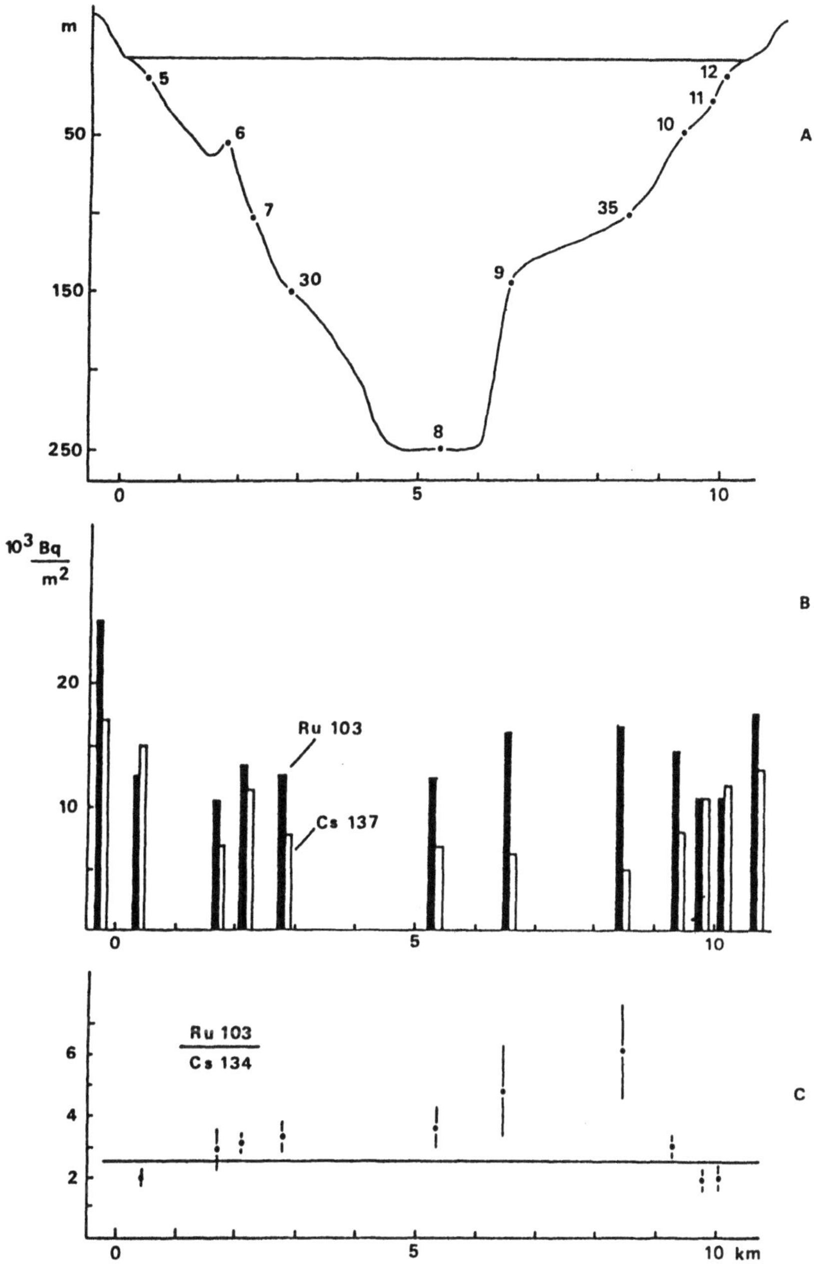


**Figure 13.5** (*a*) Numbers 5 through 35 show the locations of sediment cores taken on a transect across the deepest part of Lake Constance. (*b*) Standing crops of Cs-137 (white bars) and Ru-103 (black bars) in sediment cores. The standing crops in the soil profiles on the northern and southern shore of the lake are the bars on the extreme left and extreme right. (*c*) Ratio of the standing crops of Ru-103 and Cs-134 (decay-corrected to May 1) along this transect. The horizontal line shows the ratio of Ru-103 to Cs-134 of 2.5 at the time of deposition.

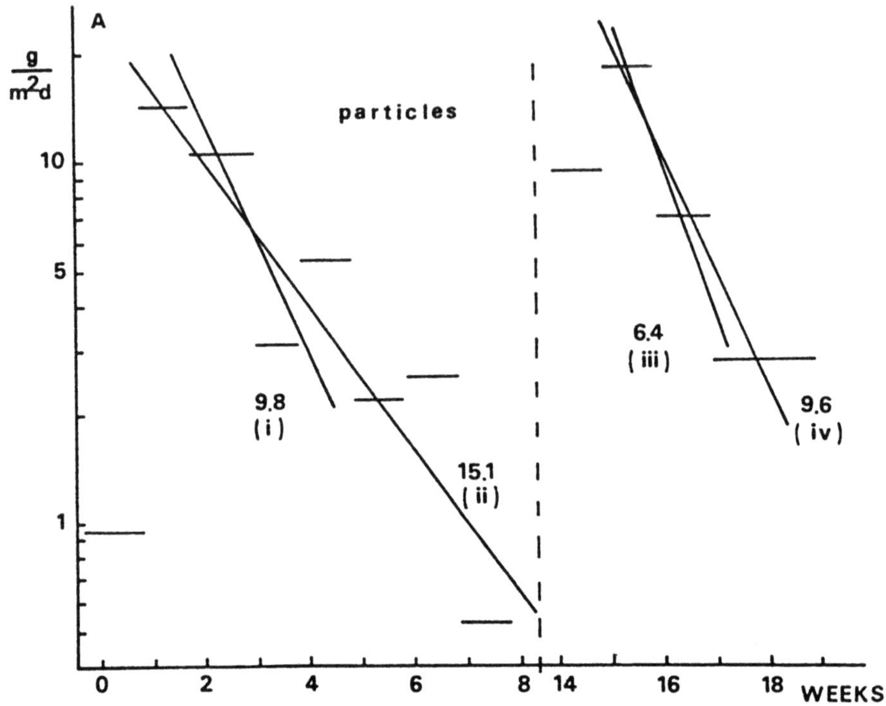


**Figure 13.6** Flux of the settling material (Figure 13.6*a*), Ru-103 (Figure 13.6*b*), and of Cs-137 (Figure 13.6*c*) into the sediment trap at 20 m as a function of time in the weeks after May 1, 1986. The residence times of particles, of Ru-103, and of Cs-137 are derived from the slope of the curves on the semi-log plot. Residence times (in days) were calculated for trap exposure periods from May 6 to May 27 (i), and from May 6 to June 24 (ii). Residence times of particles in the epilimnion were also calculated for another period of increased productivity during summer 1986 [August 12 to August 26 (iii) and August 12 to September 9 (iv)].

of radionuclides in the sediments reflects the differing supply of Cs and Ru in the northern and the southern shores of the lake, which differs by a factor of approximately 1.5 (see Figure 13.5*b*).

**The Residence Times of Cs and Ru in the Water Column** The residence time of the tracer in the system is the time until the concentration of a tracer in a system is reduced to 1/e, or 0.37, of the initial value. The residence time of particles belonging to the two peaks of settling material, which occurred in about mid-May and in mid-August, may be estimated to be 6 to 10 days (Figure 13.6). The overall residence times derived for Cs and Ru in the epilimnion during these periods of high fluxes of settling material for 12–17 days and 8–13 days, respectively. However, since only a total of 32% and 53% of the radionuclides was removed by adsorption onto particles, and, obviously, most of it was removed by some horizontal pathway directly into the shallow

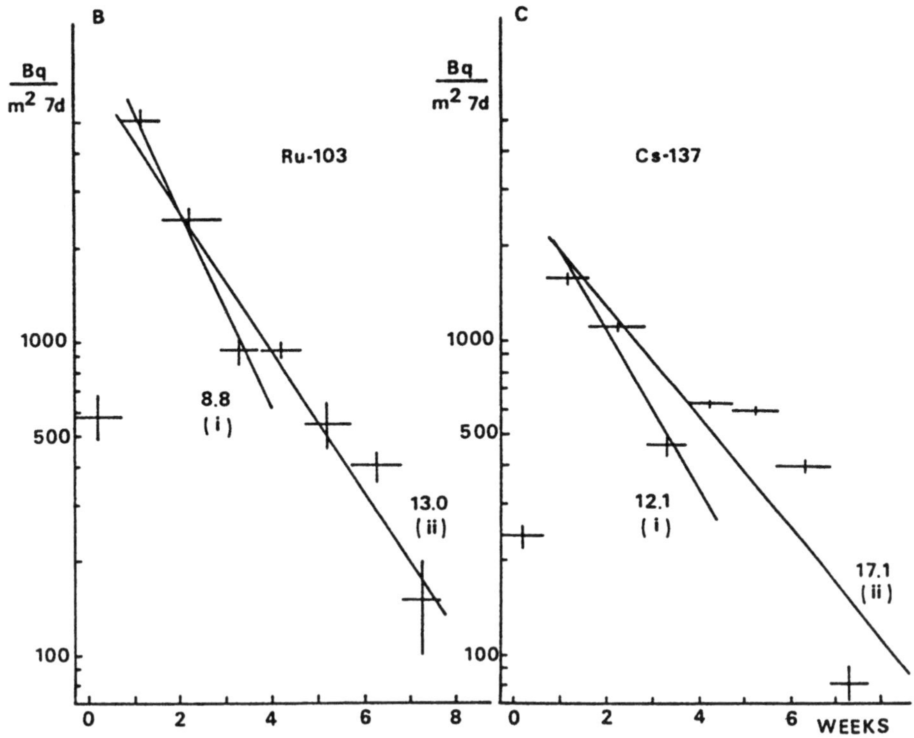


**Figure 13.6** *b* and *c*

sediments, the residence times for removal by particles are estimated to be 37–53 days for Cs and 15–24 days for Ru.

The mean epilimnetic residence times for the summer are estimated at 33–48 days (Cs) and 25–38 days (Ru) from the radiotracer concentrations in the settling material during the peaks of high fluxes in May 1986 and in August 1986, on the assumption that the settling particles had similar scavenging properties. The average settling velocity of Cs of about 50 cm/day is rather high compared with the settling velocities of 9 to 34 cm/day derived from radiotracer experiments in the "experimental lake area" of the Canadian Shield (Santschi et al., 1986). Adsorption to bottom sediments was the most effective sink of Cs in their studies.

The residence time of Cs in the hypolimnion may be calculated as 166 ± 15 days, derived from the comparison of the Cs inventory in the water column in March 1987 and July 1987. Another estimate of 155–179 days, which corresponds closely to the previous result, is based on the inventory in the water column in March 1987, which amounted to 13–17% of the deposition flux. Thus, the settling velocity of Cs in the hypolimnion of 45 cm/day is similar to the velocity evaluated in the epilimnion. This suggests that Cs-137 from the Chernobyl reactor accident was removed from both the epilimnion and the hypolimnion faster than could be expected from the results of the tracer studies in the Canadian Shield lakes by Santschi et al., 1986.

We estimate the Kd factor (distribution coefficient) for Cs in May, during the maximum of biological productivity, as $10^4$ cm$^3$/g to $3 \times 10^4$ cm$^3$/g, assuming that 7 to 20% of Cs was adsorbed on particulate matter and that the concentration of suspended matter was 9 mg/l (independent measurements of the concentration of suspended particles revealed a maximum concentration of 7.2 mg/l in mid-May (Stabel, 1986). However, this value for the Kd factor is rather high; Santschi et al. (1986) reported lower Kd values of $5 \times 10^3$ cm$^3$/g to $10^4$ cm$^3$/g in lakes with suspended particulate matter concentrations of 4 mg/l, a concentration similar to that recorded in Lake Constance. The faster removal of Cs in Lake Constance might be attributed to more reactive suspended matter in this lake.

## A Tentative Balance for Chernobyl Cs and Ru in Lake Constance

The main pathways characterizing distribution and transfer of the radiotracers Cs-137 and Ru-103 in Lake Constance are shown in Figure 13.7. In March 1987, only 15% of the initially deposited Cs and 9% of the Ru still remained in the water column. The losses presumably were due to either sedimentation or outflow.

During the first six weeks after May 1, 1986, a part of Cs-137 (approximately 32%) and of Ru-103 (53%) were removed from the epilimnion by coprecipitation with authigenically formed particles and were deposited directly on the lake bottom. By March 1987, about 75% of the Cs and 84% of the Ru of total deposition had been sedimented. A part of the Cs (≈30%) and a

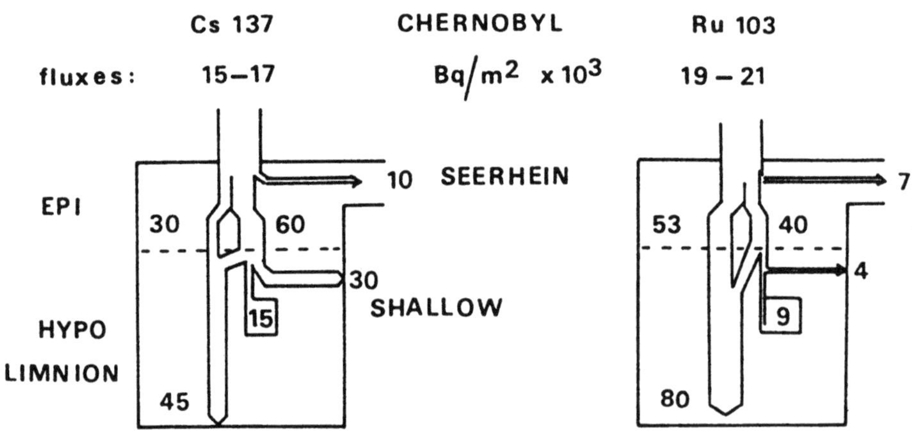


**Figure 13.7** A tentative mass-balance by March 1987 for Cs and Ru isotopes from the Chernobyl reactor accident in Lake Constance. The values in the boxes correspond to the percentage of the depositional input on the lake's surface on May 1, 1986.

smaller part of the Ru (4%) were adsorbed on the sediments in the lake littoral. We suggest that this difference in behavior relies mainly on the different affinity of Cs and Ru to particles and to their difference in residence times in the epilimnion. It also reflects the physical conditions in the lake during the spring, when mixing between epilimnion and hypolimnion is favored by a weak stratification of the water column. The outflow with the Rhine (Seerhein in Figure 13.7) was estimated at about 10% (Cs) and 7% (Ru) of total deposition. This evaluation was based on the assumptions that (1) the concentration of radioisotopes in the epilimnion decreases with their characteristic residence times, (2) the thickness of the epilimnion was 10 m, and (3) the yearly outflow with the Seerhein amounts to 10.8 km$^3$ (Wagner, 1976).

## 13.3 The tracer Pb-210 (T1/2:21.3a), a Natural and Permanent Atmospheric Component

The noble gas radon, Rn ($T_{1/2}$:3.8d), is released from soil and rocks and enters the upper layers of the troposphere and the stratosphere. Pb-210, which originates from the decay of Rn, is adsorbed on aerosols that have a typical residence time in the atmosphere of 4-7 days (Poet et al., 1972) and then are rained out or dry-deposited, generating a flux of Pb-210 on the earth's surface. The atmospheric Pb-210 deposition on continents amounts to 0.5 to 1.0 dpm/cm$^2$ year (Turekian et al., 1977). The Pb-210 flux can be determined from the standing crop of Pb-210 in soil profiles, where Pb-210 is strongly fixed in the uppermost layers rich in organic matter (Nozaki et al., 1978). There have been numerous studies on the behavior of Pb-210 in marine environments (e.g., Schell, 1977; Nozaki and Tsunogai, 1973; Bacon et al., 1979) but only a few on lakes and rivers. The studies on lakes, such as Lake Tansa (Krishnaswami et al., 1971), Lake Washington (Schell, 1974), Lake Shinji (Matsumoto, 1975), Lake Michigan (Robbins and Edgington, 1975), and, more recently, Lake Constance (Appel, 1982; Barth, 1984), showed that the main source of Pb-210 in lake sediments—in excess of Ra-226—is of atmospheric origin. In Lake Washington, the residence time of Pb-210 in the water column was estimated to be 80 days.

The average atmospheric flux at Lindau amounts to 0.99 ± 0.04 dpm/cm$^2$ yr, measured on four soil profiles, and to 0.66 dpm/cm$^2$ yr in Konstanz (Hemmerich, 1980; Appel, 1982). The difference probably reflects the difference in the precipitation patterns of 800 mm/yr in Konstanz and of 1300 mm/yr in Lindau. We evaluate the average atmospheric Pb-210 deposition in the central part of Lake Constance (average precipitation of 1000 mm/yr) at 0.78 dpm/cm$^2$ yr, assuming a linear relationship between the atmospheric depositional flux and precipitation. We will now evaluate the pathways and the residence time of Pb-210 in Lake Constance, based on monthly samplings of water and sediment traps in the central part of Lake Constance (April 1982 to November 1982).

**The Flux of Pb-210 into Sediment Traps** Sediment traps were deployed at 50 m, 90 m, 150 m and 250 m in the central part of Lake Constance. The flux of particulate matter into the traps (Figure 13.8*a*) shows a maximum of 7 to 8 $g/m^2$ d in July 1982 and minima of 1 to 1.5 $g/m^2$ d in both November and May 1982. The maximum reflects the larger supply of allochtonous material coming from the Alpenrhein and the larger amount of biological productivity during spring and summer. The mean annual average flux of sediment totals 0.098 $g/cm^2$ yr.

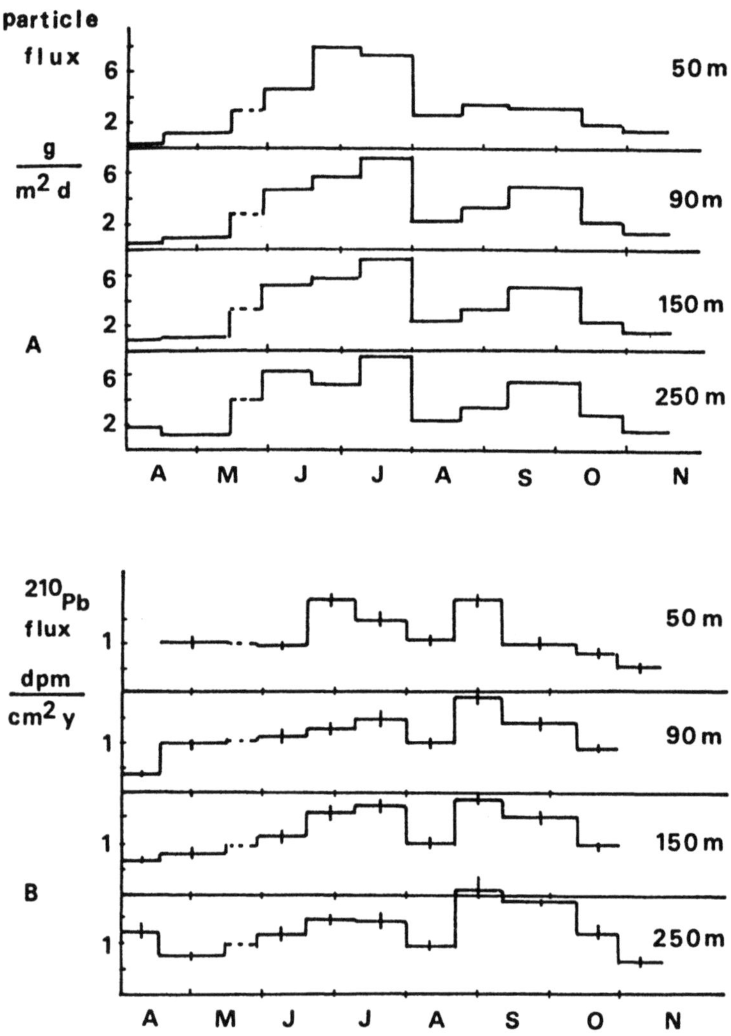


**Figure 13.8** Results from sediment traps moored at different depths in the central part of Lake Constance during the period April 1982 to November 1982 (Barth, 1984). (*a*) Depositional flux of particulate matter. (*b*) Depositional flux of Pb-210.

The depositional flux of Pb-210 averaged over the four depths amounts to 0.88 dpm/cm$^2$ yr. This mean annual flux of Pb-210 is close to the expected average flux into the lake and consists of an atmospheric component of 0.78 dpm/cm$^2$ yr and of a minor detritic component of about 0.26 dpm/cm$^2$ yr, which comes from the Alpenrhein (Dominik et al., 1981). Maxima of the Pb-210 flux up to values of 1.9 to 2.1 dpm/cm$^2$ yr were recorded in July, August, and September 1982; minima between 0.5 to 1.2 dpm/cm$^2$ yr were recorded in the winter (Figure 13.8*b*). The flux of Pb-210 increased with depth by 17% (from 50 m to 250 m, averaged over all intervals). This increase could reflect adsorption of Pb-210 on particles as they settle through the water column. However, we could not detect any increase of the Pb-210 flux with depth during four out of the 10 months. The flux of Pb-210 is apparently independent of the rate of sedimentation of particulate matter. This is shown in Figure 13.9 by the plot of the specific activity of the particulate matter against the sedimentation rate. The highest Pb-210 concentration is recorded during periods of lowest sedimentation rate; with increasing flux of the settling particles, the specific activity converges towards 2.5 dpm/g, which is the specific activity of detritic material (Dominik et al., 1981). This suggests that the larger flux of particulate matter during spring and summer does not scavenge a significant amount of Pb-210 from the hypolimnion.

**Pb-210 in the Sediments and Water Column of Lake Constance** The concentration of Pb-210 in the hypolimnetic water column was measured at

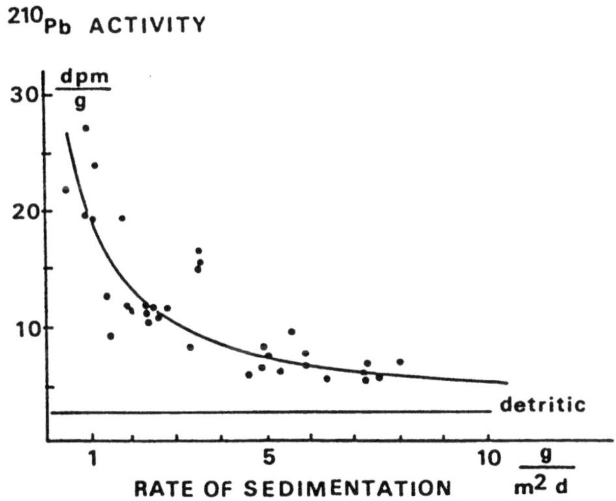


**Figure 13.9** Specific activity of particular matter as a function of the sediment rate. The sediment traps were moored in the central part of Lake Constance during the period April 1982 to November 1982 (Barth, 1984).

the depths where the sediment traps were exposed. The average annual concentration of Pb-210 for each depth is:

50 m; 0.066 ± 0.021 dpm/l
90 m; 0.086 ± 0.40 dpm/l
150 m; 0.061 ± 0.023 dpm/l
250 m; 0.066 ± 0.037 dpm/l

The concentration of Pb-210 varies from a low value of 0.04 dpm/l to a high value of 0.17 dpm/l. We assume that the higher values are caused by resuspended particulate matter in the lake and that they are not characteristic of the hypolimnion. The component of Pb-210 that is adsorbed on the particulate phase may be evaluated at 0.005 to 0.02 dpm/l, on the assumption that the specific activity of the suspended material is the same as that of the material collected with the sediment traps. This assumption seems justified because suspended material in the hypolimnion in the central part of Lake Constance (at 150 m in July 1980; Appel, 1982, separated via centrifugation) had a Pb-210 concentration of 13 ± 1 dpm/g, which is very similar to the average value for particles collected in sediment traps, and is also similar to the specific activity of the topmost section of sediments (Dominik et al., 1981). The concentration of particulate matter in the hypolimnion during 1982 was in the range of 0.5 to 0.8 mg/l (personal communication, M. Sturm). The mean, minimum, and maximum percentage of Pb-210 on particulate matter is given in Table 13.3.

The average concentrations of dissolved Pb-210, determined as (total Pb-210 minus particulate Pb-210), then are:

50 m; 0.057 ± 0.007 dpm/l
90 m; 0.079 ± 0.014 dpm/l
150 m; 0.055 ± 0.007 dpm/l
250 m; 0.057 ± 0.012 dpm/l

**Table 13.3** The amount of Pb-210 on particulate matter at various depths

| | Percentage | | | |
|---|---|---|---|---|
| Depth (m) | Mean | Minimum | Maximum | Concentration (dpm/1) |
| 50 | 12% | 5.4% | 39% | 0.0084 ± 0.0016 |
| 90 | 9% | 2.4% | 50% | 0.0080 ± 0.0016 |
| 150 | 11% | 3.9% | 22% | 0.0072 ± 0.0012 |
| 250 | 16% | 3.3% | 50% | 0.0112 ± 0.0019 |

Apparently, the concentration of dissolved Pb-210 in the water column does not increase with water depth. The deepest sample at 250 m, which was taken close to the sediment surface, has a value similar to that of the sample from a depth of 150m. This suggests that recycling of Pb-210 from the sediment into the water column is negligible. We, therefore, may evaluate an upper limit of the Pb-210 flux from the sediment back into the water column at less than 0.008 dpm/cm$^2$ year. This flux is much smaller than the annual flux of Pb-210 into the sediments of about 1 dpm/cm$^2$ yr. In the calculation of this upper limit, we used the maximum gradient between the Pb-210 concentrations at 150m and 250m water depth of $2.5 \times 10^{-6}$ dpm/l cm and a mixing coefficient in the water column of 0.1 cm$^2$/sec.

Sediment cores taken at various localities in the lake show that the deposition flux of excess Pb-210 (solely the atmospheric component) into the sediments is proportional to the accumulation rate (Figure 13.10). Some cores (e.g., SM1 and SM2) show significantly lower fluxes than the atmospheric Pb-210 flux recorded in soil profiles. Other cores (e.g., LA3, LA5, and AR1) have Pb-210 fluxes exceeding the expected flux. The different fluxes of Pb-210 and of sediment at different localities in the lake suggest that on the bottom of the lake there are some places where sediments are being accumulated preferentially and other locations where sediments are being partially eroded. On this

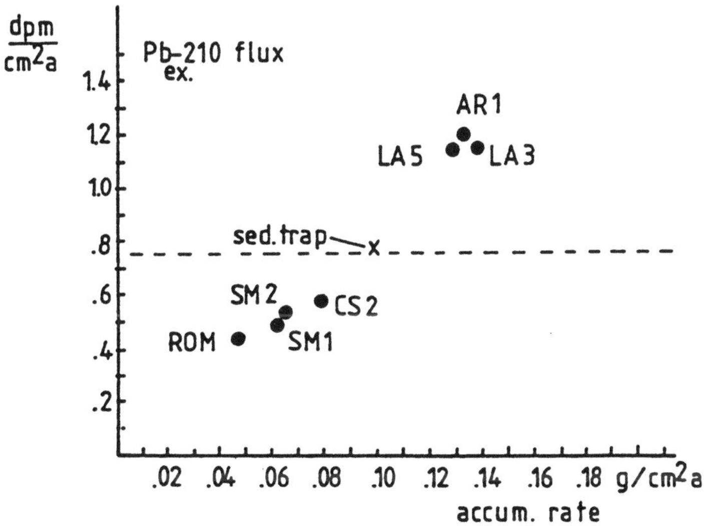


**Figure 13.10** Plot of the flux of excess Pb-210 into the sediments as recorded in cores from different localities in Lake Constance against the sediment accumulation rate. Data on cores SM1, SM2, ROM (from the central part of the lake), and AR 1, LA 3, and LA 5 (from localities closer to the northern shore, sampled at water depths of approx. 100 m) are from Dominik et al., 1981 and 1984; and Barth, 1984. The horizontal line indicates the flux of Pb-210 expected from atmospheric deposition, measured in soil profiles at localities near Konstanz and Lindau (Appel, 1982).

plot of the Pb-210 flux against sediment accumulation rate, the intersect of the best-fit line with the expected flux of 0.78 dpm/cm² yr allows us to estimate the average sediment accumulation rate in the central part of the lake at 0.095 g/cm² yr. This value comes quite close to the flux recorded in the sediment traps of 0.098 g/cm² yr and to the accumulation rate of 0.08–0.11 g/cm² yr in two cores from the central part of Lake Constance (von Gunten et al., 1987), where core recovery is assumed to be perfect.

**The Residence Time of Pb-210 in the Water Column of Lake Constance** The epilimnetic residence time may be evaluated from the ratio of the inventory in the water column and the supply rate of (atmospheric + allochtonous) Pb-210. Assuming a total Pb-210 flux of 1 dpm/cm² yr, and a concentration in the epilimnetic water column of 0.05 dpm/l (four measurements in 1980 in the range between 0.042 ± 0.0064 and 0.066 ± 0.01; Appel, 1982) the residence time in the epilimnion (of ≈20 m thickness) is estimated at approximately 36 days. However, the residence time of dissolved Pb-210 is slightly longer, because a part of Pb-210 (26%) is already adsorbed on the particles that are supplied with the Alpenrhein. The epilimnetic residence time of dissolved Pb-210 is 45 days and its settling velocity is 44 cm/day, on the assumption that the residence time of particulate matter in the epilimnion is between 6 and 15 days, as was evaluated from the sediment trap records.

The residence time of Pb-210 in the water column may be determined also from the residence time of settling material and the percentage of Pb-210 that is adsorbed on particles. In July 1980, the concentration of suspended matter was as high as 4.0 to 4.3 mg/l, and the specific Pb-210 concentration of suspended matter amounted to 6.2 ± 0.3 and 7.5 ± 0.53 dpm/g (Appel, 1982). The percentage of particulate Pb-210 may then be estimated to be between 0.025 and 0.032 dpm/l, approximately 50 to 60% of total Pb-210 in the water column. The residence time of dissolved Pb-210 (in July 1980) was therefore 10 to 30 days. This value corresponds closely with our previous estimate of 45 days, which is an average value over the summer.

The average residence time of Pb-210 in the hypolimnion (at an average depth of 80 m) may be considered to be 4.8 months (or slightly less, if we exclude the higher Pb-210 concentrations) from the ratio of the Pb-210 content of 0.05 dpm/l in the water column and from the input (or output) flux of 1 dpm/cm² yr. However, if we take into account that a large percentage (at least 80%) of the Pb-210 flux into the hypolimnion is already adsorbed on the particulate matter that settles from the epilimnion, the hypolimnetic residence time of dissolved Pb-210 ranges between 1 and 1.5 years and the settling velocity of dissolved Pb-210 in the hypolimnion ranges between 14 and 22 cm/day. This settling velocity is about three times slower than in the epilimnion, suggesting that settling particles are poor scavengers of Pb-210 in the hypolimnion. This result corroborates the findings from the sediment traps, which did not show any significant increase of the particulate flux with depth.

## 13.4 Conclusions

In the epilimnion of Lake Constance, the residence times of the Chernobyl radiotracers Ru-103 and Cs-137 are only slightly shorter than the residence time of 35–45 days obtained for Pb-210. In the hypolimnion, the residence times for Cs-137 and Pb-210 are 5.5 months and 4.8 months, respectively. Particles settling through the hypolimnion scavenged Cs-137 at the same rate as in the epilimnion. Dissolved Pb-210, in contrast, is scavenged in the hypolimnion at a significantly slower rate than in the epilimnion. The reasons for the different behavior of Cs and Pb-210 in the hypolimnion are not yet known. Our results, however, suggest that dissolved Pb-210 is being adsorbed and scavenged from the water column only on a very specific kind of particle (one originating from biological productivity?). Cs-137, on the other hand, appears to be less selective than Pb-210, as suggested by the similarity of its average settling velocity in the epilimnion and in the hypolimnion.

The standing crops of the radiotracers in the sediments reveal a fractionation in the water column between Chernobyl Ru-103 and Cs-137 due to different affinity of adsorption on particles. Finally, the standing crops of radiotracers Pb-210 and Cs-137 (from atomic bomb tests) in the sediments reveal the process of redistribution of sediment and of irregular sediment accumulation on the bottom of Lake Constance.

---

*Acknowledgements* We thank Peter Santschi for interesting discussions and suggestions. Part of this work was supported by the *Deutsche Forschungsgemeinschaft*, SFB 248.

## References

Appel, I. 1982. Das Verhalten von Pb-210 und Po-210 im limnischen Milieu am Beispiel des Bodensee-Obersees. Diplomarbeit, Institut für Umweltphysik der Universität Heidelberg, 95 pp.

Bacon, M.P., Brewer, P.G., Spencer, D.W., Murray, J.W., and Goddard, J. 1979. Lead 210, Polonium-210, manganese and iron in the Cariaco Trench. *Deep-Sea Res.* 27:119–135.

Barth, M. 1984. Blei-210 im Bodensee Obersee. Diplomarbeit, Institut für Umweltphysik der Universität Heidelberg, 55 pp.

Christian, U. 1987. Cäsium und Ruthenium im Bodensee. Diplomarbeit, Institut für Umweltphysik der Universität Heidelberg, 123 pp.

Dominik, J., Mangini, A., and Müller, G. 1981. Determination of recent deposition rates in Lake Constance with radioisotopic methods. *Sedimentology* 28:653–677.

Dominik, J., Mangini, A., and Prosi, F. 1984. Sediment rate variations and anthropogenic metal fluxes into lake Constance sediments. *Environm. Geol.* 5/3:151–157.

von Gunten, H.R., Sturm, M., Erten, H.N., Rössler, E., and Wegmüller, F. 1987. Sedimentation rates in the central Lake Constance determined with Pb-210 and Cs-137. *Schweiz. Z. Hydrol.* 49/3:275–285.

Hemmerich, R. 1980. Bestimmung des Pb-210 Flusses in Bodenprofilen. Staatsexamensarbeit, University of Heidelberg, 40 pp.

Kleiner, J., Lindner, G., Recknagel, E., and Stabel, H.H. 1984. Endogenous catalysis in the precipitation of settling material in Lake Constance investigated by PIXE. *Nucl. Inst. Meth.* B3:553–556.
Krishnaswami, S., Lal, D., Martin, J.M., and Meybeck, M. 1971. Geochronology of lake sediments. *Earth. Planet. Sci. Lett.* 11:407–414.
Matsumoto, E. 1975. Pb-210 geochronology of sediments from Lake Shinji. *Geochem. Journal* 9:167–172.
Nozaki, Y. and Tsunogai, S. 1973. A simultaneous determination of Pb-210 and Po-210 in sea water. *Anal. Chim. Acta* 64:211–214.
Nozaki, Y., De Master, D.J., Lewis, D.M., and Turekian, K.K. 1978. Atmospheric Pb-210 fluxes determined from soil profiles. *J. Geophys. Res.* 83:4047–4051.
Poet, S.E., Moore, H.E., and Martell, E.A. 1972. Pb-210, Bi-210 and Po-210 in the atmosphere: accurate ratio measurement and application to aerosol residence time determination. *J. Geophys. Res.* 77/33:6515–6527.
Robbins, J.A. and Edgington, D.N. 1975. Determination of recent sedimentation rates in Lake Michigan using Pb-210 and Cs-137. *Geochim. Cosmochim. Acta* 39:285–304.
Santschi, P.H., Nyffeler, U.P., Anderson, R.F., Schiff, S.L., O'Hara, P., and Hesslein, R.H. 1986. Response of radioactive trace metals to acid-base titrations in controlled experimental ecosystems: evaluation of transport parameters for application to whole-lake radiotracer experiments. *Can. J. Fish. Aquat. Sci.* 43:60–77.
Schell, W.R. 1974. *Sedimentation rates and mean residence times of stable Pb and Pb-210 in Lake Washington, Puget Sound estuaries and a coastal region.* Univ. of Washington, College of Fisheries, Radiat. Ecol. Lab., Seattle, Washington, USAEC, Rep. RLO-2225-118-18, 18 pp.
Schell, W.R. 1977. Concentrations, physico-chemical states and mean residence times of Pb-210 and Po-210 in marine and estuarine waters. *Geochim. Cosmochim. Acta* 41:1019–1031.
Smith, F.B. and Clark, M.J. 1986. Radionuclide deposition from the Chernobyl cloud. *Nature* 322:690–693.
Stabel, H.H. 1985. Mechanisms controlling the sedimentation sequence of various elements in prealpine lakes. In: W. Stumm (ed.). *Chemical Processes in Lakes,* J. Wiley, New York, p. 143–167.
Stabel, H.H. 1986. The role of plankton biomass in controlling fluctuations of suspended matter in Lake Constance. *Hydrobiologia* 140: 173–181.
Turekian, K.K., Nozaki, Y., and Benninger, L.K. 1977. Geochemistry of atmospheric radon and radon products. *Ann. Rev. Earth. Planet. Sci. Lett.* 5:227–255.
Wagner, G. 1976. Die Untersuchung von Sinkstoffen aus Bodenseezuflüssen. *Hydrologie* 38:191–205.

# 14

# Biological Transfer and Sedimentation of Chernobyl Radionuclides in Lake Constance

G. Lindner, M. Becker, R. Eckmann, P. Frenzel, J. Kleiner, D. Petermann-Seyboldt, W. Pfeiffer, U. Wahl, and E. Recknagel

ABSTRACT Considering the input of Chernobyl radionuclides into Lake Constance on April 30, 1986, as the starting point of an unplanned radiotracer experiment, we followed the fate of five long-lived, gamma-emitting radionuclides ($^{137}Cs$, $^{106}Ru$, $^{125}Sb$, $^{110m}Ag$, and $^{144}Ce$) in order to obtain information about their scavenging and sedimentation properties, and their transfer into and within the food-web. In particular, we studied the uptake and excretion by fish of $^{137}Cs$. With respect to the transfer of $^{137}Cs$ into fish, we also performed comparative investigations in the Schreckensee, a small shallow lake in the vicinity of Lake Constance.

The chemical speciation and the physical state of the radionuclides determine the efficiency of their scavenging and sedimentation, resulting in an enrichment of the above long-lived radionuclides in the sediments in comparison to $^{137}Cs$. The uptake of $^{137}Cs$ by planktivorous fish was considerably more rapid than by other species. For whitefish, slightly higher concentration factors and somewhat lower effective retention times than those reported in the literature were obtained. The inventory of dissolved $^{137}Cs$ in the water reflected the differences between Lake Constance and the Schreckensee with respect to their morphometry; enrichment of this isotope was considerably greater in the shallow Schreckensee than in deep Lake Constance, due to less dilution within the water column. The observed marked differences in the radioactive contamination of the fish in both lakes can be explained by the food-web structures.

With the advent of Chernobyl radionuclides in the sediments of Lake Constance in 1986, a new geochronological marker was established, allowing tracer studies of transport processes around the sediment-water interface that have a superior time resolution compared to studies using radionuclides from nuclear weapons testing fallout.

## 14.1 Introduction

As a consequence of the reactor accident at Chernobyl (USSR) on April 26, 1986, for the second time Northern Hemisphere ecosystems were exposed to widespread contamination by artificial radioactive isotopes. (The first contamination was caused by the atmospheric testing of nuclear weapons in the 1950's and early 1960's.)

The explosions in the Chernobyl nuclear power plant initiated an involuntary, worldwide "radiotracer experiment," making possible insights into the responses of different ecosystems to a sudden input pulse of radioactivity. The accident occurred at the same instant within the seasonal cycle of all ecosystems. Follow-up studies of the pathways of radiocontaminants enabled us to obtain new information on transport processes in lakes. This is because radionuclides participate in different fashions in biogeochemical cycles, depending on their lifetimes, chemical speciation, and physical states (Santschi et al., 1988).

We followed this approach at Lake Constance, a large (surface area, 539 $km^2$; mean depth, 100 m; water retention time, 4.5 years), monomictic, meso-eutrophic, prealpine lake (Elster, 1977) and, in parts, also in the Schreckensee, a small (surface area, 0.3 $km^2$; mean depth, 6 m; water renewal time, 1.3 years), glacially formed, highly eutrophic lake (Zintz, 1986) located about 40 km north of Lake Constance. Radionuclides from Chernobyl were introduced into Lake Constance by heavy rainfalls during the passage of contaminated air masses on April 30th, 1986 (Hohenemser et al., 1986a, Lindner and Recknagel, 1988). Several groups studied the residence times of radionuclides in the water (EAWAG, 1988; see Chapter 13), their removal via particulate matter (Lindner et al., 1988c; see Chapter 13), their transfer into plankton and fish (Lindner et al., 1987; Santschi et al., 1987b; Intern. Gewässerschutzkomm. Bodensee, 1987; Lindner et al., 1988b), and their deposition in sediments (Lindner et al., 1988d; see Chapter 13). From these data a mass balance of Cs and Ru radionuclides was established (see Chapter 13).

Similar investigations were also performed at other Swiss prealpine lakes (Zürichsee, Luganer See, etc.; Santschi et al., 1988; EAWAG, 1988) and in the Schreckensee (Petermann et al., 1987a; Lindner et al., 1988a; Kissling et al., 1988; Geissler et al., 1989), which suffered radionuclide input under similar conditions. A comparison of different lakes promises information about the influence of lake morphology and food-web structure on the pathways of the radionuclides.

In addition, extensive studies mainly devoted to the uptake of Cs radioisotopes by fish were performed at a number of Swedish lakes (Petersen et al., 1986; Hakanson et al., 1988). The vertical transport of Chernobyl radionuclides in seas via particulate matter was investigated for the Mediterranean (Fowler et al., 1987), the Black Sea (Büsseler et al., 1987), and the North Sea (Kempe and Nies, 1987). Recently, preliminary data from the distribution of Chernobyl radionuclides in sediments of the Masurian Lakes also has become available (Robbins, personal communication).

For seasonal cycling processes in lakes, short-lived isotopes such as $^{131}I$ ($T_{1/2}$ = 8 d) are of minor relevance, in spite of their initial abundance in the fallout. On the other hand, radionuclides with half-lives of more than 6 months, such as $^{134}Cs$ (2.06 years), $^{137}Cs$ (30.2 years), $^{106}Ru$ (1.01 years), $^{125}Sb$ (2.73 years), $^{110m}Ag$ (0.68 years), $^{144}Ce$ (0.78 years), and $^{54}Mn$ (0.86 years), could also be detected by high-resolution gamma spectroscopy. The activities of the Cs isotopes were higher than those of all other long-lived radionuclides combined. Therefore, Cs radionuclides could easily be followed in all compartments of the lake ecosystem (water, particulate matter, plankton, fish, and sediment). Their input due to Chernobyl could safely be distinguished from inputs originating from atmospheric nuclear weapon tests via their characteristic activity ratio of 0.57 on May 1st, 1986 (Hohenemser et al., 1986a). Radionuclides with lower activity ($^{106}Ru$, $^{125}Sb$, $^{110m}Ag$, $^{144}Ce$, and $^{54}Mn$) were recorded in particulate matter and sediments only. They nevertheless also provided information about the influence of chemical speciation on removal processes. Pure beta- and alpha-emitting, long-lived radionuclides, such as $^{90}Sr$ and $^{239}Pu$, were not measured. With respect to transuranium isotopes, however, $^{144}Ce$ can be looked at as a radiotracer, since we assume that it is introduced into the lake on small particles of nuclear fuel containing all nonvolatile activation products (Wahl, 1987; Lindner and Recknagel, 1988; Wahl et al., 1989).

In the following, major emphasis is put on:

1. The characterization of the fallout input
2. The uptake and excretion of $^{137}Cs$ by fish
3. The vertical transport of low-activity, gamma-emitting radionuclides via particulate matter in comparison to $^{137}Cs$
4. The horizontal and vertical distribution of Chernobyl radionuclides in sediments

Thus, in combination with other publications concerning Chernobyl radioactivity in Lake Constance (Internat. Gewässerschutzkomm. Bodensee, 1987; EAWAG, 1988; see Chapter 13), a comprehensive picture of the situation will emerge allowing a discussion of the influence of food-web interactions and lake morphometry on the fate of Chernobyl radionuclides.

## 14.2 Experimental Procedures

Most of the samples from Lake Constance (water from different depths, settling particles collected in traps at a depth of 20 m, and bottom sediments taken during 1987 and 1988) were taken from the Überlingersee, a fjord-like, northwestern bight of the lake ca. 140 m deep, which is separated from the main basin by a 60 m deep sill (see Figure 14.10*a*). Fish of different species were caught by fishermen across the whole lake, including also the smaller and shallower Untersee, whose character is more eutrophic. Plankton were collected with nets in a bay near the Isle of Mainau in the Überlingersee in 1986 and 1987. Soil samples were taken in October 1987 from 25 positions around the shores of Überlingersee in order to determine the input flux of $^{137}Cs$ to this part of Lake Constance. From the Schreckensee, water samples were taken at various depths, and fish were supplied by the local forestry agency. The $^{137}Cs$ input flux to that lake was determined from a carefully selected, large soil sample taken nearby.

Dissolved $^{137}Cs$ was extracted from water samples by addition of potassium hexacyanocobalt(II)ferrat(II) ion exchanger (Prout et al., 1965) and the total $^{137}Cs$ activity in the precipitate was measured. Plankton and fish were measured in the moist state; from the latter, heads and fins were removed and the remainder was cut into small pieces. Aliquots of particulate matter samples which had been analyzed with respect to $^{137}Cs$, $^{134}Cs$, and $^{103}Ru$ (see Chapter 13) were utilized for an extension of the isotope analysis to the less-abundant radionuclides $^{106}Ru$, $^{125}Sb$, $^{110m}Ag$, and $^{144}Ce$. These samples were centrifuged and dried at 110°C for 24 hours prior to analysis; a fraction was combusted at 520°C for 4 hours without significant losses of activity. Sediment cores of 8 cm diameter taken at several positions in the Überlingersee at depths between 100 and 140 m were separated into depth intervals of 2 cm; they were analyzed in the moist state and subsequently dried. Radionuclide analysis was performed by high-resolution gamma spectroscopy, using an intrinsic coaxial Germanium detector coupled with standard electronic signal-processing modules, including a multichannel analyzer and a personal computer for data acquisition. The samples were filled into calibrated vessels of different geometries and placed on top of the detector during measurement. All data were corrected for nuclear decay since May 1, 1986.

## 14.3 Characterization of the Radioactive Input

In contrast to the radionuclide inputs due to atmospheric nuclear weapon testing which occurred over several years, the input of Chernobyl radionuclides was restricted to few hours. Air masses, contaminated with radioactivity from Chernobyl, arrived at Lake Constance on the morning of April 30th, driven by weak northeasterly winds. In the evening of the same day, rainfalls and thunderstorms, whose precipitation was distributed rather homogeneously across the lake, caused the introduction of radioactive fallout across the water surface (Internat. Gewässerschutzkomm. Bodensee, 1987). Dry deposition was less effective by a factor of about 100 for most isotopes (Santschi et al., 1987c). The contribution from tributaries to the total radioactive input was rather limited, because, in the alpine drainage area of the Rhine, which supplies about 70% of the total water influx, the deposition of radioactivity was considerably lower than to the lake directly (Czarnecki et al., 1987). Inputs from the tributaries, therefore, will be neglected in the following considerations. At the time of the input, the lake was still only weakly stratified. As a consequence, radioactivity was immediately mixed down to depths of about 60 m (König et al., 1986; Internat. Gewässerschutzkomm. Bodensee, 1987). This was also corroborated from the shapes of the depth profiles of dissolved $^{137}Cs$ measured in autumn 1986 (EAWAG, 1988; see Chapter 13).

The isotopic composition of the fallout with respect to long-lived, gamma-emitting radionuclides is shown in Figure 14.1, which gives the gamma spectrum of cistern sediments taken about 60 km north of Lake Constance and measured in October 1986. Besides the dominant $^{137}Cs$- and $^{134}Cs$-lines, signals from $^{141}Ce$, $^{144}Ce$, $^{125}Sb$, $^{103}Ru$, $^{106}Ru$, $^{129m}Te$, $^{110m}Ag$, $^{124}Sb$, $^{95}Zr$, $^{95}Nb$, and $^{54}Mn$ could be identified. Table 14.1 gives the specific activity of areal deposition of radionuclides with a half-life higher than 0.5 years for the Überlingersee, which were obtained from the measured $^{137}Cs$ input using isotopic ratios precisely determined for a soil sample taken at Konstanz in May 1986 (Wahl, 1987). The respective proposed chemical speciation in the water, which determines the behavior within the food-web (Santschi et al. 1988), is also shown. It is the result of the physical and chemical processes that occurred during the emission from the destroyed reactor core at Chernobyl, depending on the thermodynamic stability of the respective species, and on the atmospheric transport and deposition mechanisms.

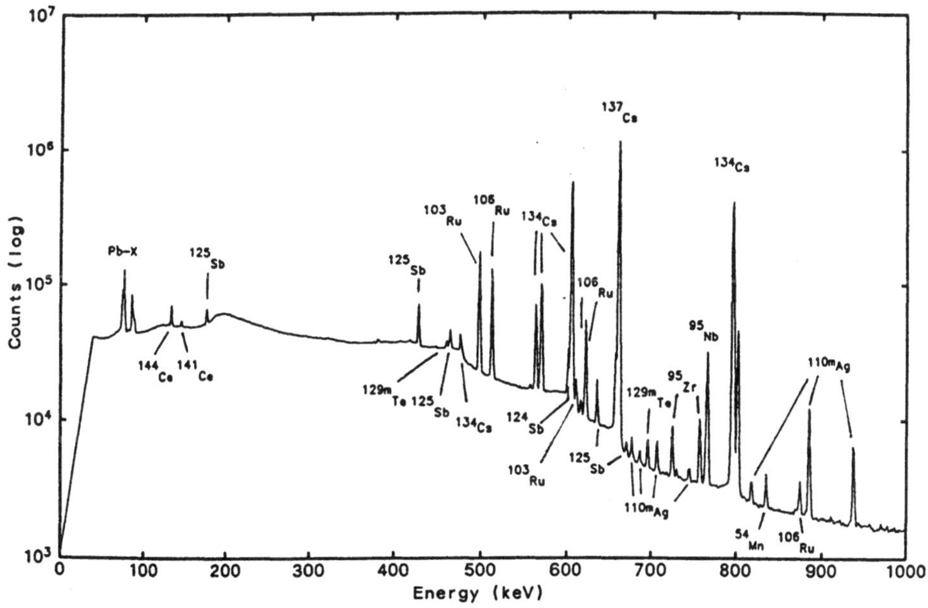


**Figure 14.1** Gamma spectrum of a cistern sediment sample taken about 60 km north of Lake Constance and measured in October 1986, which exhibits signals from long-lived, gamma-emitting radionuclides from the Chernobyl fallout.

**Table 14.1** Characterization of six long-lived, gamma-emitting radionuclides from the Chernobyl fallout

| Isotope | $^{137}Cs$ | $^{134}Cs$ | $^{106}Ru$ | $^{125}Sb$ | $^{110m}Ag$ | $^{144}Ce$ |
|---|---|---|---|---|---|---|
| Half-life (yr) | 30.2 | 2.06 | 1.01 | 2.73 | 0.68 | 0.78 |
| Emission process | Thermal evaporation from fuel | | | | Evaporation from silver metal | Ejection of fuel particles |
| Transport state | Adsorption to ambient aerosols | | | | | Fuel particles |
| Chemical speciation in water | $Cs^{+a}$ | $Cs^{+a}$ | $RuO_4^{2-a}$ | $Sb(OH)_6^{-a}$ | Unknown[b] | Fuel particles[b] |
| Transfer in Lake Constance | Scavenging by settling particles, biological uptake, adsorption at sediments | | Uptake by phytoplankton, Precipitation with calcite | | Unknown | Stokesian sinking |
| Fallout inventory (Bq/m²) | 17,500 ± 4,500 | 9,970 ± 2,560 | 4,410 ± 1,100 | 299 ± 75 | 201 ± 50 | 227 ± 57 |

[a] Soluble species.
[b] Insoluble species.

Among the radionuclides studied, the Cs isotopes are the most volatile; they were evaporated in an atomic state and thereafter were completely adsorbed by aerosols. These aerosols then were transported by moving air masses (Persson et al., 1987). In water, they dissolve as cations. $^{106}$Ru was evaporated in its oxidized form as $RuO_4$, which dissolves in an anionic, $RuO_4^{-2}$ form (Santschi et al., 1987a). A similar behavior was suggested for $^{125}$Sb, and in oxygenated waters the antimonate ion $Sb(OH)_6^-$ is proposed to be the thermodynamically stable form of Sb (Turner et al., 1981).

For $^{144}$Ce, on the other hand, thermal evaporation is rather unlikely, since temperatures were not high enough for substantial release of this nonvolatile isotope except, perhaps, during the initial few seconds after the explosive power excursions (INSAG, 1986). This isotope was found to be enriched in particles with high specific activity ("hot particles"), which were found all over Europe and which could easily be isolated from air filters, for example (Devell et al., 1986; Van der Veen et al., 1986; Devell, 1987; Wahl, 1987; Schubert and Behrend, 1987; Broda, 1987; Bode et al., 1987; Persson et al., 1987; Raunemaa et al., 1987; Weiss et al., 1987; Wahl et al., 1989; Cuddihy et al., 1989). The dominant group of these particles was identified as small fragments of nuclear fuel (Van der Veen et al., 1986; Devell, 1987; Bode et al., 1987; Schubert and Behrend, 1987; Cuddihy et al., 1989). They may have resulted from the fragmentation of the ceramic fuel material during the supercritical power excursion and were subsequently released by the explosions and by the stream of hot air emerging from the core debris (about 3% of the core material was mechanically ejected—INSAG, 1986). They lack volatile isotopes due to evaporation of the latter. There are strong indications that inclusion into nuclear fuel particles is the only form of $^{144}$Ce, since

1. The $^{144}$Ce/$^{137}$Cs activity ratio in the fallout is much higher than expected from thermal release (Wahl, 1987).
2. The activity ratio of other nonvolatile radionuclides (e.g., $^{95}$Zr, $^{95}$Nb, and $^{239}$Np) is quite similar for various samples from different sites and corresponds to that of the mature nuclear fuel before the accident (Wahl, 1987).
3. Dry deposition velocities of $^{144}$Ce are much higher than for $^{137}$Cs (Persson et al., 1987, Santschi et al., 1987c).

The latter indicated a higher specific weight of the carrier particles as compared to ambient aerosols. (These arguments will be discussed in more details in a forthcoming publication.) Since $^{144}$Ce only dissolves very slowly—with a half-life of 160 days—from fuel particles immersed into aqueous solvents (Cuddihy et al., 1989), this isotope can be regarded as being essentially insoluble, at least for the sedimentation process, which mainly occurred within a few weeks after the input event (see Section 14.5). The vertical transport of $^{144}$Ce in the water, then, is determined by the rate of sinking of $\mu$m-sized fuel particles, according to Stoke's law. Moreover, no participation in biological or chemical processes takes place, except for uptake of such particles as a whole or coagulation with other sinking material.

The radionuclide $^{110m}Ag$ was evaporated from molten neutron detectors in the reactor core made of metallic silver (Jones et al., 1986; Van Dam, 1986). Its chemical state during transport in air and after immersion in water is yet unknown, but it seems possible that it might have formed rather insoluble silver halogenides.

## 14.4 Uptake and Excretion of $^{137}Cs$ by Fish

Among the long-lived radionuclides introduced in soluble form, the Cs isotopes exhibit the highest concentration factors in freshwater fish (Poston and Klopfer, 1988). From an investigation of $^{137}Cs$ uptake and excretion by fish in Lake Constance, we attempted to obtain information on: 1) concentration factors within the biota which are characteristic for that particular lake, and 2) basis data for modelling the limnological processes controlling this transfer.

In both respects, it should be kept in mind that the uptake of dissolved Cs ions by organisms and their subsequent transfer to fish competes with the scavenging to settling particles.

The basis for this approach is the measurement of the concentration of dissolved $^{137}Cs$ in the water. Corresponding depth profiles were measured in

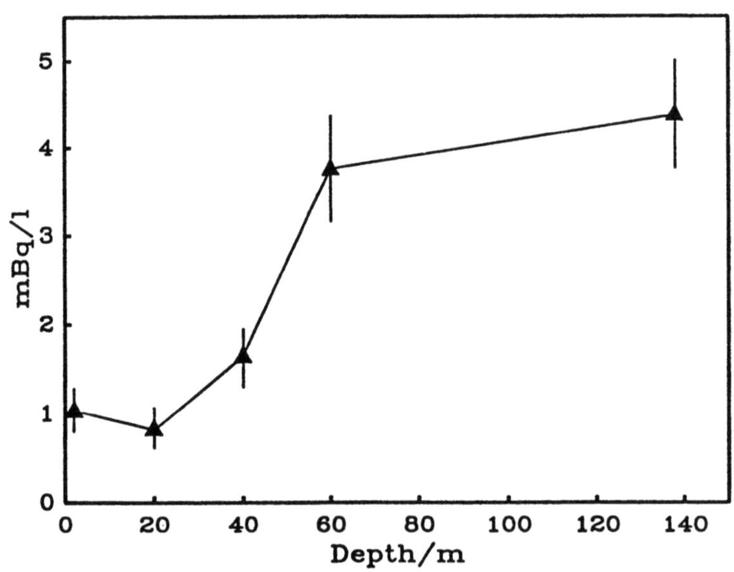


**Figure 14.2** Specific activity of dissolved $^{137}Cs$ in water from Lake Constance taken in September 1988 from different depths near the buoy in the Überlingersee (see Figure 14.10*a*).

1986 and 1987 by Mangini et al. (see Chapter 13) and by Santschi et al. (EAWAG, 1988). They reflect the seasonal dynamics of the vertical Cs transport, mainly due to scavenging. On the condition that the nuclides were initially mixed down to depths of 50 to 60 m upon introduction, the concentration maximum observed in the metalimnion during late summer of 1986 can be interpreted as the result of epilimnetic scavenging of dissolved Cs by settling particles (see Chapter 13; EAWAG, 1988). After complete homogenization of the $^{137}Cs$ profile by winter circulation, an epilimnetic minimum again appeared in the spring lasting until late summer 1987. However, no maximum occurred in the metalimnion because there had been no resupply from the lake surface (see Chapter 13; EAWAG, 1988). Similar patterns were recorded for a number of other prealpine lakes in Switzerland (EAWAG, 1988). For Lake Constance, both working groups estimated Cs residence times of about 150 to 180 d (see Chapter 13; EAWAG, 1988). We obtained a $^{137}Cs$ profile for late summer 1988 (Figure 14.2), which closely resembled those of the preceding year, although with reduced overall levels. Therefore, the epilimnetic $^{137}Cs$ minimum during summer could be expected to be a recurring phenomenon for the next decades; preliminary evaluation of the dissolved $^{137}Cs$ inventory for September 1988, however, indicates an increase in the residence time up to about 240 d. This means that a single exponential function will no longer be appropriate for the description of dissolved $^{137}Cs$ removal. Furthermore, in summer 1988, the $^{137}Cs$ activity in the epilimnion had already decreased to values below 1 Bq $m^{-3}$, which had been the typical average value prior to Chernobyl, as the result of atmospheric nuclear weapon testing (Petersen et al., 1986).

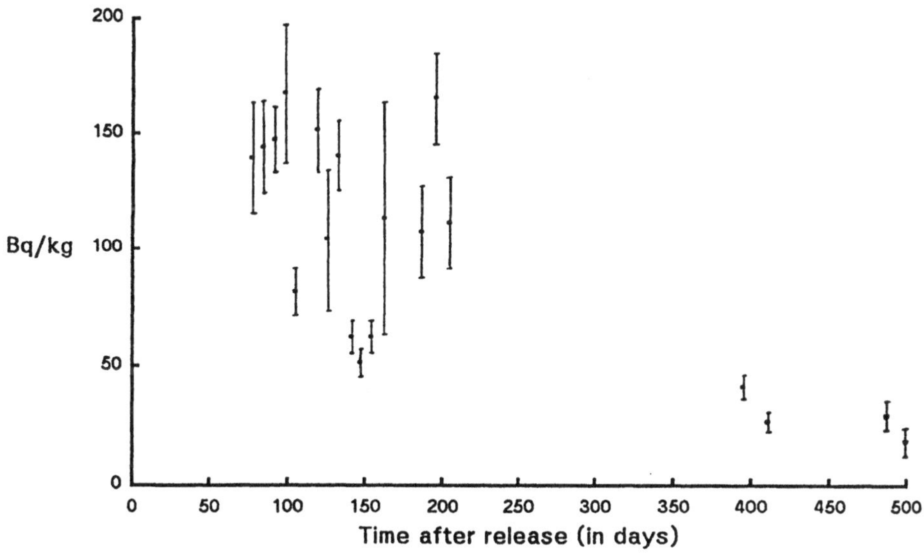


**Figure 14.3** Specific activity of $^{137}Cs$ in plankton from Lake Constance in Bq/kg dry matter as a function of time after the input event at April 30th, 1986.

Similar measurements in summer 1988 on water samples from the Schreckensee resulted in a considerably higher average $^{137}Cs$ concentration of 170 to 200 Bq/m³ (Lindner et al., 1988d). The Cs-residence time of 220 to 260 d obtained for that lake, however, is not substantially different from that of Lake Constance, in spite of fundamental differences in the limnological parameters of the Schreckensee, including lake morphology and effective water retention time (1.3 years compared to 4.5 years for Lake Constance; Zintz, 1986). The differences in dissolved $^{137}Cs$ activities are mainly the result of the differences in the magnitude of the input and the mean depths of both lakes. By multiplying the mean $^{137}Cs$ activity of 3 Bq/m³ typical for Überlingersee in 1988 by a factor of two, to account for the higher input at Schreckensee, and by 16.4, which is the ratio of the respective mean depths, one arrives at a concentration of about 100 Bq/m³, which is on the order of magnitude of

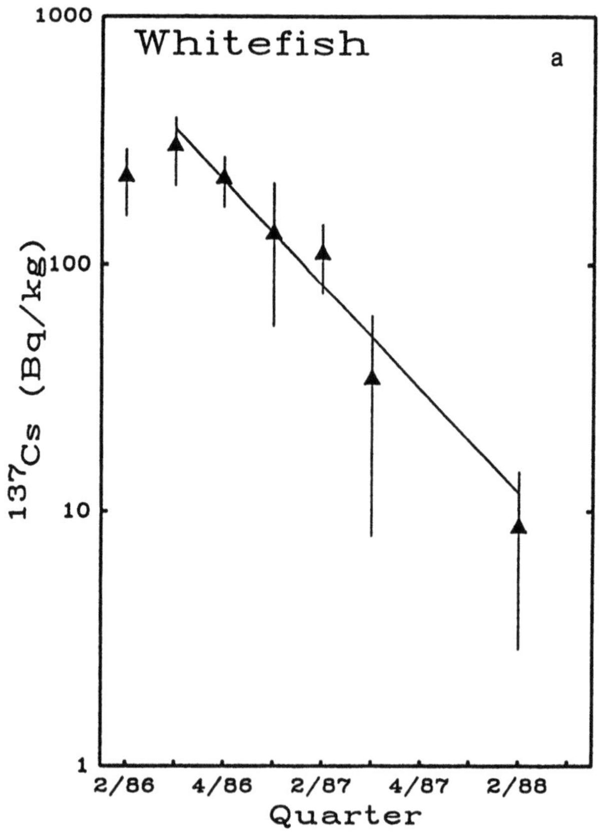


**Figure 14.4** Specific activity of $^{137}Cs$ in Bq/kg wet weight of (*a*) adult whitefish and (*b*) perches from Lake Constance after the Chernobyl input event. For each quarter, the respective mean values and standard deviations are displayed along with exponential least squares fits, from which the corresponding retention times were derived.

dissolved $^{137}Cs$ concentration in the Schreckensee in 1988. On the other hand, although there is a sharp boundary between the epilimnion and the anoxic hypolimnion at about 4 m depth in Schreckensee in the summer (Zintz, 1986), no corresponding structure could be observed for the $^{137}Cs$ profile (Lindner et al., 1988d), which is even smoother than in Lake Constance. It should be noted, however, that from other lakes, much longer $^{137}Cs$ retention times were also reported, e.g., more than 1 year for the Lago di Lugano (EAWAG, 1988).

In Lake Constance, dissolved $^{137}Cs$ was rapidly taken up by plankton, and the respective $^{137}Cs$ activities only changed within a factor of three throughout 1986 (Figure 14.3). Concentration factors of 1000 to 3000 $cm^3/g$ for 1986 and 1500 $cm^3/g$ for 1987 were obtained, in rough agreement with values given in the literature (Thomann, 1981). Comparing both years, $^{137}Cs$ activities in the plankton closely followed the pattern of dissolved $^{137}Cs$ activity in the water.

The transfer of $^{137}Cs$ to fish was rather rapid, that is, within a few weeks. Whitefish and perch, both feeding directly on plankton, showed an earlier increase in $^{137}Cs$ contamination than predatory fish like pike did (Lindner et

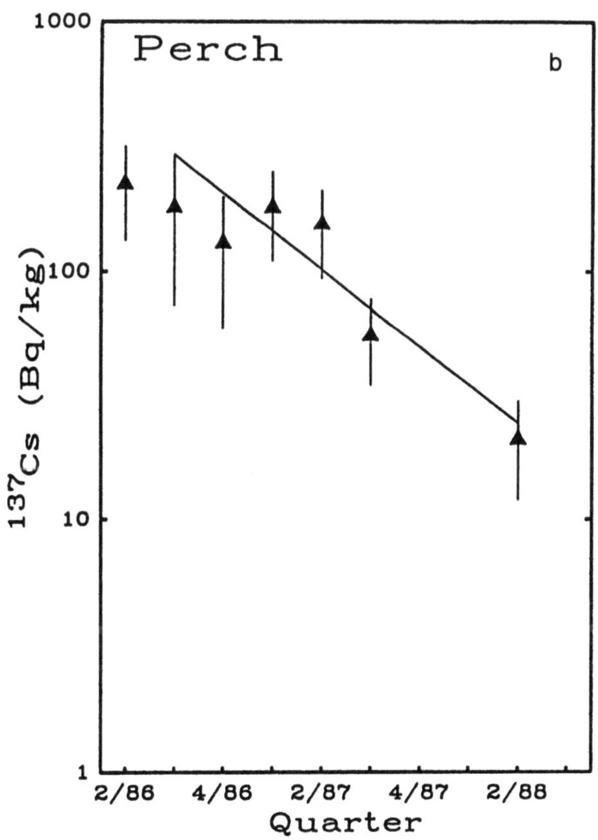


**Figure 14.4** *b*

al., 1987). A similar difference in the uptake velocity between fish species was also reported for Swedish lakes (Petersen et al., 1986). Ingestion of plankton was the main process by which fish took up $^{137}$Cs, as was shown by feeding experiments in July 1986 at the *Landesanstalt für Umweltschutz Langenargen;* young whitefish, kept in water from Lake Constance, were fed either (i) with dried uncontaminated plankton, or (ii) with plankton from the lake. Whereas, in fish from group (i), $^{137}$Cs activity remained below 30 Bq/kg, it reached 210 Bq/kg within a short time in fish from the second group (Internat. Gewässerschutzkomm. Bodensee, 1987). However, in both whitefish and perch from Lake Constance, maximum activities in $^{137}$Cs were already reached in midsummer 1986. They decreased thereafter, with time constants of 131 $\pm$ 15 and 179 $\pm$ 27 days, respectively (Figure 14.4). These values, however, should not be taken as retention times in fish individuals. Rather, they reflect the responses of the *whole* fish community and are possibly also influenced by preferential capture of certain age classes by the fishermen. These retention times are somewhat shorter than those reported for whitefish in the literature (220 $\pm$ 53 d; Kolehmainen, 1969). On the whole, the variance of specific activity within both fish communities is small as compared, e.g., to values obtained for fish from Lago di Lugano (Santschi et al., 1987a).

The significantly longer Cs-retention time in perch as compared to whitefish and the somewhat larger variability of the individual values may be related to a positive correlation between specific $^{137}$Cs activities and weights of fish, as observed in 1987 (Petermann et al., 1987b) and 1988 (Figure 14.5). A similar correlation was also reported for perch from Swedish lakes from 1987 (Hak-

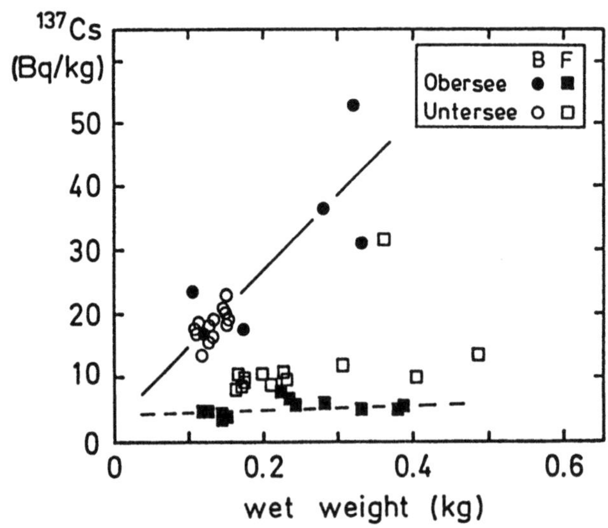


**Figure 14.5** Specific $^{137}$Cs activity in Bq/kg wet weight in perch (B) and whitefish (F) from Lake Constance (from both the Obersee and the Untersee) from summer 1988 versus fish weight (lines are guides only).

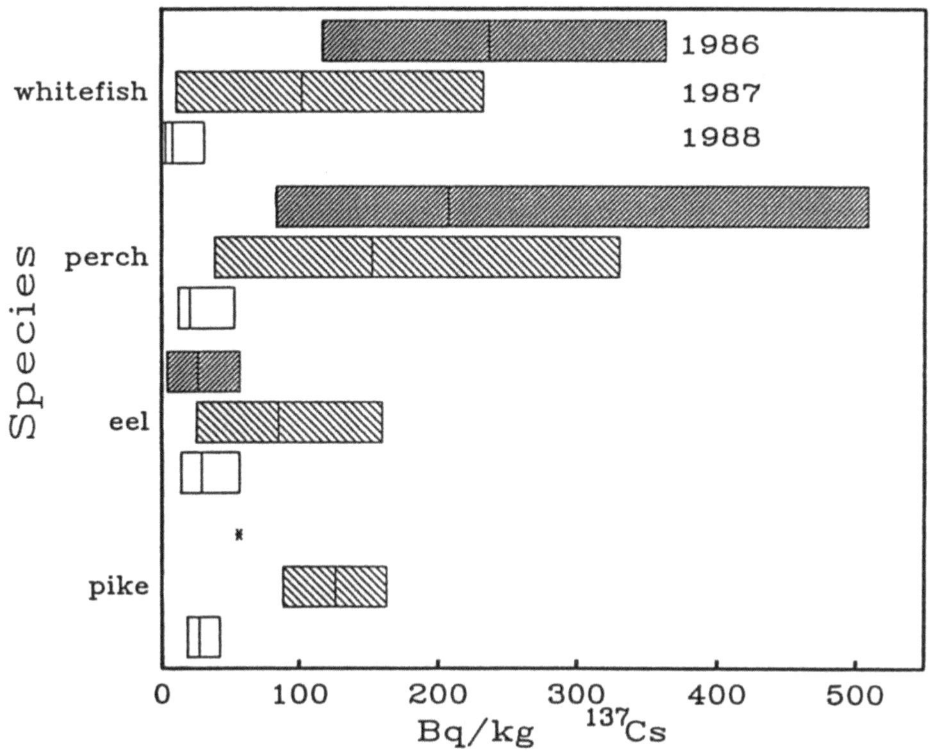


**Figure 14.6** Ranges of specific $^{137}Cs$-activity in Bq/kg wet weight for fish of four different species from Lake Constance in 1986, 1987, and 1988. The respective annual average values are indicated by a vertical line on the bar, as well as the minimum and maximum values.

anson et al., 1988) and for pike from a number of small lakes near the Schreckensee (Kissling et al., 1988). For perch, this phenomenon has tentatively been explained by shifting feeding habits which become increasingly piscivorous with age. Time constants of Cs-activity changes for predatory fish such as pike or eel are considerably longer; in Lake Constance they reached maximum $^{137}Cs$ activity as late as 1987 (Figure 14.6). Also, in the Schreckensee, a much longer retention time (more than 1 year) was obtained in pike than in perch (about 0.5 years, Geissler et al., 1989). Although it is difficult to extract concentration factors under those nonequilibrium conditions, for whitefish from Lake Constance in 1988, a concentration factor of about 3,000 was derived, which agrees reasonably well with the value of 2,000 proposed in the literature (National Council on Radiation Protection, 1985).

For herbivorous species, the three- to four-fold difference in specific $^{137}Cs$-activities between the Schreckensee (Figure 14.7) and Lake Constance (see Figure 14.6) is only slightly more than the difference in the magnitude of the

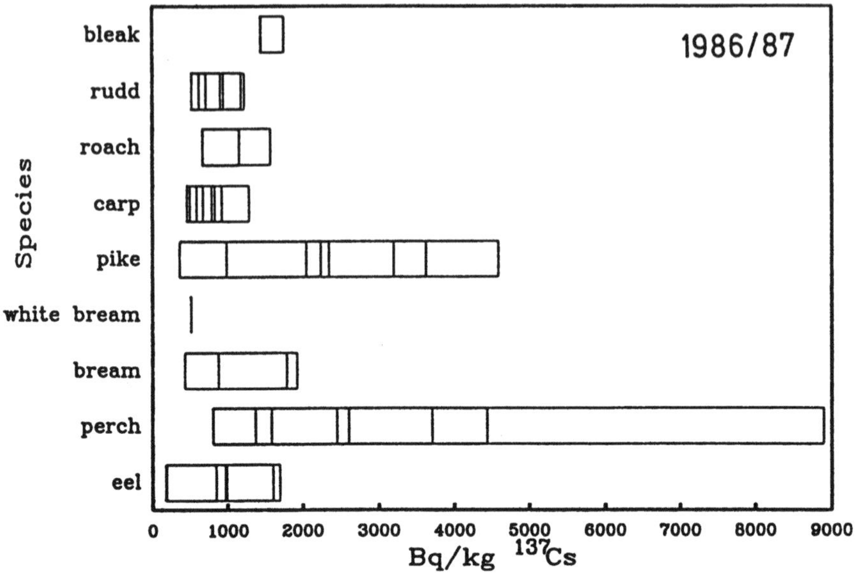


**Figure 14.7** Specific $^{137}Cs$ concentrations in Bq/kg wet weight of different fish species from the Schreckensee in 1986 and 1987. Each vertical bar corresponds to an individual sample.

inputs. This means that the initial dissolved $^{137}Cs$ concentration in the euphotic zone determined the $^{137}Cs$ uptake of these species. Perch and pike from the Schreckensee, however, accumulated $^{137}Cs$ at much higher rates than in Lake Constance. This is an indication of major differences in the food-webs with respect to predatory fish between both lakes. Whereas concentration factors for herbivorous fish from the Schreckensee in 1988 are generally lower than 2,000, they surpass 10,000 for pike.

For perch from Swedish lakes, Hakanson et al. derived a strong negative correlation between specific $^{137}Cs$ activity and the calcium-magnesium hardness of the water (Hakanson et al., 1988). By contrast, no significant influence of potassium concentrations was observed, which is in contradiction to previous statements (Petersen et al., 1986). This can be regarded as another indication of the competition between scavenging by settling particles and biological uptake.

## 14.5 Sedimentation Efficiencies of Different Radionuclide Species

Scavenging by settling particles is of general importance for the removal of contaminants introduced across the water surface (Stabel, 1985; Sigg, 1985; Eadie and Robbins, 1987). It has been shown for both Lake Constance and Lake Zürich (Santschi et al., 1988; see Chapter 13) that a peak in sedimentation flux due to autochthonous calcite precipitation occurred in May, shortly after the input event (Lindner et al., 1989), which deposited a considerable fraction of the radionuclides onto the bottom sediments. Recently it has been shown both by model calculations performed by J. Robbins and by extraction experiments with sediments, that clay mineral particles were the carriers of $^{137}Cs$ in the sedimentation process and that $^{137}Cs$ could be incorporated irreversibly by them. The sinking velocity of these particles was enhanced during the calcite precipitation event due to their nucleation of calcite crystal growth resulting in a maximum in the $^{137}Cs$ sedimentation flux in that period. The sedimentation efficiencies of Ru and Cs isotopes differed, as a result of their differing chemical speciations. This phenomenon had, in principle, already been known since the period of atmospheric nuclear weapon testing (Haberer, 1969).

We followed this approach by extending the radionuclide analysis of settling particles from Lake Constance of Mangini et al. (see Chapter 13) to the less-abundant radionuclides, $^{125}Sb$, $^{110m}Ag$, and $^{144}Ce$. Instead of looking at $^{103}Ru$, we determined the activity of the longer-lived isotope, $^{106}Ru$. Results of the cumulative radionuclide flux up to May 21, 1986, at a depth of 20 m, are given in Table 14.2. Also given are the respective isotopic inventories of sediments from the deep bottom of the Überlingersee taken in August 1987 (Figure 14.8). All values were normalized to the respective input activities. Hence, it follows that the cumulative radionuclide flux differs among these isotopes. The values for Cs and Ru agree with those given by Mangini et al. (see Chapter 13). For each radionuclide, except $^{144}Ce$, these inventories account for about 50–70% of the respective input inventories. (Details of these investigations, together with model calculations for epilimnetic scavenging, will be published elsewhere.)

We attribute the higher scavenging efficiencies of $^{106}Ru$ and $^{125}Sb$ as compared to $^{137}Cs$, to their existence in an anionic and oxidic form, in which they can act as electron acceptors and thus participate in metabolic electron transfer process by phytoplankton (Santschi et al., 1988). Support for this hypothesis

**Table 14.2** Cumulative sedimentation flux and sediment inventory from the Überlingersee for five long-lived Chernobyl radionuclides[a]

| Isotope | Cumulative radionuclide flux by settling particles[b] | Sediment inventory,[c] August 1987 |
|---|---|---|
| $^{137}Cs$ | 0.17 ± 0.02 | 0.40 ± 0.15 |
| $^{106}Ru$ | 0.36 ± 0.05 | 0.64 ± 0.23 |
| $^{125}Sb$ | 0.50 ± 0.11 | 0.67 ± 0.35 |
| $^{110m}Ag$ | 0.49 ± 0.13 | 1.25 ± 0.43 |
| $^{144}Ce$ | 1.14 ± 0.38 | 1.08 ± 0.56 |

[a] Data were normalized to the respective input inventories across the water surface.
[b] Data from particle traps at 20 m depth up to May 21, 1986.
[c] For the top layer (0–2 cm) of bottom sediment. Errors include the scattering of data from several sampling positions and the uncertainty of the input determination.

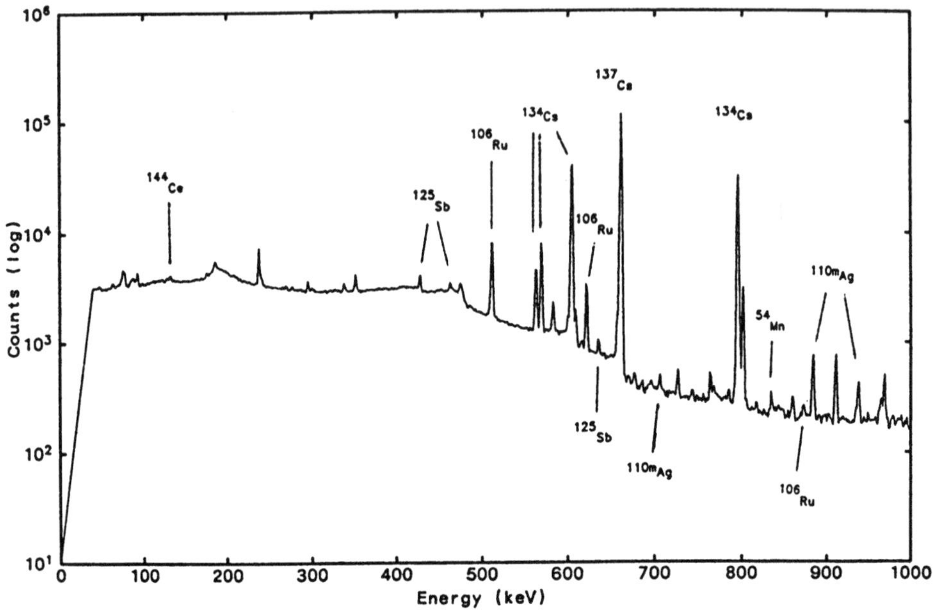


**Figure 14.8** Gamma spectrum of the sediment from the top 0 to 2 cm layer taken on August 26, 1987, at different sites from the bottom (100 to 140 m depth) of the Überlingersee. The labelled peaks indicate radionuclides from the Chernobyl fallout, whereas the unlabelled peaks can be attributed to natural radioactive isotopes in the sample and in the detector shielding.

is provided by the high accumulation rates of Ru radionuclides in sewage sludges from wastewater treatment plants in the Lake Constance area (Internat. Gewässerschutzkomm. Bodensee, 1987) and by the presence of the thermodynamically nonequilibrium species antimonite and methylantimony in rivers and oceans, which was attributed to biological activity (Andreae et al., 1981; Middelburg et al., 1988). Since Cs cations are preferentially adsorbed by clay particles (see above), the Ru/Cs isotopic ratio can be considered as an indicator for the balance between organic (phytoplankton) and inorganic (clay) contributions to the overall settling flux. In that respect, the higher Ru/Cs ratio observed for settling particles from Lake Zürich and the higher Ru scavenging efficiency correspond to the higher biomass concentration in that lake (Mäckle et al., 1987). In the same way, the isotopic Ru/Cs ratio in bottom sediments reflects the relative abundance of phytoplankton and inorganic components of settling particle flux, thus, indicating a higher contribution from phytoplankton in the western part of the Überlingersee (Figure 14.9). The high sedimentation efficiency of $^{144}Ce$ was explained by its inclusion into relatively inert nuclear fuel particles and by their sinking according to Stokes's law. This fact requires reconsideration of the interpretations given for $^{144}Ce$ precipitation in marine studies (Fowler et al., 1987; Büsseler et al., 1987; Kempe and Nies, 1987).

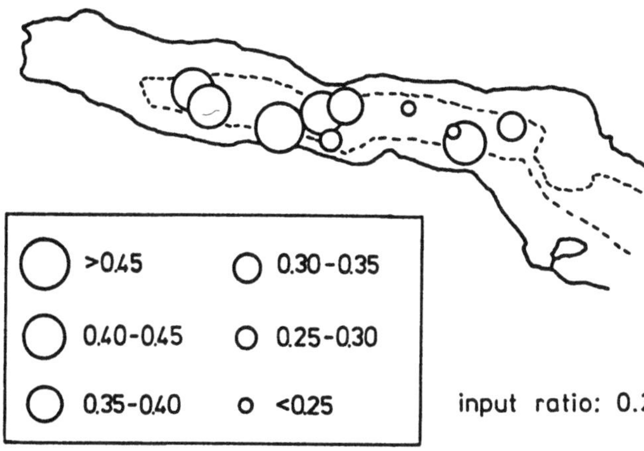


**Figure 14.9** Activity ratio of $^{106}$Ru to $^{137}$Cs, corrected for decay since May 1, 1986, in the uppermost sediment layer at various sampling positions in the Überlingersee in 1988.

## 14.6 A New Geochronological Marker in Sediments

The Chernobyl radionuclide pulse in the bottom sediment of Lake Constance was almost completed by the end of 1986. Thus, a new marker was established whose vertical extension is in the order of the thickness of the annual deposit layer, which usually reaches a few millimeters (Linder et al. 1989). The Chernobyl fallout layer promises to become an extremely sharp marker for geochronological studies. In Figure 14.10*b* this feature is shown by the $^{137}$Cs depth profiles of sediment cores taken at different sites on the deep bottom of the Überlingersee. Recent contributions from Chernobyl can clearly be distinguished from the former broad distributions of $^{137}$Cs originating from the period of nuclear weapon testing (for the latter, see also Dominik et al., 1981). Using improved coring and sectioning techniques, tracer studies of material transport within sediments will become possible with depth and time resolution superior to previous studies, which were based upon the deposits of $^{137}$Cs originating from the nuclear weapon testing (Robbins, 1982).

By gauging the depth intervals of the $^{137}$Cs peaks originating from atmospheric nuclear weapon testing and from the Chernobyl disaster, the respective average sedimentation rates for the time intervals in question can be extracted (Robbins and Edgington, 1975). At position 10, for instance, the corresponding sedimentation rates are low (0.06 g cm$^{-2}$ year), if possible resuspension is neglected, and the sediment in the top layer is rather dense, resulting in a comparatively low value for specific $^{137}$Cs activity. Near our measurement buoy, a higher sedimentation rate (0.1 g cm$^{-2}$ year) was found, which coincides with a lower sediment density and thus a higher specific $^{137}$Cs activity. In general, sedimentation rates obtained in this way agree reasonably

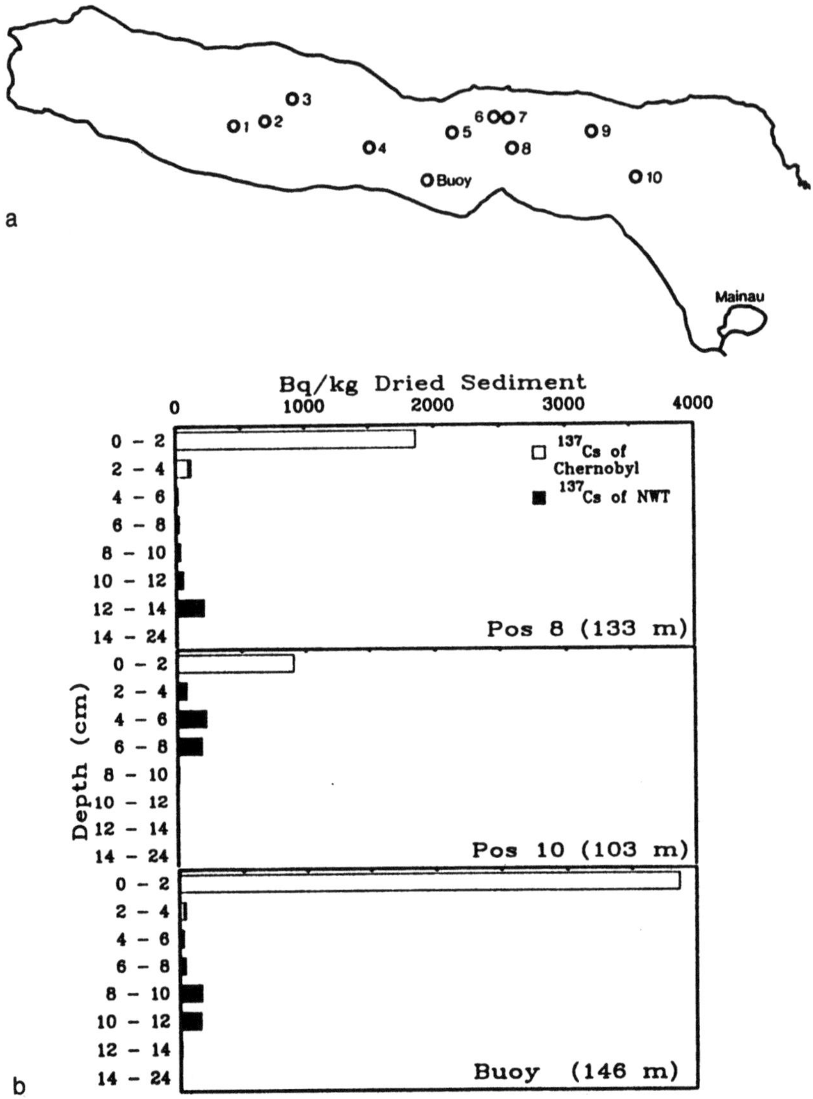


**Figure 14.10** (*a*) Sampling positions of sediment samples in the Überlingersee in 1987. The main portion of Lake Constance is located to the right. (*b*) Depth distribution of $^{137}Cs$ activity (Bq/ky dry matter) originating from Chernobyl and atmospheric nuclear weapon testing (NWT) in sediment cores taken from positions 8, 10, and at the buoy (values are given in Bq/kg dry matter).

well with those obtained by other groups for the Überlingersee (0.05 to 0.11 g/cm² yr; Stabel, 1985) and for the central part of Lake Constance (0.095 g / cm² yr—see Chapter 13; 0.08 to 0.11 g /cm² yr—von Gunten et al., 1987).

Differences in the areal inventories of bottom sediments from several positions across the Überlingersee are considerably lower than suggested by the large differences in weight-specific radioactivity. In fact, the variance of the $^{137}Cs$ inventories of 1987 sediments does not substantially exceed the variance of the input obtained from the analysis of soil samples (see Tables 14.1 and 14.2).

## 14.7 Conclusions

Perhaps the most important result of all studies of the Chernobyl fallout in Lake Constance is the observation of the rapid removal of radionuclides from the euphotic zone due to scavenging by settling particles. This process appeared to be characteristic of large, hardwater lakes. It resulted in an epilimnetic minimum of dissolved $^{137}Cs$ during the stratified period. It should be emphasized, however, that the input event happened at a point of time that might have been extremely favorable for scavenging due to the subsequent calcite precipitation which occurs regularly in the spring. The introduction of the fallout into the stratified lake after termination of calcite precipitation, e.g., in late summer or during circulation in winter, would certainly have resulted in longer retention times and different removal efficiencies and transfer coefficients for the various radionuclides. A distinct influence of the chemical speciation of different radionuclides on the scavenging efficiency could be established. $^{137}Cs$ uptake and excretion by fish in Lake Constance roughly follows the general patterns known for freshwater lakes. Compared to whitefish, time constants characterizing uptake and excretion tend to increase for species feeding on higher trophic levels. By comparison with results from the Schreckensee, the impact of the biological transfer of $^{137}Cs$ along the food-web to fish became apparent. The inventory of dissolved $^{137}Cs$ in the water, by contrast, was mainly influenced by lake morphology. The Chernobyl marker in the sediments promises information about

1. The sedimentation rates by comparison with the $^{137}Cs$ depth profile originating from atmospheric nuclear weapon testing.
2. Local sedimentation processes from the isotopic activity ratios, together with chemical information.
3. Transport processes around the sediment-water interface and within the top layers of the sediment.

Most of the work concerning scavenging and biological transfer is nearly completed. Studies of transport processes in sediments, in contrast, have just begun, keeping in mind the 30-year half-life of $^{137}Cs$.

*Acknowledgements* These investigations were supported by the Deutsche Forschungsgemeinschaft within the Special Collaborative Program (SFB 248), "Cycling of Matter in Lake Constance." Additional support was received from the Minister für Umwelt des Landes Baden-Württemberg. The cooperation of S.G. Jahn, Th. Leisner, J. Lax, K. Athanassenas, P. Brüll, H. Kühn, Dr. M. Deicher, Dr. H. Hofsäss, S. Winter, B. Landwehr, Dr. G. Scheff, S. Kissling, E. Hauser, B. Stroh, and C. Wilhelm in sampling, sample preparation, measurement, and data analysis is gratefully appreciated. In addition, we are indebted to Dr. H. Güde and Dr. H. Müller, Landesanstalt für Umweltschutz, Institut für Seenforschung (Langenargen), the Bodensee-Fischereigenossenschaft (Konstanz), the Staatliche Fischereiaufseher A. Koch and S. Grötsch (Reichenau and Uhldingen) and Forstdirektor G. Maluck (Bad Waldsee) for the provision of samples, and to Dr. L. Devell (Studsvik), Dr. L. Hakanson (Solna), and Dr. S. Luokkanen (Helsinki) for providing valuable information. Discussions with Dr. P.H. Santschi (Galveston), Dr. J.A. Robbins (Ann Arbor), Dr. A. Mangini (Heidelberg), and Dr. H.H. Stabel (Überlingen) were of help in various stages of research, and we also benefited from the critical reading of the manuscript by the editors of this volume.

## References

Andreae, M.O., Asmodé, J.F., Foster, P., and Van't dack, L. 1981. Determination of antimony (III), antimony (V) and methylantimony species in natural waters by atomic absorption spectrometry with hydride generation. *Anal. Chem.* 53:1766–1771.

Bode, M., Theile, F. W., and Metzner, R. 1987. Fallout des KKW-Unfalls Tschernobyl, Dekontamination und Partikel-Analyse. *Tagungsbericht der Jahrestagung Kerntechnik 1987 in Karlsruhe.* Deutsches Atomforum, Bonn, p. 279–282.

Broda, R. 1987. Gamma spectroscopy analysis of hot particles from the Chernobyl fallout. *Acta Physica Polonica* B 18:935–950.

Buesseler, K.O., Livingston, H.D., Honjo, S., Hay, B.J., Manganini, S.J., Degens, E., Ittekkot, V., Izdar, E., and Konuk, T., 1987. Chernobyl radionuclides in a Black Sea sediment trap. *Nature* 329:825–828.

Cuddihy, R.G., Finch, G.L., Newton, G.J., Hahn, F.F., Mewhinney, J.A., Rothenberg, S.J., and Powers, D.A. 1989. Characteristics of radioactive particles released from the Chernobyl nuclear reactors. *Environ. Sci. Technol.* 23:89–95.

Czarnecki, J., Cartier, F., Honegger, P., and Zurkinden, A. 1987. Bodenverstrahlung in der Schweiz aufgrund des Reaktorunfalles in Chernobyl. In: *Proceedings of the Symposium "Radioaktivitätsmessungen in der Schweiz nach Tschernobyl und ihre wissenschaftliche Interpretation," 1986, Bern.* Bundesamt für Gesundheitswesen, Bern, p. 93–109.

Devell, L., Tovedal, H., Bergström, U., Appelgren, A., Chyssler, J., and Andersson, L. 1986. Initial observations of fallout from reactor accident at Chernobyl. *Nature* 321:192–193.

Devell, L. 1987. Nuclide composition of Chernobyl hot particles. *Report Studsvik (NP-87/119), Nyköping,* 1987.

Dominik, J., Mangini, A., and Müller, G. 1981. Determination of recent deposition rates in Lake Constance with radioisotopic methods. *Sedimentology* 28:653–677.

Eadie, B.J. and Robbins, J.A. 1987. The role of particulate matter in the movement of contaminants in the Great Lakes. In: Hites, R.A. and Eisenreich, S.J. (eds.). *Sources and Fates of Aquatic Pollutants. Advances in Chemistry Series,* No. 216, American Chemical Society, Washington, D.C., p. 319–364.

Eidgenössische Anstalt für Wasserversorgung, Abwasserreinigung und Gewässerschutz (EAWAG). 1988. *Das Verhalten von Radionukliden in natürlichen Gewässern. Abschätzung der Gefährdung des Trinkwassers in der Schweiz im Falle eines radioaktiven Ausfalls.* (unpublished report), Dübendorf 116 p.

Elster, H.J. 1977. Der Bodensee—Bedrohung und Sanierungsmöglichkeiten eines Ökosystems. *Naturwissenschaften* 64:207–215.

Fowler, S.W., Buat-Menard, P., Yokojama, Y., Ballestra, S., Holm, E., and Van Nguyen, H. 1987. Rapid removal of Chernobyl fallout from Mediterranean surface waters by biological activity. *Nature* 329:56–68.

Geissler, A., Kissling, S., Niebuhr, J., Stroh, B., Thomas, H., Wilhelm, C., and Zibold, G. 1989. Barsche und Hechte noch erheblich belastet. *Fisch und Fang* 2/1989:40–41.

Haberer, K. 1969. *Radionuklide im Wasser.* Thiemig, München.

Hakanson, L., Andersson, T., Neumann, G., Nilsson, A., and Notter, M. 1988. *Cesium-137 in Perch in Lakes from Northern Sweden after Chernobyl—Present Situation, Relationships, Trends.* Report 3497. Statens naturvardsverk, Solna, 136 pp.

Hohenemser, C., Deicher, M., Ernst, A., Hofsäss, H., Lindner, G., and Recknagel, E. 1986a. Chernobyl: an early report. *Environment* 28:6–43.

Hohenemser, C., Deicher, M., Hofsäss, H., Lindner, G., Recknagel, E., and Budnick, J.I. 1986b. Agricultural impact of Chernobyl: a warning. *Nature* 321:817.

Internationale Gewässerschutzkommission für den Bodensee. 1987. *Die Entwicklung der Radioaktivität im Bodensee nach dem Unfall von Tschernobyl.* Report No. 36, Langenargen, 75 pp.

International Nuclear Safety Advisory Group (INSAG). 1986. *Summary Report on the Post-Accident Review Meeting on the Chernobyl Accident.* Safety series No. 75-INSAG-1. International Atomic Energy Agency, Vienna.

Jones, G.D., Forsyth, P.D., and Appleby, O.G. 1986. Observation of $^{110m}Ag$ in Chernobyl fallout. *Nature* 322:312.

Kempe, S. and Nies, H. 1987. Chernobyl nuclide record from a North Sea sediment trap. *Nature* 329:828–831.

Kissling, S., Niebuhr, J., Stroh, B., and Zibold, G. 1988. Radioaktive Kontamination von Fischen im Landkreis Ravensburg. *Fisch und Fang* 8/1988:12.

Kolehmainen, S.E. 1969. *White Oak Lakes Studies.* ORNL 4446, Oak Ridge.

König, L.A., Schüttelkopf, H., Erat, S., Fessler, H., Hempelmann, S., Maurer, K., Pimpl, M., and Radziwill, A. 1986. Der Reaktorunfall von Tschernobyl—Meßergebnisse des Kernforschungszentrums Karlsruhe. *Kernforschungszentrum Karlsruhe GmbH (KfK 4115).* Karlsruhe, 86. p.

Lindner, G., Deicher, M., Eckmann, R., Hofsäss, H., Jahn, S.G., Müller, W., Petermann, D., Pfeiffer, W., Teufel, S., Wahl, U., Winter, S., and Recknagel, E. 1987. Überregionale Aspekte der Tschernobyl-Radioaktivität im Bodensee-Gebiet. In: *Proceedings of the Symposium "Radioaktivitätsmessungen in der Schweiz nach Tschernobyl und ihre wissenschaftliche Interpretation," 1986, Bern.* Bundesamt für Gesundheitswesen, Bern, p. 312–322.

Lindner, G., Brüll, B., Athanassenas, K., Steffen, W., and Kissling, S. 1988a. Gelöste Cäsium-Radioaktivität im Schreckensee. In: *Jahresbericht Nukleare Festkörperphysik Konstanz 1988.* Universität Konstanz, p. 107–108.

Lindner, G., Jahn, S., Pfeiffer, W., Petermann-Seyboldt, D., Athanassenas, K., Brüll, P., Kühn, H., Eckmann, R., and Becker, M. 1988b. *Cäsium-Radioaktivität in Fischen aus dem Bodensee.* In: *Jahresbericht Nukleare Festkörperphysik Konstanz 1988.* Universität Konstanz, p. 105–106.

Lindner, G., Pfeiffer, W., Petermann-Seyboldt, D., Recknagel, E., Kleiner, J., and Stabel, H.H. 1988c. Die Abscheidung radioaktiver Isotope aus Tschernobyl im Bodensee durch Sinkstoffe. In: *Jahresbericht Nukleare Festkörperphysik Konstanz 1988.* Universität Konstanz, p. 98–100.

Lindner, G., Steffen, W., Pfeiffer, W., Petermann-Seyboldt, D., Brüll, P., Kühn, H., Recknagel, E., and Frenzel, P. 1988d. Radioisotope aus Tschernobyl in den Sedimenten des Bodensees. In: *Jahresbericht Nukleare Festkörperphysik Konstanz 1988.* Universität Konstanz, p. 101–104.

Lindner, G. and Recknagel, E. 1988. *Tschernobyl—Auswirkungen auf die Bodensee-Region.* Thorbecke, Sigmaringen, 126 pp.

Lindner, G., Pfeiffer, W., Wahl, U., Kleiner, J., Stabel, H.H., Frenzel, P., Robbins, J.A., Giovanoli, F., Lenhard, A., and Recknagel, E. 1989. Sedimentation of longlived radionuclides in Lake Constance. In: *Proceedings of the International Conference on Heavy Metals in the Environment,* 1989, Geneva (Vernet, J.P., ed), CEP Consultants Ltd, Edinburgh, p. 449–542.

Mäckle, H., Zimmermann, U., and Stabel, H.H. 1987. *GWF—Wasser und Abwasser* 128:544–550.

Middelburg, J.J., Hoede, D., Van der Sloot, H., Van der Weijden, C.H., and Wijkstra, J. 1988. Arsenic, antimony and vanadium in the North Atlantic Ocean. *Geochim. et Cosmochim. Acta* 52:2871–2878.

National Council on Radiation Protection and Measurements (NCRP). 1985. Radiological assessment: Predicting the transport, bioaccumulation and uptake by man of radionuclides released to the environment. *NCRP Report No. 76.* Bethesda, p. 136–156.

Persson, Ch., Rohde, H., and De Geer, L.E. 1987. The Chernobyl accident—a meteorological analysis of how radionuclides reached and were deposited in Sweden. *Ambio* 16:20–31.

Petermann, D., Jahn, S.G., Pfeiffer, W., Lindner, G., and Recknagel, E. 1987a. Cäsium-Akkumulation im Schreckensee. In: *Jahresbericht Nukleare Festkörperphysik Konstanz 1987.* Universität Konstanz, p. 133–134.

Petermann, D., Jahn, S.G., Pfeiffer, W., Lindner, G., and Recknagel, E. 1987b. Cäsium in der Nahrungskette des Bodensees. In: *Jahresbericht Nukleare Festköperphysik Konstanz 1987.* Universität Konstanz, p. 131–132.

Petersen, Jr., R.C., Landner, L., and Blanck, H. 1986. Assessment of the impact of the Chernobyl reactor accident on the biota of Swedish streams and lakes. *Ambio* 15:327–331.

Poston, T.M. and Klopfer, D.C. 1988. Concentration factors used in the assessment of radiation dose to consumers of fish: A review of 27 radionuclides. *Health Physics* 55:751–766.

Prout, W.E., Russell, E.R., and Groh, H.J. 1965. Ion exchange absorption of cesium by potassium hexacyanocobalt(II)ferrat(II). *Journal of Inorganic Nuclear Chemistry* 27:473–479.

Raunemaa, T., Lehtinen, S., Saari, H., and Kulmala, M. 1987. 2–10 $\mu$m sized hot particles in Chernobyl fallout to Finland. *J. Aerosol. Sci.* 18:693–696.

Robbins, J.A. 1982. Stratigraphic and dynamic effects of sediment reworking by Great Lakes zoobenthos. *Hydrobiologia* 92:611–622.

Robbins, J.A. and Edgington, D.N. 1975. Determination of recent sedimentation rates in Lake Michigan using $^{210}$Pb and $^{137}$Cs. *Geochim. Cosmochim. Acta* 39:285–304.

Santschi, P.F., Bollhalder, S., Farrenkothen, K., Lueck, A., Weber, C., Hermann, A., and Schuepbach, M.R. 1987a. Chemische Speziierung und Mobilität der Nuklide des Tschernobyl-Fallouts in der Umwelt. In: *Proceedings of the Symposium "Radioaktivitätmessungen in der Schweiz nach Tschernobyl und ihre wissenschaftliche Interpretation," 1986, Bern.* Bundesamt für Gesundheitswesen, Bern, p. 132–141.

Santschi, P.H., Bollhalder, S., Camani, M., Farrenkothen, K., Görlich, W., Haesler, S., Heiz, H., Lueck, A., Schuler, Ch., Sturm, M., Völkle, H., Weber, C., and Zingg, S. 1987b. Radionuklide des Tschernobyl-Fallouts in natürlichen Gewässern: Auswasch-, Verdünnungs-, Eliminierungs- und Anreicherungsprozesse. In: *Proceedings of the Symposium "Radioaktivitätsmessungen in der Schweiz nach Tschernobyl und ihre wis-*

*senschaftliche Interpretation," 1986, Bern.* Bundesamt für Gesundheitswesen, Bern, p. 323–338.

Santschi, P.H., Bollhalder, S., Farrenkothen, K., Lueck, A., Weber, C., and Zingg, S. 1987c. Messung der atmosphärischen Deposition von Tschernobyl-Radionukliden (Gesamt- und Trockendeposition) im Raume Dübendorf ZH. In: *Proceedings of the Symposium "Radioaktivitätsmessungen in der Schweiz nach Tschernobyl und ihre wissenschaftliche Interpretation," 1986, Bern.* Bundesamt für Gesundheitswesen, Bern, p. 176–187.

Santschi, P.H., Bollhalder, S., Farrenkothen, K., Lueck, A., Zingg, S., and Sturm, M. 1988. Chernobyl radionuclides in the environment: Tracers for the tight coupling of atmospheric, terrestrial, and aquatic geochemical processes. *Environ. Sci. Technol.* 22:510–516.

Schubert, P. and Behrend, U. 1987. Investigations of radioactive particles from the Chernobyl fall-out. *Radiochimica Acta* 41:149–155.

Sigg, L. 1985. Metal transfer mechanisms in lakes, the role of settling particles. In: Stumm, W. (ed.), *Chemical Processes in Lakes.* Wiley, New York, p. 283–310.

Stabel, H.H. 1985. Mechanisms controlling the sedimentation sequence of various elements in prealpine lakes. In: Stumm, W. (ed.), *Chemical Processes in Lakes.* Wiley, New York, p. 143–168.

Thomann, R.V. 1981. Equilibrium model of fate of microcontaminants in diverse aquatic food chains. *Canadian Journal of Fishery and Aquatic Sciences* 38:280–296.

Turner, D.R., Whitfield, M., and Dickson, A.G. 1981. The equilibrium speciation of dissolved components in freshwater and seawater at 25°C and 1 atm pressure. *Geochimica Cosmochimica Acta* 45:855–881.

Van Dam, H. 1986. Silver from Chernobyl. *Nature* 324:216.

Van der Veen, J., van der Wijk, A., Mook, W.G., and de Meijer, R.J. 1986. Core fragments in Chernobyl fallout. *Nature* 323:399–400.

von Gunten, H.R., Sturm, M., Erten, H.N., Rössler, E., and Wegmüller, F., 1987. Sedimentation rates in the central Lake Constance determined with Pb-210 and Cs-137. *Schweiz. Z. Hydrol.* 49:275–285.

Wahl, U. 1987. Transportmechanismen radioaktiver Isotope aus dem Tschernobyl-Fallout. Thesis, 108 p.

Wahl, U., Lindner, G., and Recknagel, E. 1989. Radioaktive Partikel im Tschernobyl-Fallout. In: Köhnlein, W., Traut, H., and Fischer, N. (eds.), *Die Wirkung niedriger Strahlendosen—biologische und medizinische Aspekte.* Springer, Berlin-Heidelberg, p. 165–176.

Weiss, W., Stockburger, H., Sartorius, H., Sittkus, A., Rath, K.H., Loosli, H., Völkle, H., and Rozanski, K. 1987. Der Einfluß von Transport-, Mischungs- und Depositionsvorgängen auf die Aktivitätskonzentration gas- und aerosolförmiger Radionuklide der bodennahen Luft in Europa. In: *Proceedings of the Symposium "Radioaktivitätsmessungen in der Schweiz nach Tschernobyl und ihre wissenschaftliche Interpretation," 1986, Bern.* Bundesamt für Gesundheitswesen, Bern, p. 32–53.

Zintz, K. 1986. *Fischereiliche Nutzung von Stillgewässern in Naturschutzgebieten.* Verlag Josef Markgraf, Langen, 531 pp.

# 15

# Lacustrine Carbonates as Geochemical Archives of Environmental Change and Biotic/Abiotic Interactions

Kerry Kelts and Michael Talbot

ABSTRACT Primary, diagenetic, and detrital carbonates are common in the sediments of large lakes. Their location, mineralogy, geochemistry, and stable isotopic composition may provide specific information on a lake's chemical and environmental history. Analyses of such archive sequences can also provide a proxy record of biotic/abiotic interactions within a lake, and supply information on the magnitude and rate of environmental change within a lacustrine system. Detrital carbonates can be used to map sources of clastic input and resedimentation processes. Surface and littoral precipitates may record seasonal productivity and temperature changes, as well as changing water composition. In isothermal settings, the carbonate shells of benthic organisms can be used to monitor variations in bottom-water isotopic composition.

Within anoxic sediments, early formed diagenetic carbonates carry carbon isotopic signatures reflecting the intensity of bacterial degradation of organic matter, chiefly via sulfate reduction or methanogenesis. Their oxygen-18 composition reflects the composition of the bottom waters during the time of host sediment accumulation.

Following an example from the Quaternary sediments of crater Lake Bosumtwi, Ghana, we illustrate how sequential isotopic changes in primary and diagenetic carbonates document periods of hydrological surplus and deficit.

## 15.1 Introduction

Lacustrine sediments may contain a high resolution history of variations in biotic/abotic interactions and other environmentally important processes in lakes. These variations are themselves dependent upon factors such as climate, basin hydrology, vegetational succession, and land use. As we become more aware of the global scale of human modification of the Earth System, it becomes increasingly important to 1) separate anthropogenic from natural environmental changes on a decadal time scale, and 2) test conceptual models of possible future feedbacks by checking predictions against environmental records of the past. Lacustrine sedimentary sequences probably represent one of the most refined archives of environmental change available. In modern systems, researchers can directly study primary production of organic matter, the microbial degradation of this and allochthonous organic matter in the water column, or during burial in oxic and anoxic sediments (Figure 15.1), as well as the distribution and accumulation of mineral matter. However, long time series are needed to understand the full consequences of biologically induced interactions between these two principal components of lacustrine sediments, and their implications in terms of the evolution of the system as a whole. There might be considerable value, for example, in having two to three centuries of seasonal productivity records accessible from core analyses. How may the signatures of these and other environmentally significant parameters be recorded in the sediments?

Sedimentary deposits include clastic mineral grains—a range of organic components derived from the lake biota—endogenic minerals precipitated from the water column, and diagenetic precipitates from sediment pore waters. Many of these may preserve useful environmental information. Geochemical signals contained in biogenic and authigenic (i.e., endogenic and diagenetic) carbonate minerals have proved particularly valuable and may readily enter the geologic record as a permanent register of environmental conditions in a lake at the time of precipitation. These signals may come from both the mineral phase of the carbonate mineral itself, which reflects salinity, and from Mg/Ca conditions within the waterbody (Müller et al., 1972); they may be either in the form of minor element concentrations (e.g., Sr/Ca) or reflected in the isotopic ratios

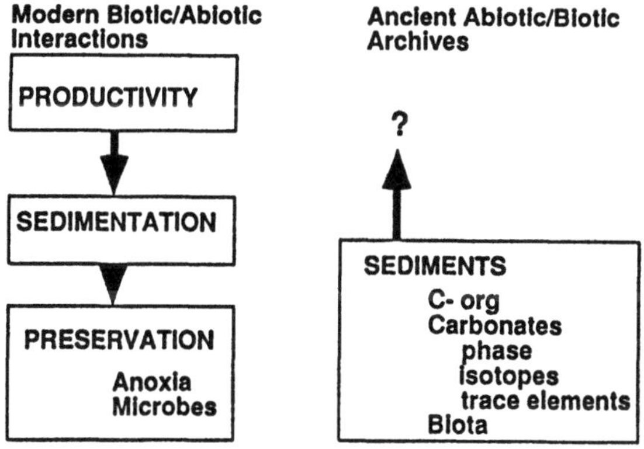


**Figure 15.1** Strategies for research in both modern and ancient large lacustrine systems. Whereas productivity and biological processes may be followed through seasonal succession in modern lakes, the sediments store only a partial record that must be deciphered from environmental signals, eg., geochemical carbonates.

($^{18}O/^{16}O$, $^{13}C/^{12}C$, $^{87}Sr/^{86}Sr$) of the carbonate mineral phases. Stuiver (1970) appears to have been among the first to suggest that the isotopic composition of lake carbonates represented potential archives of paleoclimatic information, but applications were initially hindered by uncertainties of interpretation, as well as doubts as to whether carbonates registered isotopic equilibrium conditions. In recent years, however, much progress has been made towards resolving many of these problems. Experimental studies, detailed monitoring of calcite precipitation in lakes, and the culturing of freshwater molluscs under laboratory conditions have confirmed that both biogenic and abiogenic nonmarine carbonates form in isotopic equilibrium with their growth medium (e.g., Stuiver, 1970; Fritz and Poplawski, 1974; Abell, 1982; McKenzie, 1985; Lister, in press). This means that the mineral can reliably reflect the composition and physicochemical conditions of the medium in which it was formed. The purpose of this paper is therefore to review the stable isotopic signnatures contained in typical lacustrine carbonate minerals, and to develop strategies for the interpretation of these signatures in a variety of lacustrine systems.

## 15.2 Carbonate Deposits in Large Lakes

Carbonate minerals are a common component of the sediments of many large lakes. It is a frequent misconception that catchments rich in carbonate rocks are a prerequisite for the precipitation of carbonate minerals in lakes (e.g., Siegenthaler and Eicher, 1986), but this is not so. Calcium and bicarbonate are the two most common ions released by the chemical weathering of almost all types of terrains (Livingstone, 1963); Blatt et al., 1980) and are usually present in sufficient concentrations in lake waters to allow the precipitation of some carbonate when suitable conditions are attained (Eugster and Hardie, 1978; Hardie et al., 1978). Many of the large lakes of tropical Africa, for example, periodically precipitate carbonate minerals, although in most cases their catchments consist largely of crystalline basement and volcanic rocks. That the carbonates in lakes such as these may comprise only a minor proportion of the total sediment volume in no way reduces their value as potential environmental indicators. Modern analytical techniques permit the determination of mineralogy, trace element geochemistry, and isotopic composition on extremely small samples, and, as we shall show, the environmental signatures contained in carbonate minerals from carbonate-poor lakes are just as reliable as those from systems dominated by carbonate sedimentation.

A range of carbonate deposits can accumulate in large lakes. Figure 15.2 provides a generalized summary of the principal sites of accumulation. The carbonate minerals have a variety of origins:

1. Rivers may carry clastic carbonate particles eroded from the drainage basin. These can be deposited in marginal areas, transported directly to basinal areas by turbidity underflows, or reworked to deeper water

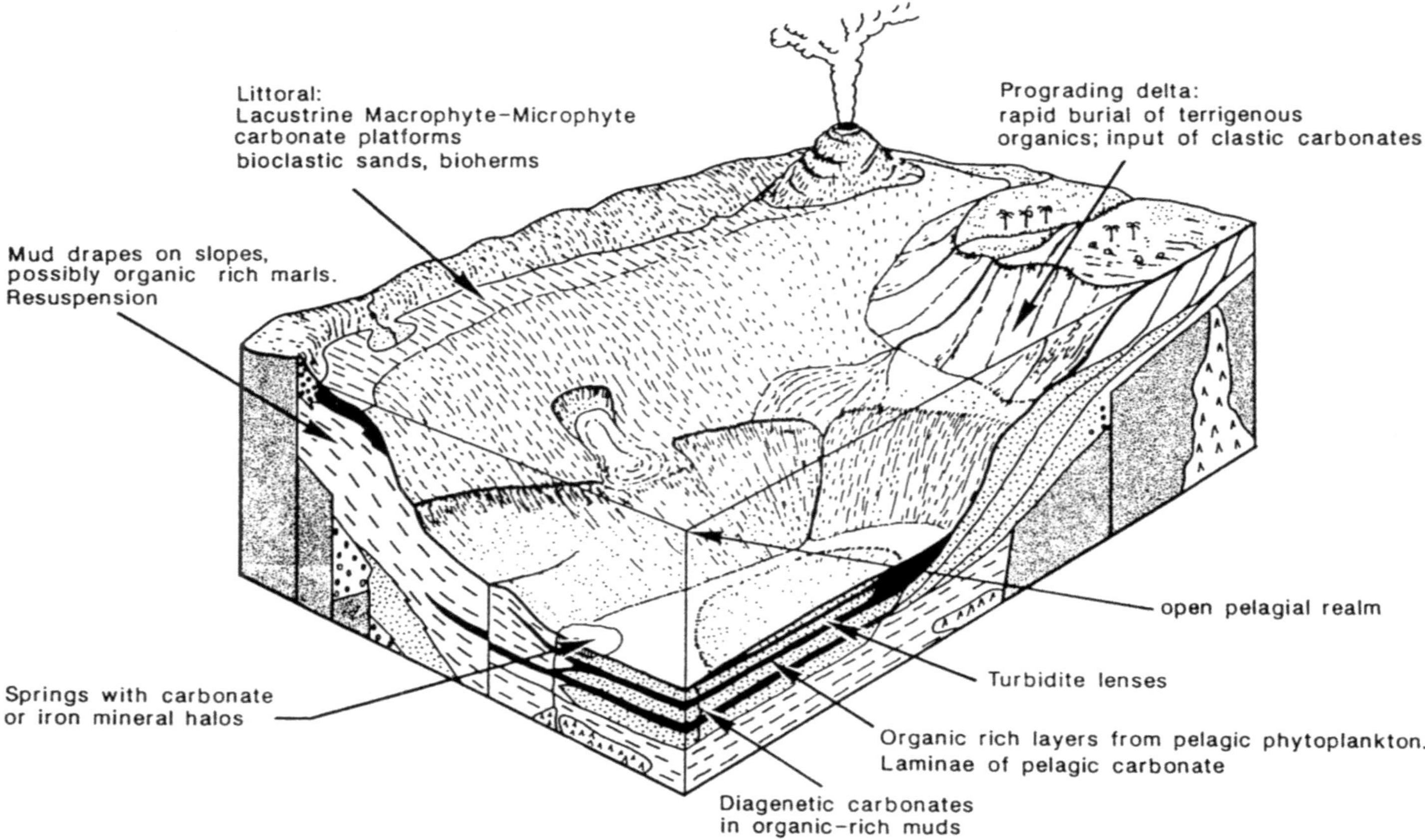


**Figure 15.2** Schematic tectonic setting of a large, deep tropical lake having carbonate facies: an example of the interplay of depositional realms and processes in large lakes such as the East African deep rift lakes.

by storm surges or slumping. In lakes adjacent to arid regions, windblown dust may also represent a significant source of detrital carbonate.

2. The littoral girdle may be the accumulation site for massive macrophyte and charophyte chalks, algal and bacterial carbonate bioherms (Dean and Fouch, 1983; Dixit, 1984; Casanova, 1986a; Cohen and Thouin, 1987), or even extensive spreads of bioclastic (chiefly molluscan) debris, as recently demonstrated from Lake Tanganyika (Cohen and Thouin, 1987).
3. If surface waters are seasonally or permanently supersaturated, the rain of crystals precipitated in the epilimnion may blanket the bottom with pelagic carbonate mud or silt (Neev and Emery, 1967; Kelts and Hsü, 1978; Hardie et al., 1978).
4. Precipitation from the upper part of the water column can also occur where inflowing river water rich in dissolved bicarbonate mixes with the lake water. Characteristic carbonate accumulations, often called chemical deltas, may be produced in this way (Hardie et al., 1978; Degens and Kurtman, 1978).
5. Carbonate minerals can precipitate from the mixing of waters around subaqueous hydrothermal vents, $CO_2$ vents, or cold groundwater springs (e.g., Scholl, 1960; Scholl and Taft, 1964; Hillaire-Marcel and Casanova, 1987; Bahrig, 1988).
6. Finally, we must be aware that the remains of carbonate-producing organisms (principally plants, molluscs, and ostracods) can occur scattered in almost any lacustrine environment.
7. Microbial processes in organic-rich muds may lead to the formation of diagenetic carbonate minerals very near the sediment/water interface (Talbot and Kelts, 1986; Bahrig, in press).

Carbonate facies types and facies distribution in large modern lakes are, in general, poorly known. Much of the carbonate is either of biological origin or a direct result of biological activity (Dean and Fouch, 1983; Eugster and Kelts, 1983), so it may be presumed that this is often reflected in the distribution of the sediments themselves (Fig. 15.3). In addition, carbonates that are apparently unrelated to any particular organism or organic activity are also widely developed in some basins. Spreads of ooid sand, for example, have been documented from the Great Salt Lake (Eardley, 1938), Lake Tanganyika (Cohen and Thouin, 1987), and Lake Urmia (Iran) (Kelts and Shahrabi, 1986); abiogenic cements may fill the pores of littoral and sublittoral sediments to form beachrocks and surface crusts (Binkley et al., 1980; Kelts and Shahrabi, 1986; Cohen and Thouin, 1987).

Mg/Ca molar ratios apparently determine which primary calcium carbonate mineral phase precipitates from a water mass (Müller et al, 1972). There

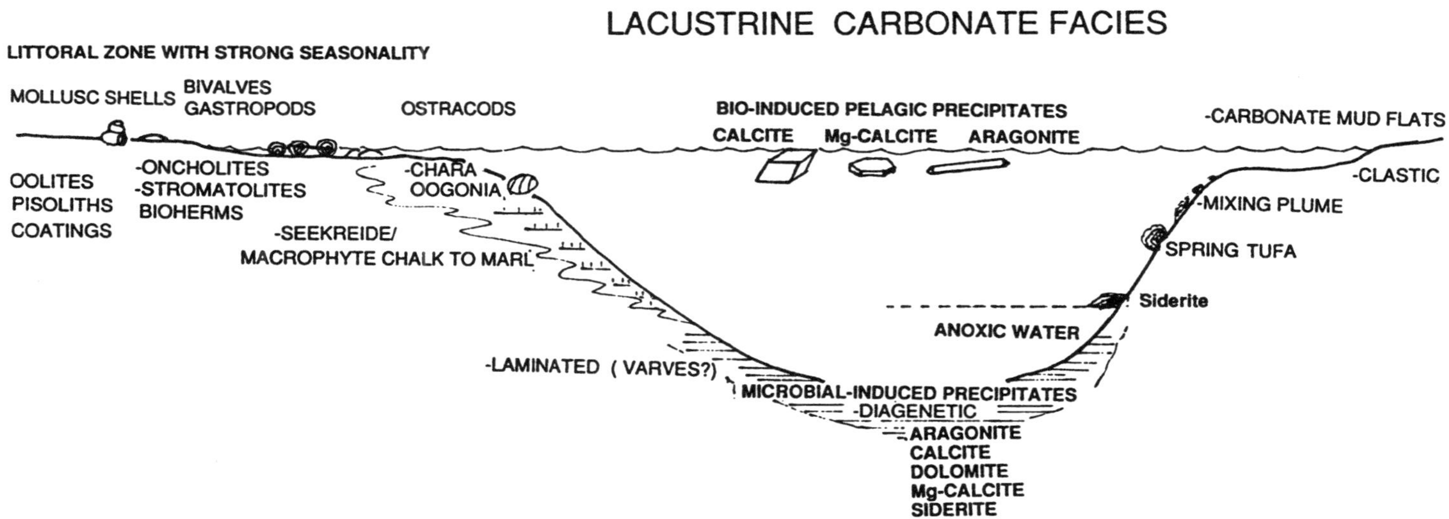


**Figure 15.3** Highly schematic overview of carbonate mineral occurrences in lake systems. Examples of sites, type, and mineral phases of common carbonates which store geochemical signals of ambient conditions.

is a general progression—from low magnesian calcite to high magnesian calcite to aragonite—as Mg/Ca increases; values higher than seven generally produce primary aragonite. Lake Tanganyika (410 mg/l TDS), for example, has a Mg/Ca of 5–8.5, and, in recent millenia, has precipitated both aragonite and high Mg-calcite (Stoffers and Hecky, 1978). Mg/Ca ratios in the fresh waters of large lakes vary, depending upon drainage basin geology, water residence time, and the history of calcium carbonate precipitation. Thus, high Mg/Ca ratios are not a direct proxy for high salinities. Some hypersaline, alkaline East African rift lakes have low Mg/Ca ratios and so precipitate low magnesian calcites (Tiercelin et al, 1987).

The symposium that this volume describes concentrated on large lakes that are fresh water today, yet most of these are able to change to closed hydrologic systems with brackish to saline waterbodies within a few tens to hundreds of years. Conversely, many of the world's great salt lakes were freshwater only a few millenia ago. For example, during the Late Quaternary, the 51,000 $km^2$, 300 m deep, freshwater Glacial Lake Bonneville reverted within a few hundred years to a 5000 $km^2$, 8 m deep Great Salt Lake, similar to today's waterbody. Many of the great African lakes have also undergone derastic water level and salinity changes during the last 20,000 years (Street-Perrott and Harrison, 1984); Talbot, 1988). Drawdown of the Great Salt Lake is recorded in only 20 cm of sediment from a piston core taken in the basin center (Eugster and Kelts, 1983; Spencer et al, 1984). One major advantage of many large lake deposits for tracing the history of environmental changes is that the rain of particles to basinal areas can accumulate undisturbed by bioturbation, bottom currents, or excessive gas ebullition. Thus, fine laminae can be preserved which allow us, in some cases, to resolve the record on an annual or even seasonal scale. Successions of laminae may then provide a continuous sedimentary record. As we shall show, stable isotope signatures preserved in lacustrine carbonates from such sequences can provide a particularly sensitive record of environmental change.

## 15.3 Stable Isotopes—An Overview

Stable isotopes are atoms in which the nucleus has the same number of protons but varying numbers of neutrons. $^{16}O$ comprises about 99.763% of the free oxygen, with about 0.199% $^{18}O$; $^{12}C$ comprises about 98.89% of the global carbon, with $^{13}C$ contributing about 1.11% (Hoefs, 1987). Mass differences of the compounds are responsible for slight variations in physico-chemical characteristics (heat capacity, vapor pressure, melting point, entropy, etc.) and thus reaction rates. Bonds with lighter isotopes break more easily, for example, than with heavier ones. Isotopic effects due to the equilibrium and kinetic processes that accompany evaporation, condensation, crystallization, melting, adsorption, and diffusion are responsible for fractionation of isotopes between various reservoirs. Some reservoirs, such as carbonate minerals, can become a permanent part of the sedimentary record. In carbonates, both oxygen and carbon isotopes are bound tightly enough to retain a record of environmental conditions at the time of mineral formation.

Stable isotope analysis of carbonates is now a routine tool in the study of sediments; details of the method are explained in numerous textbooks (Fritz and Fontes, 1980, 1986; Anderson and Arthur, 1983; McKenzie, 1985; Faure, 1986; Siegenthaler and Eicher, 1986; Hoefs, 1987). Measurements are performed in a gas-source mass spectrometer on $CO_2$ samples released by dissolution of carbonate in phosphoric acid. Ions are separated by mass, measured against known standards—e.g., PDB (*Belemnitella americana* from the Peedee Formation, South Carolina) for oxygen and carbon in carbonates, SMOW (Vienna Standard Mean Ocean Water) for oxygen in water and carbonates—and reported in a conventional delta notation (δ‰). Modern mass spectrometers can analyze as little as 0.15 mg of carbonate with a standard deviation of ca. 0.07‰. The development of laser-probe-sourced mass spectrometers will eventually permit the analysis of even smaller samples.

**Oxygen and Carbon Isotopes in Carbonates** Oxygen isotopic fractionation in carbonates is strongly temperature dependent. In sedimentary calcites, for example, a temperature change of 1°C will cause a shift of ca. 0.25‰ in the $\delta^{18}O$ of the mineral. Carbon isotopes, on the other hand, are little affected by temperature under near-surface conditions. Figure 15.4 provides a guide for possible stable isotopic signals related to various source types in lacustrine systems.

*Oxygen* In contrast to the marine reservoir, the oxygen isotopic composition of lake waters is extremely variable. Carbonates in Late Quaternary sedimentary sequences from temperate lakes typically have oxygen isotopic compositions that vary by no more than a few parts per mill (‰). These variations have generally been ascribed to the effects of temperature, either upon the isotopic composition of the rain supplied to the lake, or upon oxygen isotope fractionation during carbonate precipitation (Stuiver, 1970; Fritz et al., 1975; Mörner and Wallin, 1977; Siegenthaler and Eicher, 1986). In mid- to low-latitude lakes, however, the range in carbonate oxygen isotopic composition may be several per mil (>10)—much too great for any reasonable temperature variation—so other factors must also be involved. Normally, the source of large variations in the oxygen isotopic composition of lacustrine carbonates must be sought in the water from which mineral precipitation has occurred. The oxygen isotopic composition of a lake's water is mainly the result of the following factors:

1. The average composition of precipitation in the drainage basin, which is dependent upon latitude, continentality, altitude, and temperature of precipitation (Siegenthaler and Eicher, 1986).
2. Basin hydrology, including possible meltwater contributions.

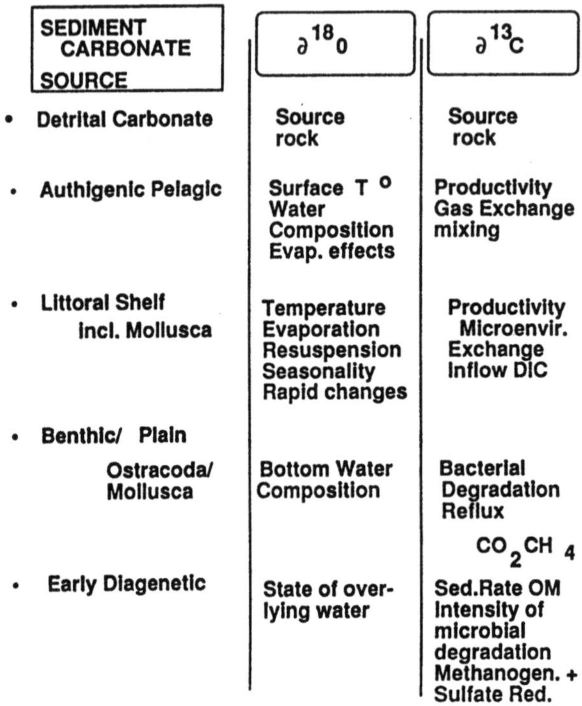


**Figure 15.4** Overview of controls on stable isotopes in lake carbonates. Types of stable isotope signals that may be expected from different lacustrine carbonate reservoirs.

3. Evaporative evolution and residence time of the water body, which are principally a function of the precipitation/evaporation balance, air humidity, and water salinity (Buchardt and Fritz, 1980; Gat, 1981; Gonfiantini, 1986).

Large, well-mixed lakes with positive water balances normally exhibit a fairly uniform isotopic composition away from points of inflow. However, significant variations may occur where waters have long residence times, are stably stratified, or where evaporation begins to have a major effect upon water composition. In Lake Tanganyika (a meromictic lake with a residence time of ca. 1000 years—Hecky, 1978), for example, the epilimnion and hypolimnion have $\delta^{18}O$ values of ca. 3.5‰ and 4.2‰, respectively (Craig, 1975). Lake Chad, a shallow lake with a negative water balance, displays a marked horizontal isotopic gradient; $\delta^{18}O$ values range from around $-3$ to $-5$‰ at points of inflow in the south, to $+12$‰ in the semi-arid north (Fontes et al., 1970). Variations through time in the relative effects of the several factors that influence the isotopic composition of lake waters can lead to differences of as much as 10–20‰ in the $\delta^{18}O$ of the precipitated carbonates. Oxygen isotopic composition of the precipitate will also change if the mineralogy changes; for example, aragonite is ca. 0.6‰ more positive (i.e., contains more $^{18}O$) than calcite formed under the same conditions, while dolomite is 3–4‰ more positive (Anderson and Arthur, 1983).

*Carbon* In lakes, the carbon isotope reservoir is a summation of the following factors:

1. $CO_2$ exchange rates with the atmosphere, which are a function of water pH, salinity, composition, total dissolved $CO_2$, etc.
2. Photosynthetic consumption of $CO_2$ (McKenzie, 1985); Schidlowski, 1987).
3. Rates of degradation and recycling of $^{13}C$-depleted aquatic and terrestrial organic matter.
4. Bacterial processes, such as sulfate reduction and methanogenesis.
5. Possible contributions from the dissolution of carbonate grains.

Because temperature has an insignificant effect on carbon isotopic fractionation in sedimentary systems, and little, if any, carbon isotope fractionation occurs through evaporative effects in fresh or brackish lakes—Stiller et al. (1985) suggest that evaporation can concentrate $^{13}C$ by 5–6‰ in hypersaline systems—carbon isotope ratios in carbonate minerals are thought to be mainly controlled by biotic/abiotic interactions. The preference of photosynthetic organisms through kinetic isotope fractionation for the $^{12}C$ atom (Deines, 1980) can lead to significant isotopic effects in the dissolved inorganic carbon (DIC) pool, from which carbonate minerals may be precipitated. These effects, which principally revolve around the photosynthetic consumption of $CO_2$ by phytoplankton to produce $^{13}C$-depleted organic matter, and the recycling of this and similarly depleted macrophyte remains (aquatic and terrestrial) through various mineralization processes, can occur in the water column or in sediment pore waters. In lakes, as in the marine system, the DIC pool will always tend towards isotopic equilibrium with the atmospheric $CO_2$ reservoir through exchange across the ater surface. This is a relatively sluggish process, however, so, in lakes having short residence times or rapid turnover, the carbon isotope system will normally be dominated by the production and degradation of organic matter. Large lakes typically have long residence times; in these, therefore, there may be a greater tendency for the DIC of open surface waters to approach isotopic equilibrium with atmospheric $CO_2$, particularly if stable stratification inhibits the recycling of $^{12}C$-rich $CO_2$, released from mineralized organic matter.

## 15.4 Carbonate Facies and Isotopic Mechanisms

In Section 15.2, we indicated that carbonates of various types can accumulate in lakes, and that accumulation occurs in one or more of several different environments. The sort of environmental information that stable isotopic studies of these carbonates may yield is primarily dependent upon the origin of the mineral (see Figure 15.4). We shall now review the major carbonate types and show how their carbon and oxygen isotopic compositions can be used to obtain paleolimnological information.

**Detrital Carbonate** Erosion of carbonate rocks in the drainage basin or resuspension of carbonate-bearing sediments from nearshore areas can cause carbonate minerals to be redistributed as clastic grains. These grains carry the isotopic signature of their parent material, and in some situations grains from different sources may become mixed, thereby producing a hybrid isotopic composition for the bulk sediment. Useful information can nevertheless be obtained from the stable isotopic study of clastic carbonates. In particular, they can be used in provenance studies as tracers of sediment supplied by a given river system. Obviously, this technique can only be applied if the drainage basin contains well-differentiated suites of carbonate rocks. Continental or marine limestones and metamorphic marbles will normally show distinctly different isotopic signatures.

No modern studies have been undertaken with this technique, but a study of Cenozoic deposits in the Ebro Basin, Spain, showed the situation in a lacustrine giant about 30 million years ago (Oberhänsli and Allen, 1987). Here, studies of stable isotope composition of bulk carbonate allowed the authors to distinguish several distinct hydrologic provinces, each reflecting the nature of the river input (duration, seasonality, catchment geology).

**Littoral Shelf** The littoral zone in lakes is characterized by photic bottom waters and unstable physical conditions, including episodic higher energy and rapid changes in temperature. $CO_2$ concentrations may also fluctuate drastically. In addition to bioinduced chalk deposits (Seekreide, lake marl), carbonate deposits may include shell sands (molluscs, ostracodes), algal oncolite gravels, charophyte silts, and other biogenic components (see Figure 15.3).

Systematic studies on the controls of stable isotope variations within such fluctuating environments are lacking, but many of the published curves of oxygen and carbon isotopes from lake marls, done for paleoclimate analyses, are taken from shallow water deposits of this sort. Mörner and Wallin (1977) used bulk oxygen isotope measurements of very shallow water *Chara* marls from a Gotland lake to estimate Holocene summer paleotemperatures. In spite of a number of inherent errors, they concluded a 1.5 to 2.0°C increase during the climatic optimum around 9,300 yrs B.P. Eicher and Siegenthaler (1976), Eicher (1987), and Siegenthaler and Eicher (1986) have systematically analyzed cores from numerous peri-alpine chalky moor deposits to document a synchronous, drastic climatic event during the Younger Dryas, 11,000 to 10,200 yrs ago. They argue that the $\delta^{18}O$ signal does not reflect a change in the temperature of formation of the calcite precipitate, but rather a decrease in the average temperature of regional precipitation and, thus, a change in the isotopic composition of inflow waters. In a third study of lake marls, Müller and Wagner (1978) have concluded that oxygen isotopic variations in primary Mg-calcite silts from Lake Balaton chiefly reflect variations in water balance. They postulate that the regional climate was driest during the period 3,000 to 5,000 yr B.P.

Algal carbonates, mainly in the form of oncolites and stromatolites or sinter "chimneys" formed around spring orifices, are a major feature of the littoral and sub-littoral zones of some lakes (Dean, 1981; Dean and Fouch, 1983; Dixit, 1984; Hillaire-Marcel and Casanova, 1987; Casanova, 1987; Cohen and Thouin, 1987), reaching their maximum development in shelf areas starved of clastic input. The structures are formed by cyanophyte or cyanophyte-bacterial associations, and the carbonate may be calcite, Mg-calcite or aragonite, apparently precipitated in isotopic equilibrium with the lake waters (Hillaire-Marcel and Casanova, 1987). Some structures display mm-scale rhythmic banding of light and dark laminae which is probably of seasonal origin. Oxygen isotope analyses of successive laminae in oncolites from Lakes Magadi, Natron, and Bogoria have been exploited by Casanova (1986b, 1987) to reconstruct seasonal variations in the rainfall regime of the Gregory Rift, East Africa, at intervals during the Quaternary.

**Pelagic Carbonate Precipitates** In temperate and mid-latitude lakes, pelagic carbonates are precipitated mainly in the summer season during periods of maximum phytoplankton productivity, when the combined effects of elevated temperature and photosynthetic consumption of $CO_2$ can bring about calcium carbonate supersaturation (Kelts and Hsü, 1978; Stabel, 1986). Little is known about the controls on primary carbonate precipitation in mid-latitude and tropical lakes. These, too, may be related to phytoplankton blooms, which in some lakes are seasonal, but in others occur more or less throughout the year (Beadle, 1981); Haberyan and Hecky, 1987; Hecky and Kling, 1987). It seems clear that a lot of organic activity is not essential for carbonate production. The surface waters of the Dead Sea contain few organisms, but are nevertheless precipitating aragonite more or less continuously (Neev and Emery, 1967).

Precipitation of carbonate minerals in the epilimnion of lakes produces a rain of crystals in which the oxygen isotopes reflect a combination of open water composition and temperature. The carbon isotopes can provide information on surface water productivity, the efficiency of internal nutrient cycling, and residence times. Sequences of pelagic carbonate precipitates yield some of the best records of environmental and limnological change in lakes.

*Geochemical Record of the Lake Bonneville Drawdown* During the Late Glacial, many basins in the western U.S.A. supported giant freshwater lakes. The most famous of these was Lake Bonneville, the Pleistocene precursor of the Great Salt Lake (Gilbert, 1890), which is estimated to have covered over 51,000 km², and to have been up to 330 m deep. Sometime around 14,000 years B.P., following a dam break, it discharged catastrophically over the Red Rock Pass into the Snake River Valley, dropping about 100 m during an 8-week flood (Jarrett and Malde, 1987). Within 1,000 years, the lake level dropped to form the present 5–10 m deep Great Salt Lake, with an area of about 5000 $km^2$.

A record of this drastic water level change is preserved in various pelagic calcium carbonate minerals (calcite, Mg-calcite, and aragonite) contained in finely laminated muds deposited in the center of the basin (Eugster and Kelts, 1983; McKenzie, in Spencer et al., 1984). These show an almost linear increase over less than 20 cm of core in both the oxygen-18 and magnesium content of the calcite. The lake waters were evidently becoming progressively more concentrated during this period of sediment accumulation. These data were used to reinterpret the classic sequence of shoreline terrace development displayed in the Great Salt Lake basin (Spencer et al., 1984). Instead of representing pauses in a downdraw of the lake, we now realize that the major terraces were formed as the lake level rose, and were then preserved due to a very rapid rate of downdraw in response to the abrupt Late Glacial climate change.

*Lake Zürich* Lake Zürich provides an example of carbonate precipitation in a peri-alpine, temperate, deep, freshwater system where evaporative effects are minor. Calcite precipitates seasonally from spring to summer at surface water temperatures of 10 to 22°C, forming carbonate-rich laminae. A clastic-dominated lamina is deposited during the autumn-winter season; couplets of these two contrasting sediment types therefore form a non-glacial varve (Kelts and Hsü, 1978). Isotopic measurements across single calcite laminae in varves of the last 80 years showed $\delta^{18}O$ variations of the order of 1‰ (Pika, 1983). Some of the temperature variation of a seasonal cycle of calcite precipitation are thus preserved in the laminae.

*Greifensee* The Greifensee is a small lake near Zürich, Switzerland, which has become eutrophic during the past century. Since 1928, varves of both calcite-rich and calcite-poor laminae have been forming in much the same manner as in Lake Zürich. Using sediment traps suspended below the epilimnion, McKenzie (1982, 1985) followed changes in the isotopic composition of the calcite rain as the precipitation season progressed. By monitoring carbon isotopic changes in the DIC at the same time, McKenzie was able to confirm a model of productivity-controlled variations in the carbon isotopic composition of the DIC-carbonate pool that had previously been hinted at by Oana and Deevey (1960) and Stiller and Hutchinson (1980). The model predicts that under conditions of stable stratification, the transfer of $^{12}C$-enriched organic matter to the hypolimnion by sinking phytoplankton remains will cause a corresponding rise in the $^{13}C$ concentration of the epilimnic DIC reservoir. Furthermore, if the system has a long response time relative to the productivity cycle, the surface DIC reservoir will become progressively more depleted in

$^{12}C$ as the seasdon of phytoplankton blooms progresses. McKenzie (1985) demonstrated that the composition of the calcite did indeed follow changes in the DIC, becoming isotopically more positive from spring through summer. Over the same period, bottom-water DIC was shown to display the reverse trend, becoming enriched in $^{12}C$, due to storage of bicarbonate liberated by mineralization of organic matter in the hypolimnion. This marked vertical isotopic gradient disappears as autumn—winter cooling causes the stratification to break down, eventually leading to complete vertical mixing of the waterbody and homogenization of the DIC pool.

McKenzie also examined a short core from the lake and showed that, over the last decades, there has been a tendency for the bulk carbon isotopic composition of each new carbonate lamina (i.e., one year's record of calcite precipitation) to be more enriched in $^{13}C$. This has been explained in terms of increasing productivity, and thus greater and greater depletion of the surface DIC pool, due to ever-increasing nutrient supply as eutrophication intensified. Since 1972, attempts to control the pollution of the Greifensee have reduced nutrient supplies and there has been a corresponding decline in productivity. This change, too, is reflected in the bulk composition of the calcite, which now shows a trend towards more negative carbon $d^{13}C$ values. Using the oxygen isotope composition of the calcites as a measure of temperature, McKenzie (1985) was also able to show that as productivity increased, the period of calcite precipitation gradually shifted to later in the season.

The pelagic calcite precipitate in the Greifensee has proved to be not only a sensitive recorder of the seasonal productivity cycle, but also of long-term changes in the productivity of the lake. Although the Greifensee is small, McKenzie (1982, 1985) suggests that the sort of DIC/carbonate isotopic relationship noted here is typical of lakes of all sizes, and even of ocean basins. A note of caution should be sounded if the Greifensee is to be used as a universal model of pelagic carbonate precipitation in productive lakes, since isotope effects not recorded in this waterbody may potentially affect the isotopic composition of the precipitate. Herczeg and Fairbanks (1987), for example, have shown that in alkaline water systems, extreme algal photosynthesis can reduce the concentration of $CO_2$ to such a degree that a thin film exchange process, via hydroxyl ions to bicarbonate, begins to operate across the air/water interface. Under these conditions, a strong kinetic isotope fractionation effect is exerted, causing marked $^{12}C$-enrichment to the surface water DIC.

*Lake Kinneret* Stiller and Kaufman (1985) summarize work on endogenic calcite in the sediments of Lake Kinneret, Israel. During summer months, phytopolankton photosynthesis induces uniform precipitation of calcite in the open waters. Isotopic analyses from three cores show that the calcites form two populations comprising minerals precipitated during the period 5,100–2,750 yr B.P. and those formed from 1,150–150 yr B.P. with transitional compositions for calcites from the intervening 1,600 yr period. Within each population, the oxygen and carbon isotopes display a covariant relationship. There is also a partial correlation between $\delta^{13}C$ and total $CaCO_3$ content. These relationships are explained in terms of climatic and productivity changes. The earlier group of calcites is thought to have formed during a warm climatic interval in a waterbody of moderate productivity, while the later group (1,150–150 yr B.P.) may indicate a cooler climate, but enhanced productivity. Greater carbonate content in the younger sediments are also thought to be a result of this increased productivity. The existence of these two populations, and the paleoclimatic information they convey, would not have been apparent without isotopic analyses.

*Lake Lisan* Probably one of the finest examples of pelagic carbonates is provided by the sediments of the Lisan Formation, which form spectacular cliffs along the Dead Sea rift valley. The finely laminated aragonitic muds accumulated in a large lake called Lake Lisan, a greatly expanded precursor of the Dead Sea (Begin et al., 1974). Katz et al. (1977) have carried out a detailed trace element and stable isotope study of the aragonite-rich laminae. Interestingly, in view of the large number of anlyses performed, no clear isotopic trends were identified, either through time, or laterally along the basin, suggesting that the surface waters of Lake Lisan were relatively homogeneous. This is rather surprising in view of the extremely elongate nature of the basin—we have already noted that Lake Chad, for example, another waterbody with essentially a single source of inflow and high rates of evaporation, displays a marked horizontal oxygen isotope gradient. A further striking feature of the Lisan carbonates is that they are 2–3‰ more enriched in $^{18}O$ than aragonite from the Dead Sea, implying that the waters of Lake Lisan were evaporatively more evolved than those of the Dead Sea. All other paleoenvironmental data (geochemical, faunal, and sedimentological) indicate the contrary, that Lake Lisan was a larger and fresher waterbody than the Dead Sea (Begin et al., 1974); Katz et al., 1977). This apparent anomaly has been explained by Gat (1984) as being due to the inhibiting effect of the Dead Sea's very high salinities on oxygen isotope exchange between brine and atmosphere. Clearly, the potential influence of such effects must be kept in mind when interpreting the isotopic record of waterbodies which may have experienced drastic variations in salinity.

**Benthic Plain Faunas** Isotopic analyses of the shells of benthic fauna (ostracods and molluscs) can provide an elegant means in some lakes to circumvent the problems posed by two unknowns: temperature of formation and water $\delta^{18}O$ composition. Following the reasoning that Shackleton (1967) applied to the ocean water–glacial ice system, Lister (1988) postulated that the deep bottom water temperature in a temperate lake may be assumed to be nearly constant. Thus, variable oxygen isotopic signals in benthic faunas through time can be used to trace compositional changes in the water sources. Lister applied this technique to bivalves and ostracods in cores from the deep plain (137 m) of Lake Zürich, as a possible tracer of meltwater inputs during deglaciation.

He was able to use a sharp, positive, step-function $\delta^{18}O$ shift of 3.7‰ at about 12,500 yr B.P. to confirm that the sudden cessation of glacial silt accumulation corresponded to an abrupt change in the source of lake inflow. A melting-regime of stagnant, isotopically light glacial ice was apparently replaced by direct catchment runoff. This isotopic shift also implies that, subsequently, the catchment was essentially ice free (Lister, 1988).

**Diagenesis near the Sediment Water Interface** Because of the lack of studies and a poor understanding of the ecology and physiology of microbial communities, isotopic effects related to processes operating at and below the sediment-water interface in lakes are not known in detail. Pore-water oxygen isotope ratios presumably reflect in some way those of the overlying lake water body. In high sedimentation rate settings, below a zone about 10-cm-thick subject to slow diffusion, pore waters may be "frozen" into the sediment and thus carry with them a record of the depositional waters.

Carbon isotopes in pore waters reflect a complex series of biotic/abiotic interactions. Depending on the supply of appropriate ions (sulfate, nitrate, Fe-oxide, Mn-oxide, $CO_2$) and whether the microenvironment is dominantly oxic, suboxic, or anoxic, bacterial processes break down organic matter by a thermodynamically determined progression of microbially mediated oxidation reactions (Hanselmann, 1986; Hesse, 1986). The destruction and creation of organic compounds, ultimately to carbon dioxide, methane, and insoluble residues, profoundly affects the carbon isotopic composition of DIC in the pore water reservoir. In many freshwater lakes, low sulfate concentrations mean that bacterial sulfate reduction is essentially completed either in the water column (of anoxic environments) or within the top few millimeters of the sediment column. In sulfate-poor systems, therefore, methanogenesis is the dominant anoxic microbial process. Bacteria produce methane which is highly enriched in $^{12}C$. Thus, within limited pore-water reservoirs, the DIC will become increasingly depleted in $^{12}C$. Thus, within limited pore-water reservoirs, the DIC will become increasingly depleted in $^{12}C$ as methanogenesis progresses (Claypool and Kaplan, 1984) Whiticar et al. 1986). The effects of this have been demonstrated in Lake Kinneret (Stiller and Magaritz, 1974) where pore waters are enriched by 10 to 20‰ with respect to the open water DIC. Similar studies have been done in Lake Washington (Kuivila and Murray, 1984) and Lake Lugano (Niessen, 1987).

Figure 15.5 provides a schematic model of DIC carbon isotopic evolution within a productive lake system and the carbon isotopic composition of associated carbonates. The recycling of carbon compounds released back into a lake during early diagenesis can affect the composition of the open-water DIC pool. An example is given by Rau (1978) who discovered organic matter with $\delta^{13}C$ of $-44$ to $-77$‰ in a freshwater lake in British Columbia. Such low values could only be produced if a major component of the lake's DIC was derived from the oxidation of methane released from the sediments. Similar arguments are applied to explain differences in the carbon isotopic composition

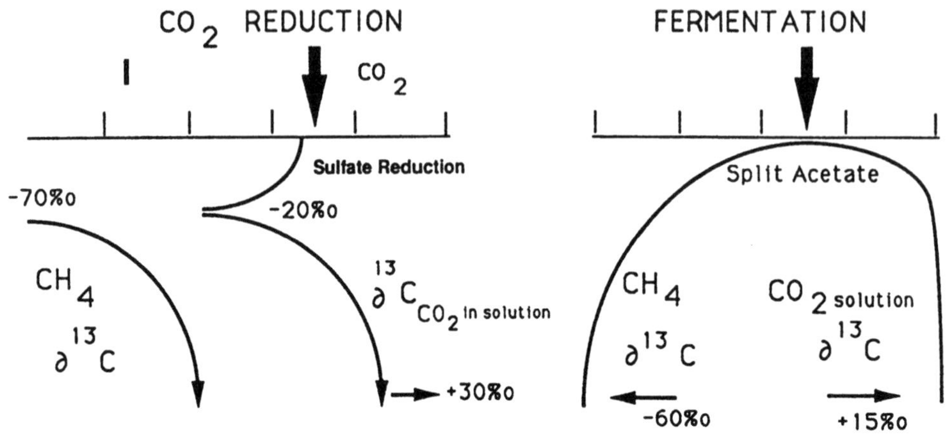


**Figure 15.5** Model of the development of the reservoir carbon isotopic signature in pore water of anoxic sediments undergoing methanogenesis for two mechanisms; $CO_2$-reduction and fermentation. Carbonate precipitated in either system will inherit the signature of the ambient pure water $\delta^{13}C$ in carbonate species. The initial pore water $\delta^{13}C$ signature in dissolved $CO_2$ and carbonate follows a mixture pathway determined by microbial reactions. $\delta^{13}C$ in methane is strongly dominated by $^{12}C$, leaving the pore water reservoir more positive.

between the oxic and anoxic water masses of Lake Tanganyika (Craig, 1975). Although known from marine shelves (e.g., Hovland et al., 1987), no lacustrine carbonates have yet been reported with the strongly negative $\delta^{13}C$ values characteristic of methane oxidation. Given the importance of methanogenesis in lacustrine environments, it can be anticipated that such minerals probably occur in some lakes. The importance of early diagenetic processes in the production of carbonates in lacustrine is discussed in Section 15.5.

**The Isotopic Signature of Local Precipitation** The previous discussion indicates that–because of the effects of residence time or of variations in temperature, evaporation rates, and salinity—large lakes may not provide a particularly good guide to the isotopic composition of the rain waters or river inflows that feed them. Data of this sort are obviously required if reconstructions of past water balances are to be attempted from the oxygen isotopic composition of lacustrine carbonates. Are there any other sources of information? A measure of the $\delta^{18}O$ composition of local precipitation seems to be available in the shells of terrestrial and swamp- and stream-dwelling freshwater gastropods. Temperature and evaporative effects also influence these environments but, with knowledge of the molluscs' habitat and careful sampling, useful information can nevertheless be obtained. Fritz and Poplawski (1974) and Abell (1982) have demonstrated that this group registers approximate isotopic equilibrium. Magaritz and Heller (1980) used land molluscs to reconstruct paleoprecipitation in Quaternary Israel, and Abell (1985) tested the method systematically against modern rainfall patterns in Africa. Results were reasonable, the gastropods displaying a $\delta^{18}O$ vs. latitude relationship parallel to that of the regional precipitation. Abell explained deviations from the expected $\delta^{18}O$ patterns in terms of local rainfall or evaporation/temperature anomalies. Freshwater gastropods provide an integrated record of their life span of 2-3 years, and thus bulk shell samples can help smooth seasonal temperature/evaporation effects. Carbon-13 values are generally depleted and more difficult to interpret environmentally. Normally, they reflect the fact that these organisms are scavengers that favor habitats where the DIC is strongly influenced by rotting plant debris and is thus carbon- 12 enriched.

## 15.5 Lake Bosumtwi: Case Study of a Carbonate-Bearing Lacustrine Sequence

Lake Bosumtwi occupies a ca. one million-year-old meteorite impact crater in southern Ghana. The almost-circular lake is closed and meromictic, with a maximum depth of 78 m and a surface area of ca. 45 km². Surface waters are alkaline (pH 9.1–9.6) and have t.d.s. of ca. 740 mg/1 (Figure 15.6). Present rainfall is ca. 1,500 mm. The lake is surrounded by forest (now much modified by agriculture), but paleobotanical and palynological evidence indicate major Late Quaternary vegetational changes (Talbot and Hall 1981; Talbot et al. 1984; Maley and Livingstone, 1983; Maley, 1986). Climatically induced changes in

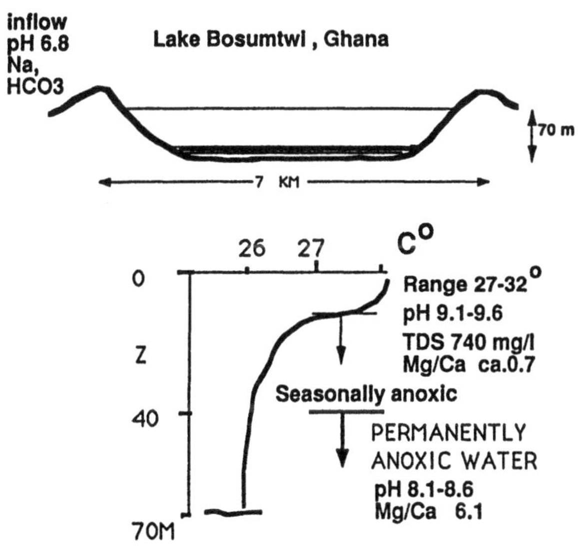


**Figure 15.6** Overview of modern physico-chemical data for Lake Bosumtwi, Ghana. The upper diagram is a cross section of the crater basin showing water level and thick, laminated, anoxic bottom sediments. The lower diagram is a schematic temperature profile with limnological data (see Talbot et al., 1984).

P/E also caused major lake level variations (Figure 15.7); present evidence suggests that the lake has been a closed waterbody for much of the past 27,500 years, overflowing only in the Early Holocene and possibly for a relatively brief period around 23,000 to 21,000 yr B.P. (Talbot et al., 1984).

Cores from the lake are composed of mixtures of clastic material and authigenic carbonates (typically $<10\%$ by weight), plus varying amounts of organic matter (typically 5–20% total organic carbon). Exposure surfaces, carrying evidence of erosion, desiccation, or pedogenesis, occur in cores from shallower water and indicate periods of lower lake level. A variety of carbonate minerals (see Figure 15.7) are present, both as discrete laminae and disseminated through the organic-rich sediments (Talbot et al., 1984; Talbot and Kelts, 1986). Crystal morphologies suggest that the carbonates are of two types: pelagic precipitates and diagenetic minerals precipitated within the sediments. The former typically occur as discrete crystals of relatively uniform size and shape, while the diagenetic minerals form crystal aggregates and micronodules. This morphological subdivision is supported by stable isotope analyses, which show that the carbonates fall into two distinct groups, defined by their carbon isotopic compositions. Primary precipitates have $\delta^{13}C$ values in the range of $-2$ to $+6‰$, while the diagentic minerals show a range of $+7$ to $+27‰$. Furthermore, the oxygen and carbon isotopic compositions of the primary minerals, despite occurring as two vertically distinct populations (see Fig. 15.7), have a strongly linear covariant trend ($r = +0.95$), while the diagenetic minerals show a much greater scatter of values. In both cases, mineralogical changes accompany changing oxygen isotopic composition.

**Oxygen Isotopic Variations** Among the primary minerals, a rise in $\delta^{18}O$ accompanies the change in mineralogy from calcite through Mg-Calcite to aragonite. Mineralogical evolution of this sort has been shown to reflect increasing salinity (Müller et al., 1972). In the case of Bosumtwi, such an interpretation is confirmed by the isotopic data, the progressively more positive $\delta^{18}O$ values indicating increasing evaporative concentration of the lake waters. Analyses of modern gastropods from local streams were used to estimate the isotopic composition of the inflow. From these it can be concluded that the $^{18}O$-enriched aragonite from the bottom of the core precipitated from a more saline (and hence lower) and isotopically more evolved waterbody than that from which the Holocene primary calcites precipitated.

A systematic change in oxygen isotopic composition also follows the diagenetic mineral sequence—Mn-siderite → calcite → Mg-calcite → dolomite. This trend follows closely that displayed by the primary phases (Figure 15.8). It appears, therefore, that the diagenetic minerals have also preserved a record of the lake's isotopic evolution. From this, we can infer that a relatively prolonged period of low water occurred in the Bosumtwi crater during the Late Pleistocene and that the dolomite-bearing sediments represent maximum evaporative lowering of the lake. Dolomite formation would have been favored by the higher Mg/Ca of the concentrated waters (Talbot and Kelts, 1986). Con-

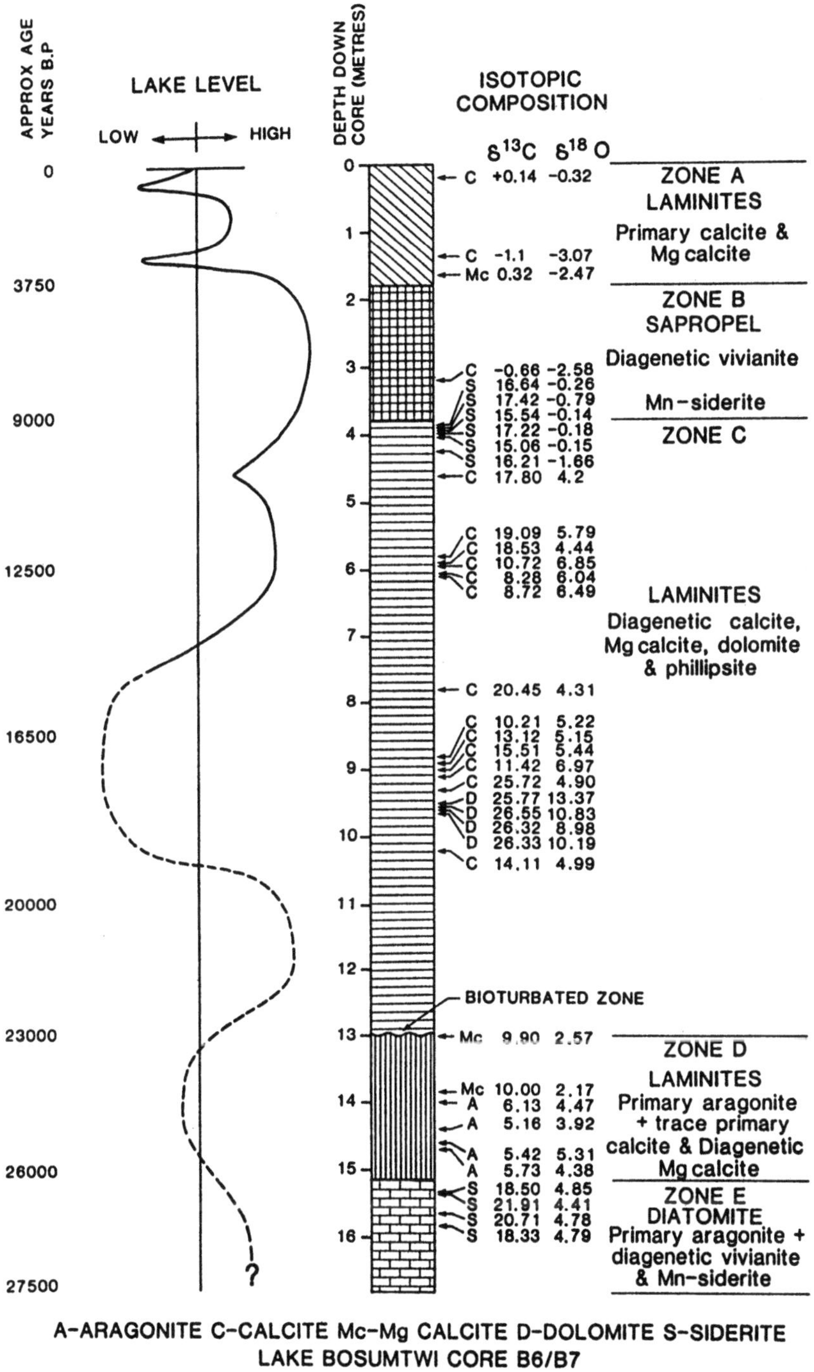


**Figure 15.7** Stratigraphy, mineralogy, and chronology of a 16 m-core (B6/B7) from offshore, Lake Bosumtwi (after Talbot and Kelts, 1986). Includes stable isotopic results for small-sample carbonate grains and derived curve of lake level variations.

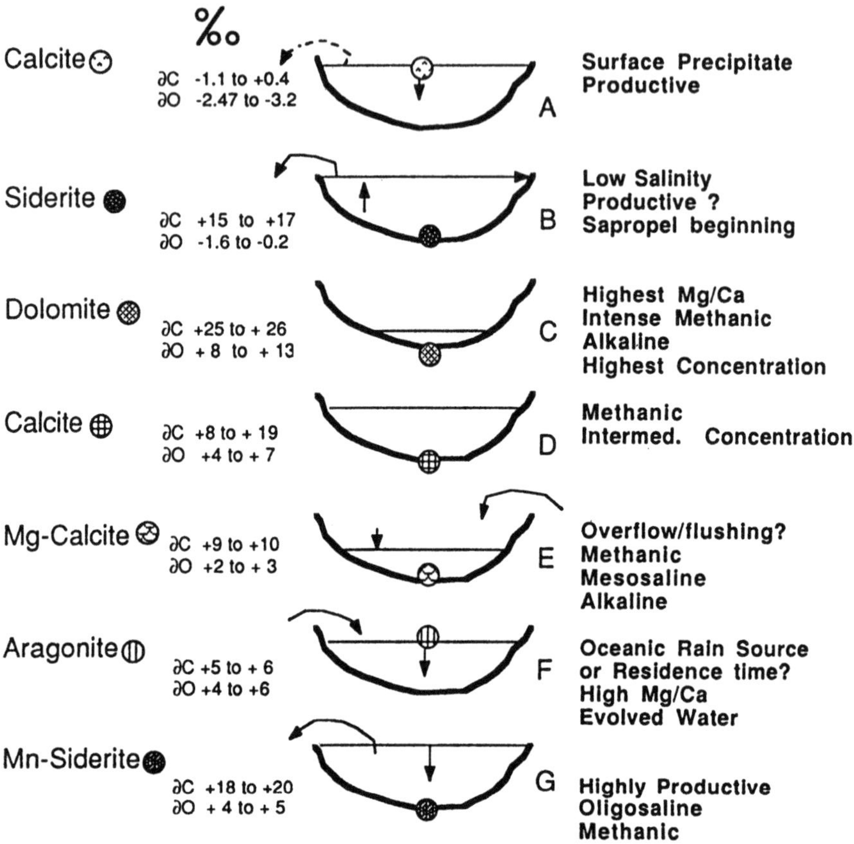


**Figure 15.8** A model of environmental interpretation, summarizing sedimentology and isotopic values for various carbonate mineral phases isolated from organic-rich sediments of Lake Bosumtwi. Column 1 gives authigenic mineral phase and signature; column 2 shows ranges of $\delta 18O$ and $\delta^{13}C$ encountered; column 3 suggests interpretations for probable site of formation of the mineral phase and assumed state of lake level in the crater; column 4 summarizes conditions interpreted from the carbonate phase found in the Bosumtwi sediment profile.

versely, the Holocene calcites and some of the siderites have more negative $\delta^{18}O$ values, indicating relatively dilute waters, and thus a rather deep lake. Independent lake level evidence confirms that for much of Early and Mid-Holocene times, Lake Bosumtwi was significantly higher than it is at present (Talbot et al., 1984; Talbot and Kelts, 1986).

**Carbon Isotopes: Primary Minerals** These show a linear trend of increasing $\delta^{13}C$ with progressive evaporative concentration of the lake water (as shown by the oxygen isotopes). The origin of this covariance, which has also been noted in other lacustrine carbonate sequences (e.g. Stiller and Kaufman,

1985; Hillaire-Marcel and Casanova, 1987; Fritz et al., 1987), is not fully understood. It probably represents a combination of different effects, including photosynthetic depletion of the DIC $^{12}C$ reservoir, a fractionation effect due to the slow degassing of $CO_2$ (Katz et al., 1977), and possible changes in the dominant dissolved carbonate ionic species with changing pH.

**Carbon Isotopes: Diagenetic Minerals** Compared with the majority of limestones (Hoefs, 1987) and most other naturally occurring carbon-bearing substances (Deines, 1980), the diagenetic carbonates from Lake Bosumtwi are strongly depleted in $\delta^{12}C$. As previously noted, bacterial methanogenesis shows a strong kinetic isotopic discrimination in favor of $^{12}C$, producing, at the same time, a bicarbonate pool enriched in $^{13}C$. In closed systems, such as the pores of fine-grained sediments, progressive depletion of the carbon reservoir during methanogenesis can lead to extreme positive isotopic values for the accompanying bicarbonate (Claypool and Kaplan, 1974). Carbonates precipitated from this bicarbonate will also be isotopically heavy. The Bosumtwi diagenetic carbonates are clearly a result of high rates of bacterial methanogenesis within the pores of rapidly deposited, organic-rich sediments (Talbot and Kelts, 1986). The evidence from Lake Bosumtwi, therefore, indicates that the carbon isotopic composition of early diagenetic carbonate phases preserves a record of the dominant microbial process responsible for organic matter degradation in the uppermost section of the sediment. Microbial activity is, in turn, determined by water chemistry; methanogenic processes dominate in sulfate-poor, neutral to alkaline systems, while sulfate reduction, with its very different $^{13}C$-depleted carbon isotopic signature characterises sulfate-rich, acid to neutral waterbodies. Thus, the carbon isotopes also provide a proxy record of lake water chemistry and some suggestion of sedimentation rates.

Both the endogenic and diagenetic carbonate minerals from Lake Bosumtwi have been shown to preserve valuable paleolimnological information, some of which could only have been obtained from the study of their stable isotope composition. Figure 15.8 schematically summarized the changing environmental conditions deduced from this diverse suite of carbonate minerals. One lesson to stress from the work at Lake Bosumtwi and some of the other studies mentioned in this review is the importance of careful sampling. In the case of the Bosumtwi sediments, great care was taken to ensure that samples came from single laminae of pelagic precipitates, and that the diagenetic phases were separated from endogenic phases. The value of the analytical data can be greatly reduced if bulk samples are taken, or material from several different stratigraphic levels is homogenized into one sample. Since there is no longer any technical reason why large samples should be analyzed, we recommend that fine-scale sampling should be a general rule in the isotopic study of lacustrine carbonates.

## 15.6 Applications to Large Lakes

Large lakes are usually of great economic importance to bordering regions, and in many cases, also represent unique ecological systems, as Hecky and Kling,

(1987) have recently emphasized with respect to the great lakes of East Africa. Nevertheless, remarkably little is known about the history of such waterbodies. Since systematic monitoring is often irregular and of variable quality, predictions about, for example, their stability, likely response to global climate change, and long-term prospects for the fishery resources of these lakes are, in almost every case, based upon inadequate data bases.

The largest lakes of tropical Africa were formed by tectonic activity; their waters all show alkaline affinities and are dominated by ions derived from the incongruent weathering of feldspathic crystalline and volcanic rocks, but in most other respects they represent an extremely diverse assemblage of waterbodies (Beadle, 1981; Livingstone and Melack, 1984; Hecky and Kling, 1987). Some are well mixed, but many have a chemocline and are meromictic (Beadle, 1981). Some, such as Lake Kivu, contain unusually high concentrations of phosphate and dissolved gases, the carbon dioxide being principally of volcanic origin, while the methane is biogenic, produced via bacterial reduction of the volcanic $CO_2$ and fermentation of sedimentary organic matter (Tietze et al., 1980). Nutrient contents in surface waters are highly variable; Lake Tanganyika, for example, receives little via influents (Hecky and Kling, 1987). Thus, phytoplankton blooms (and any associated carbonate precipitates) will be controlled by upwelling of nutrients from the anoxic hypolimnion. In spite of their low salinities, the waters of Lake Tanganyika (415 mg/l) and Lake Malawi (192 mg/l) are highly evolved isotopically, which is consistent with their long residence times (over 1,000 years in the case of Tanganyika). Hypolimnic $\delta^{18}O$ values of 4.2‰ (Tanganyika) and 2.1‰ (Malawi), contrast with values of 3.5‰ and 1.65‰ from the respective epilimnia; the former appear to be relics from an earlier cooler, but drier climatic interval (Craig, 1975; Gonfiantini et al., 1979; Gonfiantini, 1986).

Clearly, each of the large African lakes is unique, and in most of them, basinal sediments will carry the most refined record of their environmental history. Some progress has been made towards unravelling these records (Kendall, 1969; Hecky and Degens, 1973; Haberyan, 1987; Haberyan and Hecky, 1987; Talbot, 1988; Talbot and Livingstone, in press); much remains to be learned. Primary and diagenetic carbonates are known from cores from these lakes (Hecky and Degens, 1973; Stoffers and Hecky, 1978; Stoffers and Singer, 1979; Haberyan and Hecky, 1987), but have been little studied. Unexploited isotopic archives clearly exist here, and probably in many other large lakes, which could shed much light on the chemical and environmental histories of these important waterbodies

---

*Acknowledgements* We appreciate the useful comments of J. McKenzie, U. Siegenthaler, G. Lister, F. Niessen, D. Hollander, A. Losher, and E. Jansen. Research was supported by EAWAG project 27-208, the Swiss Nationalfond, the Norwegian Research Council (NAVF), and the Statoil Basic Research Fund (VISTA Project T-29415). Coring operations at Lake Bosumtwi were funded by the National Science Foundation, Washington, through a grant to Professor D.A. Livingstone, Duke University. B. Schwertfeger is thanked for help with manuscript production.

## References

Abell, P.I. 1982. Paleoclimates at Lake Turkana, Kenya, from oxygen isotope ratios of gastropod shells. *Nature* 297:321–323.

Abell, P.I. 1985. Oxygen isotope ratios in modern African gastropod shells: A data base for paleoclimatology. *Chem. Geol. (Isot. Geosci. Sect.)* 58:83–193.

Anderson, T.F. and Arthur, M.A. 1983. Stable isotopes of oxygen and carbon and their application to sedimentologic and paleoenvironmental problems. *Soc. Econ. Pal. Min. Short Course*, No. 10, 1–151.

Bahrig, B. 1988. Paleoenvironmental information from deep water siderite: Lake of Lach, W. Germany. In: K. Kelts, A Fleet, and M. Talbot, (eds.), *Lacustrine Petroleum Source Rocks*, Geol. Soc. London, Spec. Publ. 40:160–168.

Bahrig, B. 1989. Stable isotope composition of siderite as an indicator of the paleoenvironmental history of oil shale lakes. In: M. Talbot and K. Kelts (eds.), *Paleoenvironmental Signals in Lacustrine Sequences: Palaeogeogr., Palaeoclim., Palaeoecol* (in press).

Beadle, L.C. 1981. *The Inland Waters of Tropical Africa: An Introduction to Tropical Limnology.* Longman, London. 433 pp.

Begin, Z.B., Ehrlich, A., Nathan, Y. 1974. *Lake Lisan, the Pleistocene Precursor of the Dead Sea.* Bull. Geol. Survey, Israel No. 63, 30 pp.

Binkley, K.L, Wilkinson, B.H., and Owen, R.M. 1980. Vadose beachrock cementation along a southeastern Michigan marl lake. *J. Sediment. Petrol.* 50:953–962.

Blatt, H., Middleton, G., and Murray R. 1980. *Origin of Sedimentary Rocks.* Prentice-Hall, New York, 782 pp.

Buchardt, B. and Fritz, P. 1980. Environmental isotopes as environmental and climatological indicators. In: P. Fritz and J. Ch. Fontes, (eds.), *Handbook of Environmental Isotope Geochemistry: V.1, Terrestrial Environment.* Elsevier, Amsterdam, p. 473–504.

Casanova, J. 1986a. East African Rift stromatolites. In: L.E. Frostick, R. Renaut, I. Reid, and J. Tiercelin, (eds.), *Sedimentation in the African Rifts.* Geol. Soc. London, Special Publ. No. 25, p. 201–210.

Casanova, J. 1986b. La saisonnalit climatique au cours des optimum lacustres de l'est Africain et la rhythmicité de constructions stromatolitiques: Exemple du bassin Natrun-Magadi au Pliestocene terminal. In: *Changements Globaux en Afrique durant le Quaternaire—Passé-Présent-Futur.* INQUA-ASEQUA Symp., Editions de l'ORSTOM, Paris, p. 57-60.

Casanova, J. 1987. *In:* Tiercelin, J.-J. et al., *Le Demi-graben de Baringo-Bogoria, Rift Gregory, Kenya.* Bull. Centre. Rech. Expl. Prod. Elf. Aquitaine No. 11/2, p. 181–540.

Claypool, G. and Kaplan, I. 1974. The origin and distribution of methane in sediments. In: I. Kaplan (ed.), *Natural Gases in Marine Sediments.* Plenum, New York, pp. 99–139.

Cohen, A.S. and Thouin, C. 1987. Nearshore carbonate deposits in Lake Tanganyika. *Geology* 15:414–418.

Craig, H. 1975. *Lake Tanganyika Geochemical and Hydrographic Study: 1973 Expedition.* Scripps Institute of Oceanography Report 75-5, 83 pp.

Dean, W.E. 1981. *Carbonate Minerals and Organic Matter in Sediments of Modern North Temperate Hardwater Lakes.* Soc. Econ. Pal. Min. Spec. Publ. No. 31, p. 213–231.

Dean, W.E. and Fouch, T.D. 1983. Lacustrine Environment. In: P. Scholle, D. Bebout, and C. Moore (eds.), *Carbonate Depositional Environments.* Amer. Assoc. Petrol. Geol. Memoir No. 33, p. 97–130.

Degens, E.T., and Kurtman, F. 1978. *The Geology of Lake Van.* M.T.A. Press, Ankara, 97 pp.

Deines, P. 1980. The isotopic comoposition of reduced organic carbon. In: P. Fritz and J.C. Fontes (eds.), *Handbook of Environmental Isotope Geochemistry,* V. 1. Elsevier, Amsterdam, p. 329–406.

Dixit, P.C. 1984. Pleistocene lacustrine ridged oncolites from the Lake Manyara area, Tanzania, East Africa. *Sediment. Geology* 39:53–62.
Eardley, A.J. 1938. Sediments of great Salt Lake, Utah. *Bull. Am. Ass. Petrol. Geol.* 22:1305–1411.
Eicher, U. 1987. Die Spätglazialen sowie die früpostglacialen Klimaverhältnisse im Bereiche der Alpen: Sauerstoffisotopenkurven kalkhaltiger. *Geographica Helvetica* 2:99–104.
Eicher, U. and Siegenthaler, U. 1976. Palynological and oxygen isotope investigations on late-Glacial sediment cores from Swiss lakes. *Boreas* 5:109–117.
Eugster, H.P., and Hardie, L.A. 1978. Saline Lakes. In: A. Lerman (ed.) *Lakes: Chemistry, Geology, Physics*. Springer-Verlag, New York, p. 237–293.
Eugster, H.P. and Kelts, K. 1983. Lacustrine chemical sediments. In: A.S. Goudie and K. Pye (eds.), *Chemical Sediments and Geomorphology*. Academic Press, London, p. 321–368.
Faure, G. 1986. *Principles of Isotope Geology*. Wiley, New York, 486 pp.
Fontes, J.C., Maglione, G., and Roche, M.A. 1970. Elements d'hydrologie isotopique dans le bassin du lac Tchad. In: *Peaceful Uses of Atomic Energy in Africa*. IAEA, Vienna, p. 209–222.
Fritz, P., Anderson, T.W., and Leqis, C.F.M. 1975. Late Quaternary climatic trends and history of Lake Erie from stable isotope studies. *Science* 190:267–269.
Fritz, P. and Fontes, J. Ch. (eds.). 1980. *Handbook of Environmental Isotope Geochemistry: V. 1, Terrestrial Environment*. Elsevier, Amsterdam, 690 pp.
Fritz, P. and Fontes, J. Ch. (eds). 1986. *Handbook of Environmental Isotope Geochemistry: V. 2, The Terrestrial Environment B*. Elsevier, Amsterdam, 557 pp.
Fritz, P., Morgan, A.V., Eicher, U., and McAndrews, J.H. 1987. Stable isotope, fossil Coleoptera and pollen stratigraphy in Late Quaternary sediments from Ontario and New York State. *Palaeogeogr. Palaeoclim. Palaeoecol.* 58:183–202.
Fritz, P. and Poplawski, S. 1974. $^{18}O$ and $^{13}C$ in the shells of freshwater molluscs and their environment. *Earth and Planet Sci. Lett.* 24:91–98.
Gat, J.R. 1981. Lakes. In: J.R. Gat and R. Gonfiantioni (eds.), *Stable Isotope Hydrology: Deuterium and Oxygen-18 in the Water Cycle*. IAEA Technical Report No. 210, Vienna, p. 203–221.
Gat, J.R. 1984. The stable isotope composition of Dead Sea waters. *Earth Planet. Sci. Letters* 71:361–376.
Gilbert, G.K., 1890. *Lake Bonneville*. Monograph U.S. Geol. Surv., No. 1, 270 pp.
Gonfiantini, R., 1986. Environmental isotopes in lake studies. In: P. Fritz and J.C. Fontes (eds.), *Handbook of Environmental Isotope Geochemistry: V. 2, The Terrestrial Environment B*. Elsevier, Amsterdam, p. 113–168.
Gonfiantini, R., Zuppi, G.M., Eccles, D.H., and Ferro, W. 1979. Isotope investigations of Lake Malawi. In: *Isotopes in Lake Studies*. IAEA, Vienna, p. 195–203.
Haberyan, K.A. 1987. Fossil diatoms and the paleolimnology of Lake Rukwa, Tanzania. *Freshwater Biol.* 17:429–436.
Haberyan, K.A. and Hecky, R.E. 1987. The late Pleistocene and Holocene stratigraphy and paleoclimatology of Lakes Kivu and Tanganyika. *Palaeogeogr. Palaeoclim. Palaeoecol.* 61:169–197.
Hanselmann, K. 1986. Microbially mediated processes in environmental chemistry: Lake sedimenta as model systems. *Chimia* 40:146–159.
Hardie, L.A., Smoot, J.P., and Eugster, H.P. 1978. *Saline Lakes and Their Deposits: A Sedimentological Approach*. Spec. Publs. Int. Ass. Sediment. No. 2, p. 7–41.
Hecky, R.E. 1978. The Kivu-Tanganyika basin: the last 14,000 years. *Pol. Arch. Hydrobiol.* 25:159–165.
Hecky, R.E. and Degens, E.T. 1973. Late Pleistocene-Holocene chemical stratigraphy and paleolimnology of the Rift Valley lakes of central Africa. Woods Hole Oceanogr. Inst. Tech. Rep. WHOI 73-28, unpubl. ms.

Hecky, R.E., and Kling, H.J. 1987. Phytoplankton ecology of the great laeks in the rift valleys of Central Africa. *Ergebn. Limnol.* 25:197–228.

Herczeg, A. and Fairbanks, R. 1987. Anomalous carbon isotope fractionation between atmospheric $CO_2$ and dissolved inorganic carbon induced by intense photosynthesis. *Geochim. Cosmochim. Acta.* 51:895–899.

Hesse, R. 1986. Early diagenetic pore water/sediment interaction: modern offshore basins. *Geosci. Canada* 13:165–196.

Hillaire-Marcel, C. and Casanova, J. 1987. Isotopic hydrology and paleohydrology of the Magadi (Kenya) Natron (Tanzania) basin during the Late Quaternary. *Palaeogeogr. Palaeoclim. Palaeoecol.* 58:155–181.

Hoefs, J. 1987. *Stable Isotope Geochemistry.* 3rd ed. Springer–Verlag, Berlin, 241 pp.

Hovland, M., Talbot, M.R., Qvale, H., Olaussen, S., Aasberg, L. 1987. Methane-related carbonate cements in pockmarks of the North Sea. *J. Sediment. Petrol.* 57:881–892.

Jarrett, R.D., and Malde, H.E. 1987. Paleodischarge of the late Pleistocene Bonneville Flood, Snake River, Idaho, computed from new evidence. *Geol. Soc. Amer. Bull.* 99:127–134.

Katz, A., Koldny, Y. and Nissenbaum, A. 1977. The geochemical evolution of the Pleistocene Lake Lisan–Dead Sea system. *Geochim. Cosmochim. Acta* 41:1609–1626.

Kelts, K. and Hsü, K.J. 1978. Freshwater carbonate sedimentation. In: Lerman, A. (ed.), *Lakes—Chemistry, Geology, Physics.* Springer-Verlag, New York, p. 295–323.

Kelts, K., and Shahrabi. M. 1986. Holocene sedimentology of hypersaline Lake Urmia, northwestern Iran. *Palaeogeog. Palaeoclimatol. Palaeoecol.* 54:105–130.

Kendall, R.L. 1969. The ecological history of the Lake Victoria basin. *Ecol. Monogr.* 39:121–176.

Kuivila, K.M., and Murray, J.W. 1984. Organic matter diagenesis in freshwater sediments: the alkalinity and total $CO_2$ balance and methane production in the sediments of Lake Washington. *Limnol. Oceanogr.* 29:1218–1230.

Lister, G. 1988. A 15,000 year isotopic record from Lake Zürich of deglaciation and climatic change in Switzerland. *Quat. Res.* 29/2:232–242.

Lister, G. In press. Stable isotopes from lacustrine ostracoda as tracers for continental palaeoenvironments. In: J.P. Colin, P. DeDeckker, and J.P. Peypouquet (eds.), *Ostracods in Earth Sciences.* Elsevier, Amsterdam.

Livingstone, D.A. 1963. *Chemical Composition of Rivers and Lakes.* U.S. Geol. Surv. Prof. Paper, 440–G, 64 pp.

Livingstone, D.A., and Melack, J.M. 1984. Some lakes of subsaharan Africa. In: F. Taub (ed.), *Lakes and Reservoirs: Ecosystems of the World.* Elsevier, Amsterdam, 23 p., 467–497.

Magaritz, M., Heller, J. 1980. A desert migration indicator—oxygen isotopic composition of land snail shells. *Palaeogeog., Palaeoclimatol., Palaeoecol.* 32:153–162.

Maley, J. 1986. Fragmentation et reconstitution de la forêt dense humide ouest-africaine au cours du Quaternaire récent: Hypothese sur le rôle des upwellings. In: *Changements Globaux en Afrique durant le Quaternaire- Passê-Present-Futur.* INQUA-ASEQUA Symp., Editions de l'ORSTOM, Paris, p. 281–283.

Maley, J. and Livingstone, D. 1983. Extension d'un él'afement montagnard dans le sud du Ghana (Afrique de l'ouest) au Pléistocène supérieur et à l'Holocène inférieur: premières données polliniques. *Compte Rendu Acad. Sci., Ser. 2,* 296:1287–1292.

McKenzie, J.A. 1982. Carbon-13 cycle in Lake Greifen: A model for restricted ocean basins. In: S.O. Schlanger and M.B. Cita (eds.), *Nature and Origin of Cretaceous Carbon Rich Facies.* Academic Press, London, p. 197–207.

McKenzie, J.A. 1985. Carbon isotopes and productivity in the lacustrine and marine environment. In: W. Stumm (ed.), *Chemical Processes in Lakes.* Wiley, New York, p. 99–118.

Müller, G. and Wagner, F. 1978. *Holocene Carbonate Evolution in Lake Balaton (Hungary):*

*A Response to Climate and Impact of Man*. Int. Assoc. Sediment. Spec. Publ. No. 2, p. 57–81.
Müller, G., Irion, G., and Forstner, U. 1972. Formation and diagenesis of inorganic CaMg Carbonates in the lacustrine environment. *Naturwissenschaften* 59:158–164.
Mörner, N-A., and Wallin, B. 1977. A 10,000 year temperature record from Gotland, Sweden. *Palaeogeogr. Palaeoclim. Palaeoecol.* 21:113–138.
Neev, D., and Emery, K.O. 1967. *The Dead Sea: Depositional Processes and Environments of Evaporites*. Geol. Surv. Israel Bull. No. 41, 147 pp.
Niessen, F. 1987. Sedimentologische, geophysikalische und geochemische Untersuchungen zur Entstehung und Ablagerungsgeschichte des Luganersees. Dissertation, ETH-Zürich No. 8354, 180 pp.
Oana, S. and Deevey, E.S. 1960. Carbon-13 in lake waters and its possible bearing on paleolimnology. *Amer. Jour. Sci.* 258A:253–272.
Oberhänsli, H. and Allen, P.A. 1987. Stable isotopic signatures of Tertiary lake carbonates, Eastern Ebro Basion, Spain. *Palaeogeogr. Palaeoclim. Palaeoecol.* 60:59–75.
Pika, J. 1983. Zur Isotopenchemie und Mineralogie der Lacustrinen Ablagerungen im Zürichsee und im Schwarzen Meer. Dissertation, ETH-Zurich, NR6182, 232 p.
Rau, G. 1978. Carbon-13 depletion in a subalpine lake: Carbon flow implications. *Science* 210:901–902.
Schidlowski, M. 1987. Application of stable carbon isotopes to early biochemical evolution on earth. *Ann Rev. Earth Planet Sci.* 15:47–72.
Scholl, D.W. 1960. Pleistocene algal pinnacles at Searles Lake, California, *J. Sediment. Petrol.* 30:414–431.
Scholl, D.W., and Taft, W.H. 1964. Algae, contributors to the formation of calcareous tufa, Mono Lake, California. *J. Sediment. Petrol.* 34:309–319.
Shackleton, N. J. 1967. Oxygen isotope anlyses and Pleistocene temperatures reassessed. *Nature* 215:15–17.
Siegenthaler, U. and Eicher, U. 1986. Stable oxygen and carbon isotopic analyses. In: B. Berglund (ed.), *Handbook of Palaeoecology and Palaeohydrology*. J. Wiley & Sons, New York, p. 407–422.
Spencer, R.J., Baedecker, M.J., Eugster, H.P., Forester, R.M., Goldhaber, M.B., Jones, B.F., Kelts, K., McKenzie, J., Madsen, D.B., Rettig, S.L., Rubin, M., and Bowser, C.J. 1984. Great Salt Lake and precursors, Utah: the last 30,000 years. *Contrib. Mineral. Petrol.* 86:312–334.
Stabel, H.H. 1986. Calcite precipitation in Lake Constance: Chemical equilibrium, sedimentation and nucleation by algae. *Limnol. Oceanogr.* 31:1081–1093.
Stiller, M. and Hutchinson, G.E. 1980. The waters of Merom: a study of Lake Huleh BI. Stable isotopic composition of carbonates of a 54 m core: Paleoclimatic and paleotrophic implications. *Arch Hydrobiol.* 89:275–302.
Stiller, M. and Kaufman, A., 1985. Paleoclimatic trends revealed by the isotopic composition of carbonates in Lake Kinneret. *Ztschr. Gletscherkunde. Glaz.* 21:79–87.
Stiller, M. and Magaritz, M. 1974. Carbon-13 enriched carbonate in interstitital waters of Lake Kinneret sediments. *Limnol. Oceanogr.* 849–853.
Stiller, M., Rounick, J.S., and Shasha, S. 1985. Extreme carbon-isotope enrichments in evaporating brines. *Nature* 316:434–435.
Stoffers, P. and Hecky, R.E. 1978. *Late Pleistocene-Holocene Evolution of The Kivu-Tanganyika Basin*. Spec. Publs. Int. Assoc. Sediment. No. 2, p. 43–55.
Stoffers, P., and Singer, A., 1979. Clay minerals in Lake Mobutu Sese Seko (Lake Albert)—their diagenetic changes as an indicator of the paleoclimate. *Geol. Rundschau* 68:1009–1024.
Street-Perrott, F.A. and Harrison, S.P. 1984. Temporal variations in lake levels since 30,000 yr. BP—an index of the global hydrological cycle. In: J. Hansen and T. Takahashi (eds.), *Climate Processes and Climate Sensitivity*. Geophy. Monograph No. 29, Maurice Ewing Series V. 5, p. 118–129.

Stuiver, M. 1970. Oxygen and carbon isotope ratios of freshwater carbonates as climatic indicators. *J. Geophys. Res.* 75:5247–5257.
Talbot, M.R. 1988. The origins of lacustrine oil source rocks—Evidence from the lakes of tropical Africa. In: K. Kelts, A. Fleet, and M. Talbot (eds.), *Lacustrine Petroleum Source Rocks.* Geol. Soc. London, Spec. Publication, p. 26–42.
Talbot, M.R. and Hall, J.B. 1981. Further late Quaternary leaf fossils from Lake Bosumtwi, Ghana. *Paleoecology of Africa* 13:83–92.
Talbot, M.R. and Kelts, K. 1986. Primary and diagenetic carbonates in the anoxic sediments of Lake Bosumtwi, Ghana. *Geology* 14:912–916.
Talbot, M.R., Livingstone, D.A., Palmer, P.G., Maley, J., Melack, J.M., Delibrias, G., and Gulliksen, S. 1984. Preliminary results from sediment cores from Lake Bosumtwi, Ghana. *Palaeoecology of Africa* 16:173–192.
Talbot, M.R., and Livingstone, D.A. In press. Hydrogen index and carbon isotopes of lacustrine organic matter as lake level indicators. *Palaeogeog. Palaeoclimatol. Palaeoecol.*
Tiercelin, J.-J., et al., 1987. Le demi-graben de Baringo-Bogoria, Rift Gregory, *Kenya. Bull. Centr. Rech. Expl. Prod. Elf. Aquitaine* 11:181–540.
Tietze, K., Geyh, M., Müller, H., Stahl, W., and Wehner, H. 1980. The genesis of the methane in Lake Kivu, Central Africa. *Geol. Rundschau* 69/2:452–472.
Whiticar, M.J., Faber, E., and Schoell, M. 1986. Biogenic methane formation in marine and freshwater environments: $CO_2$ reduction vs acetate fermentation. *Geochim. Cosmochim. Acta* 50:693–709.

# 16

# Sedimentation and Eutrophication History of Austrian Alpine Lakes

J. Schneider, J. Röhrs, and P. Jäger

ABSTRACT The sedimentation history of Austrian prealpine lakes starts with the last ice retreat about 17,000 yr B.P. Above the Würm moraines, finely banded varves were deposited from clastic input into the arctic lakes. At 13,000 yr B. P., the autochthonous production of carbonates within the lakes from epilimnetic and benthic decalcification began as a consequence of improving climate and more intensive phytoplankton and benthic macro- and microphyte development. Within Austrian prealpine lakes the first natural eutrophication can be identified about 6,000 yr B. P. The Neolithic and the Roman colonizations had nearly no influence on these large lakes. The Bavarian colonization around 1,200 yr B. P.—with the foundation of monasteries—forest clearing, and the start of agriculture—influenced the lakes and their sediments more intensively and led to the first instance of eutrophication. During the last 40 years, fast-growing modern agriculture, increasing industrialization, construction of housing projects, and tourism, and subsequent increase of nutrient load into the lakes led to more intensive eutrophication and a growing load of heavy metals. More or less expensive restoration measures must be taken. The intensity of technical measures—lake-internal or -external measures—depend on the lake type. Within hard water lakes, carbonate precipitation takes place two times a year and is an effective scavenger for phosphorus. In contrast to soft water lakes, Austrian prealpine hardwater lakes are stable sinks for phosphorus. Reduction of external phosphorus loading is the most important restoration measure.

## 16.1 Introduction

Lakes are ecosystems which are intensively coupled with and dependent on their geological and morphological situation, on the climatic factors during their history, on the state and type of their drainage area, on the amount of sediments and nutrients washed into the waters, and on the evolution of their planktonic, nektonic, and benthonic inhabitants. Bioactivity depends primarily on the amount and type of nutrient input from the drainage area (Ohle, 1952, 1958, and 1962). Sediments are an essential part of the ecosystem.

Since humans colonized the surrounding land of lakes and their littoral zones, we notice the human impact upon lake ecosystems, e. g., in the form of eutrophication. This impact is impressively visible, e. g., by blooms of *Oscillatoria rubescens*, but also from blooms of the photosynthetic sulfur bacterium *Chromatium okenii.* For detecting both man-made and natural eutrophication, the sediments must be observed and thoroughly analyzed.

From our knowledge of a lake's eutrophication history we have much valuable information for possible conservation and restoration measures. From this point of view, paleolimnology (see Löffler, 1987) is an essential part of applied geology and applied limnology.

In some large lakes of the Austrian north-Alpine foreland that have existed since the last ice retreat (Würm), various indications of their sedimentary history and especially of their eutrophication are detectable, visible as dark layers in the sediment column, which is rich in organic material, iron sulfides, and nutrients.

## 16.2 Austrian Prealpine Lakes—Large or Small Lakes?

"By definition, large lakes have larger water volumes and greater depths than small lakes" (See Chapter 36); this is the widest and least restrictive definition. For sedimentologists the problem of defining a lake as a small or large one is considered a problem of the formation and configuration of the lake basin and of its physical properties such as exposition to winds, current distribution, mixing intensity, and sediment supply and distribution. The residence time of water within the lake depends primarily on the size and pattern of inflowing rivers and therefore cannot be regarded as a criterion for the definition of a small or large lake. Residence time of water is the most important factor for answering the question of residence time of nutrients as well as of harmful substances. Large lakes show a distinct stratification during summer and—depending on the wind—a strong holomixis during winter. This is important for the distribution of allochthonous sediments within the water body as well as for the formation of, e. g., autochthonous calcareous sediments by lake internal processes, which are controlled and triggered by the trophic conditions. The sediment supply, the type of sediments, and their distribution within the lake basin depend on climatic and hydrodynamic conditions, as well as on the relief energy in the catchment area and the relief of the lake basin. Small lakes and swamps are characterized by predominantly organic-rich sediments. In large lakes, the inorganic sediment supply is more predominant. Another important factor is that turbidity currents are characteristic features of large lakes. Turbidity currents occur especially where deltaic slopes have developed or where steep slopes between the litoral and the profundal zone occur. In all lakes described below we find turbidities.

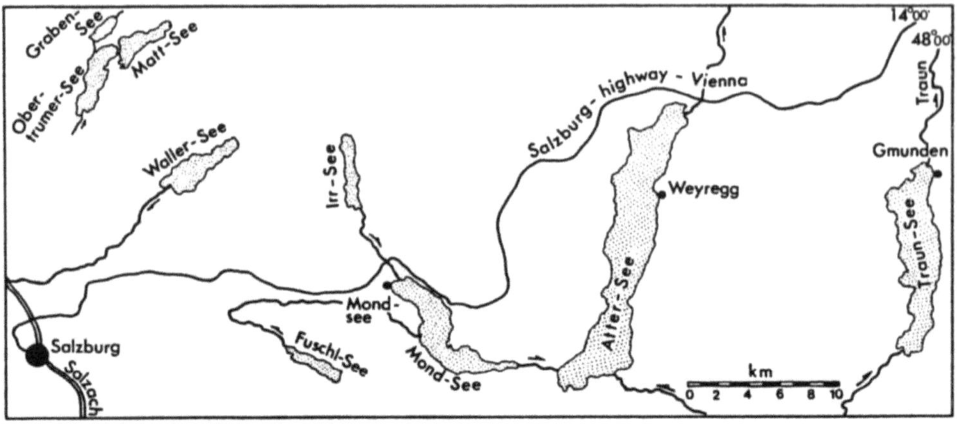


**Figure 16.1** Geographical locations of the Austrian Alpine lakes that are discussed in this chapter. Surface area in km²: Attersee, 45.6; Traunsee, 25.6; Mondsee, 14.21; Wallersee, 6.39; Obertrumersee, 4.8; Mattsee, 3.25; Grabensee, 1.26.

The Traunsee, Attersee, Mondsee, and Wallersee, Obertrumer, Matt- and Graben-See are situated north and east of Salzburg in Austria (Figure 16.1). We have investigated these lakes in detail in recent years. The general topographic features around these foreland lakes are products of the last sequence of Alpine orogenesis in the Pliocene. During Quaternary time, glaciers modified the basin morphologies, especially those of the last Würm glaciation. The overdeepened basins were filled with water and sediments during and after the last ice retreat.

All these lakes are located in the forelands of the northern Calcareous Alps. All of them are hardwater lakes. The largest lakes—Attersee, Traunsee, and Mondsee—are more influenced by allochthonous input of sediments than the Wallersee and the Trumer lakes. Yet, all these lakes have a large hypolimnion in relation to epilimnion, are intensively stratified during summer, and are holomictic because of their good exposition to winds. The ratio of profundal area to littoral area is much greater than one.

## 16.3 Principle Sediment-Supplying Processes

The sedimentation history of a lake can only be understood if the recent depositional processes are known. In general we can distinguish three main autochthonous as well as allochthonous sediment sources in these lakes:

1. Autochthonous carbonate production by biogenic epilimnetic decalcification (Minder, 1922; Ohle, 1952) over the entire lake area during summer—Especially from July to September, more or less intense photosynthesis of phytoplankton withdraws nutrients, carbon dioxide, and

bicarbonate from surface waters resulting in rise of pH and subsequent oversaturation with respect to $CaCO_3$. Carbonate then precipitates from the epilimnion in the form of tiny crystals and aggregates (e. g., Rossknecht, 1977, 1980; Kelts, 1978; Kelts and Hsü, 1978; Schröder, 1982; Schröder et al., 1983; Sturm, 1984; Lee et al., 1987). Within the Austrian prealpine lakes, this process has taken place since 13,000 yr B. P. (Behbehani et al., 1986; Schneider et al. 1987). Within eutrophic lakes this epilimnetic biogenic carbonate precipitation can eliminate a certain amount of phosphorus from the water and lead to a relatively stable deposition of P within sediments (see below and Rossknecht, 1977, Jäger and Röhrs, 1989).

2. Autochthonous carbonate production by biogenic benthic decalcification, preferably on the shallow sublittoral and the shelf areas up to a depth of about 12 m—Benthic micro-, macro-, and epiphytes are able to produce carbonate by assimilating carbon dioxide and bicarbonate (Kann, 1940; Kelts and Hsü, 1978; Schneider and Schröder, 1980; Schröder, 1982; Schneider and Schröder, 1983; Behbehani et al., 1986). This leads to the formation of "Seekreide" (lake marl) with carbonate contents of >80% (Merkt et al., 1971; Schöttle and Müller, 1968; Schäfer, 1973; Schäfer and Stapf 1978; Schröder, 1982). A $CaCO_3$ balance made in Lake Attersee in 1980/81 showed that processes no. 1 and no. 2 produce more than 11,000 t/yr of $CaCO_3$ or 250 t/km² of calcite (Schröder et al., 1983).
3. Allochthonous input into the lake from different natural and human sources (river influx, landslides, mud flows, bank erosion in the wave zone, and transport of nutrients and pollutants by rivers, surface runoff, and rain (Müller, 1979), as well as the input of industrial tailings in special cases (e. g., Behbehani et al., 1986; Schneider et al., 1986; Müller et al., 1986)—These allochthonous sediments are deposited according to lake morphology. The highest sedimentation rates are to be found in the delta areas and deep basins. The lowest sedimentation rates can be found on ridges and shelf regions.

The mineralogical character of the suspended matter of rivers draining into the lakes is determined by the different lithologies of each catchment area. Dolomite is transported from the Calcareous Alps into the prealpine lakes, whereas predominantly quartz, but also feldspar, clay minerals, and micas are derived from flysch and moraine areas.

The distribution mechanisms and autochthonous and allochthonous sediment particles in thermally stratified lakes are manifold (see Sturm and Matter, 1978; Schneider et al., 1987). We find different sedimentary features within sediment cores from the basins of the larger lakes, such as the Attersee or Traunsee, which are caused by pelagic sedimentation interrupted by the input of underflow sediments and turbidites (Behbehani et al., 1986; Müller et al., 1986; Schneider et al., 1986).

We can calculate that during the late- and postglacial periods more than 45 m of sediment accumulated in the lake basins, whereas on ridges or in shelf-regions only 1.5 to 5 m of sediment were deposited, representing the whole late- and postglacial time (Schneider et al., 1987).

The sedimentation history of late-glacial and postglacial periods can be reconstructed using sediment cores. As a typical and general example for this area, data from a core from the Attersee are shown in Figure 16.2. The Würm glaciers retreated 17,000 yr B. P. (after datings given by van Husen, 1977, 1987) at a relatively fast rate from the regions where the lakes are situated today. From cores it is possible to identify moraine material with glacially striated pebbles, overlain by finely banded varves, deposited in the arctic lake water. Drop stones are relatively common within these series (Schneider et al., 1987). Sedimentation rate is high in comparison to younger sediments.

The arctic lake existed at best for 4,000 years. Varved sediments with low carbonate contents of 30–40% are found between 17,000 and 13,000 years. After 13,000 yr B. P., the climate improved, and the stage of more intensive phytoplankton development and autochthonous biogenic carbonate production within the lakes began. As a consequence, the carbonate content increased drastically to an average of 60–70% (see Figure 16.2). As a consequence of the natural reforestation of the catchment areas by *Juniperus* (juniper) and *Betula* (birch), the sedimentation rate decreased to about one-third. The Holocene boundary (10,000 yr B. P.) cannot be identified sedimentologically. But within the larger lakes like the Attersee we are able to identify the boundary of colonization of the catchment area by *Abies* (fir) and *Fagus* (beech) around 6,000–6,500 yr B. P. and, next, the first peak of weak forest clearance (Late Neolithic) and immigration of *Carpinus* (hornbeam) about 4,700 yr B. P. (Schmidt, 1981; Behbehani et al., 1986).

A drastic non-arboreal pollen increase indicates the phase of the Bavarian colonization with the foundation of monasteries in the eighth century A. D. (Röhrs, 1986). This is the sharp border at 1,200 yr B. P. seen in Figures 16.2 and 16.3, which is indicated sedimentologically by a decrease in the carbonate content of the sediments, caused by deforestation leading to more intensive erosion rates in the catchment areas, and an increase in organic material due to the first man-made eutrophication. Within the basinal sediments, the climatic changes from 13,000 to 1,200 yr B. P.—from more humid to drier climates—are indicated by phases of more or less frequent turbidity currents.

## 16.4 Natural Eutrophication

There are many indications of former eutrophication, especially from Swiss lakes like the Baldeggersee (Niessen and Sturm, 1987) and the Lobsigensee (Ammann, 1985, 1986). Phases of natural eutrophication were identified in

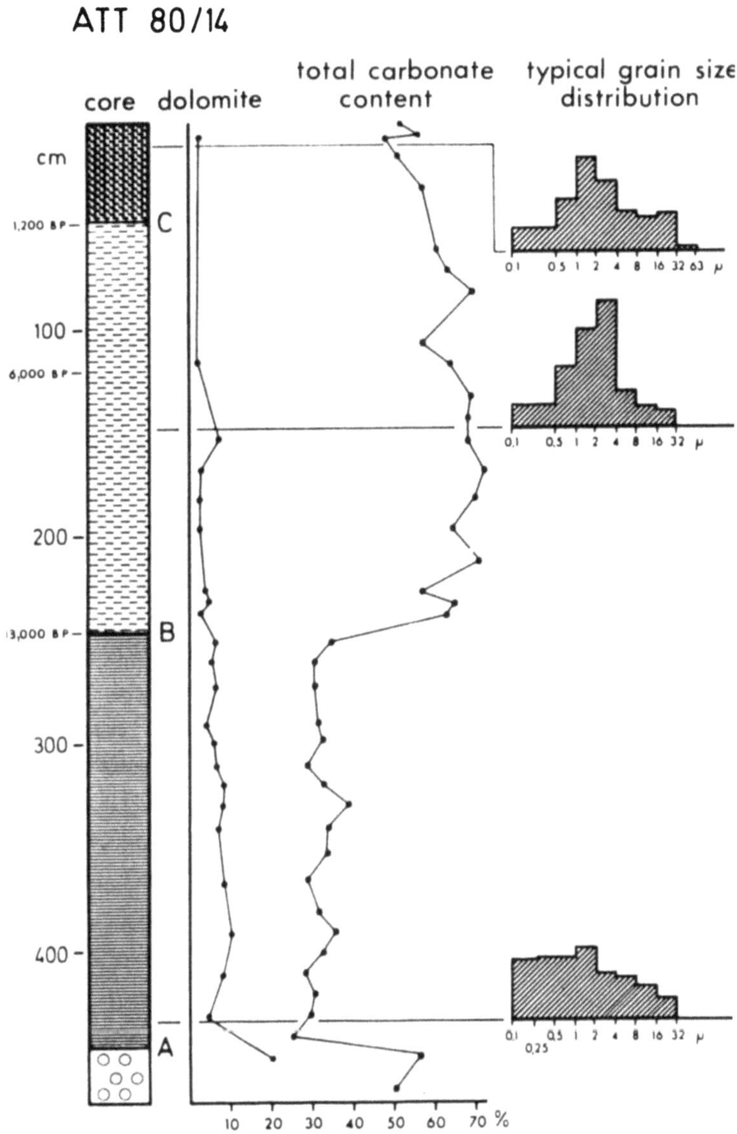


**Figure 16.2** Core ATT 80/14 from the shelf region (40 m water depth) of the Attersee showing the most important sedimentological phases and the ages of the sedimentological changes (from Behbehani et al., 1986).

these lakes, deduced from laminated anoxic sediments with higher contents of phosphorus and organic material during the Boreal and the early Older Atlanticum periods. Events of meromixis were found in lakes of Carinthia between the Bölling and the Alleröd respectively, and between the Alleröd and the

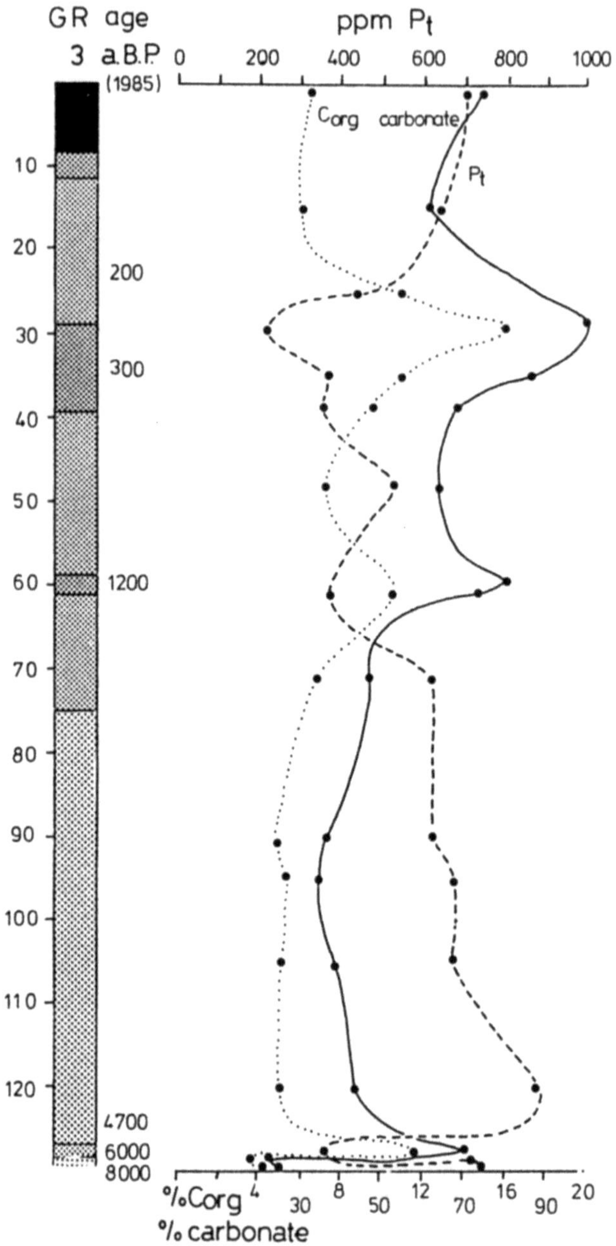


**Figure 16.3** Sedimentological and geochemical data of the core GR 3 (12 m water depth) from the Grabensee. Dark sediment layers indicate eutrophication events at 6,000 and 1,200 yr B. P., about 300 years before today, and during the last 40 years. The older eutrophication events are indicated by dark sediment layers with higher amounts of organic carbon ($C_{org}$) and total phosphorus and lower carbonate contents (after Röhrs, 1986). Core length in cm. Horizontal lines indicate sediment color changes.

Preboreal (Löffler, 1986). There also exist indications of older eutrophication phases. This was impressively shown by Züllig (1982, 1985) who worked out a "carotinoid-stratigraphy of lake sediments as a key for investigating environmental changes since the late glacial time," (Züllig, 1988). Züllig (1985, 1986) found paleopigments of *Oscillatoria rubescens* (Oscillaxanthin) within sediments of the Preboreal. This is an indication of eutrophication as a consequence of the warming trend at the beginning of the Holocene.

Natural eutrophication is especially likely to occur in small lakes which are drying up. If the epilimnion/hypolimnion ratio increases during sedimentation history, a state can be reached when the oxygen content in the hypolimnion is not great enough to oxidize and mineralize the dead organic material. The contents of organic carbon and total phosphorus ($C_{org}$ and $P_{total}$) increase, and the lake becomes a eutrophic lake and, finally, a swamp. This process took place in many small lakes within the Austrian Alpine lowlands (e. g., the small Egelseen between the Wallersee and the Trumer lakes).

Natural eutrophication indicators show the enhanced sensitivity of smaller lakes to any change in input of nutrients or of environmental conditions. The first natural eutrophication that can be observed within the sediments of the larger Austrian prealpine lakes occurred about 6,000 yr B. P., as indicated by increasing values of organic carbon, as well as of phosphate, and a contemporaneous decrease in carbonate content within the sediments (see Figure 16.3 and also Giger et al., 1984). This eutrophication was probably caused by the natural input of nutrients during a more humid climatic period with the washout of nutrients from bogs and soils. The lake recovered very quickly and normal values returned, as can be seen in Figure 16.3.

## 16.5 Man-made Eutrophication

Around 4,500 yr B. P., Neolithic man settled on the littoral zone of these lakes at a time when the lake level was 3–4 m lower than it is today (Offenberger et al., 1981; Offenberger, 1981, 1985; Behbehani, 1987). This human activity led to a local and limited load of the near-sublittoral sediments by nutrients from organic wastes and wood, which is detectable in the sediments by the increase in organic material and phosphorus. This can clearly be observed in very small lakes like the Swiss Lobsigensee (Ammann, 1986). Deforestation by the Romans can also be observed in small lakes—by an increase in phosphorus input and sedimentation rate (Ammann, 1986). The influence of Neolithic people and of Roman colonization on large lakes in the Austrian prealpine foreland, however, was very limited and therefore not detectable. Since 1,200 yr B. P., human influence on lake sedimentation has become increasingly intense, as is shown in Figure 16.3 and 16.4, indicated by dark layers, and higher amounts of phosphorus, organic carbon, and some metals.

In the small lowland lakes of Salzburg (the Trumer lakes and the Wallersee, see Figure 16.1 and Röhrs, 1986) we also found indications of eutrophication, visible as dark or black layers in the sediment column.

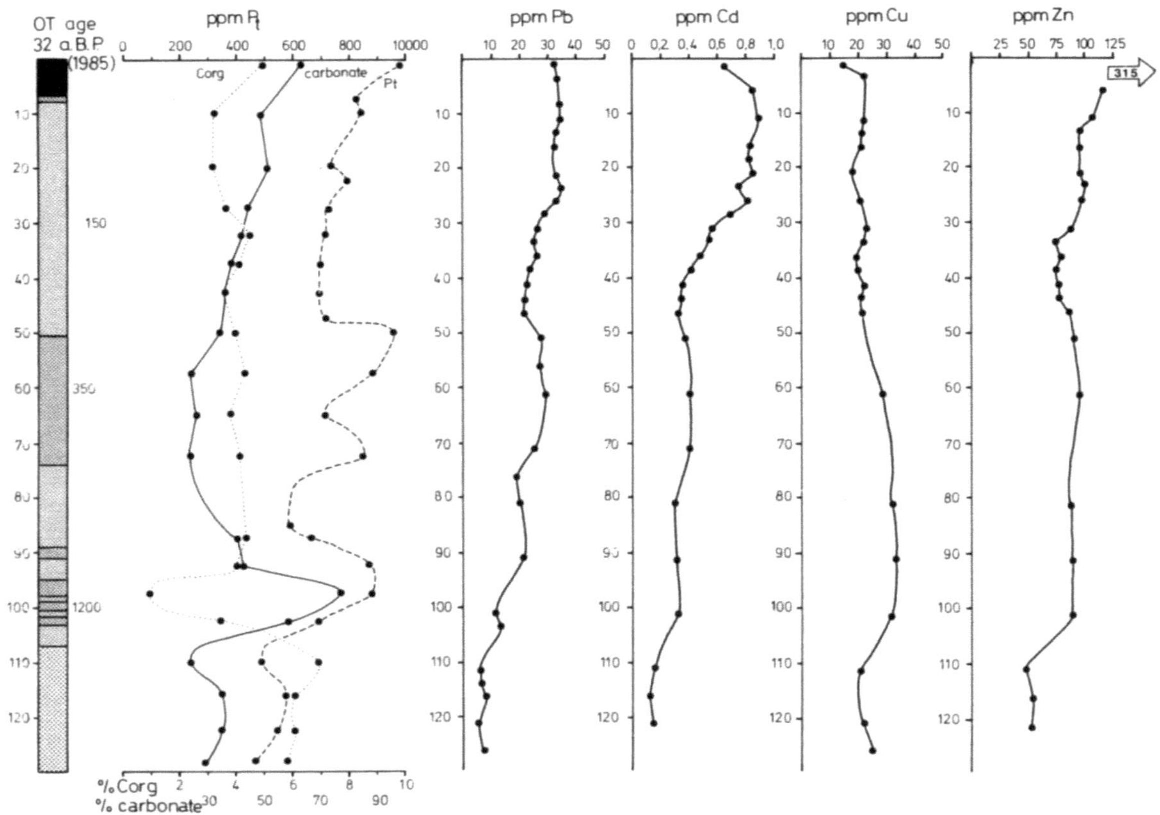


**Figure 16.4** Sedimentological and geochemical data from core OT 32 (30 m water depth) from the Obertrumersee. Eutrophication events are indicated by dark sediment layers at 1,200 and 350 yr B. P., (horizontal lines indicate sediment color changes) and during the last 40 years by higher amounts of $C_{org}$, $P_t$, and changing carbonate content (after Röhrs, 1986). Core length in cm.

The most massive input of nutrients occurred around 1,200 yr B. P., due to extensive forest clearing with subsequent soil erosion and outwash of nutrients from bog drainage. This was a consequence of the foundation of monasteries during the Bavarian colonization and the contemporaneous development of land for agriculture around the lakes. These activities influenced the entirety of the lake basins. But these events only temporarily damaged the lakes, as can be observed from the core data (see Figure 16.3). After the first massive nutrient input due to forest clearing and bog drainage the nutrients had been fixed within the sediments, eutrophication ceased; but the lakes never again reached the pre-eutrophic state because nutrient input by human activity continued.

A large brewery was constructed at Obertrumer Lake about 350 years ago. Nutrient-rich tailings were dumped directly into the lake. Contemporaneously, additional bog areas were drained, and further forest clearing occurred. However, this drastic event did not damage the lake for a long time, because carbonate sediments acted as an effective trap for nutrients (see Section 16.8).

The introduction of modern agriculture with its rising nutrient input after World War II and the subsequent increase of tourism and of housing projects without adequate treatment of waste water led to further increase, especially of phosphate input, both to lake waters and sediments. This process has been reduced recently but it still continues, in spite of the fact that many political and social measures have been taken. More scientific investigations, technical measures, and financial investments are necessary either to stop the excessive nutrient input into the lakes or to restore the lakes by internal measures (e. g., Jäger, 1986; Röhrs et al., 1986; Amt d. Salzburger Landes-Regierung, 1986; Gächter, 1987). "There is no alternative to devoting a greater percentage of our intellectual and financial resources to fresh waters; (Wetzel, 1975, p. 660).

Massive algal blooms, such as that of the cyanobacterium *Oscillatoria rubescens*, have occurred frequently over the past 20 years. Using the method of Züllig (1982, 1985, 1986), their pigments (Oscillaxanthin) and the pigments of other phytoplanktonic organisms and of bacteria have been identified in the sediments (Schultze, 1985, 1986), indicating the occasional eutrophication events.

As shown in Figure 16.3 and 16.4, the carbonate content decreased during the older eutrophication events (6,000, 1,200, and 350 yr B. P.). With the beginning of man-made eutrophication, laminated sediments occur again, as Nipkow (1920) observed in 1920 in some Swiss lakes. He was the first to correctly interpret the black and white laminations as annual lamination; those being the consequence of man-made lake eutrophication (see also Niessen and Sturm, 1987). Within the youngest eutrophicated sediments during the last 40 years, the carbonate content increases together with the organic carbon and phosphorus. Increasing growth of calcite crystals from epilimnetic decalcification as a result of phytoplankton activity and subsequent increasing saturation during eutrophication was described by Sturm et al. (1982), Kunz (1983), Kunz and Stumm (1984), and Niessen and Sturm (1987). Niessen and Sturm found, in the Swiss Baldeggersee, increasing (4–30$\mu$m) size of calcite crystals as a consequence of increasing eutrophication during the last 98 years. Diagenetic redissolution of calcite within older sediment layers can be responsible for the tinier calcite crystals (up to 20 $\mu$m) in deeper core parts (see discussion in Röhrs, 1986, p. 73–76). Röhrs found slightly corroded calcite crystals from older sediments. Another explanation could be (Sturm, 1984; Kunz and Sturm, 1984), that the modern eutrophication in the last 40 years has become more intense. Phosphates as well as organic substances act as inhibitors for the formation of calcite crystals. Enhanced photosynthesis during algal blooms uses the P-content for building organic matter, leading to a much higher calcite oversaturation of the epilimnion caused by the photosynthetic use of $CO_2$. As a consequence of the summerly phytoplankton development and P-depletion, the oversaturation is of such an order that large calcite crystals (up to 40$\mu$m) form. This also could explain why the silt content increases in sediments of the profundal zones of lakes with nearly no change in the clastic silt input (Niessen and Sturm, 1987).

Since the industrial revolution, heavy metals also have become typical anthropogenic components in the sediments. Very characteristic elements are Zn, Cd and Pb (Müller et al., 1983; Gunten et al., 1987, see also Figure 16.4). Datings of $^{137}Cs$ (Irlweck and Danielopol, 1985) designate 1952 as the beginning of the observable atmospheric testing of atomic bombs and 1963 as the year of the partial test ban treaty. According to investigations of K. Irlweck (Institute for Inorganic Chemistry of Vienna University, personal communication, 1987) about four times more radioactive $^{137}Cs$ was deposited in the Austrian prealpine lakes from the Chernobyl accident than during the entire previous period from 1952 to the present, also a significant limnogeological event.

## 16.6 Eutrophication Processes in Hard- and Soft-water Lakes

As described by Röhrs (1986), calcium carbonate within sediments of hard water lakes can act as an effective trap for phosphorus (Otsuki and Wetzel, 1972; Rossknecht, 1980) and iron hydroxides (Einsele, 1936; Mortimer, 1941, 1942; Ruttner, 1962; Syers et al., 1973; Golterman, 1984; Nürnberg, 1985).

Furthermore, phosphate can be bound to clay minerals within sediments. Within calcium-rich lakes, iron-phosphate interactions seem to be of limited importance (Golterman, 1984, p. 35). Nevertheless, within hardwater lakes bacterial mats grow on the sediment surface which can take up and store nutrients. This seems to depend on nitrate reduction (Röhrs, 1986; Boström and Petterson, 1982). The microorganisms using the nutrients (phosphate and nitrate) at the sediment/water interface can act as a barrier against the loss of nutrients from sediments to the hypolimnion (Doremus and Clesceri, 1982; Fenchel and Blackburn, 1979; Hanselmann, 1984; Gächter, 1987). We do not know enough about the ecology of microorganisms, their interactions, and the biogeochemical processes within and at the sediment/water interface (Kusnezow, 1959, p. 39–42; Golterman, 1977) to understand this phenomenon.

The complex interactions of phosphorus within the compartments of a lake must be known in order to make predictions about the P-cycle of the particular lake (see e. g., for Lake Constance: Tilzer, 1981; Stabel, 1984). Periodic input of nutrients into a hardwater lake will not necessarily be the beginning of permanent eutrophication. Lake internal eutrophication, as it is described by Ohle (1958, 1962), Psenner (1984), and Nürnberg (1988)—with an internal nutrient cycle from water to plankton to sediment and again to water—happens in hardwater lakes also but is a more characteristic feature of softwater lakes.

## 16.7 Eutrophic Softwater Lakes

Within eutrophic softwater lakes, which are usually surrounded by crystalline silicate rocks, such as the small Piburger See, the Ötztal, Tyrol (Pechlaner and Psenner, 1983; Psenner, 1984) the following processes can be summarized:

1. Normally, we find a low redox potential and, consequently, pure pyrite as the iron sulfide mineral within the sediments of eutrophic softwater lakes (Psenner, 1983, 1984). $Fe^{2+}$ (pyrite) is the stable iron phase within the sediment.
2. At the sediment/water interface under anoxic conditions, there is no stable precipitation of phosphorus phases and no essential microbial fixation of nutrients. As a result we find a phosphorus release out of the sediments into the bottom water during the stratified period. During the mixing phase the nutrients can be recycled to the surface waters where they can be available up to 90% (Nürnberg, 1984, 1988) for the plankton. This process leads to a steady re-eutrophication of a softwater lake. External as well as internal restoration measures of such softwater lakes are necessary. However, limiting nutrient input to the lake is the more important measure.

## 16.8 Eutrophic Hardwater Lakes

Within an eutrophic hardwater lake, such as the Wallersee (see Figure 16.1) (Germatsidis, 1985; Röhrs et al., 1986; Röhrs, 1986), we investigated the sedimentary history and the sedimentological processes. Two of us (Jäger and Röhrs, 1989) made a careful balance of the calcium carbonate (Ca) and phosphorus (P) cycles on the basis of an interdisciplinary study. The results of this study are shown in Figure 16.5 for carbonate and in Figure 16.6 for phosphorus (also see Amt der Salzburger Landesregierung, 1986; Jäger and Röhrs, 1989).

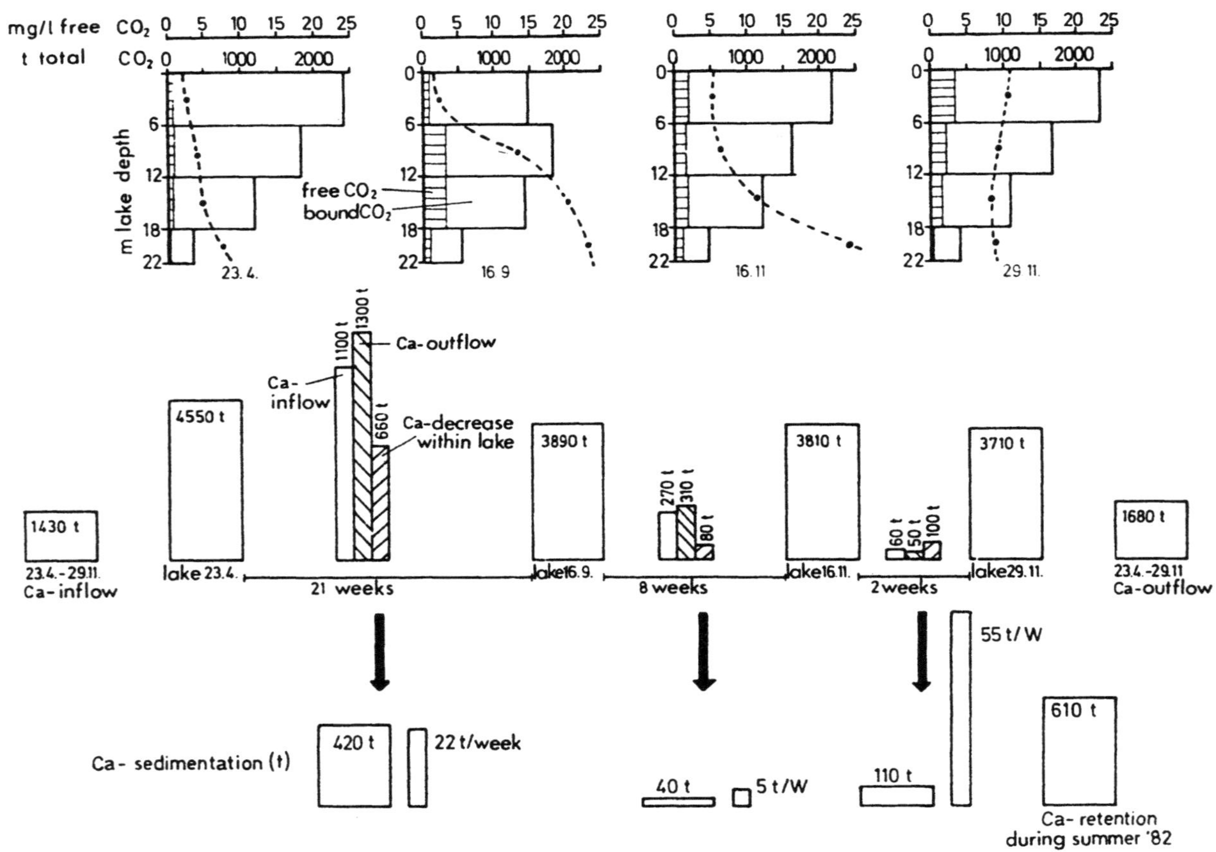


**Figure 16.5** Cycling and balance of $CO_2$ (top) and calcium (bottom) within the Wallersee from April 23, 1982 to Nov. 29, 1982 (after Jäger and Röhrs, 1989. Dashed line: Concentration of free $CO_2$ (mg/l). Horizontally hatched/white boxes: compartments enclosing loads of free/bound $CO_2$ (t/depth strata).

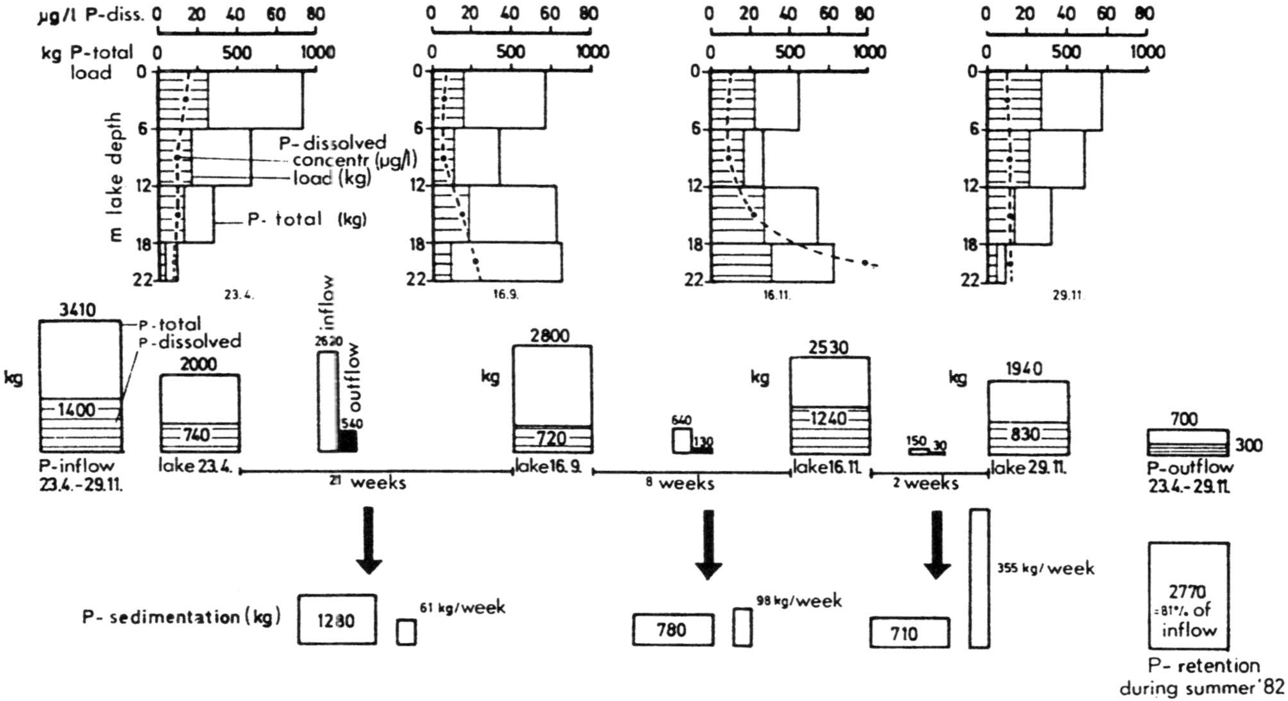


**Figure 16.6** Cycling and balance of phosphorus (P) within the Wallersee from April 23, 1982 to Nov. 29, 1982 (after Jäger and Röhrs, 1989. Dashed line: Concentration of dissolved phosphorus (µg/l). Horizontally hatched/white boxes: compartments enclosing loads of phosphorus and total phosphorus (P–total) (kg/depth strata).

**Cycles during Thermal Stratification** Wallersee begins the stratified period with a high phosphorus content within the epilimnion. Consequently, we find a massive algal bloom in spring and a subsequent P deficiency. $CO_2$ and $HCO_3^-$ are used during photosynthesis as carbon sources, and the pH rises. Therefore, epilimnetic biogenic decalcification starts. We find a scavenging of P by the precipitated calcite and an elimination out of the epilimnion of 460 t of Ca and 2.06 t of P, which is bound on and within calcite. Otsuki and Wetzel (1972) carried out experimental investigations on coprecipitation of phosphate together with calcite. These experiments showed that during photosynthetically induced calcite precipitation under high pH-conditions up to 74% of phosphate was found to be precipitated with calcite. Rates of about 30% of phosphate removal with carbonates were found experimentally by Kleiner (1987). Phosphate is mainly incorporated into calcite, rather than adsorbed on crystal surfaces. The two authors also found a removal of soluble organic substances during calcite precipitation (Otsuki and Wetzel, 1973).

Within deep waters of the Wallersee, the dead organic material is already mineralized to a greater extent. Subsequently, we find an oversaturation in $CO_2$, a pH drop, and a release of P from this organic material. At the same time calcite will be partly dissolved within the hypolimnion (see Figure 16.5, 16.11.). Consequently, P and Ca contents in deep waters increase. During core sampling, we found intensive networks of "tufted mats" built by the filamentous chemosynthetic sulfur bacterium *Beggiatoa* covering the surface of the profundal sediments. *Beggiatoa* lives under microaerophilic conditions if the $H_2S$ content is not too high (Kusnezow, 1959, p. 186–206). These *Beggiatoa* mats act as biological diffusion barriers and as nutrient traps, at the sediment/water interface (Hanselmann, 1984). Under strong anoxic conditions with extremely high $H_2S$ concentration, *Beggiatoa* disappears and this microbial mat no longer functions as a nutrient diffusion barrier. Redox measurements in cores of the Obertrumersee, as well as special analyses with SEM/EDX (Röhrs, 1986), showed that even during the stagnation phase, $Fe^{3+}$ in the form of Fe-hydroxide is the stable phase at the sediment/water interface in spite of the fact that there is no oxygen but only $H_2S$ in the deep water. Phosphorus is bound to calcite and to iron hydroxide phases. Within deeper layers of the anaerobic sediment, P can be redissolved. The question is, can this phosphorus move from the sediment surface into the bottom water as is described in laboratory experiments with softwater lake sediments, e. g., by Nürnberg (1988)? After the investigations and experiments of Röhrs (1986), a microbial mat at the sediment surface (apparently depending on the nitrate content of the bottom water) seems to be responsible for that slightly oxidizing microenvironment and thus hinder the redissolution of phosphorus. Similar observations and questions are discussed by: Ripl, 1978; Fenchel and Blackburn, 1979; Doremus and Clesceri, 1982; Herbert, 1982; and Ahlgren et al., 1988. Much more geomicrobiological investigations are necessary in the future for explaining the complex role of the different microorganisms in these processes at and within the sediment/water interface.

**Cycles during the Phase of Mixing** In autumn with the beginning of full circulation in the lake, a degassing of the surplus of $CO_2$ can be observed. The pH of the water rises, and a massive precipitation of calcite takes place because of the oversaturation of the hypolimnetic waters with respect to calcite. Phosphorus coprecipitates together with the calcite. Within one week, 55 t of Ca and 0.35 t of P were precipitated and added to the sedimentation. Therefore, phosphorus is again bound to calcite. We found a stable fixation of P in the sediment and no redissolution because the deep water is well oxygenated during the phase of mixing.

## 16.8 Conclusion

Eutrophication is not solely a modern phenomenon caused by human activity in the Austrian prealpine lakes. Former natural eutrophication events can be identified, e. g., 6,000 yr B. P. This event, as well as the events of 1,200 and 350 yr B. P., were not able to bring the described lakes into a long-term eutrophic state. The self-restoration of the hardwater lakes was possibly supported by the elimination of phosphorus through coprecipitation with calcite during epilimnetic biogenic decalcification. The results of our investigations indicate that sediments of hardwater lakes like the Wallersee may act as a stable sink for phosphorus. No extensive "self-fertilization" of such lakes takes place. Therefore, there is not necessarily a need for internal restoration measures in such lakes, which have become eutrophic due to human activity during the last decades. Gächter (1987) came to the same conclusion from observations in two Swiss lakes. However, external strategies such as limiting nutrient inputs are necessary.

*Acknowledgements* We would like to acknowledge the cooperation of many Austrian collegues from the Limnological Institute of the Austrian Academy of Sciences in Mondsee, as well as the especially good cooperation with Dr. Michael Sturm (EAWAG/ETH, Zürich) and with Dr. Jens Müller (TU Munich). We especially thank Dr. Sturm for reading the manuscript critically, for literature references, and for many valuable suggestions. We also thank Prof. Max Tilzer for critically reviewing the manuscript. The investigations were supported by grants of the Amt der Salzburger Landesregierung (Nos. 9367/8/L6602232 and 6/63-9711) and of the Deutsche Forschungsgemeinschaft (Az. Schn 16/8-12, especially 16/11-1/2). We thankfully acknowledge this support. We thank Dr. Robert B. Blodget (Illinois State University) for improving the English text.

## References

Ahlgren, I., Boström, B. and Petterson, A.-K. 1988. Seasonal variation of the microbial community in the sediments of a hypereutrophic lake. *Verh. Internat. Verein Limnol.* 23:460–461.

Ammann, B. 1985. Late glacial and Holocene environments of a lake on the Central

Swiss Plateau: Introduction and palynology. In: Lang, G. (ed.), *Swiss Lake and Mire Environments During the Last 15 000 years, Diss. Bot. No. 87,* 127–134.

Ammann, B. 1986: Litho- and palynostratigraphy at Lobsigensee: Evidences for trophic changes during the Holocene. *Hydrobiologia* 143:301–307.

Amt Der Salzburger Landesregierung (ed.) 1986. *Projekt "Vorlandseen"—Wallersee, Obertrumer See, Mattsee, Grabensee, Ergebnisband, Raumbezogene Forschung und Planung im Land Salzburg, Studien und Ergebnisse, Stud. Forsch. Salzburg* No. 2, 507 pp.

Behbehani, A.R., Müller, J., Schmidt, R., Schneider, J., Schröder, H.-G., Strackenbrock, I., and Sturm, M. 1986. Sediments and sedimentary history of Lake Attersee (Salzkammergut, Austria). *Hydrobiologia* 143:233–246.

Behbehani, A.R. 1987. Sedimentations- und Klimageschichte des Spät- und Postglazials im Bereich der Nördlichen Kalkalpen (Salzkammergutseen, Österreich) *Göttinger Arb. Geol. Paläont,* No. 34, 120 pp.

Boström, B. and Petterson, K. 1982. Different patterns of phosphorus release from lake sediments in laboratory experiments. *Hydrobiologia* 92:415–429.

Doremus, C. and Clesceri, L.S. 1982. Microbial metabolism in surface sediments and its role in the immobilization of phosphorus in oligotrophic lake sediments. *Hydrobiologia* 91:261–268.

Einsele, W. 1936 Uber die Beziehungen des Eisenkreislaufs zum Phosphatkreislauf im eutrophen See. *Arch. Hydrobiol.* 29:664–668.

Fenchel, T.M. and Blackburn, T.H. 1979. *Bacteria and Mineral Cycling.* Academic Press, London, 225 pp.

Gächter, R. 1987. Lake restoration. Why oxygenation and artificial mixing cannot substitute for a decrease in the external phosphorus loading. *Schweiz. Z. Hydrol.* 49/1:170–185.

Germatsidis, I. 1985. Die Sedimente des Wallersees. Unpubl. Diploma Thesis, Göttingen, 74 pp.

Giger, W., Sturm, M., Schaffner, C., Bonani, G., Balzer, R., Hofmann, H.J., Morenzoni, E. Nessi, M., Suter, M., and Wölfli, W. 1984. $^{14}C/^{13}C$-ratios in organic matter and hydrocarbons extracted from dated lake sediments. *Nuclear Instruments and Methods in Physics Research* B 5:394–397.

Golterman, H.L. (ed.) 1977. *Interactions between Sediments and Fresh Water—Proceedings of an Internat. Sympos. Amsterdam, Sept. 1976.* W. Jungk Publ., 473 pp.

Golterman, H.L. 1984. Sediments, modifying and equilibrating factors in the chemistry of freshwater. *Verh. Internat. Verein. Limnol.* 22:23–59.

Gunten, H.R. v., Sturm, M., Erten, H.N., Rössler, E., and Wegmüller, F. 1987. Sedimentation rates in the central Lake Constance determined with $^{210}Pb$ and $^{137}Cs$. *Schweiz. Z. Hydrol.* 49/3:275–283.

Hanselmann, K. 1984. Bakterielle Bio-Geo-Chemiker in Seesedimenten. *Neue Züricher Zeitung, Forschung und Technik,* Vol. 182, 8th August 1984, p.45.

Herbert, R.A. 1982. Nitrate dissimilation in marine and estuarine sediments. In: Nedwell, D.A. and Brown, C.M. (eds.), *Sediment Microbiology.* Academic Press, London, p. 53–71.

Husen, D. van 1977. Zur Fazies und Stratigraphie der jungleistozänen Ablagerungen im Trauntal. *Geol. Jb. B.-A.* No. 120/1, 130 pp.

Husen, D. van (ed.), 1987. Das Gebiet des Traungletschers, Oberösterreich - Eine Typregion des Würm-Glazials. *Mitt. Komm. f. Quartärforsch. d. Österr. Akad. d. Wiss.,* No. 7, 78 pp.

Irlweck, K. and Danielopol, D.L. 1985. Caesium-137 and lead-210 dating of recent sediments from Mondsee (Austria). *Hydrobiol.* 128:175–185.

Jäger, P. 1986. Chemismus der Vorlandseen und ihrer Zubringer. In: Amt d. Salzburger L.-Regierg. (ed.), *Projekt "Vorlandseen"—Wallersee, Obertrumer See, Mattsee, Grabensee, Ergebnisband, Stud. Forsch. Salzburg,* No. 2, p. 343–384.

Jäger, P. and Röhrs, J. 1989. Die Phosphorfällung über Calciumcarbonat im eutrophen

Wallersee (Salzburger Alpenvorland, Österreich. *Int. Revue d. ges. Hydrobiol.* 74 (in press).
Kann, E. 1940. Ökologische Untersuchungen an Litoralalgen ostholsteinischer Seen. *Arch. Hydrobiol.* 37:197–269.
Kelts, K. 1978. Geological and sedimentary evolution of Lake Zürich and Zug. Dissertation, ETH Zürich, No. 6146, 250 pp.
Kelts, K. and Hsü, K.J. 1978. Fresh water carbonate sedimentation. In: Lerman, A. (ed.): *Lakes–Chemistry, Geology, Physics.* Springer, Berlin, p. 295–339.
Kleiner, J. 1987. Phosphate coprecipitation with calcite: laboratory experiments modelling the phosphorus removal in hard water lakes. In: Tilzer et al. (Org. Comm.), *Functional and Structural Properties of Large Lakes–Int. Sympos. Under the Auspices of SIL, Konstanz, 13–18 Sept.* Poster, Program, and abstracts, p. 86.
Kunz, B. 1983. Heterogene Nukleierung und Kristallwachstum von $CaCO_3$ (Calcit) in natürlichen Gewässern. Dissertation, ETH Zürich No. 7355, 159 pp.
Kunz, B. and Stumm, W. 1984. Kinetik der Bildung und des Wachstums von Calciumkarbonat. *Vom Wasser* 62:279–293.
Kusnezow, S. I. 1959. *Die Rolle der Mikroorganismen im Stoffkreislauf der Seen.* VEB Deutscher Verlag der Wissenschaften, 301 pp.
Lee, C., McKenzie, J.A., and Sturm, M. 1987. Carbon isotope fractionation and changes in the flux and composition of particulate matter resulting from biological activity during a sediment trap experiment in Lake Greifen, Switzerland. *Limnol. Oceanogr.* 32/1:83–96.
Löffler, H. (ed.) 1987. Paleolimnology IV—Proceedings of the fourth international symposium in paleolimnology, held at Ossiach, Carinthia, Austria. *Dev. in Hydrobiol.*, Vol. 37, 431 pp.
Löffler, H. 1986. An early meromictic stage in Lobsigensee (Switzerland) as evidenced by ostracods and Chaoborus. *Hydrobiologia* 143:309–314.
Merkt, J., Lüttig, G., and Schneekloth, H. 1971 Vorschlag zur Gliederung und Definition der limnischen Sedimente. *Geol. Jb.* 89:607–623.
Minder, L. 1922. Über biogene Entkalkung im Zürichsee. *Verh. theor. u. angew. Limnol.* 1:20–32.
Mortimer, C.H. 1941. The exchange of dissolved substances between mud and water in lakes I. *J. Ecol.* 29:280–329.
Mortimer, C.H. 1942. The exchange of dissolved substances between mud and water in lakes II. *J. Ecol.* 30:147–201.
Müller, J., Schneider, J., and Sturm, M. 1986. Industrial tailings in Lake Traunsee (Salzkammergut, Austria). *Hydrobiologia* 143:401–405.
Müller, Ge., Dominik, J., Schröder, H.-G., and Schneider, J. 1983. Zur Chronologie anthropogener Schwermetalle in den Sedimenten des Attersees (Österreich). *Tschermaks min.-petrogr. Mitt* 31:193–206.
Müller, Gü. 1979. Grundlagendaten für Fuschlsee, Mondsee und Attersee sowie das gesamte Einzugsgebiet *Arb. Lab. Weyregg* 3:10–14.
Niessen, F. and Sturm, M. 1987: Die Sedimente des Baldeggersees (Schweiz) - Ablagerungsraum und Eutrophierungsentwicklung während der letzten 100 Jahre. *Arch. Hydrobiol.* 108:3,365–383.
Nipkow, F. 1920. Vorläufige Mitteilungen über Untersuchungen des Schlammabsatzes im Zürichsee. *Schweiz. Z. Hydrol.* 1:10–122.
Nürnberg, G.K. 1984. The prediction of internal phosphorus load in lakes with anoxic hypolimnia. *Limnol. Oceanogr.* 29/1:111–124.
Nürnberg, G.K. 1985. Availability of phosphorus upwelling from iron-rich anoxic hypolimnia. *Arch. Hydrobiol.* 104/4:459–476.
Nürnberg, G.K. 1988. Prediction of phosphorus release rates from total and reductant-soluble phosphorus in anoxic lake sediments. *Can. J. Fish. Aquat. Sci.* 45/1:453–462.

Offenberger, J. 1981. Die "Pfahlbauten" der Salzkammergutseen. In: *Das Mondseeland–Geschichte und Kultur.* O. Ö. Landesverlag, Linz, p. 295–357.
Offenberger, J., Ruttkay, E., Schmidt, R., Chondrogianni, C., Niessen, F., Schneider, J., and Stojaspal, F. 1981. Stratigraphische Untersuchungen im Bereich der neolithischen Station Weyregg I am Attersee. *Fundber. aus Österreich* 20:191–222.
Offenberger, J. 1985. Investigations of an underwater excavation and a detailed survey within the neolithic settlement of See/Mondsee. In: Danielopol, D., Schmidt, R., and Schultze, E. (eds.) *Contributions to the Paleolimnology of the Trumer lakes (Salzburg) and the Lakes Mondsee, Attersee and Traunsee (Upper Austria).* Österr. Akad. d. Wiss., Mondsee, p. 132–135.
Ohle, W. 1952. Die hypolimnische Kohlendioxyd-Akkumulation als produktionsbiologischer Indikator. *Arch. Hydrobiol.* 46:153–285.
Ohle, W. 1958. Typologische Kennzeichnung der Gewässer auf Grund ihrer Bioaktivität. *Verh. Internat. Ver. Limnol.* 13:196–211.
Ohle, W. 1962. Der Stoffhaushalt der Seen als Grundlage einer allgemeinen Stoffwechseldynamik der Gewässer. *Kieler Meeresforsch.* 18:107–120.
Otsuki. A. and Wetzel, R.G. 1972. Coprecipitation of phosphate with carbonates in a marl lake. *Limnol. Oceanogr.* 17:763–767.
Otsuki. A. and Wetzel, R.G. 1973. Interaction of yellow organic acids with calcium carbonate in freshwater. *Limnol. Oceanogr.* 18:490–493.
Pechlaner, R. and Psenner, R. 1983. Restaurierungsverlauf Piburger See II: *Nährstofflieferung aus Sedimenten als Ursache der Oligotrophierungsverzögerung.* In: *Ergebnisse des Österreichischen Eutrophierungsprogrammes 1978–1982.* Bundesministerium f. Wiss. u. Forschg., Wien p. 70–71.
Psenner, R. 1983. Die Entstehung von Pyrit in rezenten Sedimenten des Piburger Sees. *Schweiz. Z. Hydrol.* 45:219–232.
Psenner, R. 1984. Phosphorus release patterns from sediments of a meromictic mesotrophic lake (Piburger See, Austria). *Verh. Internat. Verein Limnol.* 22:219–228.
Ripl, W. 1978. Oxidation of lake sediments with nitrate—a restoration method for former recipients. Limnologiska Institutionen Lunds Universitet, Dissertation 78 B 7872, 151 pp.
Röhrs, J. 1986. Sedimentations- und Eutrophierungsgeschichte der Salzburger Vorlandseen—sedimentchemische Untersuchungen für ein Sanierungskonzept und das Problem der Phosphor-Rücklösung. Buchhandlg. Kanzler-Haseitt, München, Documenta naturae No. 31, 105 pp.
Röhrs, J., Germatsidis, I., Lindner, D., and Schneider, J. 1986. Sedimentgeologische Untersuchungen an den Trumer Seen und am Wallersee. In: Amt d. Salzburger L.-Regierg. (ed.); *Projekt "Vorlandseen"—Wallersee, Obertrumer See, Mattsee, Grabensee, Ergebnisband, Stud. Forsch. Salzburg* No. 2, p. 421–475.
Rossknecht, M. 1977. Zur autochthonen Calcitfällung im Bodensee-Obersee. *Archiv. f. Hydrobiol.* 81:35–64.
Rossknecht, M. 1980. Phosphatelimination durch autochthone Calcitfällung im Bodensee-Obersee. *Arch. f. Hydrobiol.* 88:328–344.
Ruttner, F. 1962. *Grundriß der Limnologie—Hydrobiologie des Süßwassers.* W.de Gruyter-Verlag, Berlin, 332 pp.
Schäfer, A. 1973. Zur Entstehung von Seekreide - Untersuchungen am Untersee (Bodensee).-*N. Jb. Geol. Pal., Mh.*, 4:216–230.
Schäfer A. and Stapf, K.R.G. 1978. Permian Saar-Nahe Basins and recent lake Constance (Germany): Two environments of lacustrine algal carbonates. *Spec. Publ. Int. Ass. Sediment.* 2:83–107.
Schmidt, R. 1981. Grundzüge der spät- und postglazialen Vegetations- und Klimageschichte des Salzkammergutes (Österreich) aufgrund palynologischer Untersuchungen von See- und Moorprofilen. *Mitt. Komm. f. Quartärforsch. d. Österr. Akad. d. Wiss.* No. 3, 96 pp.

Schneider, J. and Schröder, H.-G. 1980. Calcification des cyanophycees et leur contribution a la genese des sediments calcaires. In: Humbert, L. (ed.), *Cristallisation-Deformation-Dissolution des carbonates, 17./18. Nov. 1980. Bordeaux*, p. 421–428.

Schneider, J., Schröder, H.-G., and Le Campion-Alsumard, T. 1983. Algal micro-reefs—coated grains from freshwater environments. In: Peryt, T.M. (ed.), *Coated Grains*. Springer, Berlin, p. 284–298.

Schneider, J., Müller, J., and Sturm, M. 1986. Geology and sedimentary history of Lake Traunsee (Salzkammergut, Austria). *Hydrobiologia* 143:227–232.

Schneider, J., Müller, J. and Sturm, M. 1987. Die sedimentgeologische Entwicklung des Attersees und des Traunsees im Spät- und Postglazial. In: Husen, D. van (ed.), Das Gebiet des Traungletschers, Oberösterreich - Eine Typregion des Würm-Glazials. *Mitt d. Kommiss. f. Quartärforsch. d. Osterr. Akad. d. Wiss.* No. 7, p. 51–78.

Schöttle, M. and Müller, G. 1968. Recent carbonate sedimentation in the Gnadensee (Lake Constance) Germany. In: Müller, G. and Friedman, G.M. (ed.), *Recent Developments in Carbonate Sedimentology in Central Europe*. Springer, Berlin, p. 148–156.

Schröder, H.-G. 1982. Biogene benthische Entkalkung als Beitrag zur Genese limnischer Sedimente, Beispiel: Attersee (Salzkammergut, Oberösterreich). Dissertation Univ. Göttingen, private press, 178 pp.

Schröder, H.-G., Windolph, H., and Schneider, J. 1983. Bilanzierung der biogenen Karbonatproduktion eines oligotrophen Sees (Attersee, Salzkammergut, Österreich. *Arch. Hydrobiol.* 97/3:356–327.

Schultze, E. 1985. Carotenoids from selected cores of the Trumer lakes and the Mondsee (trophic development and human impact). In: Danielopol, D., Schmidt, R., and Schultze, E. (eds.), *Contributions to the Paleolimnology of the Trumer Lakes (Salzburg) and the Lakes Mondsee, Attersee and Traunsee (Upper Austria)*. Mondsee, Österr. Akad. d. Wiss, p. 52–64.

Schultze, E. 1986. Stratigraphische Pigmentanalysen aus ausgewählten Bohrkernproben der Trumer Seen. In: Amt d. Salzburger L.-Regierg. (ed.), *Projekt "Vorlandseen"—Wallersee, Obertrumer See, Mattsee, Grabensee, Ergebnisband, Stud. Forsch. Salzburg*, No. 2, p. 471–475.

Stabel, H.H. 1984. Impact of sedimentation on the phosphorus content of the euphotic zone of Lake Constance. *Verh. Internat. Verein. Limnol.* 22: 964–969.

Sturm, M. and Matter, A. 1978. Turbidites and varves in Lake Brienz (Switzerland): deposition of clastic detritus by density currents. In: Matter, A. and Tucker, M.E. (eds.), *Modern and Ancient Lake Sediments*. Spec. Publ. internat. Ass. Sediment. No. 2, p. 147–168.

Sturm, M., Zeh, U., Müller, J., Sigg, L. and Stabel, H.-H. 1982. Schwebstoffuntersuchungen im Bodensee mit Intervall-Sedimentationsfallen. *Eclogae geol. Helv.* 75/3:579–588.

Sturm, M. 1984. Suspended particles in lakes. *EAWAG-News* 16/17:3–6.

Syers, J.K., Harris, R.F. and Armstrong, D.E. 1973. Phosphate Chemistry in lake sediments. *J. Environmental Quality* 2:1–14.

Tilzer, M.M. 1981. Nährstoffbelastung und Trophiezustand des Bodensees. *Arbeitsber. Ökologie/Umwelttechnik* 5:85–110.

Wetzel, R.G. 1975. *Limnology*. W.B. Saunders Co., Philadelphia, 743 pp.

Züllig, H. 1982. Untersuchungen über die Stratigraphie von Carotinoiden im geschichteten Sediment von 19 Schweizer Seen zur Erkundung früherer Plankton-Entfaltungen. *Schweiz. Z. Hydrol.* 44:1–98.

Züllig, H. 1985. Pigmente phototropher Bakterien in Seesedimenten und ihre Bedeutung für die Seenforschung. *Schweiz. Z. Hydrol.* 47:87–126.

Züllig, H. 1986. Carotenoids from plankton and photosynthetic bacteria in sediments as indicators of trophic changes in Lake Lobsigen during the last 14 000 years. *Hydrobiologia* 143:315–319.

Züllig, H. 1988. Waren unsere Seen früher wirklich *rein*? Anzeichen von Früheutrophierung gewisser Seen im Spiegel jahrtausendealter Seeablagerungen. *Gas Wasser Abwasser* 68 (1988/1):17–32.

# Part 4

# Production, Periodicity, and Distribution of the Plankton

Deep, large lakes are essentially pelagic systems, and the plankton are their most important community. Plankters have short generation times and therefore can adjust to prevailing conditions within relatively short time spans. This applies to both the physiological state of individual organisms (within hours) and the species composition of the community as a whole (within days).

Large lakes usually are clearer and more oligotrophic than small ones. Because of the dilution of the nutrient load that occurs in large lakes, massive blooms of phytoplankton are restricted to cases of extreme eutrophication. Due to the relatively greater mixing depth of large lakes, less light is available to phototrophic organisms at any given level of surface irradiance and water transparency. There is evidence that phytoplankton are adapted to lower average irradiances in deep lakes than in shallow ones and selection in deep lakes may favor species having lower light requirements. Moreover, greater turbulence, in conjunction with lower nutrient availability, prevents cyanobacteria from predominating in the phytoplankton community and also prevents water blooms.

Large lakes have a greater tendency toward horizontal heterogeneity of environmental conditions important to phytoplankton, such as temperature, mixing regime, and nutrient concentrations. In response, patchy and qualitatively heterogeneous plankton assemblages are formed. This applies in particular to the distinction between nearshore and offshore regions.

Due to less watershed and littoral influence, internal nutrient and carbon regeneration within the pelagic zone is more intense than in small lakes. The availability of nutrient resources within the pelagial should, therefore, be more stable with time. Species selection, which to a great extent is controlled by interspecific competition for resources and predator-prey interactions, is more likely to be in equilibrium with the existing condition than in the more unstable small lakes. This should render plankton community structure in large lakes more predictable than in small lakes. A worldwide survey of phytoplankton communities in large lakes, however, thus far has revealed only a few consistent trends which suggests that our understanding of factors controlling plankton community structure in large lakes is still inadequate.

# 17

# Environmental and Physiological Control of Phytoplankton Productivity in Large Lakes

Max M. Tilzer

ABSTRACT In the vast majority of large lakes, phytoplankton are the predominant primary producers; therefore, only phytoplankton production will be discussed in this review. Primary production here is defined as the growth potential of phytoplankton, which depends on two steps—photosynthesis *and* biosynthesis.

The vertical integral of photosynthesis is controlled by both the input and the vertical gradient of radiant energy. Because of the lesser significance of inputs to the open water from both watershed and benthos, water turbidity by nonalgal material, both particulate and dissolved, should be generally less than in small and/or shallow lakes. As a consequence, phytoplankton photosynthesis and hence production at any given concentration of biomass (chlorophyll) should be enhanced, because greater proportions of the light entering the water can be harvested by photosynthetic pigments and utilized energetically thereafter. On the other hand, water column mixing, on the average, is deeper in large than in small lakes, thus leading to diminished mean water column irradiances, to which phytoplankton travelling with the circulating water masses are exposed. Using light-shade adaptation, this adverse effect can, to some degree, be compensated for. Light-shade adaptation is also capable of diminishing the effects of seasonal fluctuations in daily means of surface irradiance. Whenever the euphotic zone is thermally stratified, vertical shifts in phytoplankton light-shade adaptation also occur. This phenomenon is significantly more common in large than in the usually more turbid small lakes.

Nutrient cycling within the water body of large lakes is considerably more important than either inputs from the watershed or interactions with the littoral and lake bottom. The "microbial loop" seems to be the main pathway by which nutrients are recycled within the pelagial. During thermal stratification, major losses of nutrients from the euphotic zone occur by settling fluxes; nutrient injections to the euphotic zone are mediated mainly by internal seiches. The nutrient-dependent carrying capacity of the phytoplankton assemblage is not reached in most large lakes because of concomitant losses via respiration, herbivore grazing, and sedimentation.

The seasonality of phytoplankton biomass and productivity in large lakes is controlled by the combined effects of latitude, which governs the supply of energy;

by the trophic state, which determines the maximum biomass accumulation during summer; and by water depth, which controls the extent of respiratory losses and dilution during water column mixing.

## 17.1 Introduction

The phytoplankton community is the most widely distributed primary producer on our planet. The potential habitat for phytoplankton covers 71% of the earth's surface and has maximum vertical extent of ca. 200 m. Nevertheless, the oceans contribute only 30–50% to the global primary production (Koblentz-Mischke et al., 1970; Parsons et al., 1984; also see Chapter 3). The main reasons for this discrepancy are the constraints imposed on plant growth in the aquatic environment as compared to the terrestrial one. The most important ones are restricted nutrient availability, limited energy supply to the organisms due to the high optical density of the surrounding medium, and mixing of the water column, as well as high levels of losses. With respect to phytoplankton productivity, large and deep lakes have many similarities with oceans.

In large lakes, the relative contribution of the littoral zone to overall primary production and biomass usually is small. Therefore, this review will be restricted to phytoplankton primary productivity. The conditions for phytoplankton productivity in large lakes are described with special emphasis on factors that determine how much carbon and energy can be delivered to the subsequent food-web.

In the most general terms, *productivity* can be defined as the growth potential of a community. Biological productivity is controlled by the rate of resource input and the efficiency at which these resources are utilized (Figure 17.1). Primary production is less clearly defined. The least controversial description of what this term means is that primary production includes all processes by which autotrophic organisms increase their biomass. Applying this concept to photoautotrophic organisms, we can perceive primary production as a two-step chain of reactions: Step 1 is *photosynthesis*, by which inorganic carbon is reduced to carbohydrates by utilizing electromagnetic radiation. Step 2 is *biosynthesis*, by which the multitude of cellular components is formed. Carbohydrates are utilized as building blocks of the organic compounds synthesized and also as energy sources for nutrient uptake and inorganic anabolic reactions. From this, it follows, that only the first of the two steps of processes is "autotrophic" in the strict sense of the word, whereas the second is "heterotrophic." By adopting this broad definition of primary production, we avoid some of the dilemmas succinctly stated by Flynn (1988).

Primary production rate estimates in phytoplankton are usually based on measurements of photosynthetic rates, either by assessing carbon incorporation (Steemann-Nielsen, 1952) or oxygen evolution (Gaarder and Gran, 1927). In principle, this should enable us to predict growth rates ($\mu^*$, defined by $1/B\cdot$

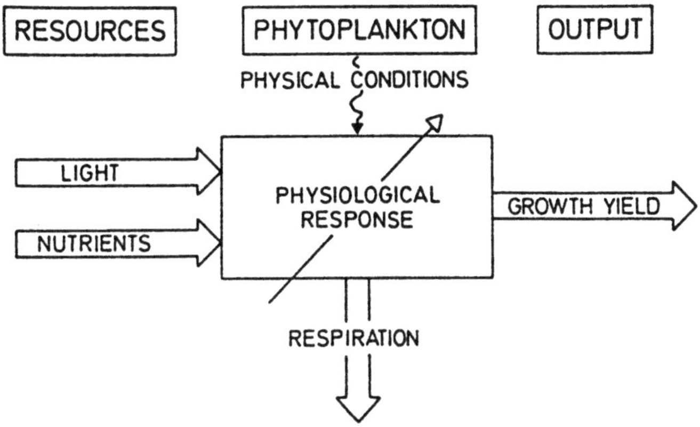


**Fig. 17.1** Phytoplankton growth yield is the net result of resource uptake and respiratory losses. The physiological response depends on the specific requirements of the organisms involved and on their adaptation to the prevailing environmental (mainly physical) conditions. Respiration consumes part of the carbohydrates initially formed by photosynthesis, in order to drive the uptake of nutrient salts from the environment and the biosynthesis of cellular components and structures. The proportions of respiratory losses are strongly influenced by the physical conditions, especially by light and temperature.

dB/dt, where B is biomass, and t is time), provided that both photosynthetic rates and biomass are expressed in the same units (e.g., carbon or chlorophyll):

$$\mu^* = \ln (1 + P/B) \, (t^{-1}) \qquad (17.1)$$

where P is the daily production rate (Peterson, 1978; Redalje and Laws, 1981; Tilzer, 1984a; Forsberg, 1985).

In reality, however, potential growth rates predicted from eq. 17.1 can never be achieved because photosynthesis is only the first step of the considerably more complex primary production proces. In fact, photosynthetic rate estimates, both by the oxygen and the radiocarbon techniques, usually yield values close to those for gross photosynthesis. The following findings support this still not universally accepted view for the radiocarbon technique, which has most commonly been applied both in large lakes and in the ocean:

1. Day-rate estimates of photosynthesis usually are based on short (e.g., 4-hr) incubations, which are extrapolated to the entire day. Linear $CO_2$ uptake during periods of less than six hours suggests negligible respiratory losses (Dring and Jewson, 1982). In the light, "dark respiration" (mitochondrial respiration in eukaryotes) is inhibited.
2. Adding up short incubations during one light period yields higher day-rate estimates than one long incubation covering the same time span (Vollenweider and Nauwerck, 1961; Tilzer et al., 1977).
3. Nighttime carbon losses usually comprise 30-50% of the previous

day's uptake and hence are probably far in excess of daytime respiration (Rhyther, 1954; Tilzer and Horne, 1979; Tilzer, unpubl. data).

In order to build up new plant biomass, some of the carbohydrates formed by photosynthesis have to be respired. Therefore, maximum possible rates of biomass increment ($\mu$) are significantly *smaller* than potential growth rates inferred from photosynthetic rates along ($\mu^*$), due to respiratory losses that are required to energetically drive biosynthesis and maintenance metabolism (designated as $\rho$):

$$\mu = \mu^* - \rho \tag{17.2}$$

(from Tilzer, 1987). Because the ratio of gross photosynthesis to respiration is variable, the potential of photosynthetic rate estimates for predicting *in situ* growth rates by using eq. 17.1 is poor (Talling, 1984; Tilzer, 1984a).

Phytoplankton production rates in the vast majority of cases are measured for the algal community as a whole, although the factors regulating the waxing and waning of phytoplankton act on the species level. This is justified if the carbon and nutrient balance of the phytoplankton as a whole is considered. If, however, the consumption of autotrophs by the "microbial loop" is studied, only the production by the small phytoplankton (usually termed "picoplankton") is relevant. In the ensuing chapter we are concerned with the contribution of the entire phytoplankton to the organic carbon pool in large lakes.

Now, the production process of phytoplankton will be discussed in view of the following *characteristics* of large lakes. [These characteristics have particular relevance for phytoplankton; however, they do not apply universally to *all* large lakes.]

- Great mean depth
- Thick, mixed upper layer during thermal summer stratification
- Nutrient upwelling during the stratified season caused by internal waves
- Relatively small nonalgal water turbidity because of reduced watershed influence and bottom sediment resuspension from the lake bottom
- Prolonged water retention time
- Prevalence of internal cycles over external inflows for the nutrient budget on a short-term basis
- Horizontal heterogeneities within the plankton community (nearshore–offshore differences and patchiness)

## 17.2 Photosynthesis Rates and Underwater Irradiance

Environmental and physiological control of phytoplankton photosynthesis can be expressed by the following equation:

$$P = I_z \cdot k_c \, \text{Chl}a \cdot 12 \, \Phi, \tag{17.3}$$

where P is the photosynthetic rate (g $cm^{-3}$ $h^{-1}$); $I_z$ is the underwater irradiance (in mol PAR $m^{-2}h^{-1}$) at the depth z, $k_c$ is the absorption cross section of the phytoplankton, normalized to chlorophyll *a* ($m^2$ $mg^{-1}$ Chl*a*) and averaged over the spectrum of PAR; and $\Phi$ is the quantum yield of photosynthesis (in mol $CO_2$ reduced/mol PAR absorbed) (Kirk, 1983). In the following, the relevance of these factors for the primary production process of large lakes will be briefly discussed.

**Exposure of Phytoplankton to Underwater Irradiance** For an individual algal cell, the photic environment is highly variable with time, due to both changes in incident irradiance and alterations in the depth of residence. The position of algal populations within the water column is controlled to a great extent by the movements of the water masses themselves, and the movements of the cells relative to the water surrounding them. Imboden (see Chapter 3) stresses the importance of water movements and spatial heterogeneities for the diversity of the community and compares the time scales of physical forcing with those of the physiological responses of the phytoplankters. In the present chapter, the physical properties of the water column are discussed mainly with respect to their influences on the amounts of radiant energy available to the phytoplankton community.

Phytoplankton photosynthesis takes places in a three-dimensional light field which is also highly variable with time. The temporal variability in the light field to which any individual algal cell is exposed, is augmented by vertical movements of the water masses with which the cell is travelling.

The *vertical gradient* of underwater irradiance with depth is described by Lambert Beer's Law:

$$I_z = I_0 \cdot e^{-Kz}, \tag{17.4a}$$

where $I_z$ and $I'_0$ are the irradiances at depth Z and immediately below the surface, respectively, and K is the vertical light attenuation coefficient. The vertical light attenuation coefficient is a function of light attenuation by both water and nonalgal material ($K_w$) and of light attenuation by phytoplankton. The latter depends on chlorophyll concentration (Chl*a*) and on the chlorophyll-specific absorption cross section of algal cells, averaged over the spectrum of photosynthetically active radiation ($k_c$ in $m^2$ $mg^{-1}$ Chl*a*):

$$K = K_w + k_c \cdot \text{Chl}a. \tag{17.4b}$$

In a mixed water column, the algae travel through an exponential light gradient. The *mean irradiance* within a water column mixed down to $Z_m'$ is

$$I = I'_0 \, (1 - e^{-Kz'}) / \, K.z' \tag{17.5a}$$

(from Riley, 1957). For mixing depths considerably greater than the euphotic depth ($Z_{eu}$, depth where underwater irradiance has decreased to 1% of the surface value and photosynthesis–respiration balance no longer is positive), eq. 17.5a can be simplified to

$$I \approx I'_0 \,/\, K.z' \qquad (17.5)$$

Riley's equation is only valid if we assume that the probability of a parcel of water to be at any given water depth within the mixed water column is the same. To my knowledge, experimental evidence for this assumption is lacking. Figure 17.2 shows the relationship between the mean light level within a water column and the mixing depth.

Large lakes are frequently characterized by enhanced mixing depths and higher water transparency, as compared to small lakes. This can lead to two opposing trends: (1) At any given water turbidity, greater mixing depth leads to diminished mean irradiances within the mixed water column to which the algal populations are exposed. (2) At any given depth of mixing, enhanced water transparency leads to increased mean water column irradiances. The relation between water transparency and mixing depth determines which of the two effects prevails in the end.

In an array of lakes having different sizes and consequently mixing depths, but comparable transparency, Hecky and Fee (1987) have observed that phytoplankton apparently were exposed to lower irradiances in the larger lakes.

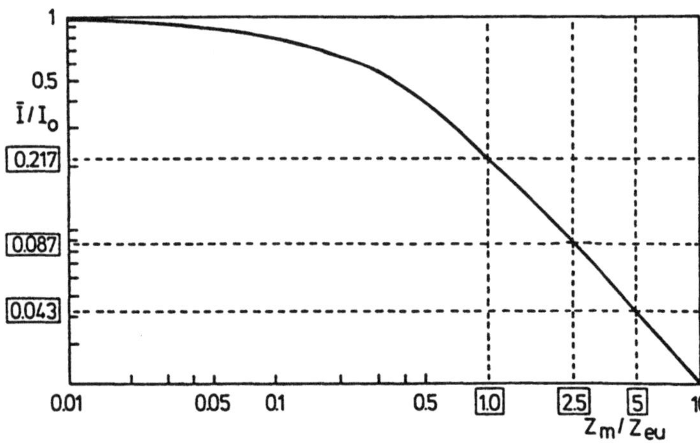


**Figure 17.2** Relationship between the mean light level within a mixed water column and mixing depth, after eq. 17.5. The mean light level of the mixed water column is expressed as fractions of the irradiance immediately below the water surface, the mixing depth is expressed as fractions or multiples of the euphotic depth (depth where irradiance is 1% of surface value). Beyond a certain mixing depth termed "critical depth of mixing," phytoplankton growth can no longer be maintained because water column respiration exceeds water column net photosynthesis. When the mixing depth equals the euphotic depth, the mean water column irradiance is 21.7% of surface irradiance. At a mixing depth of 2.5 $Z_{eu}$ it declines to 8.7%, and at a mixing depth of 5 $Z_{eu}$ to 4.3% of the surface value. The critical depth of mixing is assumed not to exceed 5 $Z_{eu}$.

In ultra-oligotrophic Lake Tahoe, by contrast, phytoplankton are exposed to considerably higher irradiances than in meso-eutrophic Lake Constance, which is of comparable size (Tilzer and Goldman, 1978; Tilzer and Beese, 1988; see Figure 17.2). In East African great lakes, different *Melosira* species occur at different mixing regimes. Their mantle pore sizes are correlated with prevailing mixing depths and, hence, with light regimes in their habitats (see Chapter 20).

**Light Harvesting and Areal Photosynthesis** Bannister, 1974, has shown theoretically and Megard et al., 1978, and Tilzer, 1983, have proven experimentally that phytoplankton photosynthesis per unit of water surface area is a function of the proportion of underwater light which is absorbed by photosynthetic pigments:

$$\pi / \psi = k_c \cdot \mathrm{Chl}a / K, \qquad (17.6)$$

where $\pi$ is the observed photosynthetic depth integral (ie., areal photosynthesis) and $\psi$ is the upper limit to areal photosynthesis that would be observed if all underwater light were absorbed by photosynthetic pigments.

Fractional light absorption can vary both with phytoplankton biomass and with background light attenuation. The upper limit to areal photosynthesis is theoretically independent of biomass, provided that chlorophyll per unit surface area exceeds the reciprocal of the absorption cross section ($1/k_c$; mg $m^{-2}$ Chl*a*). Increasing nonalgal turbidity leads to a vertical truncation of photosynthetic depth profiles, which then is an expression of an increasing degree of light limitation. Half-maximum photosynthesis is reached at chlorophyll concentrations defined by the quotient $K_w/k_c$ (Figure 17.3).

In most deep large lakes, water turbidity due to nonalgal materials can be expected to be small. Consequently, at any given level of biomass, productivity of phytoplankton should be slightly enhanced relative to smaller and/or shallow lakes. This effect should be greater in blue water than in green water because the absorption cross section of blue algal pigments is higher and increases with water depth due to the spectral absorption properties of the photosynthetic pigments (Atlas and Bannister, 1980). In contrast, light conditions in large shallow waters are particularly adverse to phytoplankton productivity because of high concentrations of resuspended material, even if algal density is high due to rich nutrients (Jewson, 1977; Dokulil, 1979). On the other hand, Kirk (1985) has demonstrated that lightly colored mineral material scatters rather than absorbs light which, therefore, remains available to the algae for photosynthesis.

**Photosynthesis-Irradiance Relationships** Within a light gradient, photosynthesis follows an optimum curve which can be explained by the dependence of quantum yields of photosynthesis on quantum flux density (Dubinsky and Berman, 1976, 1981; Bannister and Weidemann, 1984; Tilzer, 1984b). Within certain limits, algae are capable of adjusting to the prevailing light field,

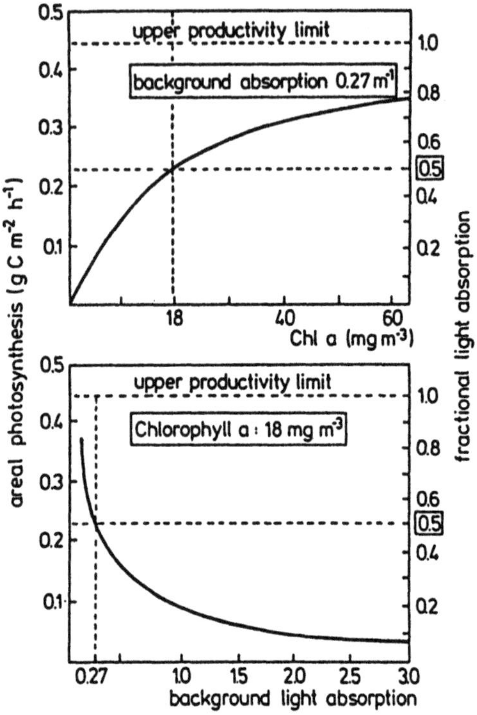


**Figure 17.3** Photosynthesis per unit surface area as a function of fractional light absorption—that is, the proportions of underwater light harvested by photosynthetic pigments (see eq. 17.6). The curves shown here are based on data from Lake Constance (Tilzer, 1983): Absorption cross section of Chl*a* ($k_c$ = 0.015 $m^2$ $mg^{-1}$ Chl*a* light-saturated assimilation number $P^Bm$ = 3.12 mg C (mg Chl*a*)$^{-1}$$h^{-1}$, optical depth of half-saturated photosynthesis (1n $I_0$ − 1n 0.5 $I_k$ = 2.08) (after Tilzer, 1983). In the upper graph, fractional light absorption and integral photosynthesis are shown as functions of chlorophyll concentration at a constant level of background light absorption (mean value for Lake Constance). At a chlorophyll concentration of ca. 18 mg $m^{-3}$, 50% of underwater light would be harvested by photosynthetic pigments. In the lower graph, fractional light absorption and areal photosynthesis at a constant chlorophyll concentration are shown as a function of background light attenuation. The lower limit of background light attenuation (0.12 $m^{-1}$) is based on calculations by Schanz (1985) for nearby Lake Zurich. The upper limit of phytoplankton productivity would be reached if all light were absorbed by photosynthetic pigments. Provided that chlorophyll per surface area exceeds $1/k_c$ (67 mg $m^{-2}$ for Lake Constance), integral photosynthesis would then be independent of pigment concentration. The higher the background light attenuation, the higher the chlorophyll concentration required to achieve a certain productivity.

by optimizing either light-harvesting and/or enzymatic reactions involved in photosynthesis (Jorgensen and Steemann Nielsen, 1968; Falkowski and Owens, 1980; Falkowski, 1980, 1981; Dubinsky et al., 1986). In low-light-adapted algae, the initial slope of the light-limited photosynthetic *versus* irradiance curve—which is a function of the maximum quantum yield, the pigment content and the light absorption cross section of the algal cells—is frequently enhanced. Light-saturated maximum photosynthesis, by contrast, is diminished. However, other patterns can also be observed, depending on the mechanism by which light-shade adaptation is accomplished (Richardson et al., 1983; also see Figure 3.12).

Increased average mixing depths in large lakes lead to photosynthesis-irradiance curves which can be interpreted as evidence of low-light adaptation (Fee and Hecky, 1987). Over the course of the year, moreover, the underwater light field undergoes more dramatic changes in deep than in shallow lakes due to greater maximum mixing depths (Talling, 1971).

Imboden (see Chapter 3) emphasizes that the response times of algae to alterations of the irradiance, to which they are exposed, changes over a wide range. Jorgensen and Steeman Nielsen (1965) had assumed that light-shade adaptation requires at least one generation time of the algae involved. Later work, however, (Falkowski and Owens, 1980; Post et al., 1985) has shown that adaptation is more rapid, but still on the order of hours.

Nevertheless, it appears well established that, during winter, phytoplankton throughout the euphotic zone are clearly low-light adapted; this is manifested by their low saturating light levels and pronounced susceptibility to surface light inhibition. This would indicate that they adapt to the *mean* irradiance within the light gradient through which they travel during the mixing process. Seasonal variations in mean daily surface irradiances can, to a significant extent, be compensated for by changing light requirements. This can be demonstrated even for relatively turbid Lake Constance, whose mixing depth should have a particularly strong influence on mean irradiances to which the algal populations are exposed (Tilzer and Beese, 1988). In clear Lake Tahoe, by contrast, the euphotic zone greatly exceeds the thickness of the epilimnion during summer.

Whenever the lake is thermally stratified, phytoplankton near the lake surface adapt to high irradiances whereas deep-living populations remain low-light adapted throughout the year. Phytoplankton collected from different water depths behave differently. This indicates that the time constant of adaptation in this case is longer than the time constant of vertical water column mixing (Tilzer and Goldman, 1978, and Figure 17.4).In shallow, continuously mixed Lough Neagh, photosynthesis-irradiance relationships are less variable seasonally and can probably be sufficiently explained by changes in water temperature (Jewson, 1976).

## 17.3 Control of Phytoplankton Productivity by Nutrients

The growth yield of phytoplankton ($\Delta$ B), as well as that of any other community, is the net result of production (P) and losses (L):

$$\Delta B = P - L. \tag{17.7}$$

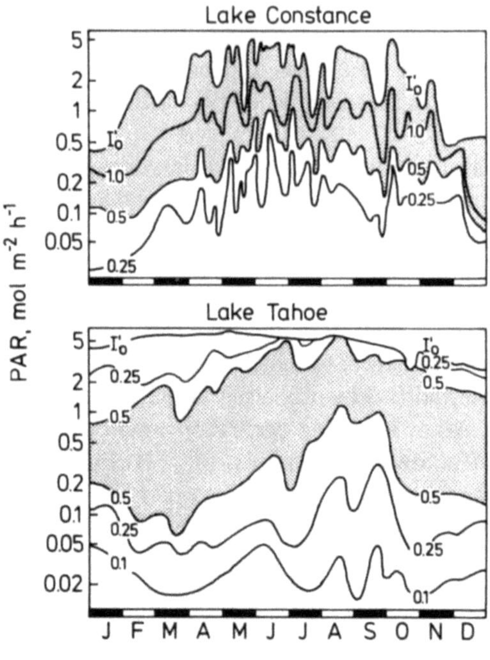


**Figure 17.4** Isopleths of relative photosynthetic rates as measured *in situ* ($A_z/A_{opt}$) plotted versus mean underwater irradiance during the incubation period (ca. 10.00–14.00 hours). Shaded areas: Photosynthetic rates exceeding 50% of the maximum profile value. Top: Lake Constance, 1981 (after data by Tilzer and Beese, 1988). Bottom: Lake Tahoe, 1974 (after data of Tilzer and Goldman, 1978). The euphotic zone as defined by the 1%-surface-light level in meso-eutrophic Lake Constance extended to a maximum depth of 21 m and was only weakly stratified between late March and early November. The 1% light depth in Lake Tahoe was at ca. 75 m; photosynthesis could be measured to a maximum depth of 105 m (0.16% light level). Between early April and late November the euphotic zone was thermally stratified with a thermocline at 5 m depth at the beginning and at ca. 35 m depth at the end of this period. Note that in Lake Constance, relative photosynthetic rates shifted to higher light levels in parallel with the surface irradiance at all depths, whereas in Lake Tahoe, deep phytoplankton showed much less variation in photosynthetic light responses. In Lake Tahoe, surface light inhibition was pronounced, especially during water column mixing.

With respect to nutrient budgets, growth can be perceived as the transition of nutrients from the ambient, mainly dissolved pool to the biomass pool. At least in theory, it should be possible to predict the maximum possible biomass that could form from the nutrient pools available in the productive water layer at the beginning of the growing season. Rhee (1978) has convincingly demonstrated that in cultures of a green alga the amount of the biomass is limited by the availability of the nutrient in shortest supply. The average elemental composition of seston as given by the *Redfield ratio* (Redfield, 1958) would then allow us to predict biomass in terms of organic carbon from the concentration of any given limiting nutrient. In reality, however, the actual accumulation of biomass also depends on the totality of losses from the newly formed biomass (Jassby and Goldman, 1974; Walsh, 1983; Tilzer, 1984a).

**Biomass Yields and Nutrient Pools** Statistical analysis has shown that, nevertheless, maximum euphotic summer concentrations of chlorophyll as a measure of phytoplankton biomass are related to maximum concentrations of growth-limiting nutrients during the preceding period of lake mixing. Because phosphorus is considered to be the main growth-limiting nutrient in most lakes, chlorophyll yield was related to the total spring phosphorus concentration (Sakamoto, 1966; Dillon and Rigler, 1974; Vollenweider, 1976; Vollenweider and Kerekes, 1980). Later work, however, has shown that even in clearly phosphorus-limited situations, N : P ratios have a bearing on total crop yields (Smith, 1979, 1982). This can be explained, at least partially, by the influence of nitrate on the rate of phosphorus release from bottom sediments (Ripl, 1976; Ripl and Lindmark, 1978; Smith, 1987). Another possible explanation could be that high N:P ratios in the environment have the potential of inhibiting the fixation of atmospheric nitrogen by cyanobacteria. Stoichiometric ratios between nutrient salts have great influence on the composition of the phytoplankton assemblage in controlling interspecific competition for nutrients (Sommer, 1989; also see Chapter 19).

Unfortunately, chlorophyll-nutrient relationships cannot be readily translated into relationships between biomass carbon and nutrients, because chlorophyll-to-carbon ratios are variable between species and, moreover, depend on the previous light history and nutrient status of the algae (Falkowski, 1980; Rhee, 1978). Much less is known about biomass yield than about chlorophyll-nutrient relationships. However, good evidence does exist that in large and/or deep lakes, at any given nutrient supply, biomass is lower than in shallow lakes (Ramberg, 1978; Ahl, 1980; Pridmore et al., 1985). Two tentative explanations can be given for this: (1) In large lakes, the water column is frequently mixed down to a greater depth. As a consequence, the algae receive less light and hence are not only light-limited to a greater extent but also have a less favorable photosynthesis-to-respiration balance. This effect becomes particularly strong when the mixing depth exceeds the euphotic depth. (2) Sedimentary losses of phytoplankton can be expected to be greater in large deep lakes than in small and/or shallow lakes, in which some of the loss due to sedimentation can be compensated for by resuspension.

In Lake Constance, maximum phytoplankton biomass is only 15–20% of what is to be expected from the Redfield ratio. The bulk of the particulate carbon pool apparently is detritus, bacteria, whose biomass approximately equals algal biomass (Simon, 1987; Simon and Tilzer, 1987), and zooplankton, both protozoans and crustaceans (Geller, 1985; also see Chapter 29). Bacteria may successfully compete with algae for inorganic nutrients such as phosphorus (see Chapter 27) and thus may be a direct cause of the fact that the full nutrient-dependent carrying capacity of phytoplankton is not reached in large lakes.

In the first half of the growing season, sedimentation from the euphotic zone is minimal. During this period, temporal variations in euphotic orthophosphate concentrations can be matched with concomitant variations in particulate organic carbon, when applying the Redfield ratio. Later in the year, however, much of the organic matter is lost from the euphotic zone by sedimentation (Stabel, 1986). Both the phosphorus content of the settling particles themselves and the adsorption of dissolved phosphorus to the sediment lead to a downward-flux of phosphorus and hence a decrease in the total phosphorus pool within the euphotic layer. Concentrations of organic matter during that time fall far short of what could be expected based on the decrease in levels of inorganic phosphate (Tilzer and Beese, 1988; and Figure 17.5).

**Phytoplankton Production and Nutrient Fluxes** Under steady-state conditions, nutrient turnover times can be calculated by

$$T_R = M / ( dM/dt), \tag{17.8}$$

where M is the nutrient pool in question and dM/dt is the rate of nutrient addition or removal, respectively (Wangersky, 1986; Klump, 1987). Applicability of this model is restricted, however, because of variable nutrient-to-carbon

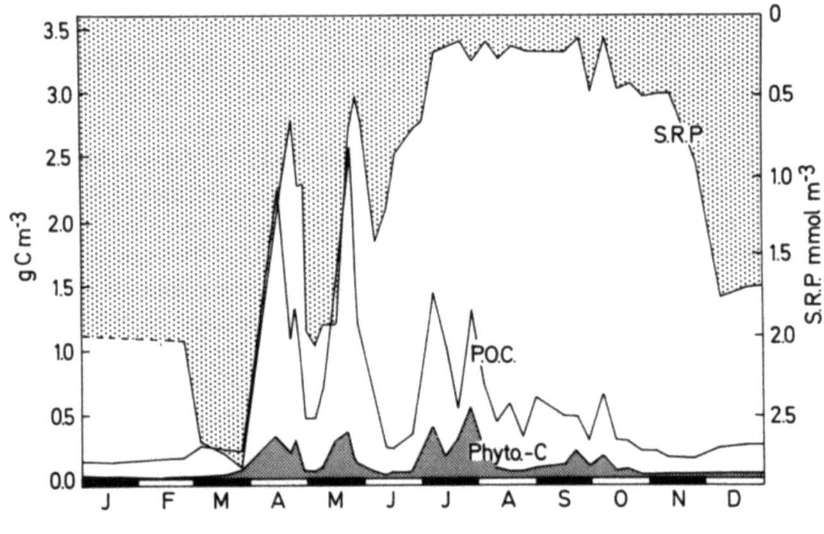


**Figure 17.5** Seasonal development of euphotic, particulate organic carbon (P.O.C.), phytoplankton carbon (shaded), and soluble, reactive phosphorus (S.R.P., stippled) in Lake Constance during 1981 (from Tilzer and Beese, 1988). The scales of particulate carbon (left, from bottom to top) and SRP (right, from top to bottom) were constructed on the basis of the Redfield ratio (106:1, atomic) in order to make them comparable. During the first half of the year, POC and SRP concentrations were reciprocal, suggesting that phosphorus moved from solution to particulate organic matter. The gap between both pools after mid-June can be explained by sedimentary carbon losses from the euphotic zone. Phytoplankton carbon comprised about 15% of the total POC; The rest was comprised of zooplankton, bacteria, and detritus.

ratios in algae, mainly due to luxury consumption (Fuhs et al., 1972; Rhee, 1978) and pure chemical reactions involved in nutrient cycles, such as nutrient adsorption to settling particles (Stabel, 1984; Stabel and Geiger, 1985). Despite this, some general features of nutrient cycles in large lakes emerge:

1. During the stratified season, nutrient infusions to the euphotic zone by upwelling should be small relative to biologically mediated *internal cycles.* Continuous estimates of net primary production should allow us to roughly estimate the time required for the consumption of the limiting nutrient pool within the euphotic zone. By comparing these predictions with observed nutrient concentration changes, we should be able to indirectly estimate the rate of nutrient recycling within the euphotic zone. If we assume that net primary production comprises ~60% of gross photosynthesis, we can estimate that the euphotic phosphorus pool in Lake Constance should be consumed within one month. In reality, however, this process requires about three months. This discrepancy can most readily be explained by nutrient regeneration (see Figure 17.5). During the period of nutrient depletion, internal nutrient regeneration within the euphotic zone is considerably accelerated. The role of the "microbial loop" in this process has been appreciated only recently (see Chapter 27). Regenerated production *sensu* Dugdale and Goering (1967) should generally be of greater significance in large than in small lakes.
2. During the stratified season, *nutrient infusions* to the euphotic zone from deep water in large lakes mainly are the result of internal waves, as well as of differential heating and cooling. In small lakes, by contrast, convective processes appear to be considerably more important (Spigel and Imberger, 1987).
3. In large deep lakes that do not mix to their bottom every year, the variable *depth of winter mixing* determines to a large extent the nutrient pool available to the phytoplankton during the ensuing growing season. In Lake Tahoe, year-to-year variations in annual primary productivity can be correlated with mixing depths (see Chapter 6).
4. The dominant process leading to losses from the euphotic zone is *sedimentation.* Although settling fluxes exhibit wide short-term variations, annual sedimentary carbon losses comprise 20–30% of net production (Bloesch et al., 1977; Eadie et al., 1984; Tilzer, 1984a; Gächter and Bloesch, 1985; Stabel, 1986). The role of settling fluxes in nutrient cycling depends to a large extent on the relative importance of nutrient sorption onto particles, versus desorption from particles. The rate of release of nutrients from the settling particles is a function of ambient concentrations of the nutrient in question and of settling times as a function of settling velocity and depth. Rates of nutrient release range from a few percent to 70% of the initial settling flux

out of the euphotic zone (Stabel and Geiger, 1985; Stabel, pers. comm.; Eadie et al., 1984). The prolonged time available for the equilibration between ambient and adsorbed nutrient pools in large lakes should to varying degrees be counterbalanced by the diminished extent of sediment resuspension from the bottom deposits.

5. *Release of nutrients from the bottom sediments* in large deep lakes is less important than in shallow and/or small lakes for three reasons: (1) the surface-to-volume ratio of the water body is smaller. (2) Near-bottom water is more likely to be oxygenated. (3) Resuspension of nutrients is considerably less important and, moreover, is restricted to layers 10–15 m above the lake bottom (Eadie et al., 1984) (Figure 17.6). Nevertheless, nutrient accumulated in near-bottom layers can represent important nutrient sources during deep mixing (Vincent, 1983). Moreover, in very deep large tropical lakes—such as Lake Titicaca and, possibly, Lake Tanganyika—anoxic bottom waters can act

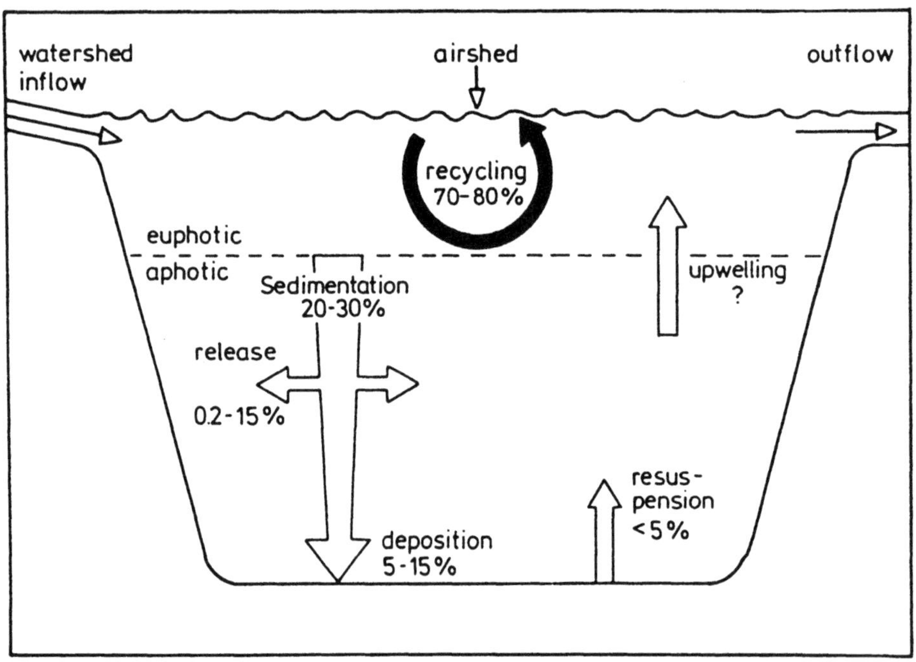


**Figure 17.6** Nutrient fluxes in large lakes. In contrast to small lakes, the vast majority of nutrients are regenerated within the euphotic zone during the growing season. Inputs from the air- and watersheds, as well as exchange with the littoral, play comparatively small roles. Sedimentary losses of nutrients are highly variable with time and can be assumed to be of a similar order of magnitude as sedimentary carbon losses. Nutrient regeneration from bottom sediments by direct release and resuspension are of minor significance in deep large lakes. The percentages are related to annual net production.

as important sources and sinks of inorganic nitrogen (Vincent et al., 1985; W.F. Vincent, personal communication).

6. *Horizontal heterogeneities* in phytoplankton distribution and composition are more common in large than in small lakes. They can be explained by differences in nutrient supply and physical conditions (mixing) of the environment. Moreover, nearshore areas are frequently more fertile and/or stratify earlier than offshore areas, provided that no thermal bar is formed. Nutrient enrichment in "neritic" areas is typical of most large lakes (Lake Constance: Lehn, 1965; Baikal, Ladoga, and Victoria: Talling, 1987, where further references are given; North American Great Lakes, Chapter 22). The effect of the spring "thermal bar" for phytoplankton development has been described for the Laurentian Great Lakes (Munawar and Munawar, 1975). In Lake Titicaca, the productivity cycle of an isolated polymictic lagoon differs sharply from the dimictic main basin (Vincent et al., 1986). Horizontal heterogeneities in phytoplankton abundance and composition can be expected whenever processes controlling phytoplankton dynamics are more rapid than horizontal exchanges of water masses (Powell et al., 1975; Chapter 3).

Figure 17.6 gives a tentative scheme of the relative importance of the above-described processes in the nutrient cycle in large lakes.

## 17.4 The Seasonal Productivity Cycle in Large Lakes

Apparent seasonal variations in production rates and biomass levels are the net results of all the variables controlling growth and losses. Over the course of the year, these factors vary and so does their relative significance in regulating the production process. There has been a long-standing dispute as to whether energy-related parameters (irradiance, temperature, mixing) or nutrient availability exert a stronger influence on the observed variability in annual primary productivity between lakes. Empirical evidence, mainly based on statistical analysis of data on a wide variety of lakes, can be used to support both views (Brylinsky and Mann, 1973; Schindler, 1978; Brylinsky, 1980).

Apart from the fact that the data used in these analyses are not necessarily comparable, the statistical treatment does not allow for the complexity of factor interactions involved in the control of the production process. Whereas growth is controlled by the resource in shortest supply or by the environmental variable that is most adverse to growth, losses are influenced by an array of factors which act simultaneously (Figure 17.7).

**The Limiting Factor Concept** The interactions of variables controlling phytoplankton growth are particularly complex because the two steps comprising the primary production process are controlled in fundamentally dif-

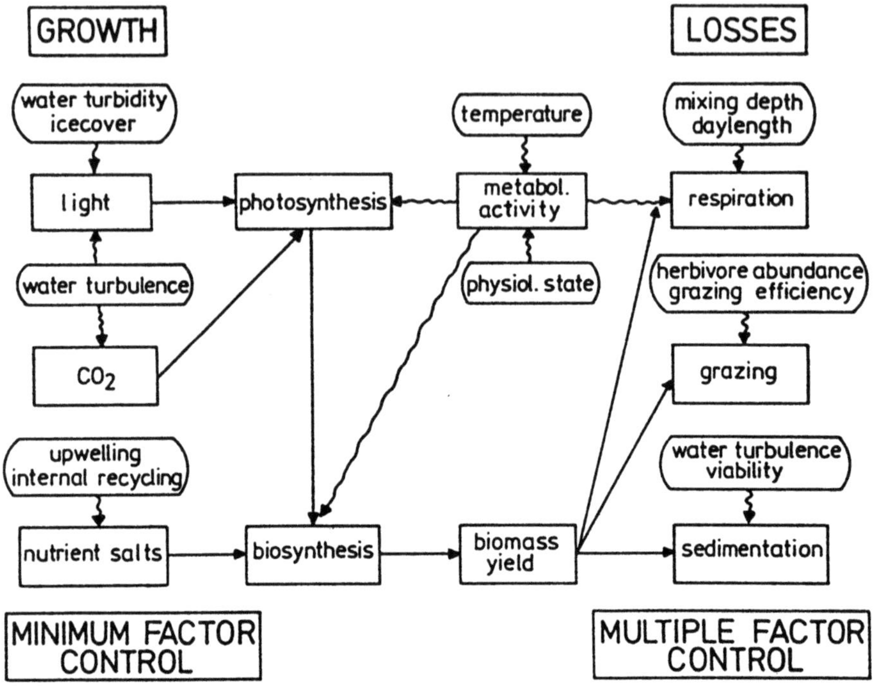


**Figure 17.7** Flow chart of the mechanisms controlling phytoplankton growth and losses in large lakes. Growth is controlled by the resource in minimum supply. Energy in the form of light controls the growth rate by driving photosynthesis, whereas nutrient salts control growth yield by supplying material for the biosynthesis of cellular components. Carbon dioxide is usually in excess and does not exert a controlling influence. Losses, by contrast, act concomitantly. The transfer efficiency of carbon and energy from the phytoplankton to the subsequent food-web depends on the contribution of respiration to overall losses.

ferent fashions. It is therefore necessary to distinguish between the control of production *rates* and the control of the *yield* of growth. The growth yield not only depends on respiratory losses required to drive biosynthesis, but also is strongly influenced by water temperature and depth of mixing. Factors controlling biomass yield are to be distinguished from factors controlling production rates.

*Yield limitation* readily allows the application of Liebig's law of the minimum: Nutrient limitation is apparent whenever one particular nutrient is being completely exhausted during the growing season, the initial pool of which determines the quantity of accumulated biomass (carrying capacity). Yield limitation is essentially control of biosynthesis.

*Rate limitation* controls the velocity at which biomass accumulates or is turned over. It is essentially controlled by factors governing photosynthesis. Rate control is mainly due to physical factors (light and temperature) and is independent of nutrients which are either progressively taken up ("new production") or recycled ("regenerated production"). The concept of rate limitation can be readily applied to the photosynthesis versus irradiance curve where a clear distinction can be made between light limitation and light saturation. Nutrient control of light-saturated photosynthesis is restricted to severe nutrient depletion, which leads to depressed nutrient cell quota (Senft, 1978). Only temperature effects on light-saturated photosynthesis are well-established (e.g., Jewson, 1976). Rate limitation of the vertical integral of photosynthesis occurs in two fashions: First, surface irradiance controls the thickness of the layer of light-saturated photosynthesis and, thus, the relative rate of photosynthetic carbon accumulation. Second, nonalgal light absorption restricts the amount of light which can be harvested by photosynthetic pigments. As already mentioned, high proportions of nonalgal light attenuation occur in systems made highly turbid by seston or color, or in clear systems with low algal densities (see Figure 17.3).

The transition in the limitation of phytoplankton growth yields by nutrient supply to control by energy can be caused by three factors: decreases in annual inputs of surface irradiance, increases in mixing depths, and increases in nonalgal turbidity. The Southern Ocean is energy-limited due to great mixing depths and an extremely short growing season (e.g., El Sayed, 1970). In impoundments, on the other hand, that are made extremely turbid by suspended fine mineral particles, primary productivity is restricted by inefficient light harvesting by phytoplankton (Grobbelaar, 1984, 1985). Thus, in both these extreme situations, biomass yields are limited by the energy supply. Loss rates due to phytoplankton respiration, then, are high relative to photosynthetic rates (cf. eq. 17.7), thus leading to diminished growth. It could be speculated that energy limitation of phytoplankton yield is conceivable in ice-covered and/or deeply mixed large lakes at high latitudes. However, to my knowledge, data are not available.

**Factor Interactions in the Seasonal Productivity Cycle** The main driving force of seasonal events in lakes, in general, is variations in energy inputs of both electromagnetic radiation and kinetic energy. Most of the seasonal changes occurring in the lacustrine environment can be explained by the interaction of irradiance and wind forcing. In this chapter, only the seasonal cycle of phytoplankton bulk biomass and productivity is discussed. The seasonality of the phytoplankton community structure is described by Pollingher in Chapter 18, with special emphasis on latitudinal effects.

We can distinguish between the primary and the secondary effects of seasonality: *Primary effects* are the seasonal variations in the light and heat inputs to the water column and the alteration of thermal stratification in mixing. Sverdrup (1953) has developed the concept of the "critical depth of mixing," beyond which phytoplankton growth is no longer possible because respiration within the mixed water column exceeds integral gross photosynthesis. The critical depth of mixing essentially depends on the daily light dose received by the phytoplankton community within the mixed water column. *Secondary effects* of seasonality are variations in concentrations of dissolved nutrients and suspended solids. Other less important and less regular seasonal driving variables are recurrent weather patterns which influence the water balance. In deep large lakes, enhanced mixing depth causes greater variations in mean irradiance within the mixed water column. In shallow large lakes, increased resuspension of particles from the bottom sediment causes wider variations in water turbidity. Both lead to greater amplitudes in the quantity of irradiance available to the phytoplankton community in large lakes as compared to small lakes. In general, effects of mixing depth on mean water column irradiance are considerably more regular than effects of sediment resuspension.

Latitude affects both the annual mean value of daily irradiances and their seasonal variability (Figure 17.8). Because the heat balance also depends on altitude, mixing regimes vary both with latitude and altitude (Hutchinson and Löffler, 1956). In the majority of lakes, the effects of solar inputs and mixing on the light climate to which the phytoplankton are exposed are multiplicative. This is particularly true in large deep lakes. Seasonal productivity cycles, therefore, are closely related to stratification and mixing.

The majority of large lakes of the world are warm-monomictic lakes (see also Chapter 18). During winter mixing, phytoplankton yield is usually light-limited. The onset of spring development frequently is coincident with the formation of thermal stratification, however weak it may be initially. Windy episodes frequently destroy thermal stratification and temporarily cause de-

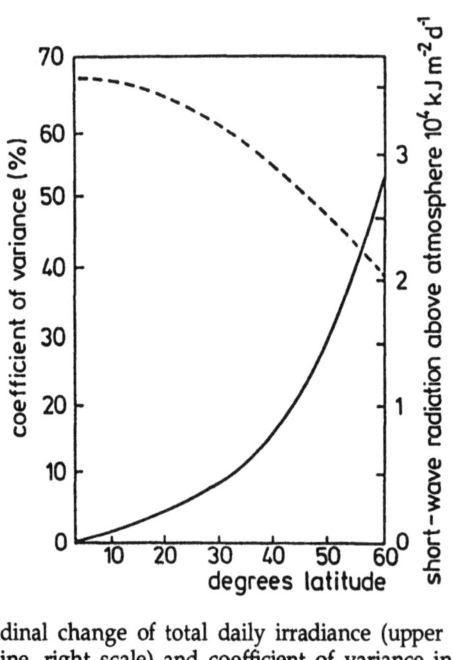


**Figure 17.8** Latitudinal change of total daily irradiance (upper boundary of the atmosphere, broken line, right scale) and coefficient of variance in percent (left scale). (Redrawn after Vincent et al., 1986, with permission).

clining biomass (Tilzer, 1984a). Riley (1957), Ramberg (1979), and Tilzer and Beese (1988) observed positive growth when the mean light doses to which the phytoplankton were exposed exceeded ca. 2 mol $m^{-2}$ PAR. In the Laurentian Great Lakes (and probably in other large lakes at higher latitude as well), the establishment of thermal stratification in spring is delayed, and spring growth of phytoplankton begins while the lake is still mixing (Schelske et al., 1986). However, beginning-nearshore thermal stratification which is separated from the main lake basin by a thermal bar (Boyce, 1974) can lead to an earlier onset of spring phytoplankton growth (Munawar and Munawar, 1975).

During winter mixing, nutrient levels are high both due to infusion from the euphotic zone and small uptake by light-limited phytoplankton. Initiation of spring growth prior to thermal stratification is only possible in extremely shade-adapted populations. In oligotrophic Lake Taupo, the annual maxima of phytoplankton biomass and productivity have been observed during winter mixing. Phytoplankton obviously are capable of effectively utilizing extremely low light doses, possibly due to efficient low light adaptation (ca. 0.9 mol $m^{-2}d^{-1}$ PAR), and thus can take advantage of enhanced nutrient levels. During summer stratification, *Melosira* populations are lost by sedimentation (Vincent, 1983, see also Chapter 18).

As soon as thermal stratification is firmly established, ample energy supply leads to the build-up of phytoplankton biomass, the magnitude of which is controlled by the available stock of the limiting nutrient species. When the limiting nutrients are exhausted, biomass can increase no further. Two other conditions can also prevent biomass growth even before the limiting nutrients are exhausted:

1. *Self-shading* by an increasing biomass should, at least in theory, at any given level of surface irradiance and mixing depth, lead to decreases in the mean irradiance within this water column. Assuming high surface irradiance (ca. 50 mol $m^{-2}d^{-1}$) and a maintenance light dose of 2 mol $m^{-2}d^{-1}$, the critical depth of mixing (where biomass accumulation would come to a halt) would be slightly above 5 $Z_{eu}$ which is the value Talling (1957, 1971) has predicted. However, neither in relatively turbid meso-eutrophic Lake Constance nor in very clear Lake Tahoe is this likely to happen; in Lake Constance, at a chlorophyll concentration of 35 mg $m^{-3}$, which is the maximum observed value in this lake (Tilzer, 1983) the critical depth of mixing would be ca. 30 m (Figure 17.9*a*) Actually, the euphotic zone is only poorly mixed. In Lake Tahoe, a mixing depth of 20 m, which is realistic in midsummer (Goldman, 1974; Tilzer and Goldman, 1978), would become critical only at a chlorophyll concentration as high as 40 mg $m^{-3}$, which is extremely unrealistic in this ultra-oligotrophic lake (Figure 17.9*b*).
2. In productive lakes with spring phytoplankton consisting of small edible phytoplankton, *excessive grazing* by crustacean zooplankton can lead to near-complete elimination of phytoplankton; this is usually

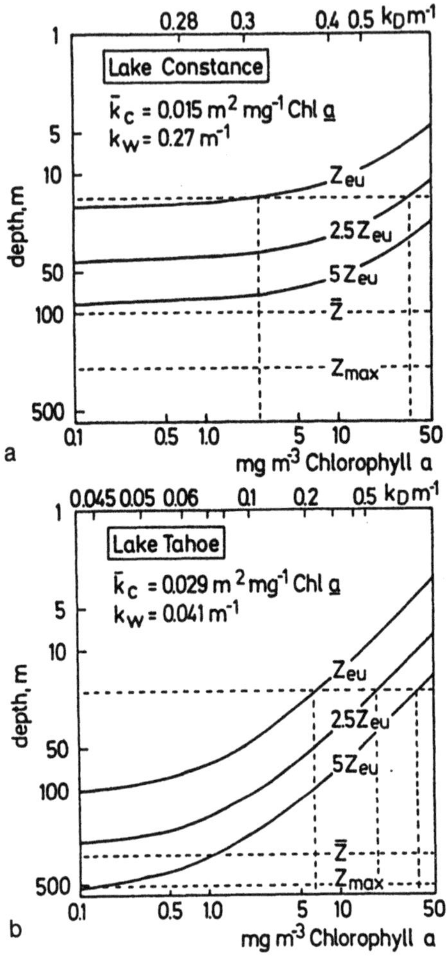


**Figure 17.9** Dependence of euphotic depth and multiples of it on chlorophyll concentration in two large lakes with vastly different optical properties. (*a*) Lake Constance is meso-eutrophic. In the absence of chlorophyll the water would be rather turbid. The absorption cross section of chlorophyll in this green lake (transmission maximum at 550 nm) is relatively small. (*b*) Lake Tahoe is ultra-oligotrophic. Background light attenuation is small. The absorption cross section of algal chlorophyll in this lake (transmission maximum at ca. 470 nm) is high. Due to these features, in Lake Constance, water transparency is less sensitive to chlorophyll concentration than in Lake Tahoe, especially at low biomass. Relationships between chlorophyll concentration and mixing depth boundary conditions, as well as mean (Z) and maximum depth ($Z_{max}$), are given as broken lines for both lakes. The upper horizontal scales give the vertical light attenuation coefficients (eq. 17.4) to be expected at the chlorophyll concentrations given on the lower scales. Optical properties of Lake Constance from Tilzer, 1983; of Lake Tahoe from Tilzer, 1978.

called the "clear-water phase" (Lampert and Schober, 1978; Sommer et al., 1986; see also Chapter 31).

During summer, the biomass development of phytoplankton in large lakes depends on their trophic state. Oligotrophic lakes which cannot support dense zooplankton populations exhibit only one major phytoplankton summer peak. During early autumn, the deepening of mixing causes a nutrient infusion into the euphotic zone. Provided that light is still sufficient, a short-lived autumnal phytoplankton maximum develops (Sommer, 1986).

Summer phytoplankton development comes to an end when the increase in nutrient levels due to thermocline erosion can no longer offset the effects of decreasing mean irradiances to which the phytoplankton are exposed. Because incident irradiance then is small, the critical depth of mixing can be reached earlier, but should be highly variable between lakes due to great difference in their optical properties. In Lake Constance, water transparency at low phytoplankton biomass is relatively insensitive to chlorophyll concentration due to high background light attenuation. The critical depth of autumnal mixing should range between 20 and 50 m, corresponding to light doses within the mixed water column of 0.9 and 2 mol $m^{-2}$ $d^{-1}$, respectively. Lake Tahoe, on the other hand, has an extremely low background light attenuation. At chlorophyll concentrations $<1$ mg $m^{-3}$ the critical depth of mixing can be assumed to be between 60 and 250 m.

The *seasonal amplitude* of primary production rates is enhanced in ice-covered, large lakes at high latitudes and/or in extremely continental climate. In Lake Baikal, the highest phytoplankton biomass is observed prior to ice-out in late spring, while maximum photosynthetic rates occur in late summer when small phytoplankton species dominate (Kozhova, 1987). Annual productivity is extremely small in high-latitude lakes due to the brevity of the growing season (Duthie et al., 1987).

Although a seasonal cycle in daily radiant energy inputs does exist in tropical lakes, the seasonality is more dependent on mixing, trade winds, and the alteration pattern of dry and wet seasons. Because these events vary less regularly than irradiance, seasonal patterns in tropical lakes are rather irregular (Talling, 1987).

Tropical large lakes are either warm-monomictic (Lake Titicaca: Richerson et al., 1977; Richerson et al., 1986), oligomictic (Lake Lanao: Lewis, 1973), or polymictic (Lake Chad: Lemoalle, 1975; Bahia de Puno of Lake Titicaca: Vincent et al., 1986). Mixing can either increase or decrease productivity. In Lake Victoria and Lake Malawi, high productivity occurs during windy periods when nutrients can be introduced to the euphotic zone and when *Melosira* dominates, which requires high water turbulence for remaining buoyant (Talling, 1965, 1966, 1987). In Lake Titicaca, mixing enhances nutrient supply from the hypolimnion (Richerson et al., 1986). On the other hand, nitrogen fixation by cyanobacteria is only possible under stratified conditions, which may provide as much as 30% of the nitrogen to this N-limited system (Vincent et al., 1986). In polymictic Lake Chad, mixing also increases nutrient influx and, hence, productivity which, due to the lake's shallowness, is not offset by decreasing availability of light. Mixed conditions prevail in Lake Chad whenever the water level is low (Lemoalle, 1975). In Lake Lanao, irregular deep mixing has an adverse effect on phytoplankton productivity (Lewis, 1974). It can be concluded that mixing causes increasing productivity whenever radiant energy supply is sufficient to allow the phytoplankton to efficiently utilize the then-abundant nutrients (Figure 17.10).

Lake George—a shallow, eutrophic, equatorial lake in Uganda—is the only well-documented example of a lake possessing a truly aseasonal phytoplankton assemblage. Although Lake George is not a true large lake by the definition applied in this book, it is mentioned here because the year-round patterns

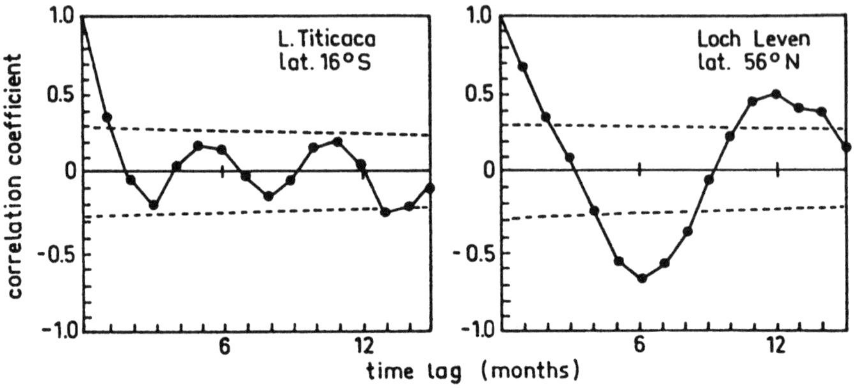


**Figure 17.10** Time series of autocorrelations (correlation coefficients, vertical scales) of primary production rates in a tropical lake (Lake Titicaca) and a northern temperate lake (Loch Leven). The near-horizontal broken lines give the envelopes within which 95% of the autocorrelations from a random series covering the time interval of each data series should fall. (Redrawn from Richerson et al., 1986, with permission).

studied so extensively in that lake should equally apply to truly large, but shallow tropical lakes. The thermal regime of Lake George is characterized by regularly recurring nocturnal mixing and diurnal stratification (Ganf and Horne, 1975). Phytoplankton productivity and biomass remain virtually constant over the course of the year (Ganf 1975, 1980) (Figure 7.11*a*).

Because fluctuations in tropical lakes are, to a great extent, controlled by factors other than the seasonal cycle of irradiance, interannual variations are frequently more pronounced than seasonal fluctuations (Talling, 1986, 1987; Vincent et al., 1986; Richerson et al., 1987). Autocorrelation analysis of time series of seasonal production rates in temperate-zone lakes usually show maximum negative correlations at a six-month time interval and maximum positive correlations after 12 months. In tropical lakes, by contrast, the patterns are irregular (Richerson et al., 1986; see also Figure 17.10).

**Factors Controlling the Amplitude of Seasonal Productivity Cycles.** Latitude, trophic state, and mean depth are the primary factors responsible for controlling the amplitude of the cycles discussed earlier.

*Latitude:* The seasonal variation of daily radiant energy supply is the single most important factor controlling the seasonal productivity cycle on a global scale. Its effect can best be examined in frequently or permanently mixed, shallow lakes where light regime and nutrient inputs are unaffected by vertical density gradients of the water column (Lake George, Lough neagh). However, within the tropics themselves, the seasonality of radiant energy inputs become decreasingly important with decreasing latitude (Vincent et al., 1986).

*Trophic state:* In monomictic, large lakes, winter productivity is usually light-limited and unaffected by nutrients. It, therefore, is equally small in oli-

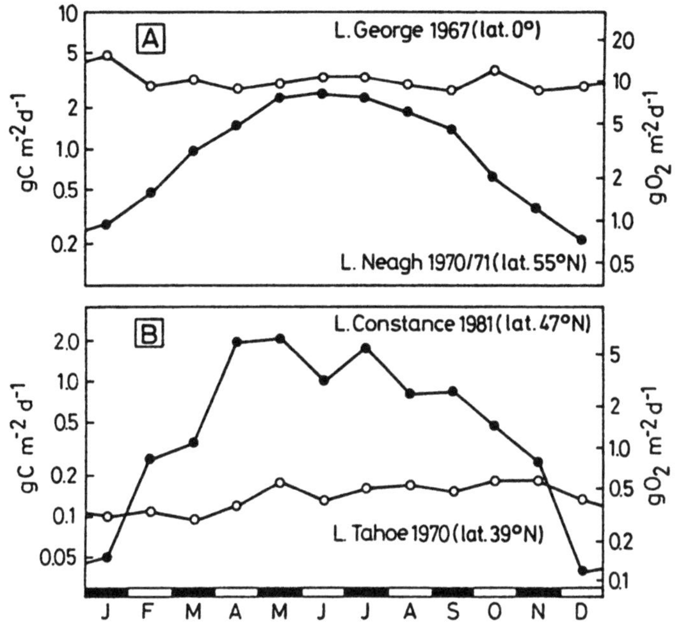


**Figure 17.11** Seasonal courses of daily production rates in four large lakes of different latitudes (given), trophic states and depth. (*a*) Lake George and Lough Neagh are both shallow, frequently or permanently mixed, and eutrophic. (*b*) Lake Constance and Lake Tahoe are both deep (mean depth 100 m and 313 m, respectively) and warm-monomictic. However, Lake Constance is meso-eutrophic, whereas Lake Tahoe is ultra-oligotrophic (from Tilzer, 1989).

gotrophic and eutrophic lakes. During summer, biomass yield and, hence, productivity are controlled by the nutrients available in the euphotic zone at the beginning of the growing season. In eutrophic lakes, the seasonal amplitude of biomass and productivity is therefore greater than in oligotrophic lakes at comparable latitudes (Lake Constance, Lake Tahoe).

*Mean depth:* Increases in mixing depth decreases phytoplankton by shifting the production/respiration balance in favor of respiration, thus, leading to a decline in the total biomass within the entire water column. In addition, phytoplankton are diluted by mixing, by which the fraction of underwater light which can be harvested and utilized by the algae is diminished with adverse effects on productivity (Lake Constance, Lough Neagh).

## 17.5 Conclusions

To end this review, the special features of large lakes with respect to phytoplankton productivity shall be briefly summarized:

1. The usually greater mean depth of large lakes, as compared to that

of small lakes, leads to a greater dilution of nutrients from the water- and airsheds and to diminished nutrient inputs from the benthos in large lakes.

2. Increased summer mixing depth of the upper layer leads to lower average light levels for the phytoplankton and, hence, to a greater extent of light limitation in large deep lakes relative to small lakes.
3. Reduced nonalgal turbidity of the water leads to higher productivity at comparable chlorophyll concentrations in large deep lakes as compared to shallow lakes, as a consequence of more efficient light harvesting.
4. Internal nutrient cycling within the euphotic zone leads to less short-term variability of the production process during the stratified season in large lakes, relative to small ones. Regenerated production should be greater relative to new production. Internal seiches provide the main mechanism by which nutrients enter the euphotic zone of large lakes during summer.

---

*Acknowledgements* This work was supported by the Deutsche Forschungsgemeinschaft within the Special Collaborative Program (SFB) 248, "*Cycling of Matter in Lake Constance.*" The numerous constructive comments of W.F. Vincent are gratefully acknowledged.

## References

Ahl, T. 1980. Eutrophication in relation to the load of pollution. *Progress in Water Technology* 12:49–61.

Atlas, D. and Bannister, T.T. 1980. Dependence of mean spectral light extinction coefficient of phytoplankton on depth, water color, and species. *Limnology and Oceanography* 25:157–159.

Bannister, T.T. 1974. Production equations in terms of chlorophyll concentration, quantum yield, and upper limit to production. *Limnology and Oceanography* 19:1–12.

Bloesch, J., Stadelmann, P., and Bührer, M. 1977. Primary production, mineralization and sedimentation in the euphotic zone of Swiss lakes. *Limnology and Oceanography* 22:511–526.

Boyce, F.M. 1974. Some aspects of Great Lakes physics of importance to biological and chemical processes. *Journal of the Fisheries Research Board of Canada* 31:689–730.

Brylinsky, M. 1980. Estimation of the productivity of lakes and reservoirs. In: LeCren, E.D., and Lowe-McConnel, R.H. (eds.), *The Functioning of Freshwater Ecosystems.* Cambridge University Press, Cambridge, p. 411–453.

Brylinsky, M. and Mann, K.H. 1973. An analysis of factors governing productivity in lakes and reservoirs. *Limnology and Oceanography* 18:1–14.

Dillon, P.J. and Rigler, F.H. 1974. The phosphorus-chlorophyll relationship in lakes. *Limnology and Oceanography* 19:767–773.

Dokulil, M. 1979. Optical properties, colour and turbidity in Neusiedlersee: The limnology of a shallow lake in Central Europe. In: Löffler, H. (ed.), *Neusiedler See, Monographiae Biologiae* No. 37. Junk, Den Haag, p. 151–167.

Dubinsky, Z. and Berman, T. 1976. Light utilization efficiencies of phytoplankton in Lake Kinneret (Sea of Galilee). *Limnology and Oceanography* 21:226–230.
Dubinsky, Z. and Berman, T. 1981. Light utilization by phytoplankton in Lake Kinneret (Israel). *Limnology and Oceanography* 26:660–670.
Dubinsky, Z., Falkowski, P.G., and Wyman, K. 1986. Light harvesting and utilization by phytoplankton. *Plant Cell Physiology* 27:1335–1349.
Dring, M.J. and Jewson, D.H. 1982. What does $^{14}C$-uptake by phytoplankton really measure? A theoretical modelling approach. *Proceedings of the Royal Society London, Series B* 214:351–368.
Dugdale, R.C. and Goering, J.J. 1967. Uptake of new and regenerated forms of nitrogen in primary productivity. *Limnology and Oceanography* 12:196–206.
Duthie, H.C. and Hart, C.J. 1987. The phytoplankton of the subarctic Canadian Great Lakes. In: Munawar, M. (ed.), *Phycology of Large Lakes in the World; Ergebnisse der Limnologie* No. 25. Schweizerbart, Stuttgart, p. 1–9.
Eadie, B.J., Chambers, R.L., Gardner, W.S., and Bell, G.L. 1984. Sediment trap studies in Lake Michigan: resuspension and chemical fluxes in the southern basin. *Journal of Great Lakes Research* 10:307–321.
El Sayed, S.Z. 1970. On the productivity of the Southern Ocean, Atlantic and Pacific sectors. In: Holdgate, M.W. (ed.), *Antarctic Ecology*. Academic Press, p. 119–135.
Falkowski, P.G. 1980. Light-shade adaptation in marine phytoplankton. In: Falkowski, P.G. (ed.), *Primary Productivity in the Sea*. Plenum Press, New York, p. 99–120.
Falkowski, P.G. 1981. Light-shade adaptation and assimilation numbers. *Journal of Plankton Research* 3:203–216.
Falkowski, P.G. and Owens, T.G. 1978. Effects of light intensity on photosynthesis and dark respiration in six species of marine phytoplankton. *Marine Biology* 45:289–295.
Fee, E.J. and Hecky, R.E. 1987. Limnological scale effects; Perspectives on Great Lake research to the year 2000. In: *Basic Issues in Great Lakes Research*, Special Report No. 123. Great Lakes Research Division, Ann Arbor, p. 70–85.
Flynn, K.J. 1988. The concept of "Primary production" in aquatic ecology. *Limnology and Oceanography* 33:1215–1216.
Fuhs, G.W., Demmerle, S.D., Canelli, E., and Chen, M. 1972. Characterization of phosphorus-limited algae (with reflections on the limiting nutrient concept). In: Likens, G.E. (ed.), *Nutrients and Eutrophication: The Limiting Nutrient Controversy*. Special Symposium American Society of Limnology and Oceanography No. 1, p. 113–133.
Gaarder, T. and Gran, H.H. 1927. Investigations of the production of plankton in the Oslo Fjord. *Rapports Proc. Verb. Conseil International d'Exploration de la Mer* 42:3–48.
Gächter, R. and Bloesch, J. 1985. Seasonal and vertical variation in the C:P ratio of suspended and settling seston in lakes. *Hydrobiologia* 128:193–200.
Ganf, G.G. 1974. Incident solar irradiance and underwater light penetration as factors controlling the chlorophyll *a* content of a shallow equatorial lake (Lake George, Uganda). *Journal of Ecology* 62:593–610.
Ganf, G.G. and Horne, A.J. 1975. Diurnal stratification, photosynthesis and nitrogen fixation in a shallow, equatorial lake (Lake George, Uganda). *Freshwater Biology* 5:13–39.
Geller, W. 1985. Production, food utilization and losses of two coexisting, ecologically different *Daphnia* species. *Archiv für Hydribiologie / Beihefte Ergebnisse der Limnologie* 21:67–79.
Goldman, C.R. 1974. *Eutrophication of Lake Tahoe Emphasizing Water Quality*. EP-66013-19-034, United States Government Printing Office, Washington, D.C.
Grobbelaar, J.U. 1984. Phytoplankton productivity in a shallow turbid impoundment, Wuras Dam. *Verhandlungen der internationalen Vereinigung für Theoretische und Angewandte Limnologie* 22:1594–1601.

Grobbelaar, J.U. 1985. Phytoplankton productivity in turbid waters. *Journal of Plankton Research* 7:653–663.
Hutchinson, G.E. and Löffler, H. 1956. The thermal classification of lakes. *Proceedings of the National Academy of Sciences* 42:84–86.
Jassby, A.D. and Goldman, C.R. 1974. Loss rates from a phytoplankton community. *Limonology and Oceanography* 21:540–547.
Jewson, D.H. 1976. The interaction of components controlling net phytoplankton photosynthesis in a well-mixed lake (Lough Neagh, Northern Ireland). *Freshwater Biology* 6:551–576.
Jewson, D.H. 1977. Light penetration in relation to phytoplankton content of the euphotic zone of Lough Neagh, N. Ireland. *Oikos* 28: 74–83.
Jorgenson, E.G. and Steemann Nielsen, E. 1965. Adaptation in plankton algae. In: Goldman, C.R. (ed.), *Primary Productivity in Aquatic Environments*. University of California Press, Berkeley, p. 39–46.
Kirk, J.T.O. 1983. *Light and Photosynthesis in Aquatic Ecosystems*. Cambridge University Press, Cambridge, 401 pp.
Kirk, J.T.O. 1985. Effects of suspensoids (turbidity) on penetration of solar radiation in aquatic ecosystems. *Hydrobiologia* 25:195–208.
Klump, J.V. 1987. Biogeochemical processes and chemical fluxes. In: *Basic Issues in Great Lakes Research*, Special Report No. 123 of the Great Lakes Research Division, Ann Arbor, p. 86–116.
Koblentz-Mishke, O.J., Volkovinski, V.V., and Kabanova, J.G. 1970. Plankton primary production of the world ocean. In: *Scientific Exploration of the South Pacific*. Standard Book No. 309-01755-6. National Academy of Science, Washington, D.C., p. 183–193.
Kozhova, O.M. 1987. Phytoplankton in Lake Baikal: Structural and functional characteristics. In: Munawar, M. (ed.), *Phycology of Large Lakes in the World; Ergebnisse der Limnologie* No. 25. Schweizerbart, Stuttgart, p. 19–37.
Lampert, W. and Schober, V. 1978. Das regelmäßige Auftreten von Algenmaximum und "Klarwasserstadium" im Bodensee als Folge von klimatischen Bedingungen und Wechselwirkungen von Phyto- und Zooplankton. *Archiv für Hydrobiologie* 82: 364–386.
Lehn, H. 1965. Ufernahe Phytoplankton-biocoenosen des Bodensees und ihre Bedeutung für das Pelagial. *Zeitschrift für Naturforschung* 206:379–382.
Lemoalle, J. 1979. *Biomasse et production phytoplanctonique du Lac Tchad (1968–1976). Relations avec les conditions due milieu*. ORSTOM. Paris, 311 pp.
Lewis, W.M., Jr. 1973. The thermal regime of Lake Lanao (Philipines) and its theoretical implications for tropical lakes. *Limnology and Oceanography* 18:200–217.
Lewis, W.M., Jr. 1974. Primary production in the plankton community of a tropical lake. *Ecological Monographs* 44:377–409.
Megard, R.O., Combs, W.S., Smith, P.D., and Knoll, A.S. 1979. Attenuation of light and daily integral rates of photosynthesis attained by planktonic algae. *Limnology and Oceanography* 24:1038–1050.
Munawar, M. und Munawar, I.F. 1975. Some observations on the growth of diatoms in Lake Ontario with emphasis on Melosira binderana Kütz. during thermal bar conditions. *Archiv für Hydrobiologie* 75:490–499.
Parsons, T.R. Takahaski, M., and Hargrave, B.T. 1984. *Biological Oceanographic Processes*. Pergamon Press, Oxford, 330 pp.
Peterson, B.J. 1978. Radiocarbon uptake: its relation to net particulate production. *Limnology and Oceanography* 23:179–184.
Post, A.F., Dubinsky, Z., Wyman, K., and Falkowski, P.G. 1985. Physiological responses of a marine planktonic diatom to transitions in growth irradiance. *Marine Ecology Progress Series* 25:141–149.
Powell, T.M., Richerson, P.J., Dillon, T.M., Aggee, B.A., Dozier, B.J., Godden, D.A., and

Myrup, L.O. 1975. Spatial scales of current speed and phytoplankton biomass fluctuations in Lake Tahoe. *Science* 189: 1088–1090.

Pridmore, R.D., Vant, W.N., and Rutherford, J.C. 1985. Chlorophyll-nutrient relationships in North Island lakes (New Zealand). *Hydrobiologia* 121:181–189.

Ramberg, L. 1978. Relations between phytoplankton and light climate in two Swedish forest lakes. *Internationale Revue der gesamten Hydrologie* 64:749–782.

Redalje, D.G. and Laws, E.A. 1981. A method for estimation phytoplankton growth rates and carbon biomass. *Marine Biology* 62:73–79.

Redfield, A.C. 1958. The biological control of chemical factors in the environment. *American Scientist* 46:250–221.

Rhee, G.Y. 1978. Effects of N:P atomic ratios and nitrogen limitation on algal growth, cell composition, and nitrate uptake. *Limnology and Oceanography* 23:10–25.

Richardson, K., Beardall, J., and Raven, J.A. 1983. Adaptation of unicellular algae to irradiance: An analysis of strategies. *New Phytologist* 93:157–191.

Richerson, P.J., Neale, P.J., Wurstbaugh, W., Alfaro, T.R., and Vincent, W.F. 1986. Patterns of temporal variation in Lake Titicaca. A high altitude tropical lake. I. Background, physical and chemical processes, and primary production. *Hydrobiologia* 138:205–220.

Richerson, P.J. Widmer, C. and Kittel, T. 1977. The Limnology of Lake Titicaca (Peru-Bolivia), a large, high altitude tropical lake. *University of California, Davis, Institute of Ecology Publication* No. 14, 78 pp.

Riley, G.A. 1957. Phytoplankton in the North Central Sargasso Sea, 1950–52. *Limnology and Oceanography* 2:252–270.

Ripl, W. 1976. Biochemical oxydation of polluted lake sediment with nitrate—a new restoration method. *Ambio* 5:132–135.

Ripl, W. and Lindmark, G. 1978. Ecosystem control by nitrogen metabolism in sediment. *Vatten* 2:135–144.

Ryther, J.H. 1954. The rate of photosynthesis to respiration in marine plankton algal and its effects upon measurement of productivity. *Deep-Sea Research* 2:134–139.

Sakamoto, M. 1966. Primary production by phytoplankton community in some Japanese lakes and its dependence on lake depth. *Archiv für Hydrobiologie* 62:1–28.

Schanz, F. 1985. Vertical light attenuation and phytoplankton development in Lake Zürich. *Limnology and Oceanography* 30:299–310.

Schelske, C.L., Stoermer, E.F., Fahnenstiel, G.L., and Haibach, M. 1986. Phosphorus enrichment. Silica utilization and biochemical Silica depletion in the Great Lakes. *Canadian Journal of Fisheries and Aquatic Sciences* 43:407–415.

Schindler, D.W. 1978. Factors regulating phytoplankton production and standing crop in the world's freshwaters. *Limnology and Oceanography* 23:478–486.

Senft, W.H. 1978. Dependence of light-saturated rates of algal photosynthesis on intracellular concentrations of phosphorus. *Limnology and Oceanography* 23:709–718.

Simon, M. 1987. Biomass and production of small and large free-living and attached bacteria in Lake Constance. *Limnology and Oceanography* 32:591–607.

Simon, M., and M.M. Tilzer. 1987. Bacterial response to seasonal changes in primary production and phytoplankton biomass in Lake Constance. *Journal of Plankton Research* 9:535–552.

Smith, V.H. 1979. Nutrient dependence of primary productivity in lakes. *Limnology and Oceanography* 24:1051–1064.

Smith, V.H. 1982. The nitrogen and phosphorus dependence of algal biomass in lakes: A theoretical and empirical analysis. *Limnology and Oceanography* 27:1101–1112.

Smith, V.H. 1987. Lake-watershed interactions: An overview of the effects of nutrients on lakes. In: *Basic Issues in Great Lakes Research,* Special Report No. 123 of the Great Lakes Research Division, Ann Arbor, p. 117–140.

Sommer, U. 1986. The periodicity of phytoplankton in Lake Constance (Bodensee) in comparison to other deep lakes of Central Europe. *Hydrobiologia* 138:1–7.

Sommer, U. (ed.). 1989. *Plankton Ecology: Succession in Plankton Communities*. Science Tech Publishers, Madison, WI., 384 pp.
Sommer, U., Gliwicz, Z.M., Lampert, W. and Duncan, A. 1986. The PEG-model of seasonal succession of planktonic events in fresh waters. *Archiv für Hydrobiologie* 106:433–471.
Spigel, R.H. and J. Imberger. 1987. Mixing processes relevant to phytoplankton dynamics in lakes. *New Zealand Journal of Marine and Freshwater Research* 21:361–377.
Stabel, H.H. 1984. Impact of sedimentation on the phosphorus content of the euphotic zone of Lake Constance. *Verhandlungen der Internationalen Vereinigung für Theoretische und Angewandte Limnologie* 22:964–969.
Stabel, H.H. 1986. The role of plankton biomass in controlling fluctuations of suspended matter in Lake Constance. *Hydrobiologia* 140:173–181.
Stabel, H.H. and Geiger, M. 1985. Phosphorus adsorption to riverine suspended matter. Implications for the P-budget of Lake Constance. *Water Research* 19:1347–1352.
Steemann Nielsen, E. 1955. The use of radioactive carbon ($^{14}C$) for measuring organic production in the sea. *Journal du Conseil International pour l'Exploration de la Mer* 18:117–140.
Strub, P.T., Powell, T., and Goldman, C.R. 1985. Climatic forcing: Effects of El Nino on a small, temperate lake. *Science* 227:55–57.
Talling, J.F. 1957. The phytoplankton population as a compound photosynthetic system. *New Phytologist* 56:133–149.
Talling, J.F. 1966. The annual cycle of stratification and phytoplankton growth in Lake Victoria (East-Africa). *Internationale Revue der gesamten Hydrobiologie* 51:545–621.
Talling, J.F. 1969. The incidence of vertical mixing, and some biological and chemical consequences in tropical African lakes. *Verhandlungen der Internationalen Vereinigung für Theoretische und Angewandte Limnologie* 17:998–1012.
Talling, J.F. 1971. The underwater light climate as a controlling factor in the production ecology of freshwater phytoplankton. *Mitteilungen der Internationalen Vereinigung für Theoretische und Angewandte Limnologie* 19:214–243.
Talling, J.F. 1986. The seasonality of phytoplankton in African lakes. *Hydrobiologia* 138:139–160.
Talling, J.F. 1987. The phytoplankton of Lake Victoria (East Africa). In Munawar, M.: (ed.), *Phycology of Large Lakes in the World; Ergebnisse der Limnologie*. Schweizerbart, Stuttgart, p. 229–256.
Tilzer, M.M. 1979. Predictions of productivity changes in Lake Tahoe at increasing phytoplankton biomass. *Verhandlungen der Internationalen Vereinigung für Theoretische und Angewandt Limnologie* 20:407–413.
Tilzer, M.M. 1983. The importance of fractional light absorption by photosynthetic pigments for phytoplankton productivity in Lake Constance. *Limnology and Oceanography* 28:833–846.
Tilzer, M.M. 1984a. Estimates of phytoplankton loss rates from daily photosynthetic rates and observed biomass changes in Lake Constance. *Journal of Plankton Research* 6:309–324.
Tilzer, M.M. 1984b. The quantum yield as a fundamental parameter controlling vertical photosynthetic profiles or phytoplankton in Lake Constance. *Archiv für Hydrobiologie, Supplementband* No. 69, *Monographische Beiträge*: 169–198.
Tilzer, M.M. 1987. Light-dependence of photosynthesis and growth in cyanobacteria: Implications for their dominance in eutrophic lakes. *New Zealand Journal of Marine and Freshwater Research*, 401–412.
Tilzer, M.M. 1989. The productivity of phytoplankton and its control by resource availability. A review. In: Kumar, P.D. (ed.), *Phycotalk*, Varanasi Hindi University, Varanasi, pp. 1–40.
Tilzer, M.M. and Beese, B. 1988. The seasonal productivity cycles and controlling factors in lake Constance. *Schweizerische Zeitschrift für Hydrologie* 50:1–39.

Tilzer, M.M. and Goldman, C.R. 1978. Importance of mixing, thermal stratification and light-adaptation for phytoplankton productivity in Lake Tahoe (California-Nevada). *Ecology* 59:810–821.

Tilzer, M.M., Hillbricht-Ilkowska, A., Kowalczewski, A., Spodmiwska, I., and Turezynska, J. 1977. Diel phytoplankton periodicity in Mikolajskie Lake, Poland, as determined by different methods in parallel. *Internationale Revue der gesamten Hydrobiologie* 62:279–289.

Tilzer, M.M. and Horne, A.J. 1979. Diel patterns of phytoplankton productivity and extracellular release in ultra-oligotrophic Lake Tahoe. *International Revue der gesamten Hydrobiologie* 62:279–289.

Vincent, W.F. 1983. Phytoplankton production and winter mixing: Contrasting effects in two oligotrophic lakes. *Journal of Ecology* 71:1–20.

Vincent, W.F., Vincent, C.L., Downes, M.T., and Richerson, P.J. 1985. Nitrate cycling in Lake Titicaca (Peru-Bolivia): the effects of high altitude tropicality. *Freshwater Biology* 15:31–42.

Vincent, W.F., Wurstbaugh, W., Neale, P.J., and Richerson, P.J. 1986. Polymixis and algae production in a tropical lake: latitudinal effects on the seasonality of photosynthesis. *Freshwater Biology 16:781–803.*

Vollenweider, R.A. 1976. Advances in defining critical loading levels for phosphorus in lake eutrophication. *Memorie del Istituto Italiano die Idrobiologia* 33:53–83.

Vollenweider, R.A. and Kerekes, J. 1980. The loading concept as basis for controlling eutrophication: Philosophy and preliminary results of the OECD Programme on eutrophication. *Progress in Water Technology* 12:5–38.

Vollenweider, R.A. and Nauwerck, A. 1961. Some observations of the C-14 method for measuring primary production. *Verhandlungen der Internationalen Vereinigung für Theoretische und Angewandt Limnologie* 14:134–149.

Walsh, J.J. 1983. Death in the sea: enigmatic phytoplankton losses. *Progress of Oceanography* 12:1–86.

Wangersky, P. 1986. Biological control of trace metal residence time and speciation: a review and synthesis. *Marine Chemistry* 18:269–297.

Westlake, D.F. 1980. Primary production. In: LeCren, E.D. and Lowe-McConnel, R. H. (eds.), *The Functioning of Freshwater Ecosystems*. Cambridge University Press, Cambridge, p. 141–246.

# 18

# Effects of Latitude on Phytoplankton Composition and Abundance in Large Lakes

Utsa Pollingher

ABSTRACT Besides the factors of the environmental conditions and the age of the lakes, latitude has major effects on the diversity and abundance of the phytoplankton in large lakes. In large lakes at high latitudes, diatoms and chrysophytes are the dominant groups. The phytoplankton of lakes located in the temperate zones are characterized by the dominance of diatoms, especially in winter and fall, and Chlorophyta, Cyanophyta, and phytoflagellates (Cryptophyta, Chrysophyta, and Dinophyta) in summer. In large lakes of the warm belt, the dominant role is taken over by Chlorophyta and Cyanophyta, and in some lakes also by diatoms. In those lakes, only a few chrysophyte species differ from those common in temperate lakes, and *Peridinium* spp. are more common than *Ceratium*. The phytoplankton biomass increases gradually from lakes at high latitudes towards the equator.

Obviously, possible changes in phytoplankton species composition, abundance, and dynamics can only be demonstrated for lakes studied over a period of several years. In Lake Michigan, for example, anthropogenic interference by decreasing the nutrient supply and increasing stocks of salmonids, has led to changes in the composition, abundance, and dynamics of the phytoplankton. Conversely, the increase of nutrients in two large, shallow lakes located in industrial zones (Lake Mälaren—mean depth, 12.8 m; and Lough Neagh—mean depth, 8.6 m) led to the development of cyanobacterial water blooms. Lake Biwa is the only ancient lake with a heavy water bloom (caused by *Uroglena americana*), which also develops as a consequence of eutrophication.

Water blooms of cyanobacteria develop during periods of stratification, low turbulence, and high temperature. The absence of real water blooms of cyanophytes and dinoflagellates in deep large lakes may be due to one of the prerequisites for their development; minimal water turbulence. A given wind velocity leads to greater turbulence per unit area in large lakes than in small ones, and as a consequence, calm periods are probably shorter in large lakes than in small water bodies. Conversely, high turbulence favors the development and maintenance of *Melosira* spp. and other heavy and large diatoms in the water column. Even in the low-latitude lakes, the diatoms and chrysophytes dominate during the mixing period.

Most of the planktonic algae are cosmopolitan forms. Among the large lakes, endemic phytoplankton species are mostly found in ancient lakes: Lakes Baikal, Biwa, and Ohrid, but were not described in the plankton of Lake Titicaca. Tropical forms of Cyanophyta and diatoms are found in the large lakes of Central Africa.

The lakes discussed here show the numerous and diverse combinations of environmental and biological factors which may affect the composition and abundance of the phytoplankton in large lakes.

## 18.1 Introduction

A comparative review of the phytoplankton composition of large lakes is a difficult enterprise. The level of knowledge varies greatly from lake to lake, and the presence or absence of a species may only reflect the number of scientists who have worked in a given area, the methods applied to count plankton, or the number and location of sampling stations. The level of knowledge depends also on the length of the time period over which the lake has been studied. Moreover, the description of the present status of a recently eutrophied lake does not enable us to compare it with a still oligotrophic lake of the same area. This is only possible if, for the former, a long-term data series is available which gives us information on the phytoplankton assemblage before eutrophication. Therefore, the data record cannot be utilized directly without allowance for the numerous causes of bias, error, and misinterpretation.

In spite of these inherent difficulties, it is clear that certain latitude-related factors deeply affect the phytoplankton communities of large lakes. These factors may have a direct effect on phytoplankton: for example, the wide seasonal variations of irradiance at various latitudes directly control photosynthesis. They may also have an indirect effect on the nutrient regime of lakes through modifications of the thermal structure.

Light and temperature are the two main factors which directly affect phytoplankton communities. As shown in Figure 18.1, at very high latitudes, algae have short growing seasons, whereas in the tropics, light is available all year long and the seasonal variations are minor.

The temperatures of the mixed layer and of the lake bottom also show wide changes with latitude. The maximum temperatures of the mixed layer vary from 28 to 30°C in the tropics, decreasing progressively to around 10°C at 70° latitude (Lewis, 1987). In contrast, the maximum temperatures of the bottom layer, at 27–28°C in the tropics, decrease rapidly with latitude and abruptly reach 4°C between 30–40° latitude (Figure 18.2).

Thermal stability is a measure of the work necessary to mix a water column. Thermal stability is very low at low latitudes, increases to a maximum value between 20° and 30° latitude, and decreases again towards higher latitudes (Serruya and Pollingher, 1983). Moreover, the Coriolis force, which deflects horizontal water currents and distorts oscillating thermoclines (Mortimer, 1974), is small-to-negligible in the tropics.

The thermal regime, in conjunction with wind frequency and with force and fetch on the water surface, governs the thickness of the mixed layer. The light gradient within a mixing epilimnion, together with surface irradiance,

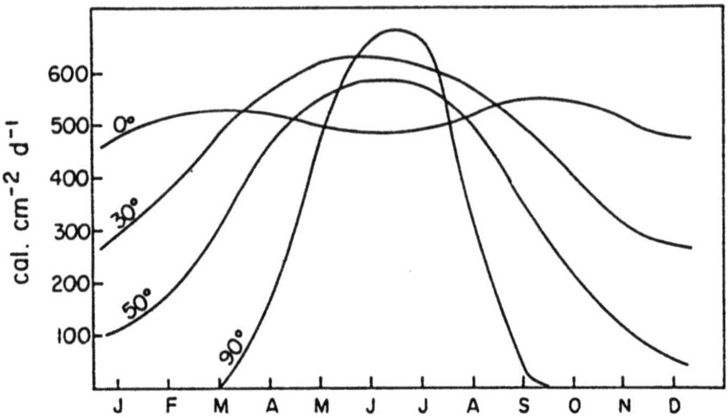


**Figure 18.1** Total solar irradiance as a function of latitude (0–90°) and month of the year.

determines the light available to a phytoplankton population in the upper layer of a lake (see Chapter 17). The sharpness of the thermocline and the eddy diffusivity caused by the shear movements along the boundary layer between the two water masses determine the rate of nutrient renewal during thermal stratification. Because the stability of thermal stratification in tropical lakes is generally less pronounced than at higher latitudes, they usually have a thicker mixed layer and a faster rate of nutrient regeneration, which ultimately leads to higher primary production and biomass.

Superimposed on all these factors that are active today is the age of a lake. Lake history introduces a time-dependent factor, which is responsible for the endemism observed in geologically ancient lakes, such as Lakes Baikal, Biwa, Ohrid, and Tanganyika.

Since the supply of energy and nutrients to the phytoplankton and some

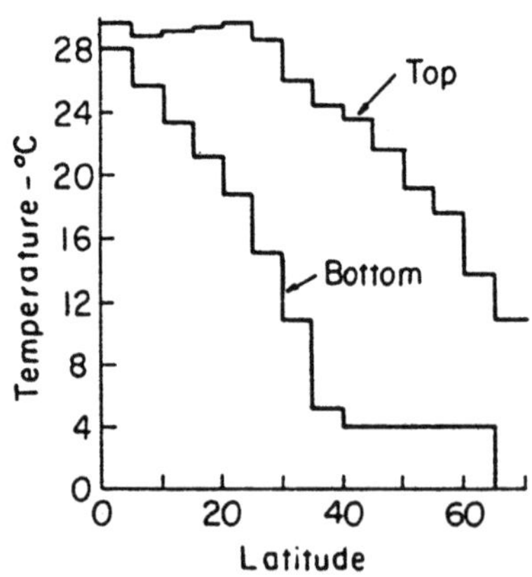


**Figure 18.2** The temperature of the bottom and of the upper mixed layer (top) as a function of latitude (reproduced from Lewis, 1987, with permission from the *Ann. Rev. Ecology and Systematics*, Vol. 18, 1987, by Annual Reviews, Inc.).

important physical and chemical mechanisms vary with latitude (Lewis, 1987; see Chapter 36), in the following discussion, three groups of large lakes are considered: the high-latitude lakes (polar and sub-polar regions, 50–70°), the mid-latitude lakes (30–50°), and the low-latitude lakes (0–30°).

It has not been my intention to cover all the lakes that have been thoroughly studied with respect to their phytoplankton composition and periodicity, but rather to select examples in which latitudinal effects could be demonstrated with particular clarity. Effects of changes in nutrient loading have led to directional changes in phytoplankton species composition and succession, which are described wherever data were available.

## 18.2 Large Lakes at High Latitudes

**Introduction** Here, only lakes whose mid-points are at a latitude greater than 50° are considered to be "high-latitude lakes." All of the high-latitude, large lakes thus defined are located on the Northern Hemisphere. Six lakes have been selected for this review. Two of them are located in the Canadian Subarctic, Great Slave Lake and Great Bear Lake, and two others belong to the hercynian crystalline shield, Lake Ladoga and Lake Baikal. Lake Baikal and Lake Mälaren, are in the temperate zone but, due to the continental climate of their surroundings, they freeze over during winter and thus have more in common with lakes located at truly high latitudes. The climate of Lough Neagh is oceanic and mild, due to the influence of the Gulf Stream. The two large northern European lakes, Lake Mälaren and Lough Neagh, have, moreover, been chosen to illustrate cases of large but shallow water bodies.

The two Canadian lakes and Lake Ladoga are subjected to extreme variations of temperature and light conditions. Surface water temperatures in these lakes never exceed 14–16°C. The lakes are ice-covered during winters and free of ice only from July to September. As thermal summer stratification is extremely weak or absent, their mixing regime can be termed cold-monomictic. The lakes have low concentrations of nutrients; $PO_4$–P concentration is less than 1 $\mu g\ l^{-1}$ in McLeod Bay and Christa Bay (Great Slave Lake) (Johnson, 1975; and Duthie and Hart, 1987). The phytoplankton are dominated by diatoms and Chlorophyta in Great Slave Lake and by diatoms and Chrysophyta in Great Bear Lake.

**Great Slave Lake and Great Bear Lake** A long-term study of the phytoplankton in Great Slave Lake (Moore, 1980a) revealed that *Asterionella formosa, Melosira islandica, Diatoma tenue* var. *elongatum*, and *Stephanodiscus astrea* dominated the biomass following the ice breakup, but in 1976 and 1977, a large population of *Chlamydomonas lapponica* developed under the ice in April, probably due to increasing light levels. The chrysophytes *Dinobryon bavaricum* and *D. cylindricum* developed high populations in June, when silica depletion by diatoms occurred. The phytoplankton biomass was very low under ice in winter and reached 130 to 340 mg $m^{-3}$ in summer (Duthie and Hart, 1987).

Moore (1980b), studying the phytoplankton of Great Bear Lake over three seasons at three different stations, found that the most common species were *Dinobryon bavaricum, D., sociales, Rhodomonas minuta, Cyclotella* sp., *Synedra acus*, and *Ankistrodesmus falcatus*. The biomass was 62 mg $m^{-3}$ in June at a water temperature of 0°C and increased to 70–91 mg $m^{-3}$ in September. Great Slave Lake exhibits considerable horizontal variation in physical and chemical properties between its various arms, and it is likely that this also applies to the phytoplankton (Duthie and Hart, 1987). The same authors suggest that the survival of phytoplankton in subarctic lakes, which are ice covered for the greater part of the year, may be due to the formation of resting stages (*M. islandica*) or of statospores (*Dinobryon*).

**Lake Ladoga** Lake Ladoga is a cold, deep lake located at the southern border of the Fenno-Scandinavian crystalline massif, U.S.S.R. Phytoplankton used to be dominated by diatoms, and algal water blooms had never been observed (Petrova, 1987). The common species of diatoms in spring were *Melosira islandica* sub. *helvetica, M. italica* var. *italica* (Ehr) Kütz, and *Melosira* spp. In summer, *Asterionella formosa* and *Tabelaria fenestrata* var. *fenestrata* dominated. Phosphorus (P) enrichment, which began in the 1960s (P loading was 0.39 g P $m^{-2}$ $yr^{-1}$; the mean concentration of total P in the water was 27 $\mu$g P $l^{-1}$), has caused alterations in the composition, abundance, and seasonal periodicity of the phytoplankton. The changes appeared first in the shallow part of the lake and in bays of the northern region, where the supply of P was maximal, and later extended over the entire lake area. The number of diatom species and their abundance increased.

Some species of diatoms (*Attheya zachariasii* and *Rhizosolenia eriensis* var. *morsa*) disappeared. In summer, the diatoms vanished completely, and a water bloom of cyanobacteria became common. In July, species of *Oscillatoria* now dominate, and in August, *Aphanizomenon flos-aquae* and *Microcystis* spp. become most prominent. The biomass has increased from a spring (May–June) maximum of 5.4 mg $l^{-1}$ wet weight in 1960–1962 to a maximum of 20.5 mg $l^{-1}$ in 1976–1979, and from a summer (July–August) maximum of 0.4 mg $l^{-1}$ in 1960–1962 to 8.2 mg $l^{-1}$ in 1976–1979. Today, Lake Ladoga is considered a meso-oligotrophic lake (Petrova, 1987).

**Lake Baikal** Lake Baikal, the most ancient lake of the world (located in Siberia), is of tectonic origin (Horie, 1961). With a maximum depth of 1,620 m, it is also the deepest lake in the world. The lake is oligotrophic, and the Secchi disk transparency reaches 40.5 m. The water is of the alkaline type with a low mineral content. The concentration of $PO_4$–P varies from 10–60 $\mu$g $l^{-1}$ (Vereskiagin, 1949—in Kozhov, 1963). The lake is frozen from mid-January to May and stratified between July and October. Circulation occurs in late June and in early December, but never reaches the deeper part of the hypolimnion (Horie, 1961). The pelagial is characterized by low temperatures with only a short period of temperatures $>$12°C at the lake surface. In the shallow littoral zone (depth <20 m), called the littoral *sor* (sor means lagoon in Russian), the water reaches up to 20°C during a relatively long period. In this zone, the concentration of nutrients is higher than in the open lake, and as a consequence, the phytoplankton are more abundant (Kozhova, 1987).

The phytoplankton community is comprised of endemic species and of cosmopolitan ones that are typical for cold, oligotrophic lakes. About 120 species inhabit the pelagic zone, of which about 8% are endemic, having been newly described from Lake Baikal (Kozhova, 1987). About 30% of the recorded species belong to the Chlorophyta, 25% to the Bacillariophyta, and 20% to the Cyanophyta.

The following species were described as endemic ones: diatoms—*Melosira baicalensis, Cyclotella baicalensis, C. minuta, Stephanodiscus skabitschevskyi*; Pyrrhophtya—*Peridinium baicalense, Gymnodinium coeruleum, G. baicalense, G. baicalense* var. *minor*; Cyanophyta—*Synechocystis limnetica*; and Chrysophyta—*Chrysosphaerella baicalensis*. Some of these species are also found in the River Angara and the alpine Lake Hubsugul, which belong to the Baikalian reef zone. With the exception of three species which are found at the littoral, all the endemic species inhabit the open lake. The dominant cosmopolitan species are: *Melosira islandica* subsp. *helvetica, Melosira binderana, Synedra acus, S. ulna* var. *danica, Nitzschia acicularis, Stephanodiscus hantzschii, Ankistrodesmus, Oocystis*, and *Scenedesmus* spp.

The Cyanophyta are represented by more than 70 species but only 30 appear in the open region and even these live in proximity to the river mouth and shallow gulfs and bays. The species occurring are: *Anabaena flos-aquae, A. lemmermannii, A. spiroides*, and *Aphanizomenon flos-aquae* (Kozhov, 1963). These algae often form water blooms in summer in coastal areas, bays, gulfs, and pre-estuarine regions. After a long period of warm calm days, the bloom extends to more open regions located near the shallow ones. Twelve species of Chrysophyta were recorded in the lake: *Mallomonas, Uroglena*, and *Dinobryon* spp. *Dinobryon cylindricum* is common in open regions of the lake, and *D. divergens* is more common in shallow ones, whereas *D. sociale* and *D. sertularia* are found in gulfs only. *Epichrysis melosirae* is an endemic chrysomonad which appears in large numbers epiphytic on the filaments of *Melosira baicalensis* (Kozhov, 1963). The Pyrrhophyta are represented by seven endemic species. Some of them, such as *Gymnodinium baicalense* and *G. coeruleum*, appear in large numbers under ice during March–April. *Peridinium baicalense* appears in spring and autumn, whereas *Ceratium hirundinella* occurs only in summer in the pelagic zone and is more common in shallow bays and pre-estuarine areas (Kozhov, 1963). Due to the contribution of the different taxonomic groups to the biomass, the Baikal plankton was considered a Bacillariophyta plankton, but in the last few years a considerable role of Pyrrhophyta was recorded, and the community can now be termed a Bacillario-Pyrrhophyta plankton (Kozhova, 1987).

The different physico-chemical conditions prevailing in both zones of the lake lead to variations in composition and abundance of the phytoplankton assemblages. The phytoplankton of the littoral zone is dominated by Bacillario-Chlorophyta assemblages, accompanied sometimes by *Anabaena* spp., and the biomass is higher than in the pelagial. In the pelagic zone, the winter under-ice water blooms of diatoms (*M. baicalensis, M. islandica* var. *helvetica*) and dinoflagellates lead to the annual maximum values of biomass. In some years, the peak of the phytoplankton biomass is 3–5 g $m^{-3}$ due to *Melosira* spp., or 100 g $m^{-3}$ when *Gymnodinium* spp. develop for a very short period (several days) when the ice breaks. The endemic species are less common in summer; only C. *minima* is dominant in late summer and in fall. In some years on warm calm days in July–August, the River Selenga's flood carries the phytoplankton (*A. lemmermannii* dominant) from the shallow waters to the offshore waters, spreading patches over the great part of the lake. In this case, the phytoplankton biomass increases. In 1979, the concentration of chlorophyll *a* in the shallow delta zone of the River Selenga reached 14 mg $m^{-3}$, and in the pelagic zones, 0.5–1.5 mg $m^{-3}$. In 1979, the concentration of chlorophyll *a* reached its maximum of 16.5 mg $m^{-3}$ under ice when *G. baicalense* dominated (biomass wet weight, 31 g $m^{-3}$). The maxima of chlorophyll *a* in 1980 were observed in mid-July and October ($>$1 mg $m^{-3}$), whereas in 1981, the maxima were recorded in February, July, August, and mid-September (i.e., in summer and fall). The mean annual chlorophyll *a* concentration was 1.3 $\pm$ 0.3 mg $m^{-3}$, and in summer for the pelagic zone, it was 0.6 mg $m^{-3}$ (0–50 m layer) (Kozhova, 1987).

Moskalenko (1972), studying the phytoplankton biomass in the offshore region (0–25 m layer) during 1964–1968, recorded a maximum biomass of 57.6 g $m^{-2}$ wet weight in spring (May ) 1968. Inter-annual changes of phytoplankton were frequently recorded in spring. A year with relatively high phytoplankton biomass appears every 2–3 years (Kozhova, 1987), mainly due to the development of *M. baicalensis, M. islandica* subsp. *helvetica,* and, sometimes, *S. binderanus*. The development of the phytoplankton in Lake Baikal depends on wind, which generates water movement, and on temperature, which determines the time of the ice-out and the warming of the water. In the shallow bays into which purified industrial effluents are flowing, no changes in the composition and abundance of phytoplankton were recorded. The increase in the water level up to 1.0–1.5 m in some shallow bays, due to the construction of a hydroelectric station on the River Angara, led to a decrease in the littoral, warm-algal assemblages and a penetration of the endemic, cold, pelagic algae—including *M. baicalensis.*

**Two Large European lakes** The two European, relatively shallow lakes are both located in the industrial zone and are affected by eutrophication, with blooms of cyanobacteria as a common response.

*Lake Mälaren* Lake Mälaren is a large (surface area, 1096 $km^2$), shallow (maximum depth, 63 m; mean depth, 12.8 m) lake located in central Sweden. The drainage area of the lake has a dense population and a developed industry. The lake is a very complicated system of waters composed of bays of different character, and the lake can be divided into five basins, each of which has its own chemical and biological characteristics (Willen, 1984, 1987). During the period 1964–1971, the lake received high loading of N and P, and as a consequence, in the polluted areas heavy blooms of cyanobacteria and diatoms developed. The P and N load were reduced in 1970–1983 and the concentration of nutrients decreased. The decrease in nutrients, as well as of the turbidity of the water, led to drastic alterations in species composition and total phytoplankton biomass. The previously blooming cyanobacteria *Microcystis* and *Anabaena* spp. nearly disappeared, and they decreased in the most polluted basins. The number of diatoms also decreased. Coenobial species generally decreased, while small and large flagellates belonging to Chrysophyta, Cryptophyta, and Dinophyta increased. The number of species increased significantly. The dinoflagellates increased tenfold in biomass per liter. The phytoplankton biomass decreased by 50% in the hypertrophic and stratified Ekoln basin but decreased less in the shallow unstratified Galten basin where the internal loading from the sediment retards the restoration. Also, in Galten basin, the number of algal species nearly doubled. This increase is especially due to unicellular large and small flagellates. The changes in species composition and the decrease of algal biomass have been interpreted as signs of improved water quality and of the reversal of eutrophication (Willen, 1987).

*Lough Neagh* Lough Neagh is the largest lake (surface area, 383 $km^2$) in the British Isles (Wood and Gibson, 1973), located at 54°36′N, 6°30′W. It is a shallow, eutrophic lake (mean depth, 8.6 m) with a homogeneously mixed water column throughout most of the year (Jewson, 1977). The phytoplankton are dominated by Cyanophyta (Wood and Gibson, 1973) which are represented by species of *Anabaena, Anabaenopsis, Oscillatoria, Microcystis, Aphanizomenon,* etc. The diatoms, which dominate in early spring, are comprised of species of *Cyclotella, Stephanodiscus,* and *Melosira* (Gibson et al., 1971). The succession of the species of cyanobacteria is correlated with the change in the concentration of nutrients.

*Oscillatoria redeckii* increases during the second half of April and declines by mid-May, when P becomes undetectable in the water. *Oscillatoria aghardhii* dominates in late summer and early autumn. Its appearance seems to be dependent on the rise of the phosphate and nitrate concentrations which precede it. *Aphanizomenon flos-aquae* dominated in July 1969 and 1970 and always appeared when nutrient levels were at their lowest (Gibson et al., 1971). The biomass of the Cyanophyta reached its highest value (*O. redeckii* dominant) of 13.4 mg $l^{-1}$ in April 1970, and its lowest of 0.6 mg $l^{-1}$ in December 1971 (Jewson, 1976).

The algal composition was similar in the open lake and the shallow (mean depth; ~2 m) Kinnega Bay, which is rich in nutrients, but since the bay is nearly isolated, the growth of the population occurs at different times from that of the open lake. The phytoplankton biomass is higher in the bay, 300 mg chl*a* $m^{-3}$, than in the open lake, 90 mg chl*a* $m^{-3}$. Jewson (1977) found that the major case of the variation of light penetration into the water was due to changes in the phytoplankton abundance, but it may also be influenced by Gelbstoff.

## 18.3 Large Lakes of the Temperate Zone

The bulk of the world's large lakes is situated in the temperate zone. Although some of them occasionally freeze over, all of them can be termed warm-monomictic. This is the major property which distinguishes the Laurentian Great Lakes from their counterparts further to the North. Most of them are of glacial origin, but two (Lake Biwa and Lake Ochrid) were formed in the Tertiary and carry numerous endemic species, even within the phytoplankton community. With only very few exceptions, the temperate-zone, large lakes of the Northern Hemisphere are heavily influenced by human activity, which is reflected in interannual differences in phytoplankton species composition and seasonal periodicity, among other factors. Lake Taupo is the only large lake in the Southern Hemisphere which has been studied extensively. The large and, as yet, pristine large lakes of Southern Chile and Argentina should also be studied systematically, both as further examples of Southern Hemisphere lakes and as paradigms of undisturbed temperate large lakes.

**Lake Superior** Lake Superior, with a surface area of 82,000 km$^2$, is the largest freshwater lake on earth. Because of its location at the edge of the Canadian Shield, it has a cold climate; due to the low population density of its watershed, man-made pollution is negligible (Matheson and Munawar, 1978). Changes in the composition of the lake water have occurred only in nearshore regions adjoining industrial centers. The mean water temperature does not rise above 6°C. Maximum surface temperature (10.6 to 16°C) is reached in mid-September. The mixing of the lake occurs from May to mid-July, and the summer stratification from July to November. The composition of the water resembles that of the rain falling on the watershed. The water is pure and clear with low concentrations of total solids. The composition of the major ions is very uniform, both vertically and horizontally, and does not present a seasonal cycle. Among the nutrients, only nitrate and reactive silica show seasonal variations. The concentration of total P is 4 mg m$^{-3}$ and that of reactive P is 1 mg m$^{-3}$ (Matheson and Munawar, 1978).

The composition, abundance, and spatial distribution of the phytoplankton was studied by Munawar and Munawar (1978) from May to November–December 1973. The phytoplankton was dominated by phytoflagellates (53%) and diatoms (38%). The diatoms had a similar composition to those of Great Slave Lake (see Section 18.2) (in Great Slave Lake only the net plankton was studied). The phytoflagellates comprised 26% Cryptomonads, 21% Chrysomonads, and 5% dinoflagellates (*Amphidinium* spp., *Gymnodinium* spp., *Ceratium hirundinella*, and *Peridinium* spp.). Among them, nanoplanktonic and picoplankton forms dominated. As in Great Slave Lake, the cyanobacteria and the green algae were never abundant. The mean phytoplankton biomass (wet weight ~0.13 g $m^{-3}$) did not show any inshore/offshore differentiation. The maximum biomass recorded was 0.19 $gm^{-3}$. Spatial differences in phytoplankton species and biomass were not as large as the differences between stratified and isothermal stations. Nalewajko and Voltolina (1986) studied the phytoplankton from May 1980 to October 1981. The phosphorus turnover time was much shorter (3.9 hrs) inshore than offshore (12.0 hrs). Moreover, the P turnover time is shorter and the phytoplankton biomass is higher in stratified water than in isothermal water. Stratified areas have a lower proportion of microalgae than isothermal ones. Growth rates of phytoplankton in stratified water were more than double the values in isothermal water. The largest differences in species composition appeared among the three seasonal groups: the proportion of diatoms was highest in May–June, green algae in July, and Chrysophytes and Cryptophytes in October.

The influence of temperature and light on the algal activity was stronger than the influence of the lake's trophic state. Algae smaller than 1 $\mu$m are important contributors to photosynthesis at low light intensities in Lake Superior. After 10 years (1983), the phytoplankton of Lake Superior was studied again (Munawar et al., 1987). Changes had occurred in the structure of the spring phytoplankton. The spring assemblages were still dominated by diatoms and phytoflagellates, but a significant increase in Chlorophyta, Chrysophyta, and Dinophyta was recorded. The total biomass was higher in 1983 than in 1973, increasing the spring and summer from 145 to 319 mg $m^{-3}$ and from 145 to 233 mg $m^{-3}$, respectively. The species composition over the whole lake continued to represent an oligotrophic and cold stenothermic community (Munawar et al., 1987). Endemic algae were not observed, but some rare species such as *Chrysolykos planctonicus* Mack, *C. skujai*, and *Stelexemonas* were recorded.

**Lake Huron** In the oligotrophic Lake Huron, the diatoms dominate the phytoplankton during all seasons with Cyanophyta and Chlorophyta as co-dominants. The most common diatoms are *Melosira granulata*, *M. italica* var. *subarctica*, and *M. islandica*, as well as *Rhizosolenia eriensis*, *Asterionella formosa*, and *Tabellaria fenestrata* (Munawar and Munawar, 1986). The phytoplankton exhibit two relatively well-developed seasonal biomass peaks: a spring maximum in June (with 86% of diatoms) and a small peak in summer (August) formed by cyanobacteria, green algae, and Cryptophyceae. At the mid-lake station, low biomass concentrations were found (3.0 to 1.7 g $m^{-3}$) (Vollenweider et al., 1974); at the northern part of the lake, the biomass was less than 1.0 g $m^{-3}$; the inshore stations of Bruce Peninsula and Saginaw Bay showed higher averages of biomass (1.5 and 8.2 g $m^{-3}$, respectively). The highest biomass was found in the eutrophic Saginaw Bay (1.6 and 17.3 g $m^{-3}$) due to a long-lasting maximum (from July until the end of September) of cyanobacteria. The annual mean phytoplankton biomass is ~0.6 g $m^{-3}$ (Munawar et al., 1987), and the authors classified the lake as an oligotrophic one, in comparison with the ultra-oligotrophic Lake Superior.

**Lake Ontario** Lake Ontario is a relatively deep, warm-monomictic lake with two major basins separated by an indistinct sill. The temperature of the water varies seasonally between 2 and 22°C. From early winter through May, the lake temperature is 4°C, and thermal stratification starts in June (Robertson and Scavia, 1984). Soluble reactive phosphorus reaches maximum values in winter and spring (~14.0 $\mu g\ P\ l^{-1}$), followed by a rapid decrease in early summer to 1 $\mu g\ P\ l^{-1}$. It increases again during mixing and fall overturn in October (Robertson and Scavia, 1984). The horizontal distribution of the nutrients is closely related to the spring thermal bar and upwelling. The distribution of nutrients in the surface waters shows a general east-to-west increase in concentration.

The basin of Lake Ontario has undergone large-scale development of population and industry. Since its volume for the dilution of polluting substances is relatively small in comparison with other large deep lakes, the effect of pollution and its reduction became more evident (Robertson and Scavia, 1984).

The phytoplankton of Lake Ontario was studied in 1970, 1975, 1979, 1982, and 1983 (Vollenwieder et al., 1974; Munawar and Munawar, 1986; Gray, 1987; Munawar et al., 1987). In winter and spring, diatoms represent 90% of the biomass. In summer and fall, the phytoplankton are more diverse, Chlorophyta and Cyanophyta predominating, with diatoms, Cryptomonads, and Chlorophyta following. Phytoflagellates represent 30% of the biomass. From January to May, diatoms dominated in both inshore and offshore regions, but the inshore region was dominated by *Melosira varians* and *M. binderana* (syn. *Stephanodiscus binderanus*; Round, 1972), whereas the offshore region was dominated by *M. islandica* var. *helvetica* (Vollenweider et al., 1974). Munawar and Munawar (1975) have studied the correlation of the ecology of *M. binderana* with a thermal bar that forms in early spring (April–May) due to the faster warming of the nearshore waters (above 4°C) than the offshore waters (which, for some time, remain at less than 4°C). *M. binderana* showed a marked decrease in abundance between nearshore and offshore areas. The highest biomass densities were observed inside the thermal bar (65% of the phytoplankton biomass). The variations in the phytoplankton abundance could be explained by differences in both nutrients and temperature. Soluble reactive phosphorus (SRP) and silica decreased concomitantly in the nearshore regions inside the thermal bar, probably due to their uptake by algae. In May, the thermal bar moved further offshore, and no difference of inshore/offshore biomass was apparent.

The diversity of the phytoplankton begins to become greater from June onwards when phytoflagellates (belonging to Chrysophyta, Cryptophyta, and Dinophyta), green algae and cyanobacteria dominated. The phytoplankton biomass in 1970 ranged from 1.4 to 5.9 $g\ m^{-3}$ (Vollenweider et al., 1974). Total phytoplankton biomass decreased considerably from 1970 to 1973 (probably correlated with the decrease in total P in response to improved waste treatment and to restrictions on phosphorus in detergents) (Robertson and Scavia, 1984) during both seasons (and especially during summer). It was accompanied by changes in the relative contribution of various taxonomic groups; Cyanophyta and Chlorophyta were reduced in 1978; diatoms and Cryptophyta increased. In the pelagic zone of the lake, which is relatively unaffected by nearshore disturbance, the biomass decreased from 3.0 g $m^{-3}$ in 1970 to 0.35 g $m^{-3}$ in 1983. Chlorophyta, diatoms, Dinophyta, and nanoplanktonic species increased in abundance between 1970 and 1983. The composition of the phytoplankton presented a mixture of species usually occurring either in oligotrophic or eutrophic environments. Lake Ontario was classified as a mesoeutrophic lake (Munawar et al., 1987).

**Lake Michigan** The phytoplankton of Lake Michigan has been extensively studied. The study of the offshore net-plankton by Parker et al. (1977) showed that diatoms are the dominant group in the lake. They contribute nearly 90% of the total biomass. A population of diatoms develops at the onset of each bloom period, and diatoms were dominant when biomass became maximal in spring (April–June) and fall. The common species were *M. islandica, M. granulata, Tabelaria flocculosa,* etc. In July, the dominance of diatoms was less pronounced, and green algae, cyanobacteria, and dinoflagellates developed. In August, nondiatom species contributed 87% to the biomass, and *C. hirundinella* was dominant. In September, diatoms began to increase, forming a second peak in October–November when they contributed 85% to the total biomass. The authors suggest that as the spring bloom of diatoms progresses, decreasing nutrient levels, especially silica concentration, slowed down the diatom growth. The development of the fall diatom bloom is regulated by the restoration of the dissolved reactive silica concentration in the euphotic zone.

Reduction in SRP-concentrations in the southeastern nearshore area (1974–1981) led to a remarkable increase in the abundance of various species of cyanobacteria during the period of lake stratification (Chang and Rossmann, 1988). The total number of cyanobacteria found in fall 1981 was twice that of 1974. The authors speculate that high phosphorus led to the uptake of dissolved silica by diatoms which then disappeared due to silica limitation. Thereafter, the major contributors to the biomass were the cyanobacteria *Anacystis incerta* and *Gomphosphaeria aponina*.

Three stages of seasonal succession were found in this region of the lake: The succession begins with a diatom-dominated assemblage due to the high availability of silica. The second stage is dominated by cyanobacteria and flagellates which do not require dissolved silica. The third stage corresponds to the breakdown of the thermal stratification and an increase of dissolved silica. As a consequence, a slight increase of diatoms occurs, but the cyanobacteria predominate. The trophic status of open waters in Lake Michigan appears to be in the upper range of oligotrophic to mesotrophic. Vollenweider et al. (1974) considered the lake to be more eutrophic than Lake Huron or Lake Superior.

Tarapchak and Stoermer (1976) classified it as mesotrophic because oligotrophic diatom assemblages have been replaced by a mesotrophic population. Silica depletion has caused pronounced changes in phytoplankton assemblages after summer stratification, diatom dominance being replaced by cyanobacteria and green algae. The shift in species composition was caused by silica depletion attributed to excessive P loading (Schelske and Stoermer, 1971). During late July and August of the years 1976 to 1978, cyanobacteria and chlorophytes constituted 50% of the total phytoplankton biomass (Bartone and Schelske, 1982). Phytoplankton dynamics in Lake Michigan were closely related to seasonal cycles of temperature and silica (Bartone and Schelske, 1982). The seasonal cycle of epilimnetic chlorophyll was unimodal, with no autumnal pulse of biomass. Parker et al. (1977) have shown that the periodicity of the blooms in Lake Michigan shows considerable annual variation.

Munawar and Munawar (1986) have studied the composition and abundance of the phytoplankton in the summer 1982. These authors found that both the spring (May) and the summer (July) biomass maxima were similar to those found in Lake Huron. Diatoms dominated the phytoplankton community during both seasons followed by phytoflagellates (Cryptophyta and Chrysophyta). In summer, Cyanophyta biomass increased compared to the spring. The dominant diatoms in spring were: *M. binderana, M. islandica, M. islandica* var. *helvetica, M. italica* var. *subarctica,* and *Stephanodiscus* spp. The dinoflagellates were a colorless form of *Gymnodinium helveticum* and *G. eurytopum.* In summer, flagellates were common; *Dinobryon, Uroglena, Cryptomonas, C. hirundinella, Peridiniopsis borgei, P. aciculiferum,* and *Peridinium gatunense.* This study has demonstrated the abundance of phytoflagellates and nanoplanktonic forms.

The phytoplankton dynamics in Lake Michigan were further studied by Fahnenstiel and Scavia (1987) from the end of the spring (May) mixing period through the mid-stratification period (July–August) from 1982 to 1984. During the middle of the stratified period, cyanobacteria and green algae contributed 32% (1982) and less than 4% (1983–1984) to the phytoplankton biomass, respectively, whereas phytoflagellates constituted 55% (1982), 91% (1983), and 68% (1984) of the total biomass. In July–August 1985, cyanobacteria constituted only 15% and the phytoflagellates 76% of the phytoplankton biomass. The authors attributed the large shift in the 1970s—from the dominance of cyanobacteria (*Anabaena flos-aquae*) and green algae to phytoflagellate dominance in the 1980s—to the large-scale ecological changes between the two sets of years; these changes were primarily the result of: (1) decrease of nutrient supply, which led to changes in P concentration—total phosphorus (TP) decreased in winter-spring; and (2) changes in food-web interactions.

The prolonged ice cover during 1977 decreased sediment resuspension, and as a consequence TP and chlorophyll *a* concentration decreased and water transparency increased. In 1983 and 1984, the increase of the population of the stocked salmonids (predators) reduced the population of the zooplanktivorous alewife (*Alosa pseudoharengus*), which allowed the development of

*Daphnia* spp., especially *Daphnia pulicaria*. The cladocerans are less selective with respect to their food than the previously dominant calanoid copepods. They reduced the concentration of trypton, cyanobacteria, and green algae; as a consequence, the water became more transparent (Scavia et al., 1986). This is an example of an anthropogenic change in the upper trophic level which led to changes in the composition and abundance of primary producers.

A hypothesis concerning the control of Lake Michigan phytoplankton dynamics was elaborated by Scavia and Fahnenstiel (1987). In early spring, at the beginning of lake circulation, the development of diatoms depends on light, temperature, and nutrients supplied by the sediments. During this period, grazing is insignificant, i.e., the spring phytoplankton dynamics are controlled "from the bottom up." During the stratification period, the epilimnion is generally nutrient (and especially phosphorus) limited. The composition and growth of the phytoplankton in summer are controlled by zooplankton nutrient regeneration and grazing. The composition and abundance of the zooplankton are controlled by the planktivorous fish, i.e., the influence on the phytoplankton dynamics is "from the top down."

**Lake Erie** Lake Erie is the shallowest (mean depth, only 17–19 m) of any of the Laurentian Great Lakes; therefore, it has the smallest volume and the largest surface-area-to-volume ratio (Robertson and Scavia, 1984). This lake is also the southernmost among the Laurentian Great Lakes. However, due to its shallowness, in a normal winter, it is the only one of the Great Lakes which usually freezes over most of its surface. The lake is divided into 3 basins.

Increased P concentrations from human activities have caused the eutrophication of Lake Erie. In Lake Erie, SRP follows a pattern similar to that in Lake Ontario—maximum concentration in winter–spring when light and vertical mixing limit the production (much of the phosphorus is in dissolved inorganic form) and a minimum during the summer (limiting concentration). Inorganic nitrogen presents low concentrations during summer, as in Lake Ontario (Robertson and Scavia, 1984). The trophic gradient increases from east to west. The inshore/offshore differentiation occurs only in the central and eastern basins.

In winter, the phytoflagellates are common, accompanied by some diatoms (*A. formosa, Diatoma, Surirella*) (Vollenweider et al., 1974). The spring phytoplankton are dominated by diatoms (*M. binderana, M. varians, M. islandica* var. *helvetica, Stephanodiscus tenuis*). During summer, cyanobacteria and dinoflagellates dominate, and in fall, Chlorophyta are also common. The phytoplankton biomass reaches high values during spring (April–May) and a second maximum in summer. Phytoflagellates play a more important role in Lake Erie than in Lake Ontario.

The abundance and succession of the phytoplankton differ in the three basins of Lake Erie. In spring, diatoms (*Stephanodiscus niagarae, M, binderana*) and phytoflagellates are more abundant, whereas in summer, flagellates dominate (from May to August), replacing the diatoms. In both periods, dinoflagellates are common (*C. hirundinella*). The inshore region presents three peaks with a maximum in summer (12 g $m^{-3}$ wet weight in July). Diatoms are the major component throughout the year. In the eastern basin, which is mesotrophic, the dinoflagellates dominate in spring and summer. The phytoplankton biomass reaches three maxima of similar magnitude (4 g $m^{-3}$) in late spring (June), in summer (July), and in later summer (September). In spring, diatoms dominate; in late spring, phytoflagellates, such as Chrysophyta, Cryptophyta, and Dinophyta; and in summer, phytoflagellates, such as cyanobacteria and green algae. Phytoflagellates are more frequent in the offshore region. The inshore region has four maxima—extra peak in fall with a greater amount of biomass than the offshore (10 g $m^{-3}$). The offshore region exhibited a well-developed pulse during the spring following the earlier inshore pulse.

**Lake Constance** Lake Constance is located on the northern fringe of the Alps within tertiary sediments (Molasse) and is of glacial origin. Due to its large watershed, which is 22 times the surface area of the lake itself (540 $km^2$), the lake—despite its great depth (mean, 100m; maximum, 254 m)—has a theoretical water retention time of only 4.5 years. The immediate surroundings of the lake are heavily populated (470 $km^{-2}$) but not industrialized. The main usages of the lake are recreation, drinking water supply (for more than $3.10^6$ people), and fishing. Until the late 1940s, the lake was oligotrophic. Due mainly to municipal waste loading, winter total phosphorus concentrations increased exponentially from undetectable levels in 1950 to values as high as 90 mg $m^{-3}$ around 1980 (Stabel, 1984). Since then, nutrient loading has decreased rapidly and, at present, winter maximum total phosphorus concentrations are below 50 mg $m^{-3}$ (Limnological Institute Konstanz, unpubl. data, see also Chapter 21).

Phytoplankton species composition, the seasonal periodicity as well as its causes, were extensively studied during the time of maximum nutrient loading. In the following, the results of these studies are briefly described. A comprehensive summary is given in Sommer (1987).

The study of Lake Constance phytoplankton can be used as an example for meso-eutrophic, temperate zone lakes. This view is confirmed by the astounding similarity of the phytoplankton succession between Lake Constance and Lake Geneva (see Chapter 31). Moreover, it has been possible to explain the similarities and differences in the seasonality of other European lakes, large and small (Sommer et al., 1986). The subsequent decline in nutrient loading has caused changes in phytoplankton species composition, relative abundance, and periodicity.

For the study period between 1979 and 1983, eight successional stages were distinguished, which shall be briefly discussed here:

1. *Rhodomonas-Stephanodiscus* stage: As soon as the lake stratified (late March-early April), small, rapidly growing species (*Rhodomonas lens, Rh. minuta, Stephanodiscus hantzschii*) dominated. Rapid reproduction,

mainly of *Daphnia hyalina* and *D. galeata*, caused the near-complete elimination of the phytoplankton by a "clear-water phase," which, at this time, recurred with great regularity, but since has been less regular and more variable in its extent (Lampert and Schober, 1978).

2. *Cryptomonas* stage: During the clear-water phase, which roughly covered the month of June, *Cryptomonas ovata* dominated. Sommer (1987) stresses that *Cryptomonas* dominance is characteristic of clear-water phases throughout the European lakes studied thus far.
3. *Pandorina-Cryptomonas* stage: After the grazing pressure had subsided, *Pandorina morum* became common, in addition to *C. ovata*. Prior to the clear-water phase, small and rapidly growing species which were highly susceptible to grazing by cladocerans dominated. Following the clear-water phase, slower-growing, larger and hence, grazer-resistant species formed the bulk of the nanophytoplankton. Sommer (1981) interpreted this shift as a transition from r- to K-selection among phytoplankton.
4. Summer diatom stage: The summer phytoplankton were consistently dominated by large diatoms such as *Fragillaria crotonensis, Tabellaria flocculosa*, and *Asterionella formosa*, until silica became depleted. In chemostat cultures using mixed natural phytoplankton as inocula, Sommer (1983) could demonstrate that competition for nutrients was the main determining factor in the succession of species during summer. Whenever silica is present, diatoms outcompete other algal groups, if the supply of SRP is short.

   Diatoms virtually disappeared in mid-summer due to silica depletion (Sommer and Stabel, 1983). As the phosphorus level decreased, diatoms persisted throughout the stratified season, apparently because the phosphorus level was now too low for the diatoms to build up sufficiently large populations that could eliminate most of the available silica.
5. *Ceratium* stage: *C. hirundinella* made a brief appearance in the phytoplankton in mid-summer but the dominance of this species has been highly variable during the study period.
6. Late summer stage: With the deepening of the upper mixed layer, nutrient infusions allowed the recurrence of larger diatoms. Other groups dominating at this time of the year were cyanobacteria (see Chapter 21) and *Ceratium*. This successional stage was less regular both in occurrence and in species composition than the other stages, although it has generally been associated with a brief biomass maximum before biomass decreased with the deepening of autumnal mixing.
7. *Mougeotia* stage: The fall phytoplankton were characterized by the dominance of filamentous algae (*Mougeotia tylespora* and *Ulothrix subtilissima*).

8. Late autumn and winter diatom stage: During the late fall and winter, the same large diatoms as those seen in early summer and, in addition, *Diatoma elongatum* and *Stephanodiscus rotula* form the bulk of the phytoplankton biomass.

The regular recurrence of most of these successional stages, as well as the apparent similarities between lakes, have strongly supported the view that the seasonal periodicity of phytoplankton can be explained in strictly deterministic terms. Whereas during winter, physical forcing and lack of radiant energy is the major controlling factor of phytoplankton biomass development, during summer intrabiocoenotic processes—such as competition for nutrients, grazing, and grazer avoidance, as well as buoyancy regulation and vertical migration—appear to be the major determinants of the changes within the community (Sommer, 1987, 1989; Sommer and Kilham, 1985).

**Lake Taupo** Lake Taupo is a deep, oligotrophic, warm-monomictic lake located on the central volcanic plateau of North Island, New Zealand (38°45′ S, 175°55′ E) (Vincent, 1983). Secchi disk transparencies vary from 11 to 21 m, and pH varies from 7 to 8. Anion concentrations are $HCO_3^- > Cl^- > SO_4^{2-}$; cations, $Na^+ > Ca^{2+} > Mg^{2+} > K^+$. The N:P ratio by weight in the input is 5.7:1 (White et al., 1980). Ammonia-nitrogen is more abundant in the epilimnion than nitrate-nitrogen. The lake is stratified in both summer (December–February) and winter (June–August). Overturn occurs in late winter (August). The phytoplankton are composed of a limited number of species. There is a continuous inoculation of Lake Taupo with *Anabaena* from Lake Rotongaio, but cyanobacteria are more common from February to July. *Melosira granulata* was present throughout the year (April 1975–March 1976), but the population reached its maximum from July to November (winter-spring). *Asterionella formosa* was another common diatom present throughout the year. The highest phytoplankton biomass was recorded in September, when *Chroococcus, Ankistrodesmus*, and *Chroomonas* were common. When the lake was mixing, *M. granulata* was the dominant form, and its wet weight biomass ranged from 60 mg $m^{-3}$ to a peak of 1160 mg $m^{-3}$ in September. During that time, the chlorophyll concentration increased by a factor of less than three. In the stratified period, the chlorophyll concentration was nearly stable.

White and Payne (1977) have found that the addition of inorganic N generally stimulated chlorophyll production during July–December, and the addition of P stimulated chlorophyll production in the other six months (January–June). Between 1979 and 1981, Vincent (1983) investigated the factors controlling the winter maximum of algal biomass in Lake Taupo. In winter (June–August), a large *Melosira* (considered to be similar to *M. baikalensis*) dominated. It was accompanied by other diatoms: *Asterionella formosa, Cyclotella meneghiniana, C. stelligera, Fragillaria crotonensis*, and *Synedra ulna*. In summer (December–February), small chlorococcaleans and flagellates (*Chroomonas* and *Dinobryon divergens*) were dominant. During the later summer, diatoms accumulated in the metalimnion. In Lake Taupo, the concentration of algal biomass and algal production reached their annual maxima during winter isothermy, when the temperature was at its annual minimum. This was due to the fact that winter circulation brought an increased supply of P and N and, moreover, prevented the large diatoms (mainly *Melosira*) and the little buoyant diatoms from settling out of the euphotic zone.

The increased availability of the limiting nutrients counterbalanced the effects of low temperature and diminished light availability. Lake Taupo constitutes an intermediate form of lake between temperate lakes, with winter minima of algal biomass and production, and tropical lakes, which present high primary production during storm-induced mixings (Vincent, 1983).

**Two Geologically Ancient Lakes** For representatives of this group, we have chosen Lake Biwa and Lake Ohrid, two ancient lakes with differing degrees of endemism.

*Lake Biwa* Lake Biwa is one of the world's oldest lakes; it is the largest lake in Japan, and supplies agricultural, industrial, and drinking water for 13,000,000 people (Nomura, 1981). The lake has "probably kept its isolated environment and stable ecological conditions for an extremely long time" (Horie, 1961). It was formerly an oligotrophic lake, but since the late 1960s, eutrophication of the lake has been developing, leading to a yearly production of a water bloom ("red tide") of *Uroglena americana*. In 1980, a eutrophication-prevention ordinance was enacted to decrease the nutrient inflow. Lake Biwa is a warm, monomictic lake with a surface area of 673.9 $km^2$ (epilimnion at 15–20 m depth). The lake has two basins; the north basin has an area of 617.6 $km^2$ and a mean depth of 44.7 m, and the sub-basin (south basin) has a surface area of 56.8 $km^2$ and a mean depth of 2.8 m. The concentrations of nutrients prior to the process of cultural eutrophication were: $PO_4$–P, 8–12 $\mu g\ l^{-1}$, $NO_3$–N, 90 $\mu g\ l^{-1}$; an $NH_4$–N, 20–30 $\mu g\ l^{-1}$.

The phytoplankton of Lake Biwa (Negoro, 1956; Wakabayashi and Ichise, 1982; Ichise and Wakabayashi, 1984; Mori and Horie, 1984; Nakanishi, 1984) is composed of a high number of Chlorophyta followed by Bacillariophyta, Cyanophyta, Chrysophyta, and Dinophyta. A few endemic species, such as *Pediastrum biwae* and *Staurastrum biwaensis*, and some rare species (called semi-endemic by Mori and Horie, 1984), such as *Melosira solida* and *Stephanodiscus carconensis*, the desmids, including *Staurastrum dorsidentiferum*, and an endemic form of *Ceratium hirundinella* have been recorded. *P. biwae*, *Staurastrum* spp., *M. solida*, and *S. carconensis* are commonly found in 200-m-long sediment core samples; this observation indicates that these species have had a long history in Lake Biwa. The species *M. solida* and *S. carconensis* developed only in the main basin (mean depth, 44.7 m) which was oligotrophic. In the years since eutrophication has been in progress, the population of *S. carconensis* seems to have been decreasing. *M. italica* and *M. granulata* are now the most common and abundant species in the very shallow sub-basin. Some species, such as

*Synura uvella, Volvox aureus,* and *Aphanocapsa,* which were dominant in the period 1935–1939 in the sub-basin, decreased.

Until 1965, the seasonal succession of the dominant phytoplankters in the main basin began with *S. carconensis* during January–March, followed by *M. solida* in April and December, *Asterionella formosa* in May, *Atteya zahariasi* in June and October, *C. hirundinella* in July, and *P. biwae* in August and September. Since 1965, *S. carconensis* and *A. zahariasi* decreased and the seasonal pattern changed. *M. solida* became dominant during January–April and November–December; *S. dorsidentiferum, Closterium aciculare,* and *P. biwae* dominate during June–October. In some years, when the desmids dominated from October to January, *M. solida* tended to decrease, and the dominant species during October–December and January–April became the desmids rather than the diatoms (1965–1966, 1971–1972). The pattern of dominant phytoplankters abundance and succession in the sub-basin varies from year to year. This fact may indicate relatively irregular seasonal changes in the physical, chemical, and biological environment. More details about the annual succession of the algal species in the northern and southern parts of the lake were described by Wakabayashi and Ichise (1982) and Ichise and Wakabayashi (1982), respectively.

A characteristic feature of the phytoplankton in Lake Biwa has been the mass development of the water bloom of *Uroglena americana* since 1977, causing deterioration of the quality of the drinking water due to fishy odors or odor of cucumbers. The water bloom now develops every year from April to June, during a period when the average daily temperature varies from 15–22°C (Mizushima and Ichise, 1983). *Uroglena* appears first in the southern part of the south basin, where the temperature increases relatively earlier than in the northern basin (Ichise and Wakabayashi, 1984). The algae extend to the northern areas of the north basin in conjunction with the increase of the water temperature. Thus, in the second half of May, *U. americana* is found all over the lake; the color of the water changes from yellowish-green to green. *U. americana* is an obligate photophagotroph (Kimura and Ishida, 1985). The wind direction and intensity probably play an important role in the water bloom's extension. At the end of June, the water temperature increases up to 22°C and higher. At this time *Uroglena* collapses rapidly. The red tide of *Uroglena* is accompanied by diatoms, green algae, and Cryptophytes.

Before the appearance of the red tide, diatoms dominate the phytoplankton, and after its disappearance, the green algae become dominant. In summer 1981 (August–September), a large development of *Anabaena macrospora* and *Microcystis aeruginosa* accompanied by a repulsive odor was recorded only in the southern shallow basin of the lake. The concentration of chlorophyll *a* in the pelagic area of the north basin now is low; its values increase gradually from the north to the south in the sub-basin. The contribution of algal cells that are less than 3 $\mu$m to the total chlorophyll *a* in the north basin increases from <5% to 45% during the late stratification period, when the larger algae decrease. This is due to an increase in the number of coccoidal cyanobacteria

(Nagata, 1986). The chlorophyll *a* concentration in the pelagic area of the main basin is relatively low (0.5–5.0 mg $m^{-3}$) and has nearly a homogeneous horizontal distribution (Nakanishi, 1984). In December 1971, when desmids increased, the concentration of chlorophyll *a* reached 20 mg $m^{-3}$. The concentration and distribution pattern of chlorophyll *a* in the sub-basin is influenced by the water movement and nutrient concentrations; it varies from 1 to 30 mg $m^{-3}$.

Lake Biwa is more influenced by human activity than Lake Baikal, but as in Lake Baikal, temperature and wind play an important role in the development and distribution of the phytoplankton.

*Lake Ohrid* Lake Ohrid is an ancient lake of tectonic origin located in the western part of the Balkan peninsula on the Yugoslavian-Albanian border. The lake has a surface area of 348 $km^2$, and a maximum and mean depth of 286 m and 145.20 m, respectively. The lake is typical for a karstic lake in that it is almost exclusively fed by very pure ground water from karstic sources. The lake can be characterized as of the subtropical, oligomictic type. Usually, thermal stratification prevails throughout the year and complete mixing only occurs in exceptionally severe winters (Stankovic, 1960). The lake water is characterized by very high transparency. During the stratified period, the trophogenic layer is about 50 m thick. Green algae and diatoms dominate the phytoplankton both in numbers of species and abundance, followed by Chrysophyta and dinoflagellates. The Cyanophyta and desmids are less important. The Chlorophyta are represented by cosmopolitan species belonging to the genera *Gloeococcus, Oocystis, Ankistrodesmus, Scenedesmus,* and *Pediastrum.* An unusual characteristic of Ohrid phytoplankton is the high number of nanoplanktonic forms (30% of the total number of species). Most of the lake production is due to the nanoplanktonic endemic species, such as *Oocystis rhomboidea, Didimogenes dubia,* and *Dispora wilhelmi,* but also to the cosmopolitan diatoms *Cyclotella ocellata, Stephanodiscus hantzschii,* and the endemic dinoflagellate *Cystodinium dominii.* Other endemic species are *Cyclotella fotti* and *C. hustedti.*

The vertical distribution of the algae during stratification is as follows: the maximum of abundance is recorded between 10 and 20–30 m depth; *C. hirundinella, Peridinium,* and *Cystodinium* are located between 0–30 m; *Gomphosphaeria lacustris* and *Merismopedia glauca* descend to 50–75 m depth; some of the Chlorophyta—*O. rhomboidea, D. dubia, Ankistrodesmus lacustris,* and *Keratococcus angalus*—descend to 75 m, but their maximum is found between 20 and 40 m. The true hypolimnetic forms are the diatoms: *C. fotti, C. hustedti, Stephanodiscus astrea,* and *Asterionella formosa,* as well as the desmid *Closterium aciculare.* Experiments (by Kozarov, 1954, cited by Stankovic, 1960) with the hypolimnetic *C. fotti* have shown that the deep population of *Cyclotella* is able to survive for a long time at minimal light intensity; the greatest photosynthetic activity was found at a depth of 25 m, but the cells seem to be able to use even the small amount of light which penetrates to a depth of 100 m. The experimental addition of inorganic and organic compounds to water samples from the upper hypolimnetic waters (50–100 m) stimulated the photosynthetic activity of the diatoms (Allen and Ocevski, 1976).

## 18.4 The Large Lakes of the Warm Belt

**The Great Lakes of Africa** The six great lakes of tropical East Africa are situated between 3°N and 15°S in latitude. Lakes Tanganyika, Malawi (or Nyasa), Albert, Kivu, and Edward lie in the western rift valley, whereas Lake Victoria is located in a shallow depression between the two ridges which border the western and eastern rift. Lake Victoria is the world's largest equatorial lake (surface area 68,000 $km^2$), but also the shallowest one among the six enumerated lakes (maximum depth, 79 m; mean depth, 40 m). Temperature and light generally have only limited influence on the development of phytoplankton in this region. However, Talling (1969) has shown that most tropical lakes do exhibit seasonal thermal cycles. Surface light is relatively constant and most of the lakes are transparent, but low mean light conditions may occur during maximum circulation due to the great depth of the mixed layer (Hecky and Kling, 1987). The deep lakes—Tanganyika (the deepest in this group), Malawi, and Kivu (mean depths of 700, 426, and 240 m, respectively)—are meromictic. Lake Edward (mean depth, 34 m) does not circulate fully every year, whereas Lake Albert (mean depth, 25 m) is polymictic (Hecky and Kling, 1987). Lake Victoria's offshore area is probably monomictic, having a full circulation from May to August and stratification during the rest of the year. Mixing and upwelling play an important role in nutrient supply in these lakes. Lakes Tanganyika, Kivu, Edward, and Albert are similar in their ionic composition and are classified (Kilham and Hecky, 1973) as "sodium-potassium-magnesium-bicarbonate waters." The calcium concentration in the East African lakes is low in comparison to other ions and is lowest in Lake Victoria. Lake Malawi is comparatively rich in calcium and sodium relative to the more northern lakes (Hecky and Kling, 1987). More details about the water chemistry of these lakes can be found in Talling and Talling (1965).

*Lake Tanganyika* The annual phytoplankton cycle of Lake Tanganyika was divided into three periods by Hecky and Kling (1981). Similar phases were observed by Talling (1966) in Lake Victoria and by Hecky and Kling (1987) in the other rift valley lakes. Lakes Tanganyika and Malawi exhibit similar phytoplankton composition and seasonal succession. The highest stability of thermal stratification in Lake Tanganyika occurs in March and April; during this period Chlorophyta and Cyanophyta predominate. In mid-February to late April 1975, the biomass varied between 50–100 mg $m^{-3}$. The onset of mixing (primary mixing period June–August) leads to a shift in the phytoplankton composition: Chrysophytes (*Chrysochromulina* and *Ochromonas*) appear and are followed by diatoms (*Nitzschia* and *Stephanodiscus*). As a consequence, a slight increase of biomass occurs. In October–November 1975, at the end of the mixing period, the phytoplankton biomass increased up to 1500 mg $m^{-3}$ (Hecky and Kling, 1981, 1987). In April-May 1975, in the plume of the Malagara River, *Glenodinium pulvisculus* was dominant. In May, the cyanobacteria were represented by *Anabaena* in the southern part of the lake and *Chroococcus* in the northern one (Hecky and Kling, 1987). Coulter (1963) has shown that in the southern part of the lake in 1961, following a period of local upwelling of nutrient-rich waters, there was a development of diatom peaks. In the northern end in 1958 in September–October, after the main rains, when superficial thermal stratification was redeveloping, *Anabaena flos-aquae* dominated (Talling, 1986). In October–November 1975, a high biomass of diatoms was found in the northern end of the lake, and a low biomass composed of unicellular chlorophytes was recorded in the southern part of the lake (Hecky and Kling, 1987). The same authors observed that diatoms generally predominated in periods of stronger vertical mixing, while cyanobacteria predominated during shallow thermal stratification. The phytoplankton biomass in Lake Tanganyika was minimal (60 mg $m^{-3}$) during the period of stratification and maximal (930 mg $m^{-3}$) at the end of the period of deep mixing. Chlorophyll *a* ranged from 0.1 to 4.5 mg $m^{-3}$ in the photic zone and from 0.2 to 20.4 mg $m^{-3}$ in surface waters (Hecky and Kling, 1981).

*Lake Malawi* The succession of the taxonomic groups in Lake Malawi is similar to that of Lake Tanganyika and also coincides with the thermal cycle. In Lake Malawi, Chrysophyta are rare, *Lyngbya* is the dominant filamentous cyanobacterium, and *Mougeotia* among the Chlorophyta. *Melosira* can be the dominant netplanktonic form during the latter part of the mixing period, as was described by Talling (1969), especially in the shallow, southern end of the lake. The biomass varied from 20 mg $m^{-3}$ in April–May to 30 mg $m^{-3}$ in October–November 1980. During January–March and June–August, the record biomass was 100–200 mg $m^{-3}$. Cyanobacteria and green algae dominated during the stratified period and diatoms during the mixing one (Hecky and Kling, 1987).

*Lakes Kivu and Edward* The phytoplankton of Lakes Kivu and Edward are similar to each other as the phytoplankton of Lakes Tanganyika and Malawi during stratification or at the onset of mixing. Both Lakes Kivu and Edward have Cyanophyta-Chlorophyta association but different dominant species. In Lake Edward, *Microcystis* and *Chroococcus* are common, whereas in Lake Kivu, the nitrogen-fixing, filamentous cyanobacteria *Lyngbya circumcreta* and *Cylindrospermopsis* sp. are common. This difference may be the result of the low concentration of nitrogen in Lake Kivu and the inflow of Lake George water, rich in cyanobacteria, in Lake Edward. The dinoflagellates *Peridinium inconspicuum* and *Gymnodinium pulvisculus* were recorded in the deepest part of the lake. *Cosmarium laeve* was prominent among the Chlorophyta. The biomass in Lake Kivu ranged from 550 to 2100 mg $m^{-3}$. Cyanophyta and Chlorophyta constitute 70–90% of the phytoplankton biomass. A high biomass was also recorded in Lake Edward, 200–300 mg $m^{-3}$. The Cyanophyta and Chlorophyta, which are the dominant groups, formed more than 80% of the phytoplankton biomass in every case (Hecky and Kling, 1987).

*Lake Albert* The phytoplankton biomass in Lake Albert (116–267 mg $m^{-3}$) was much lower than in the other lakes. Diatoms dominated the phytoplankton, representing more than 70% of the total phytoplankton biomass. Chrysophyta and Chlorophyta were subdominant. Species of *Stephanodiscus* and *Nitzschia* dominated the phytoplankton, and *Melosira* species were not observed. The explanation may be the high concentration of soluble P and the low concentration of silica (Hecky and Kling, 1987). The ecology of *Melosira* species in African great lakes has been discussed by Talling (1987) and by P. Kilham, see Chapter 20.

*Lake Victoria* Lake Victoria, like Lakes Kivu and Edward, may have higher rates of nutrient supply than Lakes Tanganyika and Malawi, because, due to their shallow nature, their mixing layers are in closer contact with the sediments. The bottom slope of Lake Victoria is very gentle, and the lake is surrounded by a wide belt of shallow inshore gulfs and bays. The shallow and semi-enclosed areas, like Pilkington and Ekunu Bays and the Kavirondo Gulf, lack any persistent thermal stratification (Talling, 1986). In deeper channels as at the open lake, a seasonal stratification occurs. The stratification period includes three phases: from September to December there is a warming of the superficial layer but isotherms are indistinct; in December-May the thermocline deepens; and from June to August the thermal gradient disappears (Talling, 1966).

The phytoplankton comprises three main groups: Chlorophyta (with a very high diversity), Bacillariophyta, and Cyanophyta. The number of flagellates is very limited; two interesting ones are a form of *Ceratium* that is not taxonomically well defined yet, called *C. brachyceros*, and *G. pulvisculus*, which was also recorded in the other African large lakes. Microflagellates are probably absent from Lake Victoria but they are replaced by very small coccoid cyanobacteria (*Aphanocapsa elachista* and *A. delicatissima*) (Talling, 1966, 1987). Other common algae present in the lake belong to the genera *Microcystis, Anabaena,* and *Anabaenopsis* with the species *A. tanganyikae, Lyngbya,* etc. Among the Chlorophyta, many net-plankters belonging to Chlorococcales (*Pediastrum, Coelastrum, Dictyosphaerium,* etc.) are found, as well as nanoplankton species (*Tetraedron, Chodatella,* etc), and desmids. The desmids are represented by species of *Staurastrum*, especially *S. leptocladum* f. *africanum* and *S. gracile* var. *nyansae*. The presence of *Mougeotia planctonica* is worth noting.

The more important contributors to the phytoplankton biomass are diatoms and cyanobacteria. The common diatoms are species of *Melosira*. *M. nyassensis* var. *victoriae* reaches its annual offshore maximum at the end of the mixing period. This species is also recorded in larger inshore channels and Mwanza Gulf. Other species of *Melosira*—*M. granulata, M. agassizzi,* and *M.*

*ambigua*—are characteristic for shallow inshore areas like Pilkington Bay and Grant Bay. They are accompanied by *Nitzschia acicularis*. The large tropical species *Surirella nyassae* and *Cymotopleura nyassae* are common in the lake. Many diatoms common in temperate lakes are lacking here. In offshore water, seasonality of the phytoplankton is dominated by a cycle involving redistribution and growth of *M. nyassensis* var. *victoriae* during mixing and its sedimentation under stratified conditions (Talling, 1986; Chapter 20). The second increase of biomass is due to *Anabaena* spp., which occurs after restratification and decreases after prolonged stratification due to a depletion of nutrients (Talling, 1986).

Both the composition and the abundance of the phytoplankton differ between the open lake and the shallow bays or channels. *Botryococcus braunii* is abundant in some bays and very scarce in the offshore waters; conversely, large diatoms are more common in the open lake, whereas the species of *Aphanocapsa* are abundant in both biotopes. The concentrations of chlorophyll *a* were higher in various inshore areas than offshore; in Kavirondo Gulf the recorded concentration was 17–21 mg $m^{-3}$ (Talling, 1965; Melack, 1979) in comparison with 1.3–4.9 mg $m^{-3}$ in surface water at an offshore station (Talling, 1965, 1966). Akiyama et al. (1977), in their 1973–1974 survey of Mwanza Gulf, found that the algae were more abundant in the gulf than in the offshore water. The concentration of *Anabaena* spp. was $6 \times 10^8$ cells $l^{-1}$ in the gulf, whereas for the open lake it was $10^6$ cells $l^{-1}$. During this period, a high concentration of nutrients was recorded. The authors suggest that nutrients such as N and P are added to the lake by inflow from the land. More details about the distribution and abundance of *Melosira* in Lake Victoria can be found in Chapter 20.

**Lake Lanao** Lake Lanao (surface area, 357 $km^2$; maximum depth, 112 m; mean depth, 60.3 m) is a low-altitude, tropical lake located on the island of Mindanao (Philippines) at 8°N and 120°E. According to recent dating, the lake has an age of 3.6 to $5.5 \times 10^6$ years (Frey, quoted after Lewis, 1978a). In Lake Lanao, circulation occurs during January and February; the lake is discontinuously stable during March and April and stratified in the rest of the year (Lewis, 1973). The epilimnion is 20–40 m thick. An unusual characteristic of this lake is the occurrence, during the stratified season of secondary and tertiary thermoclines, which split the epilimnion into an upper turbulent layer and a lower stagnant layer and last for periods of up to two months. These thermoclines are dissipated at irregular intervals by storms. This kind of mixing of water layers of different chemical properties without the disappearance of the hypolimnion is called "atelomixis" by Lewis; it allows periodic reinjection of nutrients from the deeper layers to the euphotic zone. As a consequence, the variations in the supply of nutrients in Lake Lanao are more irregular and more numerous than in other lakes, and, consequently, the phytoplankton successional episodes are also more numerous (Lewis, 1978b). Due to the deep mixing, light in Lake Lanao is the limiting factor for phytoplankton growth during the mixing period. Nutrient availability reaches maxima in September–October and in April–July. The succession of algal assemblages is a function of the two above factors (Lewis, 1978b).

The phytoplankton comprise a high number of Chlorophyta followed by Cyanophyta, Euglenophyta, Bacillariophyta, Dinophyta, Cryptophyta, and Chrysophyta (Lewis, 1978a). In terms of biomass, the diatoms are as important as the chlorophytes, followed by Cyanophyta and Cryptophyta. Diatoms appear when the mixed layer is 25–50 m thick, nutrients are available, and turbulence is high. The presence of *M. granulata* and *M. agassizzi* is correlated with high turbulence. Since *M. agassizzi* is heavier than *M. granulata*, it requires more turbulence. Cyanobacteria develop at periods of high light availability, low turbulence, and low levels of nutrients. The Chlorophyta are more abundant during the precirculation period when the radiation level is lower but the frequent atelomixis increases the concentration of nutrients. The cryptomonads generally accompany the diatoms, whereas the dinoflagellates dominate when nutrient depletion is severe and unfavorable for the growth of other algae. The annual average of phytoplankton biomass in Lake Lanao is 23.745 mg $m^{-2}$ wet weight (Lewis, 1978b).

**Lake Titicaca** Lake Titicaca is a high-altitude (3,800 m ASL), tropical lake, located in the altiplano of the Bolivian and Peruvian Andes (Lazzaro, 1981). Lake Titicaca is the only lake among the high, tropical, mountain lakes which was already formed in the Tertiary (Löffler, 1968). Its surface area is 8,372 $km^2$ with a maximum depth of 281 m (Carmouze et al., 1977). The lake is composed of 2 basins: (1) Lago Grande is monomictic (Carney et al., 1987), with a surface area of 6,489 $km^2$ and a mean depth of 134 m; it contains Puno Bay (Bahia de Puno), polymictic, which has a surface area of 552 $km^2$ and a mean depth of 14 m; and (2) Lago Pequeno, polymictic, has a surface area of 1,331 $km^2$ and a mean depth of 10 m; the deep area called Chua has a depth of 42 m. The deep part of Lago Pequeno is monomictic (Richerson et al., 1986). Lake Titicaca is characterized by a constant high tropical irradiance and a deep euphotic zone. The lake behaves physically like a low-latitude, stratified, tropical lake, despite its high elevation and lower temperature (Richerson et al., 1986). Lago Grande is stratified from October to June (temperature reaching ~14°C) (Carney et al., 1987). The lake is relatively high in solutes (TDS, 850 ppm) and chlorine 260 ppm, and has a pH of 8.6 (Richerson et al., 1977). Short- and long-term bioassays indicate that nitrogen is the principal limiting nutrient and that nitrogen ($N_2$) fixation is important for algal growth (Vincent et al., 1984, 1986; Wurtsbaugh et al., 1985; Carney et al., 1987). The concentration of nutrients in Lago Pequeno is lower than in the main basin. The most complete lists of algae are given by Lazzaro (1981) for Lago Pequeno, and by Richerson et al. (1977) and Carney et al. (1987) for the whole lake. The planktonic flora of Lake Titicaca during 1973 and 1981–82 was dominated by Chlorophyta, cyanobacteria, and diatoms. Cryptomonads were only very occasionally abundant, and dinoflagellates were common only rarely (Richerson et al.,

1986). The most abundant phytoplankton genera in L. Titicaca are: *Anabaena, Pediastrum, Coelastrum, Oocystis, Mougeotia, Closterium, Staurastrum,* and *Cyclotella*. The diatoms dominate during the period of deepest mixing, followed by Chlorophyta, cyanobacteria, and, finally, a dinoflagellate. *Anabaena spiroides* developed during the stratification period and *Cyclotella* sp. developed during the deepening of the thermocline and mixing (Richerson et al., 1986). The newly described species of *Cyclotella andina* Theriot, *Gloeotilopsis planctonica,* and *Planctonema lauterbornii* (Ulotrichale) contribute ca. 50% of the cumulative biomass. The mean concentration of biomass during 1981–82 was 511 mg $m^{-3}$, with 122 mg $m^{-3}$ minimum and 1,310 mg $m^{-3}$ maximum. The epilimnetic chlorophyll *a* concentration varies from 0.6 to 5.9 $\mu g\ l^{-1}$ (mean, 2.6 $\mu g\ l^{-1}$) (Vincent et al., 1984).

Bahia de Puno and Lago Grande present different patterns of chlorophyll *a* production and planktonic nutrient demand (Vincent et al., 1984). The two basins behave as separate entities (Vincent et al., 1986). In Bahia de Puno, green algae and/or diatoms were dominant during 1981 and 1982. Large populations of $N_2$-fixing cyanobacteria (*Nodularia harveyana, Anabaena spiroides,* and *A. affinis*) occurred during periods of high measured rates of $N_2$ fixation (November 1981 to February 1982 and October–November 1982). The period of cyanobacteria abundance coincided with a period of water column stability. $N_2$ fixation occurs in the warm season in both Bahia de Puno and Lago Grande, but their timing and magnitudes differ markedly from year to year between the two sites. From February, a shift in species composition occurred towards diatoms and green algae, and $N_2$ fixation remained detectable through May. From May to July 1982, chlorophyll *a* concentration decreased from an epilimnetic average of 4 mg $m^{-3}$ in May to 0.9 mg $m^{-3}$ at the end of June.

The phytoplankton assemblages in Puno Bay differ quantitatively, but not qualitatively from that in Lago Grande. Vincent et al. (1986) have found that mixing in Puno Bay leads to an increase in abundance of Cyanophyta and an increase in rates of $N_2$ fixation and C uptake. Tilzer (1987) discusses the problem of algal biomass rise during moderate mixing. The phytoplankton photosynthetic activity in the polymictic bay and the monomictic Lago Grande responds differently to the same environmental events. This is probably due to fact that the difference in the depth of the water column plays a more important role than temperature and light do (Vincent et al., 1986). The composition of the phytoplankton in Lago Pequeno (sub-basin) is also similar to that of Lago Grande but differs in the pattern of seasonal succession and abundance (Lazzaro, 1981; Richerson et al., 1986; Carney et al., 1987). The phytoplankton biomass varies between 0.4–1.2 g $m^{-3}$ and is composed of 60–90% chlorococcales (*Monoraphidium*) which dominate in spring and summer, followed by Zygnematales (*Mougeotia*) in fall and winter. The biomass of *Peridinium* sp. may reach 0.3 g $m^{-3}$ (especially in the deepest part of the sub-basin in spring), that of diatoms $<0.1$ g $m^{-3}$, and that of cyanobacteria is insignificant. The concentration of chlorophyll rarely exceeds 5 mg Chl $m^{-3}$ (Lazzaro, 1981).

The author classified Lago Pequeno as an oligotrophic lake. Carney et al. (1987) consider Lake Titicaca as meso-oligotrophic in terms of biomass, but eutrophic as far as primary production is concerned. The high primary production relative to biomass may be correlated with the constantly high tropical irradiance and the deep euphotic zone (20–28 m). Lake Titicaca is an old, hydrologically isolated lake. Nevertheless, the phytoplankton are composed of cosmopolitan species without any endemism.

**Lake Kariba** Lake Kariba is a man-made lake that formed after construction of a dam in 1960 along the Zambezi River, located at 17°S, 28°E between Zambia and Zimbabwe (Africa). Lake Kariba has a surface area of 5250 $km^2$; maximum depth, 120 m, and mean depth, 19.1 m (Balon and Coche, 1974). The Zambezi River provides 78% of the inflowing water, which is clear and poor in nutrients. The other tributaries are rich in nutrients and very turbid. Their inflow is high during the hot wet season (December–February), but low during the cool dry season (April–August), and ceases completely during the hot dry season (September–November). Due to the high variation of inflow and evaporation, the variation of the lake level over a year can be as much as 2–3 m (Ramberg, 1987). The lake consists of five hydrologically distinct basins. The northernmost part of the lake, Sanyati Basin, is fed by the Sanyati River, which supplies nutrient-rich water originating from agricultural land. The lake is mixed in July and August and stratified from September to June. The deepening of the thermocline starts in March. The concentrations of nutrients, especially of P (20 $\mu g\ l^{-1}\ PO_4$–P), were very high in the early phase of the lake; at present it is three to five times lower. The rainy season lasts from mid-November to mid-March—this is the period of high lake turbidity, due to the large amount of silt transported by the river, and, as a consequence, a decrease in the size of the euphotic zone occurs.

The Cyanophyta contribute more than 60% to the total phytoplankton biomass (Ramberg, 1984); two species are dominant, belonging to the genera *Cylindrospermopsis* and *Anabaena* (both $N_2$-fixing species), accompanied by *Lyngbya, Microcystis,* and *Aphanothece* species. The Bacillariophyta are represented by *Synedra ulna, S. acus,* and *Cyclotella*. The most important Dinophyta are *Peridiniopsis elpatiewskyi, P. cunningtonii, Peridinium gatunense, Ceratium hirundinella,* and *Gymnodinium*. *Cryptomonas* and *Rhodomonas* species (Cryptophyta) were more abundant in May–June. *Chrysochromulina parva* (Chrysophyceae) dominate in November. Although the Chlorophyta are represented by a large number of species, they contribute insignificantly to the total biomass. The mean phytoplankton biomass measured by Ramberg (1984) was 290 mg $m^{-3}$. The wide variations in the phytoplankton biomass during the rainy season may be due to the high turbidity of the inflowing water; perhaps, after the settling of the silt, the nutrient-rich water allows development of the algae.

Phytoplankton seasonal succession and biomass fluctuations were studied by Ramberg (1987) in the Sanyati Basin, which is the richest in nutrients among the five hydrologically distinct basins of Lake Kariba. The phytoplankton exhibit two peaks during the year. The first maximum occurs from mid-November to mid-March (730 mg $m^{-3}$ in January 1983 and 1380 mg $m^{-3}$ in January 1984). Cyanophyta contributed 70–90% to the biomass in both cases. The dominant species was *Cylindrospermopsis raciborskii,* followed by $N_2$-fixing species of *Lyngbya* and *Anabaena*. The Dinophyta were subdominant. They were abundant at the beginning of the circulation period in June when the thermocline deepened and disappeared completely during the mixing period, July–August. The second maximum occurred during the mixing period in July 1983 with a biomass of 460 mg $m^{-3}$ composed mainly of diatoms. They started increasing at the end of June, when species of *Synedra acus* and *Melosira granulata* dominated, until the end of July, when *Synedra* species, especially *S. ulna,* became common at the highest turbulence. These diatoms disappear in mid-September when thermal stratification was established; during the rest of the year, species of *Cyclotella* were common. They disappeared when mixing increased. Some species of Chlorophyta, such as *Scenedesmus, Tetraedron, Coelastrum,* and *Monoraphidium,* are present throughout the year. The mean annual phytoplankton biomass in Lake Kariba is 310 mg $m^{-3}$ wet weight, and the chlorophyll *a* content is 2.2 mg $m^{-3}$.

**The Shallow, Large Lakes of China** In China, 28 large lakes (>500 $km^2$) exist (Chang, 1987). The eight lakes located in the Pacific drainage basin are freshwater lakes, whereas the large lakes situated in the Tibet-Qinghai plateau are saline, being at the end of endorheic rivers (with no outlets). The lakes situated in the Pacific drainage basin were formed by the action of rivers. They are shallow and eutrophic. Most of the lakes in this region are polymictic and rich in nutrients, and water blooms of cyanobacteria develop, especially in summer. The lakes have a slow flushing rate. The blooms are due to *Microcystis* and *Anabaena* spp.; these species represent 80–90% of the population. The highest number of organisms were found in Lake Hongtze. Diatoms are present throughout the year. The composition and abundance of the diatoms from sheltered bays are different from those in open waters. They generally represent 10–35% of the total population. Chlorophyta are more abundant in spring and decrease in summer (10–35% of the population). Pyrrhophyta are common only in small lakes.

The largest lake situated in the Pacific drainage basin is Lake Dongting. It receives water from tributaries of the Yangtze River and also drains into it. The lake level is high during the monsoonal rain in June, July, and August, and it decreases during the dry period. The annual fluctuation in the lake level is 13.6 m, and the surface lake area varies from 2,740 to 12,000 $km^2$; the mean depth is 6.7 m and the maximum depth is 31 m. Lake Dongting is used as a storage basin for river flood waters, and during the dry period, it provides water to the river. For this reason, no water bloom develops, in spite of high nutrient loading from the catchment area. *Ceratium* is the only dinoflagellate alga found in high numbers in Lake Dongting.

Some Chinese large lakes are artificially enriched in order to increase the dominance of cyanobacteria. The cyanobacteria are then used as fertilizer to enrich the nitrogen content of the soil for agricultural purposes.

## 18.5 Conclusion–the Effects of Latitude

Besides environmental conditions and lake age, geographical latitude has apparent effects on the dominance of certain algal groups. In the large lakes of high latitude, diatoms and chrysophytes are dominant. Cyanobacteria are absent in lakes that are too cold and too low in nutrients. The phytoplankton of the lakes located in the temperate zones is characterized by the dominance of diatoms, especially in winter and fall, and by the development of Chlorophyta and Cyanophyta in summer. Diatoms during summer stratification frequently are excluded by silica limitation, in particular when high availability of other nutrients (phosphorus and nitrogen) allows the build-up of large diatom populations which lead to the depletion of silica within the euphotic zone. In large lakes of the warm belt, the dominant role is taken over by Chlorophyta and Cyanophyta and is shared in some lakes with diatoms.

In lakes which have been affected by eutrophication, such as Lake Ladoga and the shallow lakes Mälaren and Lough Neagh (located at high latitude and industrial zones), the Cyanophyta have become very abundant, to the point of forming water blooms. However, cyanophyte abundance in these lakes is restricted to periods of thermal stratification and hence of low water turbulence and elevated temperatures. Cyanophytes can be abundant in the relatively shallow large lakes—such as Lakes Erie, Kivi, Edward, Victoria, and the man-made Lake Kariba—but, massive blooms at least are virtually absent in deep large lakes. Even in the low-latitude lakes, cyanophytes appear to be restricted to periods of stable stratification. However, nitrogen-fixing species are common in situations where nutrient concentrations (in this case, nitrogen) are low. A water bloom of *Microcystis* occurs in the very shallow freshwater Lake Hongtze (China). In shallow lakes, cyanophyte populations on the lake bottom can easily reach upper layers with more favorable light conditions whenever surface heating and calm weather prevents mixing of the water column. In some cases, the abundance of cyanophytes is the result of a succession of different species, which appears to be correlated with variations in nutrient concentrations (see Section 18.2). The degree of adaptability of the species to the unstable conditions in highly turbulent water may also be of importance.

Water blooms of cyanophytes and diatoms have been recorded in protected bays, channels, and gulfs, as well as inshore and estuary regions. The absence of true water blooms of cyanobacteria and dinoflagellates in the pelagial of large lakes may be due to one of the prerequisites for their development: minimal turbulence. Taking into account the fact that the same wind intensity leads to greater turbulence per unit area in large lakes than in small ones (see Chapter 36), calm periods are probably shorter in large lakes than in small water bodies. Conversely, high turbulence favors the development and maintenance of *Melosira* spp. and other heavy and large diatoms in the water column. Even in low-latitude lakes, large and heavy diatoms, and chrysophytes dominate during the mixing period.

The only large lake with a heavy water bloom due to the chrysophyte *Uroglena americana* is the ancient lake, Lake Biwa. This bloom also develops as a consequence of nutrient increase. It first appears in the shallowest part of the lake, and the intensity and direction of the wind appears to contribute to its spreading over the whole lake.

The chrysophytes, which play an important role in lakes of high and temperate latitudes are only represented by a few different species in low-latitude large lakes (*Ochromonas, Chromulina, Bicoeca*) (Hecky and Kling, 1987).

Most of the planktonic algae are cosmopolitan forms. Even the taxonomic feature called the "cold stenotherm form" has been losing its meaning since the study of lakes from the warm belt. The flagellates *Rhodomonas minuta* and *Erkenia subaequiciliata* syn. *Chrysochromulina parva*, described by Skuja (1948) as cold stenotherm forms, were recorded in Lakes Tanganyika, Lanao, Kinneret, etc. (Serruya and Pollingher, 1983).

Endemic planktonic algae are mostly found in geologically ancient lakes. An endemic chrysophyte (*Chrysosphaerella baicalensis* Popovsk) and a cyanophyte (*Synecocystis limnetica*) were described in Lake Baikal. Endemic species of Chlorophyta were described in Lake Biwa (*Pediastrum biwae* Negro and *Staurastrum biwaensis* Hirano) and in Lake Ohrid (*Oocystis rhomboidea* Fott, *Didimogenes dubia* Fott, and *Dispora wilhelmi* Fott). Endemic species were not described in the plankton of Lake Titicaca, but the two important ulotricaleans, *Gloeotilopsis planctonica* Iyengar & Philip and *Planctonema lauterbornii* Schmidle, are not commonly found in other lakes (Carney et al., 1987). A new diatom species, *Cyclotella andina* Theriot, was found in the plankton of Lake Titicaca. Endemic planktonic diatoms were found in Lake Baikal—*Melosira baicalensis* (Meyer) Wisl., *Cyclotella baicalensis* (Meyer) Skv., *C. minuta* (Skv.) Antip., and *Stephanodiscus skabitschevsky* Popovsk—and in Lake Ohrid—*Cyclotella fottii* Hust and *C. hustedti* Jurilj. In Lake Biwa, rare species such as *Melosira solida* and *Stephanodiscus carconensis* were reported.

The dinoflagellates are represented in large lakes at temperate latitudes in winter by *Gymnodinium* species, and in summer by *Ceratium hirundinella* and species of *Peridinium, Peridiniopsis,* and *Glenodinium*. In Lake Biwa, an endemic species of *Ceratium* is found in summer. *Ceratium brachyceros* v. Daday, present in Lake Victoria, is considered a tropical form. In large low-latitude lakes, *Peridinium* spp. are more common than *Ceratium*. Endemic dinoflagellates were described in Lake Baikal—*Gymnodinium coeruleum* Antip and *G. baicalense* Antip in winter; *Peridinium baicalense* Kiss & Zwetkov in spring and fall. In Lake Ohrid, an endemic nonmotile dinoflagellate (*Cystodinium dominii* Fott) was described.

Tropical forms of cyanophytes (*Anabaenopsis raciborskyii* and *A. tanganyikae*) and diatoms (*M. agassizi, M. ikapoensis, M. nyassensis* var. *victoriae, Nitzschia nyassensis, Cymatopleura nyansae,* and *Surirella nyassae*) are found in the rift valley lakes of Central Africa. Some of them are considered to be endemic species.

The latitudinal differences in the composition of the phytoplankton are less obvious in small water bodies than in these large lakes, probably because they are more influenced by processes occurring in the watershed and usually have shorter water retention times than large lakes.

---

*Acknowledgements* Thanks are due to Dr. U. Sommer for his contribution to the description of Lake Constance phytoplankton; to Professor C. Serruya for reading the early version of the manuscript and for helpful suggestions and comments; to Professors M. Tilzer and C. Serruya for the review and editing of the manuscript; and to K. Diskin for typing the manuscript.

## References

Akiyama, T., Kajumulo, A.A., and Olsen, S. 1977. Seasonal variations of plankton and physicochemical conditions in Mwanza Gulf, Lake Victoria. *Bulletin of the Freshwater Fisheries Research Laboratory* 27:49–60.

Allen, H.L. and Ocevski, B.T. 1976. Limnological studies in a large, deep oligotrophic lake (Lake Ohrid, Yugoslavia). *Archiv für Hydrobiologie* 77:1–21.

Balon, E.K. and Coche, A.G. 1974. *Lake Kariba, A Man Made Tropical Ecosystem in Central Africa; Monographiae Biologicae* No. 24. Junk, The Hague, 767 pp.

Bartone, C.R. and Schelske, C.L. 1982. Lake-wide seasonal changes in limnological conditions in Lake Michigan in 1976. *Journal of Great Lakes Research* 8:413–427.

Carmouze, Y.P., Arce, C., and Quintanilla, J. 1977. La regulation hydrique des Lacs Titicaca et Poopo. *Cahiers ORSTOM (Serie Hydrobiologie)* 11:269–283.

Carney, H., Richardson, P.J., and Eloranta, P. 1987. Lake Titicaca (Peru/Bolivia) phytoplankton: species composition and structural comparison with other tropical and temperate lakes. *Archiv für Hydrobiologie* 110:365–385.

Chang, W.Y.B. 1987. Large lakes of China. *Journal of Great Lakes Research* 13:235–249.

Chang, W.Y.B. and Rossmann, R. 1988. Changes in the abundance of blue-green algae related to nutrient loadings in the nearshore of Lake Michigan. *Hydrobiologia* 157:271–278.

Coulter, G.W. 1963. Hydrological changes in relation to biological production in southern Lake Tanganyika. *Limnology and Oceanography* 8:463–477.

Duthie, H.C. and Hart, C.J. 1987. The phytoplankton of the subarctic Canadian Great Lakes. *Archiv für Hydrobiologie Beiheft Ergebnisse der Limnologie* 25:1–9.

Fahnenstiel, G.L. and Scavia, D. 1987. Dynamics of Lake Michigan phytoplankton: recent changes in surface and deep communities. *Canadian Journal of Fisheries and Aquatic Sciences* 44:509–514.

Gibson, C.E., Wood, R.B., Dickson, E.L., and Jewson, D.H. 1971. The succession of phytoplankton in Lough Neagh 1968–70. *Mitteilungen Internationale Vereinigung für Theoretische und Angewandte Limnologie* 19:146–160.

Gray, J.M. 1987. Differences between nearshore and offshore phytoplankton communities in Lake Ontario. *Canadian Journal of Fisheries and Aquatic Sciences* 44:2155–2163.

Hecky, R.E. and Kling, H.J. 1981. The phytoplankton and protozooplankton of the euphotic zone of Lake Tanganyika: species composition, biomass, chlorophyll content, and spatio-temporal distribution. *Limnology and Oceanography* 26:548–564.
Hecky, R.E. and Kling, H.J. 1987. Phytoplankton ecology of the Great Lakes in the rift valleys of Central Africa. *Archiv für Hydrobiologie Beiheft Ergebnisse der Limnologie* 25:197–228.
Horie, S. 1961. Paleolimnological problems of Lake Biwa-ko. *Memoirs of the College of Science, University of Kyoto, Series B (Biology)* 28:53–71.
Ichise, S. and Wakabayashi, T. 1982. The seasonal succession of plankton in the southern part of Lake Biwa. *Report Shiga Prefectural Institute of Public Health and Environmental Science* 17:78–94.
Ichise, S. and Wakabayashi, T. 1984. The distribution of red tide appeared in Lake Biwa, 1980–1983. *Report Shiga Prefectural Institute of Public Health and Environmental Science* 19:56–65.
Jewson, D.H. 1976. The interaction of components controlling net phytoplankton photosynthesis in a well-mixed lake (Lough Neagh, Northern Ireland). *Freshwater Biology* 6:551–576.
Jewson, D.H. 1977. Light penetration in relation to phytoplankton content of the euphotic zone in Lough Neagh, N. Ireland. *Oikos* 28:74–83.
Johnson, L. 1975. Physical and chemical characteristics of Great Bear Lake, Northwest Territories. *Journal of the Fisheries Research Board of Canada* 32:1971–1987.
Kilham, P. and Hecky, R.E. 1973. Fluoride: geochemical and ecological significance in East African waters and sediments. *Limnology and Oceanography* 18:932–945.
Kimura, B. and Ishida, Y. 1985. Photophagotrophy in *Uroglena americana*, Chrysophyceae. *Japanese Journal of Limnology* 46:315–318.
Kozhov, M. 1963. *Lake Baikal and Its Life: Monographiae Biologicae* No. 11. Junk, The Hague, 344 pp.
Kozova, O.M. 1987. Phytoplankton of Lake Baikal: structural and functional characteristics. *Archiv für Hydrobiologie Beihefte Ergebnisse der Limnologie* 25:19–37.
Lampert, W., and Schober, U. 1978. Das regelmässige Auftreten von Frühjahrsmaximum und "Klarwasserstadium" im Bodensee als Folge klimatischer Bedingungen und Wechselwirkungen zwischen Phyto- und Zooplankton. *Archiv für Hydrobiologie 82:364–386.*
Lazzaro, X. 1981. Biomasses, peuplements phytoplanctoniques et production primaire du lac Titicaca. *Revue Hydrobiologique Tropicale* 14:349–380.
Lewis, W.M., Jr. 1973. The thermal regime of Lake Lanao (Philippines) and its theoretical implications. *Limnology and Oceanography* 18:200–217.
Lewis, W.M., Jr. 1978a. A compositional, phytogeographical and elementary structural analysis of the phytoplankton in a tropical lake: Lake Lanao, Philippines. *Journal of Ecology* 66:213–226.
Lewis, W.M., Jr. 1978b. Dynamics and succession of the phytoplankton in a tropical lake: Lake Lanao, Philippines. *Journal of Ecology* 66:849–880.
Lewis, W.M., Jr. 1987. Tropical limnology. *Annual Review Ecological Systems* 18:159–184.
Löffler, H. 1968. Geology of the mountainous regions of the tropical Americas. *Colloquium Geographicum* 9:57–76.
Matheson, D.H. and Munawar, M. 1978. Lake Superior basin and its development. *Journal of Great Lakes Research* 4:249–263.
Mizushima, K., Ichise, S., Yagi, O., and Sudo, R. 1983. Aspects of fresh water red tide production in Lake Biwa and nutritional conditions of its lake water. *Research Data, National Institute for Environmental Studies, Japan* 24:153–163.
Moore, J.W. 1980a. Seasonal distribution of phytoplankton in Yellow Knife Bay, Great Slave Lake. *Internationale Revue der Gesamten Hydrobiologie* 65:283–293.
Moore, J.W. 1980b. Attached and planktonic algal communities in some inshore areas of Great Bear Lake. *Canadian Journal of Botany* 58:2294–2308.

Mori, S., and Horie, S. 1984. Algae. In: Horie, S. (editor), *Lake Biwa: Monographiae Biologicae* No. 54. Junk, Dordrecht, p. 291–301.
Mortimer, C.H. 1974. Lake hydrodynamics. *Mitteilungen der Internationalen Vereinigung für Theoretische und Angewandte Limnologie* 20:124–197.
Moskalenko, B.K. 1972. Biological productive system of Lake Baikal. *Verhandlungen der Internationalen Vereinigung für Theoretische und Angewandte Limnologie* 18:568–573.
Munawar, M. and Munawar, J.F. 1975. Some observations on the growth of diatoms in Lake Ontario with emphasis on *Melosira binderana* Kutz during thermal bar conditions. *Archiv für Hydrobiologie* 75:490–499.
Munawar, M. and Munawar, J.F. 1978. Phytoplankton of Lake Superior 1973. *Journal of Great Lakes Research* 4:415–442.
Munawar, M. and Munawar, J.F. 1986. The seasonality of phytoplankton in the North American Great Lakes, a comparative synthesis. *Hydrobiologia* 138:85–115.
Munawar, M., Munawar, J.F., and McCarthy, L.H. 1987. Phytoplankton ecology of large eutrophic and oligotrophic lakes of North America: Lakes Ontario and Superior. *Archiv für Hydrobiologie Beihefte Ergebnisse der Limnologie* 25:51–96.
Nagata, T. 1986. The seasonal abundance and vertical distribution of the $<3\mu m$ phytoplankton in the north basin of Lake Biva. *Ecological Research* 1:207–221.
Nakanishi, M. 1984. Phytoplankton. In: Horie, S. (editor), *Lake Biwa: Monographiae Biologicae* 54. Junk, Dordrecht, p. 281–290.
Nalewajko, C. and Voltolina, D. 1986. Effects of environmental variables on growth rates and physiological characteristics of Lake Superior phytoplankton. *Canadian Journal of Fisheries and Aquatic Sciences* 43:1163–1170.
Negoro, K. 1956. The phytoplankton of the main basin of Lake Biwa-ko. *Japanese Journal of Limnology* 18:37–46.
Nomura, K. 1981. Water quality and sediments in Lake Biwa. *Water* 23:93–95.
Parker, J.I., Conway, H.L. and Yaguchi, E.M. 1977. Seasonal periodicity of diatoms and silicon limitation in offshore Lake Michigan, 1975. *Journal of the Fisheries Research Board of Canada* 34:552–558.
Petrova, N.A. 1987. The phytoplankton of Ladoga and Onega lakes and its recent successional changes. *Archiv für Hydrobiologie Beihefte Ergebnisse der Limnologie* 25:11–18.
Ramberg, L. 1984. Phytoplankton gradients during the rainy season in Lake Kariba. *Verhandlungen der Internationalen Vereinigung für Theoretische und Angewandte Limnologie* 22:1590–1593.
Ramberg, L. 1987. Phytoplankton succession in the Sanyati basin, Lake Kariba. *Hydrobiologia* 153:193–202.
Richerson, P.J., Widmer, C., and Kittel, T. 1977. The limnology of Lake Titicaca (Peru-Bolivia), a large, high altitude tropical lake. *Institute of Ecology Publication, University of California, Davis* No. 14, 78 pp.
Richerson, P.J., Neale, P.J., Wurtsbaugh, W., Alfaro, R., and Vincent, W. 1986. Pattern of temporal variation in Lake Titicaca, a high altitude tropical lake. I. Background, physical and chemical processes and primary production. *Hydrobiologia* 138:205–220.
Robertson, A. and Scavia, D. 1984. North America Great Lakes. In: Taub, F.B. (editor), *Ecosystems of the World No. 23: Lakes and Reservoirs*. Elsevier, Amsterdam, p. 135–176.
Round, F.E. 1972. *Stephanodiscus binderanus* (Kütz) Krieger or *Melosira binderana* Kütz (Bacillariophyta, Centrales). *Phycologia* 11:109–117.
Scavia, D. and Fahnenstiel, G.L. 1987. Dynamics of Lake Michigan phytoplankton: mechanisms controlling epilimnetic communities. *Journal of Great Lakes Research* 13:103–120.
Scavia, D., Fahnenstiel, G.L., Evans, M.S., Jude, D.J., and Lehman, J.T. 1986. Influence

of salmonid predation and weather on long-term water quality trends in Lake Michigan. *Canadian Journal of Fisheries and Aquatic Sciences* 43:435–443.

Schelske, C.L. and Stoermer, E.F. 1971. Eutrophication, silica depletion and predicted changes in algal quality in Lake Michigan. *Science* 173:423–424.

Serruya, C., and Pollingher, U. 1983. *Lakes of the Warm Belt.* Cambridge University Press, New York, 569 pp.

Skuja, H. 1948. Taxonomie des Phytoplankton einiger Seen in Uppland, Schweden. *Symbolae Botanicae Upsalienses.* Vol. 9(3), Uppsala, 400 pp.

Sommer, U. 1981. The role of r- and K- selection in the succession of phytoplankton in Lake Constance. *Acta Oecologiae/Oecologia Generalis* 2:327–342.

Sommer, U. 1983. Nutrient competition between phytoplankton species in multispecies chemostat cultures. *Archiv für Hydrobiologie* 96:399–416.

Sommer, U. 1987. Factors controlling the seasonal variation in phytoplankton species composition—A case study for a deep nutrient-rich lake. In: F.E. Round and D.J. Chapman (eds.), *Progress in Phycological Research* 5:123–178. Biopress Ltd., Bristol.

Sommer, U. (ed.). 1989. *Plankton Ecology: Succession in Plankton Communities.* Science Tech Publishers, Madison, WI, 384 pp.

Sommer, U., Gliwicz, Z.M., Lampert, W., and Duncan, A. 1986. The PEG model of seasonal succession of planktonic events in freshwaters. *Archiv für Hydrobiologie* 106:433–471.

Sommer, U., and Kilham, S.S. 1985. Phytoplankton natural community experiments. A reinterpretation. *Limnology and Oceanography* 30:436–440.

Sommer, U. and Stabel, H.-H. 1983. Silicon consumption and population density changes of dominant planktonic diatoms in Lake Constance. *Journal of Ecology* 73:119–130.

Stabel, H.H. 1984. Impact of sedimentation on the phosphorus content of the euphotic zone of Lake Constance. *Verhandlungen der Internationalen Vereinigung für Theoretische und Angewandte Limnologie* 22:964–969.

Stankovic, S. 1960. *The Balkan Lake Ohrid and Its Living World: Monographiae Biologicae* No. 9. Junk, The Hague, 357 pp.

Talling, J.F. 1965. The photosynthetic activity of phytoplankton in East African lakes. *Internationale Revue der Gesamten Hydrobiologie* 50:1–32.

Talling, J.F. 1966. The annual cycle of stratification and phytoplankton growth in Lake Victoria (East Africa). *Internationale Revue der Gesamten Hydrobiologie* 51:545–621.

Talling, J.F. 1969. The incidence vertical mixing and some biological and chemical consequences in tropical African lakes. *Verhandlungen der Internationalen Vereinigung für Theoretische und Angewandte Limnologie* 17:998–1012.

Talling, J.F. 1986. The seasonality of phytoplankton in African lakes. *Hydrobiologia* 138:139–160.

Talling, J.F. 1987. The phytoplankton of Lake Victoria (East Africa). *Archiv für Hydrobiologie Beihefte Ergebnisse der Limnologie* 25:229–256.

Talling, J.F. and Talling, I.B. 1965. The chemical composition of African lake waters. *Internationale Revue der Gesamten Hydrobiologie* 50:421–463.

Tarapchak, S.J. and Stoermer, E.F. 1976. Phytoplankton of Lake Michigan. In: *Environmental Status of the Lake Michigan Region, Vol. 4.* ANL/ES-40. Argonne National Laboratory, Argonne, Illinois, p 1–211.

Tilzer, M.M. 1987. Light dependence of photosynthesis and growth in Cyanobacteria: implications for their dominance in eutrophic lakes. *New Zealand Journal of Marine and Freshwater Research* 21:401–412.

Vincent, W.F. 1983. Phytoplankton production and winter mixing: contrasting effects in two oligotrophic lakes. *Journal of Ecology* 71:1–20.

Vincent, W.F., Wurtsbaugh, W., Vincent, C.L., and Richerson, P.J. 1984. Seasonal dynamics of nutrient limitation in a tropical high altitude lake (Lake Titicaca, Peru-

Bolivia): application of physiological bioassays. *Limnology and Oceanography* 29:540–552.
Vincent, W.F., Wurtsbaugh, W., Neale, P.J., and Richerson, P.J. 1986. Polymixis and algal production in a tropical lake: latitudinal effects on the seasonality of photosynthesis. *Freshwater Biology* 16:781–803.
Vollenweider, R.A., Munawar, M., and Stadelman, P. 1974. A comparative review of phytoplankton and primary production in the Laurentian Great Lakes. *Journal of the Fisheries Research Board of Canada* 31:739–761.
Wakabayashi, T. and Ichise, S. 1982. The seasonal succession of plankton in the northern part of Lake Biwa in 1981. *Report Shiga Prefectural Institute of Public Health and Environmental Science* 17:95–103.
White, E. and Payne, G.W. 1977. Chlorophyll production in response to nutrient additions by the algae in Lake Taupo water. *New Zealand Journal of Marine and Freshwater Research* 11:501–507.
White, E., Downes, M., Gibbs, M., Kemp, L., Mackenzie, L., and Payne, G. 1980. Aspects of the physics, chemistry and phytoplankton biology of Lake Taupo. *New Zealand Journal of Marine and Freshwater Research* 14:139–148.
Willen, E. 1984. The large lakes of Sweden—Vänern, Vättern, Mälaren and Hjälmaren. In: Taub, F.B. (editor), *Ecosystems of the World No. 23: Lakes and Reservoirs*. Elsevier, Amsterdam, p. 107–134.
Willen, E. 1987. Phytoplankton and reversed eutrophication in Lake Mälaren, central Sweden, 1965–1983. *British Phycological Journal* 22:193–208.
Wood, R.B. and Gibson, C.E. 1973. Eutrophication and Lough Neagh. *Water Research* 7:173–187.
Wurtsbaugh, W.A., Vincent, W.F., Tapia, R.A., Vincent, C.L., and Richerson, P.J. 1985. Nutrient limitation of algal growth and nitrogen fixation in a tropical Alpine lake, Lake Titicaca (Peru/Bolivia). *Freshwater Biology* 15:185–195.

# 19

# Relationship of Phytoplankton and Nutrients to Stoichiometric Measures

Susan Soltau Kilham

ABSTRACT Elemental ratios can be very useful tools for characterizing phytoplankton communities in lakes. Appropriate measures of the carbon, nitrogen, silicon, and phosphorus in the various particulate fractions have generally not been collected in a systematic fashion in the past. Resource ratios and turnover rates have been shown to be important factors for controlling species composition and community structure of phytoplankton. This observation provides a mechanistic basis for the use of elemental ratios. Available nutrient data from large lakes are used to demonstrate the potential of a stoichiometric approach for understanding plankton community structure and for monitoring effects of changes in loadings of nutrients to lakes. Each of the Laurentian Great Lakes appears to have a unique resource ratio signature.

## 19.1 The Role of Elemental Ratios

The potential usefulness of elemental ratios for understanding biological processes in large lakes has been largely overlooked. For the Laurentian Great Lakes there are surprisingly few appropriate stoichiometric measurements of the major nutrient elements: carbon, nitrogen, silicon, and phosphorus (Harris, 1986). Particulate carbon (POC) and particulate silicon (PSI) are seldom measured routinely. Particulate phosphorus is also seldom measured, although total phosphorus (TP) is probably a reasonable approximation during the growing season. Particulate nitrogen fractions are sometimes measured by the Kjeldahl procedure, but, because inorganic N concentrations are usually high and more easily measured, these data are also often lacking. The reasons behind this paucity of elemental particulate data are obscure, but may be the result of assumed detrital interferences. Harris (1986) discussed this problem, but dismissed it after finding that the detrital fraction had essentially the same ratios as the living particulates. In large, deep lakes, however, the potential for detrital interferences are probably much reduced, especially in offshore regions.

I present here the argument that the elemental stoichiometry of phytoplankton particulates is a reflection of phytoplankton community structure and the processes that underlie the composition of these communities, namely resource supply ratios and turnover rates (Tilman et al., 1982, 1984, 1986; Tilman, 1982; Kilham, 1986, 1988; Sommer, 1983, 1985, 1986). Loading ratios are reflected in the elemental composition of the phytoplankton pool. Healey and Hendzel (1980) found an excellent relationship between the N:P ratios in the phytoplankton and the N:P loading ratios to the lakes. Different loading ratios lead to different phytoplankton communities because each species has unique resource requirements (Kilham and Kilham, 1984). In controlled, natural community experiments on Lake Michigan phytoplankton, I found that communities dominated by small green unicells had C:N:P ratios that averaged 40:7:1 (note: all ratios are atomic ratios), while communities dominated by araphid diatoms had C:N:P ratios that averaged 318:40:1 (Kilham, 1986).

The degree of resource limitation affects the cellular stoichiometry (Goldman, 1980). Suttle and Harrison (1986) measured the C:P and N:P ratios of

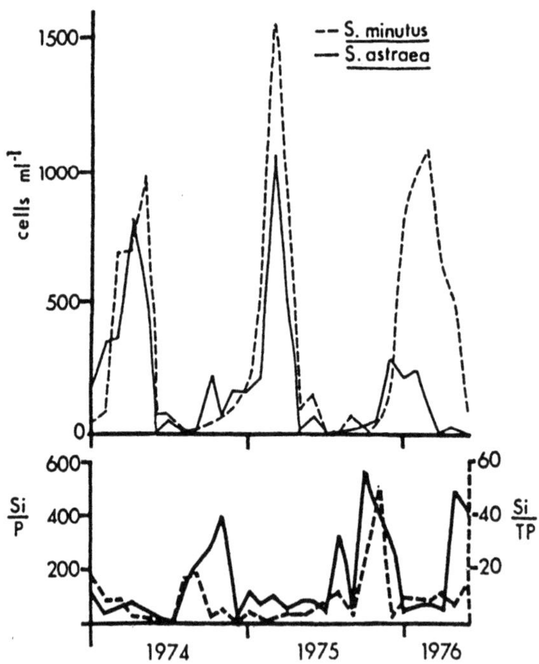


**Figure 19.1** Correspondence between high abundances of *Stephanodiscus* species and low Si:P ratios in Lake Michigan. The annual variations from 1974 to 1976 in abundance of two *Stephanodiscus* species are shown in the upper panel. The lower panel shows the atomic ratios of soluble reactive silicate (Si) to either soluble reactive phosphate (P; solid curve, lower panel) or total phosphorus (TP; dashed curve, lower panel). Data were calculated from observations by Barber and Redmond (1976) for a nearshore (3 m) station on the southwestern shore of Lake Michigan.

natural communities dominated by cyanobacteria growing with the same supply ratios but at different dilution rates (which controls the degree of nutrient limitation). They found quite different ratios across the range of dilution rates. It is even possible to make predictions about relative abundances across ratio gradients at the species level (Tilman, 1982; Tilman et al., 1984). An example of species level correlation is shown in Figure 19.1. *Stephanodiscus* species have been characterized as "low Si:P species" (Tilman et al., 1982; Kilham, 1984; Kilham et al., 1986). When the abundances of two *Stephanodiscus* species are compared with the Si:P ratio over three seasonal cycles, it can be seen that their maximum abundances are during the periods of lowest Si:P ratios. Studies such as these support the contention that elemental cellular stoichiometries of natural phytoplankton communities can reflect the type and degree of resource limitation and the relative loading rates to these populations. Many more studies are needed (especially of dominant species) to link the physiology of individual taxa with their dynamics in nature.

## 19.2 Resource Ratio Signatures

Although there is little data on the Laurentian Great Lakes to support these statements, they can at least provide support for the potential usefulness of elemental cellular stoichiometry. Table 19.1 gives examples of particulate and dissolved resource ratios for some of the lakes. The table shows that the lakes have quite different "resource ratio signatures." It can be argued that the winter/early spring dissolved fractions reflect the generally available nutrient pools for the spring phytoplankton populations and thus might serve as a useful comparative tool. I have taken data from Dobson (1985) and calculated atomic ratios from spring and summer dissolved and particulate nutrient analyses (Figure 19.2). This figure illustrates some of the problems with using inappropriate measurements for stoichiometric calculations. For the dissolved nutrients, the early spring (Mar/Apr) samples (bottom left panel) showed a consistent increasing trend in both Si:P and N:P ratios, but the summer values (Aug/Sep) (upper left) showed no pattern and were unreasonably high because of problems in measuring SRP and because the relative turnover rates of the elemental pools are so different. In the case of the particulate samples (assuming SRSi is a reasonable analog of PSi), the situation is reversed. The major signals are in the Aug/Sep samples, which again show increasing trends, while the Mar/Apr samples are generally trendless. This would be expected because the greatest proportions of the nutrient pools are in the phytoplankton particulate fractions in the late summer, and a substantial amount of the pools would be dissolved prior to the spring blooms. Yearly trends in N:P ratios for annual average data from Lake Erie (Heathcote et al., 1981) are shown in Figure 19.3. As expected, the dissolved fractions are more highly variable than the particulate fractions of both elements because a large proportion of the samples are taken in the summer when the nutrient elements are in the phytoplankton pool.

**Table 19.1** Summary of representative resource ratio signatures (by atoms) for the Laurentian Great Lakes as calculated from the reference data[a]

| | Resource ratio signature | | | | | | | |
|---|---|---|---|---|---|---|---|---|
| Location | POC | PN | SRSi | TP | IN | SRSi | SRP | Reference |
| **Lake Michigan (1976)** | | | | | | | | |
| South | 780 | 54 | 45 | 1 | 122 | 98 | 1 | Bartone and |
| North | | | 46 | 1 | 123 | 92 | 1 | Schelske, 1982 |
| **Lake Huron (1975)** | | | | | | | | |
| Early May | | | 55 | 1 | 219 | 159 | 1 | Moll et al., 1980 |
| Late May | | | 163 | 1 | 241 | 163 | 1 | |
| Aug | | | 15 | 1 | 140 | 34 | 1 | |
| Oct | | | 4.4 | 1 | 206 | 10 | 1 | |
| **Saginaw Bay (1975, Sta. 1)** | | | | | | | | |
| Early May | | | 14 | 1 | 128 | 37 | 1 | Moll et al., 1980 |
| Late May | | | 9.6 | 1 | 118 | 28 | 1 | |
| Aug | | | 29 | 1 | 104 | 71 | 1 | |
| Oct | | | 3.2 | 1 | 46 | 6.3 | 1 | |
| **Lake Erie (1979)** | | | | | | | | |
| Long Point Bay | 126 | — | 9.4 | 1 | 385 | 192 | 1 | Leach, 1981 |
| **Lake Ontario[b]** | | | | | | | | |
| | 76 | 11 | 3.6 | 1 | 143 | 42 | 1 | Dobson, 1985 |

[a] POC, particulate organic carbon; PN, particulate or Kjeldahl nitrogen; SRSi, soluble reactive silicate; TP, total phosphorus; IN, inorganic nitrogen; SRP, soluble reactive phosphate. The year in which the samples were taken is indicated in parentheses. All ratios are by atoms.
[b] Aug–Sep mean from 1974 to 1981.

The increasing trends over years for the Lake Ontario data can also be seen in other Great Lakes (Figure 19.4). The data shown are for the early spring openwater samples (except annual average data for Lake Erie) because the inorganic fractions of nitrogen and silicon were available for comparative purposes over a period of years (Dobson, 1985; Lesht and Rockwell, 1985, 1987). There are several noteworthy things about this figure. Each lake appears to have a unique elemental resource ratio signature. The generally increasing N:P and Si:P ratios may be an indication of relative success in eutrophication abatement through closer controls of P loading to the lakes. Such clear signals are often very difficult to discern when single element trends are used. Because the scales are the same for both parts of the graph, it is clear that there are differences in the N:Si ratios for the lakes (Table 19.2), Lake Huron has a ratio of 0.8; Lake Michigan, 1.0; Lake Ontario, 3.0; and Lake Erie has even higher values. This is likely to be related to the proportion of diatoms in the total phytoplankton community. Data for Lake Ontario (Dobson, 1985) indicate that the N:Si ratio has been fairly constant for the last decade. There also appears

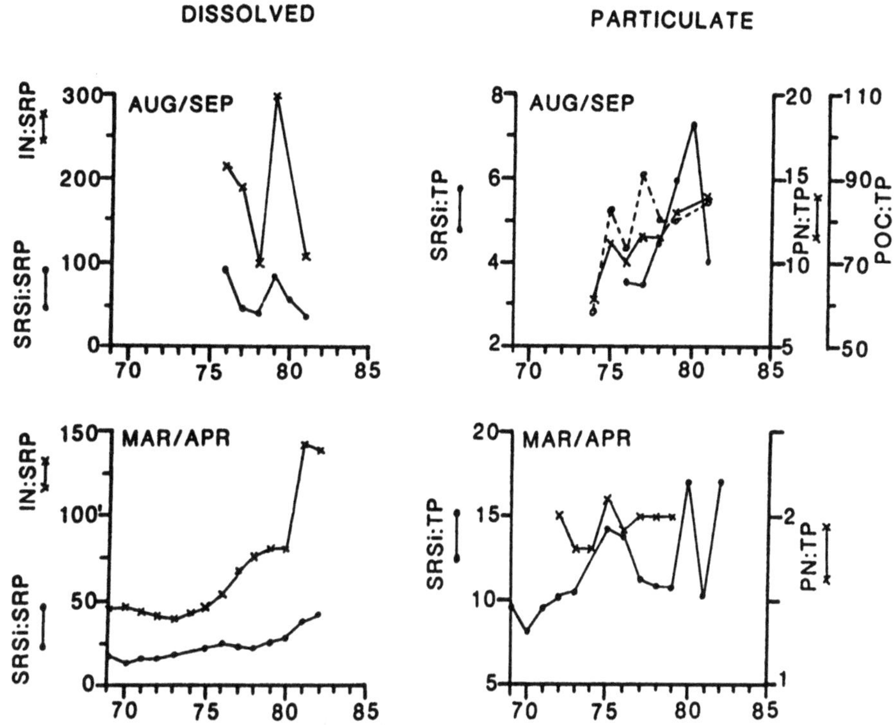


**Figure 19.2** Seasonal changes in dissolved (left panels) and particulate (right panels) elemental nutrient ratios (by atoms) for Lake Ontario. Data were calculated from Dobson (1985). Abbreviations are as given in Table 19.1.

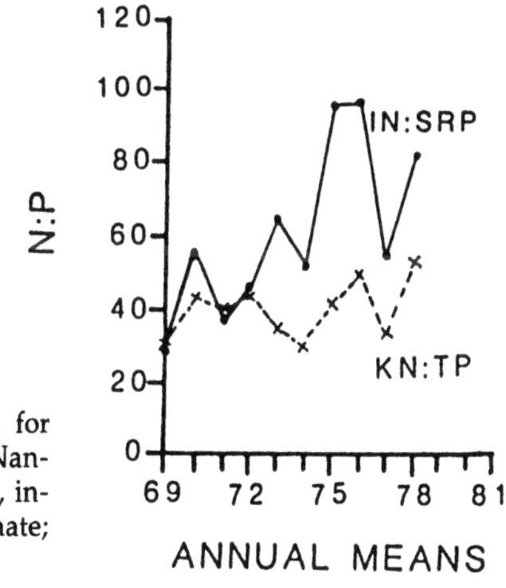


**Figure 19.3** Calculated N:P atomic ratios for mean annual nutrient data for Lake Erie at Nanticoke (data from Heathcote et al., 1981). IN, inorganic nitrogen; SRP, soluble reactive phosphate; KN, Kjeldahl nitrogen; TP, total phosphate.

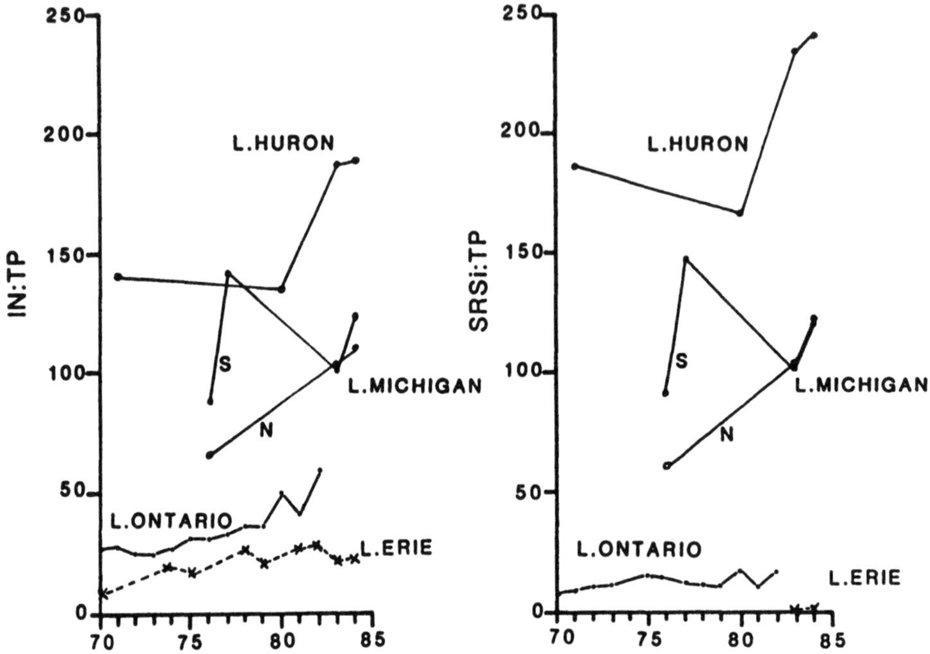


**Figure 19.4** Comparisons of resource ratio signatures of N:P (left) and Si:P (right) for four Laurentian Great Lakes (abbreviations as in Table 19.1). The data on Lakes Michigan (S, southern basin; N, northern basin) and Huron are from spring, open lake samples (Lesht and Rockwell, 1985; 1987). The data on Lake Erie are annual average values for the central basin (Lesht and Rockwell, 1985; 1987). The data on Lake Ontario are from spring, open lake samples (Dobson, 1985).

to be a clear trend in the lake resource ratio signatures reflecting changes in general trophic state. This suggestion is supported by Harris (1986) who found that C:P and N:P ratios in particulates from lakes were a function of P (the C:N ratio was constant at 8.5 by atoms). Oligotrophic lakes have C:P ratios above 193–206 and eutrophic lakes C:P ratios below 142–155.

The time of year when samples are taken and the time over which samples are averaged are important variables to consider in such a seasonally driven system as the Laurentian Great Lakes (see Figure 19.2). There is a compilation of nutrient data from several Environmental Protection Agency surveys on Lakes Michigan, Huron, and Erie for a number of seasonal cruises (Lesht and Rockwell, 1985; 1987). I have calculated the atomic ratios from their summaries of the 1983 and 1984 cruises (Table 19.3). In the spring, the ratio signatures of inorganic N (IN) to total P (TP) are highest in Lake Huron, intermediate in Lake Michigan, and lower in Lake Erie, along a trend in increasing trophy. In the summer and fall, this trend is also clear in both the dissolved and particulate N:P fractions. The Si:P ratios give the strongest signals between lakes. Seasonally within each lake, the Si:P ratios for Lakes Michigan and Huron are

**Table 19.2** Calculated N:Si atomic ratios (as inorganic nitrogen to soluble reactive silicate)

| | Location | N:Si ratio |
|---|---|---|
| **Lake Michigan (1976)** | | |
| | South; 1983 | 1.00 |
| | South; 1984 | 1.02 |
| | North; 1983 | 1.00 |
| | North; 1984 | 0.90 |
| **Lake Huron** | | |
| | 1983 | 0.80 |
| | 1984 | 0.79 |
| **Lake Erie** | | |
| | West; 1983 | 2.39 |
| | West; 1984 | 4.66 |
| | Central; 1983 | 32.7 |
| | Central; 1984 | 18.4 |
| | East; 1983 | 25.5 |
| | East; 1984 | 4.2 |
| **Lake Ontario** | | |
| | Mar/Apr 1980 | 3.0 |
| | Mar/Apr 1981 | 3.9 |
| | Mar/Apr 1982 | 3.4 |
| | Aug/Sep 1978 | 2.5 |
| | Aug/Sep 1979 | 3.7 |
| | Aug/Sep 1981 | 2.8 |

Data on Lakes Michigan, Huron, and Erie are from Lesht and Rockwell (1987), and data on Lake Ontario are from Dobson (1985).

lower in summer and higher at other times of the year, a clear indication of silica demand (Kilham, 1971). This picture is not as clear for Lake Erie, probably because there the diatoms strip the dissolved Si from the water very early in the year, making the SRSi:TP ratio highly variable.

An additional dimension where elemental cellular stoichiometry is potentially useful for understanding biological processes is in the vertical changes in ratio signatures. I have used data from one of the African Great Lakes, Lake Malawi, to illustrate this (Figure 19.5). The C:P and N:P ratios are almost always greater than the Redfield ratios for C:N:P of 106:16:1, and the peaks are associated with the major physical features (depth of euphotic zone and top of the thermocline). The TDN:TDP ratios are also high in the top 100 m. These points, along with the high Si:P ratios, indicate that phosphorus limitation of phytoplankton production is a distinct possibility (Healey and Hendzel, 1980) and that *in situ* growth rates are likely to be less than maximum (Goldman, 1980; Suttle and Harrison, 1986). Table 19.4 compares the

**Table 19.3** Seasonal averages of nutrient ratios for 1983 and 1984 at various locations in Lakes Michigan, Huron, and Erie.

| | PN:IP ratio | | | | IN:IP ratio | | | | SRSi:TP ratio | | | |
|---|---|---|---|---|---|---|---|---|---|---|---|---|
| | SPG1 | SPG2 | Summer | Fall | SPG1 | SPG2 | Summer | Fall | SPG1 | SPG2 | Summer | Fall |
| **Lake Michigan** | | | | | | | | | | | | |
| South; 1983 | | 66 | — | 56 | | 103 | 91 | 101 | | 104 | 40 | 66 |
| South; 1984 | 33 | 51 | 65 | 60 | 118 | 120 | 64 | 110 | 118 | 112 | 18 | 101 |
| North; 1983 | | 50 | 20 | 93 | | 93 | 122 | 104 | | 99 | 89 | 72 |
| North; 1984 | 37 | 46 | 92 | 31 | 71 | 95 | 73 | 82 | 77 | 99 | 35 | 75 |
| Strait; 1983 | | 65 | — | 97 | | 103 | 104 | 73 | | 88 | 91 | 105 |
| Strait; 1984 | 116 | 76 | 106 | 48 | 170 | 114 | 84 | 109 | 183 | 125 | 63 | 118 |
| **Lake Huron** | | | | | | | | | | | | |
| North; 1983 | | 74 | 69 | 94 | | 178 | 204 | 194 | | 217 | 192 | 230 |
| North; 1984 | 57 | 37 | 136 | 94 | 194 | 181 | 222 | 234 | 234 | 233 | 211 | 251 |
| South; 1983 | | 66 | 105 | 86 | | 157 | 163 | 147 | | 201 | 145 | 168 |
| South; 1984 | 75 | 61 | 100 | 30 | 200 | 132 | 186 | 197 | 248 | 172 | 133 | 222 |
| **Lake Erie** | | | | | | | | | | | | |
| West; 1983 | | 12 | 36 | 15 | | 50 | 28 | 17 | | 19 | 21 | 15 |
| West; 1984 | 24 | 12 | 47 | 34 | 77 | 34 | 61 | 52 | 17 | 5.5 | 37 | 28 |
| Central; 1983 | | 26 | 38 | 31 | | 19 | 23 | 11 | | 0.5 | 8.8 | 5.9 |
| Central; 1984 | 37 | 34 | 72 | 21 | 26 | 17 | 89 | 24 | 0.9 | 1.4 | 36 | 3.7 |
| East; 1983 | | 29 | 38 | 62 | | 53 | 34 | 39 | | 1.9 | 10 | 18 |
| East; 1984 | 19 | 38 | 45 | 40 | 31 | 32 | 28 | 44 | 7.6 | 5.6 | 7.2 | 12 |

Data were calculated from Lesht and Rockwell (1987) for 20 m depth, integrated open lake samples.

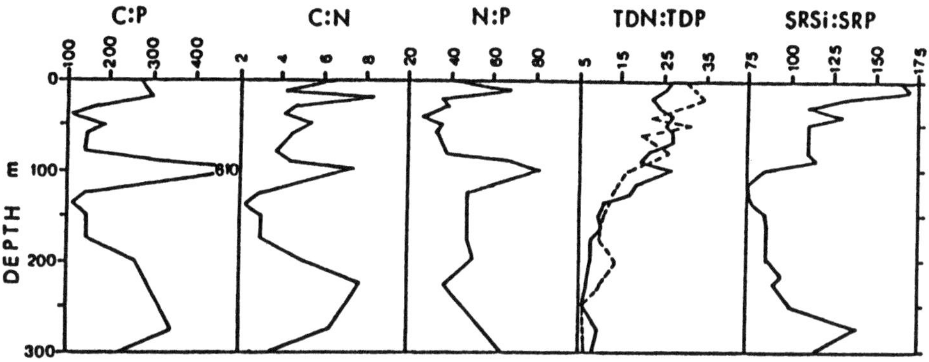


**Figure 19.5** Vertical distribution of atomic ratios of particulate (C:P, C:N, N:P) and dissolved (TDN:TDP, SRSi:SRP) nutrient elements in the top 300 m of the water column (600 m water depth) of Lake Malawi (Hecky, unpublished data). The solid lines are samples taken on 28 May, 1981; the dashed line TDN:TDP is for samples taken on 26 November 1980. TDN, total dissolved nitrogen; TDP, total dissolved phosphorus; other abbreviations are as given previously. The top of the thermocline was approximately 100 m.

**Table 19.4** Comparisons of N:P ratios in the total dissolved fractions for samples from various regions of two African Great Lakes.

| | Location | TDN:TDP ratio |
|---|---|---|
| **Lake Tanganyika** | | |
| | North | 16 |
| | Central | 40 |
| | South | 19 |
| **Lake Malawi** | | |
| | Surface | 30 |
| | Deep | 8 |

The Tanganyika values are for regional mean surface water samples from Oct/Nov 1975 (Hecky and Kling, 1981); the Malawi values are for surface and 300 m samples taken on 26 November 1980 (Hecky, unpublished data).

TDN:TDP data from Lake Malawi with data from Lake Tanganyika (Hecky and Kling, 1981). The data for central Lake Tanganyika represent a region with the highest algal chlorophyll levels compared to the other two areas. The much higher N:P ratios in this region also indicate the possibility of phosphorus limitation of phytoplankton biomass, although Hecky and Kling (1981) caution that some part of the TDN pool could be organic compounds not available to the algae.

## 19.3 Conclusion

The potential for elemental stoichiometries of phytoplankton particulates to be useful in a variety of ways is apparent. Before this potential can be realized, however, more attention will have to be paid to the collection of data on appropriate particulate fractions for elements such as C, N, Si, P (and even O, H, and S; see Vollenweider, 1985). As Azam et al. (Chapter 24) caution, bacteria have the potential to be a significant component of particulate fractions, depending on sample collection methods and lake trophic state. The time and space scales over which the samples are gathered can also have important implications for interpreting the data (Harris, 1986). To obtain the full benefits of using elemental stoichiometry, we must also improve our understanding of the relationship between resource supply ratios and a number of important biological processes, such as the type and degree of resource limitation, phytoplankton species composition (food quality), phytoplankton seasonal succession, and the rate of turnover of phytoplankton biomass. Elemental stoichiometry can also be used as a relatively easy monitoring tool for measuring the effectiveness of eutrophication abatement through control of phosphorus loading. Particulate pools are easier to quantify than dissolved pools and would be a more appropriate measure for relating the stoichiometry to phytoplankton community structure.

*Acknowledgements* I thank P. Kilham and a reviewer for making critical comments on the manuscript, and R. Kiesling for helpful discussion. R. Hecky very kindly supplied the unpublished data for Lake Malawi.

## References

Barber, B.E. and Redmond, D.G. 1976. NALCO Environmental Sciences, Operational Environmental Monitoring in Lake Michigan near Zion Station July 1975–June 1976. *Phytoplankton Study* 3:1–696.

Bartone, C.R. and Schelske, C.L. 1982. Lake-wide seasonal changes in limnological conditions in Lake Michigan in 1976. *Journal of Great Lakes Research* 8:414–427.

Dobson, H.F.H. 1985. *Lake Ontario Water Chemistry Atlas*. Scientific Series No. 139, Inland Waters Directorate, Environment Canada, Ottawa, 59 pp.

Goldman, J.C. 1980. Physiological processes, nutrient availability, and the concept of relative growth rate in marine phytoplankton ecology. In: Falkowski, P.G. (editor), *Primary Productivity in the Sea*. Plenum Press, New York, p. 179–194.

Harris, G.P. 1986. *Phytoplankton Ecology*. Chapman and Hall, New York, 384 pp.

Healey, F.P. and Hendzel, L.L. 1980. Physiological indicators of nutrient deficiency in lake phytoplankton. *Canadian Journal of Fisheries and Aquatic Sciences* 37:442–453.

Heathcote, I.W., Weiler, R.R., and Tanner, J.W. 1981. Lake Erie nearshore chemistry at Nanticoke, Ontario, 1969–1978. *Journal of Great Lakes Research* 7:130–135.

Hecky, R.E. and Kling, H.J. 1981. The phytoplankton and protozooplankton of the euphotic zone of Lake Tanganyika: Species composition, biomass, chlorophyll content, and spatio-temporal distribution. *Limnology and Oceanography* 26:548–564.

Kilham, P. 1971. A hypothesis concerning silica and the freshwater planktonic diatoms. *Limnology and Oceanography* 16:10–18.

Kilham, P., Kilham, S.S., and Hecky, R.E. 1986. Hypothesized resource relationships among African planktonic diatoms. *Limnology and Oceanography* 31:1169–1181.

Kilham, S.S. 1984. Silicon and phosphorus growth kinetics and competitive interactions between *Stephanodiscus minutus* and *Synedra* sp. *Verhandlungen Internationale Vereinigung für Theoretische und Angewandte Limnologie* 22:435–439.

Kilham, S.S. 1986. Dynamics of Lake Michigan natural phytoplankton communities in continuous cultures along a Si:P loading gradient. *Canadian Journal of Fisheries and Aquatic Sciences* 43:351–360.

Kilham, S.S. and Kilham, P. 1984. The importance of resource supply rates in determining phytoplankton community structure. In: Meyers, D.G. and Strickler, J.R. (editors), *Trophic Interactions Within Aquatic Ecosystems*. Westview Press, Boulder, p. 7–27.

Kilham, S.S. 1988. Phytoplankton responses to changes in mortality rate. *Verhandlungen Internationale Vereinigung für Theoretische und Angewandte Limnologie* 23:677–682.

Leach, J.H. 1981. Comparative limnology of inner Long Point Bay, Lake Erie, and adjacent waters of the outer bay. *Journal of Great Lakes Research* 7:123–129.

Lesht, B.M. and Rockwell, D.C. 1985. The state of the middle Great Lakes: Results of the 1983 water quality survey of Lakes Erie, Huron, and Michigan. Report ANL/ER–85–2, Environmental Research Division, Argonne National Laboratory, Argonne, 118 pp.

Lesht, B.M. and Rockwell, D.C. 1987. The state of the middle Great Lakes: Results of the 1984 water quality survey of Lakes Erie, Huron, and Michigan. Report ANL/ER—87–1, Environmental Research Division, Argonne National Laboratory, Argonne, 141 pp.

Moll, R.A., Davis, C.O., and Schelske, C.L. 1980. Phytoplankton productivity and standing crop in the vicinity of the Lake Huron—Saginaw Bay front. *Journal of great Lakes Research* 6:232–246.

Sommer, U. 1983. Nutrient competition between phytoplankton species in multispecies chemostat experiments. *Archiv für Hydrobiologie* 96:399–416.

Sommer, U. 1985. Comparison between steady state and non-steady state competition: Experiments with natural populations. *Limnology and Oceanography* 30:335–346.

Sommer, U. 1986. Phytoplankton competition along a gradient of dilution rates. *Oecologia* 68:503–506.

Suttle, C.A. and Harrison, P.J. 1986. Phosphate uptake rates of phytoplankton assemblages grown at different dilution rates in semicontinuous culture. *Canadian Journal of Fisheries and Aquatic Sciences* 43:1474–1481.

Tilman, D. 1982. *Resource Competition and Community Structure*. Princeton University Press, Princeton, 296 pp.

Tilman, G.D., Kiesling, R., Sterner, R., Kilham, S.S., and Johnson, F.A. 1986. Green, bluegreen and diatom algae: Taxonomic differences in competitive ability for phosphorus, silicon and nitrogen. *Archiv für Hydrobiologie* 106:473–485.

Tilman, D., Kilham, S.S., and Kilham, P. 1982. Phytoplankton community ecology: The role of limiting nutrients. *Annual Review of Ecology and Systematics* 13:349–372.

Tilman, G.D., Kilham, S.S., and Kilham, P. 1984. A reply to Sell, Carney, and Fahnenstiel. *Ecology* 65:328–332.

Vollenweider, R.A. 1985. Elemental and biochemical composition of plankton biomass; some comments and explorations. *Archiv für Hydrobiologie* 105: 11–29.

# 20

# Ecology of *Melosira* Species in the Great Lakes of Africa

Peter Kilham†

ABSTRACT *Melosira* species are often seasonally important components of the phytoplankton in African freshwaters. Because few experimental investigations of them have been undertaken, our knowledge of their ecology is based almost entirely on studies concerning phytoplankton distribution and seasonal succession. The ecology of *Melosira* species is of interest because these diatoms frequently occur in sediment cores that represent a historical record of more than 10,000 years. The principal limiting resources of *Melosira* species are thought to be light and phosphorus. Examination of the distribution and abundance of *Melosira* species in the Great Lakes of Africa indicates that the size of mantle pores (a species-specific characteristic) is correlated with the depth of mixing and possibly the quantity and quality of ambient light. Large-pored species, such as *Melosira nyassensis* from Lake Malawi, have 6–8 pores per 10 $\mu$m, while small-pored species that are abundant in shallower lakes, such as *M. ambigua*, have about 16–25 pores.

Even though *Melosira* species are abundant components of the plankton in lakes world-wide, knowledge of their ecology is based largely on field observations. Our analysis of *Melosira* populations from both the Great Lakes of Africa and a variety of smaller lakes indicates that resource limitation (e.g., silica, phosphorus, and light) may be common, but no experimental data are available (Kilham et al., 1986). This paper reviews the general ecology, distribution and paleoecology of *Melosira* species in the Great Lakes of Africa. In addition, arguments are presented to support the observation that the size of mantle pores (a species-specific character in *Melosira*) is correlated with the depth of mixing and possibly the quantity and quality of ambient light.

## 20.1 General Ecology

A great deal of ecological information is known about the diatoms of East and Central Africa even though these data were obtained almost entirely from

†Deceased.

studies of phytoplankton distribution (Hustedt, 1949; Kilham and Kilham, 1975; Gasse et al., 1983; Gasse, 1986) and seasonal succession (Talling, 1986, 1987; Gasse et al., 1983; Kalff and Watson, 1986; Hecky and Kling, 1987). Species of *Melosira* occur in most African freshwaters (conductivity, <1,500 $\mu S\ cm^{-1}$; alkalinity, <10 meq $l^{-1}$).

The two best known aspects of the ecology of *Melosira* species are interrelated. First, *Melosira* species are meroplanktonic organisms that enter the plankton when mixing conditions are such that they can be resuspended from the bottom sediments and maintained in the water column. Talling (1966, 1969, 1986) has shown that Lund's (1954) observations concerning the periodicity of *Melosira* species in many parts of the world are, as Lund surmised, clearly applicable to several African species. Second, *Melosira* species appear to have rather high nutrient requirements (Kilham, 1971; Kilham and Kilham, 1975). This is to be expected because they thrive in the nutrient-rich habitat that results when hypo- or metalimnetic waters are mixed throughout the water column in lakes that stratify. They also occur in many shallow, eutrophic waters that mix completely. Hutchinson (1967, p. 343) considers the freshwater species of *Melosira* to be largely or entirely meroplanktonic, but some species are more seasonal than others; in Lake Victoria, for example, offshore populations of *M. nyassensis* var. *victoriae* O. Müller exhibit greater seasonality than inshore populations of *M. granulata* (Ehrenb.) Ralfs do (Talling, 1966, 1987).

The ecology of meroplanktonic *Melosira* species contrasts markedly with that of euplanktonic species, such as *Stephanodiscus* spp., planktonic *Nitzschia* spp., and *Synedra* spp. (Table 20.1; Kilham et al., 1986). *Melosira* species are generally opportunistic organisms that are adapted to the nonequilibrium conditions that exist when wind-induced mixing permits them to enter the plankton. *Melosira* species are considered to be toward the r-selected end of the r-through K-selection continuum because they are density-independent organisms that flourish in transitional habitats (see Kilham and Kilham, 1980). They cannot maintain maximum population densities for extended periods of time in most lakes owing to changes in the environment. Euplanktonic species, on the other hand, occur when lakes are stratified and nutrients are more likely

**Table 20.1** Ecological categories associated with diatoms that divide up Si:P and Light:P ratio gradients, respectively.

| Type | Resource ratio gradient[a] Si:P | Light:P |
|---|---|---|
| Alga | Planktonic | Meroplanktonic |
| Species | More K-selected | More r-selected |
| Community | Equilibrium | Nonequilibrium |

[a] Hypothesized resource relationships among African planktonic diatoms are given in Kilham et al., 1986.

to be limiting; more equilibrium conditions prevail and euplanktonic diatoms compete for limiting resources among each other and with other algae (Tilman et al., 1982). Resource competition for nutrients would rarely occur between meroplanktonic *Melosira* species and euplanktonic species. This is because lakes often restratify, causing *Melosira* species to settle out, before nutrients (P, N, Si) are actually depleted (Petrova, 1986).

Reynolds et al. (1986) have recognized three aspects of the biology of *Melosira* that make members of this genus distinctive components of planktonic communities. First, these diatoms can adapt to the low light levels that cells experience in a well-mixed water column by increasing the amount of chlorophyll per cell and thus the light-harvesting capacity (Talling, 1957). Second, *Melosira* species have higher sinking rates than most freshwater planktonic diatoms. Consequently, they are dependent on the depth and frequency of vertical mixing such that losses resulting from sinking are at least balanced by cell division (Lund, 1959; Titman and Kilham, 1976; Reynolds, 1984). The final and perhaps most important biological characteristic of *Melosira* species is their ability to perennate on lake sediments (Lund, 1954). In other words, when *Melosira* filaments sink to the sediments their protoplasmic contents often contract into a condensed resting form. In this form, the cells are not affected by the absence of light or oxygen, and they may remain viable for months or years. Turbulent mixing then resuspends resting filaments that provide an inoculum for new populations of *Melosira*.

## 20.2 Distribution of *Melosira* Species with Depth

Water depth and, concomitantly, the depth of mixing appear to play a major role in determining the distribution of *Melosira* species in African lakes (Gasse et al., 1983). *Melosira ambigua* is the dominant planktonic diatom in many shallow lakes and in the shallow inshore waters of some of the African Great Lakes (Richardson, 1968; Gasse, 1986). In contrast, *M. nyassensis* (or varieties of it) often dominate in the offshore waters of Lakes Malawi and Victoria when stratification breaks down and the lakes are deeply mixed. The distributions of *M. granulata* and *M. agassizii* with respect to depth are less obvious, but *M. granulata* (and varieties or morphological types; see Kilham and Kilham, 1975; Gasse, 1986) is commonly found in shallow waters while *M. agassizii* occurs in both shallow lakes and as an important member of the plankton in some large lakes (e.g., Lake Victoria; Talling, 1966). Stager (1984) studied the morphology of *Melosira* species found in a sediment core from the Damba Channel, Lake Victoria, and concluded that there may be a relationship between cell morphology and environmental gradients (such as degree of mixing or depth); however, the relationship is not clear-cut. The problem, in my opinion, is that different resource requirements can obscure what may at times be a relatively straightforward relationship between cell morphology and depth-related parameters.

## 20.3 Distribution of *Melosira* Species in Large Lakes

*Melosira* species occur in all of the Great Lakes of Africa, but they are most frequently reported as dominants when deep mixing occurs in Lakes Malawi and Victoria (Lund, 1954; Talling, 1966). Considerable confusion exists concerning the distribution of *Melosira* in some of the other lakes, but recent information indicates that *Melosira* is present in all of them (Gasse, 1986; Haberyan, 1985). *M. ambigua* (a small-pored species, Figure 20.1) blooms in the shallow inshore waters of Lakes Albert and Victoria (Talling, 1966; Gasse, 1986), and it can occur in fairly high numbers in Lakes Edward and Kivu (Hustedt, 1949). *M. agassizii* and *M. granulata* (species with intermediate-sized pores) are found in Lakes Albert (Mobutu Sese Seko), Edward, Kivu, and Victoria, and perhaps in the remaining lakes. They are most abundant in somewhat deeper waters than where *M. ambigua* is dominant. In Lake Victoria, *M. agassizii* is an important component of the offshore plankton even though it is generally subdominant to *M. nyassensis* when deep mixing occurs. *M. nyassensis* and *M. nyassensis* var. *victoriae* (large-pored species) are resuspended from the inshore sediments and lower water column in Lakes Malawi and Victoria, respectively, when stratification breaks down and the lakes are mixed to great depths. *Melosira* is not commonly observed in phytoplankton samples from Lake Tanganyika, but it is undoubtedly abundant at times in the southern end of the lake (maximum mixing depth, ca. 200 m; Beadle, 1981). A relatively long sediment core obtained at a depth of ca. 440 m from Lake Tanganyika indicates that *M. argus* O. Müller (Haberyan, 1985; personal observation), which has even larger pores than *M. nyassensis*, comprised between 20–35% of the fossil diatom community for the past 2,000 years. Figure 20.2 compares the pore sizes of several different *Melosira* species (also see Figure 20.1).

The population dynamics of *M. nyassensis* (and varieties) in East Africa are quite similar to that of *Melosira* sp. (cf. *M. baicalensis*) in Lake Taupo, New Zealand. In tropical lakes, such as Lakes Victoria and Malawi, maximum populations of *Melosira* develop in offshore waters after stratification in the mixed layer breaks down owing to seasonal changes in temperature and wind velocity (Talling, 1969, 1986). Mixing events strong enough to permit the development of large populations of *Melosira* can occur several times a year in Lake Victoria (Talling, 1966). However, in the northern basin of Lake Malawi, *Melosira* blooms in some years but not in others (Hecky and Kling, 1987). In Lake Taupo, a temperate monomictic lake, *Melosira* populations are greatest during the winter when the lake circulates to ca. 60 m (Vincent, 1983). In contrast, populations of *M. baicalensis* (K. Meyer) Wisl. begin to develop in Lake Baikal, U.S.S.R., while the lake is inversely stratified in the winter and generally reach their maximum just as stratification breaks down in the early summer (Kozhov, 1963).

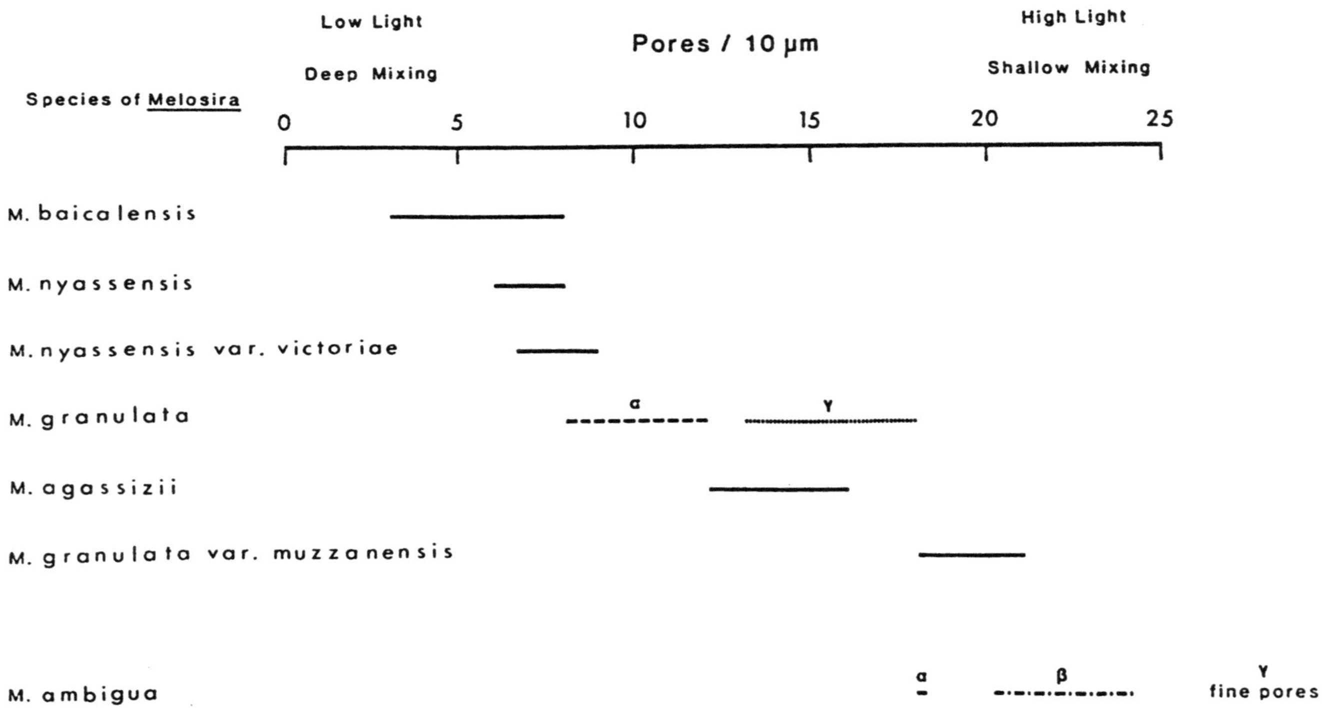


**Figure 20.1** Relative pore sizes in seven species of *Melosira*. All of these generally meroplanktonic diatoms, except *M. baicalensis*, are abundant in one or more African lakes. Data are reported as pores (areolae) per 10 µm, in keeping with the taxonomic literature. The Greek letters $\alpha$, $\beta$, and $\gamma$ are used by taxonomists as status designations. The following sources were used: Florin, 1970; Gasse, 1986; Hustedt, 1930, 1942; Müller, 1904; and Stoermer et al., 1981.

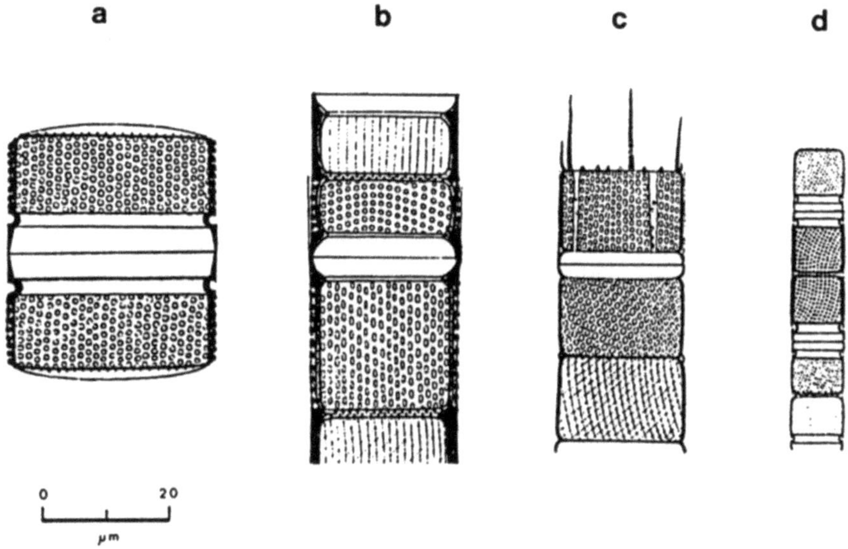


**Figure 20.2** Differences in pore size among four *Melosira* species. All of the diatoms illustrated are drawn to the same scale (Plates 3 and 4 in Müller, 1904). Identification: (*a*) *Melosira argus*, (*b*) *M. nyassensis*, (*c*) *M. granulata*, and (*d*) *M. ambigua*. All the species, except *M. granulata*, were drawn from algal samples obtained from Lake Malawi. The *M. granulata* material came from the Müggelsee, East Germany. Despite the impression given by this figure, pore size is a species-specific character and not a function of cell diameter.

## 20.4 The Ecology of *Melosira* Species in the Major Rivers of East and Central Africa

Lakes Victoria and Malawi flow, respectively, into the Nile and Zambezi Rivers, and both waterways are characterized by large populations of *Melosira*. The ecology of *Melosira* (primarily *M. granulata* and varieties) in the Nile River was reviewed in detail by Talling (1976). In this review he suggests that *Melosira* populations in both the White and Blue Niles were often limited by ambient concentrations of inorganic nitrogen. In a recent paper, Sinada and Karim (1984) suggest that *Melosira* populations in the Blue Nile are nitrogen limited, but their data indicate that phosphorus or iron may have been limiting as well. Evans (1961) found that phosphorus, but not nitrogen, stimulated growth in a water sample from Lake Victoria that was dominated by *M. nyassensis* var. *victoriae*. Reexamination of published data concerning nutrient limitation in the waters of the Nile River system indicates that neither the observational nor experimental procedures currently used to demonstrate nutrient limitation were employed when these investigations were undertaken (see Hecky and Kilham, 1988). Therefore, these results must be viewed with caution.

Despite the use of samples collected by others during earlier investigations and limited information on the physical and chemical environment, Hancock's (1979) study of diatom succession in the Zambezi River and Lake Kariba is quite informative. *M. granulata, M. granulata* var. *angustissima* O. Müller, and *M. distans* var. *alpigena* Grun. comprise the dominant diatom association in the main basins of the lake where a successional sequence is observed in response to the thermal cycle. *M. granulata* var. *angustissima* generally dominates during overturn (often March through June) and is replaced by *M. granulata* after the lake becomes homothermal and deeply mixed in May or June. During stratification (usually September to March), *Melosira* populations decrease markedly in abundance, but *M. granulata* remains dominant. Different regions of the lake show slightly different patterns. For example, M. *distans* var. *alpigena* is sometimes dominant at one station. *M. granulata* var. *angustissima* is the dominant species in the Zambezi River and when riverine conditions are found in the lake. Hancock's data indicate that *Melosira granulata*, the variety with the largest pores (Gasse, 1986), becomes dominant when deep mixing occurs. *M. granulata* var. *angustissima* is most abundant when the lake mixes to intermediate depths. Her observations are in general agreement with the relationship between pore size and mixing depth discussed above.

## 20.5 Resource Limitation

A single experiment concerning silicon-limited growth and nutrient kinetics of *Melosira granulata* from Lake Mulehe, Uganda, represents the only information available on the physiological ecology of any African planktonic diatom (S.S. Kilham, unpubl. observations). These data (Figure 20.3 and Table 20.2) indicate that *M. granulata* would be a poor competitor for silicon relative to many other planktonic diatoms (e.g., *Stephanodiscus minutus* Grun.). *Synedra* sp. (see Table 20.2) is also a poor competitor for silicon, but it is an excellent competitor for phosphorus owing to its low phosphorus requirement (Kilham, 1984).

Because the distribution of African *Melosira* species is determined to a great extent by lake depth and trophic state, we have hypothesized that the major resources involved are light (which is related to the depth of mixing and turbidity) and phosphorus (which in part determines the trophic state of a lake and turbidity: Kilham et al., 1986). We therefore proposed that most African *Melosira* species can be ranked along a Light:P resource ratio gradient such that *M. distans* > *M. ambigua* > *M. agassizii* > *M. granulata* > *M. nyassensis*. In Berra Bonita Reservoir, Brazil, Henry et al., 1985, found that phytoplankton communities dominated by *Melosira* sp. were limited by phosphorus and light.

## 20.6 Correlation between Mantle Pore Size in *Melosira* Species and Mixing Depth

One intriguing aspect of the apparent relationship between African *Melosira* species and light (or mixing depth) is the size (diameter) of the mantle pores

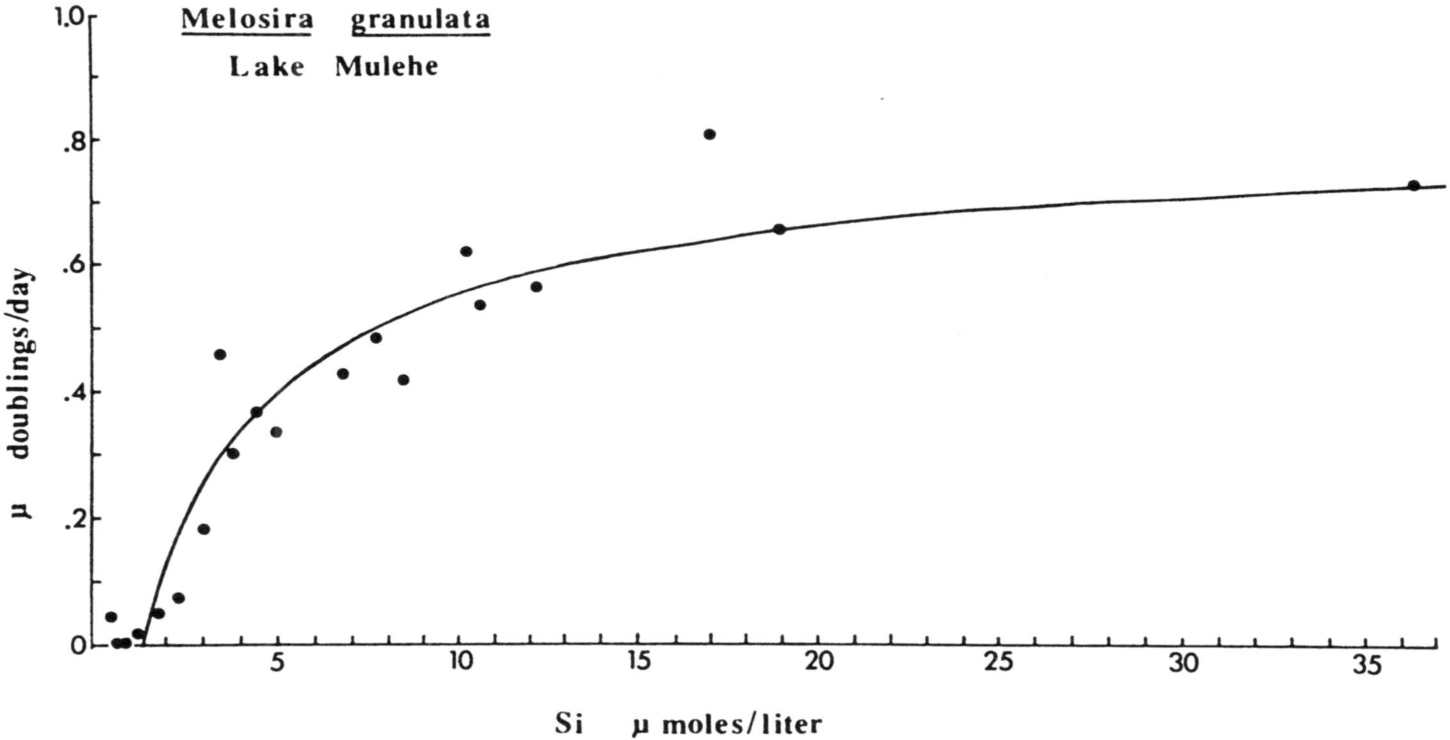


**Figure 20.3** Growth rate ($\mu$) of *Melosira granulata* clone AUX 5 as a function of silicon concentration (S.S. Kilham, unpubl. observations). Diatoms were grown at 20°C in short-term batch cultures (the line is the fitted Monod curve). Experimental procedures are given in Kilham, 1975. Growth rate in doublings/day times 0.693 equals growth rate/day.

**Table 20.2** Kinetic growth constants for *Melosira granulata* and for two species of planktonic diatoms.[a]

| | *Melosira granulata* | *Synedra* sp. | *Stephanodiscus minutus* |
|---|---|---|---|
| Source | Lake Muhele, Uganda | Lake Michigan | Lake Michigan |
| Clone no. | AUX 5 | LM SYN | LM STEPH 2-1 JM |
| $\mu_{max}$ ($d^{-1}$) | 0.55 (0.43–0.69) | 0.80 (0.58–1.0) | 0.75 (0.67–0.82) |
| $K_{Si} + Si_0$ ($\mu M$) | 5.02 (3.76–6.69) | 3.52 (1.8–6.4) | 0.88 (0.60–1.2) |
| $Si_0$ ($\mu M$) | 1.47 | 0.0 | 0.0 |
| $K_{lim}$ ($\mu M$) | 33.5[b] | 32.0 | 8.0 |
| ($SiO_2$ $mg \cdot L^{-1}$) | 2.0 | 1.9 | 0.5 |

Original data for *M. granulata* from S.S. Kilham; also see Kilham and Kilham, 1975. Experimental methods and data for *Synedra* and *S. minutus* from Kilham, 1975; and from Kilham, 1984.

[a] Monod experiments were carried out in batch culture at 20°C. Confidence intervals (95% level) are in parentheses. $\mu_{max}$, maximum growth rate; $K_{Si}$, half-saturation constant for silicon-limited growth; $Si_0$, the growth threshold; $K_{lim}$, 0.9 $\mu_{max}$.

[b] $K_{lim} + Si_0$.

(areolae) that are characteristic of each species. *Melosira* species that flourish in shallow bodies of water have small, fine pores, while species that become dominant in deep lakes have large, coarse pores (see Figures 20.1 and 20.2).

*M. baicalensis* is included in Figure 20.1 for comparative purposes because this diatom has the largest pores found in any species of *Melosira*. It is very abundant in Lake Baikal, the deepest lake in the world. Lake Baikal has a mixing depth of ca. 300 m (Kozhov, 1963), which is about 100 m deeper than the maximum mixing depths recorded in Lakes Malawi and Tanganyika (Eccles, 1974; Beadle, 1981). *M. distans* was not included in this figure because little information is available concerning its morphology and ecology (Gasse, 1986; Camburn and Kingston, 1986). Data for *M. argus* are also not included because the species has not been reported to be abundant at present in any African lake. This diatom is similar to *M. nyassensis* but has even larger pores (see Figure 20.2, 5–6 pores per 10 $\mu$m: Müller, 1904; Hustedt, 1942). Sediment cores indicate that it was often dominant during the Late Quaternary in the southern end of Lake Tanganyika (Haberyan, 1985; Haberyan and Hecky, 1987). At present, *M. argus* is reported from the sediments of Lake Malawi (Müller, 1904) and may be abundant at times in Lake Tanganyika.

Sometimes, individual *Melosira* cells are polymorphic with respect to pore size. One valve will have large pores ($\alpha$ status) while the other has small pores ($\gamma$ status). Also, a single filament can contain cells with different pore sizes (Müller, 1904; Hustedt, 1930, 1942). One can relate changes in pore size to changes in light climate even when more than one pore size is observed for a single cell or within a single filament. The proposal is that the light climate differed when the daughter cells were formed (Kilham et al., 1986).

Little (or nothing) is known about the underlying mechanism responsible for the apparent correlation between pore size and light availability. Discussions with optical physicists indicate that because the pores in diatoms are near or slightly greater than the wavelengths of visible light, defraction, scattering, and reflection are all important processes and the effects of each process are difficult to separate. Even so, it seems possible that different pore sizes will differently affect the radiant energy entering diatom cells. In the past decade it has been demonstrated that ultraviolet light (UV-B) can markedly decrease the growth rate of particular diatom species and that the susceptibility of individual species depends on the light climate where they commonly grow (Calkins, 1982; Thomson, et al., 1980; Jokiel and York, 1984). Thomson et al. (1980) found that even attached intertidal diatoms such as *Melosira nummuloides* [Dillw.] Ag. are affected (growth rates decreased by >50%) by UV-B radiation levels comparable to those observed on various parts of the globe. *M. nummuloides* is a small-pored (>25 pores/10 μm) species that is adapted to a wide range of energy conditions. Paerl (1988) has shown that a freshwater phytoplankton community that was dominated by *Melosira granulata* was also affected by UV radiation.*

If ultraviolet light is an important factor, diatoms circulating in a shallow water column would be more frequently exposed to UV light, and they would have less time in the dark to repair any damage sustained. Small pores possibly decrease the amount of UV radiation absorbed by a diatom cell as it moves in a shallow water column. Diatoms with large pores, on the other hand, may capture photons more effectively than diatoms with small pores. UV damage would still be a problem, but diatoms that are mixed to considerable depths in a large lake would spend proportionally more time in the dark where UV damage can be repaired.

Even though there are quite a few studies that suggest that exposure to UV-B can decrease the growth of individual species of diatoms, there are no data indicating that there is a relationship between susceptability to ultraviolet radiation and pore size within the freshwater species of *Melosira*, but such a relationship has probably never been looked for. Basically, the problem is that the functions of diatom pores are completely unknown and even informed speculation concerning their function is nearly impossible to find.

## 20.7 Paleoecology of *Melosira* Species in the Great Lakes of Africa

*Melosira* species were a major component of the diatom flora of most of the African Great Lakes at various times during the Late Quaternary. Lake Kivu is the only notable exception. In Lake Kivu, *Melosira* is rarely (if ever) found in cores that represent 14,000 years of accumulated sediment (Haberyan and

*Some taxonomists consider the marine and the freshwater species of *Melosira* to be in two separate genera, *Melosira* and *Aulacosira* (Simonsen, 1979).

Hecky, 1987). Also, little is known about the diatom stratigraphy of Lake Edward. Short sediment cores from Lake Edward exist, but diatom analyses have not been completed (Hecky, personal communication). *Melosira* species in Lake Victoria have occurred as subdominants to *Stephanodiscus astraea* var. *minutula* (Kütz.) Grun. almost continuously for the past 17,000 years (Stager, 1984). Between 11,700 and 8,000 yr B.P., dominance among the *Melosira* species shifted from *M. nyassensis* to *M. ambigua, M. granulata* and *M. agassizii* and then returned to *M. nyassensis* by the end of that period. Stager (1984) indicates that *M. nyassensis* again became dominant as the climate of East Africa became more seasonal and wind strengths increased. A recently analyzed core from Lake Tanganyika indicates that *Melosira* species have been an important component of the fossil diatom assemblage of that lake for most of the past 16,000 years (Haberyan, 1985; Haberyan and Hecky, 1987). About 9,900 yr B.P. there was a dramatic shift from dominance by *Stephanodiscus* spp. to dominance by *Melosira* spp. (most notably, *M. argus*). *Melosira* dominance continued until 5,200 yr B.P. when *Stephanodiscus* again increased markedly in abundance. For the past 2,000 years *Nitzschia* spp., *Stephanodiscus* spp., and *Melosira* spp. have been more or less equally abundant. Changes in the diatom flora of Lake Albert are known from a sediment core that is about 28,000 years old at its base (Harvey, 1976). *Stephanodiscus* spp. dominate the fossil diatom assemblages found throughout the core, but *Melosira* spp. are important subdominants in many regions. When diatoms are present in the core between 28,000 and 14,000 yr B.P. (there are gaps in the fossil record), *M. ambigua* and *M. granulata* are subdominant. *M. nyassensis* var. *victoriae* occurred briefly in the lake about 11,000 yr B.P. *Melosira granulata* was the most prevalent *Melosira* species found between 10,000 and 7,000 yr B.P. *Melosira* species are not found in the upper regions of the core, and *Stephanodiscus* spp. have remained dominant. Silica concentrations in the open waters of the lake are very low at present (ca. 0.04–1.1 mg $l^{-1}$: Talling, 1963) and are often within the range of silica limitation (see above). These low silica concentrations may explain in part why *Melosira* species are apparently not abundant in the offshore waters of the lake (Hecky and Kling, 1987) and why *Stephanodiscus* predominates.

## 20.8 Homeostasis in African Lakes

The fossil diatom assemblages in the sediments of several African lakes remained relatively constant for $10^2$ to $10^4$ years. This may seem like an excessively long time for any biological community to maintain its integrity, but similar time scales have also been reported for the Late Pleistocene and Holocene biota of European and North American lakes (Deevey, 1984). Lakes are believed to come into trophic equilibrium with ambient meteorological and geological conditions. This equilibrium is maintained until major environmental perturbations take place. Under certain circumstances some degree of homeostasis may result as a consequence of biotic and abiotic processes that affect nutrient cycling (Lerman and Hull 1987). K-selected diatoms (e.g., *Stephanodiscus*) may contribute to the homeostasis of a lake by utilizing nutrients as they become available. Species within this genus outcompete other diatoms by maintaining silicon concentrations at very low levels for long periods of time (Kilham, 1971; Tilman et al. 1982). On the other hand, r-selected diatoms (e.g., *Melosira*) would probably affect ambient nutrient levels only temporarily. Decreases in wind strength might eliminate *Melosira* species from the pelagic zone for extended periods.

## 20.9 Conclusions

*Melosira* species are found in all of the Great Lakes of Africa. They become dominant in offshore waters when stratification breaks down and deep mixing ensues. *Melosira* species are generally opportunistic, r-selected diatoms that are adapted to nonequilibrium conditions. The relationship between the size of mantle pores (a species-specific characteristic) and mixing depth (or light climate) is particularly clear among the *Melosira* species found in Africa.

---

*Acknowledgements* I thank Susan S. Kilham for use of her data on silicon-limited growth in *Melosira granulata* and for making critical comments on the manuscript. The comments of R.W. Genesmer, R.L. Kiesling, P.J. Kociolek, U. Sommer, and two additional reviewers were very helpful. This paper was prepared during the tenure of an Alexander von Humboldt Foundation U.S. Senior Award.

## References

Beadle, L.C. 1981. *The Inland Waters of Tropical Africa.* 2nd ed. Longman, London, 475 pp.

Calkins, J. (editor), 1982. *The Role of Solar Ultraviolet Radiation in Marine Ecosystems.* Plenum, New York, 720 pp.

Camburn, K.E. and Kingston, J.C. 1986. The genus *Melosira* from soft-water lakes with special reference to northern Michigan, Wisconsin and Minnesota. In: Smol, J.P., Davis, R.B. and Meriläinen, J. (editors), *Diatoms and Lake Acidity.* Junk Publishers, Dordrecht, p. 17–34.

Deevey, E.S., Jr. 1984. Stress, strain and stability of lacustrine ecosystems. In: Haworth, E.Y. and Lund, J.W.G. (editors), *Lake Sediments and Environmental History.* Leicester University Press, Leicester, p. 203–229.

Eccles, D.H. 1974. An outline of the physical limnology of Lake Malawi (Lake Nyasa). *Limnology and Oceanography* 19:730–742.

Evans, J.H. 1961. Growth of Lake Victoria phytoplankton in enriched cultures. *Nature* 189:417.

Florin, M.-B. 1970. The fine structure of some pelagic fresh water diatom species under the scanning electron microscope. I. *Svensk Botanisk Tidskrift* 64:51–64.

Gasse, F. 1986. *East African Diatoms (Taxonomy, Ecological Distribution). Bibliotheca Diatomologica,* Vol. 11. J. Cramer, Berlin, 201 pp.

Gasse, F., Talling, J.F., and Kilham, P. 1983. Diatom assemblages in East Africa: classification, distribution and ecology. *Revue d'Hydrobiologie Tropicale* 16:3–34.
Haberyan, K.A. 1985. The role of copepod fecal pellets in the deposition of diatoms in Lake Tanganyika. *Limnology and Oceanography* 30:1010–1032.
Haberyan, K.A. and Hecky, R.E. 1987. The late Pleistocene and Holocene stratigraphy and paleolimnology of Lakes Kivu and Tanganyika. *Palaeogeography, Palaeoclimatology, Palaeoecology* 61:169–197.
Hancock, F.D. 1979. Diatom associations and succession in Lake Kariba, South Central Africa. *Hydrobiologia* 67:33–50.
Harvey, T.J. 1976. The Paleolimnology of Lake Mobuto Sese Seko, Uganda-Zaire: The Last 28,000 Years. Ph.D. thesis, Duke University, Durham, N.C., 104 pp.
Hecky, R.E. and Kilham, P. 1988. Nutrient limitation of phytoplankton in freshwater and marine environments: A review of recent evidence for enrichment. *Limnology and Oceanography* 33:796–822.
Hecky, R.E. and Kling, H.J. 1987. Phytoplankton in the Great Lakes in the Western Rift Valley of Africa. *Archiv für Hydrobiologie Ergebnisse der Limnologie* 25:197–228.
Henry, R., Hino, K., Gentil, J.G., and Tundisi, J.G. 1985. Primary production and effects of enrichment with nitrate and phosphate on phytoplankton in the Berra Bonita Reservoir (State of Sao Paulo, Brazil). *Internationale Revue der Gesamten Hydrobiologie* 70:561–573.
Hustedt, F. 1930. Bacillariophyta (Diatomeae). In: Pascher, A. (editor), *Die Süsswasser-Flora Mitteleuropas,* Heft 10. Gustav Fischer Verlag, Jena, p. 1–466.
Hustedt, F. 1942. Diatomeen. In: Huber-Pestalozzi, G. (editor), *Das Phytoplankton des Süsswassers, Die Binnengewässer,* Band 16, Teil 2, Hälfte 2. E. Schweizerbart'sche Verlagsbuchhandlung, Stuttgart, p. 367–549.
Hustedt, F. 1949. *Süsswasser-Diatomeen. Exploration du Parc National Albert, Mission H. Damas (1935–36)* Vol. 8. Institut des Parcs Nationaux du Congo Belge, Bruxelles, 199 pp.
Hutchinson, G.E. 1967. *A Treatise on Limnology,* Vol. 2. John Wiley & Sons, New York, 1115 pp.
Jokiel, P.L. and York, R.H., Jr. 1984. Importance of ultraviolet radiation in photoinihibition of microbial growth. *Limnology and Oceanography* 29:192–199.
Kalff, J. and Watson, S. 1986. Phytoplankton and its dynamics in two tropical lakes: a tropical and temperate zone comparison. *Hydrobiologia* 138:161–176.
Kilham, P. 1971. A hypothesis concerning silica and the freshwater planktonic diatoms. *Limnology and Oceanography* 16:10–18.
Kilham, P., and Kilham, S.S. 1980. The evolutionary ecology of phytoplankton. In: Morris, I. (editor), *The Physiological Ecology of Phytoplankton.* Blackwell Scientific Publications, Oxford. p. 571–597.
Kilham, P., Kilham, S.S., and Hecky, R.E. 1986. Hypothesized resource relationships among African planktonic diatoms. *Limnology and Oceanography* 31:1169–1181.
Kilham, S.S. 1975. Kinetics of silicon-limited growth in the freshwater diatom *Asterionella formosa. Journal of Phycology* 11:396–399.
Kilham, S.S. 1984. Silicon and phosphorus growth kinetics and comparative interactions between *Stephanodiscus minutus* and *Synedra* sp. *Verhandlungen Internationale Vereinigung für Theoretische und Angewandte Limnologie* 22:435–439.
Kilham, S.S. and Kilham, P. 1975. *Melosira granulata* (Ehr.) Ralfs: morphology and ecology of a cosmopolitan freshwater diatom. *Verhandlungen Internationale Vereinigung für Theoretische und Angewandte Limnologie* 19:2716–2721.
Kozhov, M. 1963. *Lake Baikal and its Life. Monographiae Biologicae* Vol. 11. Junk Publishers, The Hague, 344 pp.
Lerman, A., and Hull, A.B. 1987. Background aspects of lake restoration: water balance, heavy metals, phosphorus homeostasis. *Schweizerische Zeitschrift für Hydrologie* 49:148–169.

Lund, J.W.G. 1954. The seasonal cycle of the planktonic diatom, *Melosira italica* (Ehr.) Kütz. subsp. *subarctica* O. Müll. *Journal of Ecology* 42:151–179.

Lund, J.W.G. 1959. Buoyancy in relation to the ecology of the freshwater phytoplankton. *British Phycological Bulletin* 7:1–17.

Müller, O. 1904. Bacillariaceen aus dem Nyassalande und einigen benachbarten Gebieten. *Engler's Botanische Jahrbücher* 34:256–301.

Paerl, H.W. 1988. Nuisance phytoplankton blooms in coastal, estuarine, and inland waters. *Limnology and Oceanography* 33:823–847.

Petrova, N.A. 1986. Seasonality of *Melosira*-plankton of the great northern lakes. *Hydrobiologia* 138:65–73.

Reynolds, C.S. 1984. *The Ecology of Freshwater Phytoplankton.* Cambridge University Press, Cambridge, 384 pp.

Reynolds, C.S., Montecino, V., Graf, M.E., and Cabrera, S. 1986. Short-term dynamics of a *Melosira* population in the plankton of an impoundment in central Chile. *Journal of Plankton Research* 8:715–740.

Richardson, J.L. 1968. Diatoms and lake typology in East and Central Africa. *Internationale Revue der Gesamten Hydrobiologie* 53:299–338.

Simonsen, R. 1979. The diatom system: Ideas on phylogeny. *Bacillaria* 2: 9–71.

Sinada, F. and Abdel Karim, A.G. 1984. A quantitative study of the phytoplankton in the Blue and White Niles at Khartoum. *Hydrobiologia* 110:47–55.

Stager, J.C. 1984. The diatom record of Lake Victoria (East Africa): the last 17,000 years. In: Mann, D.G. (editor), *Proceedings of the Seventh International Diatom Symposium, Philadelphia.* Otto Koeltz Science Publishers, Koenigstein, p. 455–476.

Stoermer, E.F., Kreis, R.G., Jr., and Sicko-Goad, L. 1981. A systematic, quantitative, and ecological comparison of *Melosira islandica* O. Müller with *M. granulata* (Ehr.) Ralfs from the Laurentian Great Lakes. *Journal of Great Lakes Research* 7:345–356.

Talling, J.F. 1957. Photosynthetic characteristics of some freshwater plankton diatoms in relation to underwater radiation. *The New Phytologist* 56:29–50.

Talling, J.F. 1963. Origin of stratification in an African rift lake. *Limnology and Oceanography* 8:68–78.

Talling, J.F. 1966. The annual cycle of stratification and phytoplankton growth in Lake Victoria (East Africa). *Internationale Revue der Gesamten Hydrobiologie* 51:545–621.

Talling, J.F. 1969. The incidence of vertical mixing, and some biological and chemical consequences, in tropical African lakes. *Verhandlungen Internationale Vereinigung für Theoretische und Angewandte Limnologie* 17:998–1012.

Talling, J.F. 1976. Phytoplankton, composition, development and productivity. In: Rzoska, J. (editor), *The Nile.* Junk, The Hague, p. 385–402.

Talling, J.F. 1986. The seasonality of phytoplankton in African lakes. *Hydrobiologia* 138:139–160.

Talling, J.F. 1987. The phytoplankton of Lake Victoria (East Africa). *Archiv für Hydrobiologie Ergebnisse der Limnologie* 25:229–256.

Thomson, B.E., Worrest, R.C., and Van Dyke, H. 1980. The growth response of an estuarine diatom (*Melosira nummuloides* [Dillw.] Ag.) to UV-B (290–320) radiation. *Estuaries* 3:69–72.

Tilman, D., Kilham, S.S. and Kilham, P. 1982. Phytoplankton community ecology: The role of limiting nutrients. *Annual Review of Ecology and Systematics* 13:349–372.

Titman (Tilman), D. and Kilham, P. 1976. Sinking in freshwater phytoplankton: Some ecological implications of cell nutrient status and physical mixing processes. *Limnology and Oceanography* 21:409–417.

Vincent, W.F. 1983. Phytoplankton production and winter mixing: Contrasting effects in two oligotrophic lakes. *Journal of Ecology* 71:1–20.

# 21

# Cyanobacteria in Large Lakes: A Case Study in Lake Constance

A. Ernst

ABSTRACT In contrast to most lakes that have undergone a process of eutrophication, the significant increase in SRP (11 to 93 $\mu g\ l^{-1}$) that took place in Lake Constance in the 1960's and 1970's was not accompanied by any noticeable increase in the cyanobacteria population; it remained a minor component of the phytoplankton assemblage. Moreover, in 1971–1972, a sudden abundance of cyanobacteria in the Untersee of Lake Constance could not be related to phosphorus modification. In 1986, a small peak of *Anabaena* was monitored for nitrate and nitrogen fixation. The results indicate that *Anabaena* has limited affinity for nitrate and has to switch to nitrogen fixation. If nitrate is present, nitrogen fixation—an energy-consuming process—does not give cyanobacteria an ecological advantage over species with greater affinity for nitrate.

Finally, it was found that, from 1900 to 1986, all the unusually high concentrations of cyanobacteria in Lake Constance corresponded to periods of extremely low water level, which are also times of low wind intensity, increased stratification, and low water turbulence. This indicates that physical factors have a profound impact on the relative abundance of cyanobacteria in Lake Constance.

## 21.1 Introduction

In lakes undergoing eutrophication, an initial increase of phytoplankton biomass is often followed by blooms of cyanobacteria. The nuisance caused by such blooms has promoted active research concerning the factors which favor the dominance of the cyanobacteria. Fuhs et al. (1972) has claimed that high P concentrations are a prerequisite to the mass development of cyanobacteria. In his world-wide survey of 22 lakes, Smith (1983, 1986) concludes that low N:P ratios give a definite advantage to cyanobacteria over other algal species, especially when light availability decreases (see also Ramberg, 1988). Other authors point to a correlation of cyanobacterial abundance, low N:P ratio, and elevated water temperature (Tilman et al., 1986; McQueen and Lean, 1987).

Increased temperature, however, may just reflect an increased thermal stability of the water body (Robarts and Zohary, 1987; Zohary and Robarts, 1989). Uehlinger (1981), Haney (1987), Burns (1987), and Lampert (1987) emphasize the absence of grazing by zooplankton as an additional factor that favors the massive development of cyanobacteria.

The different life forms of cyanobacteria comprise species with effective light-harvesting systems and other species with the ability to regulate buoyancy (Tilzer 1987; Reynolds et al., 1987). These mechanisms allow them to exploit both strongly segregated waters (light- and nutrient-rich layers separated vertically) and waters experiencing high-frequency fluctuations in vertical mixing (Reynolds, 1987).

However, although cyanobacterial blooms are common in many lakes and reservoirs, they are rare in large lakes, and, in particular, rare in Lake Constance (Mäckle et al., 1987).

In the present paper, on the basis of the long-term data series on nutrients and phytoplankton of Lake Constance, we review the factors which can explain the limited occurrence of cyanobacteria in this lake.

## 21.2 Hydrology of Lake Constance

Lake Constance has two basins: the Obersee, a large (476 km$^2$) and deep (mean depth, 100 m; maximum depth, 254 m) lake of glacial origin; and the Untersee, a small (63 km$^2$) and shallow basin (mean depth, 13.2 m; maximum depth, 46 m) which contains only 1.8% of the total water volume.

The Alpenrhein (mean discharge of 227 m$^3$ s$^{-1}$) enters the Obersee in its eastern part; then, the lake water flows into the Untersee and into the Rhine River. The maximum yield of the snow-fed Alpenrhein occurs in summer, and so does the maximum water level of the lake.

The Obersee is warm-monomictic with thermal stratification extending from mid-March to October (Kiefer, 1972). It only freezes about once every 80 years. In summer, both the flood of the Rhine and the internal seiches affect the vertical thermal profile down to 40 m (Lehn, 1963).

The Untersee often freezes over in winter and is less influenced by the Rhine flood in summer because of natural barriers (Auerbach and Rottengatter, 1960). As long ago as the 1930s, oxygen depletion, $H_2S$ formation, and phosphorus release were observed in the hypolimnion, during the summer stratification (Elster and Einsele, 1938; Frevert, 1980).

The population around Lake Constance has tripled during the last one hundred years, and intensive agriculture and tourism have developed accordingly (Zengerling, 1968). Since 1950, the Obersee has undergone accelerated eutrophication with increasing winter SRP concentrations (Wagner, 1976; Elster, 1977) (also see Figure 21.1). Maximum winter SRP concentrations of ca. 93 mg m$^{-3}$ were reached in 1980. As a result of concern for the water quality of Lake Constance, the largest reservoir of drinking water in Europe (Kiefer,

1960), an increasing fraction of the sewage entering the lake is now being sewage treated, especially to reduce the level of phosphorus. As a result, a gradual decline of winter concentrations of SRP has been observed since 1979 in the Untersee (Müller, 1984) and since 1981 in the Obersee (Mäckle et al., 1987). By 1989, winter SRP concentrations in Obersee have declined to about 50 mg $m^{-3}$ (Limnological Institute, Konstanz, unpublished data).

## 21.3 Role of Phosphorus in Cyanobacterial Growth

**Pre-Eutrophication Period: 1896–1950** In all the botanical surveys, algal productivity studies, and water monitoring programs of the last one hundred years for Lake Constance, cyanobacteria have been noted to be a minor plankton component (Table 21.1). In 1902, Schröter and Kircher reported 3 planktonic and 65 benthic cyanobacteria, one of the latter (an Oscillatoria) in the profundal and another in the littoral area. In 1921–1922, Auerbach et al. (1924) found 13 pelagic cyanobacteria. Reaching remarkable cell numbers in 1921, two of them, *Anabaena flos-aquae* and *Microcystis aeruginosa*, were classified among the 12 most abundant algal species of the lake. Zimmermann (1928) described colored filamentous and colony-forming species in the profundal down to a depth of 30 m (see also Kann and Sauer, 1982). However, in the productivity studies of the 1930s cyanobacterial biomass was negligible (Grim, 1939, 1951, 1955).

**Eutrophication Period: 1950–Present** Lehn (1973) was the first to observe a 900% increase in cyanobacterial cell numbers in the Untersee and a corresponding 600% increase in the Obersee over the period 1953–1973. On two sampling surveys, Müller (1967) reported 21 planktonic species of cyanobacteria, and Mattern (1970) identified more than 150 benthic forms, some of which exhibited high density near the sewage effluents. However, in spite of growing eutrophication of the Obersee, no further development of cyanobacteria was observed (Figure 21.1). Between 1961 and 1986, the cyanobacterial biomass generally was less than 10% of the total annual phytoplankton biomass (Bürgi, 1976; Bürgi and Lehn, 1979; Sommer, 1981, 1985; Mäckle et al., 1987; Müller, 1988). From 1965 to 1975, *Anabaena* spp. *Aphanizomenon*, and unicellular species reached their annual maximum after the August phytoplankton peak (Bürgi and Lehn, 1979; Sommer et al., 1986) as reported in the previous studies of Schröter and Kirsher (1902), Kolkwitz (1912), and Auerbach et al. (1924). In just one year, 1972, the cyanobacteria constituted approximately 40% of the summer phytoplankton biomass (Bürgi and Lehn, 1979), while in most years they remained below 20%. It has to be admitted that the numbers refer to large or colony-forming cyanobacteria only. The cyanobacterial picoplankton of Lake Constance was not included in the historical records, although it contributes significantly to primary production and cyanobacterial biomass (Weisse, 1988).

**Table 21.1** Cyanobacterial composition of the phytoplankton of Lake Constance.

| Period of study | | Cyanobacteria species observed | Reference |
|---|---|---|---|
| Obersee | Untersee | | |
| 1890–1895 | | *Microcystis, Anabaena circinalis, A. fl.-aquae, Oscillatoria,* 64 benthic species (littoral) | Schröter and Kircher, (1902) |
| | 1900–1905 | 25 benthic species (littoral) | Baumann (1911) |
| 28/8/1911 | | *Anabaena, Microcystis* (rare) | Kolkwitz (1912) |
| 1919–1923 | | *Anabaena, Microcystis,* 11 other species (rare) | Auerbach, Maerker, and Schmalz (1924) |
| | | 3 species in profundal | Zimmermann (1928) |
| 1934–1936 | | Few cyanobacteria | Grim (1939) |
| | 1934–1935 | Few cyanobacteria | Grim (1951) |
| 1938–1939 | 1950–1951 | Few cyanobacteria | Grim (1955) |
| | 1955 | Few cyanobacteria | Auerbach and Rottengatter (1960) |
| 1954–1962 | 1954–1962 | *Oscillatoria, Anabaena* | Lehn (1969) |
| 1960–1963 | | *Anabaena, Microcystis* | Elster and Motsch (1966) |
| | 1960–1961 | *Oscillatoria* | Elster and Motsch (1966) |
| 1960, 1961, 1963, 1965 | | | Findenegg (1966) |
| 1960–1962 | | $\geq$ 150 benthic species | Mattern (1970) |
| 1961–1963 | | *Anabaena* | Bürgi (1976) |
| | 1961–1963 | *Oscillatoria* | Bürgi (1976) |
| 1964–1965 | | 21 planktonic species, *Anabaena, Microcystis* | Müller (1967) |
| | 1963–1973 | *Aphanizomenon* (blooms), *Anabaena* (periodical), *Aphanothece/Aphanocapsa* | Bürgi (1977) |
| 1965–1975 | | *Aphanothece/Aphanocapsa, Anabaena* (periodical) | Bürgi and Lehn (1979) |
| 1979–1982 | | *Anabaena* | Sommer (1981, 1985) |
| 1977–1986 | | 4.4% cyanobacteria | Mäckle et al. (1987) |
| 1965–1986 | | 6.7% cyanobacteria | Müller (1988) |
| | 1976–1986 | 4.5% cyanobacteria | Müller (1988) |

In the Untersee, in addition to the regular occurrence of *Anabaena* and the sporadic appearance of unicellular species, a bloom of *Oscillatoria* was observed in the early 1960s. In October 1960, this usually rare species was found at 5 m depth (Findenegg, 1960). It reappeared in the deep zone during the summer and autumn of 1961 (Elster and Motsch, 1966; Bürgi, 1976; Lehn, 1969). Although the maximum biomass concentrations of this species did not exceed 5 g $m^{-2}$ in the pelagial (Bürgi, 1976), surface scums of *Oscillatoria* and other planktonic species were so obvious that many people enquired about this phenomenon at the local monitoring station (Mattern, 1970).

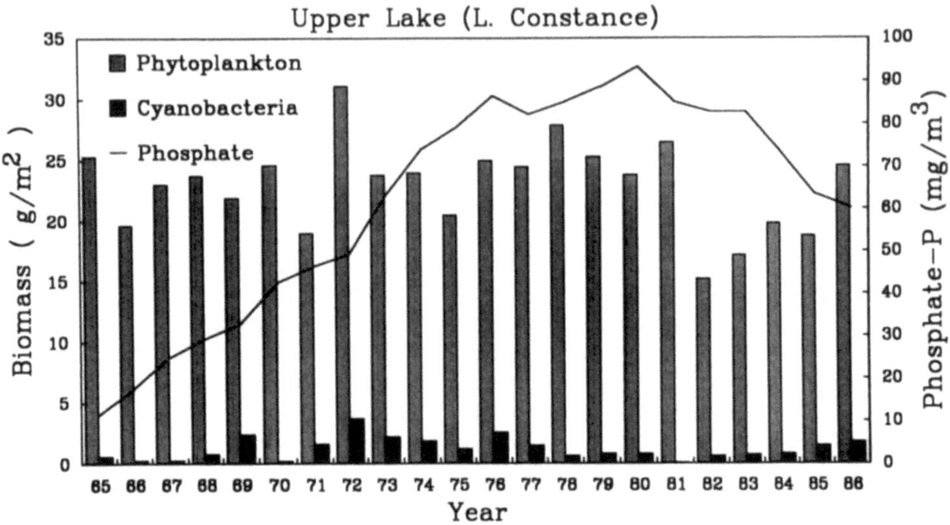


**Figure 21.1** Biomass of phytoplankton and cyanobacteria in the Obersee of Lake Constance. Picoplankton is not included. For the period 1965–1975, the data from Bürgi and Lehn (1979) integrate the water column from 0 to 50 meters. For the period 1976–1986, the data from Müller (1988) integrate a 0 to 20 m water column. Both data sets were obtained at the Fischbach-Uttwil station. Phosphate data are from Mäckle et al. (1987).

**Table 21.2** Biomass fluctuations for total cyanobacteria and for *Aphanizomenon flos-aquae* in the Untersee.

| Biomass (g m$^{-2}$) | Dec 1971 | Jul 1972 | Sep 1973 | Oct 1977 |
|---|---|---|---|---|
| Total cyanobacteria | 64 | 153 | 87 | |
| *A. flos-aquae* | 52 | 125 | 8 | 41 |

Data from Bürgi (1977) and Uehlinger (1981), integrated over a 0 to 20 m depth.

In the 1970s, unusually high concentrations of cyanobacteria, dominated by the filamentous diazotrophe *Aphanizomenon flos-aquae*, were observed at different periods (Table 21.2 and Figure 21.2) in the Untersee of Lake Constance. In conclusion, two points should be emphasized:

1. In the Obersee, the almost ninefold increase of SRP in the water over the period 1965–1980 did not trigger any mass development of cyanobacteria or even initiate an increasing trend.
2. In the Untersee, two unusual "cyanobacteria years" (1971–1972) did not correspond to any significant increase of phosphorus in comparison with the immediately preceding and following period, both very poor in cyanobacteria.

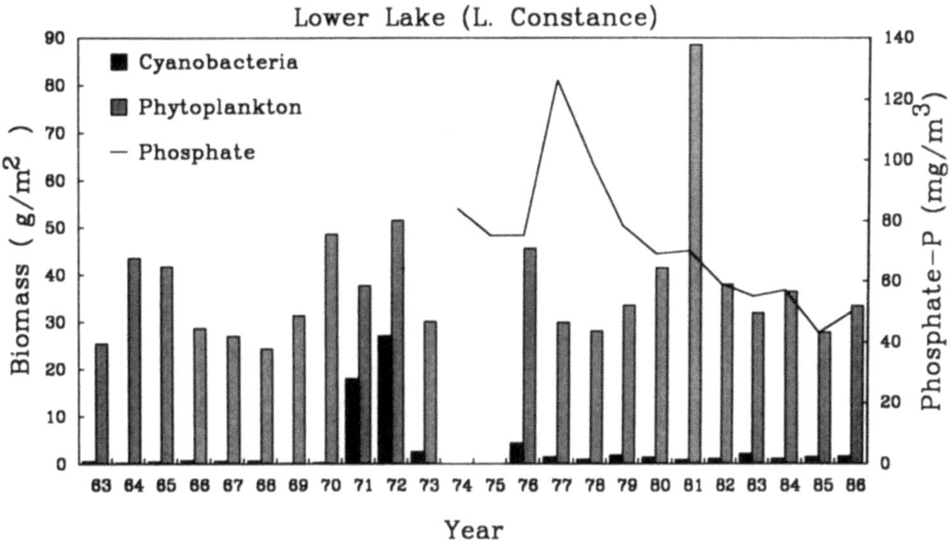


**Figure 21.2** Biomass of phytoplankton and cyanobacteria of the Untersee of Lake Constance. Data were measured at the Berlingen station (Rheinsee) for the period 1963–1973 (Bürgi, 1977) and at the Zellersee station for the period 1976–1986 (Müller, 1988). Data integrate a 0 to 20 m water column. Phosphate data (Zellersee) from Schmitz, 1974 and Müller, 1988. No data were reported for 1974 and 1975.

It seems, therefore, that in Lake Constance, phosphorus is not the main factor controlling the development of cyanobacteria.

## 21.4 Role of Nitrogen in Cyanobacterial Growth

In order to examine the effect of nitrogen limitation on the cyanobacteria of Lake Constance, we measured the concentration of $NO_3$-N in the water during a small peak of *Anabaena spiroides* in summer 1986 (Figure 21.3 *a*)*. The particulate fraction of the lake water was assayed for nitrogenase activity (NA); ammonia is not present at this period (Figure 21.3 *b*).

At maximum, *Anabaena* constituted of 7.7% of the phytoplankton biomass,

*Procedure: Samples for nitrogenase activity (3 1) were collected at 0, 1, and 2 m depth and at 7, 8, 9, and 10 m depth. The pooled samples were filtered through two layers of glass fiber filters (Schleicher and Schüll). The filters were cut into pieces and used for chlorophyll determination and nitrogenase assays. Nitrogenase assays were performed in 25-ml glass vessels closed by rubber septum stoppers in an illuminated (200 $\mu Em^{-2}\ s^{-1}$) water bath at 20°C. The vessels were partly filled with lake water; the gas phase consisted of air and 10% acetylene. Aliquots of the gas phase were analyzed after 3 and 6 h with a Carlo Erba gaschromatograph fitted with a flame ionization detector. Nitrate was determined for the pooled samples (see above), filtered through Millipore filters, Type HA, 0.45 $\mu$m, and stored at −20°C. Nitrate (and phosphate) was analyzed with a HPLC apparatus (SYCAM, Munich, F. R. G.) equipped with a suppressor column and a conductivity detector.

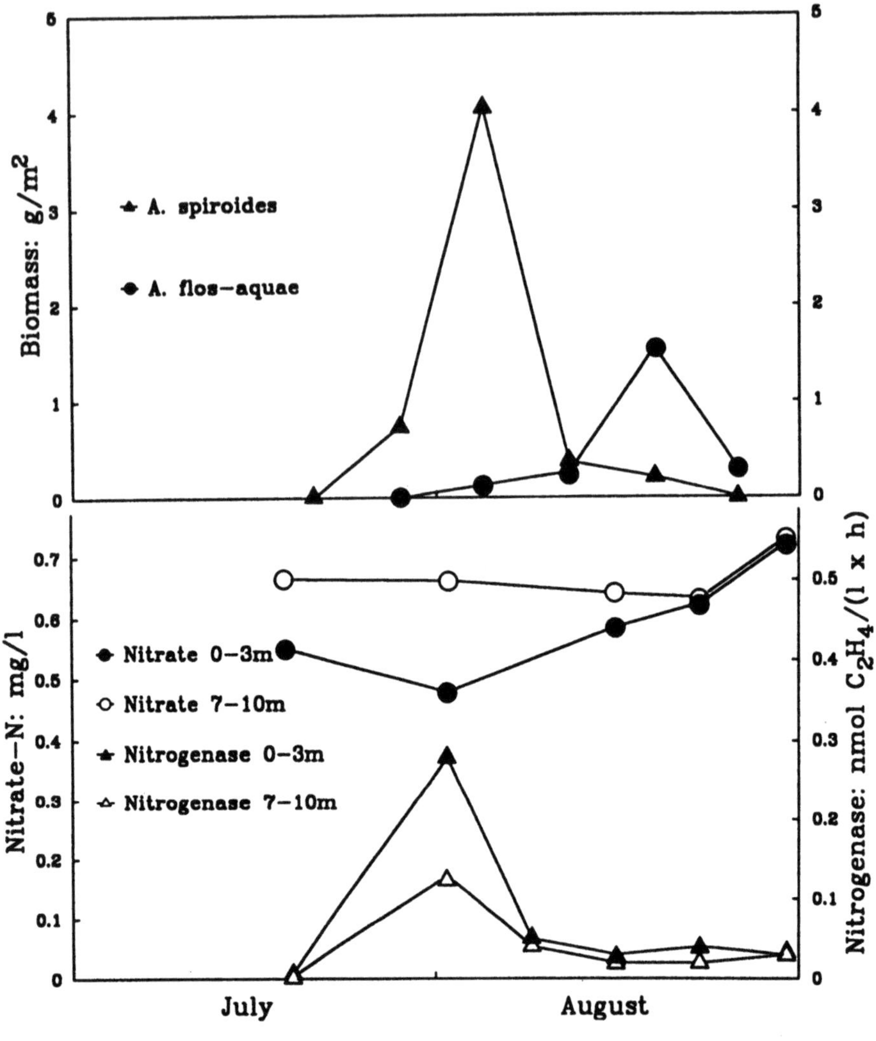


**Figure 21.3** (*a*) Biomass of cyanobacteria; (*b*) nitrogenase activity and nitrate concentration in Lake Constance in July and August 1986. Biomass samples were collected in the western Obersee and processed by C. Braunwarth (1988).

and the concentration of nitrate decreased only slightly in the surface water (0–3 m) during the development of the algal population (see Figure 21.3 *b*) but did not change in the 7–10 m depth. A low, light-induced NA was detected (see Figure 21.3 *b*) in the 0–3 m layer at the beginning of August, during the growing phase of *Anabaena*. A week later, NA had significantly diminished, along with the biomass of *Anabaena*.

The presence of NA and of heterocysts, the site of NA in *Anabaena*, in media containing relatively high concentrations of nitrate (5–10 mM) has been commonly observed in continuous culture studies (Bone, 1971; Healey, 1973; Meeks et al., 1983; Rhee and Lederman, 1983; Layzell et al., 1985). At high growth rates, the standing crop of such cultures is diminished at intermediate nitrate concentrations (150–400 $\mu g/l\ NO_3\text{-}N^{-1}$ (Elder and Parker, 1984). In this concentration range, *Anabaena* cannot use nitrate efficiently and therefore has to spend additional energy on fixing nitrogen using its functional heterocysts (Scherer et al., 1987).

Thus, when nitrate is present in the water, nitrogen fixation is not of ecological advantage for cyanobacteria. Rather, it may actually impair their competitive success against other algal species with a higher affinity for nitrate. This may explain their generally low abundance in Lake Constance, which exhibits 300–500 $\mu g/1\ NO_3\text{-}N$ in the euphotic zone during summer stratification.

## 21.5 Role of Climatological Factors

*Aphanizomenon flos-aquae*, which dominated the 1971–1972 bloom, is a nitrogen fixer with modest nutrient and light requirements (Uehlinger, 1981). What factors, then, gave such an advantage to this species in this particular period? Looking systematically at all the years with a relatively abundant biomass of cyanobacteria, we noted that they all corresponded to years of extremely low water levels (Table 21.3).

The year 1971/72 was exceptionally dry (700 mm rainfall instead of a 1000 mm average) in the lake area (Wagner et al., 1976). The water level remained low throughout the year (Kiefer, 1978) and winter overturn was incomplete (Elster, 1977). The water level record since 1871 (Kiefer, 1965, 1978) indicates that the years around 1920, the period between 1943 and 1950, the early 1960s, and the early 1970s were particularly dry and the lake was very low. All years with low water levels were characterized by summer maxima of cyanobacteria (see Table 21.3). No data were available for three dry years following World War II.

It is a definite possibility that the striking coincidence observed between low water level and cyanobacteria prosperity reflects a causal relationship. Low water levels occur in dry years with low wind velocity and are consequently associated with low water turbulence, stable stratification, and increased temperature in the upper layers of the water column. Similar conclusions have been reported by Berger and Sweers (1988) when assessing the causes for a heavy *Oscillatoria* bloom observed in 1976 in the Ijsselmeer (Netherlands).

The failure of cyanobacteria to thrive in turbulent environments may be related to an inappropriate reaction of their buoyancy regulation system and or an inability to adjust their metabolic activity rapidly enough when turbulence continually modifies light intensity. Furthermore, in presence of nitrate in a mixed surface layer, they might be outcompeted by nonbuoyant species (green algae, diatoms) able to effectively assimilate $NO_3\text{-}N$.

**Table 21.3** Relationship between low water level and cyanobacterial abundance in the Obersee[a].

| | Water level (cm) | | | |
|---|---|---|---|---|
| Year | Annual average | Summer maximum | Cyanobacterial biomass (g $m^{-2}$) | Reference |
| 1921 | 299 | 370 | | Auerbach et al. (1924) |
| 1943 | 314 | 431 | | No data |
| 1947 | 305 | 386 | | No data |
| 1949 | 289 | 377 | | No data |
| 1959 | 314 | 419 | | Fast (1960) |
| 1962 | 319 | 420 | 7.8 (0–20 m) | Elster and Motsch (1966), Bürgi (1976) |
| 1964 | 310 | 399 | | Müller (1967) |
| 1971 | 308 | 412 | 9.7 (0–50 m) | Bürgi and Lehn (1979) |
| 1972 | 299 | 390 | 24.3 (0–50 m) | Bürgi and Lehn (1979) |
| 1976 | 299 | 364 | 18.3 (0–20 m) | Müller (1988) |

[a] The 10 lowest annual average water levels observed between 1900 and 1986 were selected for this list (Kiefer, 1965, 1978, local weather monitoring station). For purposes of comparison: the average water level over a 100-year period was 345 cm (zero level: 391,766 m above sea level). The average minimum is reached in February (281 cm) and the average maximum in July (433 cm).

*Acknowledgements* The author thanks Dr. Bürgi, Dr. Mäckle, and Dr. Müller for the data sets represented in figures 21.1 and 21.2, and Dr. Braunwarth for the phytoplankton data of Figure 21.3 and is indebted to Dr. Colette Serruya for her encouraging support during preparation of this report. This work was supported by the Deutsche Forschunsgemeinschaft through the SFB 248 "Stoffhaushalt des Bodensees."

## References

Auerbach, M., Maerker, W., and Schmalz, J. 1924. Hydrographisch-biologische Bodensee-Untersuchungen. *Archiv für Hydrobiologie/Supplement* 3:597–720.

Auerbach, M. and Rottengatter, G. 1960. Untersuchungen über den Wasseraustausch der einzelnen Becken des Untersees (Bodensee) *Hydrologia* 22:598–639.

Baumann, E. 1911. Die Vegetation des Untersee (Bodensee). *Archiv für Hydrobiologie/Supplement* 1:1–324.

Berger, C. and Sweers, H.E. 1988. The IJsselmeer and its phytoplankton with special attention to the suitability of the lake as a habitat for *Oscillatoria agardhii* Gom. *Journal of Plankton Research* 10:574–599.

Braunwarth, C. 1988. Populationsdynamik natürlicher Phytoplankton-populationen: in situ Wachstums- und Verlustraten. Dissertation. Universität Konstanz.

Bone, D.H. 1971. Nitrogenase acitivity and nitrogen assimilation in *Anabaena flos-aquae* growing in continuous culture. *Archiv für Mikrobiologie* 80:234–241.

Bürgi, H.R. 1976. Die Phytoplanktonentwicklung im Bodensee in den Jahren 1961–1963. *Bericht der Internationalen Gewässerschutz-kommission Bodensee* 18:1–87.

Bürgi, H.R. 1977. Die langjährige Entwicklung des Phytoplankton im Bodensee. Teil 1, Untersee (1963–1973). *Bericht der Internationalen Gewässerschutzkommission Bodensee* 21:1–43.

Bürgi, H.R. and Lehn, H. 1979. Die langjährige Entwicklung des Phytoplanktons im Bodensee. Teil 2, Obersee (1965–1975). *Bericht der Internationalen Gewässerschutzkommission Bodensee* 23:1–86.

Burns, C.W. 1987. Insights into zooplankton—cyanobacteria interactions derived from enclosure studies. *New Zealand Journal of Marine and Environmental Research* 21:477–482.

Elder, R.G. and Parker, M. 1984. Growth response of a nitrogen fixer (*Anabaena flos-aquae*, cyanophyceae) to low nitrate. *Journal of Phycology* 20:296–301.

Elster, H.J. 1977. Der Bodensee—Bedrohung und Sanierungs-möglichkeiten eines Ökosystems. *Naturwissenschaften* 64:207–215.

Elster, H.J., and Einsele, W., 1938. Beiträge zur Kenntnis der Hydrographie des Untersees (Bodensee). *Internationale Revue der gesamten Hydrobiologie und Hydrographie* 36:241–284.

Elster, H.J. and Motsch, B. 1966. Untersuchungen über das Phytoplankton und die organische Urproduktion in einigen Seen des Hochschwarzwaldes, im Schleinsee und Bodensee. *Archiv für Hydrobiologie/Supplement* 28:291–376.

Fast, H. 1960. Neuere Untersuchungsergebnisse über den Zustand des Bodensees (westl. Obersee). *Vom Wasser, Jahrbuch für Wasserchemie und Wassergewinnungstechnik* 27:11–32.

Findenegg, I. 1966. Phytoplankton und Primärproduktion einiger ostschweizer Seen und des Bodensees. *Schweizer Zeitschrift für Hydrobiologie* 28:148–172.

Frevert, T., 1980. Dissolved oxygen dependent phosphorus release from profundal sediments of Lake Constance (Obersee). *Hydrobiologia* 74:17–28.

Fuhs, G.W., Demmerle, S.D., Canelli, E., and Chen, M. 1972. Characterization of phosphorus-limited plankton algae (with reflections on the nutrient limitation concept.) In: Likens, G.E. (ed.), Nutrients and eutrophication: the limiting nutrient controversy. *Special Symposium American Society of Limnology and Oceanography* 1:113–133.

Grim, J. 1939. Beobachtungen am Phytoplankton des Bodensees (Obersee) sowie deren rechnerische Auswertung. *Internationale Revue der Hydrobiologie* 39:13–315.

Grim. J. 1951. Ein Vergleich der Produktionsleistung des Bodensee-Untersees, des Obersees und des Schleinsees. *Abhandlungen aus der Fischerei Lfg.* 4:787–841.

Grim, J. 1955. Die chemischen und planktonischen Veränderungen des Bodensee-Obersees in den letzten 30 Jahren. *Archiv für Hydrobiologie/Supplement* 22:310–322.

Haney, J.F. 1987. Field studies on zooplankton—cyanobacteria interactions. *New Zealand Journal of Marine and Freshwater Research* 21:467–475.

Healey, F.D. 1973. Characteristics of phosphorus deficiency in *Anabaena*. *Journal of Phycology* 9:383–394.

Kann, E. and Sauer, F. 1982. Die "Rotbunte Tiefenbiocoenose" Neue Beobachtungen in österreichischen Seen und eine zusammenfassende Darstellung. *Archiv für Hydrobiologie* 95:181–195.

Kiefer, F. 1960. Der Bodensee in Gefahr! *Die Umschau in Wissenschaft und Technik* 60:65–68.

Kiefer, F. 1965. Die Wasserstände des Bodensees seit 1871. *Schriften des Vereins für Geschichte des Bodensees* 83:1–31.

Kiefer, F. 1972. *Naturkunde des Bodensees*. 2nd ed. Jan Thorbecke Verlag, Sigmaringen, F.R.G. p. 1–209.

Kiefer, F. 1978. 100 Jahre metrische Wasserstandsmessung am Bodensee. *Schriften des Vereins für Geschichte des Bodensees* 96:203–216.

Kolkwitz, R. 1912. Das Plankton des Rheinstroms von seinen Quellen bis zur Mündung. *Berichte der deutschen botanischen Gesellschaft* 30:205–226.

Lampert, W. 1987. Laboratory studies on zooplankton-cyanobacteria interactions. *New Zealand Journal of Marine and Freshwater Research* 21:483–490.

Layzell, D.B., Turpin, D.H., and Elrifi, I.R. 1985. Effect of N-source on the steady state growth and N-assimilation of P-limited *Anabaena flos-aquae. Plant Physiology* 78:739–745.

Lehn, H. 1963. Einige Frühjahrsbefunde über die Isothermen-Phytoplanktonrelation im Bodensee. *Archiv für Hydrobiologie* 59:1–25.

Lehn, H. 1969. Die Veränderungen des Phytoplanktonbestandes im Bodensee. I. Fluktuationen von Tabellaria fenestra. *Internationale Revue der gesamten Hydrobiolgie* 54:367–411.

Lehn, H. 1973. Phytoplanktonveränderungen im Bodensee und einige Folgeproblems. *Verhandlungen der Gesellschaft für Ökologie, Saarbrücken.* 24:225–234.

Mäckle, H., Zimmermann, U., and Stabel, H.-H. 1987. Phytoplankton-entwicklungen in tiefen Voralpenseen. *GWF Wasser–Abwasser* 128:544–550.

Mattern, H. 1970. Boebacht ungen über die Algenflora im Uferbereich des Bodensees (Überlinger See und Gnadensee). *Archiv für Hydrobiologie/Supplement* 37:1–163.

Meeks, J.C., Wycoff, K.L., Chapman, J.S., and Enderlin, C.S. 1983. Regulation of expression of nitrate and dinitrogen assimilation by *Anabaena* species. *Applied and Environmental Microbiology* 45:1351–1359.

McQueen, D.J., and Lean, D.R.S. 1987. Influence of water temperature and nitrogen to phosphorus ratio on the dominance of blue-green algae in Lake St. George, Ontario. *Canadian Journal of Fishery and Aquatic Science* 44, 598–604.

Müller, H. 1967. Eine neue qualitative Bestandsaufnahme des Phytoplanktons des Bodensee-Obersees mit besonderer Berücksichtigung der tychoplanktischen Diatomeen. *Archiv für Hydrobiologie/Supplement* 33:206–236.

Müller, H. 1984. Limnologischer Zustand des Bodensees. *Internationale Gewässerschutzkommission für den Bodensee* No. 11:1–69.

Müller, H. 1988. Limnologischer Zustand des Bodensee. *Internationale Gewässerschutzkommission für den Bodensee:* No. 14:1–68.

Ramberg, L. 1988. Relations between planktic blue-green algal dynamics and environmental factors in four eutrophic Swedish lakes. *Archives of Hydrobiology* 112:161–175.

Reynolds, C.S. Oliver, R.L., and Walsby, A.E. 1987. Cyanobacterial dominance: the role of buoyancy regulation in dynamic lake environments. *New Zealand Journal of Marine and Freshwater Research* 21:379–390.

Rhee, G.-Y. and Lederman, T.C. 1983. Effect of nitrogen sources on P- limited growth of *Anabaena flos-aquae. Journal of Phycology* 19:179–185.

Robarts, R.D. and Zohary, T. 1987. Temperature effects on photosynthetic capacity, respiration, and growth rates of bloom-forming cyanobacteria. *New Zealand Journal of Marine and Freshwater Research* 21:391–399.

Scherer, S., Almon, H., and Böger, P. 1988. Interaction of photosynthesis, respiration and nitrogen fixation in cyanobacteria. *Photosynthesis Research* 15, 95–114.

Schmitz, W. 1974. Limnologischer Zustand des Bodensees. *Internationale Gewässerschutzkommission für den Bodensee* No. 1:1–90.

Schröter, C. and Kircher, O. 1902. Die Vegetation des Bodensees Teil II. *Schriften des Vereins für Geschichte des Bodensees und seiner Umgebung* 31, Supplement 1:1–81.

Smith, V.H. 1983. Low nitrogen to phosphorus ratios favour dominance by blue-green algae in lake phytoplankton. *Science* 221:669–671.

Smith, V.H. 1986. Light and nutrient effects on the relative biomass of blue-green algae in lake phytoplankton. *Canadian Journal of Fishery and Aquatic Science* 43:148–153.

Sommer, U. 1981. The role of r- and k-selection in the succession of phytoplankton in Lake Constance. *Acta Oecologia/Oecol. Generale* 2:327–342.

Sommer, U. 1985. Seasonal succession of phytoplankton in Lake Constance. *BioScience* 85:351–357.

Sommer, U., Gliwicz, M.Z., Lampert, W., and Duncan, A. 1986. The PEG-model of seasonal succession of planktonic events in fresh waters. *Archive of Hydrobiology* 106:433–471.

Tilman, D., Kiesling, R., Sterner, R., Kilham, S.S., and Johnson, F.A. 1986. Green, blue-green, and diatom algae: taxonomic differences in competitive ability for phosphorus, silicon and nitrogen. *Archive for Hydrobiology* 106:473–485.

Tilzer, M.M. 1987. Light dependence of photosynthesis and growth in cyanobacteria: implications for their dominance in eutrophic lakes. *New Zealand Journal of Marine and Freshwater Research* 21:401–412.

Uehlinger, U.F. 1981. Zur Oekologie der planktonischen Blaualge *Aphanizomenon flos-aquae* in Alpenrandseen. *Schweizer Zeitschrift für Hydrologie* 43:69–88.

Wagner, G. 1976. Simulationsmodelle der Seeneutrophierung, dargestellt am Beispiel des Bodensee-Obersees. Teil 2: Simulation des Phosphorhaushalts des Bodensee-Obersees. *Archiv für Hydrobiologie* 78:1–41.

Wagner, G. Bührer, H., and Ambühl, H. 1976. Die Belastung des Bodensees mit Phosphor-, Stickstoff- und organischen Verbindungen. *Internationale Gewässerschutzkommission für den Bodensee* 17:1–55.

Weisse, T. 1988. Dynamics of autotrophic picoplankton in Lake Constance. *Journal of Phytoplankton Research* 10:1179–1188.

Zengerling, T., 1968. Zur Entwicklung der Wirtschafts- und Sozialstruktur im westlichen Bodenseegebiet. *Schriften des Verein Geschichte des Bodensees* 86:227–246.

Zimmermann, W. 1928. Algenbestände aus der Tiefenzone des Bodensees. Zur Ökologie und Soziologie der Tiefenpflanzen. *Zeitschrift für Botanik* 20:1–35.

Zohary, T. and Robarts, R.D. 1989. Diurnal mixed layers and the long-term dominance of *Microcystis aeruginosa*. *Journal of Plankton Research* 11:25–48.

# 22

# Patterns in Zooplankton Distribution and Their Causes in North American Great Lakes

K. Patalas

ABSTRACT A definition of the term "great lake" is proposed based on the mid-summer epilimnion depth. Several patterns of zooplankton distribution were discovered in North American great lakes. They were mostly a function of the following five characteristics: temperature distribution (Lake Ontario, Lake Superior, and Great Slave Lake); inflow and outflow configurations (Lake Winnipeg and Great Slave Lake); geology and drainage basin type (Lake Winnipeg and Great Slave Lake); basin morphology (Lake Erie, Lake Superior, and Great Slave L.); climate differences within the system (Lake Winnipeg). A high degree of heterogeneity was characteristic for all of the lakes but each of them developed a specific pattern determined by a specific set of external factors (climate, geology of the drainage area, hydrology) and morphology. These findings contradict some paradigms on great lakes as formulated in the preamble to this symposium. Climate (latitude and air and lake temperature) and lake morphology seem to be the primary factors controlling zooplankton abundance among large lakes. Regressions describing these relationships are presented.

We intuitively feel that processes in large lakes differ from those in small lakes. However, the nature of these differences is not yet well understood. Even the definition of a "large lake" is, and will remain, arbitrary. For the purpose of this chapter, I propose to call "large," or "great," a temperate lake in which the mid-summer epilimnion depth reaches 15–20 m. This depth approximates the asymptotic mixing depth of North American lakes (Patalas, 1984), with a fetch longer than 25 km. No deeper mixing was recorded even in lakes with fetches up to 400 km in length (Figure 22.1).

The horizontal distribution and quantity of zooplankton in lakes have always been intriguing issues. It was often tacitly assumed that zooplankton organisms with a relatively restricted mobility were rather uniformly distributed in well-mixed waters. The often-reported heterogeneous distributions of the planktonic animals can therefore be viewed as a reflection of the structural complexity of the water masses.

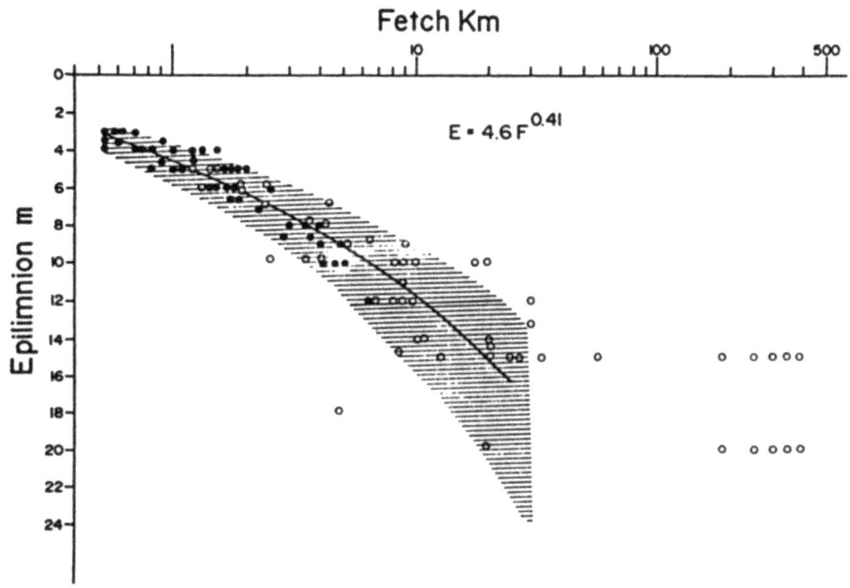


**Figure 22.1** The mid-summer epilimnion depths in lakes of different fetch in Canada (open circles) and in Poland (black dots). The regression line E equals 4.6 $F^{0.41}$ (for F = 0.5–30 km) where E is the epilimnion depth in m, F is the fetch in km. Shaded area denotes 0.95 confidence limits (Patalas, 1984).

In more-extensively investigated smaller lakes, conclusions on horizontal distribution were often a function of the size of the network of sampling stations. With a higher number of stations, a higher degree of heterogeneity was usually found. This methodological aspect was even more critical in large lakes. The effort needed to cover huge areas with a sufficient density of sampling stations was often beyond the financial capacity of the researchers who were usually associated with universities. First, in the late 1960's, due to coordinated effort undertaken in North America by the International Joint Commission (Anon., 1969), the Canada Centre for Inland Waters (Anon., 1970a, b, and c), and the Fisheries Research Board of Canada, it became possible to technically tackle the spatial structure of the Laurentian great lakes. Papers on the horizontal distribution of zooplankton began to appear (e.g., Patalas, 1972; Watson and Carpenter, 1974; Watson, 1976; Watson and Wilson, 1978). But even in these works, the planktonic distributions reported had the character of macropatterns as the distances between sampling stations were rarely shorter than 20 to 30 km.

The purpose of this paper is to present examples of distributional patterns of crustacean plankton encountered in several North American great lakes, to compare the abundance of plankton among the lakes, and to explain the causes of differences in patterns and abundances.

## 22.1 Plankton Sampling Methods

The plankton sampling in Lakes Ontario, Erie, Huron, Superior, Winnipeg, and the Great Slave Lake was carried out in cooperation with the Canada Centre for Inland Waters at Burlington and with Dr. L. Johnson and Dr. G.J. Brunskill of the Freshwater Institute in Winnipeg. Ten to 60 vertical hauls per lake, from 50 m to the surface or from the bottom to the surface in stations shallower than 50 m were taken during July or August 1967–1973. Usually, a Wisconsin plankton net, 25 cm in diameter with a mesh size of 77 $\mu$m, was used. Plankton abundance was expressed as the numbers of individuals per liter ($L^{-1}$) or per $cm^2$ area (copepod nauplii included), assuming that the filtration efficiency of the net was 100% (in fact, the efficiency is from 45–85%; later experiments indicated 63% on average). Both zooplankton abundances and mean epilimnion temperatures were averaged from 10 to 60 stations.

## 22.2 Plankton Distribution Results

There are no specific "great lake" zooplankton species. None of the 33 planktonic crustacean species found in 14 North American lakes (Patalas, 1975) are specific for large lakes. All of them are found in many other, much smaller lakes of Canada.

The one feature which may distinguish most great lakes from small ones is their highly heterogeneous horizontal distribution of plankton. The following examples will show this characteristic distribution pattern, as found in five large North American lakes.

**Lake Ontario** Lake Ontario is a relatively regularly shaped water basin, which is 19,099 $km^2$ in surface area, mean depth of 86 m, with the long axis situated from west to east (Anon., 1969). In spite of its apparent morphological simplicity, a rather prevalent pattern existed in zooplankton distribution with three to 10 times greater total plankton abundance in the eastern part than in the central and western ones (Figure 22.2). A more dramatic west-east gradient was observed for the cladoceran *Daphnia retrocurva*, ranging from 0 in the west to over 100 ind.$cm^{-2}$ in the east. The explanation of these patterns can be found in the temperature distribution which shows the same west-east trend. The prevailing, wind-generated, thermal structure of Lake Ontario shows that, on average, the epilimnion at the eastern end is twice as deep as at the western end (20 and 10 m respectively). The mean temperature of the upper 25 m layer is 5–12°C higher at the eastern end. Other causes—such as upwelling, the influx of the Niagara River, and pollutants—exert secondary influences on a more localized scale (Patalas, 1969).

**Lake Erie** Similar in size (25,821 $km^2$) to Lake Ontario but of different morphology, Lake Erie has three basins increasing in mean depth from 8 m

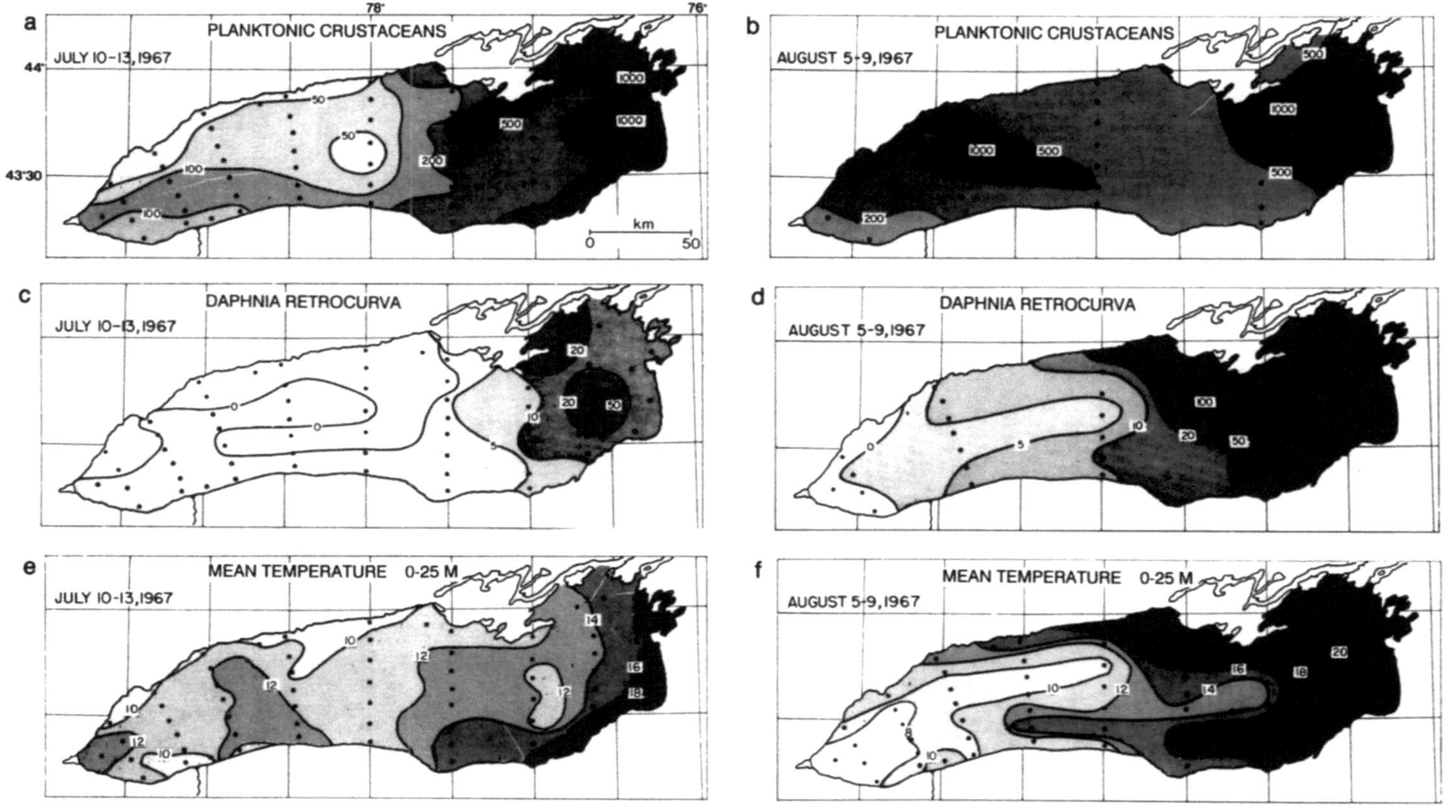


**Figure 22.2** Characteristics of Lake Ontario. Distribution of: (*a*) and (*b*) total planktonic crustaceans; (*c*) and (*d*) *Daphnia retrocurva* in Lake Ontario for July and August 1967 (indiv. $cm^{-2}$). (*e*) and (*f*) Mean temperature of the 0–25 m layer in °C (based on data from Anon., 1970a).

in the west basin, through 18 m in the central basin, to 24 m in the eastern basin (Figure 22.3). These morphological differences are reflected in the distributional patterns of three dominant cyclopids: *Acanthocyclops vernalis* occupies only the western, shallow part; *Mesocyclops edax* thrives mainly in the shallow parts of the central basin; while *Diacyclops bicuspidatus thomasi* mostly inhabits the eastern and the central basins. *Diacyclops bicuspidatus thomasi* is almost absent in the shallower western basin. In addition to this lake-morphology-related differentiation, there also appears to be a north-south trend. *Diacyclops bicuspidatus thomasi* is more abundant in the northern part of the central basin, giving way to *Mesocyclops edax* at the southern shore. This gradient is probably a reflection of temperature patterns which, due to the prevailing north-west winds, accumulate warm water at the southern shore.

**Lake Superior** Lake Superior—the largest North American lake at 82,367 $km^2$ in surface area, 148 m mean depth, and 406 m maximum depth—is the most extreme example of how morphology can affect the thermal structure of a lake and, consequently, its plankton distribution. The most prominent feature of this lake is its mid-lake, upwelling area, which coincides with the area of greatest depth (Anon., 1970c). In this area, in mid-summer, surface temperatures are less than 6–7°C, while, at the same time, water masses closer to the shore reach 16–17°C (Figure 22.4) Zooplankton patterns reflect the thermal structure. *Limnocalanus macrurus* is the most abundant species in the mid-lake, upwelling area with only a few specimens found nearshore. An opposite pattern is shown by *Diacyclops bicuspidatus thomasi,* which is absent in the upwelling area and increases in number towards shore. An extreme pattern exists for cladocerans such as *Leptodora kindtii,* present only in the warmest nearshore areas. Consequently, two different communities inhabit Lake Superior : calanoids dominating the central part, with *Limnocalanus macrurus* and *Diaptomus sicilis* most abundant; and cyclopoids dominating nearshore, with *Diacyclops bicuspidatus thomasi* most abundant. Cladocerans *Daphnia retrocurva* and *Bosmina longirostris* are also found nearshore.

**Great Slave Lake** Great Slave Lake (22,500 $km^2$ in area and 614 m max. depth) is highly complex. The main Western Basin (mean depth, 41 m) receives turbid water from sedimentary drainage resulting in Secchi transparencies in the range of 1–2 m (Figure 22.5*a*). The eastern part of the lake, McLeod and Christie Bays (mean depth 185 m), is fed by water from the Precambrian Shield, producing Secchi transparencies greater than 10 m. Mid-summer epilimnion temperatures ranged from 3.7°C in Christie Bay to 14.5°C in the Western Basin close to the inflows of the Slave and Hay rivers (Figure 22.5*g*). Maximum amounts of zooplankton were also found in the vicinity of these two river inflows, which drain sedimentary basins. Another area of high density was found in the boundary area between the two very different habitats (the "ecotone"?). Two different communities were found: in the eastern part receiving Precambrian drainage, *Limnocalanus macrurus* was the dominant component

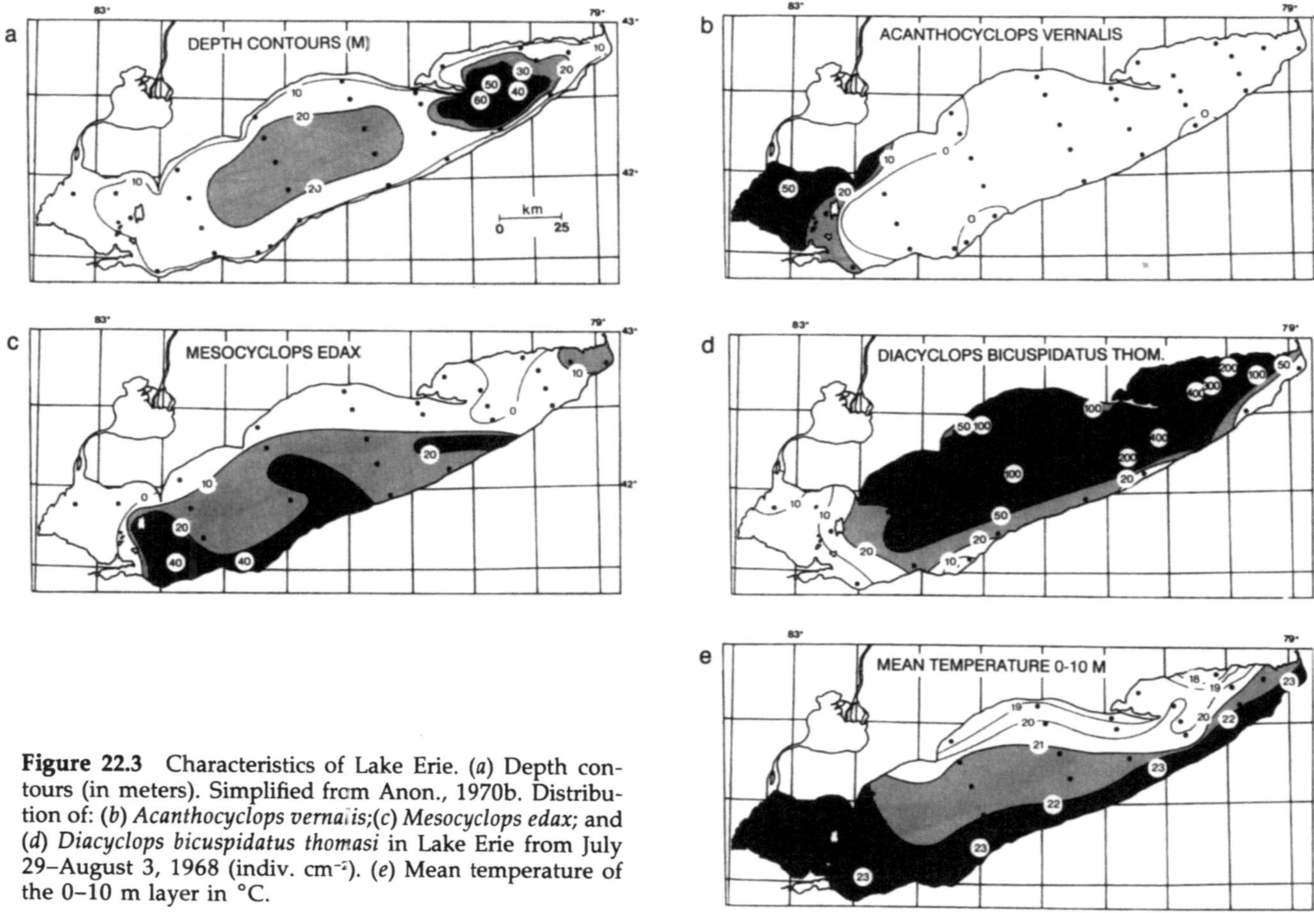


**Figure 22.3** Characteristics of Lake Erie. (*a*) Depth contours (in meters). Simplified from Anon., 1970b. Distribution of: (*b*) *Acanthocyclops vernalis*; (*c*) *Mesocyclops edax*; and (*d*) *Diacyclops bicuspidatus thomasi* in Lake Erie from July 29–August 3, 1968 (indiv. $cm^{-2}$). (*e*) Mean temperature of the 0–10 m layer in °C.

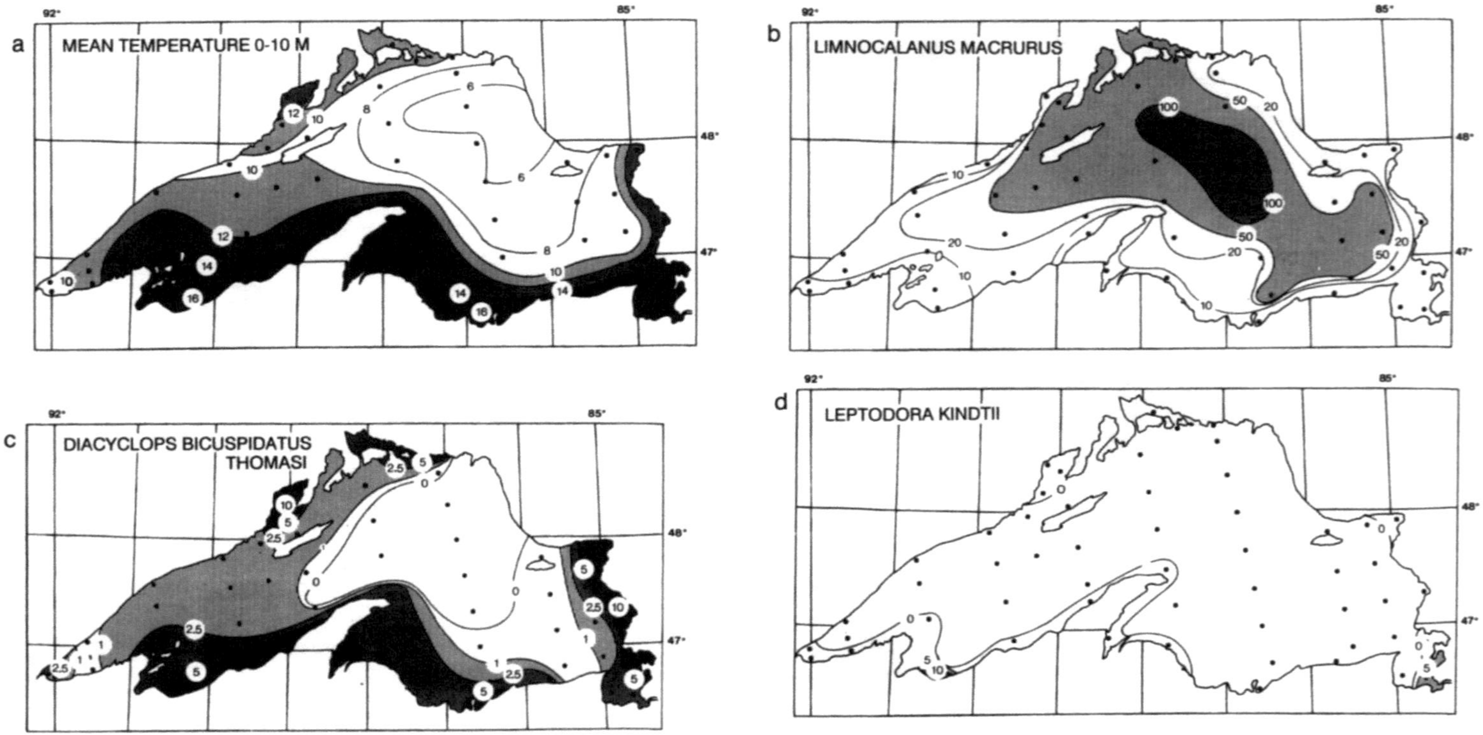


**Figure 22.4** Characteristics of Lake Superior. (*a*) Distribution of temperature August 18–28, 1968; mean temperature of the 0–10 m layer in °C. Based on data from Anon., 1970c. Distribution of selected species: (*b*) *Limnocalanus macrurus;* (*c*) *Diacyclops bicuspidatus thomasi;* and (*d*) *Leptodora kindtii.*

followed by *Diaptomus sicilis* (Patalas and Patalas, 1978). In the West Basin, *Diaptomus ashlandi* was dominant, followed by *Bosmina longirostris, Diacyclops bicuspidatus thomasi,* and *Limnocalanus macrurus. Diacyclops bicuspidatus* thomasi was present mostly in the vicinity of inflows from sedimentary drainage. Differences in the character of drainage area geology, inflow configuration, and lake morphology result in a complex temperature structure which could well explain the composition and distribution of the plankton in this lake.

**Lake Winnipeg** Lake Winnipeg has an area of 23,750 $km^2$ and a mean depth of only 12 m (Brunskill et al., 1980). It is situated on the boundary between Precambrian Shield to the east and north and sedimentary strata to the west and south. The lake is supplied by three major sources of water: a) from the Precambrian Shield via the Winnipeg River, b) from a sedimentary basin via the Red River from the south, and c) from a sedimentary basin via the Saskatchewan River from the west. This combination of water supply is reflected in the zooplankton distributional patterns.

*Diaptomus oregonensis* (Figures 22.6*a* and *b*) was absent in the area adjacent to the Winnipeg River inflow and showed distinct aggregations near the mouth of both rivers draining the sedimentary areas. *Daphnia longiremis* is evidently linked with water masses brought to the lake by the Saskatchewan River. The eastern half of the North Basin and the vicinity of Winnipeg River supplied with poorer Precambrian Shield waters were always much poorer in zooplankton than the remaining parts of the lake. These examples of distribution patterns linked with influxes from the drainage area are good indications of the significance of the external factors in this lake (Patalas, 1981).

## 22.3 Climate and Zooplankton Abundance

One of the consequences of large lake size is that diverse climatic conditions may prevail within the boundary of a large lake. For example, Lake Winnipeg, which is 436 km long, extends over 3.5° of latitudinal range, and in summer the air temperature is about 1.8°C warmer in the South Basin than the North basin. Similarly, the water temperatures of the upper 0–10 m stratum were 1.8°C higher in the south than in the north (Brunskill et al., 1980). Zooplankton abundance responded to this difference. On two out of three sampling dates, in spring and early summer, abundances in the South Basin were, on average 45% greater ($p < 0.01$) than in the North Basin, where increase in number was apparently delayed. Differences between the basins became insignificant in the autumn (Figure 22.7).

The effect of climate on zooplankton abundance is even more dramatic when lakes are compared over a wider latitudinal range. The numbers and biomass of crustaceans found in those lakes appear to be a function of water temperature. The single best predictor of mid-summer zooplankton abundance N (late July-August) is mid-summer epilimnion temperature (Figure 22.8*a*).

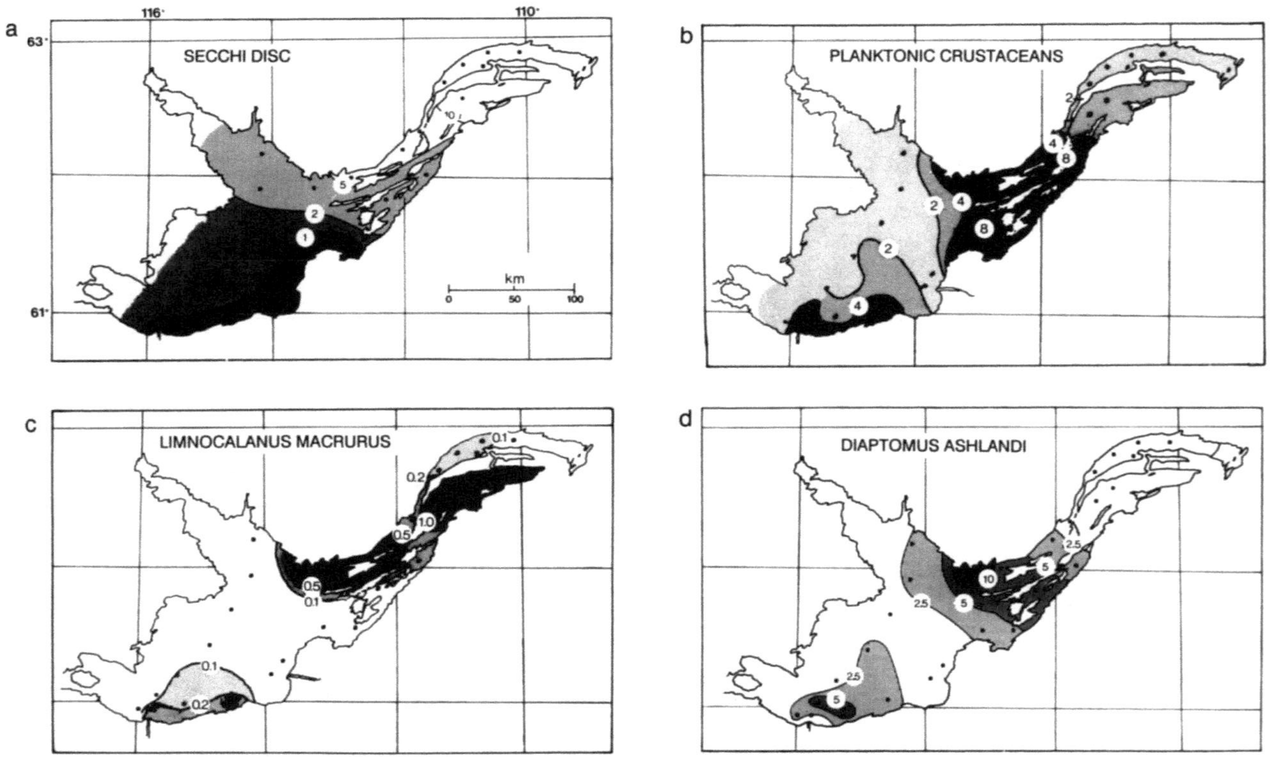


**Figure 22.5** Characteristics of Great Slave Lake. Distribution of: (*a*) Secchi disc transparency (in meters); (*b*) total planktonic crustaceans; (*c*) *Limnocalanus macrurus;* (*d*) *Diaptomus ashlandi;* (*e*) *Bosmina longirostris;* and (*f*) *Diacyclops bicuspidatus thomasi* (indiv. $L^{-1}$ in the upper 50 m). (*g*) Mean temperature of the 0–10 m layer in °C. August 1–6, 1968. (*h*) Simplified depth contours in meters.

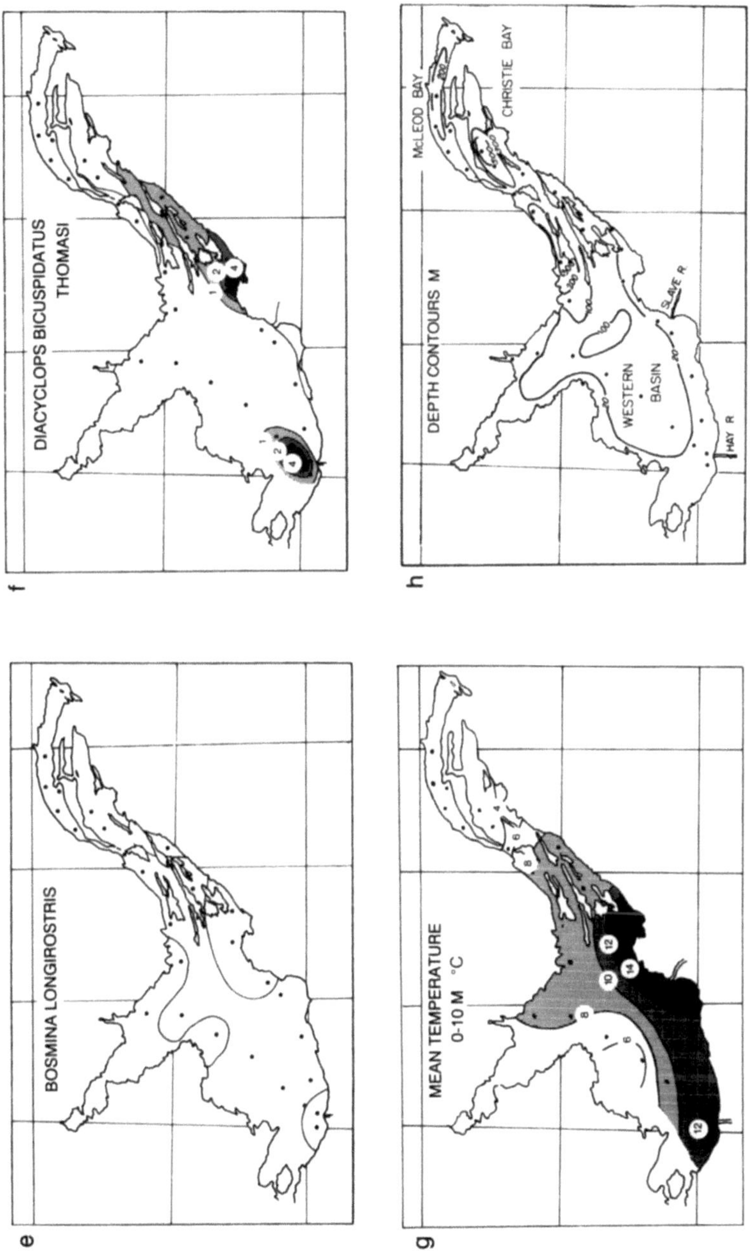


**Figure 22.5** *e* to *h*.

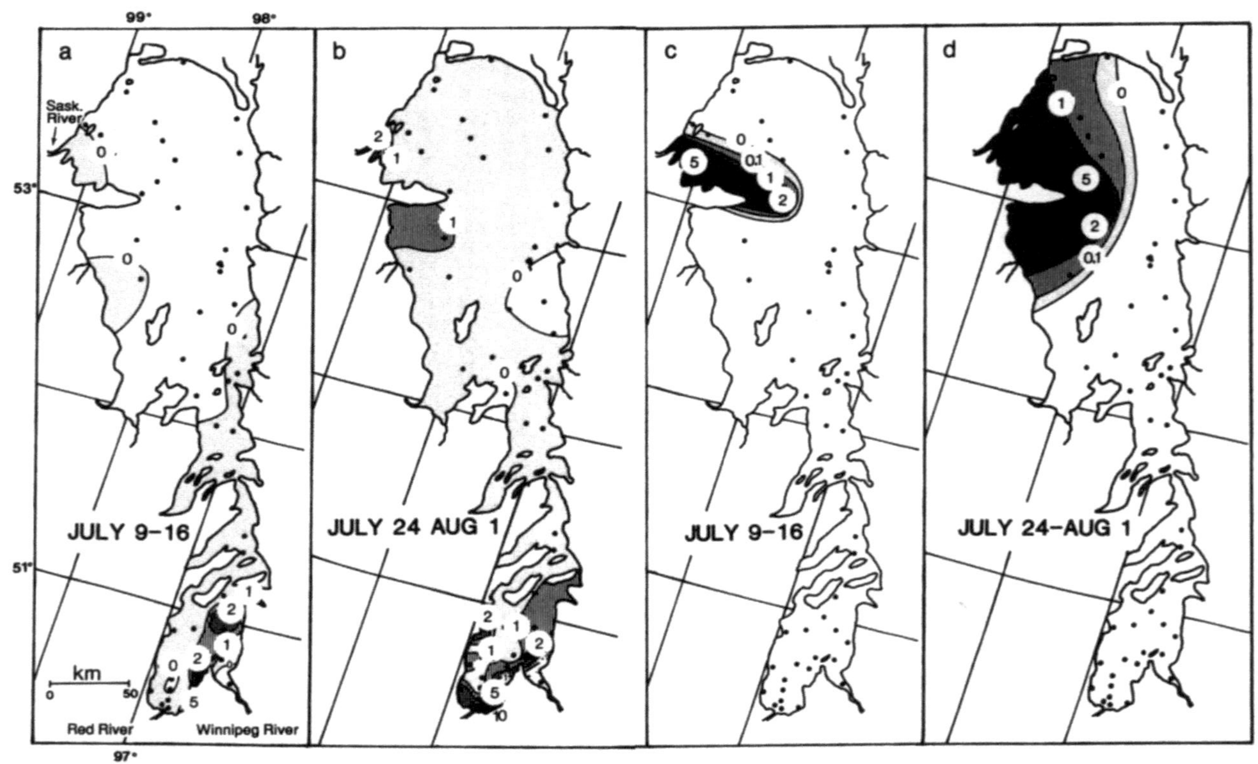


**Figure 22.6** Characteristics of Lake Winnipeg. Two different 7-day periods—July 9–16, 1969 and July 24–Aug. 1, 1969—are shown for each. Distribution of: (*a*) and (*b*) *Diaptomus oregonensis;* (*c*) and (*d*) *Daphnia longiremis;* (*e*) and (*f*) total planktonic crustaceans (indiv. $L^{-1}$). (*g*) and (*h*) Mean temperature of the 0–10 m layer in °C.

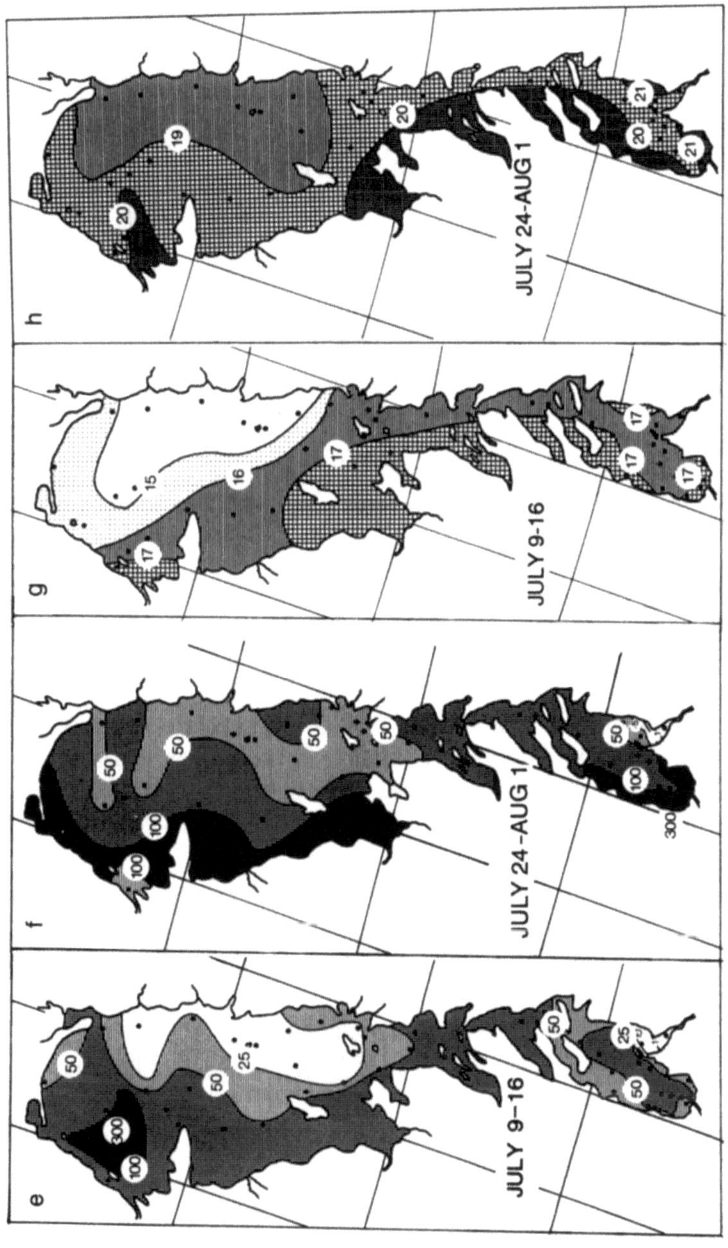


**Figure 22.6** *e* to *h*.

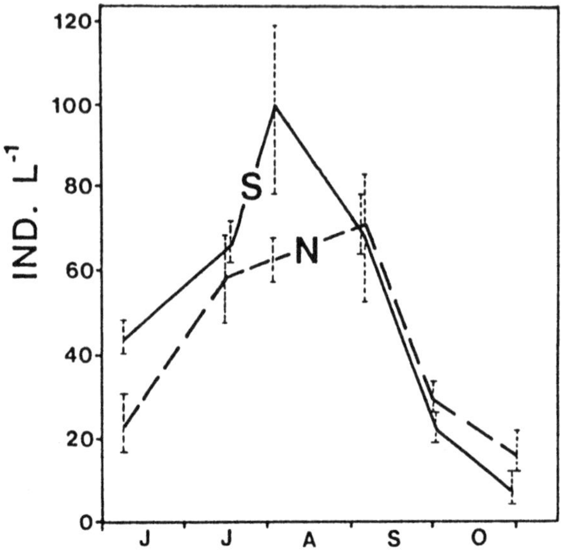


**Figure 22.7** Seasonal changes in planktonic crustacean abundance in the North (N) and South (S) Basins of Lake Winnipeg, (June–October, 1969). The vertical bars denote standard errors of the sample mean (S.E.)

$$N = 0.67e^{0.25t} \qquad (r^2 = 0.92) \tag{22.1}$$

$$B = 0.03e^{0.21t} \qquad (r^2 = 0.88) \tag{22.2}$$

where N = number of crustacean per liter, B = biomass of crustaceans (in $mg.L^{-1}$ wet weight), and t = mean epilimnion temperature in mid-summer (in °C). Epilimnion temperatures were well correlated with July mean air temperatures (a) (see Figures 22.8*b*) or with the latitude (l) and mean depth of the lake (z) (see Figure 22.8*c*).

$$t = -4.06 + 1.24a - 0.045z \qquad (r^2 = 0.93) \tag{22.3}$$

$$t = 44.13 - 0.49l - 0.065z \qquad (r^2 = 0.94) \tag{22.4}$$

Hence, if no data on epilimnion temperature are available, the following regressions can be used to predict crustacean abundance:

$$N = 0.1313e^{(0.33a - 0.008z)} \qquad (r^2 = 0.84) \tag{22.5}$$

$$N = 54176e^{(-0.13l - 0.013z)} \qquad (r^2 = 0.84) \tag{22.6}$$

## 22.4 Climate, Morphology, and Geology as Factors Defining Biotic Structure

In all the lakes described, including those with relatively uniformly shaped basins, highly heterogeneous horizontal plankton distributions prevailed. Such

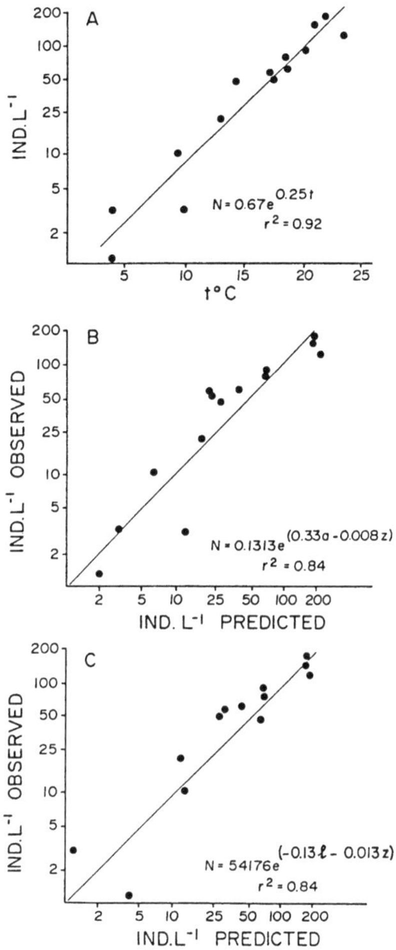


**Figure 22.8** Planktonic crustacean abundance (N) in North American large lakes as a function of: (*a*) mean epilimnion temperature in mid-summer (t); (*b*) mean air temperature in July (a) and lake mean depth (z); and (*c*) latitude (l) and lake mean depth (z).

heterogeneity was maintained throughout the summer months (Patalas, 1969) in spite of the intensive wave action characteristic of large lakes. The heterogeneity is apparently an effect of the structural complexity of large lakes, and is controlled by lake morphology, prevailing winds, upwelling, thermal bars, current patterns, configuration of inflows and outflows, etc. It is assumed that within one lake independent water masses, isolated from each other, can exist for a period of time. The water residence time within such isolated water masses can be longer than the generation times of planktonic species and populations can complete their life cycle without being affected by the communities from other water masses.

There is no general pattern of heterogeneous distribution which could apply to all the great lakes considered. Each lake develops its own specific pattern which can be explained by different combinations of factors such as climate, morphology, geology of the drainage area, and hydrology (residence time, inflow and outflow configuration). It is often difficult to separate the effects of these factors. The following examples show situations where some factors seem to have a much stronger effect on plankton distribution than others:

1. In Lake Ontario, zooplankton patterns were mostly a function of *wind-driven temperature distribution*. They could not be linked to basin morphology, which is very uniform in this lake.
2. In Lake Erie, the differences in the *depths* of the three basins and *wind-driven temperature distribution* were most important.
3. In Lake Superior, the existence of different inshore and offshore plankton communities could be linked directly with *morphology*, as well as with *morphology* and *wind-related temperature* structures (mid-lake upwelling).
4. In Lake Winnipeg, *locations* of the main *inflows* and the *geology* of their drainage area were the dominant factors. Depths and temperatures were rather uniform within the basins. *Climatic* differences between the two basins had a significant effect on plankton abundance.
5. In Great Slave Lake, all factors played an important role in plankton distribution: *geology* of the drainage basin, *morphology* of the lake, *inflow* configuration and *morphology*-related *temperature distribution*.

Climate seems to be of primary importance as an external energy input (a function of latitude and morphology), as well as a factor in controlling the thermal structure within the system. These two different aspects of climate seem to control both the abundance of plankton and its distribution within the system.

Table 22.1 provides the essential information on the physical properties and on crustacean abundance in 14 North American great lakes.

The empirical relationships shown in eqs. 22.1 and 22.2 between the measured mid-summer epilimnion temperature and the zooplankton abundance do not fully explain the mechanics. However, the high correlation coefficient ($r^2$) indicates that 92% of the variance in plankton abundance can be either directly or indirectly attributed to temperature. There is not much room left for the other factors, which, as we know, should have an effect on productivity of the system and consequently on plankton abundance. As data on phosphorus loading are not available for all those lakes, one can only make the assumption that nutrient loadings are correlated with climate in this set of lakes. This is not an unreasonable assumption, as, with the increasing latitude there is a decrease in the length of the vegetation season and also in the rate of bedrock

**Table 22.1** Physical and limnological properties and crustacean distribution in the North American Great Lakes.

| Lake | Latitude | Area ($km^2$) | Mean depth (m) | Water resid. time (yr) | Secchi (m) | Air temp.—July mean (°C) | Epil. temp.—Aug. mean (°C) | Crustacean Distribution ($IND.\cdot L^{-1}$) | Crustacean Biomass ($mg\cdot L^{-1}$)* |
|---|---|---|---|---|---|---|---|---|---|
| Erie (w.) | 41° 30′ | 3,873 | 8 | 0.2 } | 2.0 | 22.5 | 23.5 | 128 | 2.85 |
| Erie (c.) | 42° | 15,497 | 18 | } 4 | 2.9 | 22.5 | 21.7 | 184 | 3.81 |
| Erie (e.) | 42° 30′ | 6,456 | 24 | } | 3.6 | 22.5 | 21.0 | 154 | 2.88 |
| Ontario | 42° 30′ | 19,099 | 86 | 8.4 | 3.2 | 21.0 | 18.5 | 80 | 1.75 |
| Winnipeg (s.) | 51° | 4,900 | 8 | 0.3 | 0.8 | 19.0 | 20.3 | 90 | 2.20 |
| Winnipeg (n.) | 53° | 19,600 | 12 | 3.0 | 1.9 | 17.5 | 18.5 | 63 | 1.17 |
| S. Indian | 57° | 2,253 | 9 | 0.7 | 2.0 | 15.5 | 17.5 | 54 | 0.98 |
| Reindeer | 57° | 5,569 | 17 | 9.3 | 7.0 | 16.0 | 14.2 | 50 | 0.57 |
| Huron | 45° | 59,570 | 59 | 22.0 | 7.1 | 17.5 | 17.6 | 48 | 0.68 |
| La Martre | 63° 30′ | 1,707 | 11 | ~1 | 6.0 | 15.0 | 12.8 | 21 | 0.76 |
| Superior | 48° | 82,367 | 148 | 57 | 7.9 | 15.0 | 9.5 | 10 | 0.17 |
| Great Slave (w.) | 61° 30′ | 19,400 | 41 | ~6 | 2.5 | 15.0 | 10.0 | 3 | 0.07 |
| Great Slave (e.) | 62° 30′ | 3,700 | 185 | ~200 | 9.0 | 14.0 | 4.0 | 3 | 0.07 |
| Great Bear | 66° | 30,400 | 76 | 131 | 20.0 | 10.0 | 4.0 | 1 | 0.13 |

* Measured in the 0–50 m layer in mid-summer.

The limnological characteristics are mostly based on Rawson, 1960; Chandler, 1964; Brunskill, 1986; and on data from the Canada Centre for Inland Waters (Anon., 1970a, 1970b, 1970c).

weathering and mineralization of organic matter—conditions which are essential for plankton production. As the mixing depths of most of the great lakes are similar (see Figure 22.1), the epilimnion temperature reflects the total solar energy input to the system and defines the rate of metabolic processes of growth for the majority of planktonic crustaceans.

The regressions shown in eqs. 22.5 and 22.6 can be helpful in estimating an order-of-magnitude of plankton abundance in great lakes for which only data on latitude, air temperature, or mean depth are available.

In the preamble to the first circular for the conference on "Functional and Structural Properties of Large Lakes," two main characteristics of large lakes were listed:

1. Internal cycling of matter is usually of greater significance than influxes from the drainage area.
2. Processes within the pelagic zone comprise greater proportion of the overall flow of matter than interaction with the benthic zone.

With regard to the first statement, the examples of zooplankton distributions presented in this chapter provide evidence that influxes from the drainage area may have much greater significance than internally generated processes in some lakes. In Lake Winnipeg and Great Slave Lake, for example, zooplankton distributions show that areas apparently controlled by the inflowing water are significant in size. Both of these lakes are distinguished from the others by their relatively short water residence time (Lake Winnipeg—3 years and its South Basin, only 6 months; Great Slave Lake, West Basin—6 years). It can be concluded that it is not the size of the lakes but rather their water residence time which controls processes within these lakes.

A similar conclusion can be drawn regarding the second statement; the significance of processes within the pelagic zone versus the interaction with the benthic zone is defined not by the size of the lake but by its mean depth, or better, by its relative depth (the ratio of the maximum depth to the mean diameter of the lake surface area, Hutchinson, 1957). Such an interaction can be measured by the ratio of the epilimnion depth to the maximum (or mean) depth of the lake (Patalas, 1984). Lake Winnipeg's "pelagic" zone is apparently light-limited because of resuspension of bottom sediments and river-supplied seston (Healey and Hendzel, 1980; Brunskill, 1980).

Both statements—about the internal cycling of matter and the significance of the pelagic zone as characteristic features of large lakes—can be true only if long residence time and great mean depth (or relative depth) are inherent features of large lakes. These statements do not apply to all of the North American great lakes, because at least six out of 15 of them have residence times no longer than 5 months to 6 years and mean depths of less than 15 meters, i.e., less than their potential mixing depth. Their internal processes are apparently controlled by these features. The zooplankton distribution patterns well reflect the complexity and configuration of the physical properties of lakes, among which morphology and climate are the most important.

---

*Acknowledgements* I am indebted to A. Salki for the analyses of the great part of the plankton samples. Dr. G.J. Brunskill, P. Campbell, Dr. R.W. Hecky, and A. Salki provided useful criticism of the manuscript.

## References

Anon. 1969. Report to the International Joint Commission (I.J.C.) on the pollution of Lake Erie, Lake Ontario and international section of the St. Lawrence River. *International Lake Erie Water Pollution Board and International Lake Ontario St. Lawrence River Water Pollution Board*, 150 pp.

Anon. 1970a. Lake Ontario. Limnological Data Report No. 1 and 2. 1967. *Canadian Oceanographical Data Centre, Canada Centre for Inland Waters*, Burlington, Ontario, 179 pp; 212 pp.

Anon. 1970b. Lake Erie. Limnological Data Report No. 1, 1968. *Canadian Oceanographical Data Centre, Canada Centre for Inland Waters*, Burlington, Ontario, 152 pp.

Anon. 1970c. Lake Huron, Lake Superior. Limnological Data Report No. 1, 1968. *Canadian Oceanographical Data Centre, Canada Centre for Inland Waters*, Burlington, Ontario, 131 pp.

Brunskill, G.J., Elliott, S.E.M., and Campbell, P. 1980. Morphometry, hydrology and watershed data pertinent to the limnology of Lake Winnipeg. *Can. Manuscr. Rep. Fish. Aquat. Sci.* No. 1556, 32 pp.

Brunskill, G.J. 1986. Environmental features of the Mackenzie system. In: Davies, B.R. and K.F. Walker (eds.), *The Ecology of River Systems*. Junk Publishers, Dordrecht, The Netherlands, p. 435–471.

Chandler, D.C. 1964. The St. Lawrence Great Lakes. *Verh. Internat. Verein. Limnol.* 15:59–75.

Healey, F.P. and Hendzel, L.L. 1980. Physiological indicators of nutrient deficiency in lake plankton. *Can. J. Fish. Aquat. Sci.* 37:442–453.

Hutchinson, G.E. 1957. *A Treatise on Limnology*, Vol. 1. New York, J. Wiley & Sons, Inc.

Patalas, J. and Patalas, K. 1978. Spatial variation in size and reproductive cycle of Limnocalanus macrurus in a deep, subarctic lake, Great Slave Lake. *Verh. Internat. Verein. Limnol.* 20:150–158.

Patalas, K. 1969. Composition and horizontal distribution of crustacean plankton in Lake Ontario. *J. Fish. Res. Bd. Can.* 26:2135–2164.

Patalas, K. 1972. Crustacean plankton and eutrophication of St. Lawrence Great Lakes. *J. Fish. Res. Board Can.* 29:1451–1462.

Patalas, K. 1975. The crustacean plankton communities of fourteen North American great lakes. *Verh. Internat. Verein. Limnol.* 19:504–511.

Patalas, K. 1981. Spatial structure of the crustacean planktonic community in Lake Winnipeg, Canada. *Verh. Internat. Verein. Limnol.* 21:305–311.

Patalas, K. 1984. Mid-summer mixing depths of lakes of different latitudes. *Verh. Internat. Verein. Limnol.* 22:97–102.

Rawson, D.S. 1960. A limnological comparison of twelve large lakes in Northern Saskatchewan. *Limnol. Oceanogr.* 5:195–211.

Watson, N.H.F. and Carpenter, G.F. 1974. Seasonal abundance of crustacean zooplankton and net plankton biomass of lakes Huron, Erie and Ontario. *J. Fish. Res. Board Can.* 31:309–317.

Watson, N.H.F. 1976. Seasonal distribution and abundance of crustacean zooplankton in lake Erie. *J. Fish. Res. Board Can.* 33:612–621.
Watson, N.H.F. and Wilson, J.B. 1978. Crustacean zooplankton of lake Superior. *J. Great Lakes Res. Internat. Assoc. Great Lakes Res.* 4:481–496.

# 23

# Seasonal Succession of Rotifers in Large Lakes

Hans-Rainer Pauli

ABSTRACT Details of rotifer succession are known for lakes from the cold regions to the tropics and for trophic levels covering the complete spectrum of trophy. Because of its relatively small size, rotifer standing stock is usually of minor importance. Their relative significance within the energy flow of lake ecosystems, however, can be substantial and ranges from 3% to 96% of total zooplankton productivity. Temperature, food supply, competition with filter-feeding cladocerans, and predation by raptorial copepods are the main factors controlling seasonal occurrence, density, and succession of rotifers. In general, the winter rotifer communities consist of few (<5) species while the maximum number of species (>20) occurs in summer. Number of individuals per unit volume may fluctuate within three orders of magnitude. The common temporal patterns of abundance of rotifer communities in temperate lakes are: (1) low density during winter when both temperature and food levels are low, (2) exponential growth in spring when temperature and food levels increase, (3) subsequent breakdown of the rotifer community, and (4) wide fluctuations in abundance during summer and autumn. This overall temporal pattern is the result of the underlying dynamics of individual species differing either in their specific requirements for food and temperature or in their susceptibility to predators.

## 23.1 Introduction

To understand the structure and the functioning of complex ecosystems such as large lakes, it is necessary to distinguish all relevant components of the system. Rotifers are the most important soft-bodied invertebrates of the plankton (Hutchinson, 1967). They live cosmopolitanly and are found in the Antarctic as well as in the tropics. Although they are relatively small, they probably play a major role in the energy transfer in many lakes of the world. The earlier notion that they are unimportant (Nauwerk, 1963), in terms of both biomass and productivity, was rectified by a number of studies carried out in lakes ranging from extreme oligotrophy to eutrophy.

The relative importance of rotifers covers a wide range. In Lake Nakuru (Kenya) (Vareschi & Jacobs, 1985), the rotifer *Brachionus dimidiatus* reached almost 40% of the total secondary production, and in Lake Myvatn (Iceland) (Adalsteinsson, 1979), rotifer production amounted to 96% of the whole zooplankton production. In Russian and Canadian lakes of different trophic levels (Alimov et al., 1972; Winberg et al., 1972; Schindler, 1972), the relative importance of rotifers ranged from 15% to 44% of zooplankton production and from 2% to 6.6% of primary production. Similar values have been reported from Mirror Lake (Makarevicz and Likens, 1979), with 16% and 40% of zooplankton biomass and production, respectively. In Lake Oglethorpe (Orcutt and Pace, 1984), rotifers dominated during parts of the year and reached 65% to 85% of zooplankton standing crop. There are also cases where rotifers are of lesser importance. In Ovre Heimdalsvatn (Larsson, 1978), rotifers contribute 3–8% of total zooplankton production, and in Lake Lanao (Lewis, 1979) they represent only 3% of total secondary production.

Table 23.1 summarizes information on the relative importance to primary productivity of the major components of the pelagic zone of Lake Constance. The production of rotifers was estimated by summing the positive, nonzero differences of population densities between subsequent sampling days. This method measures only the "new" production and neglects the replacement productivity of populations during steady-state conditions. Hence, the minimum contribution of rotifers to the carbon cycle of Lake Constance amounts only to a portion of 1.4% of the annual gross primary production (Tilzer & Beese 1988) and to 22% of the *Daphnia* net production (Geller, 1989). According to an earlier estimate (Walz, 1986, 1987) the rotifers account, on average, only for 10% of *Daphnia* production in Lake Constance.

The seasonal succession of plankton organisms is influenced by both abiotic and biotic factors. In temperate lakes, the most important abiotic factor is the temperature. Its influence on the embryonic development time and the longevity of rotifers is well known from experimental studies (Pourriot, 1965; Pourriot and Deluzarches, 1971; Ruttner-Kolisko, 1972, 1975; Ruttner-Kolisko, 1978; Halbach, 1970; Halbach, 1973; Herzig, 1983). Furthermore, the temperature controls the time of occurrence of temperature-adapted organisms (May, 1983). The biotic factors include food supply, competition, and predator-

**Table 23.1** Relative importance of the major components of the pelagic zone of Lake Constance (0–50 m depth) to primary productivity (PPR)

| Component | Annual total (g $C \cdot m^{-2} \cdot yr^{-1}$) | Percentage of PPR | Reference |
|---|---|---|---|
| Gross PPR | 280 | 100% | Tilzer and |
| Net PPR | 224 | 80% | Breese (1988) |
| *Daphnia* productivity | 18 | 6.4% | Geller (1989) |
| *Daphnia* consumption | 36 | 12.8% | Geller (1989) |
| Rotifer productivity | 4 | 1.4% | This chapter |
| Sedimentation | 50 | 18% | Stabel (1985) |

prey relationships. Food supply generally controls the population density of consumers, but, given certain conditions, it may allow or prohibit the occurrence of specialized consumers (Pourriot, 1977; Dumont, 1977; Buikema et al., 1978; Starkweather and Bogdan, 1980; Bogdan and Gilbert, 1987). Cladoceran crustaceans are believed to be the most important competitors of rotifers (Neill, 1984). They use a similar portion of the phytoplankton species spectrum and are able to damage rotifers directly by interference competition (Gilbert and Stemberger, 1985). Common predators of rotifers are copepods (Gilbert & Williamson, 1978; Karabin, 1978; Williamson, 1983; Stemberger and Evans, 1984), some raptorial forms of rotifers, and other invertebrates.

This paper summarizes available information about the seasonal occurrence, community structure, and importance of rotifers in different types of lakes. Particular emphasis is placed on the results of a field study carried out in Lake Constance during 1984 and 1985.

## 23.2 General Course of Rotifer Abundances

The species lists of several Scandinavian and prealpine lakes are quite similar (Carlin, 1943; Pejler, 1957; Nauwerk, 1963; Wappis, 1980; Balvay and Laurent, 1981; Berner-Fankhauser, 1983). In Lake Constance, about 45 species from 20 different genera were found during the years 1984 and 1985. Figure 23.1 illustrates the seasonal course of the rotifer and phytoplankton biomass. Both components of the plankton reflect the development of the growing season which basically depends on lake stratification. The temporal dynamics of rotifer abundance in temperate lakes can be roughly separated into three phases with regard to densities and fluctuations of the population: (1) During winter and early spring, when the water is cold and food is rare, densities are low. (2) The

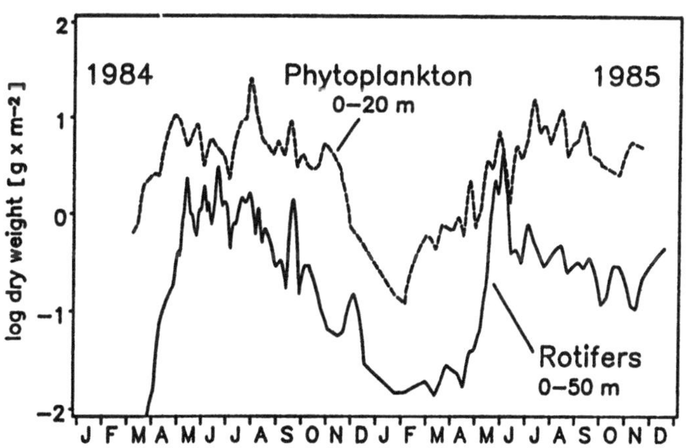


**Figure 23.1** The seasonal course of the biomass level of phytoplankton (0–20 m depth) and of rotifers (0–50 m depth) in Lake Constance in 1984 and 1985.

time of maximum increase, which results in one of the annual maxima of abundance and biomass, begins in April and lasts until May or the beginning of June. During this time, significant increases in temperature and food can be observed. Hence, two of the major factors affecting the growth of rotifers change to more suitable conditions. Some species respond with very high growth rates. This exponential growth phase is usually followed by a drastic reduction of the rotifer population. (3) The time from summer through autumn is characterized by wide fluctuations caused by the appearance of additional species and the disappearance of others. The magnitude of the fluctuations in this period may be substantially different in subsequent years. This phenomenon of different interannual productivity is common also in Lake Geneva (Balvay, 1984) and in Lago Maggiore (DeBernardi et al., 1983). Only two features seem constant: the low densities during winter and the phase of rapid growth in spring. Maximum abundances have been observed in spring and in summer, but also in autumn.

The typical pattern for the present state of Lake Geneva (Balvay, 1984) are three major growth phases—in May-June, in July–August, and in September. Annual mean densities varied within one order of magnitude ($7–60 \times 10^3$ Ind $m^{-3}$). In Lake Constance in 1984, a series of three additional growth phases could be observed. Earlier investigations of rotifers in the oligotrophic Lake Constance (Auerbach et al., 1924, 1926; Walz et al., 1987) report only one maximum during summer from investigations in the 1920's and 1930's, whereas after eutrophication began the seasonal abundance patterns found—in 1963 and 1977/78 (Walz et al., 1987)—are more comparable to that of 1984. Maximum densities in summer are also characteristic of the rotifer community in the oligotrophic Lake Stechlin (Ronneberger and Kasprzak, 1985). Hence multiple growth phases are likely to be typical for temperate nonoligotrophic lakes.

## 23.3 Growth Rates and Succession

While the development of rotifer populations in general seems to be unpredictable even for one particular lake, the successive occurrence of species is determined by specific factors with large interannual variability. Figure 23.2 shows the seasonal occurrence of growth phases of several rotifer species in 1985 in Lake Constance; Table 23.2 gives values of maximum abundances and growth rates. The growth phases of different rotifer species are clearly distinct in time, except during the period between April and June when several species are growing in parallel. *Keratella cochlearis* started growing in mid-April and was immediately followed by *Synchaeta pectinata*, *Conochilus unicornis*, and *Asplanchna priodonta*. The maximal growth of *Polyarthra dolichoptera*, *P. vulgaris*, and *Synchaeta spp.* also occurred during this period. During the summer and autumn, a series of species appeared sequentially, beginning with *Pompholyx sulcata* (July–August), *Gastropus stylifer* (August), *Ascomorpha ovalis* (August–September), and *Trichocerca porcellus* (September–October). *Notholca cau-*

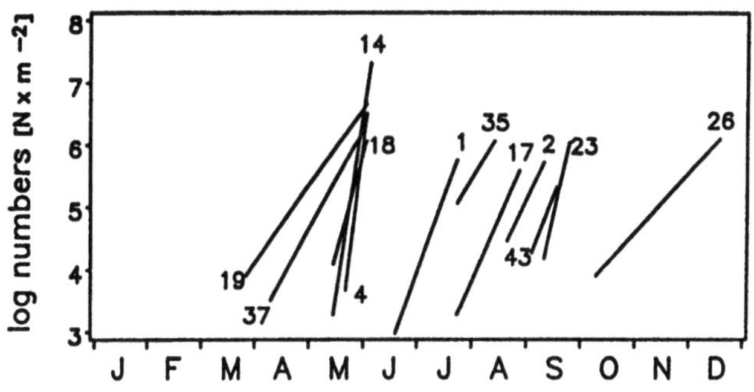


**Figure 23.2** The periods of maximal growth of 12 different rotifer species in Lake Constance in 1985. The lines connect the minimum and maximum abundance levels (logarithmic scale) during exponential growth. The species shown are: 1, *Ascomorpha ecaudis*; 2, *A. ovalis*; 3, *Asplanchna priodonta*; 14, *Conochilus unicornis*; 17, *Gastropus stylifer*; 18, *Kellicottia longispina*; 19, *Keratella cochlearis*; 23, *Lepadella ovalis*; 26, *Notholca caudata*; 35, *Pompholyx sulcata*; 37, *Synchaeta pectinata*; 43, *Trichocerca capucina*.

**Table 23.2** Maximum abundances (0–50 m depth) and maximum growth rates of 10 rotifer species in Lake Constance in 1984 and 1985

| | Maximum abundance, $A_{max}$ (N × $m^{-2}$ × $10^6$) | | Maximum growth rate, $r_{max}$ ($d^{-1}$) | |
|---|---|---|---|---|
| Rotifer species | 1984 | 1985 | 1984 | 1985 |
| *Keratella cochlearis* | 22.9 | 4.7 | .20 | .24 |
| *Keratella quadrata* | 2.1 | 3.3 | .12 | .18 |
| *Asplanchna priodonta* | 0.9 | 3.3 | .26 | .31 |
| *Conochilus unicornis* | 14.4 | 20.9 | .34 | .41 |
| *Notholca caudata* | 0.6 | 1.3 | .17 | .19 |
| *Polyarthra dolichoptera* | 6.2 | 2.1 | .23 | .23 |
| *Polyarthra vulgaris* | 5.1 | 2.1 | .18 | .17 |
| *Pompholyx sulcata* | 1.4 | 1.2 | .22 | .11 |
| *Synchaeta pectinata* | 4.2 | 1.6 | .29 | .27 |
| *Kellicottia longispina* | 2.2 | 1.2 | .14 | .16 |

*data* appeared in late October and reached its maximum abundance in December.

Not only do the phases of maximum growth differ among the rotifer species, but their seasonal abundances also show specific patterns. With respect to the duration of seasonal occurrence of the populations, perennial species and ephemeral species can be distinguished, as the populations of the latter group only exist over short periods of the year. While ephemeral species usually have only one annual population density maximum, the dynamics of the perennial species are quite different. Figure 23.3 shows some examples of both of these types; the ephemeral species shown in Figures 23.3*a* and *b* are believed

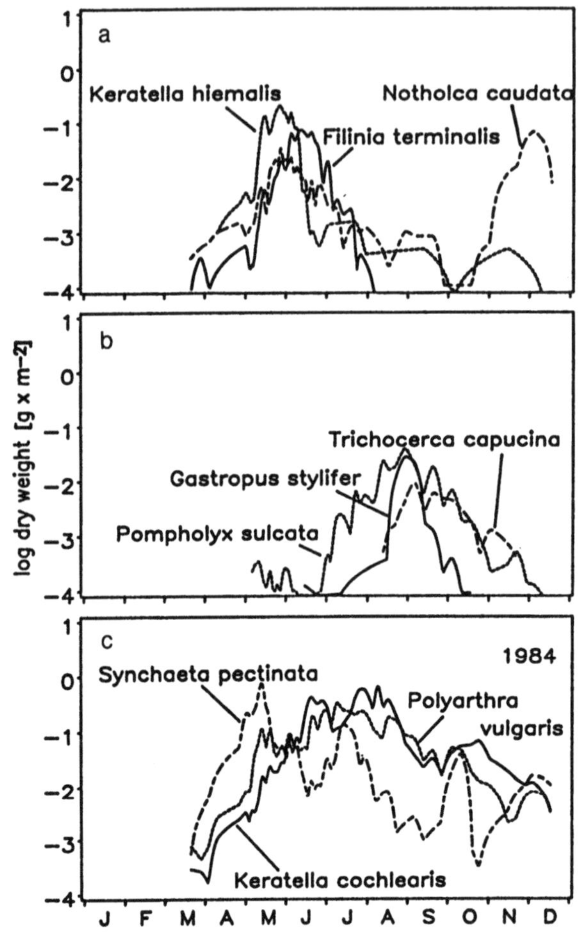


**Figure 23.3** The seasonal abundance patterns of several rotifer species (0–50 m depth, logarithmic scale) during 1984 in Lake Constance. (*a*) "Cold-stenotherm" species. (*b*) "Warm-stenotherm" species. (*c*) Perennial species.

to have specific temperature preferences and are called "stenotherms." Cold-stenotherm species include *Keratella hiemalis, Polyarthra dolichoptera, Filinia terminalis,* and *Notholca caudata;* they are only present in winter and spring, or prefer deep water layers during the summer. *Ascomorpha, Gastropus, Pompholyx,* and *Trichocerca* are typical "summer genera."

The most important perennial species (Figure 23.3c) are *Keratella cochlearis, Polyarthra vulgaris,* several *Synchaeta* species, *Asplanchna priodonta,* and *Kellicottia longispina.* They all realize their maximum growth rate of population in spring. *Keratella cochlearis* and *Polyarthra vulgaris* are able to maintain high abundances over several months, although their populations are not stable. They show wide fluctuations but they do not disappear. Similar patterns have been reported for these species from the Bielersee (Berner-Fankhauser, 1986), Lake Constance (Walz, 1987), and Lake Motalaström (Carlin, 1943). The abundance patterns of *Asplanchna priodonta* and *Synchaeta pectinata* are more dynamic. Periods of high abundances, typically two or three a year, alternate with phases of very low densities.

Successions of species within a single genus have been reported for the genera *Polyarthra* (Carlin, 1943) and *Synchaeta* (Carlin, 1943; Stemberger, 1974), where up to five different species succeeded each other in time. In Figure 23.4, the seasonal abundance patterns of three *Keratella* species, *K. cochlearis, K. hiemalis,* and *K. quadrata*, are shown. The three populations grew simultaneously, although *K. cochlearis* was able to maintain or even to increase its abundance after June, while *K. hiemalis* and *K. quadrata* disappeared—the first species quickly, the second species more slowly. The three species showed different affinities to temperature (Figure 23.5). The abundance of *K. cochlearis* was in good concordance with the seasonal course of the ambient temperature. However, the increase and decrease of *K. hiemalis* were observed at a basically constant temperature. The seasonal course of both the mean ambient tem-

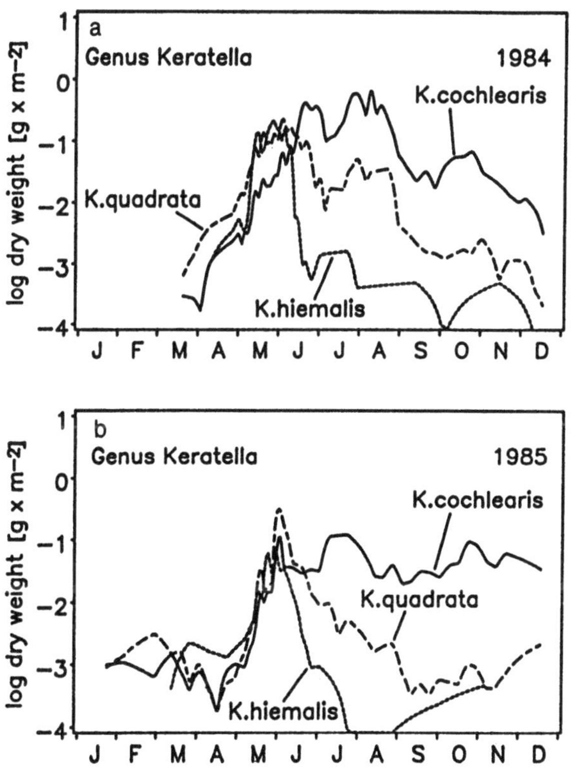


**Figure 23.4** The seasonal abundance patterns of three *Keratella* species 0–50 m depth, logarithmic scale) during (*a*) 1984 and (*b*) 1985 in Lake Constance.

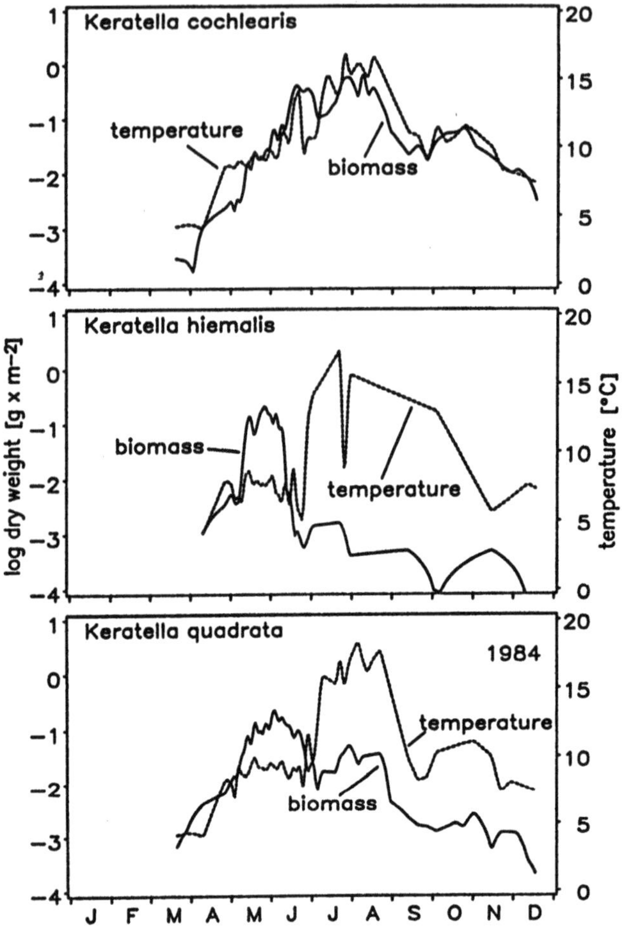


**Figure 23.5** The relationship between ambient temperature (°C) and, biomass 0–50 m depth, logarithmic scale) of three *Keratella* species during 1984 in Lake Constance.

perature and the population density of *K. quadrata* suggests an intraspecific succession of different clones with different adaption to temperature. Similar phenomena have been reported for *Euchlanis dilatata* (King, 1972), *K. cochlearis* (Hillbricht-Ilkowska et al., 1966), and *Polyarthra vulgaris* (Rey and Capblancq, 1975).

## 23.4 Community Structure

Due to the different seasonal patterns of abundance and the distinct phases of growth of rotifers, the changes observed in the community structure are rapid and substantial. Community composition can be described by two summary parameters: (1) diversity, which takes into account the number of species and their relative abundances, and (2) evenness, which describes the distribution of the relative proportions. As a rule, the number of rotifer species (the simplest measure of diversity) in temperate lakes increases from spring to summer and declines again in autumn. Similar seasonal dynamics were found using more refined measures of diversity (Margalef, 1969, 1983; Miracle, 1977; Matveeva, 1983), which consider not only the number of species but also their relative contribution to the community. Figure 23.6 shows the percentage abundances of the dominant rotifer species in Lake Constance. Indices of diversity were calculated as Shannon's "H" (1948); and indices of evenness as "A," according to Hill (1973) and Daget (1980). During winter, the number of species was low but the percentages were evenly distributed. Thus, values of H were high (Figure 23.7). During spring, several species became more abundant, and the community was dominated by either a few species or, sometimes, only one species. For example, *Synchaeta* accounted for up to 86% of total number on 24 May 1985, and *Conochilus* accounted for up to 70% on 5 June 1985. These periods were characterized by a diversity minimum. High diversity was observed during summer, when up to 20 different species coexisted. Evenness changed rapidly during this time and was not correlated with diversity, due to the high dynamics of the different species.

The seasonal change in the community structure depends on factors such as temperature, food, competition, and predation. It is one of the basic tasks of community ecology to analyze the relative strengths and the network of these interactions and factors. Correlation matrices based on population pa-

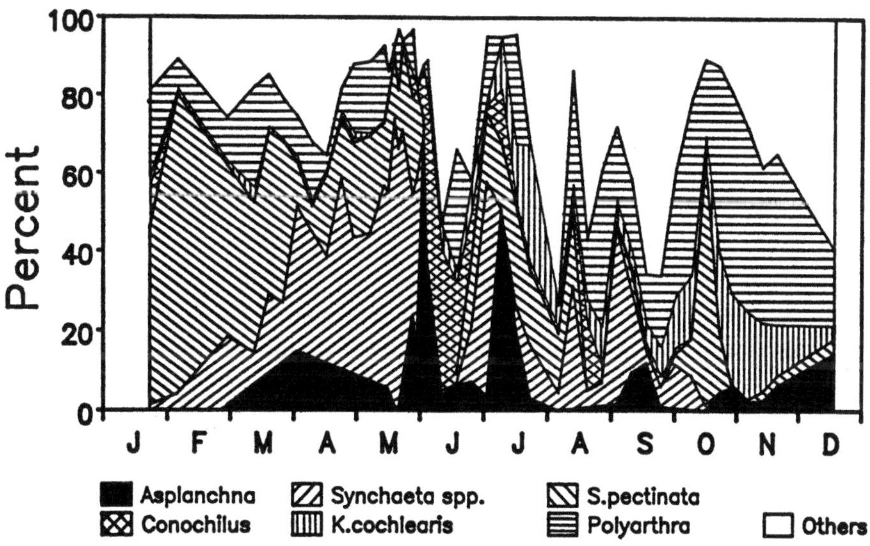


**Figure 23.6** The composition of the rotifer community in Lake Constance during 1985 (percentage abundance by biomass).

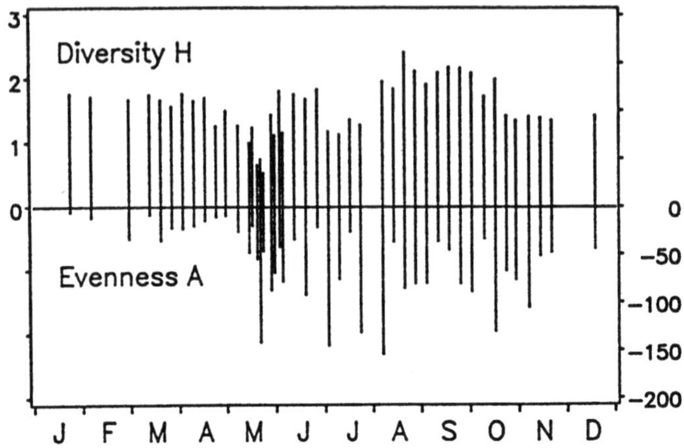


**Figure 23.7** Diversity (Shannon's "H") and evenness (Daget's "A") of the rotifer community of Lake Constance during 1985.

rameters—such as abundance (N), production (P) (Lewis, 1979), or growth rates (r) (Walz, 1986)—facilitate the analysis of the complex interactions among species within an ecosystem. Those correlations may indicate potential trophic and competitive interactions among the species of the planktonic community, but they cannot give stringent causal explanations. In an abundance (N) matrix of 60 species of rotifers, algae, and crustaceans of Lake Constance, nearly 150 significant correlations ($p<.001$) were found, but only a few of them are easy to interpret. Food specialists like *Ascomorpha* species correlated quite well ($r^2=.61-.74$) with their preferred food algae, but potential predator-prey relationships were difficult to detect, probably because of time lags and multifactorial influences, which are not considered in simple correlation analysis.

To avoid such weaknesses, more sophisticated statistical treatments have been used to analyze the community structure of rotifers. For example, factor analysis was used by Miracle (1974) and by Walz (1987). Another approach to analyze great numbers of related objects is to order them by useful descriptors (Legendre and Legendre, 1983; Gauch and Whittaker, 1981). Coincidences in time of abundances of species were used to order the sampling days, as shown in Figure 23.8. "Steinhaus coefficients" (Motyka et al., 1950) were used here as a measure of similarity; these coefficients were clustered by seven different methods, restricting the maximum possible number of clusters to five. The different methods gave similar results: there are two major clusters; the first one includes the time from January to beginning of May, and the second one includes the period from June to December. The remaining three clusters comprise only short intervals between May and July.

To get information about the community composition during the periods characterized by the observed clusters, the rotifer species were ranked according to their abundance within each cluster and then averaged over the period

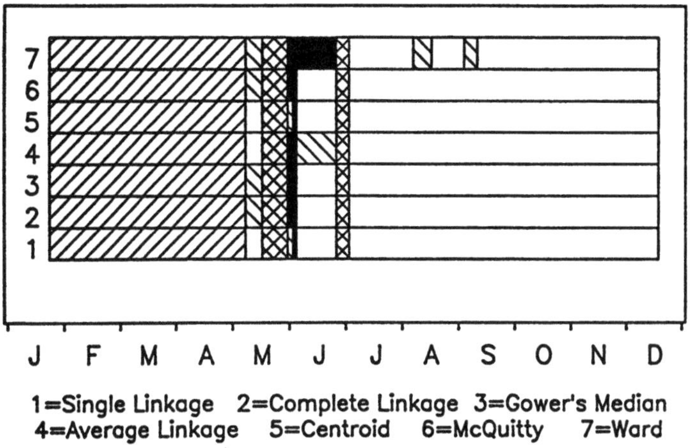


**Figure 23.8** Cluster analysis of similarities among sampling series during 1985 in Lake Constance. Steinhaus similarity coefficients were determined (Motyka et al., 1950 in Legendre and Legendre, 1983) and clustered by seven different methods, which are identified below the plot.

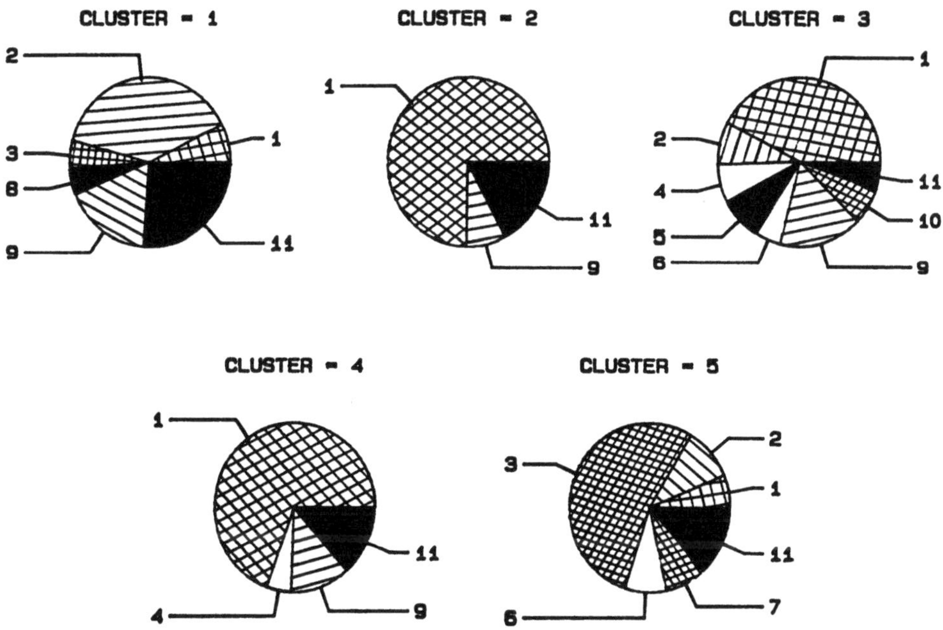


**Figure 23.9** The frequency distribution of the dominant rotifer species during the periods ordered by the clusters (from methods no. 2, 3, and 6 in Figure 23.8) during 1985 in Lake Constance. The species are: 1, *Synchaeta* spp.; 2, *Keratella cochlearis*; 3, *Conochilus unicornis*; 4, *Keratella hiemalis*; 5, *Synchaeta pectinata*; 6, *Keratella quadrata*; 7, *Asplanchna priodonta*; 8, *Pompholyx sulcata*; 9, *Polyarthra* spp.; 10, *Kellicottia longispina*; 11, others.

enclosed by each cluster. The results are shown on pie charts in Figure 23.9. Cluster no. 1 represents the main portion of the season and is dominated by *K. cochlearis* and *Polyarthra*. Cluster no. 3 encloses the first four months of the year and is dominated by *Synchaeta* and *Keratella* species. Clusters no. 2 and 4 are similar; both are dominated by small *Synchaeta* species like *S. tremula* and *S. oblonga*. The characteristic species for cluster no. 5 is *Conochilus unicornis*.

## 23.5 Problems of Rotifer Studies

Due to the cosmopolitan distribution of many rotifer species, their occurrence and seasonal succession in a particular lake seem to be mostly determined by the local situation and the environment. Apparently, there is no general rule as to which factors control rotifer succession. While some species are associated with a particular temperature, other species can withstand a wide range. Some species are highly susceptible to predation by copepods while others are able to escape or are protected by a strong lorica or by spines. The impact of food supply is also not well known. The specific requirements of food-specialists can be found easily, even from field data. The more complex mechanisms controlling the densities of generalists, which live in a large range of temperature and feed on a wide spectrum of food, are less well understood. There is some evidence for a form of regulation by competition through filter-feeding crustaceans and by predation of raptorial crustaceans.

A weak point in studies of rotifer communities is the use of heterogeneous and inadequate units to measure the density of populations. At present, there are too many different units in use—such as numbers, biovolume, fresh weight, dry weight, carbon, kcal. This makes precise comparisons between different investigations burdensome or meaningless. The use of numbers of individuals per surface area or volume unit should be restricted, for example, to calculations of diversity or similarity coefficients or to the presentation of single-species patterns. As soon as the flow of energy or the relative importance of a multi-species component within an ecosystem is considered, some mass-related unit should be used to properly compare organisms of different sizes. In fact, the range of the body size of rotifer species covers three orders of magnitude, and intraspecific variation among seasons can also be highly significant. The size differences among organisms of separate taxa or distinct trophic levels are even more important. Size measurements are easy to perform and should be included in every study dealing with biomass or productivity of rotifers. Biovolume is often used to measure the biomass of rotifers, but this unit is ambiguous and should not be used for summary presentation of different species, because the dry weight percentages among rotifer species may differ considerably, and an accurate conversion method for biovolume to organic dry weight, carbon, or energy is not yet available. These last measures of biomass are recommended for interspecies and interlake comparisons of rotifer populations and zooplankton communities.

## 23.6 Conclusions

The seasonal succession of rotifers is characterized by strong fluctuations of numbers of individuals as well as of species. These patterns are mainly influenced by water temperature, food supply, competition with other zooplankton, and predation. Temperature appears to be responsible for the seasonal occurrence of many species and, therefore, affects the diversity of the rotifer community. Food supply has two different aspects: the occurrence of those algal species which are needed by food specialists increases diversity, whereas food quantity regulates the potential abundance of other species which use a wider spectrum of food. Competition and predation affect the seasonal fluctuations of the populations' standing stocks by top-down control.

---

*Acknowledgements* This research project was supported by the Deutsche Forschungsgemeinschaft and by the Sonderforschungsbereich (Special Collaborative Program) on the "Cycling of Matter in Lake Constance" (SFB 248).

## References

Adalsteinsson, H. 1979. Zooplankton and its relation to available food in Lake Myvatn. *Oikos* 32:162–194.

Alimov, F.A., Boullion, V.V., Finogenova, N.P., Ivanova, M.B., Kuzmitskaya, N.K., Nikukina, V.N., Ozeretskovskaya, N.G., and Zharowa, T.V. 1972. Biological productivity of Lakes Krivoe and Krugloe. In: Z. Kajak and A. Hillbricht-Ilkowska (eds.), *Productivity Problems of Freshwaters*. PWN Polish Scientific Publishers, Warsaw, p. 383–404.

Auerbach, M., Märker, W., and Schmalz, J. 1924. Hydrographisch-biologische Bodensee-Untersuchungen. I. Ergebnisse der Jahre 1920-122. *Archiv für Hydrobiologie, Suppl.* 3:597–738.

Auerbach, M., Märker, W., and Schmalz, J. 1926. Hydrographisch-biologische Bodensee-Untersuchungen. II. Ergebnisse der Jahre 1923 und 1924 und Zusammenfassung 1920-1924. *Verhandlungen des naturwissenschaftlichen Vereins Karlsruhe* 30:1–128.

Balvay, G. 1984. Les rotifères. In: *CIPEL: Le Léman, Synthèse 1957-1982*. Commission internationale pour la protection des eaux du Léman contre la pollution, Lausanne, p. 304–313.

Balvay, G., and Laurent, M. 1981. Les rotifères du Léman. *Schweizerische Zeitschrift für Hydrologie.* 43:126–139.

Berner-Fankhauser, H. 1983. Abundance, dynamics and succession of planktonic rotifers in Lake Biel, Switzerland. *Hydrobiologia* 104:349–352.

Berner-Fankhauser, H. 1986. Populationsdynamik der Plankton-Rotatorien im Bielersee 1981/82. Dissertation, Universität Bern, 88 pp.

Bogdan, K.G. and Gilbert, J.J. 1987. Quantitative comparison of food niches in some freshwater zooplankton. A multi-tracer-cell approach. *Oecologia (Berlin)* 72:331–340.

Buikema, A.L., Jr., Miller, J.D., and Yongue, W.H., Jr. 1978. Effects of algae and protozoans on the dynamics of *Polyarthra vulgaris*. *Verhandlungen der Internationalen Vereinigung für Limnologie* 20:2395–2399.

Carlin, B. 1943. Die Planktonrotatorien des Motalaström. *Meddelandsen Lunds Universitet Limnol. Inst.* 5:1–255.

Daget, P. 1980. Le nombre de diversité de Hill, un concept unifacteur dans la théorie de la diversité écologique. *Acta Oecologica / Oecologia Generalis* 1:51–70.
DeBernardi, R., Giussani, G., and Origgi, I. 1983. Analisi del popolamento zooplanctonico del Lago Maggiore nel quadriennio 1979-1982. In: *Raporti su Studi e Ricerche Condotti nel Bacino del Lago Maggiore / Campagna 1982*. Istituto Italiano di Idrobiologia, Pallanza, p. 145–192.
Dumont, H.J. 1977. Biotic factors in the population dynamics of rotifers. *Archiv für Hydrobiologie Beihefte / Ergebnisse der Limnologie* 8:98–122.
Gauch, H.G. and Whittaker, R.H. 1981. Hierarchical classification of community data. *Journal of Ecology* 69:537–557.
Geller, W. 1989. The energy budget of two sympatric *Daphnia* species in Lake Constance: Productivity and energy residence times. *Oecologia (Berlin)* 78:242–250.
Gilbert, J.J. and Stemberger, R.S. 1985. Control of *Keratella* populations by interference competition from *Daphnia*. *Limnology and Oceanography* 30:180–188.
Gilbert, J.J. and Williamson, C.E. 1978. Predator-prey behavior and its effect on rotifer survival in associations of *Mesocyclops edax, Asplanchna girodi, Polyarthra vulgaris,* and *Keratella cochlearis*. *Oecologia* 37:13–22.
Halbach, U. 1970. Einfluß der Temperatur auf die Populations dynamik des planktischen Rädertieres *Brachionus calyciflorus* Pallas. *Oecologia* 4:176–207.
Halbach, U. 1973. Life table data and population dynamics of the rotifer *Brachionus calyciflorus* Pallas as influenced by periodically oscillating temperature. In: Wieser, W. (ed.), *Effects of Temperature on Ectothermic Organisms*. Springer-Verlag, Berlin, p. 217–228.
Herzig, A. 1983. Comparative studies on the relationship between temperature and duration of embryonic development of rotifers. *Hydrobiologia* 104:237–246.
Hill, M.O. 1973. Diversity and evenness: A unifying notation and its consequences. *Ecology* 45:427–432.
Hutchinson, G.E. 1967. *A Treatise on Limnology* Vol. 2. John Wiley & Sons, Inc., New York, 1115 pp.
Karabin, A. 1978. The predatory pressure of pelagic cyclopoid copepods *Mesocyclops* on small zooplankton. *Ekologia Polska* 26:241–257.
King, C.E. 1972. Adaption of rotifers to seasonal variation. *Ecology* 53:408–418.
Hillbricht-Ilkowska, A., Gliwicz, Z., and Spodniewska, I. 1966. Zooplankton production and some trophic dependencies in the pelagic zone of two masurian lakes. *Verhandlungen der Internationalen Vereinigung für Limnologie* 16:432–440.
Larsson, P. 1978. The life cycle dynamics and production of zooplankton in Ovre Heimdalsvatn. *Holarctic Ecology* 1:162–218.
Legendre, L. and Legendre, P. 1983. *Numerical Ecology*. Elsevier Scientific Publishing Company, Amsterdam, 419 pp.
Lewis, W.M., Jr. 1979. *Zooplankton Community Analysis: Studies on a Tropical System*. Springer-Verlag, New York, 163 pp.
Makarevicz, J.C. and Likens, G.E. 1979. Structure and function of the zooplankton community of Mirror Lake, New Hampshire. *Ecological Monographs* 49(1):109–127.
Margalef, R. 1969. Diversity and stability: a practical proposal and a model of interdependence. In: Diversity and stability in ecological systems. *Brookhaven Symposium in Biology* 22:25–37.
Margalef, R. 1983. *Limnologiá*. Ediciones Omega, S.A., Barcelona, 1010 pp.
Matveeva, L.K. 1983. Community structure of planktonic rotifers in a mesotrophic lake. *Hydrobiologia* 104:353–356.
May, L. 1983. Rotifer occurrence in relation to water temperature in Loch Leven, Scotland. *Hydrobiologia* 104:311–315.
Miracle, M.R. 1974. Niche structure in freshwater zooplankton: a principal components approach. *Ecology* 55:1306–1317.

Miracle, M.R. 1977. Migration, patchiness, and distribution in time and space of planktonic rotifers. *Archiv für Hydrobiologie Beihefte / Ergebnisse der Limnologie* 8:19–37.

Motyka, J., Dobrzanski, B., and Zawazki, S. 1950. Preliminary studies on meadows in the south-east of the province Lublin. *Ann. Univ. Mariae Curie-Sklodowska Sect, E. Agricultura* 5:367–447.

Nauwerk, A. 1963. Die Beziehungen zwischen Zooplankton und Phytoplankton im See Erken. *Symbolae Botanicae Upsalienses* 17:1–163.

Neill, W.E. 1984. Regulation of rotifer densities by crustacean zooplankton in an oligotrophic montane lake in British Columbia. *Oecologia (Berlin)* 61:175–181.

Orcutt, J.D. and Pace, M.L. 1984. Seasonal dynamics of rotifer and crustacean zooplankton populations in a eutrophic, monomictic lake with a note on rotifer sampling techniques. *Hydrobiologia* 119:73–80.

Pejler, B. 1957. Taxonomical and ecological studies on planktonic Rotatoria from northern Swedish Lapland. *Kunglica Svenska Vetenskapsakademiens Handlingar* 6:1–68.

Pourriot, R. 1965. Recherches sur l'ecologie des Rotifères. *Vie et Milieu Suppl.* 21:1–224.

Pourriot, R., Deluzarches, M. 1971. Recherches sur la biologie des Rotifères. II. Influence de la température sur la durée du développement embryonnaire et post-embryonnaire. *Annales de Limnologie* 7:25–52.

Pourriot, R. 1977. Food and feeding habits of Rotifers. *Archiv für Hydrobiologie Beihefte / Ergebnisse der Limnologie* 8:243–260.

Rey, J. and Capblancq, J. 1975. Dynamique des populations et production du zooplancton du lac de Port—Bielh (Pyrénées Centrales). *Annales de Limnologie* 11:157–172.

Ronneberger, D. and Kasprzak, P. 1985. The zooplankton population. In: Casper, S.J. (ed.), *Lake Stechlin: A Temperate Oligotrophic Lake.* Dr. W. Junk Publishers, The Hague, p. 243–259.

Ruttner-Kolisko, A. 1972. Der Einfluβ von Temperatur und Salzgehalt des Mediums auf Stoffwechsel- und Vermehrungsintensität von *Brachionus plicatilis* (Rotatoria). *Verhandlungen der Deutschen Zoologischen Gesellschaft* 65:89–95.

Ruttner-Kolisko, A. 1975. The influence of fluctuating temperature on plankton rotifers. A graphical model based on life data of *Hexarthra fennica* from Neusiedlersee, Austria. *Symposia Biologica Hungarica* 15:197–204.

Ruttner-Kolisko, A. 1978. Influence of fluctuating temperature on plankton rotifers. II. Laboratory experiments. *Verhandlungen der Internationalen Vereinigung für Limnologie* 20:2400–2405.

Schindler, D.W. 1972. Production of phytoplankton and zooplankton in Canadian Shield Lakes. In: Z. Kajak and A. Hillbricht-Ilkowska (eds.), *Productivity Problems of Freshwaters.* PWN Polish Scientific Publishers Warsaw, p. 383–404.

Shannon, C.E. 1948. A mathematical theory of communication. *Bell System Technical Journal* 27:379–423, 623–656.

Stabel, H.-H. 1985. Mechanisms controlling the sedimentation sequence of various elements in prealpine lakes. In: Stumm W. (ed.), *Chemical Processes in Lakes.* Wiley, New York, p. 143–167.

Starkweather, P.L. and Bogdan, K.G. 1980. Detrital feeding in natural zooplankton communities: discrimination between live and dead algal foods. *Hydrobiologia* 73:83–85.

Stemberger, R.S. 1974. Temporal and spatial distributions of planktonic rotifers in Milwaukee harbor and adjacent Lake Michigan. *Proceedings 17th Congress of Great Lakes Research*: 120–134.

Stemberger, R.S., Evans, M.S. 1984. Rotifer seasonal succession and copepod predation in Lake Michigan. *Journal of Great Lakes Research* 10:417–428.

Tilzer, M.M. and Beese, B. 1988. The seasonal productivity cycle of phytoplankton and controlling factors in Lake Constance. *Schweizerische Zeitschrift für Hydrologie* 50:1–39.

Vareschi, E. and Jacobs, J. 1985. The ecology of Lake Nakuru. VI. Synopsis of production and energy flow. *Oecologia (Berlin)* 65:412–424.
Walz, N. 1986. Wie werden Rotatorienpopulationen reguliert? *Laufener Seminarbeiträge* 2/86:80–90.
Walz, N. 1987. Stoffumsatz und Kinetik von Regulationsprozessen bei Zooplankton-Populationen. Habilitationsschrift, Ludwig-Maximilians-Universität München, 225 pp.
Walz, N., Elster, H.-J., and Mezger, M. 1987. The development of the rotifer community structure in Lake Constance during its eutrophication. *Archiv für Hydrobiologie* 74:452–487.
Wappis, E. 1980. Zur Populationsökologie des Zooplanktons des Ossiachersees. Dissertation, Universität Graz, 168 pp.
Williamson, C.E. 1983. Invertebrate predation on planktonic rotifers. *Hydrobiologia* 104:385–396.
Winberg, G.G., Babitsky, V.A., Gavrilov, S.I., Gladky, G.V., Zakharenkov, I.S., Kovalevskaya, R.Z., Mikheyeva, T.M., Nevyadomskaya, P.S., Ostapenya, A.P., Petrovich, P.G., Potaenko, J.S., and Yakushko, O.F. 1972. Biological productivity of different types of lakes. In: Z. Kajak and A. Hillbricht-Ilkowska (eds.), *Productivity Problems of Freshwaters*. PWN Polish Scientific Publishers Warsaw, p. 383–404.

# Part 5

# Food-Web Structures and Trophic Interactions

Until recently, many scientists believed that the organic matter produced by autotrophic bacteria is grazed by herbivorous zooplankton which, in turn, are preyed upon by carnivorous zooplankton and fish. In this model, bacteria were considered to be merely decomposers of organic matter. Recently, however, it has become obvious that bacteria are acutally highly active producers and that they occupy a central position within the pelagic food web.

The first section of Part 5, Chapters 24 through 27, introduces us to the fundamentals of the "microbial loop revolution": bacteria can consume up to 60–80% of the autotrophic production and are successful competitors against algae for nitrogen and phosphorus uptake. Indeed, bacterial production reaches magnitudes similar to those of phytoplankton production. The intimate relationship of planktonic bacteria with organic particles in the water and the resulting anaerobic micro-environments may play a major role in the formation of the reduced compounds found in oxygenated layers of large lakes.

The second section of Part 5, Chapters 28 and 29, deals with the bacterivorous components of the microbial food-web and their relationship with the grazing food chain. The contribution of nanoflagellates in the food-web of two lakes is described. With a growing number of trophic steps, the carbon losses through respiration increase, and the gap presently observed between autotrophic production and the heterotrophic requirements is discussed.

These new concepts have radically changed our general view of food-web interactions, a historical review of which is presented in Chapter 30. In particular, the allometric approach is compared with the species-specific approach.

In the last section of Part 5, Chapters 31 through 35, examples of interactions between phytoplankton and zooplankton and between benthos and fish, and of the effects of fish grazing on lake plankton are given. Finally, the authors of Chapters 32 to 35 demonstrate that, along a gradient of increasing lake trophy, the control of the food-web shifts from the primary resources ("bottom-up") to the consumers ("top-down").

# 24

# Bacterial Cycling of Matter in the Pelagic Zone of Aquatic Ecosystems

F. Azam, B. C. Cho, D. C. Smith, and M. Simon

ABSTRACT This paper discusses the role of bacteria in influencing the patterns of organic and inorganic matter fluxes in pelagic ecosystems of lakes and oceans. We find that the ecological and biogeochemical roles of bacteria vary greatly in response to the nutrient and trophic scenarios in the bacterium's microenvironment. We suggest that, in order to understand how bacteria influence nutrient dynamics in pelagic ecosystems, we need a conceptual framework which takes into account the interactions among bacteria, phytoplankton, and protozoa in a microenvironmental context.

## 24.1 Introduction: Stating the Problem

A fundamental goal of aquatic biology is to understand how the elements carbon, nitrogen, and phosphorus (C,N,P) cycle between the living and the non-living pools of matter, and how energy is applied to effect and pattern elemental dynamics. Until the present decade, the problem had been framed in terms of the incorporation of dissolved inorganic matter (DIM) into organic molecules by phytoplankton and the utilization of organic matter by metazoa to regenerate DIM. This perception has changed fundamentally in recent years because of the realization that bacteria and protozoa also play a significant role in elemental dynamics (Williams, 1981; Azam et al., 1983; Cole et al., 1988); consequently, there have been attempts to define and incorporate a role for the microbiota. The purpose of this paper is to discuss some mechanistic issues raised by this "new view" of the role of heterotrophic bacteria and protozoa in material cycling in aquatic ecosystems.

In the classical view, organic matter fluxes were thought to occur *within the particulate phase* along the grazing food chain: plant→herbivore→carnivore. Since bacteria can only utilize dissolved materials, one assumed that they could not be significant utilizers of particulate organic matter (POM). Given the great dilution of dissolved organic matter (DOM) in lakes and oceans, pelagic bacteria were thought to be mostly dormant (Stevenson, 1978) and to only play a minor role in the flux of materials and energy in their ecosystems. The advent of epifluorescence microscopy (Zimmermann and Meyer-Reil, 1974; Hobbie et al., 1977) revealed that bacterial abundance was 10–100 times greater than was previously believed. This meant that bacterial biomass was often a large fraction of the total plankton biomass (below). This discovery also stimulated interest in the dynamic state of bacterial biomass. New methods for measuring bacterial production (Hagström et al., 1979; Fuhrman and Azam, 1980, 1982) and its fate (Hollibaugh et al., 1980; Fenchel, 1982; Sherr et al., 1987; Riemann and Søndergaard, 1986) have now shown that bacterial production is rapid enough to generally consume 30–60% of primary production (40%; Cole et al., 1988). The abundant bacterivorous protozoa consume most bacterial production, possibly acting as a link between the microbial food web and the grazing food chain. These findings have confirmed the predictions of Pomeroy (1974) of the existence of a robust microbial food web, which is now often referred to as the "microbial loop" (Figure 24.1, and Azam et al., 1983). The microbial loop is a major route for the flow of material and energy in pelagic food webs. This new notion of the pelagic food web, developed largely on the basis of data for the coastal ocean, has been exhaustively reviewed (Williams, 1981; Ducklow, 1983; Azam et al., 1983; Sherr et al., 1988; Cole et al., 1988). The literature on lakes has been synthesized into an excellent book (Riemann and Søndergaard, 1986) and shows that the ocean systems and the lake systems are quite similar in many respects.

A corollary of the above findings is that a substantial fraction of the primary production in the euphotic zone *must pass through the dissolved phase* (since bacteria can use only DOM). This deduction represents a fundamental departure from the classical view and has implications for mechanisms of matter and energy flows and biogeochemical cycling pathways in aquatic ecosystems. In this paper, we will develop the proposition that bacteria exert a significant influence on the pathways and mechanisms of elemental fluxes in the pelagic ecosystems. In the next two sections, we will address two sets of inter-related issues (so set out below), the first regarding bacterial utilization/decomposition of organic matter and the second regarding the dual role of bacteria as producers and consumers of mineral nutrients. We will also stress the need to conceptualize and study the interactions of bacteria within their microenvironments.

Concerning the first issue, the role of bacteria as consumers of organic matter, we must ask whether the grazing food chain is inherently so leaky as to "spill" one-half of the primary production into the dissolved phase, or whether bacteria exert a "biochemical pressure" to cause DOM production from the particulate food chain? What strategies allow bacteria to perform so well in a "particulate world," and should bacteria be viewed as competitors with particle-eating metazoa and protozoa for organic matter utilization? The role of bacteria in organic matter fluxes is rendered more complex by the observation that while bacteria solubilize POM to liberate DOM, they also avidly

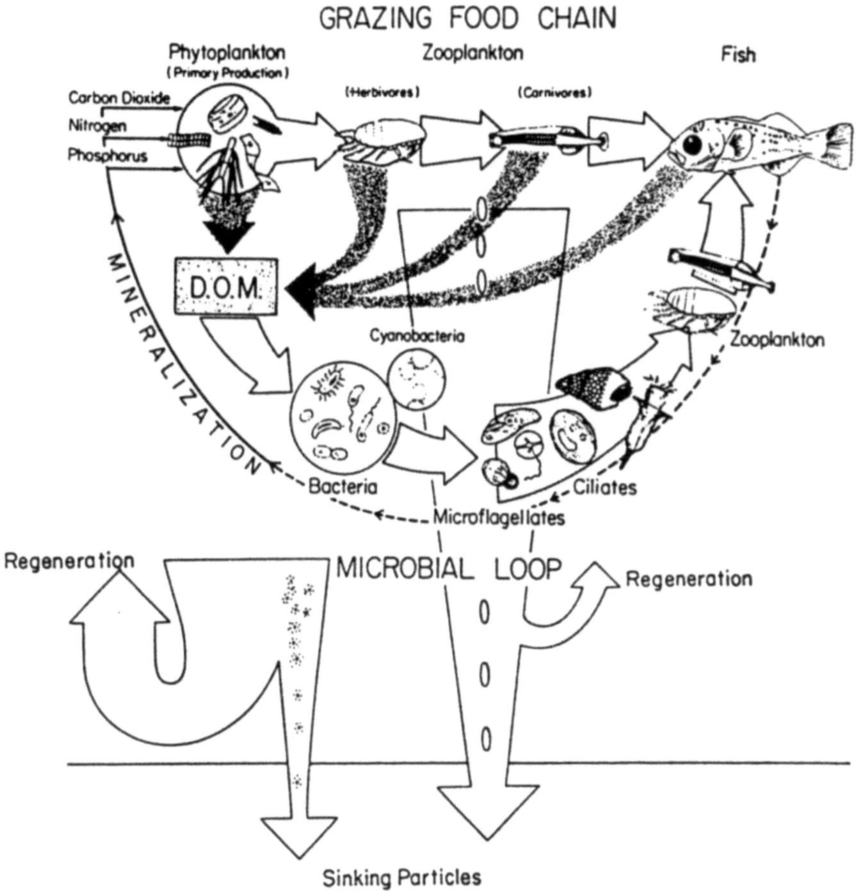


**Figure 24.1** A simplified schematic depiction of the microbial loop in aquatic ecosystems, its relationship to the classical grazing food chain, and its implications for the sinking flux of organic matter. For simplicity, mineralization by the grazing chain and DOM production by the microbial loop are not depicted.

and almost exclusively utilize DOM to form new particles (i.e., new bacterial cells).

Concerning the second issue, the role of bacteria in the dynamics of mineral nutrients: In the classical view the main role of bacteria was to mineralize the organic matter, but it now appears that they are not only mineralizers but also highly effective competitors with phytoplankton for mineral nutrients, especially in oligotrophic environments. How are the mineral fluxes into and out of bacteria regulated? Interestingly, although bacteria outcompete phytoplankton for N and P in nutrient-limited environments, they depend on phytoplankton for energy supply in the form of reduced carbon (Bratbak and Thingstad, 1985). Given these considerations, it appears that in order to understand how nutrients move within the pelagic ecosystems, we need to develop a conceptual framework which takes into account the interactions between bacteria, phytoplankton, and protozoa.

## 24.2 Bacteria as Consumers of Organic Matter

**Goals of This Research** There are two main issues concerning the role of bacteria in organic matter fluxes in pelagic ecosystems: the *magnitude* of flux and the *mechanisms* whereby bacteria interact with the sources of organic matter to achieve this flux. The proximal goal has been to measure material and energy flux *via* bacteria in the context of ecosystem dynamics and energetics. This goal has been partially achieved (see below). The eventual goal is to develop a mechanistic framework for bacteria—organic matter interactions which explains and predicts specific fluxes into bacteria in terms of the bacterium's biochemical attributes (e.g., cell surface enzymes, membrane transport systems, chemotaxis, motility). This ambitious goal is far from being achieved but, as discussed below, it is neither intractable nor dispensable.

**Magnitude of Organic Matter Flux into Bacteria** Quantification of organic matter flux into bacteria had not been possible until 1979 for lack of a suitable method. Radiotracer methodology (Parsons and Strickland, 1962; Wright and Hobbie, 1965, 1966; Azam and Holm-Hansen, 1973) showed that the pools of DOM components, such as sugars and amino acids, turn over in time frames of 10–100 h. However, the mass flux of organic matter could not be calculated for lack of methods to measure the pool sizes of various substrates. Further, bacterial uptake might be tightly coupled with substrate production, so the tracer might not completely mix with the substrate pool. An alternative approach is to use bacterial production (plus respiration) as a cumulative measure of all organic fluxes into bacteria (Hagstöm et al., 1979; Fuhrman and Azam, 1980). In pursuing this approach, several methods have been developed for measuring bacterial production (reviewed by Cho and Azam, 1988a). A method based on measuring the rate of bacterial DNA synthesis (thymidine incorporation method; Fuhrman and Azam, 1980, 1982) has been extensively used, despite on-going debate on the method's accuracy for calculating bacterial carbon production (for discussion of the method see Moriarty 1986; Cho and Azam, 1988a). Another method, based on measuring bacterial protein synthesis rate (Chin-Leo and Kirchman, 1988; Simon and Azam, 1989), compared favorably with the thymidine incorporation method in the few coastal ocean samples examined so far.

Bacterial production has now been measured in a great variety of environments in order to analyze the partitioning of the carbon flux between the microbial and the macrobial food webs: in the euphotic zone of oceans (reviewed by Cole et al., 1988) and lakes (reviewed by Riemann and Søndergaard, 1986), and in the ocean's interior (Cho and Azam, 1988). A number of studies have also reported the proportion of primary production that would be required to support the measured rate of bacterial production ($F_B$; Cho and Azam, 1988). Although $F_B$ varies considerably, it is around 0.4 in most environments (see Ducklow, 1983; Cole et al., 1988). It is noteworthy that the value of $F_B$ is independent of the trophic state of the water; there are no systematic differences in $F_B$ among oligotrophic, mesotrophic, and eutrophic environments, although the biomass ratio of bacteria and phytoplankton varies greatly with the trophic status. A question for future studies is: what mechanisms regulate the partitioning of fluxes between the microbial and macrobial food webs (as indicated by $F_B$) within such narrow limits in trophically varied environments?

Measurements of bacterial production show that organic matter flux into bacteria is one of the main pathways of material and energy flows in pelagic food webs. This also means that there must be mechanisms operating within the pelagic ecosystems which convert a substantial fraction of primary production into DOM. At present, there is little knowledge on the mechanisms of DOM production, and this is an important area for future research. In the past, the general feeling had been that, in the euphotic zone, DOM is produced by phytoplankton exudation and autolysis and by zooplankton "sloppy feeding" (see Williams, 1981). Solubilization of particulate detritus has received less attention, although it may turn out to be an important mechanism, and perhaps it may be the main mechanism of DOM production in the aphotic zone (below).

In stratified water columns, organic matter flux below the mixed layer is largely in the form of sinking particles. A central question is: do the osmotrophic bacteria manage to utilize a significant fraction of the sinking flux, or do the particle-eaters dominate the energy flow in the ocean's interior? Cho and Azam (1988) determined bacterial carbon production in mesopelagic zones of several oceanic sites and compared these with the computed sinking flux of organic matter into the mesopelagic layer. They found that $>80\%$ of the sinking flux was utilized by the mostly free-living bacteria. Thus, bacteria somehow manage to compete effectively with the particle-eaters. This also means that a large fraction of the sinking organic matter uses the following pathway: POM→DOM→bacteria. The biogeochemical implications of a large-scale phase transition of organic matter (POM→DOM) merit study. For instance, the spatial and temporal patterns of material flux in the mesopelagial may be influenced by the production of a large number of small, nonsinking particles (bacteria) at the expense of fewer, but larger, rapidly sinking particles. This area of research is ripe for study in oceans and large lakes, given the relevance of the vertical export of carbon to global climatic change.

**Mechanisms of Organic Matter Flux into Bacteria** Azam and Cho (1987) have synthesized the information on the mechanisms of bacterial utilization of organic matter in the pelagic environment. They hypothesize that bacteria compete with particle-eaters by rapidly solubilizing POM, and that bacteria also compete with less efficient osmotrophs (heterotrophic algae and protozoa) by keeping the utilizable DOM at exceedingly low concentration. Here, we discuss these aspects of bacteria–organic matter interactions and their relevance to material fluxes.

Numerous studies have shown that non-photosynthetic bacteria practically monopolize DOM, and that free-living, rather than particle-bound, bacteria are mainly responsible for uptake (Berman, 1975; Azam and Hodson, 1977; references in Azam and Cho, 1987). Fuhrman and Azam (1982) showed that most thymidine incorporation (and hence bacterial production) was also due to bacteria operationally defined as free-living. These conclusions are based on microautoradiography and on size-fractionation through sieves, and thus do not preclude the possibility that active bacteria occurred in ephemeral associations with organisms or particles rather than being absolutely free-living. (In fact, Azam and Ammerman (1984) deem it probable that free-living bacteria are able to physically interact with the sources of DOM, such as phytoplankton and detritus, in bacterial microenvironment). Significantly, these results mean that DOM has a well-defined point of entry into the pelagic foodweb, and have led to the idea that the pathway DOM→*free-living* bacteria→protozoa (a microbial loop—Pomeroy, 1974; Azam et al., 1983) is a significant pathway for material and energy flow. We note that particle-attached bacteria may also be important in some environments (e.g., some estuaries breakdown of algal blooms) and the DOM assimilated there by attached bacteria may be passed directly to metazoa feeding on particles (Azam et al., 1984 Simon, 1987).

What strategies enable bacteria to maintain a virtual monopoly on DOM? Azam and Ammerman (1984) suggest that bacteria keep utilizable DOM at low concentrations *via* a tight coupling with the sources of DOM. Biochemical strategies exist for efficient removal of two main types of compounds, those which can be directly utilized by bacteria (e.g., sugars, amino acids) and those which require extracellular hydrolysis and the uptake of hydrolysis products (e.g., proteins, organophosphorus compounds). Behavioral strategies (chemotaxis) may also enhance tight coupling by optimizing the position of the bacterium in relation to DOM source in the microenvironment.

Azam and Hodson (1981) found multiple transport systems for glucose uptake in marine bacteria. They suggested that low-flow (high affinity—low capacity) systems enable scavenging of glucose from the dilute bulk-phase while the high-flow (low affinity—high capacity) systems enable the bacterium to optimize uptake from high microevironmental concentrations in the vicinity of a source releasing glucose into seawater. The result would be a tight coupling between glucose release and uptake, thus keeping glucose concentrations in the bulk-phase at extremely low levels.

Those substrates requiring hydrolysis prior to uptake offer an intriguing challenge to bacteria which utilize such substrates in the pelagic environment. One would predict that release of hydrolytic enzymes into the bulk-phase will be energetically inefficient, so hydrolysis must be done by cell-associated enzymes. This was found for hydrolysis of dissolved protein (Hollibaugh and Azam, 1983) and ATP (Ammerman and Azam, 1985) by natural assemblages of marine bacteria. One would also predict a hydrolysis–uptake coupling, so that the hydrolysis product does not diffuse away from the cell. Evidence of hydrolysis–uptake coupling was found for both protein and ATP utilization. Interestingly, and essential to the model of P cycling which we present below, the tightness of hydrolysis–uptake coupling depended on the environmental concentration of the hydrolysis product. When P was liberated by ATP hydrolysis (by a periplasmic 5′-nucleotidase; Ammerman and Azam, 1985), the fraction of liberated P which was taken up by the cell was inversely related to the environmental P concentration. Over 50% of liberated P was taken up (100% in one instance) when the seawater P was (or was experimentally reduced to) below 30 nM; virtually all liberated P was released into seawater if P concentration was experimentally elevated to mM levels. The flexible and dynamic nature of P release explains how bacteria may act as mineralizers in some situations and competitors for P in other situations (as discussed below).

A final issue in bacteria–organic matter interaction is this: If most bacteria are free-living, then how can we justify our previous conclusion that bacteria compete with particle-eaters by rapidly solubilizing the particles. Azam and Cho (1987) hypothesize a dynamic scenario wherein colonizing bacteria channel energy into hyperproducing hydrolytic enzymes (to rapidly solubilize the particle) rather than into rapid growth. They further envision that the colonizers become "baby machines" and release their progeny in the particle's vicinity, which is enriched by DOM released by the colonizers. Simon, Alldredge, and Azam (unpubl. observation) tested aspects of this hypothesis with observations on marine snow. They found that most of protein hydrolysed from the snow was released into the water and <10% was utilized by the colonizers. They (and Jacobsen and Azam, 1984) also documented extensive progeny release consistent with the hypothesis of Azam and Cho (1987). It appears then that a colonized particle can become a source of both DOM and free-living bacteria. These preliminary observations point out that it is unrealistic to think of attached and free-living bacteria as being in separate compartments. In order to understand material and energy fluxes in the particle's microenvironment, we need to develop concepts and methods which treat particle-bacteria interaction as a dynamic system open to the environment.

## 24.3 Role of Bacteria in Dynamics of Mineral Nutrients

**Bacteria as a C, N, P Pool: Significance for Nutrient Cycling** Historically, phytoplankton and detritus have been considered the main particulate pools of C, N, P subject to nutrient liberation through the trophic activities of pelagic organisms. Are bacterial C, N, P pools significant as compared with those of phytoplankton and detritus? Recent work showed that in oligotrophic waters, bacterial carbon (and hence N, P) equalled or exceeded phytoplankton carbon (Simon and Tilzer, 1987; Cho and Azam, 1988) but the reverse was true for eutrophic samples, suggesting quite different food web structures and nutrient cycling scenarios. Producer biomass exceeded decomposer (bacteria) biomass in eutrophic waters, whereas the opposite was true for the oligotrophic waters. (It also follows that bacteria in oligotrophic waters must grow more slowly than phytoplankton, which provide the energy for bacterial growth, and this was indeed the case in the oligotrophic central North Pacific gyre where phytoplankton and bacteria doubling times were on the order of 1 and 10 d, respectively; Cho and Azam, 1987). Further, in oligotrophic waters, the food size spectrum available to the grazers will be dominated by the small, free-living bacteria, while in eutrophic waters it will be dominated by the phytoplankton cells. Thus, protozoa may be dominant grazers in oligotrophic waters and metazoa are the dominant grazers in eutrophic waters.

These observations have implications for nutrient cycling pathways and for vertical flux of biogenic production (see Figure 24.1). In the bacteria-dominated oligotrophic ecosystems, most of the N and P would occur sequestered in the nonsinking, vertically nonexportable, free-living bacteria. The reverse would be true in eutrophic waters. This could, in part, explain why oligotrophic ocean waters have low, but eutrophic ocean waters have high "new production" (i.e., the eutrophic waters vertically export a large fraction of the primary production). Further, oligotrophic waters will be "microbial-loop dominated" and, therefore, more efficient in nutrient regeneration than eutrophic waters.

**Relationship between Bacteria and Detritus** Recently, it was found that in ocean profiles, bacterial organic carbon (BOC) was 30–70% of the total POC (Cho and Azam, 1988). A high BOC/POC ratio occurred in all aphotic zone samples and also in the euphotic zone samples at oligotrophic to mesotrophic sites. Since bacterial C/N and C/P are lower than these ratios in detritus, bacteria would also represent a substantial fraction of total suspended N, P. In the context of nutrient cycling, these results indicate a need to focus on the mechanisms which liberate nutrients from bacterial cells (presumably *via* the activity of the microbial loop). A big gap in our present knowledge is that we know nothing about the structure and functioning of the detritus-based microbial loop in the deeper parts of large lakes and the oceans.

We point out in passing that bacteria may also play a role in adsorption of surface-reactive materials. In our preliminary microscopic observations, bacteria represented the largest particulate surface, so surface-reactive materials might largely adsorb onto free-living, non-sinking bacteria (Cho and Azam, 1988). Interestingly, Bacon et al. (1985) predicted, based on measured vertical flux of surface-reactive radionuclides, that there must exist in the ocean's interior a large pool of nonsinking, fine particles which adsorb radionuclides and which are swept down by rapidly sinking particles. The fine detritus of Bacon et al. (1985) may, in part, be free-living bacteria. Free-living bacteria could thus influence the sinking flux of surface-reactive materials in oceans and lakes.

**Mechanisms of Cycling of Mineral Nutrients** As stated, the classical role of net mineralizers assigned to pelagic bacteria needs revision, because while bacteria do under some conditions, release N and P from organic matter, they compete with phytoplankton for DIM when nutrients are dilute. Suttle et al. (1987) used $^{13}N$-ammonium additions to show that phytoplankton dominated ammonium uptake at high ammonium concentrations but bacteria outcompeted phytoplankton at sub-micromolar concentrations. A similar conclusion was reached by Wheeler and Kirchman (1986). These findings are consistent with the observation, discussed above, that in oligotrophic waters bacteria sequester more N and P than phytoplankton. A similar situation occurs in Lake Constance, especially during the clear-water phase, with respect to P (Simon and Tilzer, 1987). Size-fractionation experiments in the coastal ocean have also shown that the bacteria are often responsible for the uptake of 50–90% of tracer $^{32}P$ (Harrison et al., 1977; Ammerman and Azam, 1985).

If bacteria compete with phytoplankton for DIM, how can they provide mineralized nutrients for phytoplankton uptake? It is generally thought that protozoan predation on bacteria results in DIM release (Andersson et al., 1985; Goldman et al., 1987). But, in oligotrophic waters, the released DIM would diffuse to low concentrations at which bacteria would once again outcompete phytoplankton for DIM uptake. This puzzle might be solved by assuming that (1) microbial loop components interact to create microscale "hot spots" of high organism and nutrient concentrations, (2) phytoplankton attract bacteria (by exudation) to high densities, and (3) predation on "bacterial clusters" creates high DIM concentrations at which phytoplankton become effective competitors with bacteria (Azam and Cho, 1987). In the case of P, high P concentration in microbial cluster could have an "autocatalytic" effect on P release from bacteria by loosening the hydrolysis–uptake coupling (see Section 24.2).

Thus, the effects of high bacterial density in the phytoplankton's microenvironment would include: (a) high predation pressure by protozoa on clustered bacteria will mineralize the sequestered P at higher rate than in the bulk-phase; (b) DOP hydrolysis by bacteria in a high P microenvironment will loosen the hydrolysis–uptake coupling and, therefore, a larger *fraction* of the mineralized P will be released than in the bulk-phase; and (c) phytoplankton (due to their low-affinity but high-capacity uptake systems) will be able to compete with bacteria for a larger fraction of P because a high P concentration will be sustained in the microenvironment.

A central role of bacterial 5′-nucleotidase (5′-Nase) in P cycling has been suggested by the study of Ammerman and Azam (1985). However, further studies should test the generality of this role as well as the biochemical mechanism of control of hydrolysis-uptake coupling. 5′-Nase is distinct from alkaline phosphatase (APase), particularly in that 5′-Nase is not repressed by high P, while APase is repressed by high P. Thus, bacterial 5′-Nase seems designed to work in microenvironments containing both high and low P concentrations (see Ammerman and Azam, 1985, for details of the characteristics of this enzyme). Future studies should resolve the relative roles of bacterial 5′-Nase and phytoplankton APase in P dynamics in natural microbial consortia in oligotrophic lakes and oceans.

**The Significance of the Microenvironment** It is a truism that bacteria live, grow, and die in microenvironments. Hence, their role in mediating matter and energy fluxes must depend on the activities within microzones. The discussion above, as well as the arguments by Azam and Cho (1987), suggests that the solute molecules, bacteria, phytoplankton, and bacterivores are not distributed randomly with respect to each other. We note that in a "typical" euphotic zone bacteria, phytoplankton, and bacterivores are abundant enough to be within 100–1,000 μm of each other. Most of these microbes have some potential for movement. Thus, the microbial consortia are capable of spatially intimate and dynamic interactions. It is technically very difficult to demonstrate the existence of "microbial clustering." However, given their proximity and trophic interdependence, it would not be surprising if pelagic microbial components were found to cluster but it would indeed be surprising if they did not.

## 24.4 Concluding Remarks

We have emphasized that pelagic bacteria, despite their osmotrophic character, influence significantly their "particulate worlds" in lakes and oceans. Their dual ecological roles are to be "particularizers" as well as solubilizers, and as such they create dynamic scenarios of elemental fluxes in consonance with the environment's nutrient and energy status. Future studies should address the biochemical mechanisms of bacteria-organic matter coupling, to understand how organic and inorganic fluxes involving bacteria are mediated and regulated. Another challenge is to elucidate the mechanisms of DOM production in the pelagic environments, and how the dominant mechanisms vary between environments. We stress the need to study intact microbial consortia in their microenvironments, if we are to succeed in developing a comprehensive understanding of the role of bacteria in matter and energy fluxes in aquatic ecosystems. To do this, our conceptual models should take into account the microstructure of microbial communities in aquatic environments.

---

*Acknowledgements* This work was supported by an NSF grant and a DOE grant.

## References

Ammerman, J.W. and Azam, F. 1985. Bacterial 5′-nucleotidase in aquatic ecosystems: a novel mechanism of phosphorus regeneration. *Science* 227:1338–1340.

Andersson, A., Larsson, U., and Hagström, Å. 1985. Size selective grazing by a microflagellate on pelagic bacteria. *Mar. Ecol. Prog. Ser.* 23:99–106.

Azam, F. and Ammerman, J.W. 1984. In: M.J.R. Fasham (editor), *Flows of Energy and Materials in Marine Ecosystems.* Plenum Publishing, New York, p. 345–360.

Azam, F. and Cho, B.C. 1987. In: M. Fletcher (editor), *Ecology of Microbial Communities.* Cambridge University Press, Cambridge, p. 261–281.

Azam, F., Cowles, T., Banse, K., Osborne, J., Harrison, P.J., and Kennedy, C.A. 1984. Free-living pelagic bacterioplankton: Sink or link in a marine foodweb? *Eos* 65:926.

Azam, F., Fenchel, T., Field, J.G., Gray, J.S., Meyer-Reil, L.-A., and Thingstad, F. 1983. The ecological role of water-column microbes in the sea. *Mar. Ecol. Prog. Ser.* 10:257–263.

Azam, F. and Hodson, R.E. 1977. Size distribution and activity of marine microheterotrophs. *Limnol. Oceanogr.* 22:492–501.

Azam, F. and Hodson, R.E. 1981. Multiphasic kinetics for D-glucose uptake by assemblages of natural marine bacteria. *Mar. Ecol. Prog. Ser.* 6:213–222.

Azam, F. and Holm-Hansen, O. 1973. Use of tritiated substrates in the study of heterotrophy in seawater. *Mar. Biol.* 23:191–196.

Bacon, M., Huh, C., Fleer, P. and Deuser, W. 1985. Seasonality in the flux of natural radionuclides and plutonium in the deep Sargasso Sea. *Deep-Sea Res.* 32:273–286.

Berman, T. 1975. Size fractionation of natural aquatic populations associated with autotrophic and heterotrophic carbon uptake. *Mar. Biol.* 33:215–220.

Bratbak, G. and Thingstad, T. 1985. Phytoplankton-bacteria interactions: an apparent paradox? Analysis of a model ecosystem with both competition and commensalism. *Mar. Ecol Prog. Ser.* 25:23–30.

Chin-Leo, G. and Kirchman, D.L. 1988. Estimation of bacterial production in marine waters from the simultaneous incorporation of thymidine and leucine. *Appl. Environ. Microbiol.* 54:1934–1939.

Cho, B.C. and Azam, F. 1987. Significance of bacterioplankton biomass in the epipelagic and mesopelagic zone in the Pacific Ocean. *EOS* 68:1729.

Cho, B.C. and Azam, F. 1988. Major role of bacteria in biogeochemical fluxes in the ocean's interior. *Nature* 332:441–443.

Cho, B.C. and Azam, F. 1988a. Heterotrophic bacterioplankton production measurement by tritiated thymidine incorporation method. *Ergeb. Limnol.* 31:153–162.

Cole, J., Findlay, S., and Pace, M. 1988. Bacterial production in fresh and salt-water ecosystems: a cross-system overview. *Mar. Ecol. Prog. Ser.* 43:1–10.

Ducklow, H. 1983. The production and fate of bacteria in the oceans. *BioScience* 33:494–501.

Fenchel, T. 1982. Ecology of heterotrophic microflagellates. IV. Quantitative occurrence and importance as bacterial consumers. *Mar. Ecol. Prog. Ser.* 9:35–42.

Fuhrman, J.A., Ammerman, J.W., and Azam, F. 1980. Bacterioplankton in the coastal euphotic zone: Distribution, activity and possible relationships with phytoplankton. *Mar. Biol.* 60:201–207.

Fuhrman, J.A. and Azam, F. 1980. Bacterioplankton secondary production estimates for coastal waters of British Columbia, Antarctica and California. *Applied Environmental Microbiology* 39:1085–1095.

Fuhrman, J.A. and Azam, F. 1982. Thymidine incorporation as a measure of heterotrophic bacterioplankton production in marine surface waters: evaluation and field results. *Mar. Biol.* 66:109–120.

Goldman, J.G., Caron, D.A., and Dennett, M.R. 1987. Nutrient cycling in a microflagellate food chain: IV. Phytoplankton-microflagellate interactions. *Mar. Ecol. Prog. Ser.* 38:75–87.

Hagström, Å., Larsson, U., Hörstedt, P., and Normark, S. 1979. Frequency of dividing cells, a new approach to the determination of bacterial cell growth rates in aquatic environments. *Appl. Environ. Microbiol.* 37:805–812.

Harrison, W., Azam, F., Renger, E., and Eppley, R. 1977. Some experiments on phosphate assimilation by coastal marine plankton. *Mar. Biol.* 40:9–18.

Hobbie, J.E., Daley, R.J., and Jasper, S. 1977. Use of Nuclepore filters for counting bacteria by epifluorescence microscopy. *Appl. Environ. Microbiol.* 33:1225–1228.

Hollibaugh, J.T. and Azam, F. 1983. Microbial degradation of dissolved proteins in seawater. *Limnol. Oceanogr.* 28:1104–1116.

Hollibaugh, J.T., Carruthers, A., Fuhrman, J.A., and Azam, F. 1980. Cycling of organic nitrogen in marine plankton communities studied in enclosed water columns. *Mar. Biol.* 59:15–21.
Jacobsen, T. and Azam, F. 1984. The role of bacteria in copepod fecal pellet decomposition: colonization, growth rates and mineralization. *Bull. Mar. Sci.* 35:495–502.
Moriarty, D. 1986. Measurement of bacterial growth rates in aquatic systems from rates of nucleic acid synthesis. *Adv. Microb. Ecol.* 9:245–292.
Parsons, T. R. and Strickland, J. D. H. 1962. On the production of particulate organic carbon by heterotrophic processes in sea water. *Deep-Sea Res.* 8:211–222.
Pomeroy, L.R. 1974. The ocean's food web, a changing paradigm. *BioScience* 24:499–504.
Riemann, B. and Søndergaard, M. 1986. *Carbon Dynamics in Eutrophic, Temperate Lakes.* Elsevier Sci. Publishers, New York, 284 pp.
Sherr, B.F., Sherr, E.F. and Fallon, R.D. 1987. Use of monodispersed fluorescently labeled bacteria to estimate in situ protozoan bacterivory. *Appl. Environ. Microbiol.* 53:958–965.
Sherr, B.F., Sherr, E.F., Hopkinson, C.S., 1988. Trophic interactions within pelagic microbial communities. Indications of feedback regulation of carbon flow. *Hydrobiologia.* 159:19–26.
Simon. M., 1987. Biomass and production of small and large free-living and attached bacteria in Lake Constance. *Limnol. Oceanogr.* 32:591–607.
Simon, M. and Azam, F. 1989. Protein content and protein synthesis rates of planktonic marine bacteria. *Mar. Ecol. Prog. Ser.* 51:201–213.
Simon, M. and Tilzer, M. 1987. Bacterial response to seasonal changes in primary production and phytoplankton biomass in Lake Constance. *J. Plank. Res.* 9:535–552.
Stevenson, L. 1978. A case for bacterial dormancy in aquatic systems. *Microb. Ecol.* 4:127–133.
Suttle, C.A., Fuhrman, J.A., and Capone, D.G. 1987. Rapid ammonium turnover times and concentration dependent resource partitioning in planktonic communities measured using $^{13}N$. *Eos* 68:1760.
Wheeler, P.A. and Kirchman, D.L. 1986. Utilization of inorganic and organic nitrogen by bacteria in marine systems. *Limnol. Oceanogr.* 31:998–1009.
Wright, R. and Hobbie, J.E. 1965. The uptake of organic solutes in lake water. *Limnol. Oceanogr.* 10:22–28.
Wright, R. and Hobbie, J.E. 1966. Use of glucose and acetate by bacteria and algae in aquatic ecosystems. *Ecology* 47:447–464.
Williams, P.J.leB. 1981. Incorporation of microheterotrophic processes into the classical paradigm of the planktonic foodweb. *Kieler Meeresforsch* 5:1–28.
Zimmermann, R. and Meyer-Reil, L.-A. 1974. A new method for fluorescence staining of bacterial populations on membrane filters. *Kieler Meeresforsch* 30:24–26.

# 25

# Bacterial Production and the Flow of Organic Matter in Lake Constance

H. Güde

ABSTRACT During an eight-year study, the seasonal development of bacterial populations in Lake Constance showed regular patterns. In the upper water layers, high bacterial numbers (up to $10^7$ ml$^{-1}$) always occurred during summer stratification whereas minimum values (around $10^6$ ml$^{-1}$) were observed in winter. In the deep water below 30 m, however, low numbers were recorded throughout the year. Estimates of bacterial biomass and production ranged from 10 to 150 $\mu$g C l$^{-1}$ and from less than 0.1 to 4 $\mu$g C l$^{-1}$ h$^{-1}$, respectively. As suggested in this chapter, bacteria may constitute a significant fraction of planktonic biomass and production and, consequently, they may function as an important trophic link. However, attempts to quantify their contribution lead to the conclusion that bacterial net production exceeds measured primary production. The possible contribution of allochthonous sources, of horizontal inhomogeneities of biomass distribution, and of methodological restrictions are discussed as possible explanations for this inconsistency.

## 25.1 Introduction

Our knowledge of the significance of heterotrophic bacteria for the cycling of organic matter is still rather incomplete because these studies are comparatively young. Nevertheless, considerable progress has recently been made toward a better understanding of the quantitative role of bacterioplankton in the structure and function of pelagic ecosystems (also see Chapter 24). This progress has led to the currently widely accepted paradigm that bacteria convert large amounts of organic matter originating from primary production with high efficiency into bacterial biomass. Consequently, bacteria would constitute a large if not predominant part of living planktonic biomass, and bacterial production can be regarded as an important potential food source for higher trophic levels.

In such a general form, this statement may be valid for many different types of aquatic ecosystems but it can be assumed that the significance of bacterial biomass and production varies widely in different lakes. However, the data available to date are not yet sufficiently comprehensive to attribute specific patterns of bacterial developments to certain lake types. It is therefore impossible to describe specific bacteria-related patterns characteristic of large lakes.

The study presented in this paper refers to the distribution and trophic importance of bacteria in Lake Constance, a moderately large (surface 540 $km^2$), deep (maximum depth 252 m), mesotrophic lake. The hydrography, chemistry and biology of this lake have been investigated since decades (also see Chapters 5, 13, and 17). Although the investigation of bacteria has only begun recently, a comparatively large data basis is now available: Direct counts of bacteria stained with acridine orange were already performed in the late 1950s (Deufel 1967). From 1979 on, bacterial development in Lake Constance has been continuously observed (Simon and Tilzer, 1982, Güde et al., 1985; Güde, 1986, 1988, Simon and Tilzer, 1987). During these long-term observations, the seasonal and vertical distribution of bacterioplankton has been shown to exhibit regular patterns similar to those observed in other large lakes, such as Lake Michigan (Scavia and Laird, 1987), Lake Biwa (also see Chapter 28) and Lake Geneva (Dufour et al., 1988). Thus, there is some confidence that the relationships found for this lake may also be representative to a certain degree for other large deep lakes in the temperate zone.

## 25.2 Seasonal Development of Bacterial Biomass and Production

In winter, bacteria are distributed rather homogenously over the unstratified water column (Figure 25.1). Bacterial numbers are around $10^6$ $ml^{-1}$. Soon after the onset of the algal spring bloom, bacteria also start growing rapidly, reaching their first maximum of around $10^7$ $ml^{-1}$ in May. However, the increase of bacterial abundance is restricted to near-surface layers, whereas in deep waters bacterial numbers remain low throughout the year. After the spring maximum, a regular decrease of bacterial numbers is observed in June, which coincides with the clear-water phase during the maximum of daphnids (Güde 1988). During the summer bloom of phytoplankton in July, bacterial numbers increase again. Thereafter, the bacterial numbers fluctuate irregularly between 5 and 10 $\times$ $10^6$ $ml^{-1}$ until a continuous decline from October to December leads to the annual winter minimum in January. This general pattern appears to be very regular because it was consistently observed every year from 1980 through 1987 (Figure 25.2).

For Lake Constance, bacterial production was estimated by two different approaches: Simon (1984) and Simon and Tilzer (1987) calculated bacterial production on the basis of incorporation rates of amino acids. Güde et al. (1985) and Güde (1986) made thymidine incorporation measurements according to the method of Fuhrman and Azam (1980). Although the absolute values

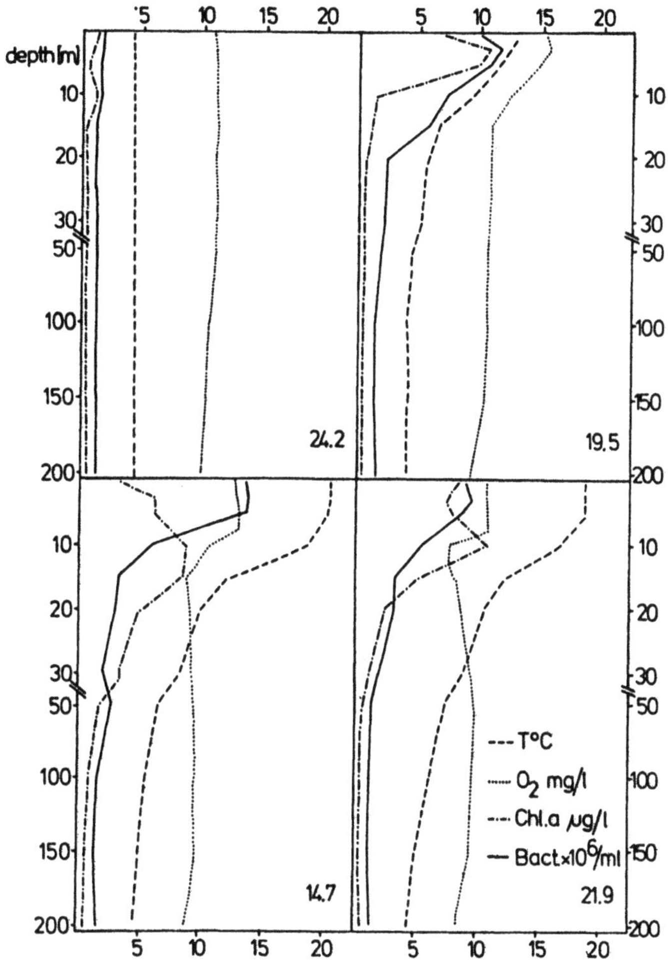


**Figure 25.1** The vertical distributions of temperature, dissolved oxygen, chlorophyll *a*, and bacteria observed in a pelagic water column of Lake Constance during each season during 1982. The observation date is shown in the lower right corner of each plot. (from Güde et al., 1985).

obtained with these approaches differed considerably in some cases, both approaches resulted in generally comparable bacterial production estimates, with values ranging from 0.1 to 4 $\mu$g C $l^{-1}$ $h^{-1}$. The observed differences may be due to fact that different water bodies were investigated and to the variety of different assumptions which must be made for both approaches. However, the general pattern of the seasonal and vertical distribution was very constant, as shown in Figure 25.3, and was comparable to the pattern for bacterial numbers (see Figure 25.2).

When integrating over the water column, a major portion of bacterial

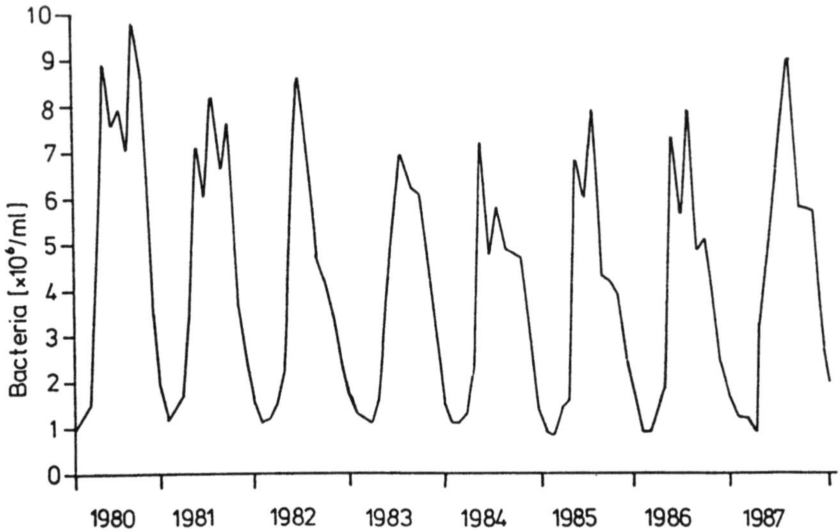


**Figure 25.2** Monthly averages of bacterial numbers observed from 1980 to 1987 in the 0 to 10 m layer at a pelagic station on Lake Constance.

biomass is always in the hypolimnion in spite of the high number of bacteria in the epilimnion, as shown in Figure 25.4. By contrast, the major fraction of bacterial productivity occurs in the epilimnion during stratification. This suggests that epilimnic bacteria, although a minor portion of total biomass, perform a major part of bacterial metabolic activities, whereas the majority of the bacteria, in the hypolimnion appears to exist in a less active state. This view is consistent with the conclusion that the bulk (70–80%) of primary production is recycled in the upper 30 m of the lake (Stabel and Tilzer, 1981).

The depth-time distribution of bacteria is the result of bacterial growth, death, and lateral and vertical advection. Although our knowledge on these processes has increased vastly during the past years, it is still insufficient to identify unequivocally the contribution of each individual process to the dynamics of the bacterial community. The available data suggest that bacterial growth is mainly supplied by organic matter originating from primary production, because high bacterial concentrations and activities are related to algal growth and are restricted to the euphotic zone. However, the link between algal and bacterial growth cannot be described by a simple stoichiometric relationship, because the correlation between algal and bacterial growth parameters usually is only weak (Güde et al., 1985; Simon and Tilzer, 1987). This suggests that the carbon flow from algae to bacteria is also controlled by other processes.

Among the loss processes from the bacterial community, neither "autolysis" nor sedimentation appear to be important (Güde, 1986). On the other hand, grazing by daphnids and hetereotrophic flagellates was considered to

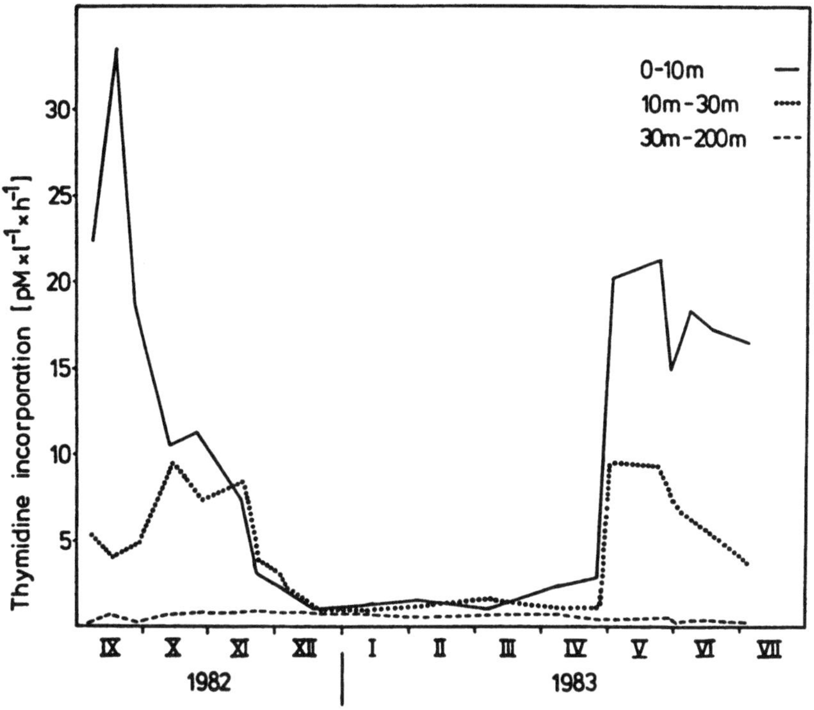


**Figure 25.3** $^3$H-thymidine incorporation rates in three different layers of the water column of a pelagic station on Lake Constance from September 1982 through July 1983 (From Güde et al. 1985).

contribute at least temporarily to significant bacterial losses, because measured or estimated grazing rates were shown to be often comparable in magnitude to estimated growth rates (Güde, 1986, 1988; also see Chapter 29). On the basis of the existing data for Lake Constance, the role of ciliates, rotifers, or phagotrophic algae as bacterial grazers can not be considered to be important. However, because the grazing activities of these organisms were less intensively studied, their contribution to bacterial losses cannot be excluded. In spite of these uncertainties, the available evidence shows that the grazing pressure exerted on the bacterial populations is also subject to strong seasonal fluctuations corresponding to the observed seasonal fluctuation of the grazer populations (Güde, 1988; Jürgens and Güde, 1989).

To what extent the bacterial distribution may be influenced by lateral transport cannot be answered for Lake Constance at present, due to the lack of pertinent studies. However, this factor is probably signnificant because the existence of horizontal inhomogeneities in the pelagic environment is obvious (see Section 25.3).

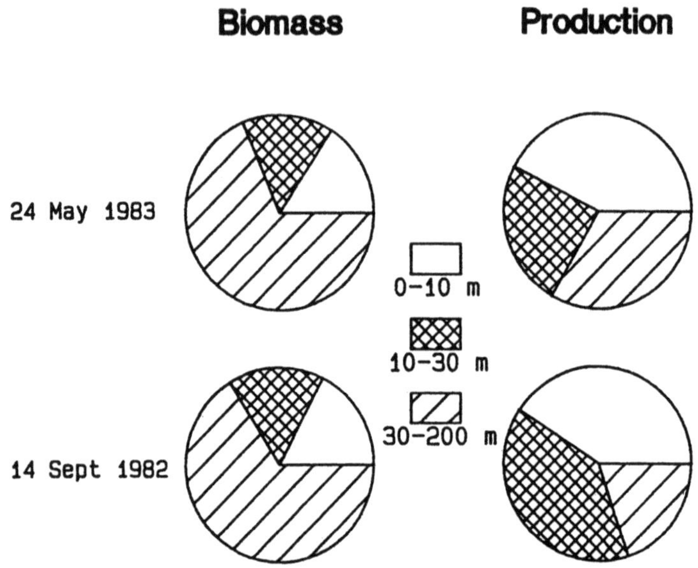


**Figure 25.4** Percentages of biomass and of production in the three different parts of the water column in two different seasons, based on estimates of total biomass and production.

## 25.3 Carbon Flow from Primary to Bacterial Net Production

Numerous attempts have been made to estimate bacterial production rates and to quantify the flow of carbon from phytoplankton to pelagic bacteria (for a review see Cole et al., 1988). Independently of the methods applied (e.g., thymidine incorporation, amino acid incorportion, frequency of dividing cells, or dilution methods), all attempts ultimately yield bacterial growth rates which thereafter can be converted into bacterial net carbon production, if the carbon content of individual bacterial cells is known. In the following, for Lake Constance, the percentage of primary production resulting in bacterial net production as estimated directly from growth rate estimates will be compared with the percentage calculated from two indirect approaches, carbon flow and biomass turnover. The results are shown in Table 25.1.

**Substrate Incorporation Approach** As has been mentioned above, bacterial production in Lake Constance has been estimated by the thymidine incorporation method (Güde et al., 1985) and by the amino acid incorporation method (Simon and Tilzer, 1987). According to these estimates, bacterial net production would correspond to 20–30% of measured primary production. These values agree favorably with the range of values reported from a comparative study, including results from different aquatic habitats (Cole et al.,

**Table 25.1** The percentage of primary production resulting in bacterial net production calculated directly from two substrate incorporation approaches and indirectly from a carbon flow and a biomass turnover approach

| Calculation method | Bacterial net production |
|---|---|
| Substrate incorporation approach | |
| Thymidine incorp. (Güde et al., 1985) | ca. 20% |
| Amino acid incorp. (Simon and Tilzer, 1984) | ca. 30% |
| Carbon flow approach[a] | |
| Assumed growth efficiency of 10% | 3–6% |
| Assumed growth efficiency of 20% | 6–12% |
| Assumed growth efficiency of 50% | 15–30% |
| Biomass turnover approach[b] | |
| Assumed turnover time of 1 d | 250–500% |
| Assumed turnover time of 3 d | 75–150% |
| Assumed turnover time of 10 d | 25–50% |
| Assumed turnover time of 40 d | 7–14% |

[a] Values are based on the carbon flow scheme illustrated in Figure 25.5. Low values refer to gross, high values to net primary production.
[b] Assuming equal bacterial and algal biomass and an algal turnover time of 1 d, and a bacterial growth efficiency of 20%. Low values refer to gross, high values to net primary production.

1988). For a critical evaluation of these results, however, it should be considered that primary production rates usually are estimated directly from the photosynthetic rates, whereas, all methods previously applied for measuring bacterial productin relied on indirect approaches. These depend largely on the reality of the theoretical assumptions involved and the conversion factors derived from them. The resulting high degree of uncertainty for bacterial production estimates is exemplified, in the case of the thymidine incorporation approach, by the fact that the reported factors for the conversion of measured thymidine incorporation to cells produced varies over a range of almost two orders of magnitude (Moriarty, 1987).

Because of this uncertainty, it may be useful to examine whether the obtained estimates of bacterial net production are consistent with energetic considerations resulting from the two indirect approaches. Both these approaches rely on measurements of bacterial growth efficiencies (ratio of increase of biomass to amount of substrate taken up). Usually, a growth efficiency of 50% is assumed, based on results from pure bacterial cultures (Calow, 1987). However, much lower growth efficiencies were obtained when bacteria were grown under conditions resembling the natural situation of pelagic environments. Values of 10–20% were reported when bacteria were grown in filtered lake water (e.g., Bjørnsen, 1986a; Tranvik and Höfle, 1987; Nagata, personal communication). Growth yields of 20–30% were obtained in chemostat studies with glucose under carbon limitation (e.g., Höfle, 1982; Güde, 1984). A similar range was estimated in a field study by Bell and Kuparinen (1984). Although it is unclear wheather these generally low growth efficiencies are mainly due to low energetic values of the natural substrates utilized or to the relative increase of demands for maintenance metabolism, the available evidence suggests that 20% rather than 50% should be regarded as a realistic value for growth efficiency.

**Carbon Flow Approach** Using this approach, one can estimate how much of the measured primary production will ultimately result in bacterial net production, when losses of primary production to compartments others than bacteria are taken into account. For this purpose, it is necessary to determine whether the $^{14}C$-primary production measurements (to which bacterial production is usually related) represent algal gross or net production. According to current views, they are much closer to gross than to net production (Dring and Jewson, 1982). This means that the amount of measured primary production potentially available for other trophic levels is reduced by about one-half due to algal respiration. Of the remaining algal net production, a further part cannot be used by bacteria due to sedimentation and zooplankton grazing. In the example of Figure 25.5, this percentage was assumed to be 20% of measured primary production (i.e., 40% of net production). This assumption is probably weighted in favor of bacteria, because estimates of sedimentation

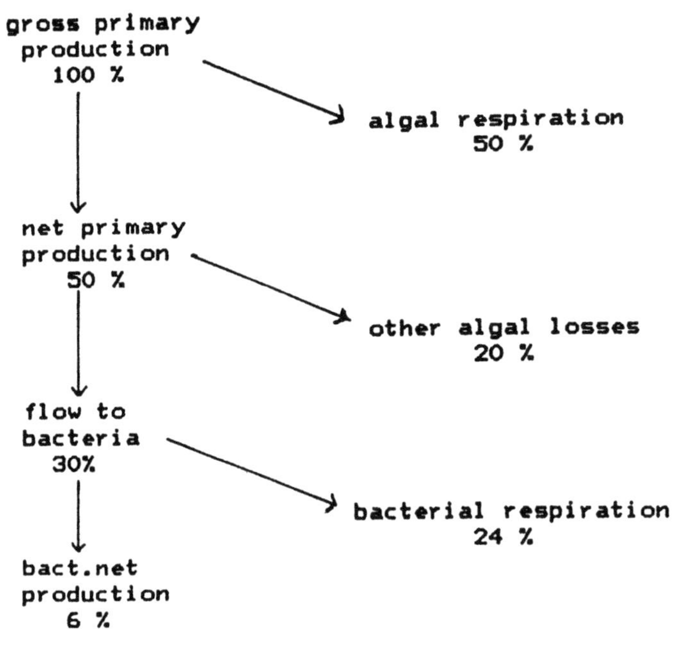


**Figure 25.5** A conceptual scheme for the flow of organic carbon from gross primary production to bacterial net production.

alone accounted for ca. 20% of measured primary production (Stabel and Tilzer, 1981). Thus, at maximum only 30% of the measured primary production (if it is really close to gross production) is available for bacterial utilization. With the assumption of a 20% growth efficiency, bacterial secondary production would ultimately amount to only 6% of measured primary production. However, even if measured primary production were close to net production, this value would, at best, increase to 12%. Because this consideration refers to only one assimilation cycle, the percentage may theoretically increase due to organic recycling during subsequent cycles (Strayer, 1988). However, this increase will be negligible again at the assumed low growth efficiencies. Thus, the high percentages obtained from the growth rate approaches (see Table 25.1) could only be explained by assuming that the measured $^{14}C$-primary productivity corresponds to net production and that bacterial growth efficiencies are high. However, they can hardly be reconciled with the apparently much more realistic assumptions of measured gross production and 20% growth efficiency. In this case, a carbon amount corresponding to 100–150% of algal gross production, or 200–300% of algal net production, would be required to explain the high production estimates resulting from the growth rate approaches.

One defect of the carbon flow approach is that usually only the particulate primary production is measured, whereas labelled primary products in the dissolved fraction are neglected. However, labelled organic substances found in the filtrate usually constitute only <10% of the measured particulate fraction (e.g., Sondergaard and Schierup, 1982). Moreover, it is suggested by experimental evidence that the direct flow from algae to bacteria via algal exudates is small (e.g., Berman and Gerber, 1980; Berman and Kaplan, 1984). In conclusion, the high production estimates obtained from the growth rate approach are not consistent with the low production values predicted by the carbon flow approach.

**Biomass Turnover Approach** This approach is based on the current knowledge of the proportions of algal to bacterial biomass and their respective turnover rates. During the past few years, evidence has been accumulating that bacterial biomass is much higher than had been suggested before, because conversion factors from biovolume to carbon were shown to considerably exceed former estimates (e.g., Bratbak, 1985; Bjørnsen, 1986b; Simon and Azam, 1988). This led to the surprising conclusion that bacterial biomass is comparable in magnitude to algal biomass in terms of carbon (also see Chapter 24). This was also shown for Lake Constance, even though a conservative factor was used for the conversion of bacterial volume to carbon (Simon and Tilzer, 1987). A further confirmation of high bacterial biomass is given by measurements of the bacterial phosphorus content by Jürgens and Güde (1990). It was shown that the bacterial size fraction (< 1 μm) represented more than 50% of the total particulate phosphorus during the entire summer. Assuming, thus, bacterial biomass to be equal to algal biomass and assuming again a growth efficiency of 20%, a carbon amount equivalent to five times the algal biomass would be required for one renewal of bacterial biomass. According to the growth rate estimates, bacterial biomass would be renewed in the range of 1–3 days (Simon, 1988). These renewal times are of the same magnitude as those estimated for algal biomass (Tilzer, 1984). This means that the carbon requirements for bacterial net production would correspond to 500% of algal net production, or 250% of gross production. As is shown in Table 25.1, much higher bacterial turnover times (i.e., lower growth rates) must be assumed in order to obtain bacterial production values which can be reconciled with those to be expected from an energetic point of view. The inevitable conclusion resulting from these considerations is that, from an energetic point of view, the current estimates of bacterial growth rates can not be explained on the basis of current estimates of primary production.

## 25.4 Possible Solutions to the Carbon Balance Problem

**Allochthonous Carbon Sources** The above-described discrepancy could be explained by the existence of alternative carbon sources of allochthonous origin. According to measurements made in the years 1978–1979, the annual load of organic carbon to Lake Constance amounted to 30,000 tons (Internationale Gewässerschutzkommission für den Bodensee 1982). During subsequent years, the load has decreased by about one-third (Wagner, personal communication). Although these allochthonous carbon inputs represent more than a negligible amount, they are considerably lower than current estimates of primary production (Tilzer, 1984; Müller, unpublished data). Moreover, it must be assumed that most of this allochthonous carbon is refractory and thus resembles the large pool of inert dissolved organic carbon in the lake. In conclusion, especially in a large lake, allochthonous carbon inputs cannot be used to explain the possibility of excessively high bacterial production rates.

**Spatial Heterogeneity** As the production estimates are related to the measurements of one central water column, horizontal inhomogeneities of primary production combined with lateral transport processes (advection) could be regarded as a further possibility to explain the discrepancy. Because it is known that primary production in the littoral zone is higher than in the pelagic environment, it is conceivable that the primary producers of the littoral zone are continuously supplying the pelagic zone with organic carbon. As shown in Figure 25.6, the horizontal distribution of primary production was indeed not homogenous, with lowest values always observed in the central station and highest values in the littoral (even though only plankton primary production but not the contribution of macrophytes or phytobenthos was measured). However, not only primary production but also bacterial production is characterized by a strong gradient from the shore to the central part of the lake (see Figure 25.6). Similar observations were made by Scavia and Laird (1987) in Lake Michigan. Thus, it seems that the descrepancy remains the same, although it

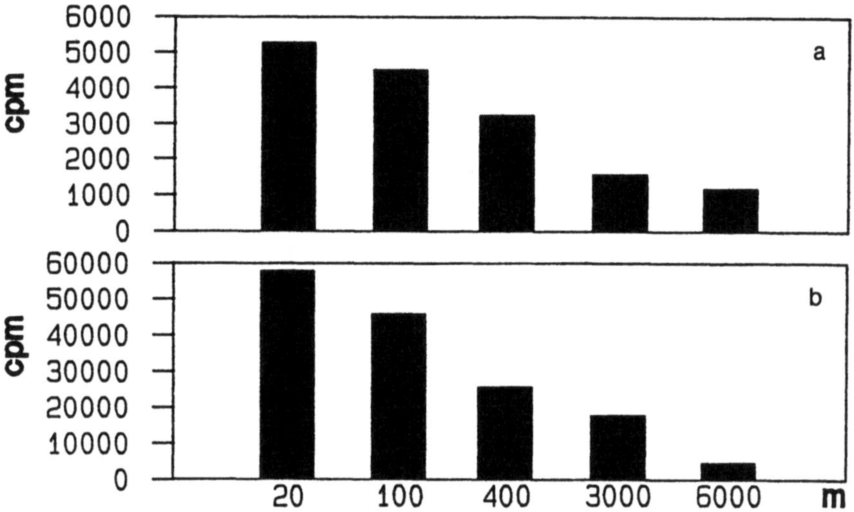


**Figure 25.6** Transsects of (*a*) primary production and of (*b*) thymidine incorporation—from the littoral to the central part of Lake Constance. Samples from the 2 m depth were compared. The horizontal scale indicates the distance from the shore.

must be emphasized that the data basis is too small by far for a complete balance of these effects. In any case, this illustrates that the error range for the carbon balance of a whole lake can be very large if data from only one water column are available.

**Methodological Problems** The remaining possibility for the explanation of the discrepancy is that bacterial production tends to be overestimated by current growth rate estimates. As has been pointed out by Güde (in press), there are indeed several lines of argumentation nourishing this suspicion: (1) As has been emphasized above, the production estimates depend largely on the conversion factors used. However, most of these production estimates rely on the unproven assumption that the conversion factors obtained in the calibration experiments are identical to those for bacteria growing in the natural evironment. Considering the morphological and physiological changes of bacterial populations frequently observed in such experiments (Ferguson et al., 1984; Güde, in press), the reality of this assumption seems questionable. (2) Interestingly, recently published conversion factors (Riemann et al., 1987; Bell, 1988) appear to converge on the theoretically expected conversion factor (Moriarty, 1987), which is at the low end of the values reported. With this factor (instead of a 3–6 fold-higher factor used by Güde et al., 1985), bacterial turnover times would increase during above 10 days most of the growing season, which would be much more consistent with the energetic considerations. (3) In most growth rate approaches, the possible occurence of biosynthesis without any resulting net growth has not been taken into account. This phenomenon, however, has been demonstrated in many studies with nongrowing, starving bacteria (Kjelleberg et al., 1987).

## 25.5 Conclusions

In summary, in Lake Constance, as in other large deep lakes in the temperate climate zone, the depth-time distribution of bacterial densities and activities is characterized by comparatively large seasonal fluctuations and a strong concentration in epilimnic waters. This pattern, together with current estimates of bacterial biomass and production, suggest a strong participation of bacteria in the flow of organic matter in Lake Constance. The bacterial secondary production resulting from this carbon flow is estimated to be no more than 30% (but probably much less) of measured primary production. Considering further energy losses within the microbial food chain, the transfer of energy to higher trophic levels via the microbial loop (see Chapter 24) appears to be of low quantitative importance. Thus, regeneration of nutrients (see Chapter 27) should be regarded as the main function of the microbial loop in Lake Constance. However, as was demonstrated by the obvious inconsistency between primary and bacterial production, further conceptual and methodological efforts are required for a more realistic quantitation of the role of bacteria in the carbon cycle.

---

*Acknowledgements* I thank M. Amann and J. Meichle for the performance of measurements of primary production and of thymidine incorporation, respectively. Further thanks go to K. Jürgens for counts of bacteria in 1987 and for numerous critical discussions. I also owe thanks to C. Serruya, M. Simon, and M. Tilzer for substantial help in the final preparation of the manuscript. This study was supported by the Deutsche Forschungsgemeinschaft (SFB 248).

## References

Bell, R.T. 1988. Importance of measuring the degree of participation of $^3$H-thymidine for calculating bacterioplankton production from rate of thymidine incorporation into DNA. *Archiv für Hydrobiologie, Beiheft Ergebnisse der Limnologie*. In press.

Bell, R.T. and Kuparinen, J. 1984. Assessing phytoplankton and bactrioplankton production during early spring in Lake Erken, Sweden. *Applied and Environmental Microbiology* 48:1221–1230.

Berman, T. and Gerber, C. 1980. Differential filtration studies of carbon flux from living algae to microheterotrophs, microplankton size distribution and respiration in Lake Kinneret. *Microbial Ecology* 6:189–198.

Berman, T. and Kaplan, B. 1984. Diffusion chamber studies of carbon flow from living algae to heterotrophic bacteria. *Hydrobiologia* 108:127–132.

Bjørnsen, P.K. 1986a. Bacterioplankton growth yield in continuous seawater cultures. *Marine Ecology Progress Series* 30:191–196.

Bjørnsen, P.K. 1986b. Automatic determination of bacterioplankton biomass by image analysis. *Applied and Environmental Microbiology* 51:1199–1204.

Bratbak, G. 1985. Bacterial biovolume and biomass estimations. *Applied and Environmental Microbiology* 49:1488–1493.
Calow, P. 1977. Conversion efficiencies in heterotrophic organisms. *Biological Reviews* 52:385–400.
Cole, J.J., Findlay, S., and Pace, M.L. 1988. Bacterial production in fresh and salt waterecosystems: a cross-system overview. *Marine Ecology Progress Series* 43:1–10.
Deufel, J. 1967. Hydrobakteriologgische Untersuchungen im Bodensee. I. Über die Vertikalverteilung der Bakterien im Pelagial. *Internationale Revue der gesamten Hydrobiologie* 52:617–626.
Dring, M.J. and Jewson, D.H. 1982. What does $^{14}C$-uptake really measure. Theoretical approach. *Proceedings of the Royal Society, London* 214:351–368.
Dufour, P., Stroffek, S., and Colon, M. 1988. Le bacterioplankton du Leman—Campagne 1987. *Rapports sur les etudes et recherches entreprises dans le bassin Lemanique, Campagne 1987*: 77–83.
Ferguson, R.L., Buckley, E.N., and Palumbo, A.V. 1984. Response of marine bacterioplankton to differential filtration and confinement. *Applied and Environmental Microbiology* 47:49–55.
Fuhrman, J.A. and Azam, F. 1980. Bacterioplankton secondary production estimates for coastal waters of British Columbia, Antarctica and California. *Applied and Environmental Microbiology* 39:1085–1095.
Güde, H. 1984. Test for validity of different radioisotope activity measurements by microbial pure and mixed cultures. *Archiv für Hydrobiologie, Beihefte Ergebnisse der Limnologie* 19:257–266.
Güde, H. 1986. Loss processes influencing growth of planktonic bacterial populations in Lake Constance. *Journal of Plankton Research* 8:795–810.
Güde, H. 1988. Influence of crustacean zooplankton on bacterial populations in Lake Constance. *Hydrobiologia* 159:63–73.
Güde, H. 1989. Bacterial net production approaching zero—A frequent phenomenon in lakes? *Archiv für Hydrobiologie Beiheft Ergebnisse der Limnologie.* In press.
Güde, H., Haibel, B. and Müller, H., 1985. Development of planktonic bacterial populations in Lake Constance (Bodensee-Obersee). *Archiv für Hydrobiologie* 105:59–77.
Höfle, M.G. 1982. Glucose uptake of *Cytophaga johnsonae* studied in batch and continuous culture. *Archives of Microbiology* 133:289–294.
Internationale Gewässerschutzkommission für den Bodensee. 1982. Die Belastung des Bodensee mit Phosphor-und Stickstoffverbindungen und Organischem Kohlenstoff im Abflussjahr 1978–79. Report no. 28.
Jürgens, K. and Güde, H. 1989. Seasonal changes in the grazing impact of phagotrophic flagellates on bacteria in Lake Constance. *Marine Microbial Food Webs.* In press.
Jürgens, K. and Güde, H. 1990. Binding and regeneration of phosphorus by aquatic bacteria and bacterivorous flagellates. *Marine Ecology Progress Series.* In press.
Kjelleberg, S., Hermansson, M., Marden, P., and Jones, G.W. 1987. The transient phase between growth and non-growth of heterotrophic bacteria with special emphasis on the marine environment. *Annual Review of Microbiology* 41:25–49.
Moriarty, D.J.W. 1987. Measurement of bacterial growth rates in aquatic systems from rates of nucleic acids synthesis. *Advances in Microbial Ecology* 6:171–191.
Nagata, T. 1987. Production rate of planktonic bacteria in the north basin of Lake Biwa. *Applied and Environmental Microbiology* 53:2872–2882.
Riemann, B., Bjørnsen, P.K., Newell, S.Y. and Fallon, R. 1987. Calculation of cell production of coastal marine bacteria based on measured incorporation of $^{3}H$-thymidine. *Limnology and Oceanography* 32:471–474.
Scavia, D. and Laird, A. 1987. Bacterioplankton in Lake Michigan: Dynamics, controls, and significance to carbon flux. *Limnology and Oceanography* 32:1017–1033.
Simon, M. 1984. Bacterial production and its relation to heterotrophic activity parameters

in Lake Constance. *Archiv für Hydrobiologie, Beihefte Ergebnisse der Limnologie* 19:131–139.
Simon, M. 1988. Growth characteristics of small and large free-living and attached bacteria in Lake Constance. *Microbial Ecology* 15:151–163.
Simon, M. and Azam, F. 1988. Protein content and protein synthesis rates of natural assemblages of marine bacteria. *Eos* 68/50:1729.
Simon, M. and Tilzer, M.M. 1982. Bacterial decay of the autumnal phytoplankton in Lake Constance (Bodensee). *Schweizerische Zeitschrift für Hydrologie* 44:192–103.
Simon, M. and Tilzer, M.M. 1987. Bacterial responses to seasonal primary production and phytoplankton biomass in Lake Constance. *Journal of Plankton Research* 9:535–552.
Stabel, H.H. and Tilzer, M.M. 1981. Nährstoffkreisläufe im Bodensee Überlinger See und ihre Beziegungen zu biologischen Umsetzungen. *Verhandlungen der Gesellschaft für Ökologie* 9:23–32.
Strayer, D. 1988. On the limits to secondary production. *Limnology and Oceanography* 33:1217–1220.
Søndergaard, M. and Schierup, H.C.H. 1982. Release of extracellular organic carbon during a diatom bloom in Lake Mosso. *Freshwater Biology* 12:313–320.
Tilzer, M.M. 1984. Estimation of phytoplankton loss rates from daily photosynthetic rates and observed biomass changes in Lake Constance. *Journal of Plankton Research* 6:309–324.
Tranvik, L.J. and Höfle, M.G. 1987. Bacterial growth on dissolved organic carbon from humic and clear waters in mixed cultures. *Applied and Environmental Microbiology* 52:684–692.

# 26

# Seasonal Changes in Methane, Hydrogen, and Carbon Monoxide Concentrations in a Large and a Small Lake

S. Schuler, B. Thebrath, and R. Conrad

ABSTRACT The water columns of Lake Constance and of the Buchensee were found to be supersaturated with respect to atmospheric $CH_4$, $H_2$, and CO. $CH_4$ concentrations were significantly lower in the large Lake Constance than in the small Buchensee. In both lakes, however, the vertical and seasonal distribution patterns indicate production of $CH_4$ during the summer in the oxic epilimnic water layers, as well as in the anoxic sediment. The distribution of CO followed a similar pattern, but $H_2$ was only produced sporadically. The origin and fate of the reduced trace gases are discussed.

## 26.1 Introduction

Methane, hydrogen, and carbon monoxide are reduced trace gases which usually are formed by anaerobic bacteria living in the sediments or in the anoxic hypolimnion of lakes (Rudd and Taylor, 1980; Zeikus, 1983; Conrad, 1988). However, these reduced trace gases also originate from oxic surface waters. It is known that large quantities of $CH_4$, $H_2$, and CO are released from oceans into the atmosphere (Seiler and Schmidt, 1974; Conrad and Seiler, 1986, 1988). Apparently, CO is mainly produced by photooxidation of dissolved organic matter, but the production processes of $CH_4$ and $H_2$ in oceanic surface water are still a matter of speculation (Rudd and Taylor, 1980; Conrad and Seiler, 1986; 1988).

In contrast to oceans, $H_2$ and CO levels in freshwater lakes have received little attention (Conrad et al., 1983a, 1983b; Dahm et al., 1983; Schütz et al., 1988), and studies of $CH_4$ have been restricted to the anoxic hypolimnion and the anoxic/oxic interface of the water column (Rudd and Taylor, 1980). The concentrations of $CH_4$, $H_2$, and CO in the oxic epilimnion of lakes are virtually unknown. The same is true for the origin, fate, and turnover of these trace gases, as well as their possible significance as dissolved energy substrates for pelagic microorganisms.

In the present study, we have assessed the seasonal distribution of dissolved $CH_4$, $H_2$, and CO in the water column of the Überlingersee of Lake Constance (the northeast arm). For comparison, we also measured the gas concentrations in the small lake, the Buchensee (located NW of Lake Constance). Whereas the Buchensee is a moderately eutrophic small lake which develops an anoxic hypolimnion during summer stratification, Lake Constance is a mesotrophic large lake and stays oxic throughout the year. The water samples were taken at different depths of the deepest stations of the lakes and analyzed by using extraction and gas chromatographic techniques with detection limits of 2–5 nl dissolved gas per liter of water (Conrad et al., 1982, 1983a).

## 26.2 Methane Concentration Patterns

The water of Lake Constance was generally supersaturated with respect to atmospheric $CH_4$ (Figure 26.1). In the surface water, $CH_4$ concentrations ranged from 0.2 to 7 $\mu l\ l^{-1}$, which is equivalent to supersaturation factors of about 2–90.

$CH_4$ concentrations were much higher in the more eutrophic Buchensee (Figure 26.2) than in Lake Constance. However, despite the different properties of these lakes, they showed a similar vertical distribution pattern of $CH_4$. In both lakes, two maxima of $CH_4$ concentrations occurred, one immediately above the sediment surface and a second one in the epilimnion. The deep maximum can be easily explained by anaerobic decomposition of organic matter and methanogenesis in the anoxic sediment, diffusion of $CH_4$ into the water column, and oxidation of $CH_4$ by methanotrophic bacteria in the oxic water layers (Rudd and Taylor, 1980; Hanson, 1980). The higher concentrations in the Buchensee as compared to Lake Constance may be due to the more anoxic hypolimnion of the small lake and its higher supply of organic matter, which favor anaerobic processes in deep water and sediment.

The second $CH_4$ maximum in the epilimnion of the lakes, however, cannot be explained so easily. In both lakes, epilimnetic $CH_4$ concentrations increased during the summer and decreased at the end of the season (see Figures 26.1 and 26.2). This indicates that $CH_4$ is produced by biological processes which show similar seasonal patterns. In case of the small Buchensee, the summer maximum may be due to increased methanogenic activity in the littoral sediments and to lateral transport into the middle of the lake. In case of Lake Constance, however, this is a very unlikely explanation, since the part of the lake where the samples were taken is surrounded by steep shores without a major littoral zone. Therefore, we have to assume that the $CH_4$ maximum in the epilimnion is due to in situ production of $CH_4$.

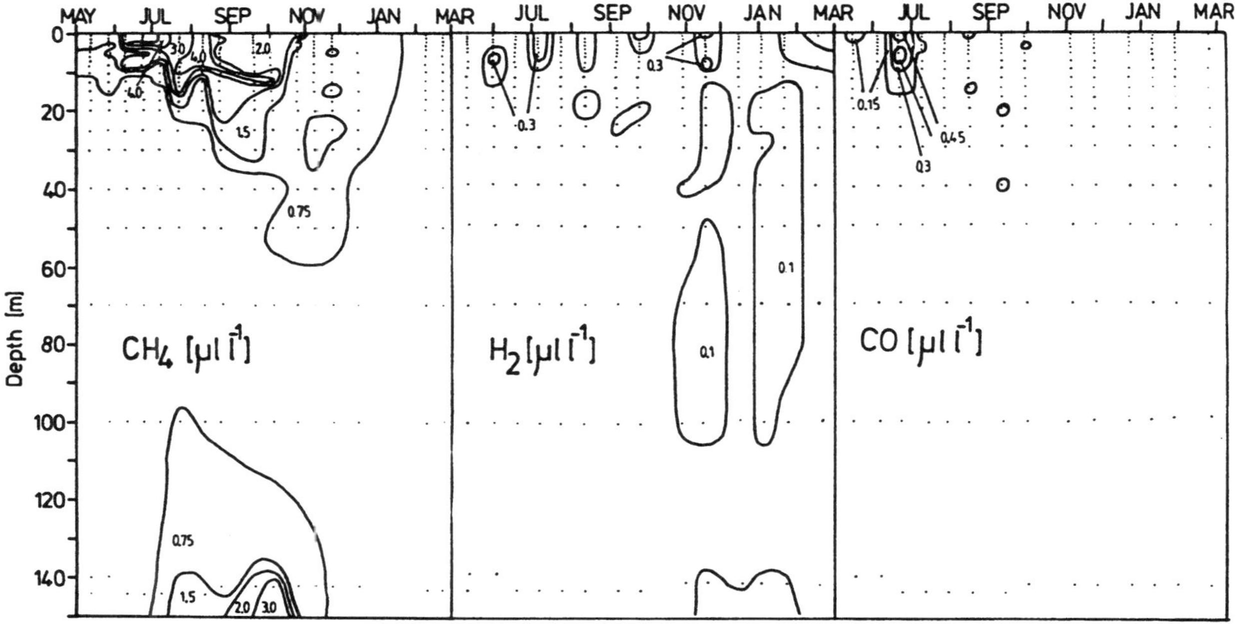


**Figure 26.1** Concentration isopleths of methane, hydrogen, and carbon monoxide in Lake Constance from May 1986 to March 1987. The numbers on the plots indicate the lower limit of concentration between two isopleths. The concentrations are given in units of *volume* of gas dissolved per liter water (μl l$^{-1}$), which are the units obtained by the analysis procedure. Concentrations in units of *mass* per liter water (μM or μg l$^{-1}$) can be calculated by correcting for water temperature.

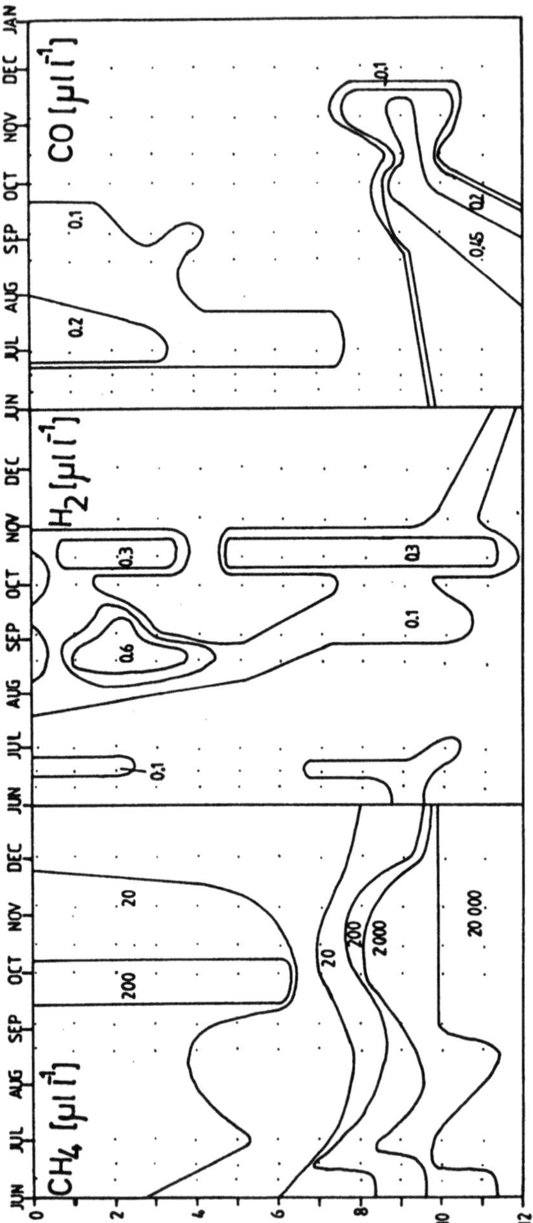


**Figure 26.2** Concentration isopleths of methane, hydrogen, and carbon monoxide in the Buchensee from June through December 1986. The numbers on the plots indicate the lower limit of concentration between two isopleths.

In-situ production of $CH_4$ has also been postulated for oceanic surface waters (Scranton and Brewer, 1977; Burke et al., 1983; Ward et al., 1987; Conrad and Seiler, 1988). In situ production of $CH_4$ in the generally oxic environment of surface waters is difficult to explain. Methanogenic bacteria, which are the only organisms that can produce $CH_4$, are strictly anaerobic. So far, a satisfactory explanation is not available, but it is assumed that the $CH_4$ is produced within anoxic microniches, such as flocs, protozoa, or the intestinal tracts of zooplankton and fish (Oremland, 1979; Traganza et al., 1979; Lilley et al., 1982; Sieburth, 1987).

## 26.3 Hydrogen Concentration Patterns

The water of Lake Constance usually was only slightly supersaturated with respect to atmospheric $H_2$, showing supersaturation factors of $>1-10$. Sporadically, however, some water layers showed supersaturations reaching a factor of 70, especially in the upper 20 m (see Figure 26.2). Highest values were observed at the end of June, but sporadic supersaturation occurred throughout the year. In the Buchensee, $H_2$ concentrations were slightly higher (see Figure 26.2), but also occurred as sporadic events. It is noteworthy that the distribution of $H_2$ followed a completely different pattern than the distribution of $CH_4$ did.

Oceanic surface and near-surface waters usually are also slightly supersaturated with respect to atmospheric $H_2$. The origin and turnover of $H_2$ in oxic water presently is not understood (Conrad and Seiler, 1986, 1988; Conrad, 1988). It has been postulated that $H_2$, as $CH_4$, might be produced within anoxic microniches by anaerobic fermenting bacteria (Lilley et al., 1982). However, the patterns of $H_2$ and $CH_4$ distribution in ocean water are completely different (Conrad and Seiler, 1988) and indicate that different processes control the turnovers of $H_2$ and $CH_4$. There is some evidence that $H_2$ might be produced by $N_2$-fixing cyanobacteria (Scranton, 1983, 1984; Conrad et al., 1983a), but final proof still requires further research.

## 26.4 Carbon Monoxide Concentration Patterns

The water of Lake Constance was generally supersaturated with respect to atmospheric CO. The CO concentrations usually were less than 0.15 $\mu l\ l^{-1}$ (supersaturation factor $< 30$) and showed a high degree of variability within this range. Only between May and August and only in the near-surface water, CO concentrations were sporadically higher reaching values of 0.6 $\mu l\ l^{-1}$ (see Figure 26.1). A similar pattern with maximum concentrations in surface and near-surface during the summer months was observed in the Buchensee (see Figure 26.2). Interestingly, this lake exhibited a second maximum in the anoxic hypolimnion, as did another small eutrophic lake (the Schleinsee) in the vicinity of Lake Constance. This result confirms earlier observations of high CO concentrations in the anoxic hypolimnion of a stratified eutrophic lake (Conrad et al., 1983b). CO production in anoxic environments may be explained by the fact that CO is produced in small quantities by methanogenic, acetogenic, and sulfate-reducing bacteria (Conrad, 1988).

It has previously been shown that photooxidation of dissolved organic matter is an important source for CO in the oxic, euphotic zone of oceans and lakes (Conrad et al., 1982, 1983b). Photooxidative CO production may thus explain the relatively high CO concentrations in the surface water. Quite by contrast, it is rather astonishing that high CO concentrations were only occasionally and not generally observed in the surface water of Lake Constance. Active CO decomposition processes by microorganisms may be the answer (see Conrad, 1988), but more research is necessary to understand the distribution pattern of CO.

## 26.5 Conclusions and Remarks

Our measurements have shown that the surface water of large Lake Constance is supersaturated with the reduced trace gases $CH_4$, $H_2$, and CO in a way similar to that of oceanic surface water (Seiler and Schmidt, 1974). Data from the small Buchensee and from other aquatic ecosystems (Conrad et al., 1983a, 1983b; Dahm et al., 1983; Schütz et al., 1988) indicate that the supersaturation of surface waters with $CH_4$, $H_2$, and CO is a general feature. For the moment, however, the origin of the reduced trace gases in the surface water is still a matter of speculation, and the producing microorganisms, the metabolic processes involved, and their regulation in situ are unknown.

A major difference between the large Lake Constance and the small Buchensee, and perhaps between large and small lakes in general, is the degree of lateral and vertical separation of the potential production zones of reduced gases; i.e., the profundal, pelagial, and littoral. In large lakes, the reduced gases produced in profundal and littoral sediments are much less likely to reach the pelagial than in small lakes, where the shore and bottom lines are much closer to the free-water zone and have a larger relative surface area. For the relatively small and eutrophic Lake Mendota, Fallon et al. (1979) have shown that during stratification almost all of the $CH_4$ released from the sediment was oxidized and assimilated by methanotrophic bacteria at the oxycline. Just 8% of the $CH_4$ entered the epilimnion and was eventually released into the atmosphere. In the Buchensee, the situation seems to be similar. In other lakes that are less eutrophic and where the oxycline is formed only within the top sediment layers, the sedimentary $CH_4$ may be completely recycled within the sediment (Heyer and Babenzien, 1985). A similar situation exists in Lake Constance, and perhaps in other large lakes, where the cycling of sedimentary $CH_4$ is restricted to the sediment and the closest hypolimnetic water. In large lakes, the additional pelagic production of reduced gases (by any process) in the epilimnion may thus play a relatively greater role than in small lakes, and it may be even more important in the ocean.

Reduced gases are energy-rich compounds which are potential substrates for methylotrophic and chemolithotrophic bacteria (Schlegel, 1974), and it has often been shown that $CH_4$, $H_2$, and CO are consumed in lake and ocean water (Rudd and Taylor, 1980; Conrad, 1988; Ward et al., 1987). The concentrations of $CH_4$, $H_2$ and CO in oxic surface water are in a range of $<0.1$ μM. These are low concentrations with respect to the $K_m$ values and the potential growth kinetics of $CH_4$-, $H_2$-, and CO-utilizing bacteria (Conrad, 1984). At present, it is not clear which aquatic microorganisms can actually utilize the trace gases (Conrad, 1988). On the other hand, the concentrations of trace gases are similar to those of individual amino acids or sugars (Vallentyne and Whittaker, 1956; Mopper and Zika, 1987). Simon (1985) determined maximum amino acid uptake rates of 30 ng C $l^{-1}$ $h^{-1}$ and turnover times of $>20$ h for Lake Constance. By comparison with the maximum $CH_4$ concentrations of 4 μg C $l^{-1}$, the microbial uptake of $CH_4$ would be as important as that of total amino acids, if microbial turnover times for $CH_4$ were in a range of $<130$ h. Future research will have to elucidate whether this is the case, and whether trace gas metabolism might be important for the productivity of bacterioplankton in Lake Constance or other large lakes.

*Acknowledgements* This work was financially supported by the Deutsche Forschungsgemeinschaft (SFB 248).

## References

Burke, R.A., Jr., Reid, D.F., Brooks, J.M., and Lavoie, D.M. 1983. Upper water column methane geochemistry in the eastern tropical North Pacific. *Limnology and Oceanography* 28:19–32.

Conrad, R. 1984. Capacity of aerobic microorganisms to utilize and grow on atmospheric trace gases ($H_2$, CO, $CH_4$). In: Klug, M.J. and Reddy, C.A. (editors), *Current Perspectives in Microbial Ecology*. American Society for Microbiology, Washington, D.C., p. 461–467.

Conrad, R. 1988. Biogeochemistry and ecophysiology of atmospheric CO and $H_2$. *Advances in Microbial Ecology* 10:231–283.

Conrad, R., Aragno, M., and Seiler, W. 1983a. Production and consumption of hydrogen in a eutrophic lake. *Applied and Environmental Microbiology* 45:502–510.

Conrad, R., Aragno, M., and Seiler, W. 1983b. Production and consumption of carbon monoxide in a eutrophic lake. *Limnology and Oceanography* 28:42–49.

Conrad, R., Seiler, W., Bunse, G., and Giehl, H. 1982. Carbon monoxide in seawater (Atlantic Ocean). *Journal of Geophysical Research* 87:8839–8852.

Conrad, R. and Seiler, W. 1986. Exchange of CO and $H_2$ between ocean and atmosphere. In: Buat-Menard (editor). *The Role of Air-Sea Exchange in Geochemical Cycling*. Reidel, Dordrecht, p. 269–282.

Conrad, R. and Seiler, W., 1988. Methane and hydrogen in seawater (Atlantic Ocean). *Deep-Sea Research* 35:1903–1917.

Dahm, C.N., Baross, J.A., Ward, A.K., Lelley, M.D., and Sedell, J.R. 1983. Initial effects of the Mount St. Helens eruption on nitrogen cycle and related chemical processes in Ryan Lake. *Applied and Environmental Microbiology* 45:1633–1645.

Fallon, R.D., Harrits, S., Hanosn, R.S., and Brock, T.D. 1979. The role of methane in internal carbon cycling in Lake Mendota during summer stratification. *Limnology and Oceanography* 25:357–360.
Hanson, R.S. 1980. Ecology and diversity of methylotrophic organisms. *Advances in Applied Microbiology* 26:3–39.
Heyer, J. and Babenzien, H.-D. 1985. Untersuchungen des Methankreislaufes in einem oligotrophen See (Stechlinsee). *Limnologica (Berlin)* 16:267–276.
Lilley, M.D., Baross, J.A. and Gordon, L.I. 1982. Dissolved hydrogen and methane in Saanich Inlet, British Columbia. *Deep-Sea Research* 29:1471–1484.
Mopper, K. and Zika, R.G. 1987. Free amino acids in marine rains: evidence for oxidation and potential role in nitrogen cycling. *Nature* 325:1–3.
Oremland, R.S. 1979. Methanogenic activity in plankton samples and fish intestines: a mechanism for in-situ methanogenesis in oceanic surface waters. *Limnology and Oceanography* 24:1136–1141.
Rudd, J.W.M. and Taylor, C.D. 1980. Methane cycling in aquatic environments. *Advances in Aquatic Microbiology*, 2:77–150.
Schlegel, H.G. 1974. Production, modification, and consumption of atmospheric trace gases by microorganisms. *Tellus* 26:11–20.
Schütz, H., Conrad, R., Goodwin, S., and Seiler, W. 1988. Emission of hydrogen from deep and shallow freshwater environments. *Biogeochemistry* 5:295–311.
Scranton, M.J. 1983. The role of the cyanobacterium *Oscillatoria (Trichodesmium) thiebautii* in the marine hydrogen cycle. *Marine Ecology Progress Series* 11:79–87.
Scranton, M.J. 1984. Hydrogen cycling in the waters near Bermuda: the role of the nitrogen fixer, *Oscillatoria thiebautii*. *Deep-Sea Research* 31:133–144.
Scranton, M.J. and Brewer, P.G. 1977. Occurrence of methane in the near-surface waters of the western subtropical North-Atlantic. *Deep-Sea Research* 24:127–138.
Seiler, W. and Schmidt, U. 1974. Dissolved nonconservative gases in seawater. In: Goldberg, E.D. (editor), *The Sea, Vol. 5: Marine Chemistry*. Wiley, New York, p. 219–243.
Sieburth, J.M. 1987. Contrary habitats for redox-specific processes: methanogenesis in oxic waters and oxidation in anoxic waters. In: Sleigh, M.A. (editor), *Microbes in the Sea*. Ellis Horwood, Chichester, p. 11–38.
Simon, M. 1985. Specific uptake rates of amino acids by attached and free-living bacteria in a mesotrophic lake. *Applied and Environmental Microbiology* 49:1254–1259.
Traganza, E.D., Swinnerton, J.W., and Cheek, C.H. 1979. Methane supersaturation and ATP-zooplankton blooms in near-surface waters of the western Mediterranean and subtropical North Atlantic Ocean. *Deep-Sea Research* 26:1237–1245.
Vallentyne, J.R. and Whittaker, J.R. 1956. On the presence of free sugars in filtered lake water. *Science* 124:1026–1027.
Ward, B.B., Kilpatrick, K.A., Novelli, P.C., and Scranton, M.I. 1987. Methane oxidation and methane fluxes in the ocean surface layer and deep anoxic waters. *Nature* 327:226–229.
Zeikus, J.G. 1983. Metabolic communication between biodegradative populations in nature. In: Slater, J.H., Whittenbury, R., and Wimpenny, J.W.T. (editors), *Microbes in Their Natural Environments*. Cambridge University Press, London, p. 423–462.

# 27

# Microbial Food-Webs and Nutrient Cycling in Lakes: Changing Perspectives

T. Berman

ABSTRACT Within the last decade, there has been a profound change in our perception of nutrient flux and trophic cycling in the oceans. Some (but not all) of the considerations leading to our present views are: 1) Frequently, most of the photosynthesis in open waters is due to picoplanktonic algae. There may have been some underestimates of net primary productivities in open oceans. In some environments, carbon fixation by bacterial autotrophs could be significant. 2) Bacteria may be actively growing on photosynthetically derived organic material, but both they and the picoplanktonic algae are probably being effectively grazed mainly by nanoplanktonic protozoans. As a result, bacterial numbers remain relatively constant although considerable material flux may still occur. 3) Heterotrophic, phagocytotic protists are efficient grazers and recyclers of nutrients. These organisms may themselves be an important link in metazoan nutrition. 4) Mixotrophic organisms able to photosynthesize or to grow as heterotrophs are sometimes common constituents of the plankton.

Thus, the simple idea of orderly trophic level transfers of carbon, phosphorus, and nitrogen from primary producers to metazoans and upward has become inadequate. As yet, the extent to which the marine paradigm can be extended to lake ecosystems is unclear, largely because of the dearth of appropriate observation and experimental evidence. This review presents some of the existing information from marine and freshwaters and suggests future areas for investigation.

*There are more things in heaven and earth, Horatio*
*Than are dreamt of in your philosophy.*

—W. Shakespeare
Elizabethan dramatist and erstwhile limnologist

## 27.1 Introduction

It is my contention that limnologists are frequently insufficiently aware of the observations made by their oceanographic colleagues. In this review, I shall note what appear to me to be some of the important recent revelations and hypotheses concerning material cycling which have been made, mainly, albeit not exclusively, in the marine environment. My thesis is that similar paradigms should be relevant in helping us to understand food-webs in large, medium-sized, or even small lakes.

A seminal observation was that "small, green, organisms" or, more often, "blue-green organisms in the pico-size range (i.e., $<2$ $\mu$m)" may be major components of the population of marine primary producers. Picophytoplankton have now been reported from almost all the oceans and seas, and from coastal waters, gulfs, and estuaries (for reviews, see Stockner and Antia, 1986; Joint, 1986). Often, a majority of the total chlorophyll and photosynthetic activity has been associated with this size fraction of the phytoplankton (e.g., Platt et al., 1983; Li et al., 1983; Takahashi and Bienfang, 1983; Berman et al., 1984). In freshwaters, there were also early reports of pico-sized ($<2$ $\mu$m-sized) organisms, for example, the "$\mu$-algae" reported in subarctic Swedish lakes by Rodhe (1955) and the *Synechococcus*-like cyanobacteria observed in Loch Leven by Bailey-Watts et al. (1968) but only recently have there been more detailed descriptions of picophytoplankton in temperate European and North American lakes. There are now well-documented cases of picophytoplankton occurring in lakes of varying size and trophic states throughout North America and northern Europe (see Stockner and Antia, 1986). As yet, little information is available for lakes in Africa, Asia, South America, and Australasia, but it would seem unlikely that these are devoid of picophytoplankton. Recently, we were somewhat surprised to discover pico-sized algal populations in Lake Kinneret, a eutrophic warm lake, despite previous indications of their absence (Berman, unpublished observation).

In general, picophytoplankton are probably more important as primary producers in marine than in freshwaters, and they tend to be relatively more dominant in oligotrophic environments. These organisms can at times be responsible for as much as 60–70% of total primary production even though their percentage contribution to total algal biomass is much lower. One suspects that the significance of picophytoplankton in freshwaters may increase in direct proportion to the numbers of limnologists who will seek them!

If picophytoplankton are responsible for a high proportion of primary production in some lakes, does this force us to reconsider the classical patterns of matter transfer in freshwater systems? Certainly, our perceptions of marine ecosystems have undergone considerable change over the last decade or so. Can pico-sized phytoplankton be effectively grazed by microzooplankton or only by nanoprotozoans? In the latter case, which appears to hold for some marine situations, bacteria and protozoans would become more important in the processing of primary produced organic compounds than the "classical" metazoan zooplankton.

## 27.2 Bacterial Production

It is now widely accepted that, in most marine systems, considerable amounts of photosynthetically fixed carbon passes into and through the bacterial biomass. Where does this carbon actually come from? Much research effort has been invested in quantifying extracellular release of organic matter by photosynthesizing algae. A recent review of this topic is given by Jones and Canon (1986). There seems to be no doubt that algae, when physiologically stressed, for example, by high light or by low nutrient levels, may release dissolved organic carbon (DOC) in amounts which can reach relatively high proportions (40–50% or more) of photosynthetically fixed carbon (Smith, 1982; Sondergaard et al., 1985). This DOC should then be readily exploited by heterotrophic bacteria (Larsson and Hagstrom, 1982). A *priori*, it would seem paradoxical that phytoplankton, especially in oligotrophic waters, should release energy-rich molecules that can stimulate the growth of bacterial competitors for a limited pool of nutrients (Bratbak and Thingstad, 1985). One possibility is that extracellular release may serve to protect the photosynthetic apparatus in the light when growth is limited by nutrients other than carbon. Thus, extracellular release, by permitting the photosynthetic mechanism to continue "ticking over," may minimize the lag time required for resumed growth when nutrients do become available (Omerod, 1983). In addition, it has been suggested that the release of such compounds can be adaptive for oceanic phytoplankton which experience temporal uncoupling of nutrient and light availability, and can serve to initiate the coevolution of complex microbial communities by patch selection (Wood and Van Valen, 1989).

Although the extracellular release is undoubtedly a factor, the relative amount of carbon which passes into the "microbial loop" from this source probably differs very considerably for different environments. It is unclear to what extent phytoplankton are physiologically stressed in nature, even in oligotrophic waters. Most studies show that actively growing algae do *not* release very large amounts (less than 10%) of recently fixed carbon (Berman and Holm-Hansen, 1974; Sharp, 1977; Falkowski et al., 1985). Obviously, one cannot generalize to all aquatic environments, but one suspects that algal biomass may be often regulated by grazing (now termed "top-down" control), rather than by nutrient ("bottom-up") limitation. If so, sloppy grazing and excretion by zooplankton grazers should be major pathways by which primarily produced, organic carbon reaches the bacteria. Where picophytoplankton dominate, the major grazers may be nano- and micro-flagellates and ciliates. In such cases, there is still little known about the fate of algal biomass or the extent of picoplankton extracellular release.

The idea of a key role for bacteria in the food-web was initially proposed mainly for marine environments (Pomeroy, 1974). This concept was placed on a firmer basis (Azam et al., 1983) when more realistic estimates of bacterial cell numbers and growth rates in aquatic systems became possible with the introduction of techniques for staining and direct counting with epifluorescence (Hobbie et al., 1977; Watson et al., 1977; Porter and Feig, 1980).

Methods based on the incorporation of radiolabelled precursors into DNA ($^{3}$H-thymidine—Fuhrman and Azam, 1982; $^{3}$H-adenine;—Karl and Winn, 1986) and protein ($^{14}$-amino acids;—Simon, 1984; Kirchman et al., 1985), as well as other approaches (e.g., frequency of dividing cells, Hagstrom et al., 1979; frequency of dividing-divided cells, Davis and Sieburth, 1984), have also given estimates of microbial productivity, even though these techniques are far from trouble-free and may be criticized. An interesting generalization is that bacterial cell numbers do not vary by much more than ~$10^5$ to ~$10^6$ cells $ml^{-1}$ in the open ocean and, perhaps, an order of magnitude more for eutrophic estuaries, coastal waters, and lakes. An indication of how important bacteria may be in freshwater systems is provided by recent studies on Lake Michigan (Scavia et al., 1986) and on Lake Constance (Simon and Tilzer, 1987). The latter investigators estimated that bacterial biomass was 11% of total particulate organic matter and, on average, equalled the phytoplankton biomass in Lake Constance. During the growing season, bacterial production was about 20% of phytoplankton production. Using autoradiography, Simon (1987) found that in Lake Constance 10%–88% of the visible bacterial cells were active, with the highest percentages occurring during the spring phytoplankton bloom. These data are compatible with the observations of Cole et al. (1988) who showed that the values determined for bacterial production correlate well with planktonic primary production and with chlorophyll *a* levels for a large number of observations from both marine and freshwater environments. Bacterial productivity averaged about 20% or 30% of primary production on a volumetric or areal basis, respectively. If this relationship is valid, it would strengthen the premise that salt and freshwater photic zone ecosystems may have identical or very similar food-webs and patterns of nutrient flux.

Sources for fuelling bacterial productivity are diverse. Most aquatic bacteria are assumed to be heterotrophic (see above), and would therefore derive both energy and carbon from reduced organic substrates. Nevertheless, a little-explored trophic mode, which may have considerable environmental importance in the ocean and possibly in lakes, is that of bacterial chemoautotrophy. (Bacterial autotrophs are, of course, at the base of a fascinating food-web in the deep oceanic hydrothermal vent ecosystems). For example, Sieburth et al. (1987) have proposed that C-1 bacterial metabolism may be significant, even in upper, well-aerated, near-surface oceanic waters. As yet, the quantitative importance of such autotrophs in euphotic zones is unknown. Presumably, chemoautotrophy could take place within naturally occurring aggregates of organic particles which could provide suitable sites for the development of anaerobic micro-environments. Are we significantly underestimating total carbon fixation in some lakes by discounting the possibility of chemoautotrophic processes?

In the open ocean and in many lakes, most bacterial metabolic activity appears to be associated with small, unattached, heterotrophic cells (Derenbach and Williams, 1974; Berman, 1975; Berman and Stiller, 1977; Azam and Hodson, 1977). Many of these bacteria seem to be growing at reasonably rapid rates (1–2 doublings $d^{-1}$, Williams, 1984). An important point to note is that these observations are based on measurements of the size partitioning of radiotracer substrate uptake and, therefore, do not necessarily reflect the size distribution of bacterial cell numbers or biomass.

In many respects, however, the bacteria associated with particles may be more versatile than unbound bacteria in the range of their metabolic activities. It has also been observed that many protozoans will flocculate bacteria in cultures (Sieburth, 1984), suggesting the possibility that bacterivorous protists may actively enhance particle aggregation in nature. Curiously though, Albright et al. (1987) found that many ciliates preferentially grazed free, rather than particle-attached, bacteria. Several studies have shown that marine "snow" particles have sites of intensive microbial activity (Silver et al., 1978; Trent et al., 1978; Shanks and Trent, 1979; Caron et al., 1982; Knauer et al., 1982). These aggregates may provide suitable microenvironments for specialized metabolic activities such as nitrate reduction or methanogenesis and may be important in the cycling of nutrients (Alldredge and Cohen, 1987). Probably, aggregates are relatively more significant in eutrophic and coastal waters than in the open ocean but further insight into their functions in both marine and freshwaters would be desirable.

I submit that we need to develop techniques for studying aggregates and particles in freshwater. Goldman (1984) has described such particles as "life-support systems" for a host of microorganisms and has pointed out that firm attachment to such floating islands may not be necessary or even desirable for motile microorganisms. Indeed, evolution may have led to the proliferation of small, fast-swimming organisms such as nano- and microflagellates and ciliates which can move rapidly from "oasis" to "oasis" in an "aquatic desert." (Alternatively, these organisms might "hop" onto particles which are sinking out of the water column.) Chemotaxis, which seems to be a common attribute of many marine microorganisms, may be a key factor in permitting these organisms to sense the location of nutrients (Azam and Ammerman, 1984).

Consideration of the micro-niches which may be provided by aquatic particles leads us to the question of patchiness in the environment. Goldman (1984) succinctly reviews some of the fluid-mechanical considerations pertaining to the retention of high levels of nutrients in the vicinity of micro-patches, and the relation between the swimming capabilities of aquatic microorganisms and their ability to exploit a diffusing patch of nutrients. Goldman (1984) concludes that nutrient uptake by algae from nutrient micropatches in oligotrophic aquatic environments can be diffusion limited. Nevertheless, there is some experimental evidence that phytoplankton cells can and do exploit transient point sources of nutrients such as might be generated by zooplankton excretion. Lehman and Scavia (1982) found that algae incubated in the presence of $^{32}P$-labelled zooplankton became labelled either in a Poisson or non-Poisson manner depending on whether the experimental flasks were stirred (homogeneous environment) or not (possibly micropatchy environment). Recently, Lehman (1987) has observed that algal cells taken from a lake environment also showed a non-Poisson distribution of intracellular P-content, indicating that, in nature, some cells encountered and exploited micro-regions of higher

P concentrations than others. There are also suggestions that bacteria tend to cluster close to point sources of nutrients such as phytoplankton cells or detrital particles (Bell and Mitchell, 1972; Azam and Ammerman, 1984) but this phenomenon may be restricted to certain more stable zones of reduced turbulence such as exist at thermoclines (Mitchell et al., 1985).

In addition to the microscale inhomogeneities caused by the presence of micro-aggregates or planktonic organisms, it appears that small-scale (meters) patchiness of plankton is a regular feature of many water bodies (Steele, 1978). For example, in a recent study by Atkinson et al. (1987) the presence of oxygen patches (approximately 100–200 m long by 1–2 m deep) was demonstrated in a stratified river system using current drogues and a rapid-response $O_2$ electrode mounted on a CTD probe. These authors suggested that the patches were caused by discrete photosynthetic and respiratory activities which generated different ambient concentrations of oxygen. The existence of meso-scale patchiness in stratified water bodies will require more careful consideration of problems of scale, not only in practical terms of sampling strategies and instrumentation, but also for theoretical considerations of ecosystem structure and function.

There is still considerable confusion concerning the fate in aquatic systems of the carbon, nitrogen, and phosphorus which are synthesized into bacterial cells. Despite earlier assertions (Stevenson, 1978; Hobbie, 1979), the consensus is that most bacteria are metabolically active and that protozoan grazing is mainly responsible for determining cell numbers. Does the microbial loop act as a carbon sink by continuously releasing organic carbon as respired $CO_2$? If there are more than one or two transfer steps through the microbial loop before carbon reaches the level of metazoan zooplankton, then this should indeed be the case. Ducklow et al. (1986) concluded from an experimental study of an enclosed water column that the microbial loop was acting as a carbon sink in Loch Ewe, but Sherr et al. (1987) have disputed their conclusions. In contrast, Sherr et al. (1986) and Turner et al. (1988) have suggested that protists (both ciliates and flagellates) grazing on picophytoplankton and bacteria do appear to transfer carbon to metazoan zooplankton. It seems reasonable to assume that there is no one generalization that will hold for all aquatic environments. However, for marine systems, a relatively greater flow of carbon probably occurs through the microbial loop in oligotrophic than in productive waters. Often, but not always, oligotrophic systems also tend to have a higher percentage of picoplankton. Do lakes also fall into this pattern?

## 27.3 Nanoprotist Grazing

In addition to their role as processors of bacterial picophytoplankton and detrital carbon, the nanoprotists may be central to the recycling of nutrients, such as nitrogen and phosphorus. By virtue of their grazing activities, these nanoprotists can release N (mainly as $NH_4$) and P nutrients, and probably organic compounds also (Sherr et al., 1983; Andersson et al., 1985; Andersen et al., 1986; Berman et al., 1987). The significance of nanoprotists as remineralizers depends upon the degree to which N and P are retained or released by the grazers. An elegant series of studies using both batch and continuous cultures of bacteria, cyanobacteria, and a nanoprotist (Goldman and Caron, 1985; Goldman et al., 1985, 1987; Andersen et al., 1986) has shown that the extent of recycling of $NH_4$ and orthophosphate (i.e., the efficiency of mineralization) by the nanoflagellates was greatly dependent on the C:N:P ratios of the bacterial source. Release of reasonably high amounts of $NH_4$ and orthophosphate during exponential growth of the nanoprotists was observed only when bacteria or cyanobacteria with relatively low C:N or C:P ratios were provided as food for the nanoprotists. Güde (1985) showed that phagotrophic nanozooflagellates were necessary for high mineralization rates of N and P in a freshwater ecosystem. Moreover, even if the recycling efficiency of nanoprotists is low, the fact that N and P are nicely packaged in nanoprotist cells, which can subsequently be grazed by metazoan zooplankters (Turner et al., 1988), may be an effective way of passing these nutrients to the larger members of the food-web. At least for oligotrophic systems, Berman et al. (1987) have demonstrated that nanoprotists can enhance nutrient flux to microalgae by grazing on intact bacterial cells.

Sherr et al. (1988) have proposed that nanoprotozoans may have a major cybernetic function in material flux in aquatic environments because of their rapid growth and grazing rates, and their high release rates of ammonia and phosphorus. Larger prozozoans, such as dinoflagellates (Kimor, 1981), tintinnids (Lessard and Swift, 1985; Heinbokel and Beers, 1979), and metazoan zooplankters (Gophen et al., 1974; Peterson et al., 1978), also graze bacteria, but presumably do not have as significant an impact as nanoprotists. The cropping of bacterial populations by heterotrophic protists may act to keep these populations more active physiologically, and there is also some evidence that specific kinds of microorganisms may be selected for as a result of nanoprotist grazing. For example, Sherr et al. (1982) showed that bacterial decomposition of dead *Peridinium* cells was accelerated in the presence of nanoflagellates. In addition, when nanoflagellates were present, there was a specific enhancement of the bacterial breakdown of dinoflagellate polysaccharide theca but no increase in the rate of degradation of cell protoplasm. The inference was that the nanoprotist grazers were selectively stimulating a population of bacteria which was capable of hydrolising material with a high carbohydrate content.

Although most research attention has been directed recently at the colorless heterotrophic nanoprotists, there is now renewed interest in the possible ecological roles of the pigmented phagotrophs (Porter et al., 1985). In addition to being protozoans, many of these belong to the common *algal* phylum, Chrysophyceae. The almost-ubiquitous presence of chlorophyll-bearing protozoa in water samples has long been observed. However, only recently have investigators begun to consider the impact of planktonic algae as grazers of bacterial (Bird and Kalff, 1986, 1987) and, presumably, detrital particles in lakes.

Where does the mixotrophic way of life fit into the overall ecological pattern? Estep et al. (1986) found that more than 50% of nanoflagellate chrysophytes in samples from an Atlantic transect contained chloroplasts, and that such organisms could be maintained for extended periods in rice-grain cultures. Obviously, it would be highly advantageous for pelagic organisms subjected to varying light climates to be able to survive by consuming bacteria or other organic particles when photosynthesis is impaired. Furthermore, some of these creatures (e.g., *Ochromonas*—Aaronson and Baker, 1959) are also able to utilize dissolved organic carbon substrates for energy and growth. Apparently, such organisms are not "aware" of the neat ordering of trophic levels made by classical ecologists.

Somewhat surprisingly, there seem to be few data on diurnal fluctuations of bacterial and protozoan activity and cell numbers, or of ambient concentrations of inorganic and organic nutrients in lakes. Marine studies suggest that nutrient flux patterns may vary considerably over a 24-hour period (Burney et al., 1982). Not only photosynthesis but also bacterial growth (Sieracki and Sieburth, 1986) may be affected by sunlight. In a poetic but somewhat speculative passage, Sieburth (1984) conjectures that many bacterial metabolic processes may be inhibited by daylight.

## 27.4 Nutrient Cycles

How should we now view matter flux and nutrient cycling in lakes? Consider, for example, the status of our knowledge about phosphorus flux in lake ecosystems (Tarapchack and Nalewajko, 1986). It is generally accepted that orthophosphate-P ($P_i$) is the major form of this nutrient utilized by primary producers and other microplankton. However, it has also been known for a long time that algae and bacteria can also obtain phosphorus from organic P sources (Chu, 1946; Harvey, 1953; Provasoli, 1958), and that dissolved organic P (DOP) often constitutes a significant proportion of the total dissolved phosphorus. Bacteria (and presumably picoplankton) appear to have higher affinities and more rapid specific uptake rates for $P_i$ than larger-celled phytoplankton under most, but not all, circumstances (Rhee, 1973; Berman, 1985; Currie et al., 1986). Based on $^{32}P_i$ uptake experiments in the laboratory and in natural lake waters, Currie and Kalff (1984a, 1984b) have suggested that bacteria in oligotrophic situations outcompete algae for $P_i$ and that phytoplankton may utilize P from DOP released from the bacteria. In most cases, bacteria appear to be net sinks for P, although it has been shown that P is both rapidly taken up and released by algae and bacteria (Berman and Skyring, 1979). Also, in respect to the utilization of DOP, there is presently no direct evidence that phytoplankton have higher affinity for organic phosphorus sources. On the contrary, some recent experiments in Lake Kinneret showed that the addition of organic phosphorus sources tended to spare the $P_i$ requirement of microorganisms smaller than 3 $\mu$m (mainly bacteria) and, conversely, to stimulate $P_i$ uptake into $>3$ $\mu$m organisms (mainly algae). This implied that bacteria (and perhaps picophytoplankton) were usually capable of exploiting the added organo-phosphorus substrates more effectively than the larger organisms, and the smaller organisms then had an increased opportunity to take up $P_i$ from the environment (Berman, 1985).

After the incorporation of P into bacterial cells, the recycling of this nutrient is probably mainly dependent on bacterivorous protist grazing and subsequent mineralization. In oligotrophic waters, bacteria may act as scavengers of low concentrations of P (and N) which are first taken up by the protist grazers, and then by the grazers of the protists. Major unanswered questions concern the fate of the protozoa and the efficiency whereby the nutrients originating in bacterial cells are either transferred or released again into the environment. Also unknown are the sources of DOP in most aquatic environments, especially the labile fractions thereof. In brief, P cycling between the biotic components in lake trophogenic layers has become more, rather than less, confusing with the recognition of an active and multi-faceted protist involvement in aquatic food-webs.

This, of course, applies to aquatic nutrient cycling in general. Clearly, Nature is more complex than our simple paradigms. One model of the current view is shown in Figure 27.1 (from Sherr et al., 1987). The dotted lines in this figure indicate some problematical areas. Is there a significant microbial loop (or perhaps, more appropriately, microbial maze) in lakes, whereby bacteria growing rapidly on the products of primary production are grazed mainly by protozoans, some of which also photosynthesize and/or use dissolved organic material heterotrophically? What is the fate of the carbon, nitrogen, and phosphorus in protozoan cells? Do metazoan zooplankton and larval fish grow well on nanoprotists? The recycling of nitrogen and phosphorus incorporated into bacterial cell material certainly seems to depend upon the grazing activities of the nanoprotozoans. If metazoan zooplankton feed on the protists of the microbial loop to any significant amount, then perhaps even fish yields and population composition may be strongly affected by the efficiencies with which carbon, nitrogen, and phosphorus cycle through this pathway.

There are, of course, obvious differences between the patterns of nutrient flux in oceans and those in lakes. "New production" in the sea is generally dependent on upwelling events which entrain $NO_3$ into the euphotic zone. The algae responsible for "new production" are usually larger than the phytoplankters growing on nutrients regenerated within the photic layer. In lakes, analogous inputs of nutrients leading to "new production" may come not only from upwelling but also from allochotonous sources. Nevertheless, there is no *a priori* reason to believe that material flux in the photic zones of large- or medium-sized lakes should be much different from that in the near-surface waters in the ocean. In many ways, lakes present easier and perhaps more attractive objects for research on basic questions relating to food web dynamics

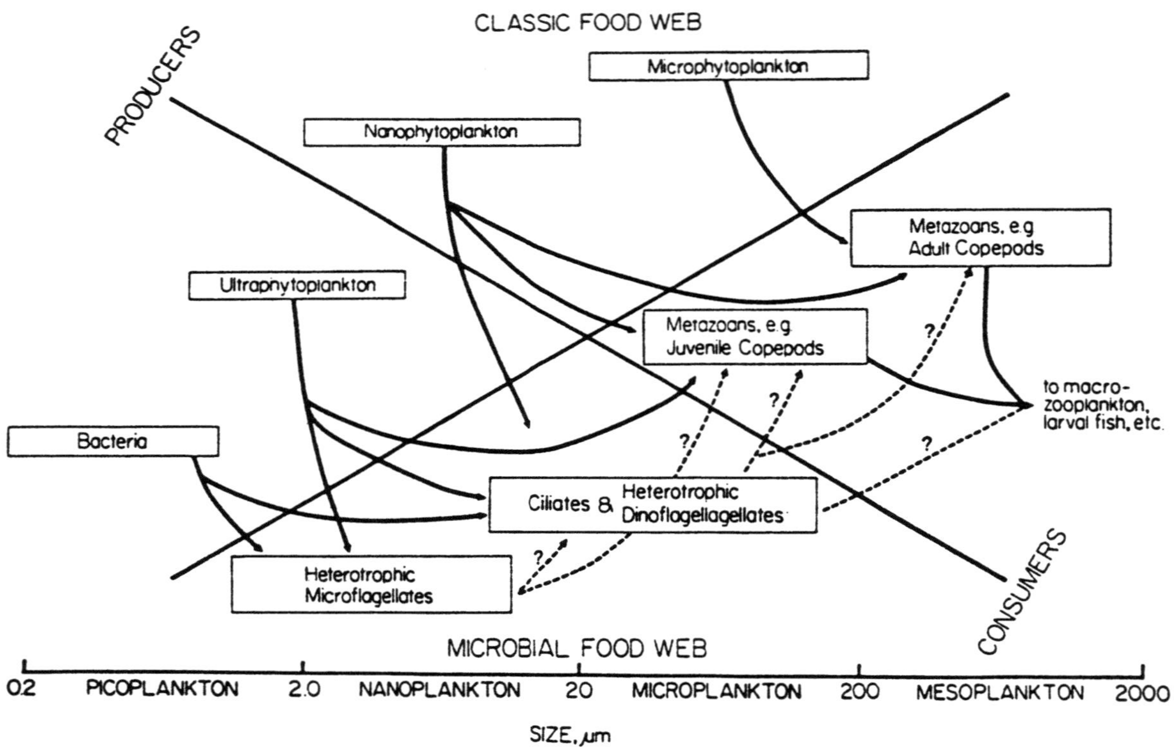


**Figure 27.1** The relationship between the classic food-web and the microbial food-web: diagram of the major groups of autotrophic and heterotrophic organisms at the base of marine pelagic food-webs. Classes of organisms are grouped in two different ways: 1) Producers of biomass from dissolved organic carbon are above the diagonal line sloping up to the right; consumers of this biomass are below the line. 2) Organisms considered to be part of the "microbial loop" are below the diagonal line sloping down to the right, and organisms considered to be in the "classic" phytoplankton-copepod-fish food-web are above this line. The relative-size ranges of each group of organisms, from 0.2 to 2,000 $\mu m$, are shown in relation to the logarithmic scale at the bottom. Trophic links for which there is fairly good evidence are shown as solid arrows: Postulated trophic pathways are shown as dotted arrows. (By permission of Sherr, E.B. et al., 1987).

in aquatic environments. Clarification of some of the points briefly noted in this paper should keep us busy for some time to come.

---

*Acknowledgments* This research was supported in part by Grant no. 87-00006 of the United States–Israel Binational Foundation (BSF), Jerusalem, Israel; and by the Office of the Israel Water Commission. A contribution of Israel Oceanographic and Limnological Research.

## References

Aaronson, S. and H. Baker. 1959. A comparative biochemical study of two species of *Ochromonas*. *J. Protozool.* 6:282–284.
Albright, L.J., E.B. Sherr, B.F. Sherr, and R.D. Fallon. 1987. Grazing of ciliated protozoa on free and particle-attached bacteria. *Mar. Ecol. Progr. Ser.* 38:125–129.
Alldredge, A.L. and Y. Cohen. 1987. Can microscale chemical patches persist in the sea? Microelectrode study of marine snow, fecal pellets. *Science* 235:689–691.
Andersen, O.K., J.C. Goldman, D.A. Caron, and M.R. Dennett. 1986. Nutrient cycling in a microflagellate food chain: III. Phosphorus dynamics. *Mar. Ecol. Progr. Ser.* 31:47–55.
Andersson, A., C. Lee, F. Azam, and A. Hagstrom. 1985. Release of amino acids and inorganic nutrients by heterotrophic marine microflagellates. *Mar. Ecol. Prog. Ser.* 23:99—106.
Atkinson, M.J., T. Berman, B.R. Allanson, and J. Imberger. 1987. Fine-scale oxygen variability in a stratified estuary: patchiness in aquatic environments. *Mar. Ecol. Progr. Ser.* 36:1–10.
Azam, F. and J.W. Ammerman. 1984. Cycling of organic matter by bacterioplankton in pelagic marine ecosystems: microenvironmental considerations. In: M.J. Fasham (ed.), *Flows of Energy and Materials in Marine Ecosystems: Theory and Practice.* Plenum, London, p. 345–360.
Azam, F. and R.E. Hodson. 1977. Size distribution and activity of marine microheterotrophs. *Limnol. Oceanogr.* 22:492–501.
Azam, F., T. Fenchel, J.G. Field, J.S. Gray, L.A. Meyer-Reil, and F. Thingstad. 1983. The ecological role of water-column microbes in the sea. *Mar. Ecol. Prog. Ser.* 10:257–263.
Bailey-Watts, A.E., M.E. Bindloss, and J.H. Belcher. 1968. Freshwater primary production by a blue-green alga of bacterial size. *Nature* 220:1344–1345.
Bell, W.H. and R. Mitchell. 1972. Chemotactic and growth responses of marine bacteria to algal extracellular products. *Biol. Bull.* 143:265–277.
Berman, T. 1975. Size fractionation of natural aquatic populations associated with autotrophic and heterotrophic carbon uptake. *Mar. Biol.* 33:215–220.
Berman, T. 1985. Uptake of $^{32}P$ orthophosphate by algae and bacteria in Lake Kinneret. *J. Plank. Res.* 7:71–84.
Berman, T. and O. Holm-Hansen. 1974. Release of photoassimilated carbon as dissolved organic matter by marine phytoplankton. *Mar. Biol.* 28:305–310.
Berman, T. and G.W. Skyring. 1979. Phosphorus cycling in aquatic microorganisms studied by phased uptake of $^{33}P$ and $^{32}P$. *Current Microbiol.* 2:47–49.
Berman, T. and M. Stiller. 1977. Simultaneous measurement of phosphorus and carbon uptake in Lake Kinneret. *Microbial. Ecol.* 3:279–299.
Berman, T., D.W. Townsend, S.Z. Sayed, C.C. Trees, and Y. Azov. 1984. Optical transparency, chlorophyll and primary productivity in the Eastern Mediteranean near the Israeli Coast. *Oceanologia Acta.* 7:367–372.

Berman, T., M. Nawrocki, G.T. Taylor, and D.M. Karl. 1987. Nutrient flux between bacteria, bacterivorous nanoplanktonic protists and algae. *Mar. Microbial Food Webs* 2:69–82.
Bird, D.F. and J. Kalff. 1986. Bacterial grazing by planktonic lake algae. *Science* 231:493–495.
Bird, D.F. and J. Kalff. 1987. Algal phagotrophy: Regulating factors and importance relative to photosynthesis in *Dinobryon* (Chrysophyceae). *Limnol. Oceanog.* 32:277–284.
Bratbak, G. and T.F. Thingstad. 1985. Phytoplankton-bacteria interactions: an apparent paradox? Analysis of a model ecosystem with both competition and commensalism. *Mar. Ecol. Progr. Ser.* 25:23–30.
Burney, C.M., P.G. Davis, K.M. Johnson, and J.McN. Sieburth. 1982. Diel relationships of microbial trophic groups and in situ dissolved carbohydrate dynamics in the Caribbean Sea. *Mar. Biol.* 67:311–322.
Caron, D.A., P.G. Davies, L.P. Madin, and J.McN. Sieburth. 1982. Heterotrophic bacteria and bacterivorous protozoa in oceanic macroaggregates. *Science* 218:795–797.
Chu, S.P. 1946. Utilization of organic phosphorus by phytoplankton. *J. Mar. Biol. Assoc. U.K.* 26:285–295.
Cole, J.J., M.L. Pace, and S. Findley. 1988. Bacterial production in fresh and salt water ecosystems: A cross system overview. *Mar. Ecol. Prog. Ser.* 43:1–10.
Currie, D.J. and J. Kalff. 1984a. The relative importance of bacterioplankton and phytoplankton in phosphorus uptake in freshwater. *Limnol. Oceanogr.* 29:311–321.
Currie, D.J. and J. Kalff. 1984b. A comparison of the abilities of freshwater algae and bacteria to acquire and retain phosphorus. *Limnol. Oceanogr.* 29:298–310.
Currie, D.J., E. Bentzen, and J. Kalff. 1986. Does algal-bacterial phosphorus partitioning vary among lakes? A comparative study of orthophosphate uptake and alkaline phosphatase activity in freshwater. *Can. J. Fish. Aquat. Sci.* 43:311–318.
Davis, P.G. and J.McN. Sieburth. 1984. Estuarine and oceanic microflagellate predation of actively growing bacteria: estimation by frequency of dividing-divided bacteria. *Mar. Ecol. Prog. Ser.* 19:237–246.
Derenbach, J.B. and P.J.LeB. Williams. 1974. Autotrophic and bacterial production: fractionation of plankton populations by differential filtration of samples from the English Channel. *Mar. Biol.* 25:263–269.
Ducklow, H.W., D.A. Purdie, P.J.Leb. Williams, and J.M. Davies. 1986. Bacterioplankton: a sink for carbon in a coastal marine plankton community. *Science* 232:865–867.
Estep, K.W., P.G. Davis, M.D. Keller, and J.McN. Sieburth. 1986. How important are oceanic algal nanoflagellates in bacterivory? *Limnol. Oceanogr.* 31:646–650.
Falkowski, P.G., Z. Dubinsky, and K. Wyman. 1985. Growth irradiance relationships in phytoplankton. *Limnol. Oceanogr.* 30:311–321.
Fuhrman, J.A. and F. Azam. 1982. Thymidine incorporation as a measure of heterotrophic bacterioplankton production in marine surface waters: Evaluation and field results. *Mar. Biol.* 66:109–120.
Goldman, J.C. 1984. Conceptual role for microaggregates in pelagic waters. *Bull. Mar. Sci.* 35:462–476.
Goldman, J.C. and D.A. Caron. 1985. Experimental studies on an omnivorous microflagellate: implications for grazing and nutrient regeneration in the marine microbial food chain. *Deep Sea Res.* 32:899–915.
Goldman, J.C., D.A. Caron, O.K. Andersen, and M.R. Dennett. 1985. Nutrient cycling in a microflagellate food chain: I. Nitrogen dynamics. *Mar. Ecol. Prog. Ser.* 24:231–242.
Goldman, J.G., D.A. Caron, and M.R. Dennett. 1987. Nutrient cycling in a microflagellate food chain: IV. Phytoplankton-microflagellate interactions. *Mar. Ecol. Prog. Ser.* 38:75–87.

Gophen, M., B.Z. Cavari, and T. Berman. 1974. Zooplankton feeding on differentially labelled algae and bacteria. *Nature* 247:393–394.

Güde, H. 1985. Influence of phagotrophic processes on the regeneration of nutrients in two-stage continuous culture systems. *Microb. Ecol.* 11:193–204.

Hagstrom, A., U. Larsson, P. Horsted, and S. Normark. 1979. Frequency of dividing cells, a new approach to the determination of bacterial production. *App. Environ. Microbiol.* 37:805-812.

Harvey, H.W. 1953. Note on the absorption of organic phosphorus compounds by *Nitzschia closterium* in the dark. *J. Mar. Biol. Assoc. U.K.* 31:475–476.

Heinbokel, J.F. and J.R. Beers. 1979. Studies on the functional role of tintinids in the Southern Californian Bight. III. Grazing impact of natural assemblages. *Mar. Biol.* 52:23–32.

Hobbie, J.E. 1979. Activity and bacterial biomass. *Arch. Hydrobiol. Beih. Ergebn. Limnol.* 12:59–63.

Hobbie, J.E., R.J. Daley, and S. Jasper. 1977. Use of Nuclepore filters for counting bacteria by fluorescence microscopy. *Appl. Envir. Microbiol.* 33:1225–1228.

Iturriaga, R. and B.G. Mitchell. 1986. Chroococcoid cyanobacteria: a significant component in the food web dynamics of the open ocean. *Mar. Ecol. Prog. Ser.* 28:291–297.

Joint, I.R. 1986. Physiological ecology of picoplankton in various oceanographic provinces. *Can. J. fish. Aquat. Sci.* 214:287–309.

Jones, A.K. and R.C. Cannon. 1986. The release of microalgal photosynthate and associated bacterial uptake and heterotrophic growth. *Br. Phycol. J.* 21:341–358.

Karl, D.M. and C.D. Winn. 1986. Does adenine incorporation into nucleic acids measure total microbial production? A response to comments by Fuhrman et al. *Limnol. Oceanogr.* 31:1384–1394.

Kimor, B. 1981. The role of phagotrophic dinoflagellates in marine ecosystems. *Kieler Meeresforsch. Sonderh.* 5:164–173.

Kirchman, D.L., K.E. Nus, and R.E. Hodson. 1985. Leucine incorporation and its potential as a measure of protein synthesis by bacteria in natural aquatic systems. *App. Environ. Microbiol.* 49:599–607.

Knauer, G.A., D. Hebel, and F. Cipriano. 1982. Marine snow: major site of primary production in coastal waters. *Nature* 300:630–631.

Lancelot, C. 1979. Gross excretion rates of natural marine phytoplankton and heterotrophic uptake for excreted products in the southern North Sea, as determined by short-term kinetics. *Mar. Ecol. Prog. Ser.* 1:179–186.

Larsson, U. and A. Hagstrom. 1982. Fractionated phytoplankton primary production, exudate release and bacterial production in a Baltic eutrophication gradient. *Mar. Biol.* 67:57–70.

Lehman, J.T. 1987. Microscale nutrient patchiness in plankton communities. *ASLO Abstracts* p. 46.

Lehman, J.T. and D. Scavia. 1982. Microscale patchiness of nutrients in plankton communities. *Science* 216:729–730.

Lessard, E.J. and E. Swift. 1985. Species-specific grazing rates of heterotrophic dinoflagellates in oceanic waters, measured with a dual-label radioisotope technique. *Mar. Biol.* 87:289–296.

Li, W.K.W., D.V. Subba Rao, W.G. Harrison, J.C. Smith, J.J. Cullen, B. Irwin, and T. Platt. 1983. Autotrophic picoplankton in the tropical ocean. *Science* 219:292–295.

Mitchell, J.G., O. Akira, and J.A. Fuhrman. 1985. Microzones surrounding phytoplankton form the basis for a stratified marine microbial ecosystem. *Nature* 316:58–59.

Omerod, J.G. 1983. The carbon cycle in aquatic ecosystems. In: J.H. Slater, R. Whittenbury, and J.W.T. Wimpenney (eds.), *Microbes in Their Natural Environments*. Soc. Gen. Micro. Symp. No. 34. Cambridge, Cambridge Univ. Press, p. 461–482.

Peterson, B.J., J.E. Hobbie, and J.F. Haney. 1978. *Daphnia* grazing on natural bacteria. *Limnol. Oceanogr.* 23:1039–1044.
Platt, T., D.V. Subba Rao, and B. Irwin. 1983. Photosynthesis of picoplankton in the oligotrophic ocean. *Nature* 301:702–704.
Pomeroy, L.R. 1974. The ocean's food web, a changing paradigm. *Bioscience* 24:499–504.
Porter, K.G. and Y.S. Feig. 1980. The use of DAPI for identifying and counting aquatic microflora. *Limnol. Oceanogr.* 25:943–948.
Porter, K.G., E.B. Sherr, B.F. Sherr, M. Pace, and R.W. Sonders. 1985. Protozoa in planktonic food webs. *J. Protozool.* 32:402–415.
Provasoli, L. 1958. Nutrition and ecology of protozoa and algae. *Ann. Rev. Microbiol.* 12:279–308.
Rhee, G.-Y. 1973. A continuous-culture study of phosphate uptake, growth rate, and polyphosphate in *Scenedesmus* spp. *J. Phycol.* 9:495–506.
Rodhe, W. 1955. Productivity: can planktonic production proceed during winter darkness in subarctic lakes? *Verh. Int. Ver. Limnol.* 12:117–122.
Scavia, D., G.A. Laird, and G.L. Fahnenstiel. 1986. Production of planktonic bacteria in Lake Michigan. *Limnol. Oceanogr.* 31:612–626.
Shanks, A.L. and J.D. Trent. 1979. Marine snow: microscale nutrient patches. *Limnol. Oceanogr.* 24:850–854.
Sharp, J.H. 1977. Excretion of organic matter by marine phytoplankton: do healthy cells do it? *Limnol. Oceanogr.* 22:381–399.
Sherr, B.F., E.B. Sherr, and T. Berman. 1982. Decomposition of organic detritus: a selective role for microflagellate protozoa. *Limnol. Oceanogr.* 27:765–769.
Sherr, B.F., E.B. Sherr, and T. Berman. 1983. Grazing, growth, and ammonium excretion rates of a heterotrophic microflagellates fed with four species of bacteria. *Appl. Environ. Microbiol.* 45:1196–1200.
Sherr, B.F., E.B. Sherr, and C.S. Hopkinson. 1988. Trophic interactions within pelagic microbial communities: indications of feedback regulation of carbon flow. *Hydrobiologia* 159:19–26.
Sherr, E.B., B.F. Sherr, and G.-A. Paffenhofer. 1986. Phagotrophic protozoa as food for metazoans: a "missing" trophic link in marine pelagic food webs? *Mar. Microb. Food Webs.* 1:61–80.
Sherr, E.B., B.F. Sherr, and L.J. Albright. 1987. Bacteria: link or sink. *Science* 235:88.
Sieburth, J.McN. 1984. Protozoan bacterivory in pelagic marine waters. In: J.E. Hobbie and P.J.LeB. Williams (eds.), *Heterotrophic Activity in the Sea.* Plenum Press, New York, p. 405–444.
Sieburth, J.McN., P.W. Johnson, M.A. Eberhardt, M.E. Sieraci, M. Lidstrom, and D. Laux. 1987. The first methane-oxidizing bacterium from the upper mixing layer of the deep ocean: *Methylomonas pelagica* sp. nov. *Curr. Microbiol.* 14:285–293,
Sieracki, M.E. and J.McN. Sieburth. 1986. Sunlight-induced growth delay of planktonic marine bacteria in filtered seawater. *Mar. Ecol. Prog. Ser.* 33:19–27.
Silver, M.W., A.L. Shanks, and J.D. Trent. 1978. Marine snow: microplankton habitat and source of small-scale patchiness in pelagic populations. *Science* 201:371–373.
Simon, M. 1984. Bacterial production and its relation to heterotrophic activity parameters in Lake Constance. *Ergeb. Limnol.* 19:131–139.
Simon, M. 1987. Growth characteristics of small and large free-living and attached bacteria in Lake Constance. *Limnol. Oceanogr.* 32:591–607.
Simon, M. and M.M. Tilzer. 1987. Bacterial response to seasonal changes in primary production and phytoplankton biomass in Lake Constance. *J. Plank. Res.* 9:535–552.
Smith, R.E.H. 1982. The estimation of phytoplankton production and excretion by carbon-14. *Mar. Biol. Lett.* 3:325–334.
Sondergaard, M., B. Rieman, and N.O.G. Jorgensen. 1985. Extracellular organic carbon (EOC) released by phytoplankton and bacterial production. *Oikos* 45:323–332.

Steele, J.H. 1978. Some comments on plankton patches. In: Steele, J.H. (ed.), *Spatial Patterns in Plankton Communities.* Plenum Press, New York, p. 1–21.

Stevenson, L.H. 1978. A case for bacterial dormancy in aquatic systems. *Microbiol. Ecol.* 4:127–133.

Stockner, J.G. and N.J. Antia. 1986. Algal picoplankton from marine and freshwater ecosystems: a multidisciplinary perspective. *Can J. Fish. Aquat. Sci.* 43:2472–2503.

Takahashi, M. and B.K. Bienfang. 1983. Size structure of phytoplankton biomass and photosynthesis in subtropical Hawaiian waters. *Mar. Biol.* 76:203–211.

Tarapchak, S.J. and C. Nalewajko. 1986. Synopsis: phosphorus-plankton dynamics symposium. *Can. J. Fish. Aquat. Sci.* 43:416–419.

Trent, J.D., A.L. Shanks, and M.W. Silver. 1978. *In situ* laboratory measurements on macroscopic aggregates in Monteray Bay, California. *Limnol. Oceanogr.* 23:626–635.

Turner, J.T., P.A. Tester, and R.L. Ferguson. 1988. The marine cladoceran *Penila avirostris* and the "microbial loop" of pelagic food webs. *Limnol. Oceanogr.* 33:245–255.

Watson, S.W., T.J. Novitsky, H.L. Quinby, and F.W. Valois. 1977. Determination of bacterial number and biomass in marine environments. *Appl. Environ. Microbiol.* 33:940–954.

Williams, P.J.LeB. 1984. Bacterial production in the marine food chain: the emperor's new suit of clothes? In: M.R. Fasham (ed.), *Flow of Energy and Materials in Marine Ecosystems—Theory and Practice.* Plenum Press. London, p. 271–299.

Wood, A.M. and L.M. Van Valen. 1989. Paradox lost? On the release of energy rich compounds by algae. *Marine Microbial Food Webs.* (In Press).

# 28

# Contribution of Picoplankton to the Grazer Food Chain of Lake Biwa

Toshi Nagata

ABSTRACT Recent research has revealed that minute planktonic organisms, called picoplankton, represent significant biomass and production within pelagic systems. In lake environments, bacterial net production is at least 10–40% of primary production, and up to 50–90% of primary production is attributable to picophytoplankton. Although there are some possible loss factors of picoplankton (sinking, lysis, and grazing), evidence has been accumulated that grazing, especially that by phagotrophic protists, is most significant. It is likely that the "microbial loop" operates actively in large lakes. Dissolved organic carbon produced through several planktonic processes, and a fraction of particulate organic carbon that is not edible for zooplankton enter into the grazer food chain through the microbial loop. It is difficult, however, to determine the efficiency of carbon transfer through the loop, because less information is available about the exact number of trophic steps within the loop and the carbon transfer efficiency of each step.

## 28.1 Introduction

According to the traditional view of planktonic food webs in marine systems (Steele, 1974) and mesotrophic and oligotrophic lakes (Gliwicz, 1969), a major proportion of primary production is directly consumed by zooplankton, and consequently, the algal-zooplankton food chain is the key linkage among planktonic trophic interactions. Now we need to modify this view. Based on recent improvements in epifluorescence microscopy and tracer techniques, it has become apparent that bacteria and picophytoplankton (0.2–2 $\mu$m) significantly contribute to planktonic biomass and production (Fuhrman and Azam, 1982; Stockner and Antia, 1986), and that phagotrophic protists play a pivotal role both as consumers of picoplankton and as prey of zooplankton (Fenchel, 1982a, b; Sherr et al., 1986; Turner et al., 1988). The revised model of the marine food web, which has recently been proposed, incorporates these microbial components (Azam et al., 1983). This model has become accepted as an appropriate food web scheme for lakes of various sizes and trophic status. However, the structure of the microbial loop seems to be more complex than was supposed, and the functional role of the loop in pelagic systems is still being debated (sink vs. link argument; Ducklow et al., 1986; Sherr et al., 1986).

In this chapter, I review recent microbial loop studies in lakes. I mainly consider production of picoplankton and the fate of the carbon that is produced. A case study in Lake Biwa, a large lake in Japan, is also reviewed and its relevance to other large lakes is discussed.

## 28.2 Bacterial Production in Lakes

Bacterial production seems to increase with increasing lake trophic status (Bell et al., 1983; Nagata, 1987). It is difficult, however, to generalize the relation between lake trophic status and percentage contribution of bacterial production to primary production. Recent research has revealed that bacterial net production is at least 10–40% of primary production in eutrophic lakes (Lovell and Konopka, 1985) and mesotrophic and oligotrophic large lakes (Güde et al., 1985; Nagata, 1987; Scavia and Laird, 1987; Simon and Tilzer, 1987). Considering that the gross growth efficiency of bacteria is 20–50% (Pomeroy and Wiebe, 1988), the above value implies that 20– >100% of primary production passes to bacteria in lakes of various trophic status. The larger value may be attributable in part to underestimation of primary production and to allochthonous organic input. However, based on careful examinations, Søndergaard et al. (1988) showed that bacterial gross production accounted for 33–75% of primary production in an oligotrophic lake.

What is the underlying mechanism of this large flux? One of the well-documented processes is release of extracellular organic carbon (EOC) by algae and consumption of EOC by bacteria. Although methodological problems of measurement of the EOC release were claimed, it seems to be valid at present that this process is one of the important mechanisms of the carbon flux from phytoplankton to bacteria (Cole et al., 1982; Søndergaard et al., 1988). Another potentially important process is associated with activities of zooplankton and protozoans. Herbivorous zooplankton loose a significant portion of the carbon of collected algal cells as dissolved organic carbon (DOC) through "sloppy feeding", excretion, or both (Lampert, 1978; Riemann et al., 1986). This DOC includes amino acids (Riemann et al., 1986) and may be a good bacterial substrate. Protozoans are also known to excrete dissolved free amino acids (Andersson et al., 1985). Although DOC excretion by protozoans might provide significant flux of carbon, because they are active components of planktonic communities (Fenchel, 1987), the significance of this flux in lakes has not yet been examined. In addition to the above processes, release of DOC through lysis (autolysis or lysis by viruses) of algae may also be related to the carbon flux from phytoplankton to bacteria. The relative importance of the above processes in pelagic systems has been poorly understood. Based on enclosure experiments in lakes, Riemann and Søndergaard (1986) suggested that there are three different situations in which the single major process responsible for the algal-bacterial flux is EOC, algal lysis, and zooplankton activity, respectively. Apparently, complicated interactions among several trophic levels of a planktonic community are related to the flux of carbon through bacteria.

Data on seasonal change of bacterial biomass and production in the pelagic area of large lakes may be helpful for understanding the internal regulation factors of bacterial production, because this system is relatively free from sporadic, allochthonous effects such as rainfall (Nagata, 1987). The field data (Güde et al., 1985; Scavia and Laird, 1987) suggest that temperature is the most important regulation factor of bacterial production under low temperature regime (<10°C). Under a medium to high temperature regime (10–30°C), however, factors other than temperature seem to be much more important. It has been reported in some large lakes that the seasonal maxima of bacterial abundance and production are found when phytoplankton biomass is decreasing or when it is lowest (Nagata, 1987; Scavia and Laird, 1987; Simon and Tilzer, 1987; Figure 28.1). Nagata (1987) observed that bacterial abundances were negatively correlated with chlorophyll *a* in Lake Biwa during warm periods of 1986. These results support the idea (Fuhrman et al., 1980) that DOC released from senescent algae, or that released through zooplankton and protozoan activity is a more significant bacterial carbon source than that released from actively growing, healthy algae.

An important question that is related to bacterial carbon cycling in lakes is whether bacteria attach to larger particles. Although it has been reported that unattached small bacteria (<1 $\mu$m) are major bacterial producers in most marine environments (Azam et al. 1983), this thesis is under debate for lakes. In most mesotrophic and oligotrophic large lakes, as well as in the marine environment, unattached bacteria seem to be the most significant bacterial producers. Nagata (1987) has shown that, throughout warm periods in Lake Biwa, more than 90% of bacterial production is always attributable to the bacterial fraction less than 1 $\mu$m. Contrary to this, Pedrós-Alió and Brock (1983) suggested that production of attached bacteria is significant in eutrophic Lake Mendota. As pointed out by Güde (1988), methodological problems inherent in the use of radiolabeled glucose or amino acids, which can be taken up by algae, might have led to overestimation of production of attached bacteria in earlier studies. However, by using $^3$H-thymidine, which is specifically incorporated by bacteria (Fuhrman and Azam, 1982; Bern, 1985), Bell et al. (1983) and Lovell and Konopka (1985) reported that up to 30–50% of bacterial production was attributable to the fraction of attached bacteria in eutrophic lakes. Although more systematic work is needed, it seems that the production of attached bacteria becomes significant when a cyanobacterial bloom occurs. Dense colonization of bacteria in the slime of colonial cyanobacteria may be related to the prevalence of attached bacteria.

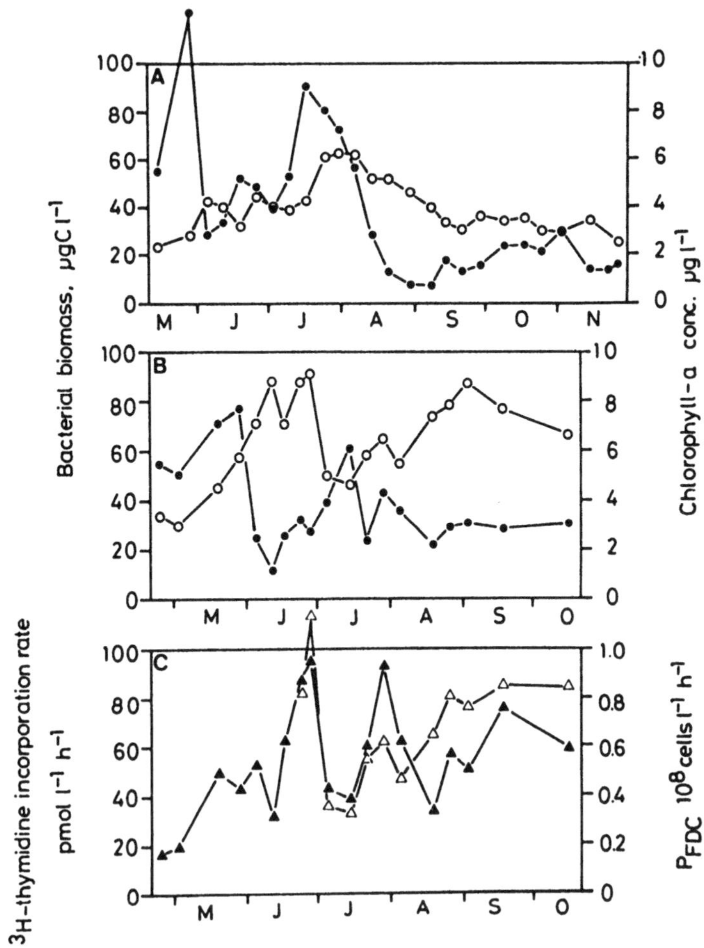


**Figure 28.1** (*a* 1984; *b*, 1986). Seasonal changes of bacterial biomass (○) and chl *a* concentration (●) in Lake Biwa. Biomass was estimated from cell count (acridine-orange direct count), assuming a cell contains 11 fg of C (Nagata, 1986a, 1988b). The average values for the water column (0 to 10 m) are presented. (*c*) $^3$H-thymidine incorporation rate (▲) into cold TCA-insoluble fraction measured by the method of Fuhrman and Azam (1982). $P_{FDC}$(△) is the bacterial growth rate estimated by the FDC method (Hagström et al., 1979). The regression of FDC vs. the instantaneous growth rate, $\mu$ was obtained from bottle incubation experiments using Lake Biwa samples ($\ln \mu = 0.33$ FDC $- 5.3$, Nagata, 1987). The average values for the water column (0 − 10 m) are presented. Data from Nagata (1987, 1988b).

## 28.3 Picophytoplankton Abundance and Production in Lakes

Picophytoplankton have been overlooked in conventional limnological studies on phytoplankton. They are hardly observable by traditional Utermöhl technique, because the cells settle too slowly, and the discrimination between pigmented and nonpigmented cells is very difficult. Recent application of epifluorescence microscopy (Waterbury et al., 1979; Nagata, 1986b) first made it possible to estimate cell density and biomass of picophytoplankton in a relatively simple manner. By this method, the ubiquitousness of procaryotic and eucaryotic picophytoplankton has become recognized in both marine and freshwater ecosystems. Moreover, measurements of photosynthetic rate and pigment concentration in size-fractionated samples have increasingly revealed that a good proportion of biomass and primary production is attributable to picophytoplankton in diverse environments. Ecology and physiology of this "new" plankton component have recently been reviewed comprehensively (Stockner an Antia, 1986). I here comment on some characteristics of the occurrence of picophytoplankton in lakes and its significance in carbon cycling.

In lakes, as well as in marine environments, unicellular chroococcoid cyanobacteria are one of the most ubiquitous picophytoplankton. Chroococcoid cyanobacteria have been observed in lakes of various trophic status (Bailey-Watts et al., 1968; Craig, 1984; Caron et al., 1985; Fahnenstiel et al., 1986; Nagata, 1986b, 1988b). Cell densities in the lakes range from $10^2$ to $10^6$ cells $ml^{-1}$, and within a lake, the density fluctuates seasonally and vertically by three to four orders of magnitude (Caron et al., 1985; Nagata, 1986b, 1988b). This large variation is contrasted with the relatively small variation in bacterial density within a lake (fluctuation is less than an order of magnitude; Nagata, 1987, 1988b), and among lakes of various trophic status ($10^5$–$10^7$ cells $ml^{-1}$), suggesting that chroococcoid cyanobacteria are a relatively unstable picoplanktonic carbon pool in lakes. Cyanobacterial density is usually less than 5% of bacterial density, whereas biomass of cyanobacteria makes up 30–40% of bacterial biomass (Nagata, 1988b).

Figure 28.2 shows seasonal changes in cell density of chroococcoid cyanobacteria in Lake Biwa. The seasonal dynamics can be characterized by rapid increase during midsummer, from approximately $10^3$ cells $ml^{-1}$ up to $10^5$ cells $ml^{-1}$. By contrast, total chl-a concentration decreased during this period because of a decrease in larger phytoplankton ($>3$ $\mu$m). Consequently, the percentage contribution of picoplanktonic chl-a (i.e., $<3$ $\mu$m fraction) to total chl-a increased from less than 5% to 45%. Caron et al. (1985) have also reported a midsummer maximum of chroococcoid cyanobacterial abundance in Lake Ontario. They thought that temperature was related to the seasonal change of the cyanobacteria. Because our knowledge of seasonal dynamics of chroococcoid cyanobacteria in lakes is still poor, we cannot decide at present whether a midsummer peak of chroococcoid cyanobacteria is a general phenomenon in various lakes. Further research is needed to clarify the effects of light, temperature, nutrients, and grazing on chroococcoid cyanobacteria.

Although eucaryotic picophytoplankton have been described in both marine and freshwater environments (Stockner and Antia, 1986), the relative contributions of procaryotes and eucaryotes to total picophytoplankton biomass has rarely been studied in lakes. In Lake Biwa, I measured volume biomass

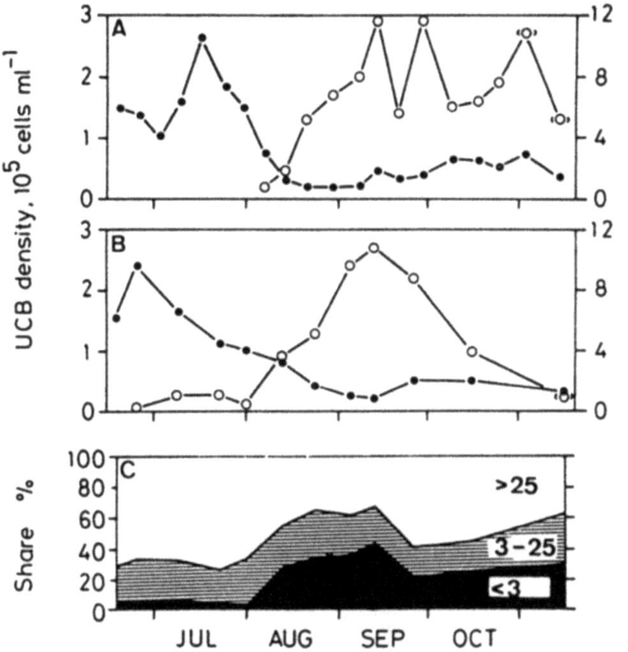


**Figure. 28.2** Seasonal changes in the density of unicellular chroococcoid cyanobacteria (UCB) (○) and in chl *a* concentration (●) in Lake Biwa (*a*, 1984; *b*, 1985). Water samples were collected at a depth of 5 m (the points in parentheses are from a depth of 10 m). (*c*) Relative contribution of the size-fractionated (<3 µm, 3–25 µm, >25 µm) chl *a* concentration in 1985 (5 m depth). Data from Nagata (1986b, 1988b).

of phytoplankton passing through 3 µm-pore size Nuclepore filter by the FITC staining technique of epifluorescence microscopy (Nagata, 1986b). I observed small flagellates (3–4 µm) and minute greenlike cells (2–3 µm) in this fraction. However, the most significant biomass was usually attributable to chroococcoid cyanobacteria from July to December 1985.

The relative contribution of picophytoplankton to total primary production has usually been estimated by the size-fractionation method using Nuclepore filters (pore size, 1–3 µm; Stockner and Antia, 1986). By this method, Fahnenstiel et al. (1986) have reported that approximately 50% of primary production is attributable to picophytoplankton (<3 µm fraction) throughout the year in Lake Superior. Craig (1984) observed that the percentage contribution of <3 µm fraction was 24–97% to primary production in a small oligotrophic lake in Canada. From a review of the literature, Stockner and Antia (1986) suggested that the contribution of picophytoplankton production tends to be higher in oligotrophic lakes than eutrophic lakes. However, as indicated by Stockner and Antia, a chroococcoid cyanobacterial "bloom" has been reported in a eutrophic lake, Loch Leven, Scotland (Bailey-Watts et al., 1968).

## 28.4 Fate of Picoplankton and Food Web Implications of the Microbial Loop

We here discuss three possible pathways of picoplankton loss in pelagic systems of lakes. These pathways are sinking, lysis, and grazing.

Because picoplankton are small, single free-living cells of picoplankton hardly sink in natural waters. However, if they are associated with larger particles, they would be lost by sedimentation. Pedrós-Alió and Brock (1983) have proposed that a significant proportion of bacteria in Lake Mendota attach to larger particles and they are lost from the pelagic system through sedimentation. As mentioned above, this process of bacterial loss seems to be significant only in eutrophic lakes with colonial cyanobacterial blooms. In mesotrophic and oligotrophic large lakes, in which unattached bacterial production prevails, it is unlikely that sedimentation is an important bacterial loss factor. Association of picophytoplankton with larger particles has been observed in marine environments (Johnson et al., 1982; Silver et al., 1986). However, using sediment traps, Silver et al. (1986) estimated that daily sinking losses of algal picoplankton represent less than 1% of the standing stock of these cells in the euphotic zone of the sea. Their result suggests that sinking is not an important loss factor of picophytoplankton in pelagic systems.

We know little about lysis of picoplankton. Autolysis is a complex phenomenon that involves cellular disintegration accompanied by various enzymatic processes. Some environmental stress (e.g., deficiency of nutrients, toxic substances, allelopathy) may cause autolysis of picoplankton. Klein and Alexander (1986) suggested that antibacterial toxic substances, which inhibit some bacterial species, exist in natural lake water. However, the effect of toxins on planktonic bacterial populations and whether they cause autolysis of picoplankton have not yet been examined. Lysis of bacteria might also be caused by viruses and bdellovibrios. Although occurrence of bacteriophages (Torrella and Morita, 1974) and bdellovibrios (Shilo, 1984) in seawater has been reported, it has not yet been clarified whether they are indigenous to these environments or are temporal immigrants with polluted water. According to the experimental work by Wiggins and Alexander (1985), and theoretical consideration by Rittenberg (1979), neither bacteriophages nor bdellovibrios can proliferate in unpolluted marine and lake environments, because host cell density is too low. However, if picoplankton distribute heterogeneously in waters, host cell density in "micropatches" may become higher than average density. I consider that our present knowledge on ecology of lysis is too poor to conclude that there is no impact by bdellovibrios and viruses on picoplankton in lakes.

A variety of planktonic suspension feeders, including crustaceans, rotifers, ciliates, and flagellates are potential picoplanktivores in lakes. Ecological studies of feeding have revealed that several cladocerans (Peterson et al., 1978; Nagata and Okamoto, 1988) and rotifers (Bogdan et al., 1980) crop bacteria. It has also been reported that cladocerans feed on chroococcoid cyanobacteria (Brendelberger, 1985). The efficiency of picoplankton collection compared to nanoplankton collection depends on grazer species. Some species collect at low efficiency while others collect highly efficiently. For cladocerans, this efficiency may be related to the mesh size of the filtering combs (Geller and Müller, 1981; Brendelberger, 1985). In contrast to cladocerans, freshwater copepods are, in general, less efficient bacterial feeders (Nagata and Okamoto, 1988). We know little about distribution and feeding ecology of ciliates in lakes. Although Fenchel (1980) suggested, based on bioenergetics of ciliates, that bacterivorous ciliates have difficulty living in pelagic environments, where bacterial density is low, recent marine work (Sherr and Sherr, 1987) has revealed that small, planktonic ciliates can consume bacteria more efficiently than Fenchel's estimation. Sherr and Sherr (1987) estimated that ciliates are major bacterial consumers in estuaries. Quantitative assessment of ciliate grazing on picoplankton has not yet been done in lakes. Heterotrophic flagellates, which are ubiquitous planktonic components in lakes (Nagata, 1988a, b), are potentially important picoplanktivores. It has been documented (Fenchel, 1982a, b; Nagata, 1988b) that heterotrophic flagellates crop bacteria and algal picoplankton efficiently, they can grow at the picoplankton biomass that is found in pelagic systems, and they often become the major picoplankton consumers in pelagic environments. Recently, phagotrophy of pigmented flagellates has received attention in the context of picoplankton loss factor (Porter, 1988). Most important mixotrophic algae in lakes seem to be chrysomonads (Kimura and Ishida, 1985; Bird and Kalff, 1986), although mixotrophy of other taxa has also been reported (Porter, 1988). The relative contribution of each grazer to total picoplankton loss in a given system has rarely been examined. Based on enclosure experiments conducted in a eutrophic lake in Denmark, Riemann (1985) found that flagellates are the dominant bacterial grazers in enclosures with fish, whereas macrozooplankters become the major bacterial grazers in enclosures without fish.

Figure 28.3 schematically summarizes the possible pathways of pelagic picoplankton loss. It is apparent that if lysis or sinking are major bacterial loss factors, picoplankton carbon cannot reach higher trophic levels. On the other

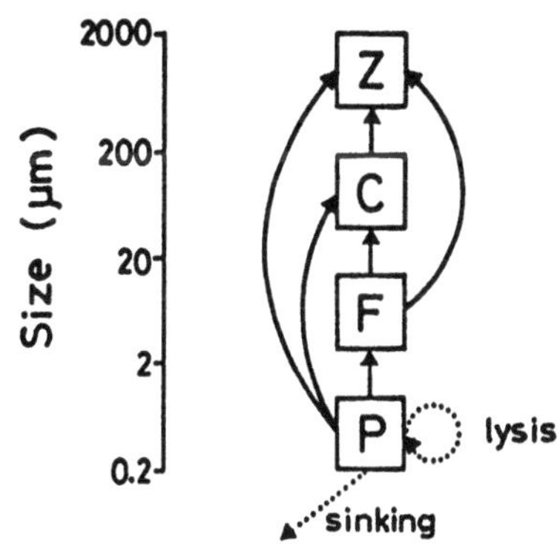


**Figure 28.3** A schematic representation of the possible fate of picoplankton in pelagic systems. Each compartment corresponds to planktonic organisms of an approximate size scale: P, picoplankton; F, flagellates; C, ciliates; Z, macrozooplankton. The solid arrows indicate the transfer of energy through the food chain, while the dotted arrows indicate the dissipation of energy from the system through the lysis or sinking of picoplankton.

hand, if picoplankton enters the grazers' linkage, they would provide a base for food webs. The efficiency of the transfer from picoplankton to macrozooplankton, however, depends on number of trophic steps and growth efficiency. Assuming gross growth efficiency of protozoans to be 30–40% (Nagata, 1988b), 9–16% of picoplankton net production can be passed to macrozooplankton through the picoplankton-flagellates-ciliates chain, which has been supposed to be the major pathway of the microbial loop (Azam et al., 1983). If one step is bypassed (eg., picoplankton-flagellates-zooplankton or picoplankton-ciliates-zooplankton), a situation that seems to be realistic as discussed above and by Sherr et al. (1986) and Turner et al. (1988), 30–40% of picoplankton net production is available to macrozooplankton. The most efficient pathway seems to be direct consumption of picoplankton by macrozooplankton. In lakes, where daphnids often dominate, this linkage might be significant. However, if assimilation efficiency of bacteria and cyanobacteria by daphnids is much lower than that of eucaryotic algae and protozoans, this direct pathway would not act as an efficient carbon link.

Some investigators (Ducklow et al., 1986; Pomeroy and Wiebe, 1988) have stressed the importance of the microbial loop as an energy sink rather than link, because respiratory loss of carbon through the loop is significant (see Figure 28.3). It is obvious that if the carbon of algae edible to zooplankton enters the microbial loop, the carbon transfer efficiency is considerably reduced. Pomeroy and Wiebe (1988) argued that metazoans and microbes compete for particulate organic carbon in the sea. However, we must recall that primary producers may escape direct consumption by zooplankton by a shift in species composition to dominance by the large fraction ($>30$ $\mu$m), picoplankton fraction, or gelatinous colonial types that pass through the gut of zooplankton in intact form (Porter, 1977; Johnson et al., 1982). Moreover, as discussed above, loss of particulate carbon as DOC is a general and significant phenomenon in planktonic communities. The particulate and dissolved organic carbon, to which zooplankton cannot get access, goes back to the grazing food chain only through the operation of the microbial loop.

## 28.5 The Microbial Loop in Lake Biwa

Finally, I briefly consider a microbial loop study conducted in the pelagic area of Lake Biwa. Lake Biwa is a large monomictic lake in Japan. The major water basin of this lake is the north one, with a surface area of 618 km$^2$ and mean water depth of 46.5 m. The recent trophic state of this water basin is mesotrophic. Based on biomass and production, it has been suggested that bacteria and algal picoplankton are important components in Lake Biwa (Nagata, 1986a, b; 1987; Table 28.1). Nagata (1987) estimated that net bacterial production corresponded to at least 30% of primary production in this system. Heterotrophic flagellates have been identified as major picoplankton consumers (Nagata, 1988b). Both biomass of flagellates and feeding experiments indicate that

**Table 28.1** Summary of the quantitative properties of epilimnetic picoplankton in Lake Biwa

| Major group | Density[a] (cells $ml^{-1}$) | Biomass[a] ($\mu g$ C $l^{-1}$) | Contribution to TPOC[b] | Contribution to total chl *a* conc.[c] | Net production rate ($\mu g$ C $l^{-1}$ $d^{-1}$) | Contribution to primary production rate[g] |
|---|---|---|---|---|---|---|
| Nonpigmented bacteria | 2–9 × $10^6$ | 30–90 | 6–13% | | 5–59 (mean 28)[d]<br>9–33 (mean 18)[e]<br>5–25 (mean 17)[f] | ca. 30% |
| Pigmented unicellular chroococcoid cyanobacteria | 0.0–4 × $10^5$ | <1–50 | <1–12% | <5–45% | Not determined | Not determined |

[a] Ranges of the data obtained during warm seasons between 1984 and 1986 (from Nagata, 1986a, 1986b, 1987, and 1988b).
[b] TPOC; total particulate organic carbon (from Nagata, 1986a).
[c] From Nagata (1986b).
[d] By the $^3$H-thymidine incorporation method (Fuhrman and Azam, 1982). The incorporation rate was converted to production rate using a factor of 2 × $10^6$ cells per pmol of thymidine. Data were obtained between 24 April and 15 October of 1986 (n=57) (from Nagata, 1987).
[e] By the FDC method (Hagström et al., 1979). FDC values were converted to the instantaneous growth rate using the formula given in the legend of Figure 28.1. The bacterial production rate was estimated by multiplying the instantaneous growth rate by bacterial biomass. Data were obtained between 23 June and 15 October of 1986 (n=36) (from Nagata, 1987).
[f] By the bottle incubation method (Wright and Coffin, 1984). Bacterial production rate was estimated from the changes of cell numbers in grazer-eliminated lake waters (1 $\mu$m filtrate). Data were obtained from June to September, 1984 and 1985 (n=7) (from Nagata, 1988b).
[g] The primary production rate was estimated by the $^{13}$C-method from May to October, 1983 (Tsuda and Nakanishi, unpublished observations). Bacterial production in the water column (0 − 10 m) was compared with the primary production in the same water basin.

a major part of bacterial daily production is usually consumed by flagellates (Nagata, 1988b). Although *Daphnia longispina*, a predominant herbivorous zooplankter, collect bacteria highly efficiently (Nagata and Okamoto, 1988), their grazing is significant only during the periods of temporal maximum abundance (Nagata, unpublished). These data suggest that the pathway of the microbial loop that is mediated by protozoans (Figure 28.3), is active, and that picoplankton is a potentially important basis of the pelagic food web in Lake Biwa.

Our tentative conclusion about the microbial loop in Lake Biwa agrees well with results recently obtained in Lake Michigan (Scavia and Laird, 1987), Lake Ontario (Caron et al., 1985), and Lake Constance (Güde, 1986; Simon and Tilzer, 1987). Contribution of the microbial loop to the grazer food chain in large lakes may be significant. However, we know little about the regulation factor of picoplankton production in these systems. Moreover, our discussion on the carbon transfer efficiency (see Figure 28.3) is severely limited because we lack information about the fate of flagellates, the ecological role of ciliates, and assimilation efficiency of picoplankton and protozoans by macrozooplankters. Are heterotrophic flagellates important food resources for macrozooplankters? Do ciliates consume mainly picoplankton or nanoplankton? By what factor(s) are the dynamics and production of protozoans controlled? We have just reached the recognition that the pelagic food web should be analyzed from a more comprehensive viewpoint than has previously been done. Many aspects of ecology and natural history of planktonic microbes in pelagic environments remain to be clarified.

---

*Acknowledgements* I thank Y. Tezuka, M. Nakanishi, T. Narita, K. Okamoto, K. Kawabata, and other researchers at the Otsu Hydrobiological Station, Otsu, Japan, for helpful advice, criticism, and encouragement. I also thank the Yoshida Foundation of Science and Technology for financial support for my attendance at Large Lakes Symposium. This study is partly supported by a grant for Scientific Research No. 61790227 from the Ministry of Education, Science and Culture of Japan.

## References

Andersson, A., Lee, C., Azam, F., and Hagström, Å. 1985. Release of amino acids and inorganic nutrients by heterotrophic marine microflagellates. *Marine Ecology Progress Series* 23:99–106.

Azam, F., Fenchel, T., Field, J.G., Gray, J.S., Meyer-Reil, L.-A., and Thingstad, F. 1983. The ecological role of water-column microbes in the sea. *Marine Ecology Progress Series* 10:257–263.

Bailey-Watts, A.E., Bindloss, M.E., and Belcher, J.H. 1968. Freshwater primary production by a blue-green alga of bacterial size. *Nature* 220:1344–1345.

Bell, R.T., Ahlgren, G.M., and Ahlgren, I. 1983. Estimating bacterioplankton production by measuring [$^3$H]thymidine incorporation in a eutrophic Swedish lake. *Applied and Environmental Microbiology* 45:1709–1721.

Bern, L. 1985. Autoradiographic studies of [methyl-$^3$H]thymidine incorporation in a cyanobacterium (*Microcystis wesenbergii*)-bacterium association and in selected algae and bacteria. *Applied and Environmental Microbiology* 49:232–233.

Bird, D.F., and Kalff, J. 1986. Bacterial grazing by planktonic lake algae. *Science* 231:493–495.

Bogdan, K.G., Gilbert, J.J., and Starkweather, P.L. 1980. *In situ* clearance rates of planktonic rotifers. *Hydrobiologia* 73:73–77.

Brendelberger, H. 1985. Filter mesh-size and retention efficiency for small particles: comparative studies with Cladocera. *Archiv für Hydrobiology: Beiheft, Ergebnisse der Limnologie* 21:135–146.

Caron, D.A., Pick, F.R., and Lean, D.R.S. 1985. Chroococcoid cyanobacteria in Lake Ontario: Vertical and seasonal distributions during 1982. *Journal of Phycology* 21:171–175.

Cole, J.J., Likens, G.E., and Strayer, D.L. 1982. Photosynthetically produced dissolved organic carbon: An important carbon source for planktonic bacteria. *Limnology and Oceanography* 27:1080–1090.

Craig, S.R. 1984. Productivity of algal picoplankton in a small meromictic lake. *Verhandlungen der internationalen Vereinung für Theoretische und Angewandte Limnologie* 22:351–354.

Ducklow, H.W., Purdie, D.A., Williams, P.J.LeB., and Davies, J.M. 1986. Bacterioplankton: A sink for carbon in a coastal marine plankton community. *Science* 232:865–867.

Fahnenstiel, G.L., Sicko-Goad, L., Scavia, D., and Stoermer, E.F. 1986. Importance of picoplankton in Lake Superior. *Canadian Journal of Fisheries and Aquatic Science* 43:235–240.

Fenchel, T. 1980. Relation between particle size selection and clearance in suspension-feeding ciliates. *Limnology and Oceanography* 25:733–738.

Fenchel, T. 1982a. Ecology of heterotrophic microflagellates. II. Bioenergetics and growth. *Marine Ecology Progress Series* 8:225–231.

Fenchel, T. 1982b. Ecology of heterotrophic microflagellates. IV. Quantitative occurrence and importance as bacterial consumers. *Marine Ecology Progress Series* 9:35–42.

Fenchel, T. 1987. *Ecology of Protozoa: The Biology of Free-Living Phagotrophic Protists.* Science Tech Publishers, Madison, WI, 197pp.

Fuhrman, J.A., Ammerman, J.W., and Azam, F. 1980. Bacterioplankton in the coastal euphotic zone: Distribution, activity and possible relationships with phytoplankton. *Marine Biology* 60:201–207.

Fuhrman, J.A., and Azam, F. 1982. Thymidine incorporation as a measure of heterotrophic bacterioplankton production in marine surface waters: Evaluation and field results. *Marine Biology* 66:109–120.

Geller, W., and Müller, H. 1981. The filtration apparatus of Cladocera: filter mesh-sizes and their implications on food selectivity. *Oecologia* 49:316–321.

Gliwicz, Z.M. 1969. Studies on the feeding of pelagic zooplankton in lake with varying trophy. *Ekologia Polska Seria A* 17:663–708.

Güde, H. 1986. Loss processes influencing growth of planktonic bacterial populations in Lake Constance. *Journal of Plankton Research* 8:795–810.

Güde, H. 1988. Incorporation of $^{14}C$-glucose, $^{14}C$-amino acids and $^{3}H$-thymidine by different size fractions of aquatic microorganisms. *Archiv für Hydrobiologie Beiheft, Ergebnisse der Limnologie* 31:61–69.

Güde, H., Haibel, B., and Müller, H. 1985. Development of planktonic bacterial populations in a water column of Lake Constance (Bodensee-Obersee). *Archiv für Hydrobiology* 105:59–77.

Hagström, Å., Larsson, U., Hörstedt, P., and Normark, S. 1979. Frequency of dividing cells, a new approach to the determination of bacterial growth rates in aquatic environments. *Applied and Environmental Microbiology* 37:805–812.

Hobbie, J.E., Daley, R.J., and Jasper, S. 1977. Use of Nuclepore filters for counting bacteria by fluorescence microscopy. *Applied and Environmental Microbiology* 33:1225–1228.

Johnson, P.W., Xu, Huai-Shu, and Sieburth, J.McN. 1982. The utilization of chroococcoid cyanobacteria by marine protozooplankters but not by calanoid copepods. *Annales de l'institut océanographique*, Paris, 58:297–308.
Kimura, B., and Ishida, Y. 1985. Photophagotrophy in *Uroglena americana*, Chrysophyceae. *Japanese Journal of Limnology* 46:315–318.
Klein, T.M., and Alexander, M. 1986. Bacterial inhibitors in lake water. *Applied and Environmental Microbiology* 52:114–118.
Lampert, W. 1978. Release of dissolved organic carbon by grazing zooplankton. *Limnology and Oceanography* 23:831–834.
Lovell, C.R., and Konopka, A. 1985. Primary and bacterial production in two dimictic Indiana lakes. *Applied and Environmental Microbiology* 49:485–491.
Nagata, T. 1986a. Carbon and nitrogen content of natural planktonic bacteria. *Applied and Environmental Microbiology* 52:28–32.
Nagata, T. 1986b. The seasonal abundance and vertical distribution of the $<3$ $\mu$m-phytoplankton in the north basin of Lake Biwa. *Ecological Research* 1:207–221.
Nagata, T. 1987. Production rate of planktonic bacteria in the north basin of Lake Biwa, Japan. *Applied and Environmental Microbiology* 53:2872–2882.
Nagata, T. 1988a. Seasonal abundance, grazing impacts on bacteria and vertical distribution of heterotrophic microflagellates in the south basin of Lake Biwa. *Japanese Journal of Limnology* 49:167–174.
Nagata, T. 1988b. The microflagellate-picoplankton food linkage in the water column of Lake Biwa. *Limnology and Oceanography* 33:504–517.
Nagata, T., and Okamoto, K. 1988. Filtering rates on natural bacteria by *Daphnia longispina and Eodiaptomus japonicus* in Lake Biwa. *Journal of Plankton Research* 10:835–850.
Pedrós-Alió, C., and Brock, T.D. 1983. The importance of attachment to particles for planktonic bacteria. *Archiv für Hydrobiology.* 98:354–379.
Peterson, B.J., Hobbie, J.E., and Haney, J.F. 1978. *Daphnia* grazing on natural bacteria. *Limnology and Oceanography* 23:1039–1044.
Pomeroy, L.R., and Wiebe, W.J. 1988. Energetics of microbial food webs. *Hydrobiologia* 159:7–18.
Porter, K.G. 1977. The plant-animal interface in freshwater ecosystems. *American Scientist* 65:159–170.
Porter, K.G. 1988. Phagotrophic phytoflagellates in microbial food webs. *Hydrobiologia* 159:89–97.
Riemann, B. 1985. Potential importance of fish predation and zooplankton grazing on natural populations of freshwater bacteria. *Applied and Environmental Microbiology* 50:187–193.
Riemann, B., Jørgensen, N.O.G., Lampert, W., and Fuhrman, J.A. 1986. Zooplankton induced changes in dissolved free amino acids and in production rates of freshwater bacteria. *Microbial Ecology* 12:247–258.
Riemann, B., and Søndergaard, M. 1986. Regulation of bacterial secondary production in two eutrophic lakes and in experimental enclosures. *Journal of Plankton Research* 8:519–536.
Rittenberg, S.C. 1979. Bdellovibrio: A model of biological interactions in nutrient-impoverished environments? *In* Shilo, M. (editor), *Strategies on Microbial Life in Extreme Environments.* Verlag Chemie, New York, p. 305–322.
Scavia, D., and Laird, G.A. 1987. Bacterioplankton in Lake Michigan: Dynamics, controls, and significance to carbon flux. *Limnology and Oceanography* 32:1017–1033.
Sherr, E.B., and Sherr, B.F. 1987. High rates of consumption of bacteria by pelagic ciliates. *Nature* 325:710–711.
Sherr, E.B., Sherr, B.F., and Paffenhöfer, G.-A. 1986. Phagotrophic protozoa as food for metazoans: A "missing" trophic link in marine pelagic food webs? *Marine Microbial Food Webs* 1:61–80.

Shilo, M. 1984. *Bdellovibrio* as a predator. *In* Klug, M.J., and Reddy, C.A. (editors), *Current Perspectives in Microbial Ecology*. American Society of Microbiology, Washington D.C., p. 334–339.

Silver, M.W., Gowning, M.M., and Davoll, R.J. 1986. The association of photosynthetic picoplankton and ultraplankton with pelagic detritus through the water column (0–2000 m). *In* Platt, T. and Li, W.K.W. (editors), *Photosynthetic Picoplankton*, Department of Fisheries and Oceans, Ottawa, pp. 311–341.

Simon, M., and Tilzer, M.M. 1987. Bacterial response to seasonal changes in primary production and phytoplankton biomass in Lake Constance. *Journal of Plankton Research* 9:535–552.

Søndergaard, M. et al. 1988. Pelagic food web processes in an oligotrophic lake. *Hydrobiologia* 164:271–286.

Steele, J.H. 1974. *The Structure of Marine Ecosystems*. Harvard Univ. Press, Massachusetts. 128pp.

Stockner, J.G., and Antia, N.J. 1986. Algal picoplankton from marine and freshwater ecosystems: A multidisciplinary perspective. *Canadian Journal of Fisheries and Aquatic Science* 43:2472–2503.

Torrella, F., and Morita, R.Y. 1979. Evidence by electron micrographs for a high incidence of bacteriophage particles in the waters by Yaquina Bay, Oregon: Ecological and taxonomical implications. *Applied and Environmental Microbiology* 37:774–778.

Turner, J.T., Tester, P.A., and Ferguson, R.L. 1988. The marine cladoceran *Penilia avirostris* and the "microbial loop" of pelagic food webs. *Limnology and Oceanography* 33:245–255.

Waterbury, J.B., Watson, S.W., Guillard, R.R.L., and Brand, L.E. 1979. Widespread occurrence of a unicellular, marine, planktonic, cyanobacterium. *Nature* 277:293–294.

Wiggins, B.A., and Alexander, M. 1985. Minimum bacterial density for bacteriophage replication: Implications for significance of bacteriophages in natural ecosystems. *Applied and Environmental Microbiology* 49:19–23.

Wright, R.T., and Coffin, R.B. 1984. Measuring microzooplankton grazing on planktonic marine bacteria by its impact on bacterial production. *Microbial Ecology* 10:137–149.

# 29

# Significance of Heterotrophic Nanoflagellates and Ciliates in Large Lakes: Evidence from Lake Constance

Thomas Weisse and Helga Müller

ABSTRACT This paper presents a case study on the dynamics of heterotrophic nanoflagellates (HNF) and ciliates in Lake Constance illustrating the relationships between bacteria and protozoa in a large lake. The results will be discussed in the context of our present knowledge of the existence and functioning of a highly dynamic microbial food web ("microbial loop") in pelagic systems.

In Lake Constance, bacterial production is mainly cropped by grazing of HNF which occur in concentrations of $10^2$–$10^4$ cells $ml^{-1}$. HNF less than 10 $\mu m$ in size ingest between 10 and 100 bacteria $HNF^{-1}h^{-1}$. The size spectrum shows pronounced seasonal variations. Although the smallest HNF (<2 $\mu m$) dominate in numbers, their contribution to total HNF biomass is negligible. A biomass peak at 3 to 4 $\mu m$ is obvious throughout most of the year. In epipelagic waters, HNF typically double every 20–50 hours. Thus, growth rates of HNF are only slightly lower than those of pelagic bacteria.

HNF are grazed upon by nano- and microciliates. Total ciliate concentrations range from 1–100 cells $ml^{-1}$, and the orders Prostomatida and Oligotrichida numerically dominate the population. Although these ciliates are known to feed on small algae, their ability to efficiently crop bacteria is under debate. Further experimental evidence is needed in order to estimate the impact of pelagic ciliate grazing on bacteria and autotrophic picoplankton in lakes.

## 29.1 Introduction

The importance of the microbial food web in pelagic systems has been recognized over the past decade (Williams, 1981; Pomeroy, 1984). A considerable percentage of the organic material initially produced by primary production of phytoplankton is channeled through a 'microbial loop' (Azam et al., 1983) consisting of pelagic bacteria, autotrophic pico- and nanoplankton, and phagotrophic protozoa (for reviews see Porter et al., 1985; Sherr, B.F. and Sherr, 1984; Sherr, E.B. et al., 1986). Among planktonic protozoa, heterotrophic nanoflagellates (HNF) are currently considered the major consumers of unattached bacteria (Fenchel, 1982c; Sherr et al., 1984). HNF in the size range of less than 2.0 $\mu$m to 20 $\mu$m are a heterogeneous assemblage including both nonpigmented species within four orders of Phytomastigophorea (Cryptomonadida, Dinoflagellida, Euglenida, and Chrysomonadida) and two orders of Zoomastigophorea (Choanoflagellida and Kinetoplastida), which are the zooflagellates sensu stricto (Levine et al., 1980; Fenchel, 1982a). Pigmented and nonpigmented forms are often closely related taxonomically and physiologically. Thus, it is not surprising that bacterivory has also been found occasionally among chloroplast-containing chrysomonads both in the sea (Estep et al., 1986) and in lakes (Bird and Kalff, 1986, 1987). However, whether bacterivory of autotrophic flagellates occurs only sporadically or has a permanent grazing impact on bacteria has to be further investigated.

Whereas HNF have been quantified in a number of recent marine studies, (summarized by Porter et al., 1985) information on HNF concentrations in lakes is scarce (Bloem and Bär-Gilissen, 1989). In situ growth and grazing rates of HNF and their importance in the energy transfer within planktonic lake systems are virtually unknown.

By contrast, the quantitative importance of pelagic ciliates in lakes is well documented. Numbers, biomass, and species composition have been studied at various locations (for review, see Porter et al., 1985; Fenchel, 1987; Beaver and Crisman, 1989). Their role in the pelagic food web, however, is still controversial. According to Fenchel (1980a, b), ciliate bacterivory in open waters is of minor importance, whereas Rivier et al. (1986) and Sherr and Sherr (1987) conclude from their experiments that the grazing impact of ciliates on pelagic bacteria might be highly significant.

This chapter presents results of an investigation of both HNF and ciliates in the pelagic zone of Lake Constance. The objective is to provide a first evaluation of their ecological role in large lakes, particularly their vertical and seasonal distribution, biomass, and feeding impact on pelagic bacteria.

## 29.2 Vertical and Seasonal Distribution of HNF and Ciliates

The distribution of HNF and ciliates has been investigated at the site of maximum water depth (146 m) in Überlinger See, the northwest bay of Lake Constance, from 23 March through 16 November, 1987, in weekly to biweekly intervals. Samples were taken with a water sampler in five 4-m depth intervals in the upper 20 m, so that the entire water column from the surface to 20 m was covered, and from 50 m and 140 m depth. HNF were fixed with Formalin (final concentration 1.5% vol.) and counted using DAPI staining and epifluorescence microscopy according to Porter and Feig (1980). Ciliates were fixed with Lugol's solution and counted by the Utermöhl technique (Utermöhl, 1958). In addition, protargol-stained permanent slides were prepared from $HgCl_2$-fixed material for further taxonomic identification of ciliates and measuring of cell dimensions (Müller, 1989). Ciliate taxonomy followed the scheme of Corliss (1979); identifications were based on Kahl (1930–35) and Foissner and Wilbert (1979).

In early spring, HNF were more or less evenly distributed in the entire water column from surface to bottom (Figure 29.1). After the onset of thermal stratification in late April, HNF abundances were always higher in the epipelagic zone than in the deep water. The highest concentrations, $8.1 \times 10^3$ cells $ml^{-1}$, were recorded at 4 to 8 m in mid-July. During summer, HNF were always slightly more abundant at this depth interval than closer to the surface. Thus, HNF might tend to be concentrated in the thermocline. However, due to the unusually bad weather conditions in 1987, thermal stratification of the water column was less pronounced in this year compared to previous years. Therefore, the vertical distribution of HNF was almost uniform at the end of July, when surface temperature had dropped from 21.2 to 13.3 °C within 2 weeks. Close to the bottom, variation of HNF numbers was comparatively small, with the values below $3 \times 10^3$ cells $ml^{-1}$.

Figure 29.2 shows the seasonal and vertical distribution of ciliates. From March through November, cell concentrations in the uppermost 4 m were from 2 to 100 cells $ml^{-1}$, and the maximum was observed on 27 April, simultaneously with the first peak of HNF abundance. Numbers recorded from the 4 to 8 m

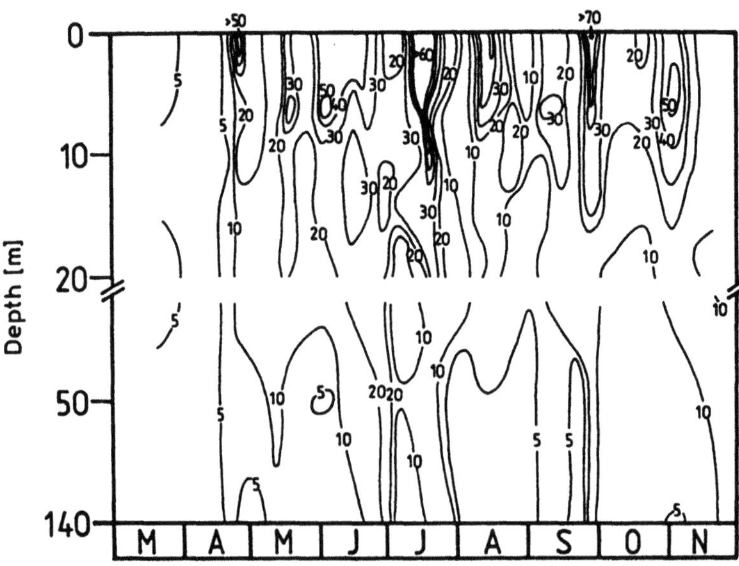


**Figure 29.1** Vertical distribution of heterotrophic nanoflagellates in Lake Constance ($10^2$ cells $ml^{-1}$).

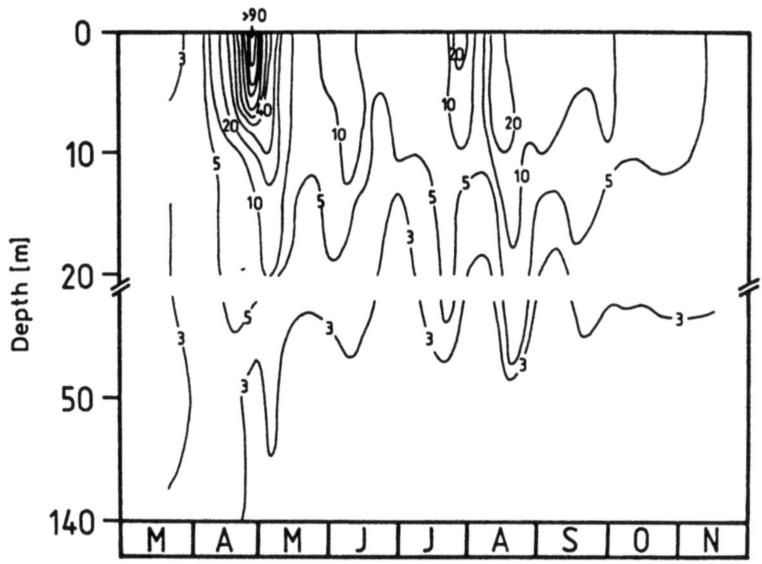


**Figure 29.2** Vertical distribution of ciliates in Lake Constance (cells $ml^{-1}$).

depth interval were, in general, similar or lower than those recorded close to the surface. Mean numbers at 8 to 20 m ranged from 2 to 20 cells $ml^{-1}$. However, because integrated samples from 8 to 20 m were counted from June onwards, the decline of ciliate numbers over the 8–12 m depth interval can only roughly by estimated from the diagram shown in Fig. 2. As in the case of HNF, ciliate numbers were low at 50 and 140 m depths, ranging from 1 to 4 cells $ml^{-1}$.

## 29.3 Abundance, Biomass, and Species Diversity

In a mixed sample from the upper 8 m of the water column, HNF were enumerated in 3 different size classes, $<2$ $\mu m$, 2 to 5 $\mu m$, and 5 to 10 $\mu m$. Cells larger than 10 $\mu m$ were found occasionally, however, they occurred in numbers too low to be quantified with the epifluorescence counting procedure.

Total cell counts of HNF varied between $0.2 \times 10^3$ cells $ml^{-1}$ in April and $7.1 \times 10^3$ cells $ml^{-1}$ in mid-July (Figure 29.3a). Similar concentrations were found in Obersee, the large main basin of Lake Constance, by Güde (1986). He reported epilimnetic abundances of up to $6 \times 10^3$ HNF $ml^{-1}$ during lake stratification (May to November) and $10^2$ cells $ml^{-1}$ in winter. These values were within the range of data reported from other lakes. Boraas et al. (1987) found in Lake Michigan HNF concentrations of 3 to $5 \times 10^3$ cells $ml^{-1}$ during stratification and 0.2 to $0.3 \times 10^3$ cells $ml^{-1}$ in winter. In central Lake Superior, Fahnenstiel et al. (1986) reported protozoan numbers of about $0.5 \times 10^3$ cells $ml^{-1}$. However, this was only a single record in late September. In Lake Vechten, The Netherlands, epilimnetic HNF concentrations amounted to 0.6 to 7.4 $\times$

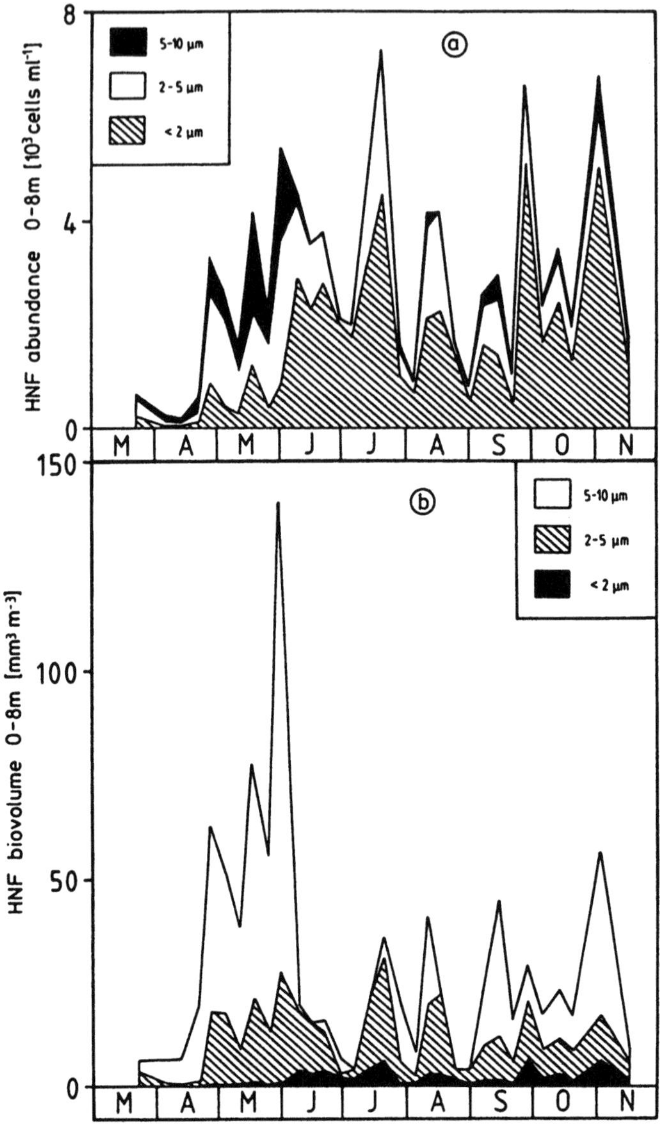


**Figure 29.3** (*a*) Size-specific abundance and (*b*) biovolume of heterotrophic nanoflagellates in the upper 8 m of Lake Constance.

$10^3$ cells $ml^{-1}$ (Bloem et al., 1986; Bloem and Bär-Gilissen, 1989). Nagata (see Chapter 28) also reported HNF concentrations in the order of $10^2$ to $10^3$ cells $ml^{-1}$ in Lake Biwa.

The range of HNF concentrations in lakes is in agreement with results obtained from different marine areas. Values from nine out of ten studies summarized by Porter et al. (1985) fall in the range of 0.1 to $9.5 \times 10^3$ HNF ml$^{-1}$. Thus, HNF cell numbers in various planktonic systems ranging from oligotrophic to eutrophic vary by two orders of magnitude.

Throughout most of the year, the smallest HNF with a longest linear dimension below 2 μm dominated in numbers in Lake Constance (see Figure 29.3a). The percentage of these small cells is highest in summer and lowest in winter and spring. On the annual average, 48.1% of all HNF fall into the smallest size class, 38.8% in the second smallest, and 13.1% in the third smallest size class. However, cell numbers are only a poor indicator of the ecological importance of a specific planktonic taxon or group of organisms. Therefore, cell counts were converted to biomass, assuming average geometrical cell shapes and dimensions for each size class, and a specific gravity of 1.07. In terms of biomass, the smallest HNF are negligible, and 2 to 10-μm-sized cells determine the seasonal fluctuation of the overall HNF biomass (Figure

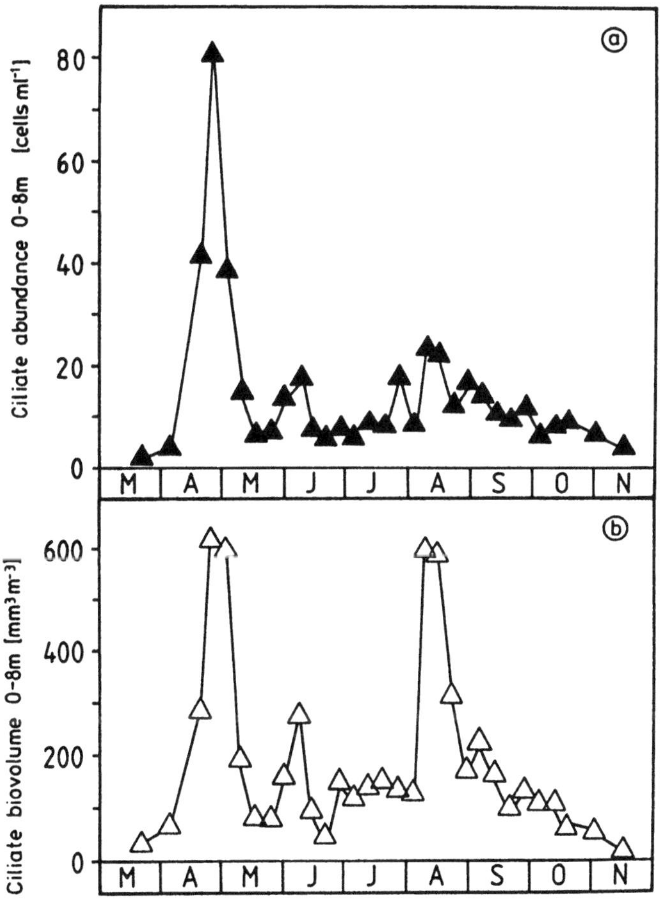


**Figure 29.4** (*a*) Abundance and (*b*) biovolume of ciliates in the upper 8 m of Lake Constance.

29.3b). In spring, HNF biomass began to increase simultaneously with increases in bacteria and phytoplankton but the maximum was reached only several weeks later at the end of May. The decline in June led to a second peak in bacterial biomass in early July, which was again followed by an increase in HNF biomass. The mutual relationship between HNF and bacteria points to a predator-prey relationship, as previously assumed by Güde (1986) for Lake Constance. The biomass of HNF typically amounts to 10 to 20% of the bacterial biomass throughout most of the year. In May and June, however, HNF biomass equals that of free-living bacteria. This estimate is higher than the one for Lake Biwa (5 to 10%), which has been derived from literature data (see Chapter 28).

Mean cell concentrations and biovolumes of ciliates in the upper 8 m of Lake Constance are shown in Figure 29.4. Numbers ranged from 2 to 80 cells $ml^{-1}$, biovolumes from 25 to 620 $mm^3$ $m^{-3}$. These findings are in agreement with data reported from other mesotrophic lakes (Beaver and Crisman, 1982; Pace, 1985; Taylor and Heynen, 1987). A shift in species composition from smaller to larger forms was observed from spring to late summer. Mean cell volumes on 27 April (spring maximum) and 11 August (summer maximum) were 7600 $\mu m^3$ and 25100 $\mu m^3$, respectively.

**Table 29.1** Ciliate taxa identified in the pelagic waters of Lake Constance, March to October 1987.

| Order | Genus |
|---|---|
| Prostomatida | *Pseudobalanion* |
| | *Urotricha* |
| Haptorida | *Lacrymaria* |
| | *Lagenophrya* |
| | *Paradileptus* |
| | *Didinium* |
| | *Askenasia* |
| Suctorida | *Staurophrya* |
| Scuticociliatida | *Cyclidium* |
| | *Histiobalantium* |
| Peritrichida | *Vorticella* |
| | *Epistylis* |
| | *Vaginicola* |
| Heterotrichida | *Stentor* |
| Oligotrichida | *Halteria* |
| | *Strombidium* |
| | *Strobilidium* |
| | *Codonella* |
| | *Tintinnidium* |

A list of ciliate taxa identified in Lake Constance is given in Table 29.1. Some genera, e.g., *Urotricha* and *Strobilidium*, were represented by several species differing widely in size. Others, such as *Pseudobalanion*, were uniform populations of single species. This nanociliate of the order Prostomatida has been identified recently as *Pseudobalanion planctonicum*, nov. gen., nov. spec. by Foissner et al. (1990). In the uppermost 8 m, *Pseudobalanion planctonicum* (mean cell diameter 15 $\mu m$, mean cell volume 1300 $\mu m^3$) and a small *Strobilidium* sp. (mean cell diameter 25 $\mu m$, mean cell volume 6500 $\mu m^3$) (Figure 29.5) were the numerically dominant forms. Together, they accounted for 40% to 90% of total ciliate numbers throughout most of the year. Scuticociliates, in contrast, were of minor importance in the epipelagic waters (Figure 29.6).

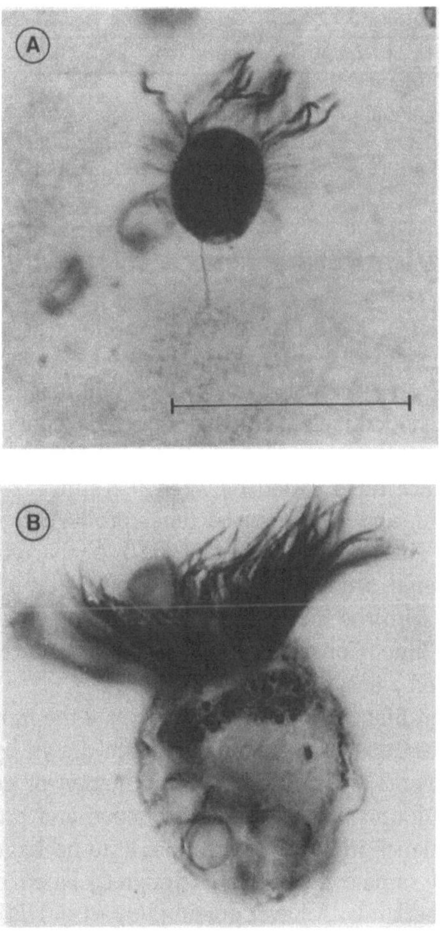


**Figure 29.5** (*a*) *Pseudobalanion planctonicum* and (*b*) *Strobilidium* sp. with ingested *Stephanodiscus hantzschii*; scale bar 25 $\mu m$, protargol stain.

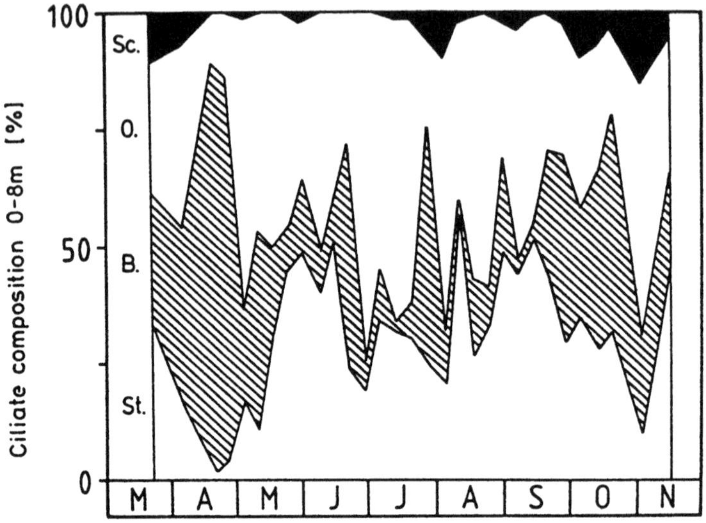


**Figure 29.6** Percentage abundance of different ciliate taxa in the upper 8 m of Lake Constance: Oligotrichs of <35 μm, predominantly *Strobilidium* sp. (St); *Pseudobalanion planctonicum*, 10–20 μm (B); Scuticociliates (Sc); and others (O).

## 29.4 Grazing Impact on Pelagic Bacteria

The grazing impact of HNF on free-living bacteria as measured by in situ experiments (see Section 29.5) has been reported elsewhere in detail (Weisse, 1989b). Therefore, only the main conclusions will be summarized here.

HNF have been identified as the major bacterial consumers in Lake Constance. An average HNF ingests between 10 and 100 bacteria per hour. Corresponding clearance rates varied between 0.2 and 122.8 nl $HNF^{-1}$ $h^{-1}$. On the annual average, about 50% of bacterial production is removed through grazing by HNF. The grazing impact of HNF on bacteria is highest in late spring when bacterial production is almost completely consumed. The high grazing pressure, with the increasing HNF standing stocks caused food limitation of HNF in June. Consequently, HNF biomass decreased, and bacterial populations reached a second peak.

The conclusion that HNF <12 μm in size are the most important grazers on bacteria in Lake Constance has already been drawn by Güde (1986), who conducted experiments with differential elimination of grazers. Sanders and Porter (1986) also found that bacterial production and protozoan grazing are in approximate balance in Lake Oglethorpe, a 30-ha-large man-made lake in Georgia, USA. The same conclusion was drawn by Bloem et al. (1989) for Lake Vechten, The Netherlands. A lower grazing impact of HNF on pelagic bacteria in lakes has been assumed by Nagata (Chapter 28). He estimated that HNF in Lake Biwa crop 10 to 20% of bacteria per day. Grazing rates similar to those measured in Lake Constance were also obtained in Lake Michigan (Scavia and

Laird, 1987). Ingestion and clearance rates reported from both lakes (Seale et al. ,1987; Weisse, 1989b) are well within the range of data obtained from different marine areas and from laboratory cultures (for review see Sherr, B.F. et al., 1986).

As far as we currently know, ciliate bacterivory in the epipelagic waters of Lake Constance is small compared to the grazing impact of HNF on bacteria. Based on information from the literature (Kahl, 1930–1935; Fenchel, 1980a, b; Stoecker et al., 1986; Skogstad et al., 1987) and own observations (see Figure 29.5b), we assume that the dominant ciliates in the upper 20 m mainly feed on particles in the size range of small algae. Bacterial feeders such as *Cyclidium* and other scuticociliates occur in low numbers in the epipelagic waters, whereas they reach higher percentages in the deep water (140 m) and thus might have a significant grazing impact on bacteria close to the lake bottom (Müller et al. ,1989b). The in situ grazing experiments with natural and nano- and microplankton also pointed to low bacterial feeding among epilimnetic ciliates (Weisse, 1989b).

The grazing impact of larger crustacean zooplankton on pelagic bacteria is also low in Lake Constance (Güde, 1986, 1988).

## 29.5 In Situ Growth Rates of HNF and Fate of HNF Production

Growth rates of HNF and grazing pressure on HNF were measured under in situ conditions using a dilution technique (modified after Landry and Hasset, 1982) and dialysis bags. Natural microplankton populations were diluted in ratios of 1:0, 1:1, and 1:2 of unfiltered to filtered lake water. Experiments were conducted at a 3 m depth and lasted for 24 hours (for details see Weisse, 1989b). Net growth ($\mu$), i.e., the change between initial ($N_o$) and final ($N_t$) population densities with time [$\mu = (\ln N_t - \ln N_o)/t$] in undiluted treatments, showed pronounced seasonal variations. With the exception of two high positive values measured in March and June 1987, there was a close relationship between measured net growth rates and observed changes in HNF standing stocks in spring and summer 1987. Negative net growth rates prevented an increase of HNF biomass (in April) or caused a subsequent decrease (in May/June). Positive rates were followed by increasing standing stocks (in May and July) (see Figure 29.3). Negative growth rates measured in autumn 1986 (Weisse, 1989b) also agreed well with decreasing population densities in the water column. Thus, the experimental results reflected population dynamics of HNF at the sampling location, although the time of incubation (24 hours) was short compared to the sampling interval.

Observed changes in population net growth rates ($\mu$) are the results of two antagonistic processes: intrinsic growth measured as gross growth rate (k) and grazing (g) as the major loss factor ($\mu = k - g$). Other loss factors, such as sedimentation or autolytic cell disintegration, were prevented by the experimental setup or have been neglected. The dilution experiments revealed growth and grazing rates reported in Figure 29.7. Both gross growth rates and grazing pressure were highest in spring. Although there was a considerable scattering, most values were in the range of 0.010 to 0.030 $h^{-1}$ in summer.

From gross growth rates, generation times (G) of HNF were calculated according to $G = \ln 2/k$. Generation times varied between 16 and 119 hours. As the majority of values fell in the range of 20 to 50 hours, HNF typically doubled once in 1 to 2 days. Similar doubling times have been found in laboratory cultures (Fenchel, 1982b, Rivier et al., 1986). Thus, HNF populations reproduce about half as fast as pelagic bacteria, which have generation times mostly between 5 and 46 hours during lake stratification (Simon, 1988; Weisse, unpubl.).

Considering the fate of HNF production, we assume that HNF are mainly cropped by ciliate feeding. As larger zooplankton were excluded from the grazing experiments by prefiltration through 200 μm mesh gauze, and other microzooplankton, such as rotifers and developmental stages of crustacea, were of minor importance, the measured grazing impact on HNF (Figure 29.8) had

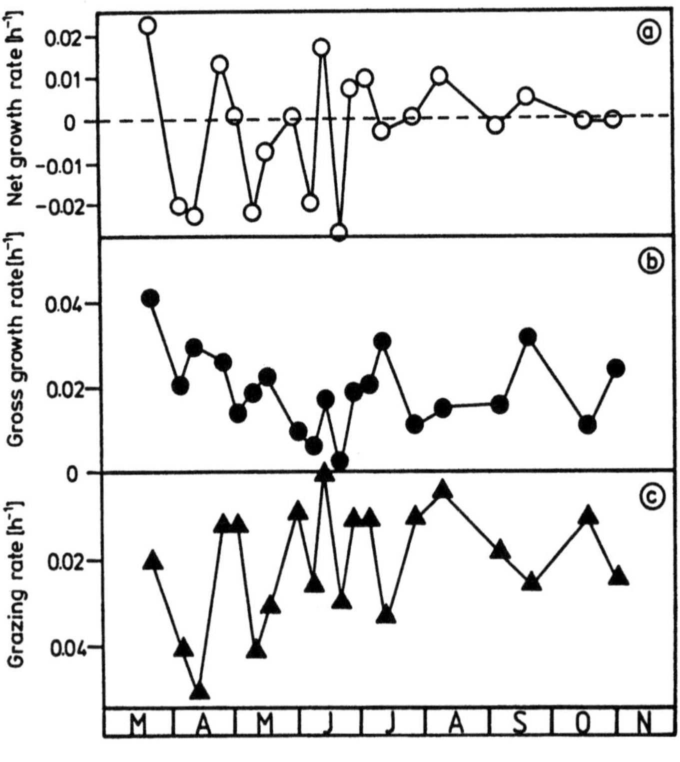


**Figure 29.7** (*a*) Net growth rate μ, (*b*) gross growth rate k of heterotrophic nanoflagellates, and (*c*) grazing impact g on heterotrophic nanoflagellates measured in the in situ experiments in Lake Constance.

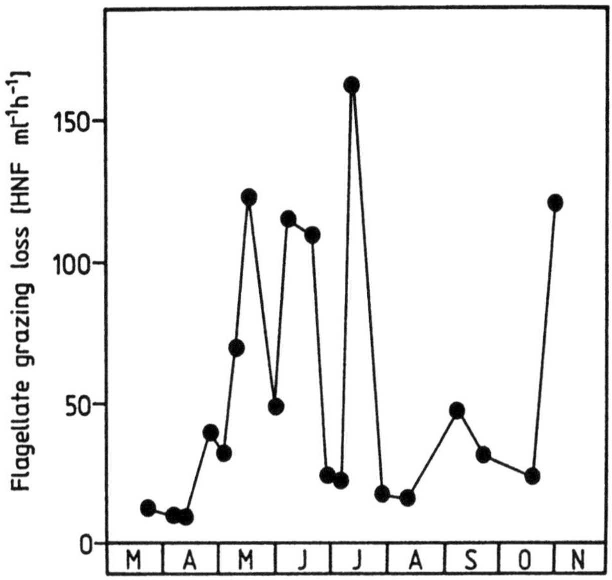


**Figure 29.8** Grazing impact on HNF populations in Lake Constance (0 to 8 m).

to be primarily caused by ciliate feeding. Accordingly, the energy transfer within the microbial loop of Lake Constance is mainly channelled through a linear food chain consisting of bacteria, HNF, and ciliates, which are themselves grazed by planktonic crustacea. Similar food chains have been found in marine waters (Rassoulzadegan and Sheldon, 1986; Sheldon et al., 1986; Sheldon and Rassoulzadegan, 1987; Weisse, 1989a). However, it is well known that ciliates also feed on phytoplankton. Furthermore, it must be considered that crustacean zooplankton might directly graze HNF and, to a lesser extent, pelagic bacteria (Güde, 1988). Therefore, the microbial loop and the classic pelagic food chain via phytoplankton, herbivorous zooplankton, and fish are ultimately linked and not mutually exclusive. Further effort is needed to clarify the obvious seasonal shifts that we observe in the relative contribution of these two major pathways to the overall energy transfer in the pelagic waters of large lakes.

## 29.6 Conclusions

HNF and ciliates occur in Lake Constance in concentrations similar to those in other large lakes and in marine waters. HNF concentrations are about three orders of magnitude lower than those of free-living bacteria and two orders of magnitude higher than ciliate concentrations. During lake stratification, protozoans are concentrated in epilimnetic waters. Although the smaller HNF are numerically most abundant, the biomass shows a consistent peak in medium-sized HNF between 3 and 6 $\mu$m. Experimental evidence, as well as trophic considerations, strongly suggest that HNF primarily feed on free-living bacteria and control bacterial population dynamics to a great extent. Rapid growth rates enable HNF to keep track with their bacterial prey. Except for food supply, HNF population dynamics are regulated mainly by grazing impact through nano- and microciliates.

The functional role of HNF in the planktonic food web primarily consists in transferring bacterial production to large protozoans such as ciliates. At the present state of knowledge, it is unknown whether other food sources in the size range of picoplankton, such as coccoid cyanobacteria, which are abundant in lakes (Craig, 1984; Caron et al., 1985; Fahnenstiel et al., 1986; Weisse, 1988; Weisse and Schweizer, 1990), and detritus particles, are also consumed efficiently. All we currently know about the microbial loop points to great structural and functional similarities between freshwater and marine pelagic systems. In order to assess the contribution of the microbial food web to the overall energy transfer, the fate of protozoan production has to be determined more accurately. So far, HNF have been defined operationally. Further effort is needed to characterize the heterogeneous assemblage of HNF taxonomically and ecologically. The same holds for the nutritional mode of ciliates, which has to be investigated experimentally.

---

*Acknowledgements* This study was supported by Deutsche Forschungsgemeinschaft within the Sonderforschungsbereich 248, "Cycling of Matter in Lake Constance." The technical assistance of G. Baldringer and A. Schöne is gratefully acknowledged. J. Barica and W. Geller provided helpful comments on the manuscript.

## References

Azam, F., Fenchel, T., Field, J.G., Gray, J.S., Meyer-Reil, L.A., and Thingstad, F. 1983. The ecological role of water-column microbes in the sea. *Marine Ecology Progress Series* 10:257–263.

Beaver, J.R. and Crisman, T.L. 1982. The trophic response of ciliated protozoa in freshwater lakes. *Limnology and Oceanography* 26:822–830.

Beaver, J.R. and Crisman, T.L. 1989. The role of ciliated protozoa in pelagic freshwater ecosystems. *Microbial Ecology* 17:111–136.

Bird, D.F. and Kalff, J. 1986. Bacterial grazing by planktonic algae. *Science* 231:493–495.

Bird, D.F. and Kalff, J. 1987. Algal phagotrophy: Regulating factors and importance relative to photosynthesis in *Dinobryon* (Chrysophyceae). *Limnology and Oceanography* 32:277–284.

Bloem, J. and Bär-Gilissen, J.M.B. 1989. Bacterial activity and protozoan grazing potential in a stratified lake. *Limnology and Oceanography* 34:297–307.

Bloem, J., Bär-Gilissen, M.J.B., and Cappenberg, T.E. 1986. Fixation, counting, and manipulation of heterotrophic and nanoflagellates. *Applied and Environmental Microbiology* 52:1266–1272.

Bloem, J., Ellenbroek, F.M., Bär-Gilissen, M.J.B., and Cappenberg, T.E. 1989. Protozoan grazing and bacterial production in stratified Lake Vechten estimated with fluorescently labeled bacteria and by thymidine incorporation. *Applied and Environmental Microbiology* 55:1787–1795.

Boraas, M.E., Nealson, K., Remsen, C.C., and Seale, D.B. 1987. Phagotrophic flagellates in Lake Michigan, in 30th Conference on Great Lakes Research, IAGLR, Ann Arbor, May 11–14, 1987 (mimeo).

Caron, D.A., Pick, F.R., and Lean, D.R.S. 1985. Chroococcoid cyanobacteria in Lake Ontario: vertical and seasonal distributions during 1982. *Journal of Phycology* 21:171–175.

Corliss, J.O. 1979. *The Ciliated Protozoa: Characterization, Classification and Guide to the Literature*. 2nd ed. Pergamon Press, London, 455 pp.

Craig, S.R. 1984. Productivity of algal picoplankton in a small meromictic lake. *Verhandlungen der Internationalen Vereinigung für Theoretische und Angewandte Limnologie* 22:351–354.

Estep, K.W., Davis, P.G., Keller, M.D., and Sieburth, J.McN. 1986. How important are oceanic algal nanoflagellates in bacterivory? *Limnology and Oceanography* 31:646–650.

Fahnenstiel, G.L., Sicko-Goad, L., Scavia, D., and Stoermer, E.F. 1986. Importance of picoplankton in Lake Superior. *Canadian Journal of Fisheries and Aquatic Sciences* 43:235–240.

Fenchel, T. 1980a. Suspension feeding in ciliated protozoa. Functional response and particle size selection. *Microbial Ecology* 6:1–11.

Fenchel, T. 1980b. Suspension feeding in ciliated protozoa. Feeding rates and their ecological significance. *Microbial Ecology* 6:13–25.

Fenchel, T. 1982a. Ecology of heterotrophic microflagellates. I. Some important forms and their functional morphology. *Marine Ecology Progress Series* 8:211–223.

Fenchel, T. 1982b. Ecology of heterotrophic microflagellates. II. Bioenergetics and growth. *Marine Ecology Progress Series* 8:225–231.

Fenchel, T. 1982c. Ecology of heterotrophic flagellates. IV. Quantitative occurrence and importance as bacterial consumers. *Marine Ecology Progress Series* 9:35–42.

Fenchel, T. 1987. *Ecology of Protozoa: The Biology of Free-Living Phagotrophic Protists*. Science Tech. Publ., Madison, WI, 197 pp.

Foissner, W., and Wilbert, N. 1979. Morphologie, Infraciliatur und Ökologie der limnischen Tintinnina: Tintinnidium fluviatile (Stein), Tintinnidium pusillum (Entz), Tintinnopsis cylindrata (Daday) und Codonella cratera (Leidy) (Ciliophora, Polyhymenophora). *Journal of Protozoology* 26:90–103.

Foissner, W., Oleksiv, I., and Müller, H. 1990. Morphology and infraciliature of some ciliates (Protozoa: Ciliophora) from stagnant waters. *Archiv für Protistenkunde* (in press).

Güde, H. 1986. Loss processes influencing growth of planktonic bacterial populations in Lake Constance. *Journal of Plankton Research* 8:795–810.

Güde, H. 1988. Direct and indirect influences of crustacean zooplankton on bacterioplankton of Lake Constance. *Hydrobiologia* 159:63–73.

Kahl, A. 1930–1935. Urtiere oder Protozoa. I: Wimpertiere oder Ciliata. In Dahl, F. (editor). *Die Tierwelt Deutschlands*. G. Fischer, Jena. parts 18, 21, 25, 30, pp. 1–886.

Landry, M.R. and Hasset, R.P. 1982. Estimating the grazing impact of marine microzooplankton. *Marine Biology* 67:283–288.

Levine, N.D. et al. 1980. A newly revised classification of the Protozoa. *Journal of Protozoology* 27:37–58.

Müller, H. 1989. The relative importance of different ciliate taxa in the pelagic food web of Lake Constance. *Microbial Ecology* 18: (in press).

Müller, H., Geller, W., and Schöne, A. 1990. Pelagic ciliates in Lake Constance: Comparison of epilimnion and hypolimnion. *Verhandlungen der Internationalen Vereinigung für Theoretische und Angewandte Limnologie* 24: (in press).

Nagata, T. 1988. Bacterial production and their contribution to the grazer food chain of Lake Biwa. This volume.

Pace, M.L. 1985. An empirical analysis of zooplankton community size structure across lake trophic gradients. *Limnology and Oceanography* 31:45–55.
Pomeroy, L.R. 1984. Significance of microorganisms in carbon and energy flow in marine ecosystems. In Klug, M.J., and Reddy, C.A. (ed.), *Current Perspectives in Microbial Ecology*, American Society for Microbiology, Washington, D.C., p. 405–411.
Porter, K.R. and Feig, Y.S. 1980. The use of DAPI for identifying and counting aquatic microflora. *Limnology and Oceanography* 25:943–948.
Porter, K.G., Sherr, E.B., Sherr, B.F., Pace, M., and Sanders, R.W. 1985. Protozoa in planktonic food webs. *Journal of Protozoology* 32:409–415.
Rassoulzadegan, F. and Sheldon, R.W. 1986. Predator-prey interactions of nanozooplankton and bacteria in an oligotrophic marine environment. *Limnology and Oceanography* 31:1010–1021.
Rivier, A., Brownlee, D.C., Sheldon, R.W., and Rassoulzadegan, F. 1986. Growth of microzooplankton: a comparative study of bacterivorous zooflagellates and ciliates. *Marine and Microbial Food Webs* 1:51–60.
Sanders, R.W. and Porter, K.G. 1986. Use of metabolic inhibitors to estimate protozooplankton grazing and bacterial production in a monomictic eutrophic lake with an anaerobic hypolimnion. *Applied and Environmental Microbiology* 52:101–107.
Scavia, D. and Laird, G. 1987. Lake Michigan bacteria dynamics and production, in 30th Conference on Great Lakes Research, IAGLR, Ann Arbor, May 11–14, 1987 (mimeo).
Seale, D.B., Bleakley, B., Boraas, M.E., and Nealson, K. 1987. Use of a bioluminescent bacterium to obtain clearance rates for Lake Michigan phagotrophic flagellates, in 30th Conference on Great Lakes Research, IAGLR, Ann Arbor, May 11–14, 1987 (mimeo).
Sheldon, R.W., Nival, P., and Rassoulzadegan, F. 1986. An experimental investigation of a flagellate-ciliate-copepod food chain with some observations relevant to the linear biomass hypothesis. *Limnology and Oceanography* 31:184–188.
Sheldon, R.W. and Rassoulzadegan, F. 1987. A method for measuring plankton production by particle counting. *Marine Microbial Food Webs* 2:29–44.
Sherr, B.F. and Sherr, E.B. 1984. Role of heterotrophic protozoa in carbon and energy flow in aquatic ecosystems. In Klug, M.J., and Reddy, C.A. (ed.), *Current Perspectives in Microbial Ecology*, American Society for Microbiology, Washington, D.C., p. 412–423.
Sherr, B.F., Sherr, E.B., Andrew, T.L., Fallon, R.D., and Newell, S.Y. 1986. Trophic interactions between heterotrophic protozoa and bacterioplankton in estuarine water analyzed with selective metabolic inhibitors. *Marine Ecology Progress Series* 32:169–179.
Sherr, B.F., Sherr, E.B., and Newell, S.Y. 1984. Abundance and productivity of heterotrophic nanoplankton in Georgia coastal waters. *Journal of Plankton Research* 6:195–202.
Sherr, E.B. and Sherr, B.F. 1987. High rates of consumption of bacteria by pelagic ciliates. *Nature* 325:710–711.
Sherr, E.B., Sherr, B.F., and Paffenhöfer, G.-A. 1986. Phagotrophic protozoa as food for metazoans: a 'missing' trophic link in marine pelagic food webs? *Marine Microbial Food Webs* 1:61–80.
Simon, M. 1988. Growth characteristics of small and large free-living and attached bacteria in Lake Constance. *Microbial Ecology* 15:151–163.
Skogstad, A., Granskog, L., and Klaveness, D. 1987. Growth of freshwater ciliates offered planktonic algae as food. *Journal of Plankton Research* 9:503–512.
Stoecker, D.K., Cucci, T.L., Hulburth, E.M., and Yentsch, C.M. 1986. Selective feeding by Balanion sp., Ciliata, Balanoidae, on phytoplankton that best support its growth. *Journal of Experimental Marine Biology and Ecology* 95:113–130.
Taylor, W.D. and Heynen, M.L. 1987. Seasonal and Vertical Distribution of Ciliophora in Lake Ontario. *Canadian Journal of Fisheries and Aquatic Sciences* 44:2185–2191.

Utermöhl, H. 1958. Zur Vervollkomnung der quantitativen Phytoplankton-Methodik. *Mitteilungen der IVL* 9:1–38.

Weisse, T. 1988. Dynamics of autotrophic picoplankton in Lake Constance. *Journal of Plankton Research* 10:1179–1188.

Weisse, T. 1989a. The microbial loop in the Red Sea: dynamics of pelagic bacteria and heterotrophic nanoflagellates. *Marine Ecology Progress Series* 55:241–250.

Weisse, T. 1989b. Trophic interactions among heterotrophic microplankton, nanoplankton, and bacteria in Lake Constance (FRG). *Hydrobiologia Dev. Hydrobiol.* (in press).

Weisse, T. and Schweizer, A. 1990. Seasonal and interannual variation of autotrophic picoplankton in a large prealpine lake (Lake Constance). *Verhandlungen der internationalen Vereinigung für Theoretische und Angewandte Limnologie* 24: (in press).

Williams, P.J.leB. 1981. Incorporation of microheterotrophic processes into the classical paradigm of the planktonic food web. *Kieler Meeresforschungen,* Sonderheft 5:1–28.

# 30

# On Pelagic Food Web Interactions in Large Water Bodies

Karl Banse

ABSTRACT Principles and observations relevant to the understanding of abundance (concentration) of the common species of freshwater and marine plankton are reviewed, beginning with A.C. Hardy's food web centered round North Sea herring, R.L. Lindeman's pyramid of energy, and G.A. Riley's model of the seasonal plankton cycle on Georges Bank. Isotope enrichment data suggest that preferred strands persist in food webs for appreciable periods, so that J.D. Isaac's unstructured webs may not be the rule, at least not among organisms of the size of copepods and larger. Because neither Hardy's nor Lindeman's approaches are workable, although for different reasons, it is suggested that future field and laboratory work might be organized increasingly around mathematical models. The models should combine allometric rules with species-specific information, rather than be based on pure biomass spectra, and should not rely on fixed growth or ecological efficiencies for the species. As these models of abundance go beyond current multispecies management approaches in fisheries, they obviously are long-term goals; moreover, in large water bodies, they will run into validation problems. For the time being, therefore, replicated experiments in enclosed brackish or freshwater mesocosms seem to be promising tools for studying food web interactions, including population dynamics and mass balances, in complex webs, even though such experiments present formidable problems of scaling.

## 30.1 Introduction

One hundred years ago, V. Hensen (1887) published his pace-setting paper, "On the Determination of the Plankton, i.e., the Plants and Animals Drifting in the Sea," that introduced to ecology the counting of organisms collected under a known area or in a known volume. The reason for these studies was his belief that successful management of fisheries would be possible only when the productivity of the sea, i.e., the production of organic matter by plants and animals, was known and understood. Hensen, as the physiologist of the Faculty of Medicine at the University of Kiel, thought correctly in terms of rates and aimed ultimately at an understanding of the metabolism, or a physiology, of the sea as a whole. Yet, the time was not ripe for rigorous, process-oriented work. Even now, one hundred years later, we are unable, by using our understanding of the various links instead of applying empirically based transfer efficiencies, to calculate production of carnivores from the temporal distribution or the rate of production of phytoplankton. However, much of Hensen's original rationale still pertains.

This paper will review major advances in quantitative investigations of food webs in large water bodies, be they marine or freshwater; highlight principles and concepts; and comment on questions that need to be addressed before the goal of a solid biological understanding of large water bodies is attained. The treatment will center on the pelagic realm, including plankton and nekton. The premise and presumption of the paper are that, in the years and decades to come, the explanation and prediction of abundance (concentration) of, at least, the common pelagic species will be the main issues, rather than the explanation of presence and absence of species. Studies aimed at such understanding will have to concentrate on the rates of gain and loss that result in a particular abundance or biomass. Hopefully, they will also elucidate the year-to-year variations in abundance of the common species already in focus in fisheries.

Because my experience is with the marine realm, I will refer more often to marine than freshwater examples, such that the limnological literature may appear to be slighted. However, it is ironic that I conclude the review by suggesting that major progress in understanding pelagic food web interactions in the seas and lakes can be expected from work in small volumes of brackish or freshwater, thus slighting marine as well as large lake interests.

## 30.2 Development of the Current Understanding

Several points will be made in this section about the development of our understanding since Hensen's studies, beginning with concepts introduced or formalized by the benchmark papers by Hardy (1924), Lindeman (1942), and Riley (1946). First however, large water bodies will be defined and the omission of the benthos will be justified. The physical definition of size is relatively straightforward, i.e., a basin is large if it has appreciable horizontal motion, and the effect of the earth's rotation becomes important. At intermediate latitudes, the Coriolis force is not negligible in basins larger than about 5 km in diameter (Harris, 1986, p. 53). Considered from a chemical or biological viewpoint, however, the domain of interest is an offshore region where the fluxes from internal cycling are larger than the horizontal or vertical fluxes. On the premise that most mass flux or cycling per unit volume and time will proceed through the smallest organisms, with realized generation times of about a day or a little less, large water bodies may be defined here as well-stratified and at least 1 to 10 km in diameter. This range estimate is based on scaling arguments by Wroblewski (1983) and literature values for concentrations of plankton in freshwater, eddy diffusivity, Michaelis-Menten kinetics underlying algal growth, and grazing losses. By implication, the bottom of such large systems (lakes, or seas on broad continental shelves) is negligible for offshore processes on this short time scale. Obviously, for organisms with generation times appreciably longer than 1 d, basins would need to be larger, the diameter increasing roughly with the square root of the turnover or generation time. To complicate matters further, the horizontal eddy diffusivity coefficient in a given basin may be time dependent, not only because of direct effects of changing winds on the center of the basin but also through the occasional breakup of coastal fronts (e.g., the 'thermal bar' in Harris 1986). Moreover, in large lakes, for intervals larger than microbial generation times, advection tends to be more important for the distribution of material than diffusion (Imboden and Schwarzenbach, 1985), so that the scaling used here is quite approximate. Finally, on a seasonal basis (during winter), the bottom effects certainly will be felt, except in meromictic lakes. Thus, there is much grey area regarding the applicability of the attribute "large" to lakes. Similarly, the treatment of a seasonally stratifying shelf sea may neglect the sea bed during the summer, but must take the benthos into account during the other seasons.

Hardy (1924; Figure 30.1) illustrated the feeding relations of the planktivorous North Sea herring for ten months in each of two years studied. He was not, however, in a position to weigh the importance of the ties, in this respect remaining in Hensen's situation. Note that for the animals, a food web, rather than a number of trophic levels, was depicted. However, Wyatt (1976) pointed out that, at a particular time, the web leading to adult herring is actually far less complex than shown, because, for example, in the North Sea they feed during two or three spring months almost exclusively on juvenile sand eels (*Ammodytes* sp.), neglecting the other nine illustrated food items.

Can the herring data be generalized? Simple webs or actual chains should lead to biomagnification, whereas a food web where, within the constraints in size, everyone eats everything (the so-called unstructured food web, Isaacs, 1972; Fasham, 1981), should not. Studies of the Cs/K ratios in muscle tissue of pelagic invertebrates, sharks, fishes, and mammals off California, as well as in the eastern tropical Pacific (Mearns et al., 1981) did not bear out earlier data from the Gulf of California, which had suggested an unstructured web. Isotope ratios plotted against trophic position deduced from stomach contents of the same specimens showed roughly the same increase as expected from the mean residence times of the two elements in tissues; this was in contrast to lesser biomagnification in benthic food webs off California (see Thomann, 1981 about freshwater data, and the critique by Vézina, 1986).

Biomagnification was observed also for $^{13}C/^{12}C$ and $^{15}N/^{14}N$ ratios in pelagic animals. The increase of ratios between groups of species assigned to adjoining trophic levels often was not greatly different from that found in laboratory feeding experiments (carbon: Rau et al., 1983; Fry and Sherr, 1984;

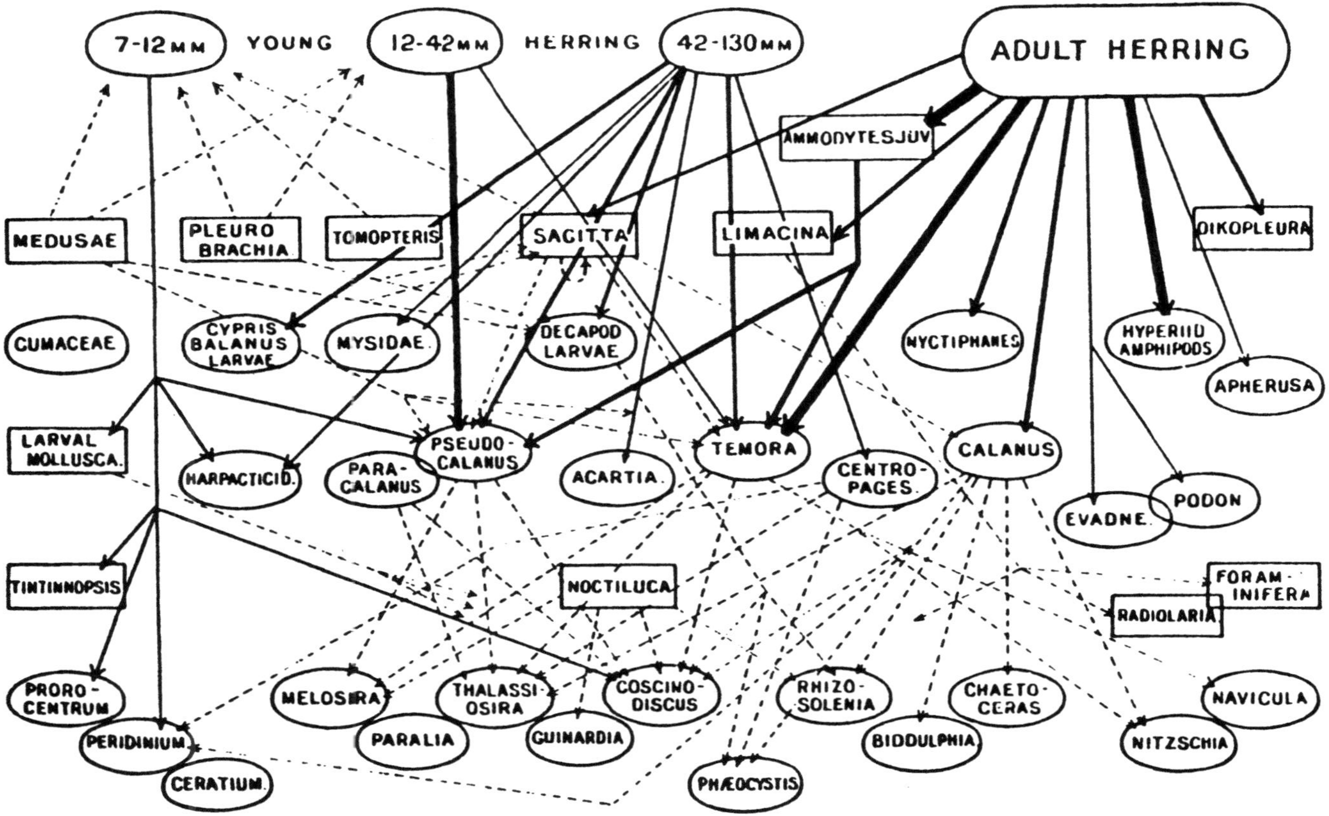


**Figure 30.1** Overview of the feeding relations of young and adult herring in the North Sea. Full and broken lines based on research by A.C. Hardy and others, respectively. Note that bacteria, nanophyto- and nanozooplankton, and juveniles of invertebrates are omitted, and that the food of many of the herring's prey organisms has not been indicated. (Reprinted from Hardy, 1924).

nitrogen: Pang and Nriagu, 1977; Minagawa and Wada, 1984; Mullin et al., 1984). Excepting the seston misnamed as phytoplankton, copepods were the smallest class considered in the previously cited papers. It is possible, however, that different size ranges of organisms follow different rules. Using $^{32}P$ in lakewater enclosures with organisms from bacteria up to small copepods, Vézina (1986) found that uptake, metabolism, and turnover of the element in the various size fractions was better described by assuming an unstructured web. Presently, we lack natural history data to judge such differences if they prove to be general. Although suspension-feeding crustaceans can differentiate among, and select from particles of quite similar sizes during feeding (Huntley et al., 1983; DeMott, 1986), the question for isotope enrichment and food web structure is how often the animals either avail themselves of this capability or simply track the natural spectrum of food (cf. Boyd, 1976, for copepods; Fryer, 1987 for cladocerans). Small protozoa, which may bump into and then ingest particles one by one, could on that account be quite selective (Stoecker et al., 1981, 1986; Skogstad et al., 1987); but then, microflagellates also may indiscriminately ingest microspheres (Sieracki et al., 1987), as do copepods. In conclusion, rather simple webs or even food chains may pertain for many plankton and nekton species during most periods, whereas the appearance of a complex web may be due to our temporal (and often, geographical) averaging. The difficulty will be to predict the time of switching to another principal food item.

In addition to illustrating a food web rather than a food chain, Hardy (1924) depicted both the larval and juvenile herring occupying places in the web that differ from those of the adults because of the great ontogenetic size range of the species. A similar change of mass during development holds for most invertebrates, and therefore, the small and large stages need to be discriminated as their functional role (e.g., suspension versus raptorial feeding) or even habitat (plankton versus benthos) may change with age. Because we lack information on many of these features of natural history and the applicable rates, Hardy's, and by implication, Hensen's, conceptual model are at present unable to make useful quantitative statements about the pelagic realm.

Whereas Hardy (1924) had presented the naturalist's realistic picture of a food web, Lindeman (1942) tilted to almost total abstraction with a pyramid of energy. He formalized earlier notions of organisms acting on trophic levels into the so-called "trophic-dynamic" concept and at the same time (following unpublished notes of G.E. Hutchinson) cast the concept in terms of flow of energy rather than of organic matter. In contrast to the then prevailing ecological classifications, he grouped organisms within an ecosystem as [primary] producers, primary consumers, secondary consumers, etc., with the classification admittedly becoming looser for members of the higher levels. Each level depended on the production by the level below it and degraded less energy than the one below it so that a pyramid of energy flow resulted (Figure 30.2).

The "endearing simplicity" of the trophic-level concept (Ivlev, 1945, thereby intending to criticize Hutchinson more than Lindeman) made the idea

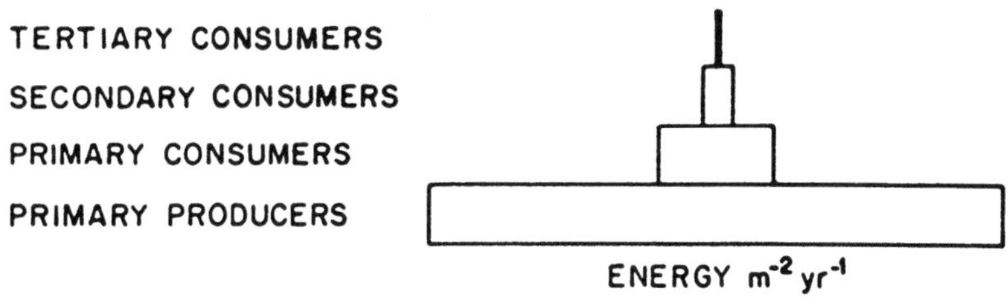


**Figure 30.2** A pyramid of energy flow (arbitrary units) in a community, with the degraders not designated. (Based on Lindeman, 1942).

easy to grasp and enter textbooks as well as minds, in spite of Lindeman (1942) himself having illustrated the bypassing of trophic levels and of early formal criticism. Ivlev (1945) and later authors noted the common difficulty, or ambiguity, of assigning a species to a particular trophic level. In the plankton, for example, an adult, "herbivorous", i.e., suspension-feeding, copepod also might consume suspension-feeding ciliates and its own offspring, as well as larvae of its predator. It is easily shown that relatively little animal matter added to the principal plant food greatly lowers the efficiency (copepod production/total plant production) of this food web. Therefore, attempts to calculate, for example, world fish production, using essentially the trophic-dynamic scheme (Ryther, 1969) have not met with kind reception (Alverson et al., 1970). The most cogent criticism against the approach, however, was levied in another context by Murdoch (1966; see also Rigler, 1975; Cousins, 1985b). Because many animals and the facultatively heterotrophic algae (Bird and Kalff, 1986) cannot be assigned to one trophic level, these levels in an ecosystem (e.g., the primary carnivores as a group) are abstractions and do not possess measurable properties, in contrast to the constituent populations. Therefore, a theory, meant to make predictions which is based on this concept, is not amenable to rigorous testing, hence cannot be disproven, and thus should not be employed (for Lindeman's lasting contribution, see Cook, 1977).

The almost contemporary paper by Riley (1946) on the phytoplankton of Georges Bank, off New England, cast existing concepts of phytoplankton dynamics and new experimental and field data into a differential equation that described the seasonal rate of algal population change (in units of C) as the balance between the environmentally influenced photosynthetic rate and rates of respiration and grazing. The calculations suggested that grazing is the principal loss term during long periods of the year, a lesson too often neglected even at the present time—the population dynamics of phytoplankton cannot be understood from light, nutrients, sinking, and mixing alone. The theoretical section containing the model was preceded by a statistical treatment, using correlations between observations and environmental factors to demonstrate that this approach is quite limited in making progress toward a broadly applicable theory. The model, relatively simple (and transparent!), as it may appear to us now, was subsequently found to be broadly applicable. For example, with only slight changes of the values of coefficients, it reproduced differences between two successive years in the shallow harbor of Pusan, Korea (Riley and van Arx, 1949; see also Riley, 1947, on the zooplankton of Georges Bank). The importance of Riley's (1946) paper to the development of plankton ecology can hardly be overstated (see introductory essays in Wroblewski, 1982). The quantitative understanding of abundances by biomass or specimen numbers of a species or larger category, or of the temporal change of size structure in a cohort, require mathematical models.

The four decades following Riley's (1946) paper, of course, experienced the introduction of better methods for determining concentrations, especially for microbes, gelatinous zooplankton, and fish biomass, provided numerous rate measurements for the laboratory and field, and yielded values for the production by field populations, especially among freshwater plankton, aquatic benthos, and on land. The rate measurements supplied the data for establishing a number of allometric relations of rates (y) with body mass (x) of the form $y = a\ x^b$ or $y/x = a\ x^{b-1}$, which contributed to the development of food web theory based on body size or mass rather than taxonomic relation (see further below). These four decades also witnessed the creation of many process models, a development that accelerated after the advent of computers, which facilitated numerical calculations and experiments.

Allometric rules entered food web studies by process models directed at mass balances for more or less specific sites and by providing the basis for the development of general ecological theory. The models of marine ecosystem structure by Steele (1974) and on the size structure of plankton communities by Steele and Frost (1977) may serve as examples for general (i.e., broadly applicable) process models. Both hold in principle for freshwater systems and will be referred to later. Neither these nor the following models included spatial components, in contrast to the fishery model by Laevastu and Larkins (1981) to be discussed later. An example for a realistic (i.e., directed at particular sites) freshwater model is the food web representation for the pelagic domain of Lake Michigan by Canale et al. (1976), comprising eleven abiotic and fourteen biotic compartments, which in part define actual species and their attributes. Similar marine models are the sequence of descriptions of tropical pelagic ecosystems by M. Ye. Vinogradov and coworkers (Vinogradov et al., 1976) and the idealized temperate shelf ecosystem by Pace et al. (1984). The latter model comprises three abiotic and fourteen biotic compartments, each of the latter involving broad groups like "nano- (phyto-) plankton" or "mucus net feeders", which are distinguished by function, body mass, and habitat. If the mucus net feeders are omitted, the model should be applicable to large lakes. The seventy-three fluxes among compartments are modelled by differential equations, specifying, for instance, food preference or uptake, and the coefficients are taken from the literature. The model, which also calculates concentrations in, as well as production by, compartments, is used to reproduce time- and depth-averaged pictures for exploring seasonal cycles in four shelf systems. Essentially, the correctness of the model is evaluated by its general agreement with published data, as did Vinogradov and colleagues, rather than by a statistical test. Among the specific results of Vinogradov et al. (1976) and Pace et al. (1984) is the finding that the ecological efficiencies (rates of production/input for a compartment) that were an issue of great concern to the trophic-dynamic school are highly variable between simulations for all compartments (note that compartments do not equal trophic levels).

This variability among efficiencies is known from laboratory experiments (Winberg, 1971: Figure 5.9), annual budgets in field populations (Banse 1979), and trophic-level summaries (see Morgan, 1980, Table 6.3, for zooplankton data obtained with presumably uniform methods). The trophic-dynamic literature notes that the ecological efficiency of the invertebrate-plankton-predators level tends to be higher than that of the suspension feeders, but the trend is far from having any statistical significance (herein checked for Morgan's table; see also Humphrey's, 1979; Brylinski, 1980, Table 9.13). Because of this variability, it is inadvisable to estimate fluxes from fixed values of, for example, net growth efficiencies for energy budgets. Conversely, efficiencies, which are ratios among rates or fluxes should not be derived by comparing compartments (i.e., concentrations). Steele and Mullin (1977) showed by a simple set of differential equations describing the rates of change of concentrations that conclusions by Blackburn (1973) about changes of food chain efficiency with altered nutrient enrichment, mainly based on correlations and ratios between biomasses on successive trophic levels, were incorrect.

The development of general ecological theory in the context of this paper was largely reviewed by Platt (1985). The present focus of effort is directed at understanding the rules about, and the reasons for, the distribution of biomass over classes of body (cell) mass in a community, the so-called biomass spectrum (Peters, 1983; Platt, 1985). In the pelagic realm of the open sea, biomass tends to decline only slightly over logarithmically increasing classes of body mass and even less so when based on wet weights or volumes rather than C as mass measures (Rodriquez and Mullin, 1986), and is without peaks except when there are blooms of algae. The underlying mechanism is the regularity of the dependence on body mass of (i) individual growth rate and population production and of (ii) turnover or generation time. The same kind of spectrum pertains also to lakes (Sprules and Munawar, 1986) and marine coastal waters, but there appear to be more irregularities than in the open sea. As an example of a generalization that may be based on mass (size) spectra, Sprules and Munawar (1986) noted the trend for eutrophic temperate lakes to have over an order of magnitude more specimens of large zooplankton supported by about the same amount of small algae. In a spectrum where concentrations are normalized by dividing by the body mass of the particular mass class, thus plotting approximate specimen numbers on body mass, C.S. Elton's pyramid of numbers (many small, few large animals) is quantified (Rodriquez and Mullin, 1986).

Setting out from such mass spectra, largely taken from situations close to temporal equilibrium, Silvert and Platt (1980) formulated a general, nonlinear, time-dependent equation defining the exchange of energy in an ecosystem described by the body mass of its organisms rather than trophic levels. It allows feedbacks, such as reproduction (i.e., production of small particles from large ones) but, so far, a general time-dependent solution is not known (W. Silvert, personal communication). Also, the predicted biomass spectrum apparently has not been checked by field observations for the large size classes, i.e., the fishes and squids, for want of assessment methods. However, from very preliminary data it appears to me the old notion of a small fish/zooplankton equilibrium ratio (as, e.g., in Blackburn, 1981, Table 1, from midwater trawls), which is mirrored by the biomass spectrum theory, may not be correct, but the biomass in the mass classes for fish may approach or even exceed that of their prey. Examples are the comparison of midwater trawl-derived micronekton concentrations with about one hundred-fold higher acoustical assessment off northwest Africa (Blackburn, 1977; Mathisen et al., 1978), and new, not as dramatic observations, off California (P. Smith, La Jolla, California, personal communication).

Dickie et al. (1987), who reviewed recent developments in the formulation of body mass dependence, emphasized that, whereas mass-specific physiological processes scale by the usual $-0.25$ power of body mass, mass-specific processes and rates on the population level, such as predation or production scale by a more negative exponent (e.g., $-0.37$ expected for cumulative annual respiration/mean biomass [observed, $-0.35$, Banse, 1979]). Dickie et al. (1987) suggested that an ecological scaling factor, which is additional to that known from physiology, is related to the relative sizes and distributions of predator and prey. This greater exponent becomes apparent when less than the entire animal kingdom is considered in one equation and smaller taxa, such as fishes, or ecological groups, such as meiofauna, are being studied. McGurk's (1987) and Gulland's (1987) papers on mortality rates in larval plankton also proposed a change of coefficients and exponents upon using smaller (and, in my view, more biological) groupings. Incorporation of information of this kind in theories like that of Silvert and Platt (1980) may predict an equilibrium biomass of fishes and other large cold-blooded animals in pelagic ecosystems that is appreciable relative to that of the smaller size classes.

All of the above discussion neglects the role of size- (not mass-) selective predation on species composition and production, and the structuring of the food web from the top down. Limnologists have contributed most of the knowledge about predation effects on specific composition of aquatic habitats. The principal reason, I believe, is that sites and populations can be revisited. Although the interplay of body mass and growth efficiency was alluded to from the outset of work on predation (for example, Brooks and Dodson, 1965), the effect, if any, on production rates and biomass achieved by an assemblage was rarely addressed beyond noting that large cladocerans tended to enhance water clarity by reducing algal concentrations. However, the very first set of papers in the genre (Hrbáček, 1958; 1962) already reported that the presence or absence of at least partially planktivorous fish did not significantly affect the mean biomass of the cladocerans collected. More recent examples of theoretical and experimental work on predation effects among pelagic producers and consumers were presented by Lynch (1979), Kerfoot and DeMott (1984), Vanni (1986, 1987), and Walters et al. (1987). Other articles treating at least plankton biomass as being dependent on fish predation were mentioned by Carpenter et al. (1985; see also Carpenter et al., 1987), and the indirect effects of fish and other predation on plankton were discussed in several papers in Kerfoot and Sih (1987; see also review by Lazzaro, 1987). A formal treatment of such experiments in terms of the biomass spectrum theory might be interesting.

The sketched approaches centered around allometry have provided quantitative theory rather than mere concepts about food web interactions. Moreover, they are cast in terms of measurable variables. So far, however, only general comparisons with field observations have been made, and it is yet to be seen how far we can go without incorporating information on at least the dominant species in a particular community. The views about future prospects vary. Steel and Frost (1977) ended their large paper that modelled the size structure of phytoplankton and planktonic suspension feeders by surmising that "prediction of certain features of population structure for parts of the food web may be a more attainable objective of theory and more practically important, than attempts to predict biomass." Dickie et al. (1987), on the other hand, expressed guarded optimism that calculation of average fish production from zooplankton production may be justifiable without assigning position in the food web, when the calculation is based on understanding of body mass relations and the associated ecological efficiencies. In fact, they seem to suggest that the calculation would not require detailed information on species composition. Also, in contrast to Steele and Frost (1977), B.W. Frost (personal communication) now believes that estimates of biomass and abundance, rather than merely stage composition, are needed and will be attainable in the not-too-distant future.

## 30.3 Remarks on the Current Understanding

Where do we stand, more than 100 years after Hensen's (1887), more than 60 years after Hardy's (1924), and 40 years after Riley's (1946; 1947) studies? On the one hand, we still are making fundamental discoveries and rediscoveries, e.g., of the very small autotrophs, the so-called picoplankton; the heterotrophic microbial loop; and the marine gelatinous zooplankton, and do not know how much fish biomass abounds in the sea and even in many lakes. On the other hand, we have developed quite comprehensive models on the functioning of ecosystems that are in some agreement with reality, and we are attempting multispecies management models in fisheries. Being somewhat on the sidelines both of field research and mathematical modeling, I am not in a position to provide an even reasonable comprehensive review of the current understanding, quite apart from the space constraint. Instead, I will offer remarks on some approaches, toward the goal of better explaining and later predicting abundance and biomass.

Because it was indicated earlier that both Hensen's (or Hardy's), and Lindeman's conceptual models are unworkable, a middle course is to be looked for. In my view, it should follow Riley's example of constructing process models, but drive toward more realism by allowing for size structure and the recognition of some species. Comprehensive ecosystem models of this kind may by themselves have limited explanatory power (Pielou, 1981). However, they seem to be the only way to summarize our knowledge rigorously; they also can pinpoint important processes, as well as data gaps, and thus guide further research. In any case, because very large ecosystem models of the type sketched below are some time away, I will suggest that intermediate steps in experimental field work might be taken now by using small and biologically simple systems, such as small lakes or artificial fresh or marine enclosures (mesocosms). Here, functional relations can be investigated now, and existing models and submodels can be tested for accuracy by mass balances, before the issues posed by large spatial and temporal scales are tackled.

It is worth noting for the reader with a practical bent, who may need to make estimates of animal production now, that the precision of biomass or production values calculated even by excellent ecosystem models will for several reasons always be limited, for example, because of the uncertainty about the initial state of the system and the resulting uncertainty of prediction, which is analogous to numerical weather forecasting. The precision will probably not be better than the empirical relations already at hand, such as the multiple regression estimates for secondary production in temperate lakes (Brylinski, 1980) or the correlation of fish yield with primary production data in tropical lakes (Melack, 1976). The coefficients of determination ($r^2$) of the regressions in these papers account for about half of the variance. The difference between correlation now and accurate modeling later, however, will be the ability of the latter approach to extrapolate to new conditions, by having incorporated true understanding.

In my view, the structure-caused variability in food webs mentioned earlier will diminish the applicability of the theory developed around the biomass spectrum. Additional conceptual doubts about any but a very general food web theory, based almost purely on mass (size) and allometry, arise from consideration of (i) the role of certain functional groups, (ii) questions about the physiological universality of allometric relations, and (iii) species-specific differences when they pertain to dominant forms. First, examples of functional relations needed in realistic (as opposed to general) models of marine pelagic food webs are the millimeter-long larvaceans (appendicularians) and some of the centimeter-long salps that employ mucus nets and feed on bacteria, thus short-circuiting the normal microbial loop from bacteria to zooflagellates to ciliates, etc. Similarly, baleen whales feed on relatively very small prey, such as large copepods, bypassing the small fish feeding on these copepods and being eaten by large fish or squids that are the prey of toothed whales. These peculiarities can, of course, be built into an allometric model but deprive it of its generality. Second, over a broad range of body mass that is central for zooplankton ecology, size alone is a poor predictor of rates of respiration and of intrinsic growth among major taxonomic groups. The intrinsic growth rate among the freshwater cladocerans is independent of adult body mass (Banse, 1982a), again reducing the generality of the available theory. Further, within functional groups, such as "crustacean suspension feeders," unrelated species of similar body mass may exhibit quite different thresholds and rates, e.g., a *Eudiaptomus* sp. and a *Daphnia* sp. studied by Lampert and Muck (1985). Third, Frost (1980) noted that for two dominant, related marine pelagic copepods (*Pseudocalanus* sp., *Calanus* sp.), with very similar feeding habits, accurate modeling of their relative abundance depended on using some characteristics that did not follow allometric rules, such as an increase of specific growth rate with adult body size. Thus, for introducing some realism into quantitative formulations of our ecological understanding, it will be necessary to combine functional groupings, such as guilds, or even species, with the allometric equations required among these groups as well as within, to account for growth, etc.

In view of these difficulties and the need to test our theories on real systems, models with some built-in realism are needed—body mass alone is not sufficient. Paraphrasing Cousins (1985a; see also Cousins, 1985b, for a more quantitative treatment), we must think of pyramids again, with many small and few large specimens in the community. These pyramids can be neither as abstract as Lindeman's (1942) nor as realistic as Hardy (1924) would have constructed them, and incorporate actual species, or at least functionally defined, relatively small groups (about the latter, cf. Sprules and Holtby, 1979; Bahr, 1982). The new models must combine mass (energy) flow with population dynamics and incorporate the functionally oriented approaches and aspects of community structure, thereby recognizing the functional limitations affecting the processes as well as the structural, and sometimes spatial, constraints affecting the populations (cf. O'Neill et al., 1986).

A recent overview of large freshwater models was provided by Straškaba and Gnauck (1985). A number of marine and freshwater models incorporating a few species already exist. For multispecies fisheries management, the first steps have been taken in the direction of large-scale, realistic ecosystem models of the marine pelagic domain that demand physiological and ecological understanding of abundance, biomass, and production on the species level. For example, Andersen and Ursin (1977) published a biologically involved model for the North Sea with eleven fish species and nine other biotic compartments but without spatial components. Laevastu and Larkins' (1981) model for parts of the Gulf of Alaska and the Bering Sea, with thirteen fish species or groups and seven other biotic compartments, was biologically less involved (for example, growth coefficients were fixed) but it allowed geographic differences and horizontal migrations. Comments on the multispecies management aspect of such efforts were provided by Larkin (1984). More broadly speaking, the fisheries experience demonstrates the enormous difficulties ahead for large ecosystem models that incorporate species-related information: Even in a well-studied region like the North Sea, the basic reasons for the often huge fluctuations in abundance of common species (year-class variability) are not yet predictable, and therefore are not fully understood (cf. Steele, 1984; Beddington, 1986), in spite of essentially the entire geographic range of the commercially utilized species being sampled every year with comparable methods, and in spite of fishes of the temperate zone being among the best-studied of all cold-blooded aquatic animals in regard to ontogeny, physiology, population dynamics, etc. To realize the magnitude of the task, one may contemplate the interaction of fish predation with an actual suite of zooplankton and algae in freshwater (referred to earlier under the subject of size-selective predation), the trends in the pattern of algal succession on a gradient from oligotrophic to eutrophic deep lakes (Sommer, 1986), or the marine model by Steele and Frost (1977), and then combine the modeling of these processes with spatial gradients of species distributions, caused by predation, between shallow and deep parts of a lake (for example, Jakobsen and Johnsen, 1987).

Of course, one has to be practical about the idea of creating a comprehensive ecosystem model that includes some taxonomy of groups smaller than fishes. By and large, we do not have the physiological data for tackling the marine plankton at the specific level, although it has been repeatedly stated that more and better rate measurements are of paramount importance for better modeling (for example, Platt, 1981; Longhurst, 1984); rate measurements are needed to connect model compartments and also as tools for observational strategy (Harris, 1986; O'Neill et al., 1986). We may even lack field observations, such as those on seasonal aspects of vertical distribution of ordinary net-collected zooplankton (Longhurst and Williams, 1979). Further, there may be holistic properties of ecosystems, such as resilience or multiple equilibria, that may or may not be amenable to the reductionist approach of the present mathematical models (cf. Paloheimo et al., 1984, for lakes; Allen, 1985, and O'Neill et al., 1986, for general treatments). Finally, there are mundane problems, such as the difficulties of rigorous validation, that need to be overcome. On the positive side, our colleagues in physical limnology and oceanography, with their more easily processed observations and encouraged by advancing theory and superfast computers, design very fine-gridded circulation models that will provide the environmental underpinning for ecosystem models with spatial components.

Before addressing the validation issue, a point that has been stated before by others bears repeating. The comprehensive ecosystem models involving species will be very large, so that their intricacies most likely can be appreciated fully only by those who work with them. In contrast, a small, transparent model can do much to illuminate principles to a broad audience. Recent marine examples are papers by Steele and Henderson (1984) on long-term, seemingly aperiodic fluctuations of fish stocks and by Evans and Parslow (1985) on the minimum conditions needed for the occurrence or absence of the phytoplankton spring bloom in large water bodies. Authors of large models, therefore, owe their readers a sensitivity analysis, as otherwise, the effect of the known variability of a forcing variable is hard or even impossible to guess.

Turning to the validation problem, mathematical models in effect present hypotheses and must, hence, meet the criterion, raised in connection with the trophic-dynamic approach, of being refutable at least in principle. The straightforward test of one of the present models of plankton dynamics that do not incorporate algal species is to measure the initial state, such as the chlorophyll concentration, then calculate a future state on the basis of environmental data, and compare prediction with new observations. Usually, the comparison is qualitative, and general agreement is looked for in timing and amplitude of the occurrence of a variable. For example, Steele (1974) utilized the ability to create three generations per year of an idealized, "herbivorous" copepod as an essential feature for the validation, rather than the capacity to reproduce faithfully the copepod biomass. By applying such models to different regimes and continuing to find reasonable agreement, confidence in the validity of the constructs is enhanced (cf. the mentioned marine work by Vinogradov et al., 1976, and Pace et al., 1984; Jørgensen et al., 1986, for a freshwater eutrophication model). Even so, there may be doubt as to whether a particular, accurate balance (for example, of biomass) results from both calculated production and predation rates being either too high or too low, with only the difference being accurate; direct measurements of in situ rates of production or mortality may not be possible for animals of millimeter- or centimeter size or are inherently not feasible because the entity chosen for a compartment, such as small, planktonic predators, as a whole may not possess measurable properties that yield such rates. In large water bodies, additional difficulties arise, such as the uncertainty about the initial state, obtained from sampling patchy distributions, that limits the extent and quality of the prediction. Moreover, horizontal advection, plus patchiness of distributions of chlorophyll, etc. will present an almost unsurmountable problem for deciding which deviation between prediction and observations of the future state proves failure of the model (Banse, 1982b). Therefore, some of the large models in principle might not be directly testable by biomass or abundance in large water bodies.

Testing, however, might be possible in enclosures or large aquaria. The cited models did not include a stochastic component to simulate the role of patchiness in large water bodies, but were essentially of a deterministic nature. Large, stirred aquaria may be regarded as deterministic, living models of the open water, in which entire mathematical models or their microbial and copepod components could, and should, be usefully tested. For example, one might borrow from the essentially physiological model of Steele and Frost (1977) that investigated the food web (size) structure in the photic zone during temperate spring-to-summer cycles and involved an assemblage of twenty size (mass) classes of phytoplankton (in place of species) facing three herbivorous copepods. For example, filtered water in a controlled environment might be stocked with four spineless species (strains) of single-celled algae of the same shape but differing size and three suspension-feeding copepods, with unambiguous plankton nets being substituted for their predators. Such gnotobiotic aquaria would have the four following advantages over field experiments: (i) replication, normally missing in the field, is easy; (ii) advection is excluded and the same specimens are being revisited; (iii) mass balances comprising the dissolved and the particular phases permit budgets (the terms must add up to 100%); (iv) plant physiology and animal feeding will operate on the same algal cell diameters; and (v) beyond the immediate purpose of testing the Steele and Frost model, the addition of isotopes allows determination of specific activities and opens additional avenues of attack. Control over the quality (shape) of the algae would be the key difference between aquarium and field manipulations [Note that in the field, physiological size of the cells, such as the common colonial diatom genera *Chaetoceros* or *Asterionella*, and their entity size as perceived by the grazers, are nearly independent of each other, a point not considered by Steele and Frost (1977).] With natural water in floating plastic enclosures (mesocosms) instead of gnotobiotic aquaria, which allow the replicated manipulation of larger volumes and hence of larger animals, neither the initial biological state nor the subsequent environmental conditions of irradiance and temperature are easily controlled; otherwise, the same advantages accrue over traditional field work. In any case, paraphrasing Banse (1982b), it remains to be shown that we in fact can make the necessary measurements and model large "beakers" and bags. If we cannot—and we do not know—how can we model and really understand bays, let alone big lakes or the oceans?

## 30.4 Freshwater and Salt Water—Small and Large Water Bodies

Are the suggestions in the preceding account, which was biased toward the marine experience, applicable to oceans and lakes alike, and was the glossing over of the issues of spatial scale justified? The brief answer to the first part of the question is that most processes and rules of concern to production biology seem to be common to the aquatic domain in spite of the radical taxonomic differences between the two major realms. This fact has allowed general treatments largely independent of the specific assemblages. However, the end of the journey is in sight; if we are to build pyramids that incorporate species, these more realistic models cannot remain general. In response to the second part of the question, large spatial size does add new rules which, however, we might want to circumvent at the present time.

Some obvious differences between fresh- and saltwater do not seem to affect greatly the overall trophic interactions. For example, nitrogen seems to be the nutrient element usually controlling primary production in the sea (see, however, Zentara and Kamykowski, 1977); phosphorus usually assumes this role in freshwater. The contrast between N and P control may lead to somewhat different chemical composition of algae from nutrient-depleted habitats but will not remove the controlling role of nutrients on production. The more frequent occurrence in lakes of odd N/P ratios, due to local variability of source waters, introduces more variety into algal species competition, and thus dominance, than is known from the sea (Kilham and Kilham, 1984), and in turn may alter palatability of algae to animals, thus affecting the composition of the animal community.

The importance of ratios of nutrient supply is an example of structuring the food web from the bottom up, while the cited references on size-selective predation demonstrate the opposite. Bottom-up and top-down structuring also imply principal control of community interactions by resource limitation or predation. I have the impression that limnologists favor top-down control of pelagic food webs, whereas oceanographers more often tend toward bottom-up control (for the latter, see Koslow, 1983) but, in either case, it is not quite clear what is fact and what is artifact of approach. Similarly, Lampert (1985) noted a tendency among limnologists to tackle food chain dynamics, such as food limitation and competition in different ways than their marine colleagues. Possibly, the divergence of approaches and the underlying conceptual models concerning bottom-up or top-down control result also from the differences in the main features of seasonal succession between eutrophic and oligotrophic lakes (Sommer et al., 1986). The former type of lake is studied experimentally more often, whereas the latter, with a perhaps different underlying balance of mechanisms, appears to be more similar to the sea with respect to the seasonal trends, but has not influenced our views about control mechanisms as much as the eutrophic lake. Because, however, small eutrophic impoundments and basins as large as Lake Michigan have responded very similarly to fish predation, the relative roles of trophic state and scale (size) on bottom-up versus top-down are not yet clear. Moreover, as noted by Dodson (1979), studies merely of presence or absence of species, which are so common, may yield different conclusions from those focused on dominance; dominance is the more important issue for understanding fluxes and population dynamics. For the marine realm, the single largest puzzle in this context seems to be the North Sea ecosystem, where a recent great reduction of herring stock has not resulted in the changes predicted by a model based on the top-down freshwater paradigm (Koslow, 1983). As remarked by this author, among the possible reasons for the difference may be substitution of herring by sandeels (*Ammodytes*) or attributes of physical scale. As far as we know, the abundance of gelatinous ambush predators, which are absent in fresh water, has not changed. In my view, the prospect of an early understanding of changes of abundances in the marine realm in general are dim if we cannot explain change in the North Sea, probably the best-studied truly marine region.

Salt has an huge impact on taxa; major phyla are missing entirely (or almost so) from freshwater, but this seems to have little effect on principles of food web interactions in the pelagic realm of large water bodies. Apparently, nothing that insects do in freshwater is not done by some marine group, except the export of organic matter by imagines from the water when they move to the land. Conversely, the function of marine mucus net feeders, such as planktonic tunicates and pteropods, seems to be performed, in part, by bacterivorous cladocerans in freshwater plankton, with the limitation that their small size excludes the lack of size-selective feeding exhibited by the large members of the mentioned marine groups. However, of greater import for the structure of food webs may be scarcity of planktonic ambush predators in open lakes, especially of the entangling type, whereas in the sea, medusae, siphonophores ctenophores, and chaetognaths (the later being ambush-raptorial, similar to *Chaoborus* larvae in freshwater) contribute an appreciable fraction to the zooplankton biomass. Although they may comprise a few ten percent (Longhurst, 1985), the gelatinous organisms may even be under-represented in most reports because of the difficulty of their collection and preservation. The consequence of ambushing for patterns of size-selective predation was discussed by Greene (1985; 1986). Also largely absent form freshwater are the planktivorous clupeids, often the dominant fish in pelagic food webs of temperate seas and subtropical upwelling areas. There are limnic exceptions, besides the alewife in northeastern North America, such as two dominant clupeids in the pelagic domain of Lake Tanganyika, or the Lake Kinneret "sardine" (a cyprinid). Hecky (1984), in his discussion of the roles of lake age and colonization routes in the structuring of the pelagic foodweb in East African lakes, stressed the great importance of an efficient planktivore for pelagic production in Lake Tanganyika. More generally, Hecky remarked on the role, for the size structure, of what I may call "historic accidents," such as isolation and evolution, which may have lead to differing ecological efficiencies, especially with respect to fish yield available to man. Elsewhere in low-latitude lakes, a number, large relative to temperate lakes, of phytoplankton-feeding fishes not only compete with "herbivorous" zooplankton but also, similar to the marine mucus feeders and baleen whales, short-circuit the rules of the biomass spectrum. Thus, they introduce extra strands into the usual lacustrine food web (see also Nilssen, 1984). Finally, absent from fresh water is the biome producing the highest marine fish yields, the subtropical upwelling regions. As far as is known, though, its principal features rest on the temporal and spatial aspects of hydrography rather than on biological peculiarities not found anywhere else. In summary, realistic food webs will differ within both the fresh or salty realms, but general theory should not differ between them.

Important for realistic models is the low number of animal species in the open pelagic domain of freshwater at all latitudes, even within groups like copepods, in contrast to the situations in the marine pelagic domain (Lehman, 1988). This poverty also holds for the few old lakes where the benthos and fish fauna often evolved numerous endemisms. Further, the latitudinal trends of species richness differ, increasing toward higher latitudes in lakes (according to Lehman; remaining about equal according to Lewis, 1987), but increasing toward the tropics in the sea. No explanation seems to be at hand for the fact that the freshwater situation is opposed also to the terrestrial trend. As a consequence of the lower species numbers, however, the building of food web pyramids that include dominant species should be easier in lakes than in the sea.

Turning to the issue of spatial scale, the obvious difference between small and large lakes is the absence of littoral influences on the pelagic domain in terms of nutrient fluxes and immigration or advection of species and specimens over short intervals of days to weeks. During the periods of stratification, the lake bottom also can be neglected. In addition, however, the open pelagic realm of large water bodies is not horizontally well mixed over the same temporal intervals. There will be phytoplankton patches, the causes depending on the spatial scale, as reviewed by Legendre and Demers (1984) and Harris (1986). Such patches may be accompanied, and partially caused by physiological differences, as exemplified by a study of the chlorophyll-specific photosynthetic rate in a marine basin of a few kilometers diameter (Platt and Filion, 1973). Even on the temporal scale of copepod life cycles, the open pelagic realm of large lakes may not be uniform, as shown by biometric differences observed by Baldi et al. (1949) over distances of several kilometers in Lago Maggiore. Small water bodies may more easily be visualized as being sufficiently mixed in the horizontal dimension but it behooves the investigator to check whether the forces that usually provide horizontal uniformity on the small scale are operating; the spatial genetic heterogeneity of *Daphnia* sp. in a 0.8 ha V-shaped pond (Weider, 1985) is astonishing. As reviewed by Legendre and Demers (1984), the consequences of investigating water bodies of different size, with horizontal unevenness and the accompanying temporal scales of biota and environment, is that the proper scaling-up of the observational effort is not a trivial matter (cf. Platt, 1981). Neither is the question of how to translate results from small to large lakes, as was emphasized by E.J. Fee at the Large Lakes symposium.

In conclusion, where might research on interactions in pelagic food webs intent on incorporating dominant species, be most profitably done in the near future? The premise underlying my answer is that most processes in the fresh- and saltwater pelagic domains occur in both, perhaps with the exception of the role of the large marine invertebrates preying by ambush. In my opinion, lakes are the choice because of the greatly lowered specific diversity at least at the level of the metazoans. Because of the drastic effect of lowered salinity on diversity, the Baltic and Black seas—those large meromictic lakes—also qualify. Further, because large water bodies are horizontally not well mixed, resulting in sampling and validation problems discussed earlier, the choice should be stratified small, and preferably deep, replicated water bodies, such as kettle lakes or artificial enclosures. Such relatively small volumes have the additional, great advantage that the same water and virtually the same specimens can be revisited but may be spatially or temporally too limited for studying the role of fishes or medusae. To make rapid progress, the experiments should be guided by quantitative hypotheses (mathematical models) formulated beforehand. With the validation issue so important, it might well be that much advance in understanding food web interactions in large water bodies will come from field work in small, natural freshwater or artificial fresh- or brackishwater lakes.

---

*Acknowledgements* I wish to thank M.M. Tilzer for the invitation to attend the stimulating symposium on Functional and Structural Properties of Large Lakes, as well as numerous colleagues for their suggestions and criticism. The preparation of the paper was partially supported by grant No. NAWG-1007 from the National Aeronautics and Space Administration. D.P. Henry and S. Neuer improved on the style of the paper. Contribution No. 1844 from the School of Oceanography, University of Washington, Seattle.

## References

Allen, P.M. 1985. Ecology, thermodynamics, and self-organization: towards a new understanding of complexity. *Canadian Bulletin of Fisheries and Aquatic Sciences* 213:3–26.

Alverson, D.L., Longhurst, A.R., and Gulland, J.A. 1970. How much food from the sea? *Science* 168:503–505.

Andersen, K.P. and Ursin, E. 1977. A multispecies extension to the Beverton and Holt theory of fishing, with accounts of phosphorus circulation and primary production. *Meddelelser fra Danmarks Fiskeri- og Havundersøgelser* N.S. 7:319–435.

Bahr, L.M., Jr. 1982. Functional taxonomy: an immodest proposal. *Ecological Modelling* 15:211–233.

Baldi, E., Cavalli, L.L., Pirocchi, L., and Tonolli, V. 1949. L'isolamento delle popolazioni di *Mixodiaptomus laciniatus* Lill. del Lago Maggiore e i suoi nuovi problemi. *Memorie dell' Istituto Italiano di Idrobiologia "Dott. Marco de Marchi"* 5:295–305.

Banse, K. 1979. On weight dependence of net growth efficiency and specific respiration rates among field populations of invertebrates. *Oecologia* (Berlin) 38:111–126.

Banse, K. 1982a. Mass-scaled rates of respiration and intrinsic growth in very small invertebrates. *Marine Ecology Progress Series* 9:281–297.

Banse, K. 1982b. Experimental marine ecosystem enclosures in a historical perspective. *In* Grice, G.D. and Reeve, M.R. (ed.), *Marine Mesocosms*. Springer-Verlag, New York, p. 11–24.

Beddington, J.R. 1986. Shifts in resource populations in large marine ecosystems. *In* Sherman, K. and Alexander, L.M. (ed.), *Variability and Management of Large Marine Ecosystems*. American Association for the Advancement of Science, Washington, D.C., p. 9–18.

Bird, D.F. and Kalff, J. 1986. Bacterial grazing by planktonic lake algae. *Science* 231:493–495.

Blackburn, M. 1973. Regressions between biological oceanographic measurements in the eastern tropical Pacific and their significance to ecological efficiency. *Limnology and Oceanography* 18:552–563.

Blackburn, M. 1977. Studies on pelagic animal biomasses. *In* Andersen, N.R. and Zahuranec, B.J. (ed.), *Oceanic Sound Scattering Prediction*. Plenum Press, New York, p. 283–299.

Blackburn, M. 1981. Low latitude gyral regions. *In* Longhurst, A.R. (ed.), *Analysis of Marine Ecosystems*. Academic Press, London, p. 3–29.

Boyd, C.M. 1976. Selection of particle sizes by filter-feeding copepods: A plea for reason. *Limnology and Oceanography* 21:175–180.

Brooks, J.L. and Dodson, S.I. 1965. Predation, body size and composition of plankton. *Science* 150:28–35.

Brylinski, M. 1980. Estimating the productivity of lakes and reservoirs. *In* Le Cren, E.D. and Lowe-McConnell, R.H. (ed.), *The Functioning of Freshwater Ecosystems.* Cambridge University Press, London, p. 411–453.

Canale, R.P., DePalma, L.M., and Vogel, A.H. 1976. A plankton-based food web model for Lake Michigan. *In* Canale, R.P. (ed.), *Modeling Biochemical Processes in Aquatic Ecosystems.* Ann Arbor Science Publishers, Ann Arbor, Michigan, p. 33–74.

Carpenter, S.R., Kitchell, J.F., and Hodgson, J.R. 1985. Cascading trophic interactions and lake productivity. *BioScience* 35:634–639.

Carpenter, S.R., Kitchell, J.F., Hodgson, J.R., Cochran, P.A., Elser, J.J., Elser, M.M., Lodge, D.M., Kretchmer, D., He, X., and von Ende, C.N. 1987. Regulation of lake primary productivity by food web structure. *Ecology* 68:1863–1876.

Cook, R.E. 1977. Raymond Lindeman and the trophic-dynamic concept in ecology. *Science* 198:22–26.

Cousins, S.H. 1985a. Ecologists build pyramids again. *New Scientist* 107 (1463):50–54.

Cousins, S.H. 1985b. The trophic continuum in marine ecosystems: structure and equations for a predictive model. *Canadian Bulletin of Fisheries and Aquatic Sciences* 213:76–93.

DeMott, W.R. 1986. The role of taste in food selection by freshwater zooplankton. *Oecologia* (Berlin) 69:334–340.

Dickie, L.M., Kerr, S.R., and Boudreau, P.R. 1987. Size-dependent processes underlying regularities in ecosystem structure. *Ecological Monographs* 57:233–250.

Dodson, S.I. 1979. Body size patterns in arctic and temperate zooplankton. *Limnology and Oceanography* 24:940–949.

Evans, G.T. and Parslow, J.S. 1985. A model of annual plankton cycles. *Biological Oceanography* 3:327–347.

Fasham, M.J. 1981. Analytic food-web models. *In* Platt, T., Mann, K.H., and Ulanowicz, R.E. (ed.), *Mathematical Models in Biological Oceanography.* Monographs on Oceanographic Methodology (Unesco) 7, Paris, p. 54–64.

Frost, B.W. 1980. The inadequacy of body size as an indicator of niches in the zooplankton. *In* Kerfoot, W.C. (ed.), *Evolution and Ecology of Zooplankton Communities.* University Press of New England, Hanover, N.H., p. 742–753.

Fry, B. and Sherr, E.B. 1984. $\delta\,^{13}C$ measurements as indicators of carbon flow in marine and freshwater ecosystems. *Contributions in Marine Science* 27:13–47.

Fryer, G. 1987. The feeding mechanisms of the Daphniidae (Crustacea: Cladocera): recent suggestions and neglected considerations. *Journal of Plankton Research* 9:419–432.

Greene, C.H. 1985. Planktivore functional groups and patterns of prey selection in pelagic communities. *Journal of Plankton Research* 7:35–40.

Greene, C.H. 1986. Patterns of prey selection: implications of predator foraging tactics. *The American Naturalist* 128:824–839.

Gulland, J.A. 1987. Natural mortality and size. *Marine Ecology Progress Series* 39:197–199.

Hardy, A.C. 1924. The herring in relation to its animate environment. Part I. The food and feedings habits of the herring with special reference to the east coast of England. *Fishery Investigations* (London), series II, vol. VII(3):1–53.

Harris, G.P. 1986. *Phytoplankton Ecology.* Chapman and Hall, London. 384 pp.

Hecky, R.E. 1984. African lakes and their trophic efficiencies: a temporal perspective. *In* Meyers, D.G. and Strickler, J.R. (ed.), *Trophic Interactions within Aquatic Ecosystems.* American Association for the Advancement of Science, Washington, D.C., AAAS Selected Symposium 85, p. 405–408.

Hensen, V. 1887. Ueber die Bestimmung des Plankton's oder des im Meere treibenden Materials an Pflanzen und Tieren. *Fünfter Bericht der Kommission zur Untersuchung der deutschen Meere in Kiel für die Jahre 1882 bis 1886,* XII-XVI. Jahrg.:1–107.

Hrbáček, J. 1958. Typologie und Produktivität der teichartigen Gewässer. *Verhandlungen der Internationalen Vereinigung für Theoretische und Angewandte Limnologie* 13:394–399.

Hrbáček, J. 1962. Species composition and the amount of the zooplankton in relation to the fish stock. *Rospravy Československé Akademi Věd, Řada Matematických a Přírodních Véd* 72(10):116 pp.

Humphreys, W.F. 1979. Production and respiration in animal populations. *Journal of Animal Ecology* 48:427–453.

Huntley, M.E., Barthel, K.-G., and Star, J.L. 1983. Particle rejection by *Calanus pacificus*: discrimination between similarly sized particles. *Marine Biology*: 74:151–160.

Imboden, D.M. and Schwarzenbach, R.P. 1985. Spatial and temporal distribution of chemical substances in lakes: modeling concept. *In* Stumm, W. (ed.), *Chemical Processes in Lakes*. Wiley, New York, p. 1–30.

Isaacs, J.D. 1972. Unstructured marine food webs and "pollutant analogues". *Fishery Bulletin U.S.* 70:1053–1059.

Ivlev, V.S. 1945. [The biological productivity of waters.] *Uspekhi Sovremennoi Biologii* 19:98–120. (In Russian. Translation by Ricker, W.E., *Journal of the Fisheries Research Board of Canada* 23:1727–1753, 1966.)

Jakobsen, P.J. and Johnsen, G.H. 1987. The influence of predation on horizontal distribution of zooplankton species. *Freshwater Biology* 17:501–507.

Jørgensen, S.E., Kamp-Nielsen, L., and Jørgensen, L.A. 1986. Examination of the generality of eutrophication models. *Ecological Modelling* 32:251–266.

Kerfoot, W.C. and DeMott, W.R. 1984. Food web dynamics: dependent chains and vaulting. *In* Meyers, D.G. and Strickler, J.R (ed.), *Trophic Interactions within Aquatic Ecosystems*. American Association for the Advancement of Science, Washington, D.C., AAAS Selected Symposium 85, p. 347–382.

Kerfoot, W.C. and Sih, A. (ed.) 1987. *Predation: Direct and Indirect Impacts on Aquatic Communities*. University Press of New England, Hanover, New Hampshire. 386 pp.

Kilham, S.S. and Kilham, P. 1984. The importance of resource supply rates in determining phytoplankton community structure. *In* Meyers, D.G. and Strickler, J.R. (ed.), *Trophic Interactions within Aquatic Ecosystems*. American Association for the Advancement of Science, Washington, D.C., AAAS Selected Symposium 85, p. 7–27.

Koslow, J.A. 1983. Zooplankton community structure in the North Sea and Northeast Atlantic: development and test of a biological model. *Canadian Journal of Fisheries and Aquatic Sciences* 40:1912–1924.

Laevastu, T. and Larkins, H.A. 1981. *Marine Fisheries Ecosystem. Its Quantitative Evaluation and Morphogenesis*. Fishing News Books, Fournhan, England. 159 pp.

Lampert, W. 1985. Preface. *In* Lampert, W. (ed.), *Food Limitation and the Structure of Zooplankton Communities*. Archiv für Hydrobiologie, Beiheft 21, p. v–vii.

Lampert, W. and Muck, P. 1985. Multiple aspects of food limitation in zooplankton communities: the *Daphnia–Eudiaptomus* example. *In* Lampert, W. (ed.), *Food Limitation and the Structure of Zooplankton Communities*. Archiv für Hydrobiologie, Beiheft 21, p. 311–322.

Larkin, P.A. 1984. Strategies for multispecies management. *In* May, R.M. (ed.), *Exploitation of Marine Communities*, (Dahlem Konferenzen, 1984). Springer-Verlag, Berlin, p. 287–301.

Lazzaro, X. 1987. A review of planktivorous fishes: Their evolution, feeding behaviours, selectivities, and impacts. *Hydrobiologia* 146:97–167.

Legendre, L. and Demers, S. 1984. Towards dynamic biological oceanography and limnology. *Canadian Journal of Fisheries and Aquatic Sciences* 41:2–19.

Lehman, J.T. 1988. Ecological principles affecting community structure and secondary production by zooplankton in marine and freshwater environments. *Limnology and Oceanography* 33:931–945.

Lewis, W.M., Jr. 1987. Tropical limnology. *Annual Review of Ecology and Systematics* 18:159–184.

Lindeman, R.L. 1942. The trophic-dynamic aspect of ecology. *Ecology* 23:399–418.
Longhurst, A. 1984. Importance of measuring rates and fluxes in marine ecosystems. *In* Fasham, M.J.R. (ed.), *Flows of Energy and Materials in Marine Ecosystems*. Plenum Press, New York, p. 3–22.
Longhurst, A.R. 1985. The structure and evolution of plankton communities. *Progress in Oceanography* 15:1–35.
Longhurst, A. and Williams, R. 1979. Materials for plankton modelling: Vertical distribution of Atlantic zooplankton in summer. *Journal of Plankton Research* 1:1–28.
Lynch, M. 1979. Predation, competition, and zooplankton community structure: An experimental study. *Limnology and Oceanography* 24:253–272.
Mathisen, O.A., Thorne, R.E., Trumble, R.J., and Blackburn, M. 1978. Food consumption of pelagic fish in an upwelling area. *In* Boje, R. and Tomczak, M. (ed.), *Upwelling Ecosystems*. Springer-Verlag, Berlin, p. 111–123.
McGurk, M.D. 1987. Natural mortality and spatial patchiness revisited. *Marine Ecology Progress Series* 39:201–206.
Mearns, A.J., Young, D.R., Olson, R.J., and Schafer, H.A. 1981. Trophic structure and the cesium—potassium ratio in pelagic ecosystems. *California Cooperative Oceanic Fisheries Investigations Reports* 22:99–110.
Melack, J. 1976. Primary productivity and fish yields in tropical lakes. *Transactions of the American Fisheries Society* 105:575–580.
Minagawa, M. and Wada, E. 1984. Stepwise enrichment of $^{15}$N along food chains: further evidence and the relation between $\delta$ $^{15}$N and animal age. *Geochimica et Cosmochimica Acta* 48:1135–1140.
Morgan, N.C. 1980. Secondary production. *In* Le Cren, E.D. and Lowe-McConnell, R.H. (ed.), *The Functioning of Freshwater Ecosystems*. Cambridge University Press, London, p. 247–340.
Mullin, M.M., Rau, G.H., and Eppley, R.W. 1984. Stable nitrogen isotopes in zooplankton: some geographic and temporal variations in the North Pacific. *Limnology and Oceanography* 29:1267–1273.
Murdoch, W.M. 1966. "Community structure, population control, and competition"—a critique. *The American Naturalist* 100:219–226.
Nilssen, J.P. 1984. Tropical lakes—functional ecology and future development: The need for a process-orientated approach. *Hydrobiologia* 113:231–242.
O'Neill, R.V., DeAngelis, D.L., Waide, J.B., and Allen, T.F.H. 1986. *A Hierarchical Concept of Ecosystems*. Princeton University Press, Princeton, New Jersey. 252 pp.
Pace, M.L., Glasser, J.E., and Pomeroy, L.R. 1984. A simulation analysis of continental shelf food webs. *Marine Biology* 82:47–63.
Paloheimo, J.E., Zimmerman, A.P., Sprules, W.G., and Gates, M.A. 1984. The structure of aquatic ecosystems and its dependence on environmental variables. *In* Fasham, M.J.R. (ed.), *Flows of Energy and Materials in Marine Ecosystems*. Plenum, New York, p. 85–104.
Pang, P.C. and Nriagu, J.O. 1977. Isotopic variations of the nitrogen in Lake Superior. *Geochimica et Cosmochimica Acta* 41:811–814.
Peters, R.H. 1981. *The Ecological Implications of Body Size*. Cambridge University Press, Cambridge. 329 pp.
Pielou, E.C. 1981. The usefulness of ecological models: a stock-taking. *The Quarterly Review of Biology* 56:17–31.
Platt, T. 1981. Thinking in terms of scale: introduction to dimensional analysis. *In* Platt, T., Mann, K.H., and Ulanowicz, R.E. (ed.), *Mathematical Models in Biological Oceanography*. Monographs on Oceanographic Methodology (Unesco) 7, p. 112–121.
Platt, T. 1981. General conclusions. *In* Platt, T., Mann, H., and Ulanowicz, R.E. (ed.), *Mathematical Models in Biological Oceanography*. Monographs on Oceanographic Methodology (Unesco) 7, p. 138–139.

Platt, T. 1985. Structure of the marine ecosystem: its allometric basis. *Canadian Bulletin of Fisheries and Aquatic Sciences* 213:55–64.
Platt, T. and Filion, C. 1973. Spatial variability of the productivity:biomass ratio for phytoplankton in a small marine basin. *Limnology and Oceanography* 18:743–749.
Rau, G.H., Mearns, A.J., Young, D.R., Olson, R.J., Schafer, H.A., and Kaplan, I.R. 1983. Animal $^{13}C/^{12}C$ correlates with trophic level in pelagic food webs. *Ecology* 64:1314–1318.
Rigler, F.H. 1975. The concept of energy flow and nutrient flow between trophic levels. *In* van Dobben, W.H. and Lowe-McConnell, R.H. (ed.), *Unifying Concepts in Ecology.* Junk, The Hague, p. 15–26.
Riley, G.A. 1946. Factors controlling phytoplankton populations on Georges Bank. *Journal of Marine Research* 6:54–73.
Riley, G.A. 1947. A theoretical analysis of the zooplankton of Georges Bank. *Journal of Marine Research* 6:104–113.
Riley, G.A. and von Arx, R. 1949. Theoretical analysis of seasonal changes in the phytoplankton of Husan Harbor, Korea. *Journal of Marine Research* 8:60–72.
Rodriguez, J. and Mullin, M.M. 1986. Relation between biomass and body weight of plankton in a steady state oceanic ecosystem. *Limnology and Oceanography* 31:361–370.
Ryther, J.H. 1969. Photosynthesis and fish production in the sea. *Science* 166:72–76.
Sieracki, M.E., Haas, L.W., Caron, D.A., and Lessard, E.J. 1987. Effect of fixation on particle retention by microflagellates: underestimation of grazing rates. *Marine Ecology Progress Series* 38:251–258.
Silvert, W. and Platt, T. 1980. Dynamic energy-flow model of the particle size distribution in pelagic ecosystems. *In* Kerfoot, W.C. (ed.), *Evolution and Ecology of Zooplankton Communities.* University Press of New England, Hanover, N.H., p. 754–763.
Skogstad, A., Granskog, L., and Klaveness, D. 1987. Growth of freshwater ciliates offered planktonic algae as food. *Journal of Plankton Research* 9:503–512.
Sommer, U. 1986. The periodicity of phytoplankton in Lake Constance (Bodensee) in comparison to other deep lakes of central Europe. *Hydrobiologia* 138:1–7.
Sommer, U. Gliwicz, Z.M., Lampert, W., and Duncan, A. 1986. The PEG-model of seasonal succession of planktonic events in freshwaters. *Archiv für Hydrobiologie* 106:433–471.
Sprules, W.G. and Holtby, L.B. 1979. Body size and feeding ecology as alternatives to taxonomy for the study of limnetic zooplankton community structure. *Journal of the Fisheries Research Board of Canada* 36:1354–1363.
Sprules, W.G. and Munawar, M. 1986. Plankton size spectra in relation to ecosystem productivity, size, and perturbation. *Canadian Journal of Fisheries and Aquatic Sciences* 43:1789–1794.
Steele, J.H. 1974. *The Structure of Marine Ecosystems.* Harvard University Press, Cambridge, MA, 128 pp.
Steele, J.H. 1984. Kinds of variability and uncertainty affecting fisheries. *In* May, R.M. (ed.), *Exploitation of Marine Communities,* (Dahlem Konferenzen, 1984). Springer-Verlag, Berlin, p. 245–262.
Steele, J.H. and Frost, B.W. 1977. The structure of plankton communities. *Philosophical Transactions of the Royal Society of London* B 280:485–534.
Steele, J.H. and Henderson, E.W. 1984. Modeling long-term fluctuations in fish stocks. *Science* 224:985–987.
Steele, J.H. and Mullin, M.M. 1977. Zooplankton dynamics. *In* Goldberg, E.D., McCave, I.N., O'Brien, J.J., and Steele, J.H. (ed.), *The Sea, Vol. 6, Marine Modelling.* Wiley/Interscience, New York, p. 857–890.
Stoecker, D.K., Cucci, T.L., Hulburt, E.M., and Yentsch, C.M. 1986. Selective feeding by *Balanion* sp. (Ciliata: Balanionidae) on phytoplankton that best support its growth. *Journal of Experimental Marine Biology and Ecology* 95:113–130.

Stoecker, D., Guillard, R.R.L., and Kavee, R.M. 1981. Selective predation by *Favella ehrenbergii* (Tintinnia) on and among dinoflagellates. *Biological Bulletin* 160:136–145.

Straškraba, M. and Gnauck, A.H. 1985. *Freshwater Ecosystems. Modelling and Simulation.* Elsevier, Amsterdam, 309 pp.

Thomann, R.V. 1981. Equilibrium model of fate of microcontaminants in diverse aquatic food chains. *Canadian Journal of Fisheries and Aquatic Sciences* 38:280–296.

Vanni, M.J. 1986. Fish predation and zooplankton demography: indirect effects. *Ecology* 67:337–354.

Vanni, M.J. 1987. Effects of food availability and fish predation on a zooplankton community. *Ecological Monographs* 57:61–88.

Vézina, A.F. 1986. Body size and mass flow in freshwater plankton: models and tests. *Journal of Plankton Research* 8:939–956.

Vinogradov, M.Ye., Shushkina, E.A., and Kukina, I.N. 1976. Functional characteristics of a planktonic community in an equatorial upwelling region. *Oceanology* 16:67–76 (translation from *Okeanologiya* 16, 1976).

Walters, C.J., Krause, E., Neill, W.E., and Northcote, T.G. 1987. Equilibrium models for seasonal dynamics of plankton biomass in four oligotrophic lakes. *Canadian Journal of Fisheries and Aquatic Sciences* 44:1002–1017.

Weider, L.J. 1985. Spatial and temporal genetic heterogeniety in a natural *Daphnia* population. *Journal of Plankton Research* 7:101–123.

Winberg, G.G. (ed.) 1971. *Methods for the Estimation of Production of Aquatic Animals.* Academic Press, New York, 161 pp.

Wroblewski, J.S. (ed.) 1982. *Selected Works of Gordon A. Riley.* Dalhousie University, Halifax, N.S., 489 pp.

Wroblewski, J.S. 1983. The role of modeling in biological oceanography. *Ocean Science and Engineering* 8:245–285.

Wyatt, T. 1976. Food chains in the sea. *In* Cushing, D.H. and Walsh, J.J. (ed.), *The Ecology of the Seas.* Saunders, Philadelphia, p. 341–358.

Zentara, S.-J. and Kamykowski, D. 1977. Latitudinal relationships among temperature and selected plant nutrients along the west coast of North and South America. *Journal of Marine Research* 35:321–337.

# 31

# Lake Trophic Status and the Development of the Clear-Water Phase in Lake Geneva

Gérard Balvay, Meg Gawler, and Jean Pierre Pelletier

ABSTRACT A brief period of very clear water following the spring phytoplankton bloom has been observed in many mesotrophic and eutrophic lakes. For Lake Geneva, data on water transparency from Forel (1895) demonstrate that the phenomenon of a clear-water phase did not exist when the lake was still oligotrophic in the late nineteenth century. Comparable Secchi measurements over the last three decades indicate that the clear-water phase developed along with increasing eutrophication as a result—not of clearer water in June—but of significantly lower transparencies at the time of the spring bloom, which is usually dominated by fast-growing, nanoplanktonic algae. Generally in Lake Geneva, following this spring bloom, the numbers of cyclopoid copepods decline strongly, and the *Daphnia* populations explode. Severe Si limitation in May was observed in 9 out of the last 14 years. Although the sharp reduction in algal standing crop at this time of year may begin by the loss of diatoms due to nutrient stress, the development of a dramatic clear-water phase is mediated primarily by grazing, and in particular, by the grazing of *Daphnia*.

## 31.1 The Clear-Water Phase

The term "clear-water phase" was first coined in German ("Klarwasser-Stadium") by Uhlmann (1961) to describe the disappearance of algae, due to a mass development of daphnids following a food pulse in hypereutrophic ponds. Lampert and Schober (1978) later showed that this brief, but spectacular period of very clear water following the spring phytoplankton bloom also occurred in Lake Constance, when a "window" for the daphnids is opened under conditions of increased water temperatures, abundant food, and reduced predation pressure, as the spring pulse of *Cyclops vicinus* declines. The clear-water phase ends several weeks later when the daphnids' food resources are exhausted, and predation pressures resume. This phenomenon has been observed in many lakes, and seems to be a predictable event in mesotrophic and eutrophic systems.

In comparing three Swiss lakes, Bürgi et al., (1985) observed several overgrazing periods per year in the eutrophic lakes Hallwilersee and Greifensee, but only one period of overgrazing in mesotrophic Lake Lucerne. In all cases, however, a period of clear water occurred only during the algal collapse of May and June. Similarly, in a comparison of nine deep lakes of central Europe, Sommer (1986) noted that not only the swiftness of seasonal changes, but also their amplitude, increase with increasing trophic status. Although a spring phytoplankton bloom occurred in all nine lakes, only the eutrophic lakes produced a clear-water phase.

During the course of eutrophication of Lake Constance, a lake very similar to Lake Geneva, Secchi measurements reveal a clear-water phase as early as 1961 (Lehn, 1968). Phytoplankton data for Lake Constance (1953 to 1972) show that a regular clear-water phase began to be observed during the 5-year period beginning in 1963 (Lehn, 1973), when the $PO_4$ holomixis level surpassed 20 $\mu g$ P $l^{-1}$ (Tilzer et al., 1985). The regular occurrence of the clear-water phase in Lake Constance from 1965 to 1975 is illustrated in detail by Bürgi and Lehn (1979).

In the present chapter, we examine the recent development of the clear-water phase in Lake Geneva in relation to changes in lake trophic status and zooplankton community structure, based on lake monitoring data collected over the last three decades for the "Commission Internationale pour la Protection des Eaux du Léman". Despite the highly variable composition of the plankton, seasonal events in Lake Geneva generally concur with the predictions of the PEG-model (Sommer et al., 1986), particularly with respect to the sequential depletion of nutrients, P, Si and N, and the diatom-cryptophyte-chlorophyte-dinoflagellate sequence in the phytoplankton (Gawler et al., 1988).

## 31.2 Long-term Changes in Transparency

Numerous Secchi disk measurements performed during the end of the last century in Lake Geneva allow long-term studies for transparency evolution. This lake (also named Lake Léman), summarily described in Table 31.1, is the largest volume of freshwater in western Europe, exclusive of the Baltic-Scandinavian region. Formerly oligotrophic, this lake has been subject to accelerated eutrophication, with an increase by nearly an order of magnitude in the annual mean concentration of total phosphorus, from 10 $\mu g$ $l^{-1}$ in the late 1950s to over 90 $\mu g$ $l^{-1}$ in the late 1970s (Monod et al., 1984). Orthophosphate available at the start of the growing season now fluctuates between 65 and 80 $\mu g$ P $l^{-1}$.

In Lake Geneva, transparency is negatively correlated with chlorophyll concentrations and thus gives a rough indication of phytoplankton abundance; about 80% of the variation in transparency is explained by chlorophyll concentration (Balvay et al., 1984). Similar observations have been made for Lake

**Table 31.1** Limnological characteristics of Lake Geneva

| | |
|---|---|
| Geographic coordinates | N 46°27′, E 6°32′ |
| Altitude | 372 m |
| Area | 582 $km^2$ |
| Watershed area (exclusive of lake) | 7400 $km^2$ |
| Maximum depth | 309 m |
| Mean depth | 152 m |
| Length | 72 km |
| Maximum width | 14 km |
| Shoreline | 167 km |
| Volume | 89 $km^3$ |
| Mean outflow | 236 $m^3\ s^{-1}$ |
| Theoretical retention time | 12 years |
| Mixis type | monomictic |
| pH | 7.5 to 9 |

Constance by Tilzer (1988) in a detailed discussion on this topic. Early measurements of Secchi depth by Forel (1895) demonstrate that, in the late nineteenth century, seasonal variations of transparency, from approximately 15 m in winter to 6 m in summer, were quite regular, following a bell-shaped curve with the annual minimum in July (Figure 31.3). The seasonal pattern of recent years presents a distinct contrast, with two minima, one very pronounced in May and the other in late summer, separated by a June clear-water phase. Secchi transparency, now typically about 2 m during the vernal phytoplankton peak in May, often increases to over 10 m a few weeks later (Gawler et al., 1988). Likewise, spring chlorophyll concentrations of 10 to 40 $\mu g\ l^{-1}$ drop during the clear-water phase to only 1 to 2 $\mu g\ l^{-1}$, values which are now often lower than those found in mid-winter.

## 31.3 Recent Evolution in Transparency and Composition of the Plankton

**Methods** The data presented herein are based on samples collected at the only pelagic station that has been sampled continuously since 1959—the central station, midway between Evian and Lausanne. Sampling frequency was once a month from 1959 to 1980, and twice a month starting in 1981. Zooplankton has been sampled since 1959, using a net 35 cm in diameter, mesh size 200 $\mu m$, from a depth of 50 m, and phytoplankton (0 to 10 m) has been sampled since 1974 with a Pelletier integrating sampler, and counted by the inverted microscope method. Both zooplankton and phytoplankton counts are converted to fresh weight by length-weight regressions and cell volume estimates, respectively (Bottrell et al., 1976; Balvay et al., 1984). This study is based on year-to-year changes from the data of the minimum observed transparency

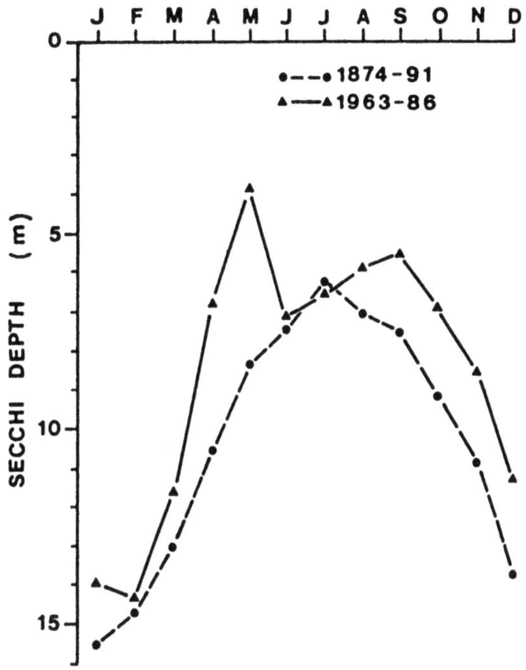


**Figure 31.1** Monthly mean Secchi transparency at the time of Forel (●—●), and from 1963 to 1986 (▲–▲).

(spring algal bloom) to the date of the summer maximum observed transparency (clear-water phase). Minimum and maximum transparency will be referred to in the text as "May" and "June" for the sake of convenience. Minimum Secchi depth was always recorded in May, with the exception of 1959 and 1961, when it occurred in June. Maximum transparency was generally observed in June, but sometimes in July (1959, 1961, 1984, 1986). Generally, in order to observe a distinct clear-water phase, a sampling frequency of at least once a week is required (Lampert et al., 1986). Thus, for the years of this study when a clear-water phase is not apparent, it should be remembered that it was not necessarily absent – it simply may have not been detected.

**Transparency** It is remarkable that, despite the low sampling frequency for these data collected for the purposes of lake monitoring, a distinct increase in transparency, expressed by the ratio max/min Secchi depth $> 1.5$, can be observed for 15 out of 22 years based on monthly samples (1959 to 1987), and for 6 out of 7 years with bimonthly samples (1981 to 1987) (Figure 31.2). The evolution of Secchi transparency from 1959 to 1987 seems to comprise four phases: 1959 to 1962, a period of little enrichment, with relatively high transparency during the spring bloom (more than 5.5 m), and no apparent clear-water phase (ratio $< 1.3$); 1963 to 1974, a long period of increasing

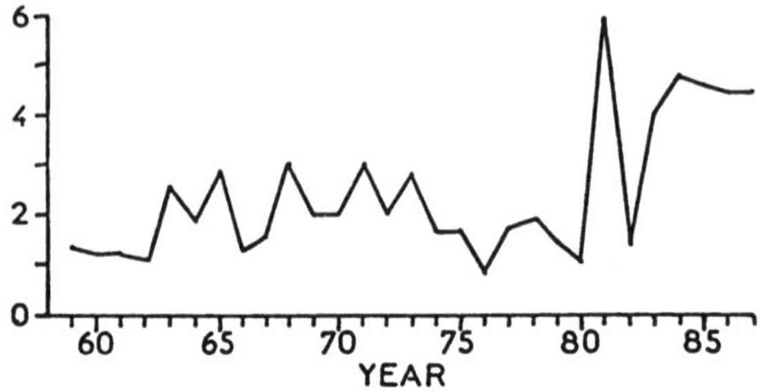


**Figure 31.2** Change in ratio of the max/min Secchi depth from 1959 to 1987.

eutrophication, with a reduction in minimum spring transparency (2.8 to 5.2 m) and frequent occurrence of a clear-water phase (ratio > 1.5 except for 1966); 1975 to 1980, a period of maximum phosphorus loading, slight decrease in minimum transparency (2.5 to 4.8 m), and little evidence of a clear-water phase (ratio ≈ 1 for 1976 and 1980); 1981 to 1987 (with higher sampling frequency), a period of further reductions in minimum transparency (1.5 to 2.3 m) and a distinct clear-water phase (ratio > 4, except for 1982). Although the increasing spring water temperature influences the growth of both phytoplankton and zooplankton populations, analysis of the thermal regime in the upper 20 m of the water column showed no apparent relationship with the magnitude of the change in transparency from May to June for each year.

**Phytoplankton** The changes in phytoplankton from May to June (Figure 31.3) include the periods from 1975 to 1980 with an occasional clear-water phase, and the period from 1981 to 1987 with a distinct clear-water phase, except in 1986. It is evident that phytoplankton biomass is often strongly depressed between May and June. Furthermore, the disappearance of the highly grazeable fraction, the ultraplankton (length <5 μm) and nanoplankton (5 to 20 μm), is responsible for the decline of the algal standing crop at this time of year. In fact, maximum transparency often occurs slightly after the June chlorophyll minimum because of the change in the size structure of the phytoplankton community, due to selective losses of smaller particles that diffuse light to a greater extent than do larger particles.

Interesting patterns are also revealed in the changes in the taxonomic classes of the phytoplankton. The spring bloom biocoenosis is usually dominated by the Cryptophyceae, accompanied by diatoms (which have often peaked slightly earlier) and/or chlorophytes. In the clear-water phase, however, diatoms are almost entirely absent, and the cryptophytes and green algae are generally reduced in biomass. Little change is seen from May to June in the occasional abundance of dinoflagellates. A high biomass of Cryptophyceae during the spring bloom is generally followed by a particularly strong clear-water phase (see years 1981 and 1983 to 1987 in Figure 31.2). These same crytophytes (particularly *Rhodomonas minuta*) have been found to be the pre-

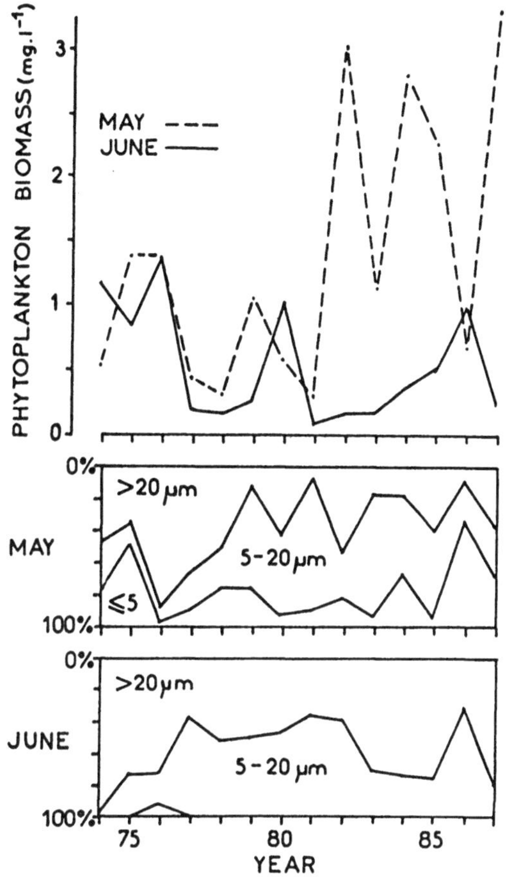


**Figure 31.3** Biomass (fresh weight) and size structure of the phytoplankton biocoenosis in May and June from 1974 to 1987.

ferred food for herbivorous zooplankton in Lake Constance (Knisely and Geller, 1986). Although the low algal biomass of the clear-water phase is usually dominated by the Cryptophyceae, this pattern is not as uniform as stated by Sommer (1986); in Lake Geneva, Chlorophyceae can also be the dominant group during this period.

**Nutrients** As shown in Figure 31.4, levels of inorganic N ($NO_3$ + $NO_2$ + $NH_4$) in the trophic zone varied little from the spring bloom to the clear-water phase, whereas concentrations of $PO_4$ decreased (although sometimes only slightly) from May to June in every year from 1974 to 1987, particularly during the period 1975 to 1980. The lack of a net increase in inorganic N and P in June from 1981 to 1987, despite very low algal standing crops and reduced rates of primary production, shows that no recycling of nutrients by zooplankton excretion could be detected during this period of presumably high grazing

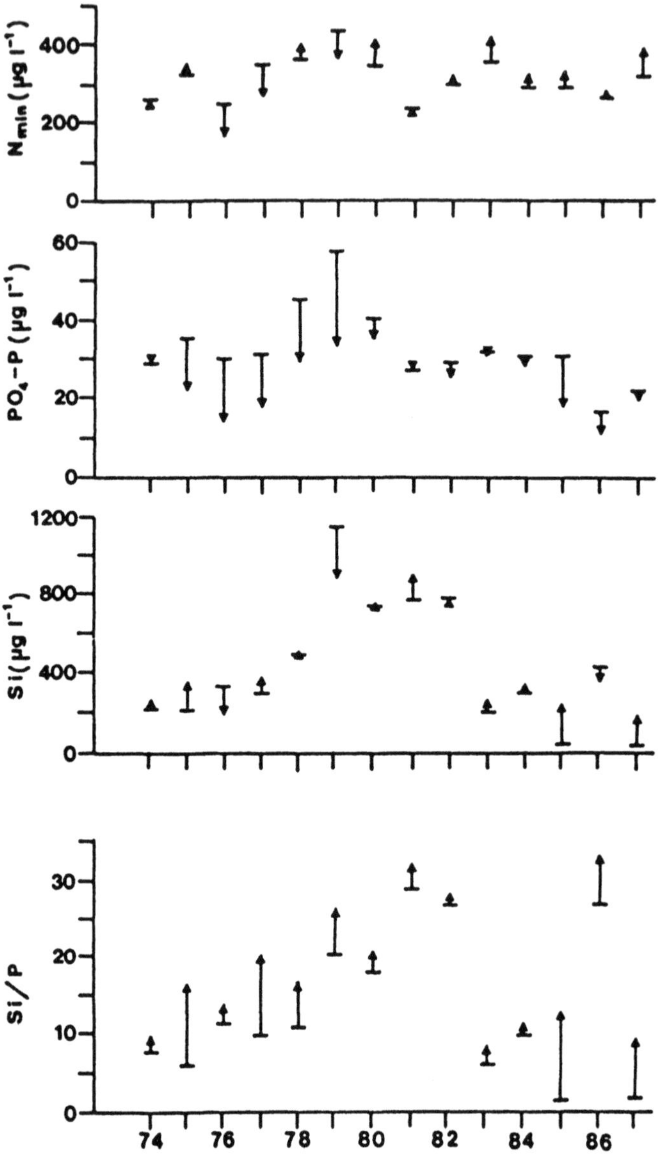


**Figure 31.4** Change in nutrient concentrations (weighted mean, 0–10 m), and Si/P ratio from May (bar) to June (triangle) from 1974 to 1987.

activity. Si also remained little changed from May to June. However, the atomic ratio Si/P increased systematically in June, undoubtedly as a result of the nearly total disappearance of diatoms, with continued uptake of P by other algae. At the time of the spring bloom, the Si/P ratio in several years had already fallen below 10 — a threshold considered critical for the survival of most diatoms (Sommer, 1985). Although *Stephanodiscus* is known to be a very good competitor under limiting silicon conditions (Kilham, 1984), Gawler et al. (1987) have shown that, as a result of early Si limitation, the nanoplaktonic diatoms dominant in 1985, *Stephanodiscus minutula* and *S. alpinus*, sedimented below the euphotic zone in late May, when measured grazing intensity was still negligible. In this case, the loss of phytoplankton leading to the clear-water phase was initiated by nutrient limitation, and then exacerbated by grazing pressure. In the present study, particularly severe Si limitation occurred in May 1985 and 1987 during an exceptionally strong spring diatom pulse (Figure 31.4).

**Zooplankton** Total zooplankton biomass, although variable, has increased substantially over the years for both May and June (Figure 31.5). This is primarily due to an increase over time in the biomass of *Cyclops* spp. during the spring algal bloom, and in the biomass of *Daphnia* spp. during the period of maximum transparency. The relationship of the increase in annual mean

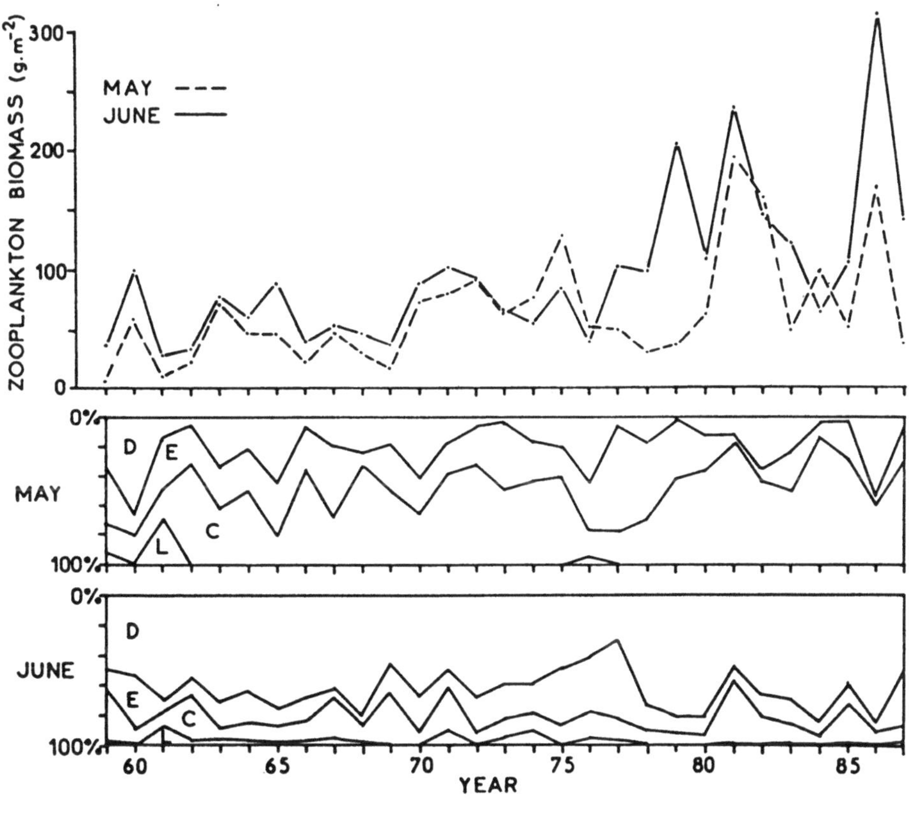


**Figure 31.5** Biomass (fresh weight) and composition of the zooplankton biocoenosis in May and June from 1959 to 1987. C = *Cyclops* spp., D = *Daphnia* + *Bosmina*, E = *Eudiaptomus*, L = *Leptodora* + *Bythotrephes*.

zooplankton biovolume to increasing phosphorus concentrations and primary production has been described by Balvay (1984) and Balvay and Pelletier (1988).

The composition of the zooplankton community changes markedly from May to June: the predatory cladocerans, *Leptodora kindtii* and *Bythotrephes longimanus,* begin to develop; the biomass of the cyclopoid copepods, *C. vicinus* and *C. prealpinus,* is greatly reduced; the calanoid *Eudiaptomus gracilis* may increase or decrease slightly, and the biomass of *Daphnia* + *Bosmina* increases dramatically, especially in the later years. Out of the five years with abundances of *Daphnia* in June lower than 20 g fr wt $m^2$, (i.e. 1959, 1961, 1962, 1969, 1976 in Figures 31.2 and 31.5), only the year 1969 had an apparent clear-water phase. Lampert (1985) provides estimates of the "critical" concentration of zooplankton necessary to initiate a clear-water phase in several lakes. For Lake Constance, the lake most similar to Lake Geneva, this critical value is 2.0 to 2.4 g dry wt $m^{-2}$, i.e., 20 to 24 g fr wt $m^{-2}$ when assuming a 10/1 ratio of fresh to dry weight (Balvay, 1987).

The correlation coefficients for log-normalized values of water transparency in May and June as a function of the biomass of various zooplankton groups are given in Table 31.2. The correlation is improved by the elimination first of carnivores, then of all "herbivorous" copepods, and finally of *Bosmina.* The correlation of *Daphnia* biomass (X) with Secchi transparency (Y) in May and June, despite considerable scattering of the points, is highly significant ($P < 0.001$); the best fit is

$$\log Y = 0.203 \log X + 0.491 \ (R = 0.589, N = 58).$$

**Table 31.2** Zooplankton/transparency correlation coefficients[9]

| Zooplankton type | r |
|---|---|
| Total zooplankton | 0.026 |
| Total herbivorous zooplankton | 0.077 |
| Herbivorous calanoids | 0.394 |
| *Daphnia* spp. | 0.589 |

[9]X = log biomass (g $m^{-2}$); Y = log Secchi transparency (m); N = 58.

## 31.4 General Trends

In Figure 31.6 data from these 29 years are condensed into 5 periods, chosen arbitrarily with respect to transparency values. Concentrations of total P increased sharply over the first 22 years of this study, and have tapered off slightly since then. Zooplankton biomass has also grown with increasing eutrophication, but surprisingly, the mean zooplankton biomass in June increased in recent years despite lower mean levels of P. Similarly, spring food resources seem to have increased continuously past the point of maximum P concentrations, as indicated by the steadily decreasing transparency during the spring

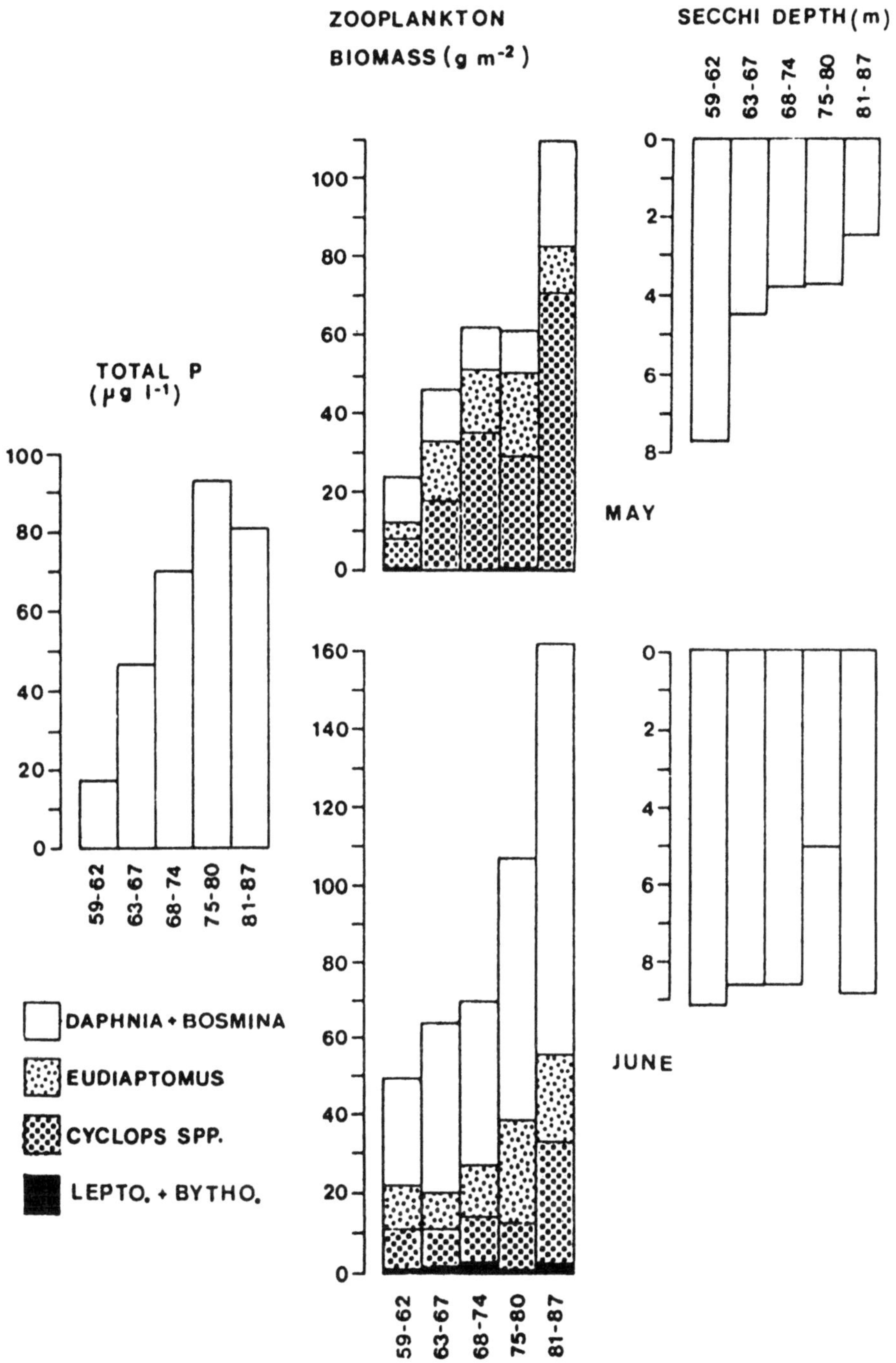


**Figure 31.6** Long-term trends in annual mean concentration of total P over the entire water column, trends for May and June of zooplankton fresh weight biomass and community structure and Secchi transparency.

phytoplankton bloom. The most dramatic change over time has been the surge in the biomass of *Daphnia* spp., probably in response to the continued enrichment of appropriate food resources.

One period (1975 to 1980) does not agree with this general trend; this is the period of maximum eutrophication with little or no evidence of a clear-water phase. The most unusual year on record in terms of climatic events and phytoplankton composition and dynamics was 1976. Not only was there no decline from May to June, but the entire growing season was dominated by a very large biomass of dinoflagellates that fluctuated dramatically; this was also a year of exceptionally low *Daphnia* abundance.

With the exception of the period 1975 to 1980, it is evident from Figure 31.6 that maximum Secchi depth in June has changed little over the years. Thus we can conclude that the historical development of the clear-water phase is not due to clearer water in June, but results from an increasing change in transparency at this time due to greatly reduced Secchi depth during the spring bloom. The ensemble of these data show that this rapid change in transparency from May to June is mediated primarily by the grazing activities of *Daphnia*. Although zooplankton grazing is widely recognized as the most important factor responsible for the decline of the spring phytoplankton, nutrient limitation can sometimes initiate the clear-water phase: e.g., P in Lake Constance (Geller, 1980), Si in Lake Geneva (Gawler et al., 1987).

---

*Acknowledgements* We wish to thank: P. Blanc, who provided the data on nutrient chemistry as well as logistical support; J.-C. Druart, who enumerated the phytoplankton samples; J. Barraud and J.-P. Moille for help with the drawings; the sampling crew, particularly P. Chifflet; and W. Lampert and W. Geller for their pertinent comments on the manuscript. This study was supported by the "Commission Internationale pour la Protection des Eaux du Léman" (CIPEL) and the "Institut National de la Recherche Agronomique" (INRA).

## References

Balvay, G. 1984. Premières données sur l'évolution quantitative des entomostracés planctoniques dans le lac Léman. *Schweizerische Zeitschrift für Hydrologie* 46:224–229.

Balvay, G. 1987. Equivalence entre quelques paramètres estimatifs de l'abondance du zooplancton total. *Schweizerische Zeitschrift für Hydrologie* 49:75–84.

Balvay, G., Druart, J.-C., Pelletier, J., Pongratz, E., and Revaclier, R. 1984. Plancton, pp. 261–313 in *Le Léman: Synthèse des Travaux de la Commission Internationale pour la Protection des Eaux du Léman contre la Pollution, 1957–1982*, Secrétariat de la Commission internationale, Lausanne.

Balvay, G. and Pelletier, J.P. 1988. Evolution quantitative du plancton durant la phase d'eutrophisation accélérée du lac Léman (Suisse-France). *Verhandlungen Internationale Vereinigung für Theoretische und Angewandte Limnologie* 23:474–481.

Bottrell, H.H., Duncan, A., Gliwicz, Z.M., Grygierek, E., Herzig, A., Hillbricht-Ilkowska, A., Kurasawa, H., Larsson, P., and Weglenska, T. 1976. A review of some problems in zooplankton production studies. *Norwegian Journal of Zoology* 24:419–456.

Bürgi, H.R. and Lehn, H. 1979. Die Langjährige Entwicklung des Phytoplanktons im Bodensee (1965–1975). *Int. Gewasser, für den Bodensee*, Rep. 23.

Bürgi, H.R., Weber, P., and Bachmann, H. 1985. Seasonal variations in the trophic structure of phyto- and zooplankton communities in lakes in different trophic states. *Schweizerische Zeitschrift für Hydrologie* 47:197–224.

Forel, F.-A. 1895. *Le Léman: Monographie Limnologique*, Vol. 2. F. Rouge, Lausanne.

Gawler, M., Blanc, P., Druart, J.C., and Pelletier, J.P. 1987. Dynamique de quelques populations majeures du phytoplancton printanier du Lac Léman en relation avec le broutage et les sels nutritifs. *Actes du Colloque National du CNRS*, Biologie des Populations, Université Claude Bernard 1986, p. 412–419.

Gawler, M., Balvay, G., Blanc, P., Druart, J.-C., and Pelletier, J.P. 1988. Plankton ecology of Lake Geneva: a test of the PEG-model. *Archiv für Hydrobiologie* 114(2):161–174.

Geller, W. 1980. Stabile Zeitmuster in der Planktonsukzession des Bodensees (Überlinger See). *Verhandlungen der Gessellschaft für ökologie* 8:373–382.

Kilham, S.S. 1984. Silicon and Phosphorus growth kinetics and competitive interactions between *Stephanodiscus minutus* and *Synedra* sp. *Verhandlungen Internationale Vereinigung Limnologie* 22:435–439.

Knisely, K. and Geller, W. 1986. Selective feeding of four zooplankton species on natural lake phytoplankton. *Oecologia* 69:86–94.

Lampert, W. 1985. The role of zooplankton: an attempt to quantify grazing, pp. 54–62 in *Lakes Pollution and Recovery*, Proceedings of the International Congress of the European Water Pollution Control Association, Rome, 1985.

Lampert, W., Fleckner, W., Rai, H., and Taylor, B.E. 1986. Phytoplankton control by grazing zooplankton: a study on the spring clear-water phase. *Limnology and Oceanography* 31:478–490.

Lampert, W. and Schober, U. 1978. Das regelmässige Auftreten von Frühjahrs-Algenmaximum und "Klarwasserstadium" im Bodensee als Folge von klimatischen Bedingungen und Wechselwirkungen zwischen Phyto- und Zooplankton. *Archiv für Hydrobiologie* 82:364–386.

Lehn, H. 1968. Sichttiefen im Überlinger See (Bodensee) 1953–1962. *Schweizerische Zeitschrift für Hydrologie* 30:67–74.

Lehn, H. 1973. Phytoplanktonänderungen im Bodensee und einige Folgeprobleme. *Verhandlungen der Gesellschaft für ökologie*: 225–235.

Monod, R., Blanc, P. and Corvi, C. 1984. Les formes du phosphore, pp. 121–140 in *Le Léman: Synthèse des Travaux de la Commission Internationale pour la Protection des Eaux du Léman contre la Pollution, 1957–1982*, Secrétariat de la Commission internationale, Lausanne.

Sommer, U. 1985. Comparison between steady state and nonsteady state competition: experiments with natural phytoplankton. *Limnology and Oceanography* 30:335–346.

Sommer, U. 1986. The periodicity of phytoplankton in Lake Constance (Bodensee) in comparison to other deep lakes of central Europe. *Hydrobiologia* 138:1–7.

Sommer, U., Gliwicz, Z.M., Lampert, W., and Duncan, A. 1986. The PEG-model of seasonal succession of planktonic events in fresh waters. *Archiv für Hydrobiologie* 106:433–471.

Tilzer, M.M. 1988. Secchi disk-chlorophyll relationships in a lake with highly variable phytoplankton biomass. *Hydrobiologia* 162:163–171.

Tilzer, M.M., Sommer, U., Geller, W., Eckmann, R., and Stabel, H.-H. 1985. Untersuchungen zum Stoffhaushalt in der Freiwasserzone des Bodensees/Uberlinger See. *Sonderdruck aus "gwf-wasser/abwasser" 126*, 9:461–466.

Uhlmann, D. 1961. Über den Einfluss von Planktonorganismen auf ihr Milieu. *Int. Rev. ges. Hydrobiol.* 46:115–129.

# 32

# Impact of Zooplankton Grazing on Phytoplankton Along a Trophic Gradient

S. Richman, D. K. Branstrator, and M. Huber-Villegas

ABSTRACT Green Bay is the largest embayment of the North American great lake, Lake Michigan, the sixth-largest lake in the world. The bay is 190 km long, 37 km in mean width, and has an area of approximately 4,000 $km^2$. Its most significant tributary is the Fox River, which has a high mean flow rate (125 $m^3$/sec) and a heavy load of dissolved and particulate materials. It enters Green Bay at its extreme southern end. The transport and dispersion of these substances result in a gradient of trophic conditions ranging from a hypereutrophic extreme southern part to oligo-mesotrophic characteristics in the northern bay. These conditions have a major impact on the abundance and species composition of the zooplankton community and its ability to control phytoplankton species composition and biomass. Species comparisons show that the composition of the zooplankton shifts from small-sized cladocerans and cyclopoid copepods in the lower and mid-bay regions to larger-sized *Daphnia* and calanoid copepod species in the upper bay. Furthermore, zooplankton/phytoplankton biovolume ratios and zooplankton fecundity in relation to phytoplankton biovolume are significantly greater in the upper bay region as compared to lower and mid-bay regions. These differences are related to the effectiveness of *in situ* zooplantkton community grazing, which shows that larger herbivorous zooplankton and higher phytoplankton food quality in the upper bay result in the highest grazing rates and phytoplankton control.

## 32.1 Introduction

Green Bay, the largest embayment of Lake Michigan, stretches 190 km long in a southwest to northeast direction and has an average width of 37 km. Its volume is approximately 70 $km^3$ with the shallowest depths (2–3 m) in the southern bay and the deepest (40–50 m) in the northern bay near Washington Island (Bertrand et al., 1976). Surrounding rich farmland is responsible for heavy nutrient loading which enters the bay through eight major tributaries. The heaviest loading comes through the Fox River, located at the extreme southern end of the bay, with a mean flow rate of one hundred twenty five $m^3$/sec (Gannon, 1974). Discharge from municipal treatment plants and paper industries along the river contributes to annual average loadings of 3,000 kg/day of phosphate and 2,000 kg/day of ammonia-nitrogen into southern Green Bay (Sager and Wiersma, 1972). This heavy nutrient loading, coupled with a shallow depth of 7 m and strong river currents, has created an extremely productive, homothermous, turbid environment at the lower-bay sampling station (Figure 32.1, station no. 2).

Just north of station 2 is a natural shoal extending eastward from the land extension on the west bank. This shoal restricts about 80% of the Fox River nutrient and sediment inputs from travelling further north in the bay (Ahrnsbrak and Ragotzkie, 1970). For this reason, station 3, just north of the shoal, which was also sampled in this study, is less productive than station 2 (see Figure 32.1). The vulnerability of this shallow area to continual surface currents and moderate mixing with Fox River water, however, has created an environment far more productive than those in the northern portions of the bay.

The middle bay station (no. 14) is also affected by the Fox River currents, which run northward along the east bank and deposit Fox River sediments and nutrients up to 40 km north of the outlet (Gannon, 1974). This loading has increased productivity in this deep station to unusually high levels for a mesotrophic-type environment. The middle bay also mixes indirectly with Lake Michigan through the Sturgeon Bay canal. Summer stratification is common at a depth of 10–15 m.

The upper station (no. 30) is continually mixed with oligotrophic water from northern Lake Michigan. This station is essentially unaffected by Fox River nutrient loading. Deep summer stratification is characteristic.

Order-of-magnitude differences in total phosphorus (TP) and other physical/chemical parameters relating to trophic conditions are shown for the sampling for the sampling areas of this study in Table 32.1 after Richman et al. (1984).

A dramatic decrease in phytoplankton particle concentrations ranging from 10,000–23,000 particles/ml in the lower bay to 1,000 particles/ml in the upper bay was also observed in the summer of 1982 (Richman et al., 1984). *In situ* measurements of primary productivity support this decrease in phytoplankton concentration. Sager et al. (1984) measured $^{14}C$ primary productivity and found a dramatic decrease from 1382 down to 7 mg C/$m^3$/day across the trophic gradient from the lower to the upper bay during the 1982 study; similar differences were evident in the present study.

Shifts in species composition of the phytoplankton assemblage over this gradient have been studied extensively (Gannon, 1974; Stoermer and Stevenson, 1979; Richman et al., 1984). The lower bay phytoplankton assemblage is dominated by the diatom *Melosira* and by the cyanobacteria *Anabaena, Aphanizomenon,* and *Microcystis,* all good indicators of eutrophic conditions. The diatoms *Fragilaria* and *Stephanodiscus,* common eutrophic-mesoptrophic species, are highly concentrated in the lower and middle bay. Dominant species

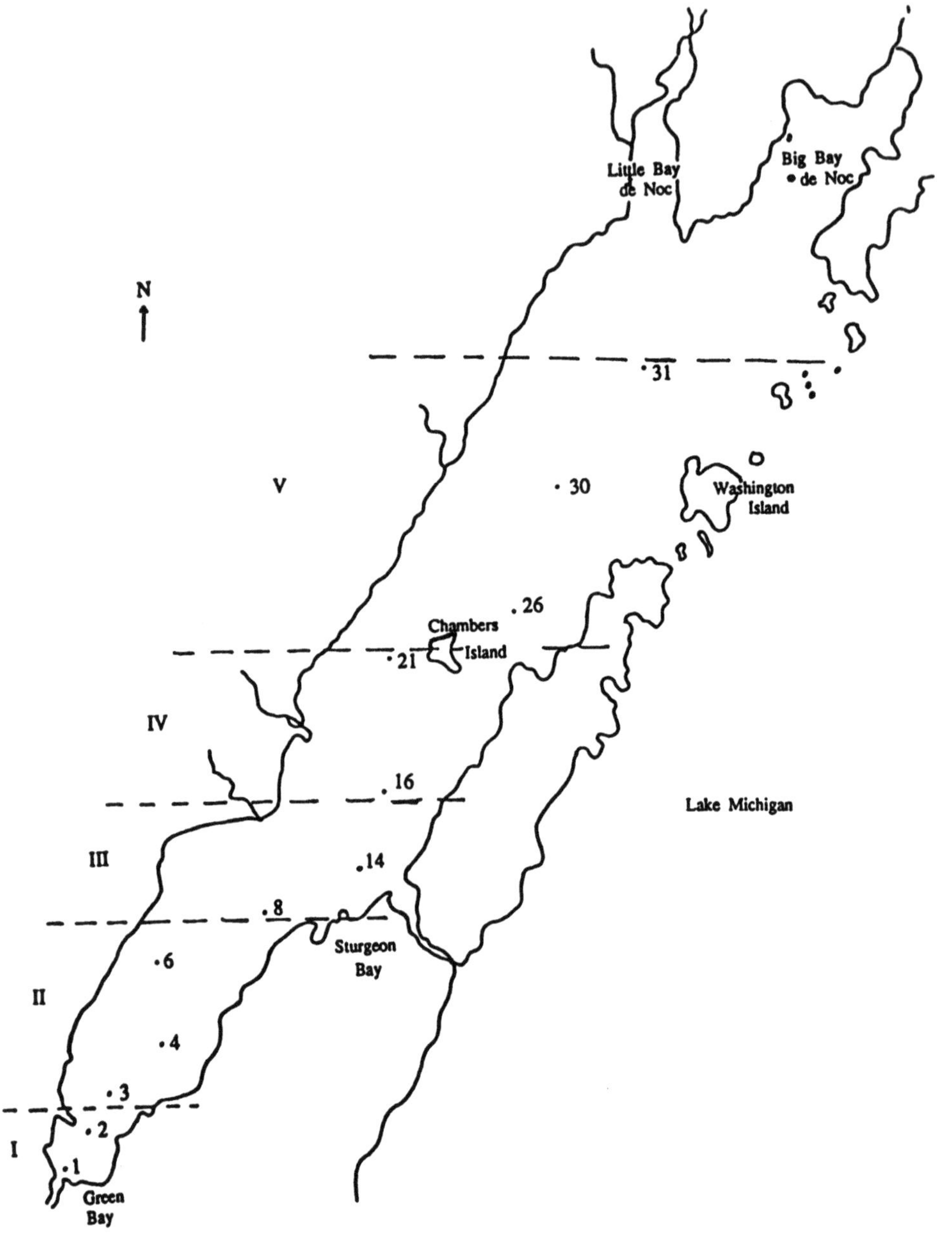


**Figure 32.1** The locations of sampling stations in Green Bay are indicated by the numbered dots: station no. 2—lower bay angle light, no. 3—lower bay entrance light, no. 14—middle bay, no. 30–upper bay. The Fox River enters the bay at the bottom left. The cities of Green Bay and Sturgeon Bay are also indicated. The regions of Green Bay are shown by the roman numerals at the left.

**Table 32.1** Mean trophic characteristics of Green Bay during summer, 1982.

| Region* (1982) | TP (μg/l) | Chl. *a* (μg/l) | Secchi disc (m) | Mean depth (m) |
|---|---|---|---|---|
| I | 190 | 49.7 | 0.5 | 2.9 |
| II | 76 | 8.4 | 2.0 | 7.6 |
| III | 45 | 5.8 | 2.8 | 8.9 |
| V | 27 | 2.1 | 5.6 | 29.0 |

Source: Richman et al. (1984).
* The regions are indicated on the left side of Figure 32.1.

in the middle and upper bay include *Ceratium* and *Dinobryon*, the latter a good indicator of oligo-mesotrophic conditions. Prevalent diatoms in the upper bay, *Asterionella* and *Tabellaria*, are indicative of meso-oligotrophic conditions and suggest that mixing with Lake Michigan is common there. Associated with changes in species composition is a marked shift in the average size of individual phytoplankton. The lower bay supports a broad range of particle sizes—from approximately 4–60 μm in volumetric-converted spherical equivalent diameters compared to the upper bay, which is characterized by a particle range of 7–25 μm throughout the summer.

Earlier studies show that the species composition of the zooplankton community along the bay also shifts according to changing trophic conditions. The small cladocerans of the Bosminidae and Chydoridae groups, *Daphnia retrocurva*, *Ceriodaphnia*, and cyclopoid copepods constituted the major portion of the assemblage in the lower bay (Gannon, 1974; Richman et al., 1984a). The dominance of small cladocerans here may be related to their ability to effectively feed on small particles (3–5 μm), abundant in the lower bay, or on periphyton associated with the filamentous species (Gliwicz, 1969; Gliwicz and Hillbricht-Ilkowska, 1972). The larger cladocerans of the Daphnidae group and calanoid copepods, on the other hand, increased to greater proportions of the assemblage in the upper bay (Gannon, 1974; Richman et al., 1984a).

This study further examines the diverse zooplankton communities ranging from microfilter feeders and sedimentators in the southern eutrophic waters to macrofiltrators and size-selective feeders in the northern oligotrophic waters. Being the primary consumers of phytoplankton primary production, zooplankton communities represent a crucial link between primary production and the carnivorous food web. Carpenter and Kitchell (1984) have shown that shifts in zooplankton biomass and/or size structure can alter primary productivity by factors of several hundred. Although productivity rates in Green Bay have been related to allochthonous nutrient supplies (Sager et al., 1984), the impact of the zooplankton community on primary production must also be considered. McQueen and Post (1986) predicted that in oligotrophic environments, consumer control will be more significant and zooplankton will regulate primary production to a greater extent than in eutrophic environments. Thus, it is expected that shifts in trophic conditions will be closely related to shifts in zooplankton grazing.

The coupled occurrence in Green Bay between a decreasing phytoplankton standing stock and changes in zooplankton species composition suggests a difference may exist in the grazing effectiveness of these zooplankton communities. The potential impacts of Green Bay *Daphnia* and calanoid copepod grazing capabilities have been studied through laboratory feeding experiments (Richman et al., 1980; Richman and Dodson, 1983). These results suggest the importance of the interaction between food quality and feeding effectiveness on the establishment of trophic characteristics in the bay. By performing *in situ* feeding experiments on zooplankton communities in Green Bay, a direct measure of the extent to which zooplankton control primary productivity and overall trophic structure can be obtained.

This study measures *in situ* zooplankton grazing rates in lower, middle, and upper areas of Green Bay, representative of eutrophic, mesotrophic, and oligotrophic regions. The study relates cladoceran egg production and the size and species composition of the zooplankton and phytoplankton to this grazing activity in order to assess the effectiveness of grazing on phytoplankton control under conditions of varying trophy.

## 32.2 Methods

Field samples were collected during five cruises on Green Bay in the summer of 1986: June 30–July 6, July 14–16, August 5–9, August 18–22, and August 25. On August 18 and 25, the lower station was moved from station 3 to station 2 (see Figure 32.1), about 8 km closer to the Fox River outlet.

At each station, zooplankton were collected at six depths by an ABS Robusta (201) electric pump using a 80-μm net, pumping approximately 30–35 liters of water at each depth. The samples were immediately preserved in a 4% formalin-sucrose solution and brought to the laboratory for later examination. The sucrose was used at a concentration of 40 g/l to prevent hypertonicity and to reduce carapace distortion and egg loss by preserved organisms (Haney and Hall, 1973).

Zooplankton species identification and measurements were done using both binocular-compound and stereo-dissecting microscopes. At least two subsamples were obtained for each sample using a 2-ml Hensen-Stempel pipet and 2-ml counting chambers. When necessary, the samples were concentrated until at least 30–50 animals could be obtained in each 2-ml subsample. Typical standard deviations of replicate subsamples were approximately 10–15% of the mean.

All the Cladocera in each subsample were identified, counted, and measured. A large number of length measurements (over 6,300) provided data for size-frequency distributions for a comparison of key Cladocera species relative to station location. *Daphnia* were measured from the front of the head to the base of the spine. Bosminids and chydorids were measured from the posterior to the anterior end, excluding mucro in bosminids. The number of eggs or embryos per adult female found in the brood chamber were also recorded. The copepods were identified and counted, and subsamples from each station were measured for body length, excluding caudal rami.

Length measurements for all microcrustacea were converted to biovolume using formulae specified for particular geometric solids corresponding to each genus, adapted from Bottrell et al. (1976). Phytoplankton biovolumes were measured using a Model B Coulter Counter interfaced to a Commodore Pet 2001 microcomputer. Particles were analyzed using two aperture tubes of 140- and 200-$\mu$m diameter to determine the size spectra of the phytoplankton over a range of 18–180,000$\mu m^3$, equivalent to 3–75$\mu$m in spherical equivalent diameters (Richman et al., 1984). Separate sets of phytoplankton biovolume data, determined with each aperture tube, were meshed to obtain continuous distributions and to determine total phytoplankton biovolume per unit of lake water.

Feeding experiments were carried out according to the method of Lampert and Taylor (1985). Unialgal cultures of *Chamydomonas* (mean diameter, 4–6$\mu$m) were grown in a constant temperature room at 18°C. Two days before the field experiments, 150$\mu$C of $^{14}C$-$NaHCO_3$ were added to 150 ml of algal culture, placed in bright light on a shaker table. The day before each cruise, the culture was concentrated by centrifugation, washed several times with nonradioactive water and resuspended to obtain the desired concentration.

Day grazing tests were performed between 10 a.m. and 3 p.m., while night grazing tests were initiated between ½ to 1 hour after sunset and were run between 10 p.m. and 2 a.m. The same six depths mentioned above were sampled through the water column at each station. Grazing experiments were performed using a 3-liter feeding chamber, modified from Haney's (1971) original design: the food-releasing vial was mounted to the inside of the chamber wall and was physically released by the closing of the top door. Before each test the food-releasing vial was filled with 3 ml of radioactive algae, resulting in an added food concentration of approximately 150 cells/ml. With such a low concentration addition, there was little chance of altering normal grazing activity. The trap was lowered to the desired depth and closed by messenger. After approximately 7 min., it was raised and about 25–50 ml of liquid was removed from the bottom drain for filtering. The remainder of the water was run through an 80-$\mu$m plankton mesh, fitted in a plexiglass cylinder, to collect the zooplankton. The mesh was submersed in two successive baths of soda water to anesthetize the zooplankton and to remove any radioactive algae not ingested. Zooplankton were washed from the filter and collected on a 0.45-$\mu$m Millipore filter which was then placed in a 20-ml plastic scintillation vial. The previously extracted 25–50-ml sample was processed by filtering replicate 2-ml subsamples through separate 0.45-$\mu$m Millipore filters. These filters were placed in replicate scintillation vials and used to determine phytoplankton radioactivity. The 2 ml of filtrate remaining from each run was pipetted into separate scintillation vials and used to correct for dissolved radioactivity. Each scintillation vial was filled with 15 ml of a cocktail solvent (Scinti Verse Bio-HP, Fisher) and counted on a Beckman LS-230 Liquid Scintillation System. The values determined for each vial were used in the following equation, adapted from Lampert and Taylor (1985), to determine the community grazing rate:

$$G = \frac{\text{R(animals}}{[\text{R(water)} - \text{R(filtrate)}]/\text{v}} \cdot \frac{60}{\text{t} \cdot \text{V}}$$

where:

G = community grazing rate (in milliliters per liter per hour),
R = measured radioactivity (in counts per minute),
v = volume of the water samples (in milliliters),
t = feeding time (in minutes), and
V = volume of the chamber (in liters).

## 32.3 Results

For analyzing the impact of zooplankton community feeding on trophic relationships in Green Bay, the large number of species of zooplankton was divided into four groups, consisting of calanoid copepods, cyclopoid copepods, species of *Daphnia*, and cladocerans of the Bosminidae and Chydoridae groups. Percentage composition of these groups in terms of$\mu m^3$ per unit volume shows the relative biovolume importance of each in the lower, middle, and upper regions of Green Bay (Figure 32.2). In the lower bay, the major proportion of zooplankton biovolume consists of small cyclopoid copepods, primarily *Acanthocyclops vernalis*, and small cladocerans, including *Bosmina longirostris*, *Eubosmina coregoni*, and *Chydorus sphaericus*. These groups make up 52–87% of the lower bay zooplankton biovolume. In the middle bay, a similar composition is present with these groups representing 70% or more of the total biovolume. They are also present in the upper bay but represent a smaller proportion of the total community, between 44–65% (see Figure 32.2). Also, they are present in larger size within species to be discussed later, and are represented by larger species as well, such as *Diacyclops thomasi* and *Mesocyclops edax*.

The cladocerans—*Daphnia retrocurva*, *Daphnia galeata mendotae*, and *Ceriodaphnia* sp.—and the calanoid copepods are the major herbivores. While *D. retrocurva* and *Ceriodaphnia* show a definite preference for the lower bay, both lower and middle regions show a considerably smaller overall percentage of Daphnidae and calanoid copepod biovolume (13–29%) than does the upper station (33–55%), with the exception of the Aug, 9 sample when *D. retrocurva* was very high in the lower station (see Figure 32.2*b*). This general trend suggests a shift to a more effective grazing population in the upper region.

The greater grazing effectiveness of the zooplankton community in the upper station is further suggested by an examination of shifts in the size dis-

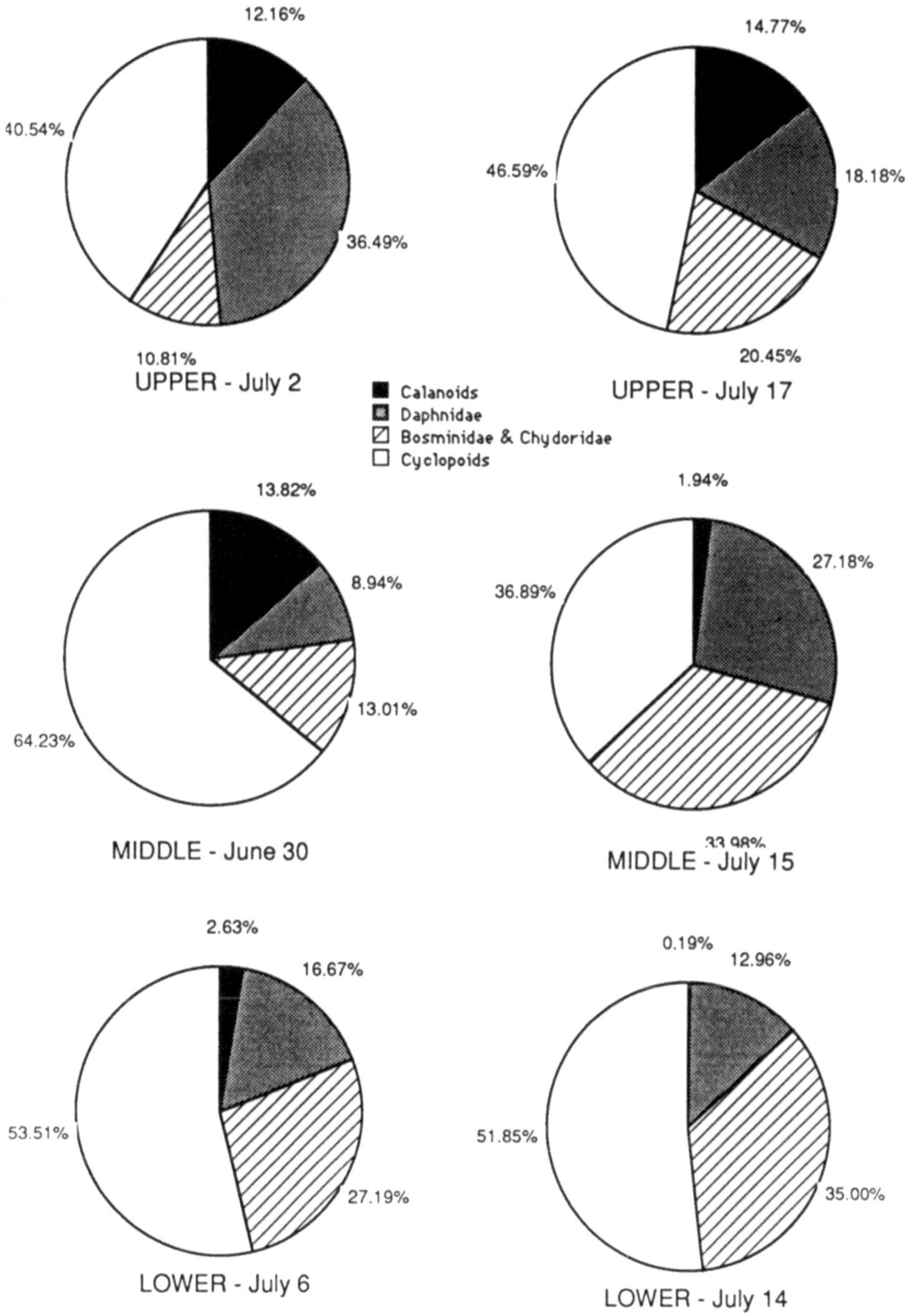


**Figure 32.2** The percentage composition of the average water column concentrations of Green Bay copepods and cladocerans based on biovolume ($\mu m^3$/liter).

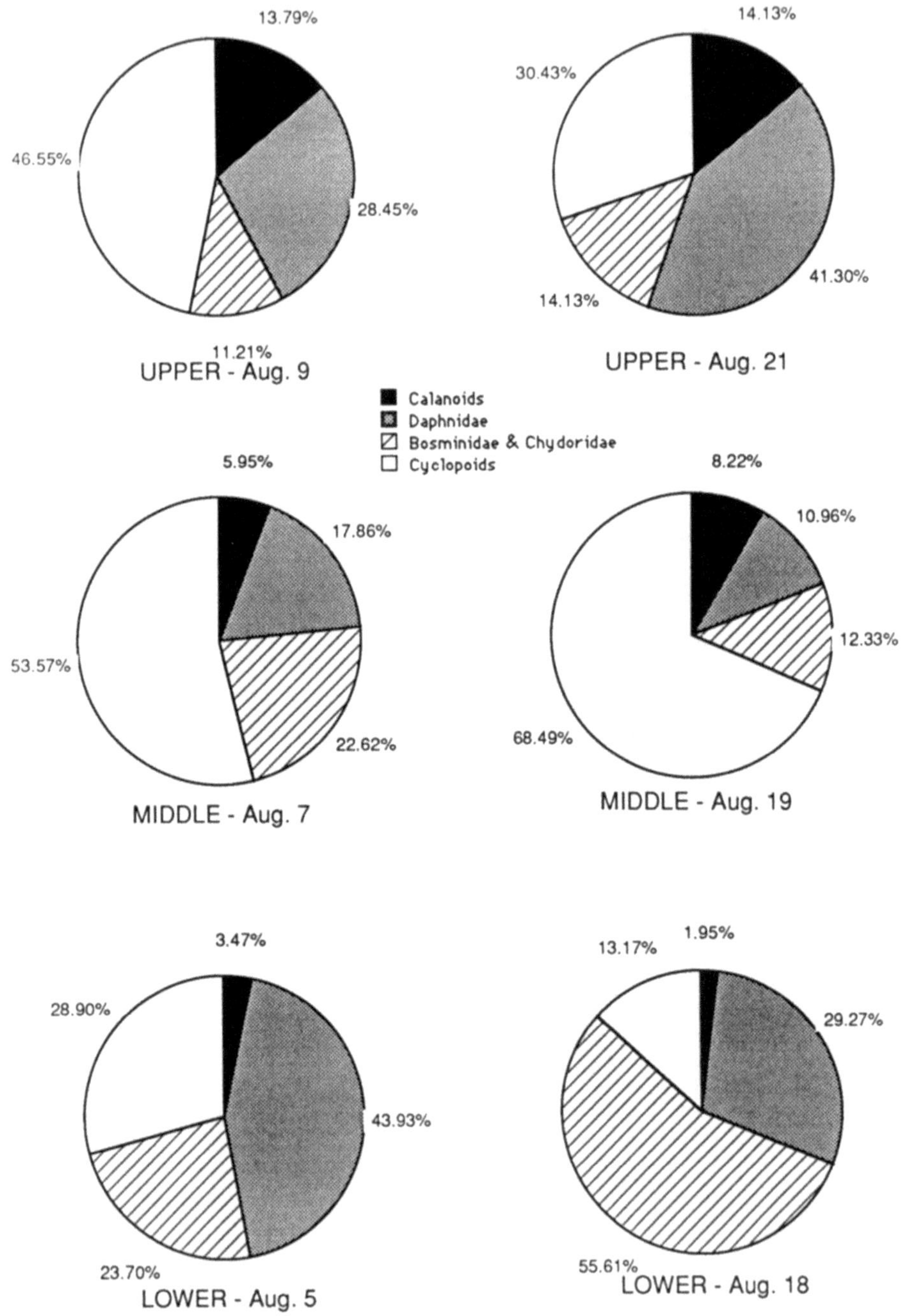


**Figure 32.2** Continued

**Table 32.2** Mean and standard deviation for cladoceran body lengths at three stations in Green Bay during summer, 1986.

| | Body length (mm) | | |
|---|---|---|---|
| Species | Lower station | Middle station | Upper station |
| *Bosmina longirostris* | 0.27 ± 0.05 | 0.29 ± 0.05 | 0.31 ± 0.06[a] |
| *Eubosmina coregoni* | 0.35 ± 0.07 | 0.37 ± 0.07 | 0.39 ± 0.09[a] |
| *Chydorus sphaericus* | 0.25 ± 0.04 | 0.23 ± 0.04 | 0.24 ± 0.04 |
| *Ceriodaphnia sp.* | 0.36 ± 0.08 | 0.35 ± 0.08 | 0.43 ± 0.10[b] |
| *Daphnia galeata mendotae* | 0.68 ± 0.18 | 0.67 ± 0.20 | 0.84 ± 0.25[b] |
| *Daphnia retrocurva* | 0.70 ± 0.20 | 0.73 ± 0.15 | 0.88 ± 0.19[b] |

[a] Upper station is significantly different with respect to lower station (Student's $t$-test, $P < .05$).
[b] Upper station is significantly different with respect to lower and middle stations.

tribution of the six cladoceran species (Table 32.2, Figure 32.3). While there are no significant differences in the size distribution of the six major species between the lower and middle station, the mean size of all species, except *Chydorus sphaericus*, is significantly larger in the upper station when compared to the lower station (P<01), and the three *Daphnia* species (see Figure 32.3*b*) are significantly larger in the upper station when compared to the middle station as well. In addition, the distributions clearly show increases in the size of the largest animals in the upper station for the three *Daphnia* species, as well as for *B. longirostris* and *E. coregoni* (see Figure 32.2*a*).

Comparisons of the vertical distributions of phytoplankton and zooplankton biovolume and grazing rates show the extent to which these parameters are coupled to one another (Figures 32.4 and 32.5). Several trends are apparent in these data. First, a direct relationship between zooplankton community grazing (ml/1/hr) and zooplankton concentration is shown (see Figure 32.5). This relationship is particularly noticeable on all sampling dates in the upper station but is also apparent in the middle station on July 15, 0–3 m; Aug 7, 0–10 m; Aug. 19, 3–8 m; and in the lower station on July 6, 1–2 m and 6 m; July 14, 2 m; and Aug. 18 at 5 m. A second trend is the increasing zooplankton biovolume concentration relative to phytoplankton concentration in the upper bay as compared to middle and lower bay regions (see Figure 32.4 and Table 32.3). On all sampling dates (except on Aug. 5 at the lower station, when *D. retrocurva* was in great abundance) this trend is indicated by the fact that the zooplankton/phytoplankton ratio is consistently greater than one in the upper station while the ratio is less than one in the middle and lower bay regions. These results, summarized in Table 32.3, suggest that in the upper bay, conversion of phytoplankton to zooplankton is significantly greater than in the middle and lower bay. Related to this trend is the cladoceran egg ratio (no. of eggs/adult female) relative to the phytoplankton concentration. While the number of eggs per female in the water column decreases as phytoplankton de-

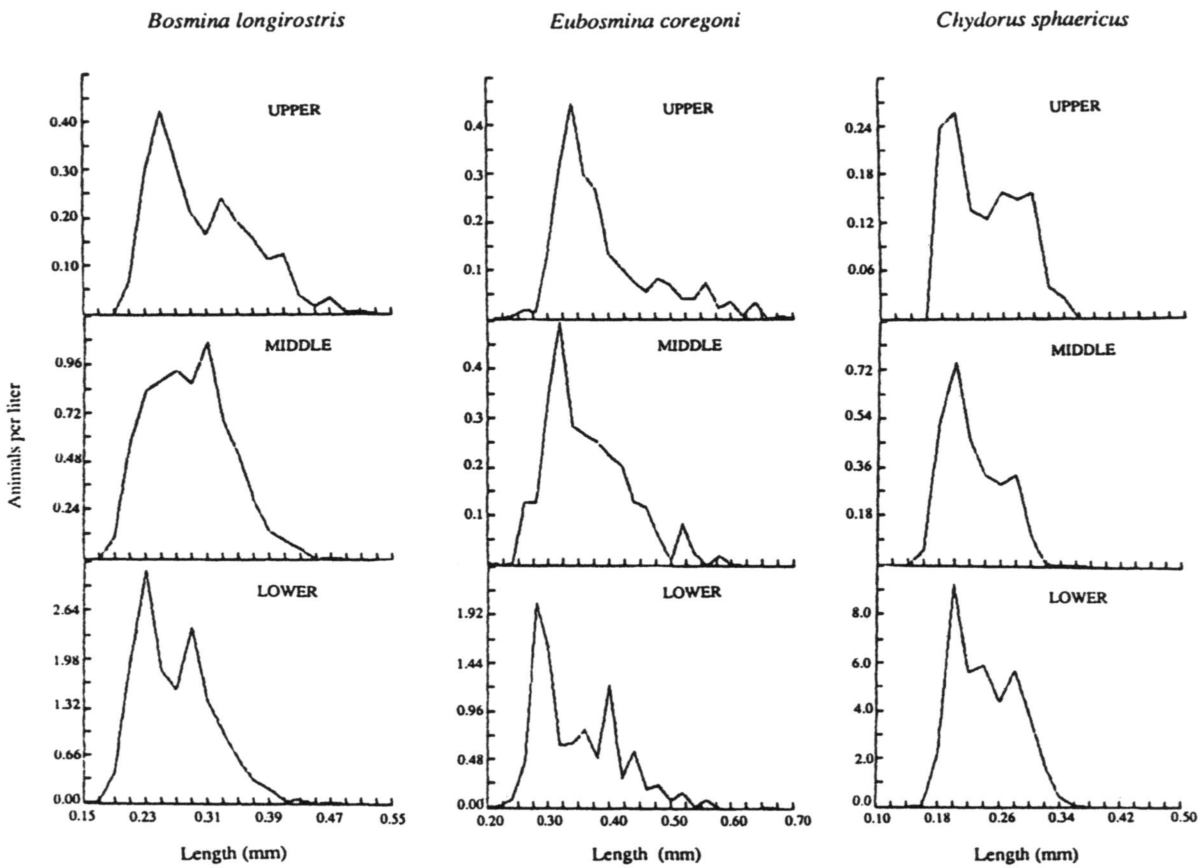


**Figure 32.3a** The mean size-frequency distributions for the six cladoceran species at the upper, middle, and lower regions of Green Bay.

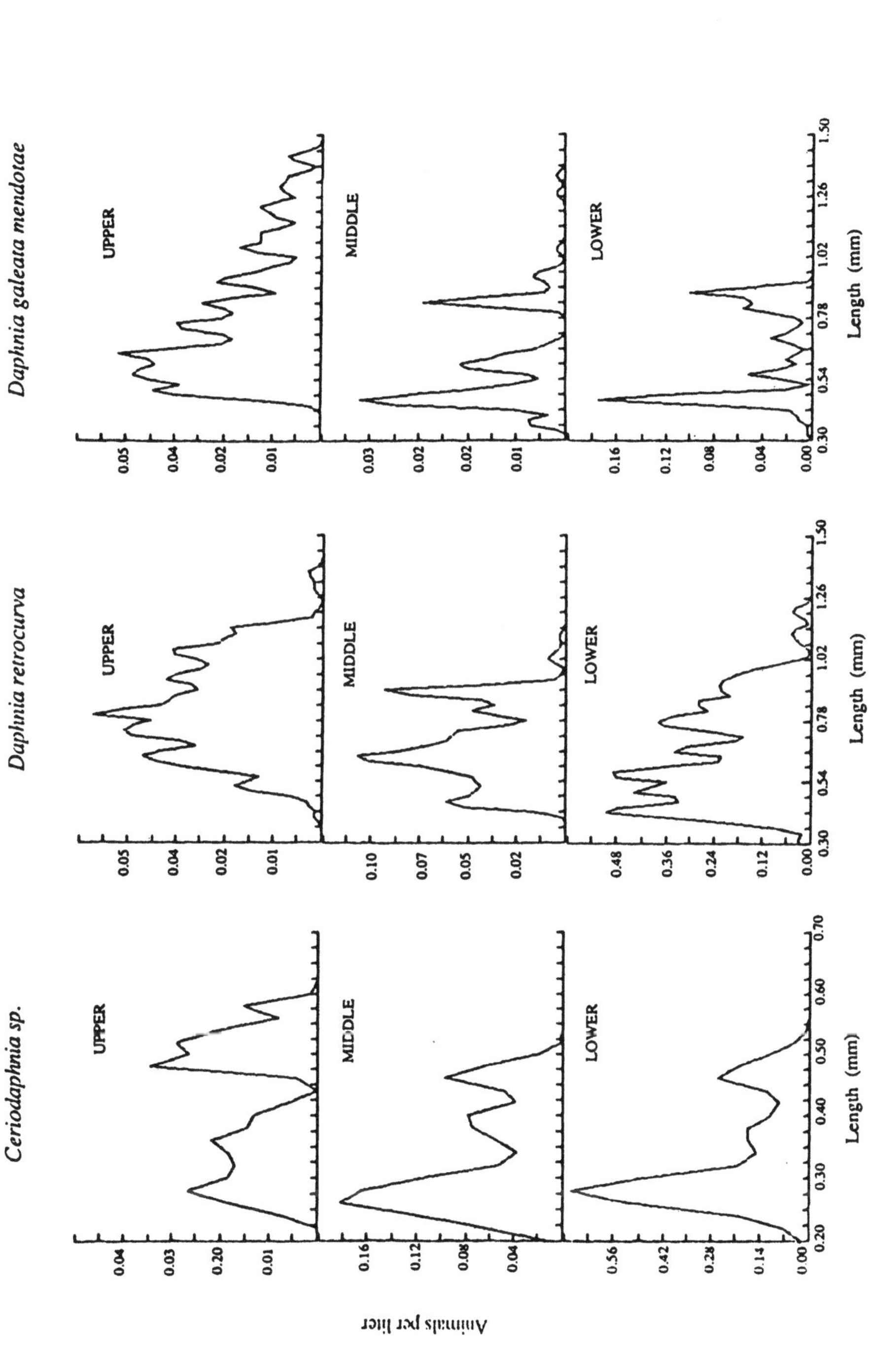
Ceriodaphnia sp.
Daphnia retrocurva
Daphnia galeata mendotae
UPPER
MIDDLE
LOWER
Animals per liter
Length (mm)

**Figure 32.3b**

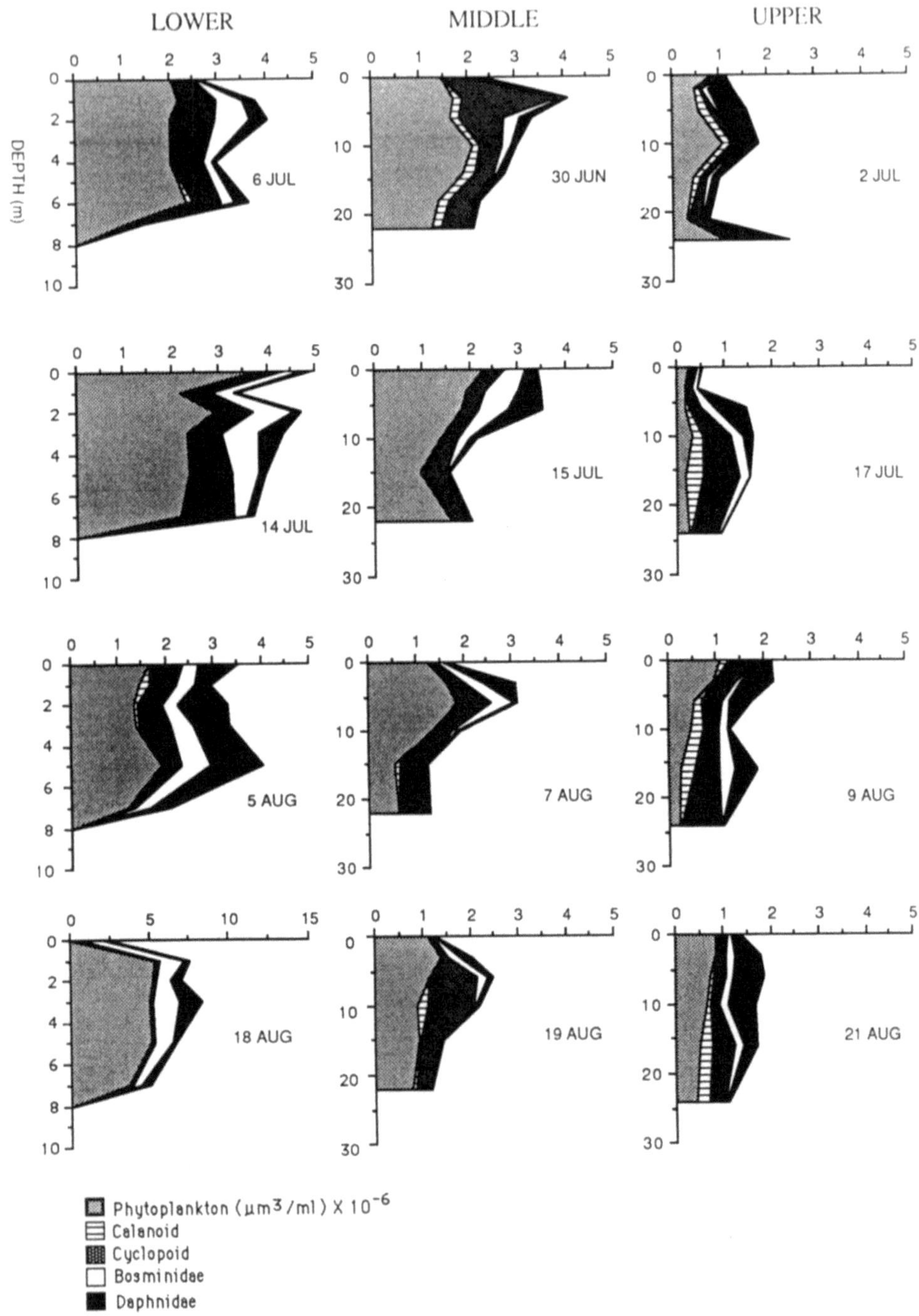


**Figure 32.4** The daytime vertical distributions of phytoplankton and zooplankton biovolume at the three regions of Green Bay.

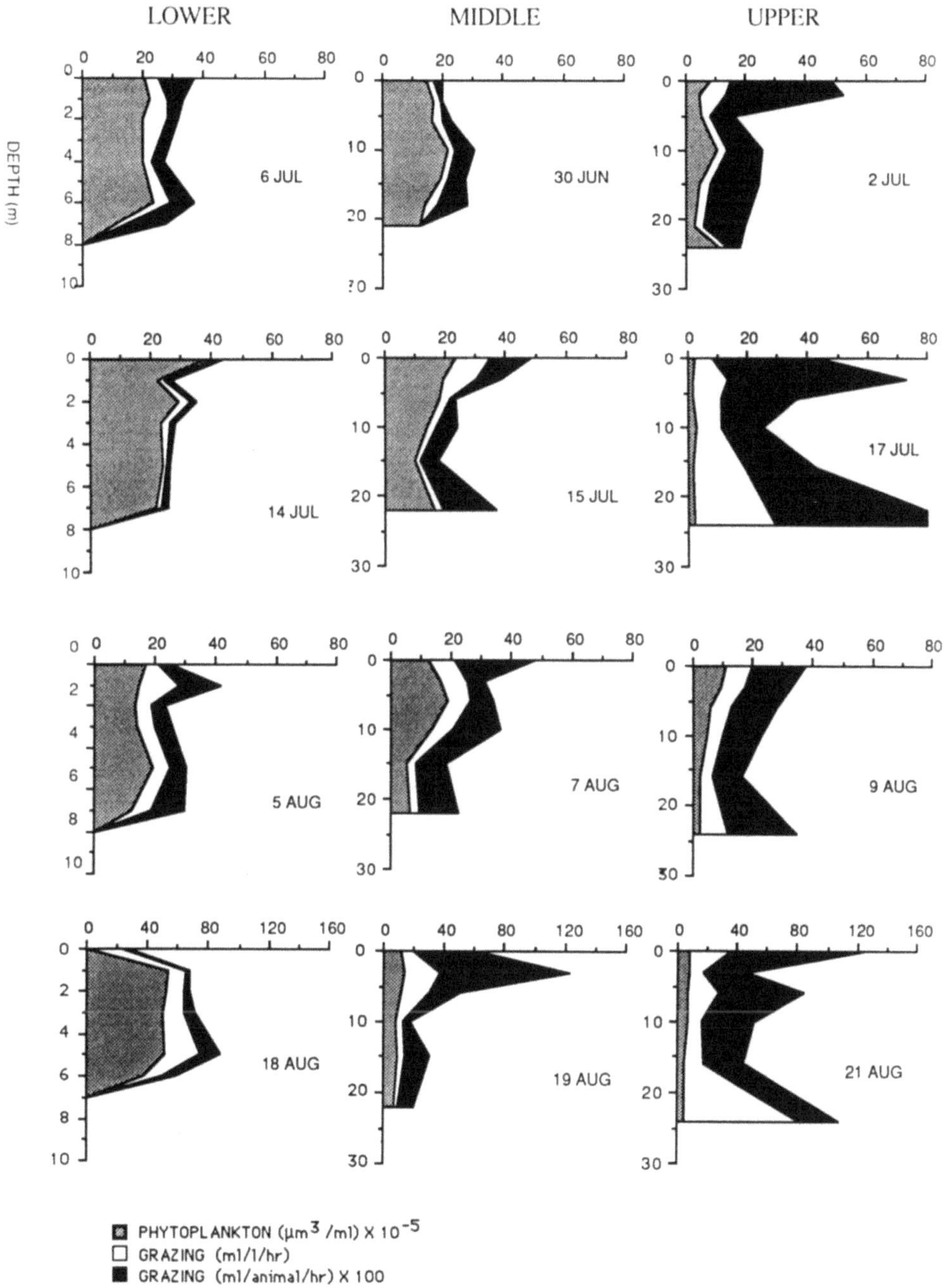


**Figure 32.5** The daytime vertical distributions of phytoplankton biovolume and of zooplankton *in situ* grazing rates, based on *Chlamydomonas* as the indicator food, at the three regions of Green Bay.

creases from lower to upper bay, the number of eggs/female/phytoplankton increases and is significantly greater in the upper station, further suggesting more effective conversion of food to zooplankton in upper bay waters (see Table 32.3).

*In situ* grazing rate on a ml per liter per unit time basis is a measure of the effectiveness of the zooplankton community in the removal of naturally occurring phytoplankton Table 32.4, also see Figure 32.5. These values divided by 10 express the community grazing rate as the percentage of the water volume cleared per unit time. Grazing in these terms shows a trend of increasing percent removal of food from lower to upper bay. This trend is depicted during the July 14–17 and Aug. 18–21 day-sampling periods. Vertical and diel differences in grazing rates within a station are related to biovolume concentrations of the zooplankton, as has already been noted. Overall summer percentages for day and night feeding show that the upper bay zooplankton community removes 26% of the phytoplankton per day as compared to 17.4% and 15.4% for the lower and middle bay respectively. This point is particularly interesting considering the decrease in phytoplankton concentration in the upper bay (see Figure 32.5 and Table 32.4). Grazing expressed on an individual

**Table 32.3** Mean water column zooplankton standing stock per unit of phytoplankton, cladoceran egg ratios, and egg ratios per unit of phytoplankton in three regions of Green Bay.

| Date | Station | Zooplankton ÷ phytoplankton | Egg ratio (no. eggs/female) | Egg ratio ÷ phytoplankton ($\mu m^3 \times 10^{-7}$/ml) |
|---|---|---|---|---|
| July 6 | Lower | 0.61 | 0.71 | 3.65 |
| June 30 | Middle | 0.76 | 0.44 | 2.61 |
| July 2 | Upper | 1.25[a] | 0.31 | 5.04 |
| July 14 | Lower | 0.53 | 0.43 | 1.63 |
| July 15 | Middle | 0.66 | 0.43 | 2.56 |
| July 17 | Upper | 4.28[a] | 0.26 | 11.82 |
| Aug. 5 | Lower | 1.14 | 0.75 | 4.97 |
| Aug. 7 | Middle | 0.78 | 0.79 | 6.52 |
| Aug. 9 | Upper | 2.91[a] | 0.23 | 3.92 |
| Aug. 18 | Lower | 0.43 | 1.45 | 2.94 |
| Aug. 19 | Middle | 0.73 | 0.64 | 6.22 |
| Aug. 21 | Upper | 1.47[a] | 0.37 | 5.52 |
| Aug. 25 | Lower | 0.73 | 0.90 | 1.56 |
| Summer average | Lower | 0.69 | 0.85 | 2.95 |
| | Middle | 0.73 | 0.58 | 4.48 |
| | Upper | 2.48[a] | 0.29[a] | 6.58[b] |

[a] Upper station is significantly different with respect to lower and middle stations.
[b] Upper station is significantly different with respect to lower station (Student's *t*-test, $P < .05$).

**Table 32.4** Phytoplankton standing stock removed per day by zooplankton during day and night feeding in three regions of Green Bay.

| | | Phytoplankton decrease (%/day)* | | |
|---|---|---|---|---|
| Date | Feeding time | Lower station | Middle station | Upper station |
| June 30–July 6 | Day | 12.1 | 7.2 | 11.3 |
| | Night | 13.4 | 9.0 | — |
| July 14–16 | Day | 6.2 | 14.0 | 31.7 |
| | Night | 10.1 | 13.2 | 14.8 |
| Aug. 5–9 | Day | 16.4 | 14.7 | 17.4 |
| | Night | 25.9 | 24.2 | 18.7 |
| Aug. 18–22 | Day | 38.0 | 25.2 | 62.9 |
| Summer average | Day and night | 17.4 | 15.4 | 26.1 |

* Based on *Chlamydomonas* as the indicator food.

**Table 32.5** *In situ* day and night grazing rates at the three regions of Green Bay.

| | | Grazing rate | | |
|---|---|---|---|---|
| Date | Feeding time | Lower station | Middle station | Upper station |
| June 30–July 6 | Day | 0.08 | 0.05 | 0.19[b] |
| | Night | 0.06 | 0.05 | — |
| July 14–16 | Day | 0.02 | 0.09 | 0.37[b] |
| | Night | 0.03 | 0.16 | 0.12[c] |
| Aug. 5–9 | Day | 0.07 | 0.13 | 0.16[c] |
| | Night | 0.10 | 0.12 | 0.25[b] |
| Aug. 18–22 | Day | 0.07 | 0.30 | 0.44[c] |
| Summer average | Day and night | 0.06 | 0.13 | 0.26[b] |

[a] Based on *Chlamydomonas* as the indicator food.
[b] Upper station is significantly different with respect to lower and middle stations.
[c] Upper station is significantly different with respect to lower station (Student's $t$-test, $P < .05$).

basis (ml per animal per time) shows a dramatic increase from lower to upper bay waters for both the day and night periods (Table 32.5, also see Figure 32.5). Although there are smaller concentrations of zooplankton in the upper waters, individual grazing rates clearly show the increased effectiveness of larger species of *Daphnia* and calanoid copepods.

## 32.4 Discussion

Our results show that the species composition of zooplankton communities in Green Bay are substantially different in tthe three regions studied. These differences, due to a host of parameters—including physicochemical factors, phytoplankton composition, and planktivore predation—result in important differences in community grazing rates and in the ability of zooplankton to control phytoplankton production.

The two dominant cladoceran species by biovolume in the lower bay consisted of *Bosmina longirostris* and *Eubosmina coregoni* (see Figure 32.2 and 32.4). The small size of these invertebrates reduces the size of particles which they can easily ingest. It has been reported that in eutrophic conditions, microfiltrators ingest, on the average, 5–15% phytoplankton, 10–20% detritus, and 70–80% bacteria (Sprules, 1980; Gliwicz, 1969). The diet of *Bosmina*, nevertheless, has been found to be quite flexible. DeMott and Kerfoot (1982) show the *Bosmina's* selectivity between *Chlamydomonas* and *Aerobacter* is variable and depends mainly on previous feeding history and relative food abundance. Bogdan and Gilbert (1982) find this fluctuation is mostly dependent on temperature and food concentration. In either case, the eutrophic conditions of the lower bay are prime for the grazing of *Bosmina*. The rich benthos contains an abundance of bacteria and detritus, in addition to a broad size-range of phytoplankton, which is continually mixed due to shallow depths and strong surface currents. *Bosmina*, which are commonly found concentrated in littoral zones and warm benthic environments, show a fairly even vertical distribution in the lower bay.

In the middle bay, there is a clumping of bosminids above the thermocline, which occurs between 10–15 m. Mesotrophic conditions here support a less abundant stock of bacteria and detritus, but water currents running along the eastern bank of the bay from the Fox River provide a high supply of nutrients to this area. In the upper bay there is a near absence of Bosminidae. Their filtering efficiency drops markedly in this oligotrophic environment where food particles are larger and more dispersed.

Within the bosminid group, *Bosmina longirostris* was frequently more abundant in the lower bay while *Eubosmina coregoni* was more common in the upper bay. The middle bay supported equal populations of the two species. This shift in species can be explained by the impact of planktivore grazing in the southern half of Green Bay. Approximately one-half of the alewife landed commercially in the Lake Michigan basin are caught in Green Bay (Gannon, 1972). Since 1972, the alewife population has been decimated by salmonid feeding, but recent increased populations of lake perch, another planktivore, have been reported by commercial fishermen. These planktivores, which forage visually, have been shown by Stenson (1976) to prefer the larger of the two species, *Eubosmina coregoni*, which may explain its scarcity in the lower bay. In the middle and upper bay, where planktivore grazing pressure is decreased, *E. coregoni* is more abundant and is even able to outcompete *B. longirostris*, given its preference for meso-oligotrophic conditions (Beeton, 1965).

On August 18, the lower bay station was moved closer to the mouth of the Fox River, from station 3 to station 2 (see Figure 32.1), and a dramatic change in zooplankton species composition was observed. The new dominating member was *Chydorus sphaericus*. This change was associated with a bloom of cyanobacteria which occurred simultaneously. *C. sphaericus* does not feed on cyanobacteria, but rather attaches to their larger filaments in the open water. Presumably, it feeds more profitably there, in addition to feeding on bacteria growing on the filament surfaces (Gliwicz, 1977). Haney (1973) observed *Chydorus* cells attached to and swimming to and from the cyanobacterium *Anabaena*.

It is well known that the Daphnidae graze more effectively on the nannoplankton assemblage than do the Bosminidae (Gliwicz and Hillbricht-Ilkowska, 1972; Porter and Orcutt, 1980; Sprules, 1980; Brendelberger, 1985). Larger species such as *Daphnia galeata mendotae* clearly prefer upper bay water where they feed more effectively on high-quality food (Richman and Dodson, 1983).Even though food is limited there, it is likely that nutrients are released in feeding and excretion, allowing for fast algal turnover (Lehman, 1980). Many studies have been done to understand adaptations to limited food conditions; Tillmann and Lampert's (1984) studies on the effect of food limitation on differently sized *Daphnia* show that under food-limited conditions, the smaller sizes are favored, since they require less food to reproduce, and they conclude that "the smallest species was the best competitor." Contrary to Tillmann and Lampert's ideas, Gerritsen and Kou (1985) present a model in which foraging activity favors the larger species when food is limited in the pelagic environment. The results of their study demonstrate that a more oligotrophic environment sustains larger species than does a eutrophic one. The model is based on the swimming speed of cladocerans of different sizes. The swimming speed of the smaller cladocerans can be as large as 30-body-lengths or more per second, while the larger ones can swim less than one-body-length per second. In terms of distance covered per body movement, the larger forms swim faster, since to cover any distance a large cladoceran will need to move its body length fewer times than a smaller one. Therefore, in addition to the energy expenditure caused by higher swimming rates by the smaller species, particle-encountering rates will also be lower in these species than in larger ones. The 1986 Green Bay zooplankton samples indicate that indeed the large species are found in the more oligotrophic upper bay.

The size distributions of the six cladoceran species studied show that the smaller species had relatively little or no change in the size shift from the lower to the upper bay. *Chydorus sphaericus*, being the smallest, showed essentially no size shift. This cladoceran is too small to be seen by fish predators. *Bosmina longirostris*, the second-smallest cladoceran sampled, shows a slight size increase from the lower to the upper bay (see Table 32.2 and Figure 32.3). This shift is larger for *Eubosmina coregoni*, which is a larger species than *B. longirostris*. These smaller crustaceans are more common prey to invertebrate predators, such as the copepods, *Leptodora*, and *Chaoborus*. Invertebrate predation generally does not depend on the size of the prey, but rather on their shape. Small limnetic crustaceans undergo cyclomorphic effects to alter their morphology, becoming unrecognizable to their predator upon capture, and thus are released unharmed (Dodson, 1974; Kerfoot and Peterson, 1980; Black, 1980; Black and Hairston, 1983).

The larger species show very significant size shifts from the lower to the upper bay. The size shift from the lower to the middle bay is not as significant as the shift from the middle to the upper bay. *Daphnia* populations are particularly affected by heavy predation (Hall, 1964). The most impressive shift in size is shown by *Daphnia galeata mendotae,* the largest of the *Daphnia* species collected. Planktivorous fish predation in the lower bay, coupled with the problems that larger filter-feeders encounter while trying to filter in waters where cyanobacteria are dominant (Porter and Orcutt, 1980; Richman and Dodson, 1983; Gliwicz, 1985), keeps the larger members of this species from developing in this part of the bay.

The increasing concentration trend of the Daphnidae from lower to upper bay is also evident with the calanoid copepods, which adds to the increasing effectiveness of grazing control of phytoplankton in the upper bay. Their size-selective feeding behavior is well documented (Richman et al., 1980; Price and Paffenhöfer, 1985). In an oligotrophic environment the calanoids may encounter more food per unit of energy expended than do similarly sized cladocera with their size-selective raptorial feeding behavior (Price and Paffenhöfer, 1985).

Unlike the other groups, the cyclopoid copepods are more or less evenly distributed throughout the bay. Their ability to feed on both phytoplankton and zooplankton and their resourceful grazing strategy of feeding on high energy food, result in a predominance of these forms. The cyclopoids did feed on the indicator food, $^{14}$C-*Chlamydomonas,* and it is clear that they contribute to phytoplankton control and, therefore, are included in the community grazing analysis.

While there are substantial decreases in the number of zooplankters from lower to upper bay, the zooplankton biovolume remains quite constant, the result of the presence of a large number of smaller forms in the lower and middle bay, and a small specimen number of larger species in the upper bay. Grazing rates, therefore, should be assessed in relation to zooplankton biovolume. Larger animals consume more phytoplankton to sustain their size and subsequently have a greater impact on food resources and total community grazing. Lampert and Taylor (1985) found that zooplankton grazing rates, measured in ml/1/hr, closely followed zooplankton biomass. Similar correlations are found in Green Bay where peaks in the vertical distribution of zooplankton biovolume are commonly associated with higher grazing rates (see Figures 32.4 and 32.5).

While community grazing is more or less constant from station to station, as is total zooplankton biovolume, there is a trend of increasing grazing from lower to upper bay. This trend is particularly evident in the July 14–17 and Aug. 18–21 results (see Figure 32.5 and Table 32.4). Similar grazing rates are evident at night and no consistent pattern of vertical migration was evident. Water-column day and night averages of community grazing shows that the lower and middle bay community removes 15–17% of the phytoplankton standing stock per day, whereas the upper bay community removes an average of 26%. This difference might play an important role in the decreased phytoplankton standing stock in the upper bay and in the increased proportion of zooplankton to phytoplankton and cladoceran egg/female/phytoplankton standing stock there (see Table 32.3). The composition of the zooplankton in the upper bay makes for a more effective grazing community.

This effectiveness is evident when grazing rates are compared on an individual basis. From the lower to the upper bay there is a dramatic increase in the effectiveness of individual grazers at removing limnetic phytoplankton (see Figure 32.5 and Table 32.5). This increase is directly related to the feeding strategies of the predominant zooplankters in the different areas of the bay. In the lower bay, bosminids mainly remove the smaller phytoplankton species. The presence of scattered calanoids and prevalent *Daphnia* in the middle bay results in a greater average grazing rate per individual than values in the lower bay. The abundance of calanoids and *Daphnia galeata mendotae* in the upper bay is most responsible for the effectiveness of this assemblage throughout the summer.

The lower bay is the most eutrophic region of Green Bay. The zooplankton community here, dominated by cyclopoids, *Bosmina, Chydorus,* and, to a lesser extent by *Daphnia,* is faced with extremely high phytoplankton productivity. The less-effective grazing ability of these zooplankton, compared to those in the middle and upper stations, is apparent. In the lower bay, the ratios of zooplankton biovolume to phytoplankton biovolume varies between 0.43–1.14. Microfiltrators here feed mainly on smaller phytoplankton, while the bulk of primary phytoplankton production remains unconsumed. These algae settle and are metabolized through benthic and bacterial food webs. Strong currents and wave action continually resuspend settled material, which increases turbidity in the lower bay. This material increases the small particulate forage base for zooplankters, but subsequently decreases the overall quality of food. Particulate matter in the lower bay contains a large proportion of detrital material. Zooplankton may ingest this material, but assimilation rates are likely to drop, and grazing effectiveness thus declines. In this eutrophic environment, a majority of the phytoplankton crop is not transferred through the pelagic food web. Instability in the food web is a result of increased primary production, but the inefficiency of energy transfer in the pelagial is a result of ineffective zooplankton grazing. Zooplankton here clear only 17.4% of available phytoplankton biovolume per day. This means the time required to convert a given stock of algae to zooplankton biomass is 5.7 days.

Physical parameters in the middle bay, including a deep thermocline between 10–15 m and a continual flushing with Lake Michigan waters via the Sturgeon Bay canal, suggest this water should be mesotrophic in nature. High zooplankton and phytoplankton biovolume measurements, however, show that this station is biologically more comparable to the lower bay. Phytoplankton biovolume is relatively high because of extended nutrient loading from the lower bay. The zooplankton assemblage, composed of cyclopoids, *Bosmina,*

*Daphnia*, and, to a lesser extent, of calanoid copepods, is similar to the lower bay in its ability to control the algae. The ratio of zooplankton to phytoplankton biovolume is similar to the lower bay (0.43–0.90), and the time for the grazers to turn over the phytoplankton is about the same (6.5 days).

The upper bay is a meso-oligotrophic environment marked by a deep thermocline. The small phytoplankton stock is a result of a depleted nutrient base and more effective grazing activity. Zooplankton to phytoplankton biovolume ratios are 1.25–4.28, indicative of an extremely efficient zooplankton assemblage. The zooplankton convert a given stock of algae to equivalent zooplankton biovolume every 3.8 days, or 33% faster than in the lower bay. This region is highly efficient at passing energy through trophic levels and thus supports the largest assemblage of zooplankton per unit of food available. With minimal allochthonous nutrient loading in the upper bay, it is unusual that the phytoplankton community there is not reduced to an even-smaller density over the summer. Lehman (1980) has shown that food excretion and egestion frequently creates concentrated nutrient pools from which nitrogen and phosphorus are rapidly reassimilated by algae. The raptorial mode of copepod grazing also produces free, nutrient-rich, organic fragments. An abundance of raptorial grazers in the upper bay may thus induce large nutrient pools and the relatively high phytoplankton turnover rate needed to sustain the zooplankton biomass.

---

*Acknowledgements* This investigation was supported by the University of Wisconsin Sea Grant College Program under an institutional grant from the National Oceanic and Atmospheric Administration, U.S. Department of Commerce. We thank Kathryn Keating, Joseph Krupka, and Dr. P.E. Sager for technical assistance in sampling and analysis. The authors are also grateful for the detailed comments of two anonymous reviewers.

## References

Ahrnsbrak, W.F., and R.A. Ragotzkie. 1970. Mixing processes in Green Bay. *International Association Great Lakes Research* 13:880–890.

Beeton, A.M. 1965. Eutrophication of the St. Lawrence Great Lakes. *Limnology and Oceanography* 10:240–254.

Bertrand, G., J. Lang, and J. Ross. 1976. The Green Bay Watershed, Past/Present/Future. *University of Wisconsin Sea Grant College Program, Technical Report No. 229.*

Black, R.W. 1980. The nature and causes of cyclomorphosis in a species of the *Bosmina longirostris. Ecology,* 61:1122–1132.

Black, R.W. and N.G. Hairston, Jr. 1983. Cyclomorphosis in *Eubosmina longispina* in a small North American pond. *Hydrobiologia* 102:61–67.

Bogdan, K.G. and J.J. Gilbert. 1982. Seasonal patterns of feeding by natural populations of *Keratella, Polyarthra,* and *Bosmina;* Clearance rates, selectivities, and contributions to community grazing. *Limnology and Oceanography* 27:918–934.

Bottrell, H.H., A. Duncan, Z.M. Gliwicz, E. Grygierek, A. Herzig, A. Hillbricht-Ilkowska, H. Kurasawa, P. Larsson, and T. Weglenska. 1976. A review of some problems in zooplankton production studies. *Norwegian Journal of Zoology* 24:419–456.

Brendelberger, H. 1985. Filter mesh-size and retention efficiency for small particles: comparative studies with cladocera. In: Lampert, W. (ed.). *Food Limitation and Structure of Zooplankton Communities. Archiv für Hydrobiologie Beiheft Ergebnisse der Limnologie* 21:135–146.

Carpenter, S.R. and J.F. Kitchell. 1984. Plankton community structure and limnetic primary production. *American Naturalist* 124:159–172.

DeMott, W.R. and W.C. Kerfoot. 1982. Competition among cladocerans: nature of interaction between *Bosmina* and *Daphnia. Ecology* 63:1949–1966.

Dodson, S.I. 1974. Adaptive change in plankton morphology in response to size-selective predation: a new hypothesis of cyclomorphosis. *Limnology and Oceanography* 19:721–729.

Gannon, J.E. 1972. Effects of eutrophication and fish predation on recent changes in zooplankton crustacea species composition in Lake Michigan. *Transactions American Microscopic Society* 91:82–84.

Gannon, J.E. 1974. The crustacean zooplankton of Green Bay, Lake Michigan. *International Association Great Lakes Research* 17:28–51.

Gerritsen, J. and J. Kou. 1985. Food limitation and body size. In: Lampert, W. (ed.). *Food Limitation and the Structure of Zooplankton Communities. Archiv für Hydrobiologie Beiheft Ergebnisse der Limnologie* 21:173–184.

Gliwicz, Z.M. 1969. Studies on the feeding of pelagic zooplankton in lakes with varying trophy. *Ekologia Polska-Seria A,* 17:665–708.

Gliwicz, Z.M. 1977. Food size selection and seasonal succession of filter feeding zooplankton in an eutrophic lake. *Ekologia Polska-Seria A,* 25:179–225.

Gliwicz, Z.M. 1985. Predation or food limitation: an ultimate reason for extinction of planktonic cladoceran species. In: Lampert, W. (ed.) *Food Limitation and the Structure of Zooplankton Communities Archiv für Hydobiologie Beiheft Ergebnisse der Limnologie* 21:419–430.

Gliwicz, Z.M. and A. Hillbricht-Ilkowska. 1972. Efficiency of the utilization of nannoplankton primary production by communities of filter feeding animals measured *in situ. Internationale Vereinigung für Theoretische und Angewandte Limnologie* 18:197–203.

Hall, D.J. 1964. An experimental approach to the dynamics of a natural population of *Daphnia galeata mendotae. Ecology* 45:94–112.

Haney, J.F. 1971. An *in situ* method for the measurement of zooplankton grazing rates. *Limnology and Oceanography* 16:970–977.

Haney, J.F. 1973. An *in situ* examination of the grazing activities of natural zooplankton communities. In: Lampert, W. (ed.). *Archiv für Hydrobiologie Beiheft Ergebnisse der Limnologie* 72:87–132.

Haney, J.F. and D.J. Hall. 1973. Sugar-coated *Daphnia*: A preservation technique for Caldocera. *Limnology and Oceanography* 18:331–333.

Kerfoot, W.C. and C. Peterson. 1980. Predatory copepods and *Bosmina:* replacement cycles and further influences of predation upon prey reproduction. *Ecology* 61:417–431.

Lampert, W. and B.A. Taylor. 1985. Zooplankton grazing in a eutrophic lake: implications of diel vertical migration. *Ecology* 66:68–82.

Lehman, J.T. 1980. Nutrient recycling as an interface between algae and grazers in freshwater communities. *American Society of Limnology and Oceanography Special Symposium* 3:251–263. Univ. Press of New England, Hanover, New Hampshire.

McQueen, D.J. and J.R. Post. 1986. Trophic relationships in freshwater pelagic ecosystems. *Canadian Journal of Fisheries and Aquatic Science* 43:1571–1581.

Porter, K.G. and J.D. Orcutt, Jr. 1980. Nutritional adequacy, manageability, and toxicity as factors that determine the food quality of green and blue-green algae for *Daphnia. American Society of Limnology and Oceanography Special Symposium* 3:268–281. New England.

Price, H.J. and G.A. Paffenhöfer. 1985. Perception of food availability by calanoid copepods. In: Lampert, W. ed. Food Limitation and the Structure of Zooplankton Communities. *Archiv für Hydrobiologie Beiheft Ergebnisse der Limnologie.* 21:115–124.

Richman, S., S.A. Bohon, and S.E. Robbins. 1980. Grazing interactions among freshwater calanoid copepods. *American Society of Limnology and Oceanography Special Symposium,* 3:219–233. New England.

Richman, S. and S.I. Dodson, 1983. The effect of food quality on feeding and respiration by *Daphnia* and *Diaptomus. Limnology and Oceanography* 28:948–956.

Richman, S., P.E. Sager, G. Banta, T.R. Harvey, and B. T. DeStasio. 1984. Phytoplankton standing stock, size distribution, species composition and productivity along a trophic gradient in Green Bay, Lake Michigan. *Internationale Vereinigung für Theoretisch und Angewandte Limnologie* 22:460–469.

Richman, S., M.D. Bailiff, L.J. Mackey, and D. Bolgrien. 1984a. Zooplankton standing stock, species composition and size distribution along a trophic gradient in Green Bay, Lake Michigan. *Internationale Vereinigung für Theoretisch und Angewandte Limnologie* 22:475–487.

Sager, P.E. and J.H. Wiersma. 1972. Nutrient discharges to Geeen Bay, Lake Michigan from the lower Fox River. *International Association Great Lakes Research* 15:132–148.

Sager, P.E., G. Banta, and J. Kirk. 1984. The relation between areal and volumetric expressions of $^{14}C$ productivity in Green Bay, Lake Michigan. *Internationale Vereinigung für Theoretisch und Angewandte Limnologie* 22:470–474.

Spurles, G.W. 1980. Zoogeographic patterns in the size structure of zooplankton communities, with possible applications to lake ecosystem modeling and management. *American Society of Limnology and Oceanography Special Symposium,* 3:642–656. New England.

Stenson, J.A.E. 1976. Significance of predator influence on composition of *Bosmina* spp. populations. *Limnology and Oceanography* 21:814–822.

Stoermer, E.F. and R.J. Stevenson. 1979. Green Bay phytoplankton composition, abundance, and distribution. *U.S. Environmental Protection Agency, Region 5, Chicago, Illinois, Report No. 905/3-79-002.* 104 pp.

Tillmann, U. and W. Lampert. 1984. Competitive ability of differently sized *Daphnia* species: an experimental test. *Journal of Freshwater Ecology* 2:311–323.

# 33

# Strength of Zooplankton-Phytoplankton Coupling in Relation to Lake Trophic State

Heath J. Carney and James J. Elser

ABSTRACT We examine the interactions of zooplankton and phytoplankton in lakes representing a broad range of environments and trophic states. Comparisons focus on the grazing impact and the regeneration rate of the most limiting nutrient during summer stratification in several large lakes: Lake Michigan (USA), Lake Titicaca (Peru), and Lake Tahoe (USA). These lakes range over an order of magnitude in plankton biomass and productivity, and in all of them copepods are the dominant crustaceans. In Lake Michigan (2.2 mg $m^{-3}$ chl *a* biomass and 2.86 mg $C\text{-}m^{-3}\cdot h^{-1}$ primary productivity), zooplankton have a very significant grazing impact on phytoplankton, and they can supply the majority of phytoplankton phosphate demand through regeneration. Daily phosphate demand is much greater than the available ambient concentration, so turnover of the available phosphate pool is quite rapid. Lake Titicaca has similar phytoplankton biomass (0.95 mg $m^{-3}$ chl. *a*) and productivity (3.5 mg $C\text{-}m^{-3}\cdot h^{-1}$), yet zooplankton concentrations are lower. Thus, zooplankton have less impact, though size-selective grazing may be significant, and nutrient regeneration may reach 20 percent of phytoplankton N demand. Plankton biomass (0.23 mg $m^{-3}$ chl. *a*) and productivity (0.25 mg $C\text{-}m^{-3}\cdot h^{-1}$) are much lower in Lake Tahoe than the other two lakes, and the zooplankton-phytoplankton coupling is much weaker. Grazing by zooplankton is two orders of magnitude lower than phytoplankton growth, and zooplankton regeneration appears to be negligible in relation to other fluxes and to the available concentrations of limiting nitrogen.

These examples indicate that there is substantial variability in strength of zooplankton-phytoplankton interactions. For these three large lakes the strength of coupling increases with increasing trophic state. Data from these lakes in combination with smaller and more eutrophic lakes demonstrate that nutrient regeneration and grazing by crustaceans peak at intermediate trophic levels. Thus, zooplankton-phytoplankton coupling is strongest in meso-oligotrophic to meso-eutrophic lakes, and less important in both oligotrophic and very eutrophic lakes.

The potential impact that herbivores may have on phytoplankton biomass and species composition has long been appreciated (Porter, 1977). Thus, methods have been developed to control algal blooms through manipulations of the zooplankton (Shapiro, 1979). More recently, the significance of zooplankton as a source of limiting nutrients to phytoplankton has been studied (e.g., Lehman, 1980; Sterner, 1986). Also, it is now clear that food limitation of zooplankton is widespread in lacustrine systems (Lampert, 1985), and competition among zooplankton for resources has been demonstrated (Smith and Cooper, 1982; DeMott, 1983). Together, these studies indicate the critical role that coupling of zooplankton and phytoplankton may have on the ultimate effect of perturbations which may begin below (e.g., nutrient addition or reduction), above (e.g., fish manipulation), or simultaneously at more than one point.

Despite this recent work, relatively little is known about which factors control the strength of zooplankton-phytoplankton interactions. For example, does the degree of coupling change as a lake becomes eutrophic? Does lake size or morphometry influence the nature of zooplankton-phytoplankton interactions? This paper examines zooplankton-phytoplankton coupling in large lakes that represent a broad range of environments and trophic states. We focus on the impact of crustacean zooplankton grazing and nutrient regeneration in these systems. (Other potentially important components of the interaction, such as the zooplankton response to changing algal resources, are not considered here.) We find that there is substantial variability in the extent of coupling between macrozooplankton and phytoplankton in large lakes. This is demonstrated by a detailed examination of three large lakes (Lakes Michigan, Titicaca, and Tahoe), and additional examples. We then suggest that the variation within large lakes is part of a broader trend, in which zooplankton-phytoplankton interactions first become stronger and then diminish in going from ultraoligotrophy to hypereutrophy.

## 33.1 General Background

The obvious direct effect of herbivorous zooplankton on phytoplankton is grazing mortality, which can reduce phytoplankton density. However, recent considerations have emphasized the positive effects of zooplankton on phytoplankton, specifically the stimulation of algal growth by zooplankton regeneration of limiting nutrients. This results in high turnover rates for both the dissolved and algal nutrient pools (Lehman, 1984; Sterner, 1986). Zooplankton nutrient regeneration is a complex function of zooplankton size and species composition (Peters and Downing, 1984; Taylor, 1984), the quantity and quality of food sources (Lehman, 1984), as well as various environmental factors. Likewise, the response of the phytoplankton community to recycled nutrients may depend on ambient nutrient levels, environmental conditions, and various species-specific factors (see below). Thus, the degree of coupling between the activities of herbivorous zooplankton and the growth of phytoplankton may vary considerably both between lake systems and seasonally within a given system. This complicates the straightforward prediction of effects of both food web manipulations and changes in nutrient regimes of any given lake.

The complex responses of phytoplankton communities to experimental manipulation of zooplankton have been clearly demonstrated in a number of recent studies (Lehman and Sandgren, 1985; Bergquist and Carpenter, 1986; Elser et al., 1987 and 1988). Different algal species respond differently to zooplankton grazing: net growth rates of some species decline with increasing grazing while others are consistently stimulated. Other algal species achieve highest net growth rates at intermediate grazing levels. The intimate interactions between grazers and certain algal species are most clearly demonstrated by algal species which survive gut passage and grow more rapidly after exposure to the high-nutrient gut environment (Porter, 1973 and 1975). Vertical migration behavior of zooplankton adds another dimension to phytoplankton-zooplankton coupling (Lampert, 1987).

A number of investigators have attempted to assess the strength of zooplankton-phytoplankton coupling by estimating total nutrient regeneration by zooplankton and comparing the estimate to algal nutrient demand. The percentage of algal demand satisfied by zooplankton nutrient recycling in lakes covers a wide range (5–400% at any given time; Axler et al., 1981; Barlow and Bishop, 1965; Devol, 1979; Jacobsen, 1978; Liao and Lean, 1978), and one can generally conclude that high productivity during the nutrient-limited, summer stratified period in many lakes is largely the result of intensive nutrient regeneration by zooplankton and subsequent uptake by phytoplankton with high nutrient affinities. However, the lack of studies of this sort in the most oligotrophic systems precludes a more general assessment of the relationship between planktonic coupling and lake trophic status. In addition, the biomass and size distribution of a lake's zooplankton community will contribute to the degree to which recycling can keep pace with algal demand; Hargrave and Geen (1986) attributed the large discrepancy between the percentage of algal demand satisfied by zooplankton recycling in their two study systems (20% vs. 200%) to the unknown contribution of microzooplankton. The relative role of microzooplankton (small noncrustacean) vs. macrozooplankton in regenerating limiting nutrients requires further evaluation.

The importance of phytoplankton-zooplankton coupling in maintaining high rates of productivity may be greater in large lakes than in smaller lakes. In lakes of substantial size, a large portion of the pelagic zone will be relatively unaffected by nutrient inflows from streams and rivers and by nutrient loading from littoral processes (wave action, macrophyte effects, litter inputs, etc.). In the Laurentian Great Lakes, the importance of zooplankton nutrient recycling has been emphasized (Ferrante and Parker, 1978; Korstad, 1983; Scavia, 1979). In the largest lakes, pelagic coupling may resemble the situation in open ocean surface waters, where the distinction between new and regenerated production has long been recognized (Dugdale and Gering, 1967) and where close interactions between producers and microzooplankton may maintain algal growth rates near maximum despite vanishingly low nutrient concentrations (McCarthy and Goldman, 1979, but see Jackson, 1980).

Important questions remain in evaluating the interactions between phytoplankton and zooplankton in large lakes. Most fundamentally, what is the grazing impact and what proportion of algal nutrient demand can be satisfied by zooplankton recycling in some of the large lakes of the world? Is the degree of coupling a function of lake trophic status? Given the dependence of both quantitative and qualitative aspects of zooplankton regeneration of limiting nutrients on the species composition and size distribution of the zooplankton (Elser et al., 1988), how might pelagic coupling be influenced by food web changes in large lakes heavily impacted by man?

## 33.2 Zooplankton-Phytoplankton Coupling in Three Large Lakes

**Methods** Data for a comparison of three large lakes were collected primarily from recent field and laboratory studies. Lake Michigan (4,920 $km^3$, volume; 85 m, mean depth), one of the Laurentian Great Lakes, is located in the midwestern USA (44° 00′N, 87° 00′W) at an elevation of 176 m. Most fieldwork was carried out near Grand Haven during 1982 (Carney, 1982, unpublished data). Tropical Lake Titicaca (827 $km^3$, volume; 107 m, mean depth) is situated at the border of Peru and Bolivia (15° 48′S, 69° 24′W) at an elevation of 3,810 m. Data were collected during a 1981–82 field project (see Richerson et al., 1986, and Vincent et al., 1984, for general project; also Carney, 1984; Pawley and Alfaro, 1984; and Carney et al., 1987). The report of Richerson et al. (1977) was also consulted. Lake Tahoe (124 $km^3$, volume; 249 m, mean depth) is located along the border of California and Nevada, USA (39° 06′N, 120° 02′W) at an elevation of 1,899 m. Fieldwork was completed during 1985 (Carney, 1987; Carney et al., 1988) and the long-term data were also reviewed (Goldman, 1974, annual monitoring reports and database). Details for methods are provided in the above references.

For each lake, comparable data on basic features of phytoplankton and zooplankton were collected (see Tables 33.1, 33.2, and 33.3). Data were from the field studies mentioned above unless otherwise noted. In addition, flow diagrams were constructed for the most significant limiting nutrient in each lake (see Figures 33.1, 33.2, and 33.3). These provide concentrations (mg $m^{-3}$) for phytoplankton and the soluble available fraction of the nutrient, and fluxes (mg $m^{-3}$ $d^{-1}$) for grazing, zooplankton nutrient regeneration, and phytoplankton demand. The flux estimates were based on measurements of zooplankton filtration, nutrient regeneration, and actual nutrient uptake and productivity, respectively. In all these lakes there is substantial interannual variability and change during the summer stratification period. Average summertime values are provided here for comparison rather than as precise estimates for each lake. A number of the above studies have discussed some aspect of zooplankton-phytoplankton coupling for a given system, and this paper is consistent with their results and interpretations.

Comparable data on other systems were obtained through a computer search and additional literature review. We found the relevant information to be quite heterogeneous (different types of data reported by different studies, and different units used for similar information), so relatively few other lakes could be included for statistical analyses (see Table 33.4). The extent of zooplankton-phytoplankton coupling (operationally defined as the percent of algal nutrient demand supplied by zooplankton) was related quantitatively to lake trophic state with least square polynomial regressions of these data (See Figure 33.5). For these analyses a Statview 512+ statistical package and a Macintosh microcomputer were used.

**Lake Michigan** The algal biomass and primary productivity of this lake indicate that it is meso-oligotrophic (Table 33.1). There is a clear seasonal succession from diatoms in the spring to flagellates, cyanobacteria, and diatoms during stratification. Some of the major reasons for this are silica depletion and sinking of diatoms (Schelske and Stoermer, 1971), and species-specific grazing by herbivores. During recent decades, copepods of the genera *Diaptomus* and *Cyclops* have been dominant (Evans et al., 1980; Scavia et al., 1986).

Phosphorus is a key limiting nutrient, and a plankton flow diagram for this element indicates that there is rapid flux between compartments (Figure 33.1), with most phosphorus residing in the particulate fraction at any given time. Soluble reactive phosphate values are quite low ($<0.10$ mg m$^{-3}$ P) according to modified Rigler assays (Carney, unpublished data), and they are still low according to standard chemical assays (0.4–1.0 mg m$^{-3}$ P). These values are only slightly lower than those measured by Tarapchak et al. (1982) in a careful study of orthophosphorus during 1977. Phytoplankton biomass is proportionately quite high, and so are grazing and nutrient regeneration rates. Thus, the turnover time of the available phosphate pool is quite rapid, with

**Table 33.1** Characteristics of Lake Michigan phytoplankton and zooplankton in the epilimnion during the summer.

| Dominant phytoplankton | Chlorophyll biomass (mg m$^{-3}$ chl. *a*) | Productivity (mg C m$^{-3}$ h$^{-1}$) |
|---|---|---|
| Flagellates | | |
| Cyanobacteria (*Anabaena* spp.) and green algae | 2.2 | 2.86 |
| *Fragilaria crotonensis* | | |
| *Asterionella formosa* | | |
| **Dominant zooplankton** | | **Individuals (m$^{-3}$)** |
| *Diaptomus* spp. | | 12422 |
| *Cyclops* spp. | | 12193 |
| *Bosmina* spp. | | 11790 |
| *Daphnia* spp. | | 2053 |

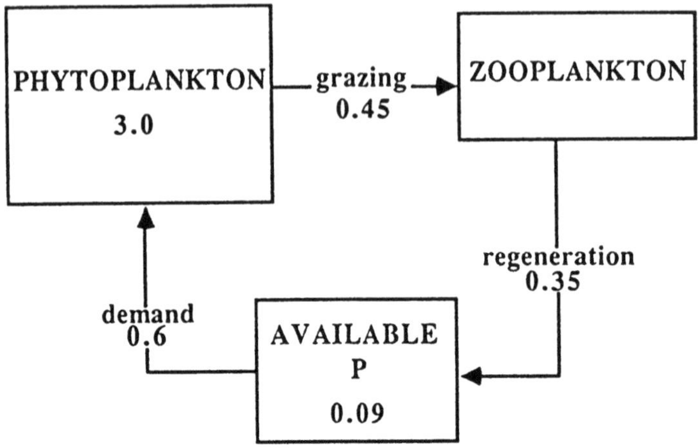


**Figure 33.1** A plankton flow diagram for phosphorus in Lake Michigan (summertime epilimnion). Figures 33.1, 33.2, and 33.3 have the following units: mg $m^{-3}$ of P or N for concentrations in boxes, and mg $m^{-3}$ $d^{-1}$ for fluxes.

fluxes dominated by phytoplankton uptake and zooplankton regeneration. During 1982, this strong coupling of phytoplankton and zooplankton was confirmed with semi-continuous bioassays. When zooplankton regeneration was eliminated by filtering out zooplankton, phytoplankton populations declined rapidly because of washout. However, when the estimated phosphate input due to zooplankton regeneration was added (0.35 mg $m^{-3}$ P), the control biomass and growth levels remained relatively constant. Similar planktonic phosphorus cycling has been found experimentally for Lake Huron (Korstad, 1983), through modelling for Lake Ontario (Scavia, 1979), and in more recent Lake Michigan studies (Scavia and Fahnenstiel, 1987). In Lake Michigan, zooplankton and phytoplankton may have become even more strongly coupled in recent years. A reduction in the planktivorous alewife (*Alosa pseudoharengus*) has allowed an increase in large *Daphnia*, particularly *Daphnia pulicaria*. The increased grazing by these zooplankton species is considered the major reason for the lower chl. *a* content and the greater transparency of Lake Michigan during the past several years (Scavia et al., 1986).

**Lake Titicaca** Phytoplankton biomass and productivity values for Lake Titicaca and Lake Michigan are similar (Table 33.2). The productivity/biomass ratio is higher in Lake Titicaca since it is meso-oligotrophic according to biomass, yet well within the range for temperate eutrophic lakes in productivity (Carney et al., 1987). As in Lake Michigan, there is a gradual decline in diatoms with stratification, but seasonality is less pronounced. Green algae such as *Gloeotilopsis planctonica* and *Oocystis* spp. are abundant throughout the year. Zooplankton densities, dominated by the endemic *Boeckella titicacae* and other copepods, are considerably lower than in Lake Michigan.

**Table 33.2** Characteristics of Lake Titicaca phytoplankton and zooplankton in the epilimnion during the summer.

| Dominant phytoplankton | Chlorophyll biomass (mg $m^{-3}$ chl. *a*) | Productivity (mg C $m^{-3}$ $h^{-1}$) |
|---|---|---|
| *Gloeotilopsis planctonica* | | |
| *Planctonema lauterbornii* | | |
| *Oocystis borgei* | 0.95 | 3.5 |
| *Cyclotella* spp. | | |
| Small *Stephanodiscus* | | |
| *Anabaena* spp. | | |
| Small green cells | | |
| **Dominant zooplankton** | | **Individuals ($m^{-3}$)** |
| *Boeckella titicacae* | | 2320 |
| *Microcylops leptopus* | | 198 |
| *Boeckella occidentalis* | | 40 |
| *Bosmina hagmanni* | | 29 |

In Lake Titicaca, nitrogen is the key limiting nutrient (Vincent et al., 1984; Carney, 1984), and the plankton flow diagram for this element (Figure 33.2) is interesting to compare to the diagram for Lake Michigan (see Figure 33.1). In Lake Titicaca, the available fraction of the most limiting nutrient is quite low in relation to phytoplankton biomass, as in Lake Michigan. However, according to direct estimates of grazing (Haney and Trout, 1985) and nutrient regeneration (Pawley and Alfaro, 1984), fluxes of these processes are relatively low in comparison with both ambient concentrations and phytoplankton de-

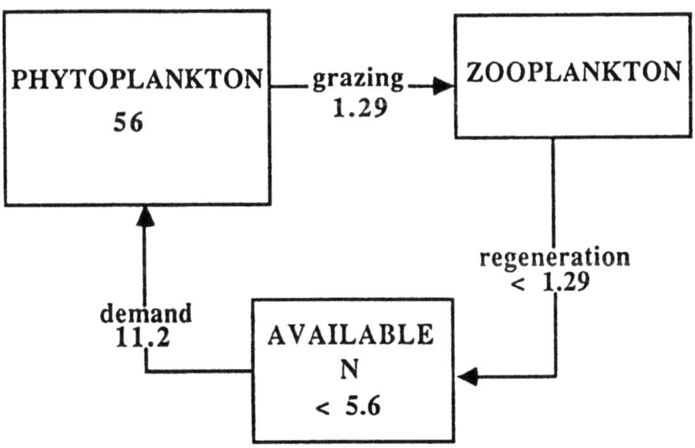


**Figure 33.2** A plankton flow diagram for nitrogen in Lake Titicaca (summertime epilimnion). The units are the same as in Figure 33.1.

mand. There is substantial variability in these processes; they vary from an almost negligible to a minor, though significant, impact on phytoplankton biomass, nitrogen concentrations, and phytoplankton demand. Long-term, semi-continuous bioassays here were also consistent with the above studies (Carney, 1984). In contrast to Lake Michigan, when zooplankton were filtered out and lakewater alone was added to control treatments, biomass and growth levels were maintained for at least two weeks without washout. Since demand was still high in relation to the available concentrations of nitrate and ammonium, other mechanisms, such as bacterial regeneration, mineralization of dissolved organic nitrogen, or direct uptake of that nitrogen fraction, must have been operating.

Zooplankton-phytoplankton coupling in tropical lakes such as Lake Titicaca may be affected by factors which vary with latitude. It is beyond the scope of this paper to examine these factors in detail, but we can briefly mention some potentially important ones. Lewis (1987) has identified three primary causes of distinctive properties of tropical lakes: high annual irradiance, low variation in irradiance, and Coriolis effects. These factors have important effects on latitudinal gradients of surface water temperature, water column stability, net primary production, nutrient regeneration rate, and oxygen level. The effect which is most relevant to this paper is nutrient regeneration. Lewis reasons that there is more efficient recycling of nutrients in the tropics because (a) decomposition is more rapid at a higher temperature, and (b) there is more frequent mixing, which resupplies nutrients from the hypolimnion to the epilimnion. These factors may help explain why zooplankton nutrient regeneration is relatively unimportant in Lake Titicaca. Hecky (1984) also considers the potential significance of time and developmental history, factors which Lewis explicitly declines to consider. Most temperate lakes are of glacial origin, while most tropical lakes originated in other ways. The large ancient tropical lakes, including Lake Titicaca and the African rift lakes, originated before the Pleistocene ice ages. Hecky proposes that in large, ancient systems, such as Lake Tanganyika, highly efficient food webs have an opportunity to evolve. Because tropical systems differ in the above ways from temperate ones, they constitute natural experiments with important treatment effects which should provide valuable comparative data for understanding zooplankton-phytoplankton interactions.

**Lake Tahoe** This lake is clearly ultraoligotrophic, so phytoplankton biomass and productivity are much lower than in both Lake Michigan and Lake Titicaca (Table 33.3). In contrast to these two systems, diatoms remain dominant during the summer, as is common in oligotrophic systems (Carney et al., 1988). Two copepods, *Diaptomus tyrrelli* and *Epischura nevadensis*, dominate throughout the year at concentrations much lower than zooplankton densities in the other lakes.

Both nitrogen and phosphorus are important limiting nutrients in Lake Tahoe (Goldman, 1986; Carney et al., 1988), and the plankton flow diagram

**Table 33.3** Characteristics of Lake Tahoe phytoplankton and zooplankton in the epilimnion during the summer.

| Dominant phytoplankton | Chlorophyll biomass (mg $m^{-3}$ chl. *a*) | Productivity (mg C $m^{-3}$ $h^{-1}$) |
|---|---|---|
| *Cyclotella* spp. | | |
| *Synedra radians* | 0.23 | 0.25 |
| Cryptomonads | | |
| Small coccoids, flagellates | | |
| **Dominant zooplankton** | | **Individuals ($m^{-3}$)** |
| *Diaptomus tyrrelli* | | 527 |
| *Epischura nevadensis* | | 27 |

is given in terms of nitrogen here (Figure 33.3) since more careful work has been done on this nutrient. Here the concentration of phytoplankton is relatively low in relation to measured levels of available nitrate. Since crustacean populations are also quite low, their grazing impact is almost insignificant. Indeed, during 1985, a selectively grazed diatom, *Cyclotella glomerata,* still became dominant during the summer. In agreement with grazing estimates, deletion of zooplankton in 100-L bags had no significant effect on the phytoplankton (Carney et al., 1988). Nutrient regeneration has not been measured in this lake, but it must be less than the nitrogen ingested through grazing. Thus, regeneration of nutrients by zooplankton must contribute very little per day to the available nitrogen, and it also contributes very little to phytoplankton demand. In contrast to the other two lakes, concentrations of the particulate and dissolved compartments are relatively high in relation to fluxes.

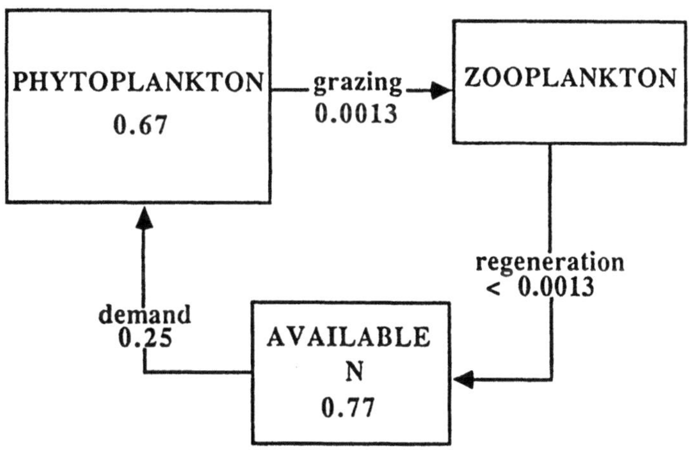


**Figure 33.3** A plankton flow diagram for nitrogen in Lake Tahoe (summertime epilimnion). The units are the same as in Figure 33.1.

The above results suggest a relatively straightforward dependence of phytoplankton growth on nutrient concentration, zooplankton growth on phytoplankton concentration, and little feedback from zooplankton back to nutrients or phytoplankton. This simple "bottom-up" picture could be altered by several factors. First, the Lake Tahoe "micro-food web" of bacteria, flagellates, and protozoans that seems important (Carney et al., 1988) is relatively unstudied and may influence nutrient cycling in significant ways. Microscale patchiness may also have a significant effect on all processes and compartments discussed here. Finally, the available information, especially chemical measurements, may need to be refined. For example, available nitrogen, measured here chemically as nitrate, may be lower if measured biologically. On the other hand, available N may be higher, since other potentially available sources of nitrogen (e.g., ammonium, urea, dissolved organic nitrogen) may exist. These factors indicate the special need for the best possible techniques in analyzing oligotrophic systems.

## 33.3 Discussion and Conclusion

The above results clearly demonstrate substantial variance in the coupling of zooplankton and phytoplankton in a variety of lakes. One possible explanation, consistent with results from both the three lakes studied above and the other literature we have reviewed, is that macrozooplankton (including crustaceans) have a greater impact in relatively eutrophic freshwater environments, while microzooplankton and other factors play a larger role in oligotrophic lakes and oceans. Lakes Tahoe, Titicaca, and Michigan are part of a much broader spectrum of systems in which, we suggest, there is a major shift in importance from microzooplankton to macrozooplankton with increasing nutrient inputs (Figure 33.4). This is correlated with a gradient from large systems to small lakes and ponds. Food limitation for macrozooplankton would logically be most severe in large oligotrophic systems. In hypereutrophic systems the impact of macrozooplankton once again declines because of blooms of large, inedible cyanobacteria and green algae. Additionally, zooplankton nutrient recycling may be of little importance in such hypereutrophic systems, as light availability may be a more important factor limiting continued algal growth (Barica and Mur, 1980).

Support for this general model was found when we considered our results in the context of previous accounts. We combined the results from Lake Tahoe, Lake Titicaca, and Lake Michigan with results from previous studies which assessed the percentage of algal demand satisfied by zooplankton recycling in more productive systems. The combined studies are summarized in Table 33.4. The criteria for selection of studies were that: (1) studies involved freshwater systems of at least moderate size (i.e., no ponds); (2) acceptable estimates of algal demand and zooplankton supply of nutrients were either supplied by the authors or could be derived from data presented; and (3) some measure of the

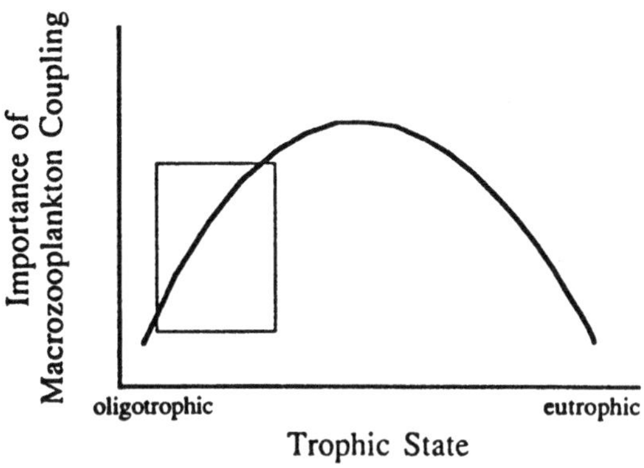


**Figure 33.4** A diagram of the proposed relationship between lake trophic status and the strength of coupling between phytoplankton and macrozooplankton. The box indicates the range of trophic states encompassed by lakes Michigan, Titicaca, and Tahoe.

**Table 33.4** Relative trophic rank, chlorophyll concentration, and estimated percentages of phytoplankton nutrient demand satisfied by zooplankton nutrient recycling (ZPER) in eight lakes.

| Lake | Rank[a] | Chlorophyll ($mg\ m^{-3}$)[b] | ZPER (%) | Reference |
|---|---|---|---|---|
| Tahoe | 1 | 0.23 | <0.5 | This study |
| Titicaca | 2 | 0.95 | <11.5 | This study |
| Michigan | 3 | 2.2 | 58 | This study |
| Castle | 4 | 2.3* | 32 | Axler et al. (1981) |
| Washington | 5 | 4.1* | 50 | Devol (1979) |
| Cayuga | 6 | 8.0 | 78.5 | Barlow and Bishop (1965) |
| Bay of Quinte | 7 | 44.0* | 24 | Liao and Lean (1978) |
| Frederiksborg Slottso | 8 | 330.0* | <5 | Jacobsen (1978) |

[a] Rank indicates the relative trophic ranking of a lake based on its average chlorophyll concentration.

[b] The average chlorophyll concentration during the study period (which was estimated by particulate N:chl, particulate P:chl, or particulate C:chl ratios when chlorophyll was not measured in a given study, indicated by *).

lake trophic status was supplied, that is, when chlorophyll concentration was not given, values of particulate C, N, or P were converted to chlorophyll equivalents using ratios published in Harris (1986). A significant unimodal relationship existed between percent algal demand supplied by zooplankton in a lake (ZPER) and its relative trophic rank (Figure 33.5A; $p < 0.05$, $R^2 = 0.72$) and between ln(ZPER) and ln(chlorophyll[CHL]) (Figure 33.5B; $p < 0.001$, $R^2$

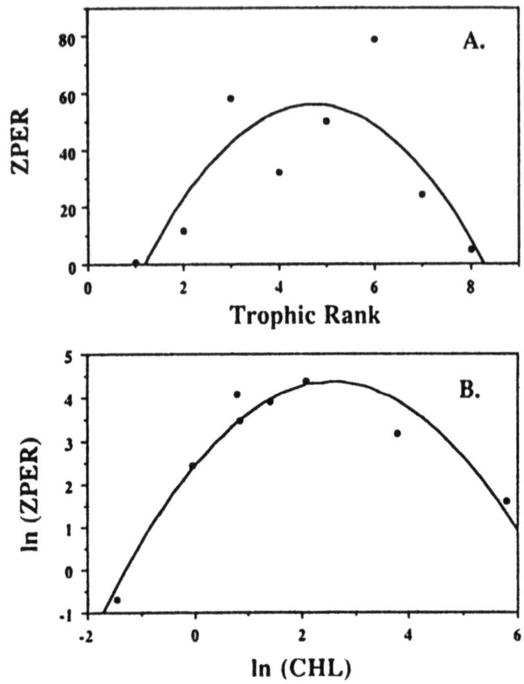


**Figure 33.5** (A) The percentage of algal demand supplied by zooplankton (ZPER) as a function of lake trophic rank. (B) The natural log of ZPER as a function of the natural log of chlorophyll (CHL). For both plots, the data used are shown in Table 33.4, and the curves shown are the best fit to the data using least-square quadratic polynomial equations.

= 0.94). The quadratic polynomial equation for the curve shown in Figure 33.5B was:

$$\ln(\text{ZPER}) = 2.434 + 1.50[\ln(\text{CHL})] - 0.293[\ln(\text{CHL})]^2$$

While a greater number of studies would clearly enhance this analysis, particularly in both highly oligotrophic and highly eutrophic lakes, the unimodal relationship remained significant when either the lowest biomass (Tahoe) or the highest biomass (Frederiksborg Slottso) lakes were withheld from the analysis.

These patterns are consistent with other surveys of zooplankton-phytoplankton interactions across lake trophic gradients, notably those of Gliwicz (1973, 1975) and Geller and Müller (1981). Gliwicz found that, for nine Polish lakes along a trophic gradient, the grazing impact of filter-feeding zooplankton first increased as the biomass and productivity of relatively small, edible nannophytoplankton increased. The range of trophic states here is represented by Lake Tahoe at one extreme and Lake Michigan at the other. Then, with increasing trophic status, there was a gradual decline in grazing impact as the proportion of large, inedible net phytoplankton increased, and bacteria assumed a greater fraction of the zooplankton diet. This is commonly the case for smaller eutrophic lakes and for bays of larger systems. Ejsmont-Karabin (1983) reported lowest phosphorus turnover times (resulting from estimated zooplankton excretion rates) in lakes of intermediate trophic status in a study of 42 Polish lakes. In reviewing zooplankton feeding in a broad range of lakes, Geller and Müller (1981) found the following pattern: (a) in oligotrophic systems, dominance of "macrofiltrator" zooplankton which feed on edible phytoplankton, and (b) a greater proportion of cladocerans with finer filter meshes which can collect bacteria in more eutrophic systems. Lake Tahoe is clearly dominated by "macrofiltrator" copepods, and Lake Michigan has a greater proportion of smaller fine-mesh cladocerans. The above studies combine experiments with field observations, and more experimental field investigations of the nature of phytoplankton-zooplankton interactions across a trophic gradient are clearly needed.

The degree of coupling between zooplankton and phytoplankton observed in these systems must also be considered in the context of the changes in their zooplankton communities which have resulted from major food web alterations imposed by man. In Lake Michigan, the introduction, establishment, and maintenance of exotic piscivorous and planktivorous fish species have caused pronounced changes in the species composition of the zooplankton community (Evans, 1986; Scavia et al., 1986; Kitchell and Carpenter, 1987). Introduction of the opossum shrimp *Mysis relicta* to provide forage for introduced kokanee salmon in Lake Tahoe resulted in the elimination of large-bodied cladocerans from the lake (Richards et al., 1975). In Lake Titicaca, introduction of exotic fish species has markedly altered the fish community of the lake (Alfaro et al., 1982; Vaux et al., 1988) and impacts on the zooplankton are likely but have not been thoroughly evaluated. Thus, in these systems, the present strength of phytoplankton-zooplankton coupling may reflect the influence of planktivory in structuring the current zooplankton community, and may well have been considerably different in times when planktonic structure was unaltered by food web disruptions. Given that large lakes are particularly subject to the impacts of exploitation and introduction of fish species, a complete consideration of the nature of phytoplankton-zooplankton interactions in large lakes should consider both the present and past dynamics of the pelagic food web.

The questions we have addressed in this paper clearly need much more study. Comparative studies which clarify the roles of microzooplankton and macrozooplankton in nutrient cycling along trophic gradients are needed. Alternative pathways, such as bacterial remineralization, purely chemical transformations, and algal death and lysis, also need to be investigated and compared quantitatively.

---

*Acknowledgements* Field support for the Lake Michigan work was provided by the Great Lakes Environmental Research Laboratory (Ann Arbor, Michigan). Studies at Lake Titicaca were funded primarily by National Science Foundation grant

DEB-7921933 to Peter J. Richerson, and also by Sigma Xi and the University of California. The Lake Tahoe study was supported by National Science Foundation dissertation grant BSR-8501872 to Heath J. Carney, Sigma Xi, the Jastro-Shields fund, and the University of California. This paper was written while both authors were supported by University of California graduate fellowships. Fatimah Mateen assisted with the manuscript preparation. Walter Geller and Z. Maciej Gliwicz provided thoughtful reviews which improved this paper. We thank Charles R. Goldman for his help during manuscript preparation and Max M. Tilzer for his substantial organizational and editorial efforts.

## References

Alfaro, R., E. Bustamante, J. Torres, H. Trevino. and W. Wurtsbaugh. 1982. *La Pesqueria en el Lago Titicaca (Peru): Presente y Futuro.* Rome: FAO FI:1DP/PER/76/022. 59 pp.

Axler, R.P., G. Redfield, and C.R. Goldman. 1981. The importance of regenerated nitrogen to phytoplankton productivity in a subalpine lake. *Ecology* 62:345–354.

Barica, J. and L.R. Mur (eds). 1980. Hypertrophic ecosystems. *Developments in Hydrobiology* 2:348 pp.

Barlow, J.P. and J.W. Bishop. 1965. Phosphate regeneration by zooplankton in Cayuga Lake. *Limnology and Oceanography* 10 (suppl.):R15-R24.

Bergquist, A.M. and S.R. Carpenter. 1986. Limnetic herbivory: effects on phytoplankton populations and primary production. *Ecology* 67:1351–1360.

Carney, H.J. 1982. Nutrient-Phytoplankton Dynamics in Lake Michigan: Importance of Physical Processes and Trophic Interactions. Paper presented at American Geophysical Union/American Society of Limnology and Oceanography Winter Meeting, San Francisco, California.

Carney, H.J. 1984. Productivity, population growth and physiological responses to nutrient enrichments by phytoplankton of Lake Titicaca, Peru-Bolivia. *Internationale Vereinigung für Theoretische und Angewandte Limnologie: Verhandlungen* 22:1253–1257.

Carney, H.J. 1987. Field tests of interspecific resource-based competition among phytoplankton. *Proceedings of the National Academy of Sciences,* USA 84:4148–4150.

Carney, H.J., P.J. Richerson, and P. Eloranta. 1987. Lake Titicaca (Peru/Bolivia) phytoplankton: species composition and structural comparison with other tropical and temperate lakes. *Archiv für Hydrobiologie* 110(3):365–385.

Carney, H.J., P.J. Richerson, C.R. Goldman, and R.C. Richards. 1988. Seasonal phytoplankton demographic processes and experiments on interspecific competition. *Ecology* 69:664–678.

DeMott, W.R. 1983. Seasonal succession in a natural *Daphnia* assemblage. *Ecological Monographs* 53:321–340.

Devol, A.H. 1979. Zooplankton respiration and its relation to plankton dynamics in two lakes of contrasting trophic state. *Limnology and Oceanography* 24:893–905.

Dugdale, R.C. and J.J. Goering. 1967. Uptake of new and regenerated forms of nitrogen in primary productivity. *Limnology and Oceanography* 12:196–206.

Ejsmont-Karabin, J. 1983. Ecological characteristics of lakes in north-eastern Poland. VIII. Role of nutrient regeneration by planktonic rotifers and crustaceans in 42 lakes. *Ekologia Polska* 31:411–427.

Elser, J.J., N.J. Goff, N. MacKay, M.M. Elser, A. St. Amand, and S.R. Carpenter. 1987.

Species-specific algal responses to zooplankton: Experimental and field observations in three north temperate lakes. *Journal of Plankton Research* 9:699–717.
Elser, J.J., M.M. Elser, N. MacKay, and S.R. Carpenter. 1988. Zooplankton-mediated transitions between N and P limited algal growth in three north temperate lakes. *Limnology and Oceanography* 33:1–14.
Evans, M.S. 1986. Recent major declines in zooplankton populations in the inshore region of Lake Michigan: probable causes and implications. *Canadian Journal of Fisheries and Aquatic Sciences* 43:154–159.
Evans, M.S., B.E. Hawkins, and D.W. Sell. 1980. Seasonal features of zooplankton assemblages in the nearshore area of southeastern Lake Michigan. *Journal of Great Lakes Research* 6:275–289.
Ferrante, J.C., and J.I. Parker. 1978. The influence of planktonic and benthic crustaceans on silicon cycling in Lake Michigan, U.S.A. *Internationale Vereinigung für Theoretishe und Angewandte Limnologie: Verhandlungen* 20:324–328.
Geller, W., and H. Müller. 1981. The filtration apparatus of Cladocera: filter mesh-sizes and their implications on food selectivity. *Oecologia* 49:316–321.
Gliwicz, Z.M. 1973. Importance of zooplankton for plankton system functioning in more and less eutrophic lakes. Symposium, "Eutropheirung und Gewässerschutz," Biol. Gesellschaft der DDR, Schloss Reinhardsbrunn, 16–16 d.
Gilwicz, Z.M. 1975. Effect of zooplankton grazing on photosynthetic activity and composition of phytoplankton. *Internationale Vereinigung für Theoretische und Angewandte Limnologie: Verhandlungen* 19:1490–1497.
Goldman, C.R. 1974. Eutrophication of Lake Tahoe emphasizing water quality. U.S. Environmental Protection Agency Ecological Research Series (EPA-660/3-74-034), Corvallis, Oregon, USA.
Goldman, C.R. 1986. Lake Tahoe: an oligotrophic lake's response to nutrient loading. Lake Pollution and Recovery: Proceedings International Congress, European Water Pollution Control Association, Rome, 15–18 April 1985, p. 249–254.
Haney, J.F. and M.A. Trout. 1985. Size selective grazing by zooplankton in Lake Titicaca. *Archiv für Hydrobiologie, Ergebnisse der Limnologie* 21:147–160.
Hargrave, B.T. and G.H. Geen. 1968. Phosphorus excretion by zooplankton. *Limnology and Oceanography* 13:332–342.
Harris, G.P. 1986. *Phytoplankton Ecology: Structure, Function, and Fluctuation.* Chapman and Hall, New York. 384 p.
Hecky, R.E. 1984. African lakes and their trophic efficiencies: a temporal perspective. In: *Trophic Dynamics of Aquatic Ecosystems* (D.G. Meyers and J.R. Strickler, ed.). AAAS Special Symposium, Westview Press, Boulder, p. 405–448.
Jackson, G.A. 1980. Phytoplankton growth and zooplankton grazing in oligotrophic oceans. *Nature* 284:439–441.
Jacobsen, O.S. 1978. Nitrogen and phosphorus dynamics in eutrophic Frederiksborg Slotsso. *Internationale Vereinigung für Theoretishe und Angewandte Limnologie: Verhandlungen* 20:696–701.
Kitchell, J.F. and S.R. Carpenter. 1987. Piscivores, planktivores, fossils, and phorbins. In: *Predation: Direct and Indirect Effects on Aquatic Communities* (W.C. Kerfoot and A. Sih, eds.). University Press of New England, Hanover, N.H., p. 132–146.
Korstad, J. 1983. Nutrient regeneration by zooplankton in southern Lake Huron. *Journal of Great Lakes Research* 9:374–388.
Lampert, W. (ed.). 1985. Symposium, Food limitation and the structure of zooplankton communities. *Archiv für Hydrobiologie, Ergebnisse der Limnologie* 21.
Lampert, W. 1987. Vertical migration of freshwater zooplankton: indirect effects of vertebrate predators on algal communities. In: *Predation: Direct and Indirect Effects on Aquatic Communities* (W.C. Kerfoot and A. Sih, eds.). University Press of New England, Hanover, N.H., p. 291–299.
Lehman, J.T. 1980. Nutrient recycling as an interface between algae and grazers in

freshwater communities. In: *Evolution and Ecology of Zooplankton Communities* (W.C. Kerfoot, ed.). University Press, New England, p. 251–263.
Lehman, J.T. 1984. Grazing, nutrient release, and their impacts on the structure of phytoplankton communities. In: *Trophic Dynamics of Aquatic Ecosystems* (D.G. Meyers and J.R. Strickler, eds.). AAAS Special Symposium, Westview Press, Boulder, p. 49–72.
Lehman, J.T. and C.D. Sandgren. 1985. Species-specific rates of growth and grazing loss among freshwater algae. *Limnology and Oceanography* 21:501–516.
Lewis, W.M. 1987. Tropical limnology. *Annual Review of Ecology and Systematics* 18:159–184.
Liao, C.F.-H. and D.R.S. Lean. 1978. Nitrogen transformations within the trophogenic zone of lakes. *Journal of the Fisheries Research Board Canada* 35:1102–1108.
McCarthy, J.J. and J.C. Goldman. 1979. Nitrogenous nutrition of marine phytoplankton in nutrient-depleted waters. *Science* 203:670–672.
Pawley, A.L. and R. Alfaro. 1984. Zooplankton in the Lake Titicaca ecosystem: the importance of regenerated nitrogen to phytoplankton productivity. *Internationale Vereinigung für Theoretische und Angewandte Limnologie: Verhandlungen* 22:1258–1263.
Peters, R.H. and J.A. Downing. 1984. Empirical analysis of zooplankton filtering and feeding rates. *Limnology and Oceanography* 29:763–784.
Porter, K.G. 1973. Selective grazing and differential digestion of algae by zooplankton. *Nature* 204:179–180.
Porter, K.G. 1975. Viable gut passage of gelatinous green algae ingested by *Daphnia. Internationale Vereinigung für Theoretische und Angewandte Limnologie: Verhandlungen* 19:2840–2850.
Porter, K.G. 1977. The plant-animal interface in freshwater ecosystems. *American Scientist* 65:159–170.
Richards, R.C., C.R. Goldman, T.C. Frantz. and R. Wickwire. 1975. Where have all the *Daphnia* gone? The decline of a major cladoceran in Lake Tahoe, California-Nevada. *Internationale Vereinigung für Theoretische und Angewandte Limnologie: Verhandlungen* 19:835–842.
Richerson, P.J., C. Widmer, and T. Kittel. 1977. *The Limnology of Lake Titicaca (Peru-Bolivia), a Large, High Altitude Tropical Lake.* Inst. of Ecology, Univ. of California, Davis, Publ. no. 14, 78 pp.
Richerson, P.J., P.J. Neale, W. Wurtsbaugh, R. Alfaro, and W. Vincent. 1986. Patterns of temporal variation in Lake Titicaca. A high altitude tropical lake. I. Background, physical and chemical processes, and primary production. *Hydrobiologia* 138:205–220.
Scavia, D. 1979. Examination of phosphorus cycling and control of phytoplankton dynamics in Lake Ontario with an ecological model. *Journal of the Fisheries Research Board Canada* 36:1336–1346.
Scavia, D., G.L. Fahnenstiel, M.S. Evans, D.J. Jude, and J.T. Lehman. 1986. Influence of salmonid predation and weather on long-term water quality trends in Lake Michigan. *Canadian Journal of Fisheries and Aquatic Sciences* 43:435–443.
Scavia, D., and G.L. Fahnenstiel. 1987. Dynamics of Lake Michigan phytoplankton: mechanisms controlling epilimnetic communities. *Journal of Great Lakes Research* 13:103–120.
Schelske, C.L., and E.F. Stoermer. 1971. Eutrophication, silica depletion, and predicted changes in algal quality in Lake Michigan. *Science* 173:423–424.
Shapiro, J. 1979. The importance of trophic-level interactions to the abundance and species composition of algae in lakes. In: *Developments in Hydrobiology* (J. Barica and L.R. Mur, eds.) 2:105–116.
Smith, D.W. and S.D. Cooper. 1982. Competition among cladocera. *Ecology* 63:1004–1015.

Sterner, R.W. 1986. Herbivores' direct and indirect effects on algal populations. *Science* 231:605–607.

Tarapchak, S.J., S.M. Bigelow, and C. Rubitschun. 1982. Overestimation of orthophosphorus concentrations in surface waters of southern Lake Michigan: effects of acid and ammonium molybdate. *Canadian Journal of Fisheries and Aquatic Sciences* 39:296–304.

Taylor, W.D. 1984. Phosphorus flux through epilimnetic zooplankton from Lake Ontario: relationship with body size and significance to phytoplankton. *Canadian Journal of Fisheries and Aquatic Sciences* 41:1702–1712.

Vaux, P., W. Wurtsbaugh, H. Trevino, L. Marino, E. Bustamante, J. Torres, P. Richerson, and R. Alfaro. 1988 Ecology of the pelagic fishes of Lake Titicaca (Peru-Bolivia). *Biotropica* 20:220–229.

Vincent, W., W. Wurtsbaugh, C.L. Vincent, and P.J. Richerson. 1984. Seasonal dynamics of nutrient limitation in a tropical high-altitude lake (Lake Titicaca, Peru-Bolivia): application of physiological bioassays. *Limnology and Oceanography* 29:540–552.

# 34

# *Pontoporeia hoyi*—a Direct Trophic Link between Spring Diatoms and Fish in Lake Michigan

Wayne S. Gardner, Michael A. Quigley, Gary L. Fahnenstiel, Donald Scavia, and William A. Frez

ABSTRACT Several lines of evidence suggest that the Lake Michigan benthic amphipod, *Pontoporeia hoyi* (an important fish prey in large, temperate, low-nutrient lakes), may obtain a large portion of its annual energy directly from the spring diatom bloom: 1) Energetic considerations suggest that *P. hoyi* must assimilate a large fraction of energy from incoming organic material, but that summer input rates are not sufficient to support observed annual production of *P. hoyi*. 2) The weight-specific lipid content of *P. hoyi* at some locations in Lake Michigan doubles within a few weeks after the spring diatom bloom. 3) Lipids accumulate in *P. hoyi* primarily as the storage products, triglycerides. 4) *P. hoyi* feeds intermittently and can survive for months without food. 5) The dominant spring diatom in Lake Michigan, *Melosira*, is not significantly cropped by zooplankton and settles rapidly through the water column in the spring and early summer. 6) After the spring diatom bloom, the phytoplankton changes to a dominance of flagellates that are mostly eaten by pelagic zooplankton and therefore largely unavailable to benthic organisms.

The ability of *P. hoyi* to rapidly accumulate and store energy from spring diatom blooms may help explain why this amphipod thrives in many temperate, oligotrophic/mesotrophic lakes. This apparently direct trophic linkage between spring diatoms and *P. hoyi* is energetically important because it involves a minimum of trophic energy loss between primary production and fish.

*Pontoporeia hoyi*, the dominant benthic invertebrate in the upper Great Lakes, constitutes about 65% of the benthic-invertebrate biomass in southern Lake Michigan (Nalepa et al., 1985). This amphipod is a desirable fish prey that is consumed by various developmental stages of most Great Lakes fish species (Wells, 1980). Despite the importance of *P. hoyi* to fish, pathways between primary producers and *P. hoyi* have not previously been detailed. *P. hoyi* has generally been considered to be a detritivore that prefers small particles enriched with organic material or bacteria (Marzolf, 1965). Food availability was reported to be the critical factor limiting growth of *Pontoporeia* in Lake Maalaren, Sweden (Wiederholm, 1973; Johnson and Wierderholm, 1989). Studies in the Baltic Sea have linked spring phytoplankton sedimentation with subsequent production of *Pontoporeia* species (Cederwall, 1977), but linkage between sedimented phytoplankton blooms and subsequent *P. hoyi* growth has not been established as definitively in the North American Great Lakes. In this paper, we summarize some of our recent Lake Michigan experiments showing that energy may be transferred quite directly from spring diatoms to *P. hoyi* and suggest that *P. hoyi* exploits seasonal inputs of high-quality food in the Great Lakes.

## 34.1 Areal Comparisons of Sediment-Trap Inputs to Annual *P. hoyi* Production

Initially, we compare inputs of organic material, estimated by sediment-trap collections, to areal *P. hoyi* metabolic rate processes (biomass-production and excretion rates). Net particulate organic carbon (POC) inputs into the hypolimnion of Lake Michigan were estimated for 1980 (Eadie et al., 1984) from the mass flux and organic carbon content of sediment trap materials collected below the epilimnion during the period of summer stratification. These materials were collected in traps suspended at 35-m depth at four stations (offshore from Grand Haven, Michigan) with total depths ranging from 83 to 102 m (Eadie et al., 1984). Sediment-trap collections made during the period of stratification were thought to provide a better estimate of "freshly produced" organic particles than estimates made during holomixis because they did not contain significant amounts of "old" resuspended particles from the hypolimnion. Using this approach, average inputs of "new" organic carbon at four offshore stations in southern Lake Michigan were estimated to be 63 $\pm$ 36 mg $m^{-2}$ $d^{-1}$ (Eadie et al., 1984), or 23 g $m^{-2}$ $y^{-1}$ if summer rates were extrapolated over the annual cycle.

*P. hoyi* production in Lake Michigan was calculated from mean areal *P. hoyi* biomass estimates and production/biomass (P/B) ratios reported for the organism (Lubner, 1979; Winnell and White, 1984). Mean areal *P. hoyi* biomass [2.76 g AFDW (ash-free dry weight) $m^{-2}$] was determined from a 40-station grid sampled in 1980–1981 (Nalepa et al., 1985) by summing areal-weighted biomass estimates from various regions (Gardner et al., 1985a). Mean areal biomass was multiplied by an annual P/B ratio (about 1.2 for *P. hoyi* in offshore Lake Michigan; Lubner, 1979; Winnell and White, 1984) to calculate the mean productivity of *P. hoyi* in southern Lake Michigan (3.3 g AFDW $m^{-2}$ $y^{-1}$). Assimilation efficiencies have not been estimated, but based on a growth-efficiency range of 0.3 to 0.6 for *P. hoyi* (Johnson and Brinkhurst, 1971), this calculation suggests that *P. hoyi* must assimilate at least between 5.5 to 11 g

$m^{-2}$ $y^{-1}$ of organic detritus (Gardner et al., 1985a). More recent work indicates that Ponar grab samples underestimate actual *P. hoyi* biomass by a factor of 1.5 (Nalepa et al., 1988). With this correction, the calculated minimum requirements of *P. hoyi* would be 8.2 to 16.5 g POM (particulate organic matter) $m^{-2}$ $y^{-1}$ (or 4.1 to 8.2 g C $m^{-2}$ $y^{-1}$, assuming a carbon to biomass ratio of 0.5). This food requirement equals 18 to 36% of the organic C input as estimated above from summer sediment-trap accumulation rates of offshore material (Eadie et al., 1984). If the summer sediment-trap input rates provide a reasonable estimate of the annual supply of new available material to the benthos, this calculation suggests that a relatively high percentage of freshly settled organic material must be assimilated by *P. hoyi* and other benthic invertebrates (Gardner et al., 1985a). However, nearly complete assimilation of ingested material probably is not a good assumption, particularly if the food quality is low, as may be expected for benthic-detrital material. In addition, if settling rates are slow, a portion of the particles (measured at near-thermocline depth) would probably be eaten or decomposed in the water column before becoming available to the benthic invertebrates. Thus the "observed" level of *P. hoyi* productivity appears higher than expected if estimated summer input rates are assumed to represent maximum available food supply.

Similar comparisons were made by comparing areal *P. hoyi* nitrogen regeneration rates, as measured by ammonium excretion, to areal inputs of particulate organic nitrogen. Seasonal studies of ammonium excretion by *P. hoyi* in southeastern Lake Michigan (Gauvin et al., 1989; Gardner et al., 1987) indicate that ammonium excretion rates of *P. hoyi* range from about 1 to 4 ng atom nitrogen (mg AFDW)$^{-1}$ $h^{-1}$.* Assuming this range of excretion rates, and an areal *P. hoyi* biomass of 4.1 g AFDW $m^{-2}$ (see above), the mean areal excretion rate for *P. hoyi* would range between 0.5 and 2.0 g N $m^{-2}$ $y^{-1}$ as compared to a Kjeldahl nitrogen input rate of 2.3 g N $m^{-2}$ $y^{-2}$, as extrapolated from summer sediment-trap data (Eadie et al., 1984). If approximately as much nitrogen is assimilated for growth as is excreted by *P. hoyi* (N growth efficiency = 0.5), the total nitrogen requirement would be 1 to 4 g N $m^{-2}$ $y^{-1}$. This calculation indicates that incoming particulate organic nitrogen, as measured in the summer, may not be sufficient to account for nitrogen assimilation by *P. hoyi* and other benthos. An explanation for these apparent inconsistencies between estimated nutritional inputs and *P. hoyi* production and excretion rates is that annual rates of food supply were underestimated by extrapolating summer input rates over the whole season. Recent seasonal sediment-trap studies indicate that the fluxes of "potentially available" nutrient indicators (AFDW, organic nitrogen, total lipids, biogenic silica) peak in the spring (Gardner et al., 1989).

*As would be expected, these rates are much lower than those reported for warm-water pelagic species, e.g., 35–83 ng atom nitrogen (mg dry weight)$^{-1}$ $h^{-1}$ for *Daphnia* (Lehman, 1980), or for other benthic macroinvertebrates held at relatively high temperatures, e.g., 5–14 ng atom nitrogen (mg AFDW)$^{-1}$ $h^{-1}$ for chironomids and tubificids held at 22°C (Gardner et al., 1983).

## 34.2 Biochemical/Physiological Studies of *P. hoyi*

Insights about seasonal trends in the transfer of energy to *P. hoyi* and about the possible role of spring diatoms in this exchange were gained by examining seasonal changes in lipid content (Gardner et al., 1985a), lipid composition (Gardner et al., 1985b), and feeding patterns of *P. hoyi* (Quigley, 1988).

In the early spring, lipids constituted about 20 to 25% of the AFDW of *P. hoyi*, but within a period of weeks, mean lipid content of the animals dramatically increased to 40 to 50% of AFDW (Figure 34.1). This increase in lipid content corresponded in time to expected inputs of diatoms from the spring bloom and suggests a likely direct link between the diatoms and *P. hoyi*, as observed in the Baltic Sea for *P. affinis* (Cederwall, 1977). Similar patterns of lipid accumulation were observed in subsequent years (Gauvin et al., 1989; Landrum, 1988). A more recent comparison of alkane ratios (C-17:C-29) in seasonally collected sediment trap materials indicated that fresh phytoplankton constituted a higher proportion of the trap materials during the period immediately preceding the observed increases in *P. hoyi* lipid content than during other parts of the year (Gardner et al., 1989).

To differentiate the effects of incoming food on lipid content from other season-dependent factors, *P. hoyi*, collected in spring, were divided into two groups and incubated in the laboratory. One group was held for 80 days in filtered lake water whereas the other group was held in lake sediments with no fresh additions of particles. The lipid content of constituent animals was determined over time and compared to that of animals collected in the field over the same interval. The results (Figure 34.2) showed pronounced differences among the three groups. The mean lipid content of animals held in filtered lake water decreased during the incubation period of about 80 days to about 12% of animal dry weight. Lipids in animals held in lake sediments remained relatively constant over the first 40 days of incubation and then decreased relative to dry weight. Thus, in contrast to the lake animals, neither

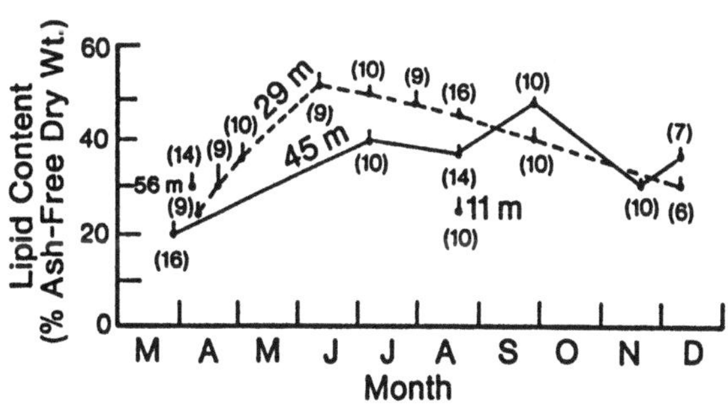


**Figure 34.1** Total lipid content, relative to AFDW, in *Pontoporeia hoyi* sampled seasonally from southeastern Lake Michigan in 1984. Range bars = standard errors. Numbers in parentheses represent the number of replicate animals analyzed. (Modified from Gardner et al., 1985a.)

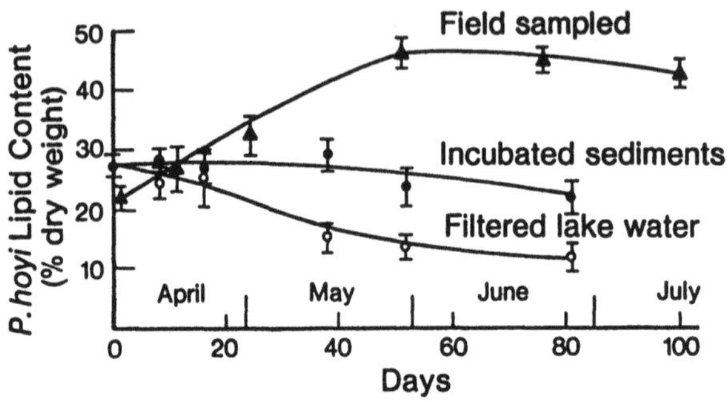


**Figure 34.2** Lipid content, relative to dry weight, in *P. hoyi* samples collected directly from Lake Michigan and those incubated in the laboratory in the presence and absence of natural sediments. *P. hoyi* samples analyzed directly after collection were taken from a site directly offshore from Grand Haven, Michigan, at a water depth of 24 m. Incubated animals were collected offshore from South Haven, Michigan at a depth of 56 m in April 1984. Range bars = standard errors.

laboratory group showed a springtime increase in lipid content. Similar results were observed in a subsequent study (Gauvin et al., 1989). Although other factors, such as possible differences in feeding activities between laboratory and field animals, cannot be completely disregarded, *in situ P. hoyi* populations received inputs of sedimenting spring diatoms, while laboratory populations did not. The differences in lipid accumulation among the groups support the hypothesis that *P. hoyi* may accumulate lipids from settling diatoms.

Further insights about the use of lipids for energy storage by *P. hoyi* were gained by examining lipid-class composition. Class analysis of lipids in *P. hoyi* collected from Lake Michigan in August 1984 (Gardner et al., 1985b) indicated that triglycerides are the dominant lipid component in *P. hoyi*. Triglycerides are accumulated by organisms specifically for the purpose of storing energy (Lee, 1975). Although we did not examine lipid compositional changes in the seasonal study, more recent laboratory experiments indicate that triglycerides are the main lipid class reflecting observed changes in total lipids in *P. hoyi* (unpublished data).

Examination of *P. hoyi* gut fullness (Quigley, 1988) indicated that Lake Michigan *P. hoyi* populations intensify their feeding in spring. Highest mean percent gut fullness was 59% in March and declined through the summer to approximately 40%. The incidence of animals with empty and partially filled guts also suggested that *P. hoyi* feed intermittently rather than in the continuous mode commonly noted among amphipods. In general, guts of continuously feeding animals remain constantly full (Moore, 1977; Marchant and Hynes, 1981). *P. hoyi's* intermittent feeding habits are unusual, but not surprising, given their ability to accumulate and store lipids. In fact, lipid storage may afford *P. hoyi* the option of exploiting high-quality food when it becomes available (i.e., following sedimentation of the spring diatom bloom). At other times, the animals may rely on stored lipids to satisfy normal metabolic energy requirements, while feeding at a reduced rate. *P. hoyi*'s lipid-storage capability may, in part, explain the amphipod's widespread distribution throughout the upper Great Lakes (Cook and Johnson, 1974) and other environments where high-quality food resources are often limited. Further insights into sources of "high-quality" food for *P. hoyi* were gained by seasonal investigations of epilimnetic phytoplankton and settling particles in Lake Michigan.

## 34.3 Seasonal Changes in the Composition of Particles Settling to the Hypolimnion

The above studies suggest that *P. hoyi* could obtain a substantial portion of its nutrition in spring before stratification. Phytoplankton dominance in Lake Michigan shifts from diatoms in the spring to cryptophytes and other small flagellates in the late summer (Figure 34.3; Fahnenstiel and Scavia 1987b). The large, heavy diatoms, i.e., the *Melosira* spp. that often are present in Lake Michigan during spring (Holland and Beeton, 1972; Fahnenstiel and Scavia, 1987b), may become directly available to *P. hoyi* either through turbulent mixing during isothermal periods, or by rapidly sinking through the water column.

Unfortunately, during spring when the lakes are not stratified, inputs of "new" organic particles from the epilimnion cannot be readily distinguished from those of resuspended organic particles, if comparisons are based only on carbon or AFDW analysis. The problem of quantifying spring inputs of "new" POC was circumvented by enumerating algae in Lugel's preserved sediment trap samples and converting the results to POC concentrations based on size measurement (Fahnenstiel and Scavia, 1987a). In a comprehensive study of the dynamics of phytoplankton at an offshore station (100 m depth) in Lake Michigan, both zooplankton grazing and particle settling rates were evaluated seasonally as processes for removing phytoplankton and other particles from the epilimnion (Figure 34.4; Scavia and Fahnenstiel, 1987). The algal loss rates due to sedimentation from the epilimnion decreases from 5–12 ug C $l^{-1}$ $d^{-1}$ in early spring to $<0.5$ ug C $l^{-1}$ $d^{-1}$ during summer stratification (Figure 34.5). Algal losses in the spring were strongly dominated by diatoms that constituted $>95\%$ of the settling algae. Losses of settling phytoplankton were much smaller in the summer than in the spring, but diatoms still constituted from 23 to 62% of the total phytoplankton collected in the traps. Diatoms constituted only 5% of the total phytoplankton in the epilimnion during late summer, but settled out more than the other species (Scavia and Fahnenstiel, 1987). Epilimnetic zooplankton grazing on algae followed a seasonal pattern inversely related to that of algal settling into the sediment traps; grazing losses were lowest in the spring but increased as the season progressed (Scavia and Fahnenstiel, 1987).

The percentage of total POC accounted for by algae in the sediment-trap materials decreased from 16 to 95% in the spring to 0.4 to 2% during summer stratification. Thus, the quality as well as quantity of settling POC changed

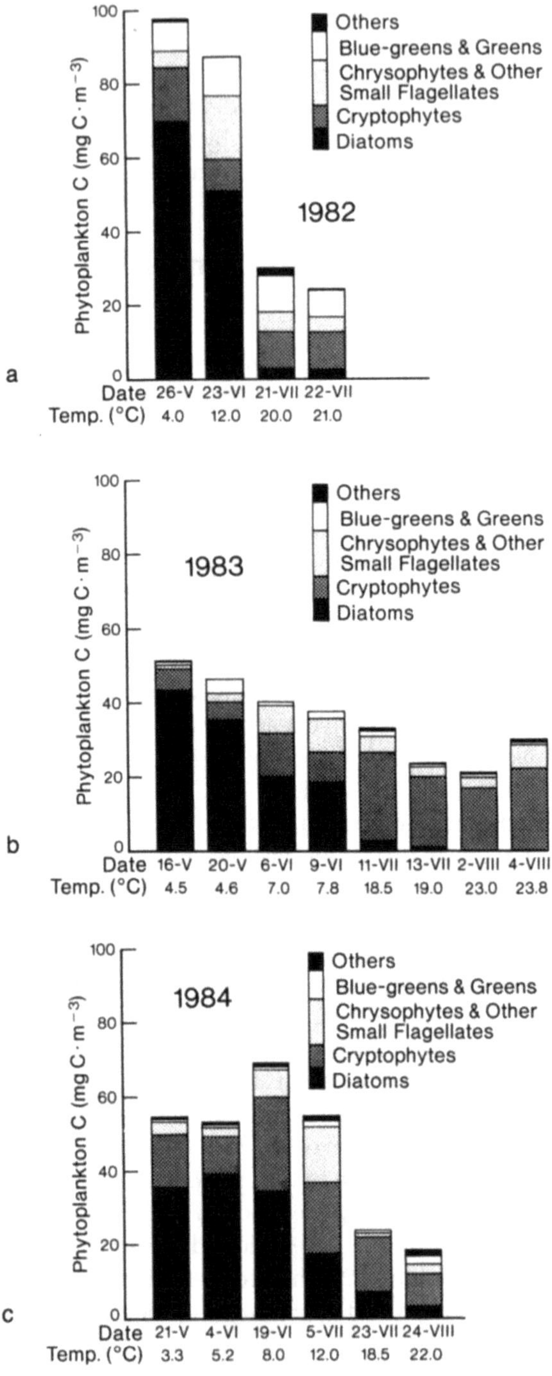


**Figure 34.3** Phytoplankton composition in the epilimnion at various dates and temperatures in (*a*) 1982, (*b*) 1983, and (*c*) 1984. (From Fahnenstiel and Scavia, 1987b.)

a
b
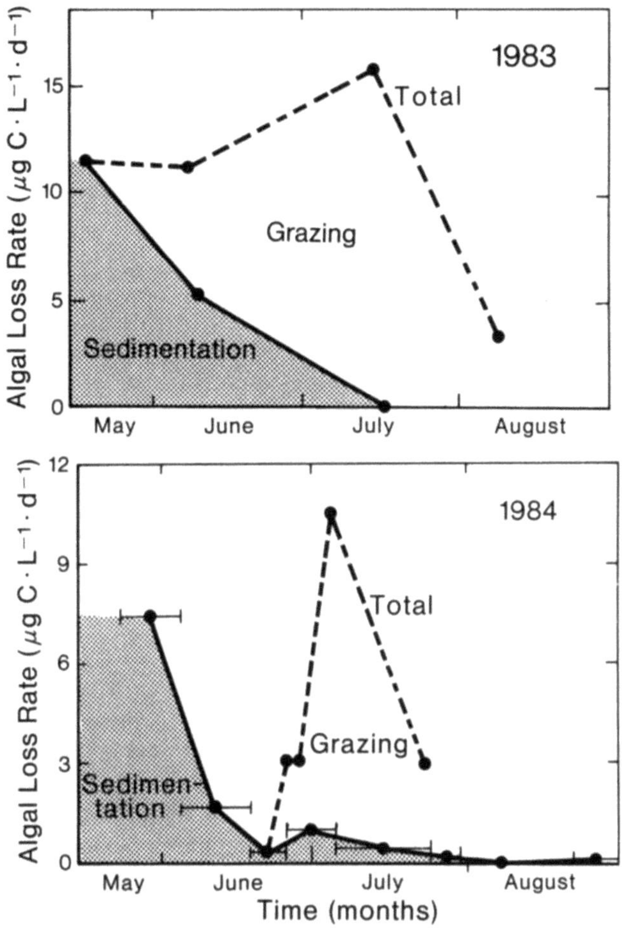


**Figure 34.4** Partitioning of total loss rate (grazing plus sinking) during the field seasons of (*a*) 1983 and (*b*) 1984. (From Scavia and Fahnenstiel, 1987.)

substantially with season. During stratification, a substantial portion of the settling POC was detritus (e.g., zooplankton fecal pellets or residues of dead algal cells) rather than living phytoplankton. Feces from pelagic zooplankton and other detrital particles would be less nutritious than fresh phytoplankton because part of the labile nutrients would already be removed. The potential capacity of the diatom phytoplankton bloom to support *P. hoyi* production in Lake Michigan can be roughly estimated from the primary production and phytoplankton species-composition data of Fahnenstiel and Scavia (1987a; 1987b). Integration of *in situ* measurements of primary production over the euphotic zone of Lake Michigan indicated that primary production rates were in the range of 300 to 420 mg C $m^{-2}$ $d^{-1}$ during the spring diatom bloom (May measurements; Fahnenstiel and Scavia, 1987a). If (1) the diatoms are assumed

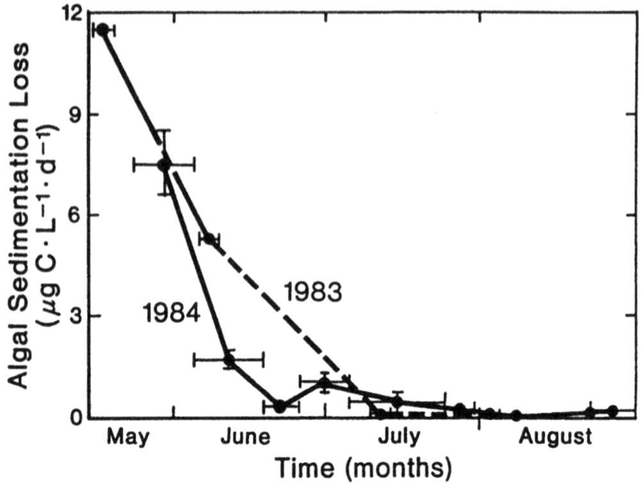


**Figure 34.5** Sedimentation loss rates from algal counts of trap collections in 1983 and 1984. Horizontal bar lengths represent deployment duration; vertical bars are mean ± standard errors, from multiple deployments in 1984. (From Scavia and Fahnenstiel, 1987.)

to constitute about 73% of spring phytoplankton (Fahnenstiel and Scavia, 1987b), (2) the duration of the bloom is about 60 days (Holland 1969), and (3) sedimentation is the dominant mechanism removing spring diatoms from the water column (Scavia and Fahnenstiel, 1987), then potential annual input of diatom carbon to the hypolimnion from the spring bloom would be 13 to 18 g C m$^{-2}$. This production would be adequate to support annual *P. hoyi* production (4.1 to 8.2 g organic C m$^{-2}$) as calculated above. Recent calculations (Laird et al., 1988) suggest that the above estimate for diatom production may be too high because diatoms have slower growth rates than most of the other species of epilimnetic phytoplankton found in the spring bloom. However, even with this more conservative approach for estimating diatom production, spring diatom production would still supply as much as one-third of the energy required to support annual *P. hoyi* production.

All of the above calculations must, of course, be interpreted with caution because of the potential errors in extrapolating results from a relatively small number of field measurements, made in different investigations, to more general interpretations. Despite such precautions, they do suggest that the spring diatom bloom is an important source of food for *P. hoyi*.

## 34.3 Summary and Discussion

In agreement with previous productivity results on *Pontoporeia affinis* in the Baltic Sea, *P. hoyi* appears to have developed a feeding strategy that enables it to thrive in low-nutrient, temperate aquatic systems such as the upper Laurentian Great Lakes, according to the following scenario (diagrammed in Figure 34.6): During spring holomixis when silica is available and temperature and light conditions are appropriate, large diatoms are abundant in these lakes. These diatoms are not grazed significantly by pelagic zooplankton because of their large size and fast sinking rates. In addition, concentrations of zooplankton are relatively low in the spring. The large size and high density of the spring diatoms cause them to settle out of the pelagic zone into the hypolimnion where they are directly (or indirectly) available to *P. hoyi*. This "seasonally available" energy is accumulated by *P. hoyi* in the spring as lipids, and, these lipids are used during the rest of the season when both the quantity and quality of incoming available food is low. As the summer stratification progresses, silica limitation causes the diatoms to decline and dominance to switch to other phytoplankton species (Schelske and Stoermer, 1971; Fahnenstiel and Scavia, 1987b) that tend to be grazed more efficiently by zooplankton. As a result, the composition of the dominant settling particles changes from diatoms to zooplankton feces and other detrital residues with low nutritive value.

The apparently direct transfer of nutrients/energy from large diatoms to *P. hoyi*, that is, in turn, eaten directly by fish, has interesting and important implications in comparative energy-transfer routes in Great Lakes food-web dynamics. The large size of spring diatoms compared to other phytoplankton, and of *P. hoyi* relative to most pelagic zooplankton, should make this energy-transfer route more efficient for fish production than would be true for traditional pelagic or microbial food webs. Only one intermediate trophic step may be required to convert photosynthetically fixed energy into a form (*P. hoyi*) sufficiently large to be readily eaten by fish. In contrast, energy originating from the microbial food web would usually require more trophic transfers with

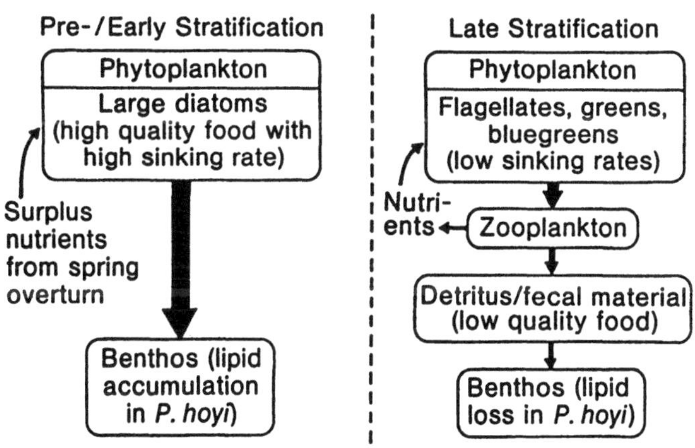


**Figure 34.6** Schematic diagram of a conceptual model of the predominant processes for energy flow from phytoplankton to the benthos during (left) the pre/early stratification period (right) the late stratification periods, in southern Lake Michigan.

associated energy losses before a small portion of the energy can be incorporated into upper trophic levels (Ducklow et al., 1986). The classical food web, including intermediate-sized phytoplankton and zooplankton, falls between these two extremes. Another factor that appears to make *P. hoyi* an efficient food-web intermediate is its ability to accumulate high levels of lipids. Lipid energy is transferred more efficiently through the food web than is energy from other organic nutrients (Stewart et al., 1983).

The seasonal, direct transfer of biochemical energy from pelagic phytoplankton to benthic macroinvertebrates, which in turn serve as fish-prey, has not been studied extensively for a broad spectrum of lakes of different size and nutritional status. However, certain lake characteristics appear to be necessary for this phenomenon to be ecologically important:

1. Abundant oxygen and low temperature in hypolimnetic waters
2. Development of algal bloom(s) (e.g., a spring bloom of large diatoms) with high sedimentation loss rates and minimal euphotic grazing losses, and
3. A dominance of macroinvertebrate populations that can use seasonal nutrient inputs and are desirable prey for fish.

Low temperature is an important characteristic because animals with low metabolic rates (such as *P. hoyi;* Gauvin, et al., 1989) can survive with minimal "new" food inputs during much of the seasonal cycle and also because the rate of bacterial decomposition of settling materials is relatively low. Low-nutrient lake status is important because the hypolimnetic waters of eutrophic lakes tend to become anaerobic during stratification due to the oxygen demand of organic matter resulting from high primary production rates in the water column. Seasonal or episodic inputs of nutrients to the euphotic zone, such as occurs in temperate lakes during spring turnover, provide a means for phytoplankton blooms to occur in otherwise low-nutrient waters. As mentioned above, large diatoms such as *Melosira* in these blooms are not quantitatively grazed by pelagic zooplankton because of their size and relatively short residence times, relative to zooplankton regeneration times, in the pelagic zone. On the other hand, phytoplankton that do not readily settle to the sediments would be prone to ingestion or decomposition in the water column. Finally, conditions, like those mentioned above, must be appropriate for the development of high populations of macroinvertebrates, such as *P. hoyi, M. relicta,* or possibly Chironomids, that can effectively cope with infrequent inputs of high-quality food and provide a food base for fish. These conditions occur in several large, monomictic and dimictic lakes and some relatively small, oligotrophic lakes but would not apply to warm-water lakes or highly eutrophic ones.

---

*Acknowledgements* We thank E.A. Cichocki for measuring lipids in *P. hoyi* and B.J. Eadie, T. Nalepa, and H. Vanderploeg for helpful suggestions on the manuscript.

# References


Cederwall, H. 1977. Annual macrofauna production of a soft bottom in the northern Baltic proper. *In*: B.F. Keenen, P. OCeidigh, and P.J.S. Boaden (eds.), *Biology of Benthic Organisms*. Pergamon Press, Ltd., Oxford, p. 155–164.

Cook, D.G. and M.G. Johnson. 1974. Benthic macroinvertebrates of the St. Lawrence Great Lakes. *J. Fish. Res. Board Can.* 31:763–782.

Ducklow, H.W., D.A. Purdie, P.J.LeB. Williams, and J.M. Davies. 1986. Bacterioplankton: A sink for carbon in a coastal marine plankton community. *Science* 232:865–868.

Eadie, B.J., R.L. Chambers, W.S. Gardner, and G.L. Bell. 1984. Sediment trap studies in Lake Michigan: resuspension and chemical fluxes in the southern basin. *J. Great Lakes Res.* 10:307–321.

Fahnenstiel, G.L. and D. Scavia. 1987a. Dynamics of Lake Michigan phytoplankton: primary production and growth. *Can. J. Fish. and Aquat. Sci.* 44:499–508.

Fahnenstiel, G.L. and D. Scavia. 1987b. Dynamics of Lake Michigan phytoplankton: recent changes in surface and deep communities. *Can. J. Fish. and Aquat. Sci.* 44:509–514.

Gardner, W.S., T.F. Nalepa, D.R. Slavens, and G.A. Laird. 1983. Patterns and rates of nitrogen release by benthic *Chironomidae* and *Oligochaeta*. *Can. J. Fish. and Aquat. Sci.* 40:259–266.

Gardner, W.S., T.F. Nalepa, W.A. Freż, E.A. Cichocki, and P.F. Landrum. 1985a. Seasonal patterns in lipid content of Lake Michigan macroinvertebrates. *Can. J. Fish. and Aquat. Sci.* 42:1827–1832.

Gardner, W.S., W.A. Frez, E.A. Cichocki, and C.C. Parrish. 1985b. Micromethod for lipid analysis in aquatic invertebrates. *Limnol. Oceanogr.* 30:1099–1105.

Gardner, W.S., T.F. Nalepa, and J.M. Malczyk. 1987. Nitrogen mineralization and denitrification in Lake Michigan sediments. *Limnol. Oceanogr.* 32.1226–1238.

Gardner, W.S., B.J. Eadie, J.F. Chandler, C.C. Parrish, and J.M. Malczyk. 1989. Mass flux and "nutritional composition" of settling epilimnetic particles in Lake Michigan. *Can. J. Fish. and Aquat. Sci.* 46:1118–1124.

Gauvin, J.M., W.S. Gardner, and M.A. Quigley. 1989. The effects of food removal on nutrient release rates and lipid content of Lake Michigan *Pontoporeia hoyi*. *Can. J. Fish. and Aquat. Sci.* 46:1125–1130.

Holland, R.E. 1969. Seasonal fluctuations of Lake Michigan diatoms. *Limnol. Oceanogr.* 14:423–436.

Holland, R.E. and A.M. Beeton. 1972. Significance to eutrophication differences in nutrients and diatoms in Lake Michigan. *Limnol. Oceanogr.* 17:88–96.

Johnson, M.G. and R.O. Brinkhurst. 1971. Production of benthic macroinvertebrates of the Bay of Quinte and Lake Ontario. *J. Fish. Res. Board Can.* 28:1699–1714.

Johnson, R.K. and T. Wiederholm. 1989. Long-term growth oscillations of *Pontoporeia affinis* Lindstrom (Crustacea: Amphipoda) in Lake Malaren. *Hydrobiologia* 175:183–194.

Laird, G.A., D. Scavia, G.L. Fahnenstiel, L.A. Strong, and G.A. Lang. 1988. Dyamics of Lake Michigan: relationship to nitrogen and silica fluxes. *Can. J. Fish. Aquat. Sci.* 45:1459–1466.

Landrum, P.F. 1988. Toxicokinetics or organic xenobiotics in the amphipod *Pontoporeia hoyi*: role of physiological and environmental variables. *Aquat. Toxicology* 12: 245–271.

Lee, R.F. 1975. Lipids of arctic phytoplankton. *Comp. Biochem. Physiol.* 51B:263–266.

Lehman, J.T. 1980. Release and cycling of nutrients between planktonic algae and herbivores. *Limnol. Oceanogr.* 25:620–732.

Lubner, J.F. 1979. Population dynamics and production of the relict amphipod *Pontoporeia hoyi* at several Lake Michigan stations. Ph.D. thesis, University of Wisconsin, Milwaukee, WI., 98 pp.

Marchant, R. and H.B.N. Hynes. 1981. Field estimates of feeding rate for *Gammarus pseudolimnaeus* (Crustacea: Amphipoda) in the Credit River, Ontario. *Freshwater Biology* 11:27–36.
Marzolf, G.R. 1965. Substrate relations of the burrowing amphipod *Pontoporeia affinis* in Lake Michigan. *Ecology* 46:579–592.
Moore, J.W. 1977. Some aspects of the feeding biology of benthic herbivores. *Hydrobiologia* 53:139–146.
Nalepa, T.F. 1987. Long-term changes in the macrobenthos of southern Lake Michigan. *Can. J. Fish. Aquat. Sci.* 4:515–524.
Nalepa, T.F., M.A. Quigley, K. Childs, J. Gauvin, T. Heatlie, M. Parker, and L. Vanover. 1985. Macrobenthos of Southern Lake Michigan 1980–81. NOAA Data Report, Environmental Research Laboratories, Great Lakes Environmental Research Laboratory, Ann Arbor, MI., 273 pp.
Nalepa, T.F., M.A. Quigley, and R.W. Ziegler. 1988. Sampling efficiency of the Ponar grab in two different benthic environments. *J. Great Lakes Res.* 14:89–93.
Quigley, M.A. 1988. Gut fullness of the deposit-feeding amphipod, *Pontoporeia hoyi*, in southeastern Lake Michigan. *J. Great Lakes Res.* 14:178–187.
Scavia, D. and G.L. Fahnenstiel. 1987. Dynamics of Lake Michigan phytoplankton: Mechanisms controlling epilimnetic communities. *J. Great Lakes Res.* 13:103–120.
Schelske, C.L. and E.F. Stoermer. 1971. Eutrophication, silica depletion, and predicted changes in algal quantity in Lake Michigan. *Science* 17:423–424.
Stewart, D.J., D. Weininger, D.V. Rottiers, and T.A. Edsall. 1983. An energetics model for lake trout, *Salvelinus mamaycush*: Application to the Lake Michigan population. *Can. J. Fish. Aquat. Sci.* 40:681–698.
Wells, L. 1980. Food of alewives, yellow perch, spottail shiners, trout-perch, and slimy and fourhorn sculpins in southeastern Lake Michigan. U.S. Fish and Wildl. Serv. Tech. Paper no. 98.
Wiederholm, T. 1973. On the life cycle of *Pontoporeia affinis* (Crustacea Amphipoda) in Lake Malaren. *Zoon* 1:147–151.
Winnell, M.H., and S.S. White. 1984. Ecology of shallow and deep water populations of *Pontoporeia hoyi* (Smith) (Amphipoda) in Lake Michigan. *Freshwater Invertebr. Biol.* 3:118–138.

# 35

# Fish Predation Effects on Plankton Community Structure in Large Lakes: Lake Kinneret

M. Gophen and S. Serruya

ABSTRACT The analysis of the long-term (1969–1985) data record for Lake Kinneret (Israel) indicated a reduction in zooplankton biomass and the enhancement of nanoplankton, together with the lowering of TP/TN ratios in the epilimnion. The zooplankton community structure was modified by a reduction in the number of large cladocerans and copepods relative to small organisms. Small animals like copepod nauplii and *Keratella* spp. also declined in recent years. Long-term fisheries management and stocking in the Kinneret caused an intensification of predation pressure on zooplankton, resulting in reduction of their densities. Predation on zooplankton was enhanced by both particulate- and filter-feeding fishes. Consequently, nanoplankton densities were enhanced mostly in the summer-fall period when the fish predation pressure is heaviest. An increase in total P and a decrease in total N in the epilimnion of Lake Kinneret occurred at the same time as the nanoplankton and zooplankton changes. It is suggested that phytoplankton enhancement was affected by alterations in both nutrients and zooplankton.

## 35.1 Introduction

Plankton community structure in lakes and reservoirs can be altered by human control of fish populations (Shapiro et al., 1975). This alteration may be the result primarily of top-down, or consumer-controlled, forces (Carpenter et al., 1985, 1988; McQueen et al., 1986; Benndorf, 1987; Threlkeld, 1987). The complexity of the ecosystem response and the combined role of bottom-up or producer-controlled forces (nutrients) and/or nutrient-mediated effects of planktivorous fish on plankton community structure were considered by Carpenter (1988), McQueen and Lean (1987), and Threlkeld (1987).

The classic approach to lake ecosystem structure has been the food-chain concept, in which energy or carbon is considered to pass from phytoplankton to zooplankton to fish. Unfortunately, fish as top consumers were poorly considered as a component that may strongly affect water quality, while classical limnology gave inadequate consideration to fishery biology (Rigler, 1982).

Pioneering studies on fish zooplanktivory and its effect on algal communities were carried out by Hrbacek and his team during the late 1950's and early 1960's (Hrbacek et al., 1961). Shortly thereafter, Brooks and Dodson (1965) and Hall et al., (1970) also documented the effect of zooplanktivorous fish predation on lower trophic levels.

Effects of high-trophic-level consumers (fish) on water quality in natural and artificial water bodies were presented by Hurlbert et al. (1972), Zaret and Pain (1973), Shapiro et al. (1975), Anderson et al. (1978), Hrbacek et al. (1978), and Stenson et al. (1978).

Responses of phytoplankton and zooplankton to a reduction or increase of planktivorous fish predation pressure and the effect of piscivory—as well as the nutrients and fish nutrient-mediated effects on phytoplankton, bacteria, microzooplankton, and picoplankton—were analyzed during the 1980's by Kitchell et al. (1982), Shapiro et al. (1982), Benndorf et al. (1984), Drenner et al. (1984) Hansson et al. (1984) Goad (1984), Reinertsen and Olsen (1984), McQueen et al. (1986), Threlkeld (1987), Vinyard et al. (1988), Carpenter (1988), and Stockner (1988).

Major studies of natural fish effects on aquatic communities were carried out in small lakes, ponds, reservoirs, outdoor tanks, and enclosures (Anderson et al., 1978; Benndorf et al., 1984; Reinertsen and Olsen, 1984; Shapiro and Wright, 1984; Stenson et al., 1978; Hrbacek et al., 1961; Leventer, 1981; Drenner, et al., 1984; 1987; Goad, 1984; Tatrai and Istvanovics, 1986; Vinyard et al., 1988; Threlkeld, 1987). However, fish effects in natural, large water bodies were poorly documented. These few studies were mostly the result of ecological changes or disasters caused by modifications of fish populations in large lakes (Barel et al., 1985; Gophen, 1987, 1988; Gophen et al., 1983; Zaret and Pain, 1973; Scavia et al., 1986). One additional aspect which has not been thoroughly studied is the long-term effect of fish on the aquatic environment.

Among the 24 fish species in Lake Kinneret, 15 are commercially fished, four (two sardines and two tristramellids) are endemic to the lake, three (mullets: *Mugil cephalus* and *Liza ramada*; silver carp, *Hypophthalmichthys molitrix*) are exotics, and one, *Tilapia aureus* (*Oreochromis aureus*, or blue tilapia) was rare prior to its introduction. The most abundant species are the zooplanktivorous bleaks (*Mirogrex* sp. and *Acanthobrama* sp.). The majority of the Kinneret fishes are planktivores, although several are benthophagous. The level of piscivory is low (Gophen, 1987).

The lake has been stocked with fingerlings since the early 1960's. From 1960 to 1987, 123 million fingerlings [37% *Sarotherodon galilaeus*, 26% *Tilapia aureus*, 11% silver carp (*H. molitrix*), and 26% mullets (*M. cephalus* and *L. ramada*)] were planted in the lake. Although there are differences in major food sources, most of the commercial species consume zooplankton, several of them during all seasons, and others mostly in the summer-fall period (Table 35.1). Fish predation pressure on zooplankton is heaviest during the summer-fall

**Table 35.1** Percentage of zooplankton biomass in the gut content of the major commercial fishes of Lake Kinneret[a]

| | Zooplankton biomass (%) | |
|---|---|---|
| Species | Summer–fall | Winter–spring |
| *Sarotherodon galilaeus* | 10 | 6 |
| *Oreochromis aureus* | 35 | 41 |
| Silver carp | 59 | 9 |
| *Tristramella sacra* | 24 | 19 |
| *Tristramella simonis* | 32 | 11 |
| *Tilapia zillii* | 42 | 42 |
| *Mirogrex terraesanctae* | 90–100 | 90–100 |

Source: Gophen, 1987a.

[a] Preliminary observations also indicated a high level of zooplanktivory by grey mullets (Gophen, unpublished data).

period and is produced by both filter- and particulate-feeder fishes. Zooplanktivory by larval and adult sardines and the increase of this zooplanktivorous fish stock biomass in the lake after the 1960's were documented by Gophen and Landau (1977), Gophen and Scharf (1981), Landau et al. (1987), and Gophen and Threlkeld (1988). Landau (1984) indicated an annual increase of 3–7% of the stock biomass of sardines in Lake Kinneret during 1968–1981.

Partial replacement of the native *Sarotherodon galilaeus* by stocked *Tilapia aureus* and its effect on suppression of zooplankton was documented by Gophen et al. (1983), Drenner et al. (1987), and Vinyard et al. (1988). Zooplanktivory by fingerlings of cichlids was described by Drenner et al. (1982) and Gophen (1980).

Intensification of fishing effort (boat days, improvements of technological and fishery operations) was mostly aimed at high-value species (cichlids, mullets, and silver carp) (Gophen et. al., 1983), resulting in enhancement of low-value fish populations, especially sardines. Fishing effort on sardines declined in recent years and sardines feed mainly on zooplankton.

As a consequence of these modifications the level of fish predation pressure on zooplankton increased, followed by several alterations in plankton community structure. Independently, nutrient (P and N) contents also showed temporal changes. In this paper, we describe temporal trends in plankton communities and analyze the effect of fish predation (top-down forces) on zooplankton. The role of nutrient impact (bottom-up forces) is also considered.

## 35.2 Methods

Zooplankton and phytoplankton data were collected by Gophen and U. Pollingher, respectively (Annual Reports of the Kinneret Limnological Laboratory,

**Table 35.2** Correlation coefficients of regressions between variables of zooplankton density (no./l), phytoplankton biomass, and years (1969–85).

| | Correlation coefficient[b] | | | | | | |
|---|---|---|---|---|---|---|---|
| | Year | Pyr. | Chl. | Cya. | Bac. | Nonpyr. | Total |
| Year | – | NS | 0.64* | NS | NS | 0.68** | 0.62* |
| Copepoda biomass ($g/m^2$) | −0.73** | NS | −0.85** | −0.54* | NS | −0.85** | −0.59* |
| Copepoda eggs/ female | NS | NS | NS | NS | NS | NS | NS |
| Adult copepoda | −0.84** | NS | NS | −0.62* | NS | NS | NS |
| Copepodite | NS | NS | NS | NS | NS | NS | NS |
| Nauplius | −0.50* | NS | NS | −0.63* | NS | NS | NS |
| Cladocera biomass ($g/m^2$) | −0.67* | NS | NS | −0.53* | NS | −0.55* | NS |
| *Diaphanosoma* | NS | NS | NS | NS | NS | NS | NS |
| *Diaphanosoma* S/L[a] | 0.77** | NS | 0.67* | NS | NS | 0.67* | 0.71* |
| *Ceriodaphnia* | NS | NS | NS | NS | NS | NS | NS |
| *Ceriodaphnia* S/L[a] | 0.66* | NS | 0.61* | NS | NS | NS | 0.68* |
| *Bosmina* | −0.60* | NS | NS | NS | NS | −0.54* | NS |
| *Bosmina* S/L[a] | 0.79* | NS | 0.69* | NS | NS | 0.67* | 0.76* |
| Rotifera biomass ($g/m^2$) | −0.66* | NS | −0.64* | NS | NS | −0.56* | −0.70** |
| *Keratella* | −0.74** | NS | NS | NS | NS | NS | −0.70** |
| Large rotifers | NS | NS | NS | NS | NS | NS | NS |
| Small rotifers | NS | NS | −0.57* | NS | NS | NS | −0.73** |

NS, not significant
* $P < 0.05$
** $P < 0.01$
[a] S/L, the ratio between small (S) and large (L) animals.
[b] Pyr., Pyrrhophyta; Chl., Chlorophyta; Cya., Cyanophyta; Bac., Bacillariophyta; Nonpyr., Nonpyrrhophyta; Total, total phytoplankton.

1969–1985). Nutrient (total P and total N) mass contents (tons) of the lake epilimnion (0–16 m depth) were calculated from routinely measured concentrations and water layer volumes (Serruya, S., Kinneret database—unpublished observations). The raw data were annually averaged over years.

Correlation coefficients (r) and probabilities ($p$) of linear regressions for each variable versus time (years) and between variables were calculated; the results are presented in Table 35.2 for zooplankton groups, genera, and size classes vs. years and phytoplankton groups. In Figure 35.1, results presented for zooplankton, phytoplankton, nutrients, and years.

## 35.3 Results

**Zooplankton** The biomass of total zooplankton, Copepoda, Cladocera, and Rotifera, and the production of Copepoda and Cladocera significantly ($p < 0.01$) declined from 1989 to 1985 (Figures 35.2 and 35.3; also see Figure

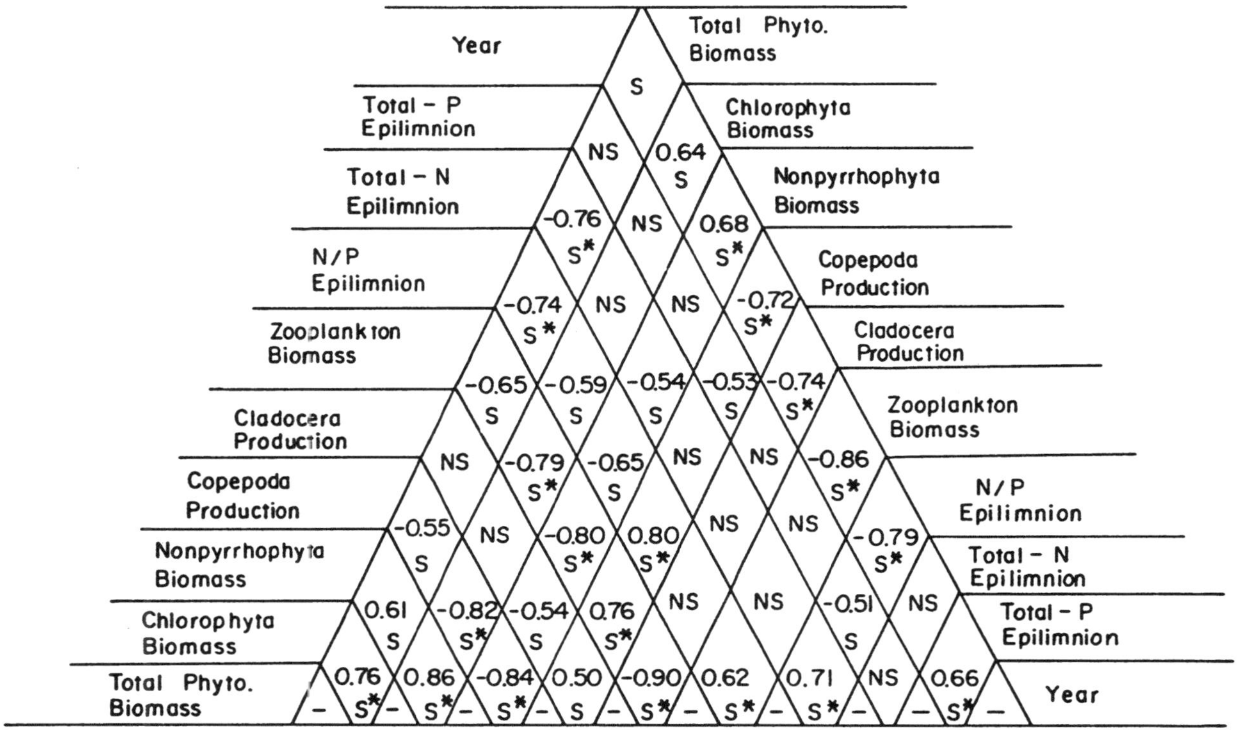


**Figure 35.1** Correlation coefficients of linear regressions between variables of zooplankton (1969–85), phytoplankton (1972–85), nutrients (1969–85), and years. S* = $p<0.01$; S = $p<0.05$; NS = not significant.

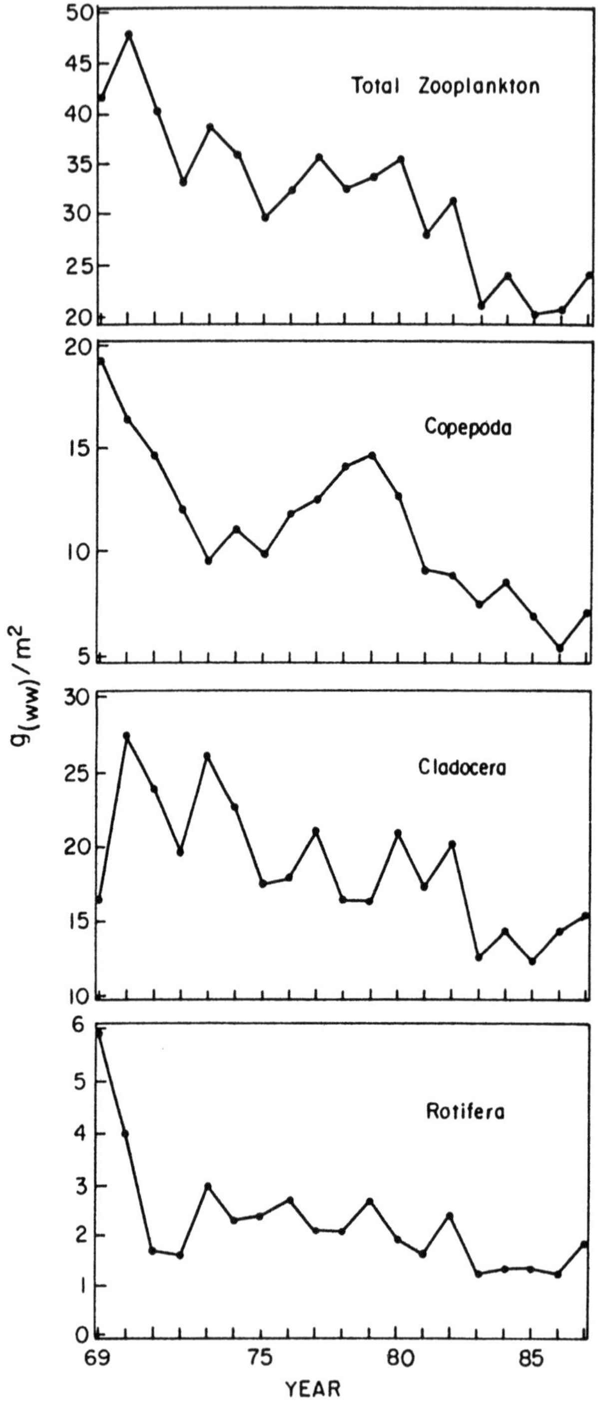


**Figure 35.2** Annual averages of the biomass of total zooplankton, Copepoda, Cladocera, and Rotifera, in Lake Kinneret during 1969–1987.

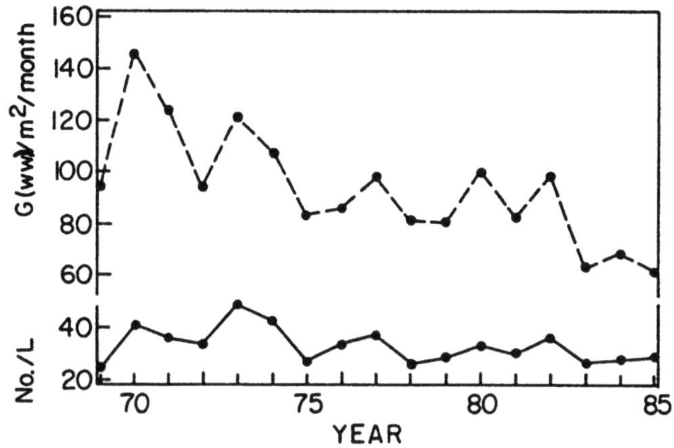


**Figure 35.3** Annual averages of cladoceran densities and production in Lake Kinneret during 1969–1985.

35.1). Densities of adult copepods and *Keratella* ($p < 0.01$), and of nauplii and *Bosmina* ($p < 0.05$) were significantly reduced during this period (see Table 35.2 and Figure 35.3; Figure 35.4). The ratios of small to large *Diaphanosoma, Bosmina* ($p < 0.01$), and *Ceriodaphnia* ($p < 0.05$) were positively correlated with time (1975–1985), i.e., densities of small cladocerans relative to large ones increased (Figure 35.5).

The egg ratio of Copepoda (eggs/female), densities of copepodite stages, *Diaphanosoma, Ceriodaphnia,* and large and small rotifer groups all showed insignificant trends with time (1969–1985) (see Figures 33.3 and 33.4 and Table 35.2).

Densities of cladocerans did not change significantly but the organisms became smaller and thus, biomass and production declined. Densities of the large-bodied *Mesocyclops* sp. declined relative to the smaller, cyclopoid *Thermocyclops* sp. (Figure 35.6).

**Phytoplankton** The biomass concentrations of nonpyrrhophyta ($p < 0.01$), Chlorophyta, and total phytoplankton ($p < 0.05$) increased in Lake Kinneret during 1972–1985 (Figure 35.7; also see Figure 35.1). On the other hand, the biomass concentrations of Pyrrhophyta, Cyanophyta, and Bacillariophyta did not change significantly (see Table 35.2).

**Zooplankton-Phytoplankton Correlations** The biomass levels of Copepoda and Rotifera were negatively correlated with phytoplankton variables, except with Pyrrhophyta and Bacillariophyta, and, except Rotifera with Cyanophyta (see Table 35.2). Cladocera biomass was inversely correlated with Cyanophyta and nonpyrrhophyta (see Figure 35.1 and Table 35.2).

Copepoda egg ratio and population densities of copepodites, *Diaphano-*

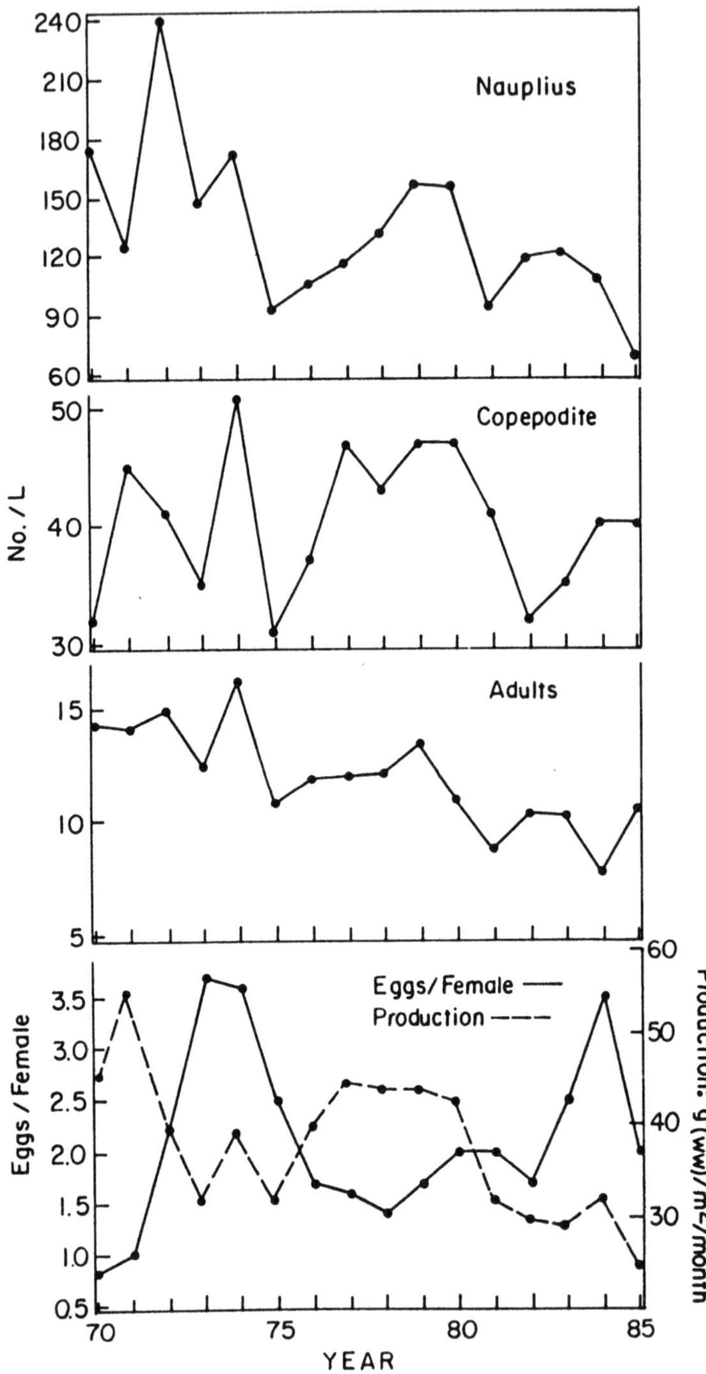


**Figure 35.4** Annual average densities of the three life stages of Copepoda: nauplius, copepodite, and adult. Bottom: Rates of eggs to females and copepod production in Lake Kinneret during 1970–1985.

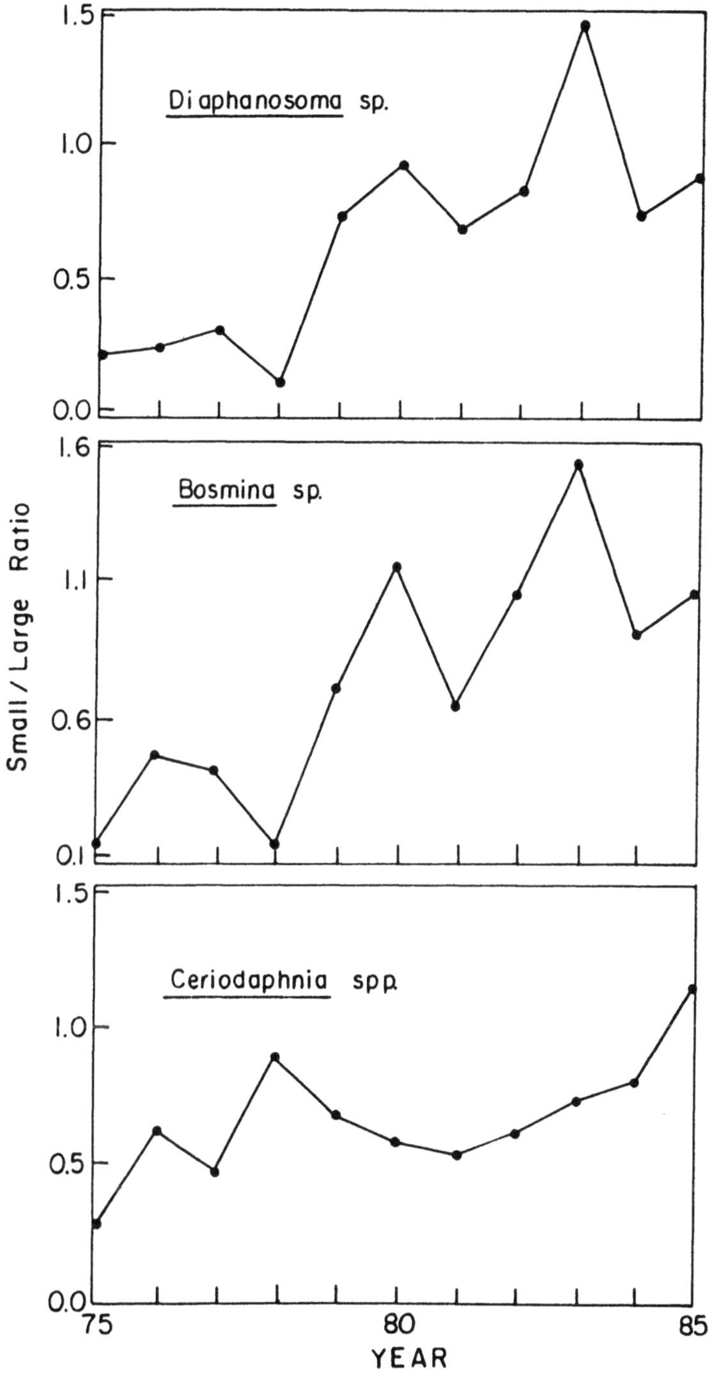


**Figure 35.5** Annual averages of the ratio of small to large individuals for *Diaphanosoma* sp. *Bosmina* sp., and *Ceriodaphnia* spp. in Lake Kinneret during 1975–1985.

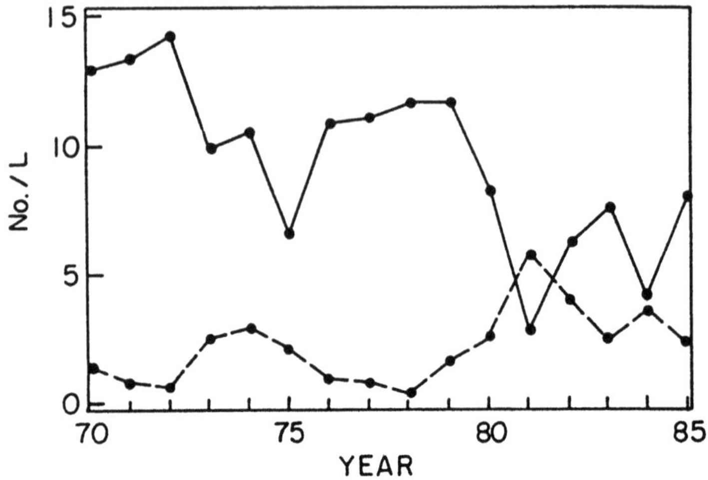


**Figure 35.6** Annual averages of the densities of adults of *Mesocyclops* sp. (———) and *Thermocyclops* sp. (----) in Lake Kinneret during 1970–1985.

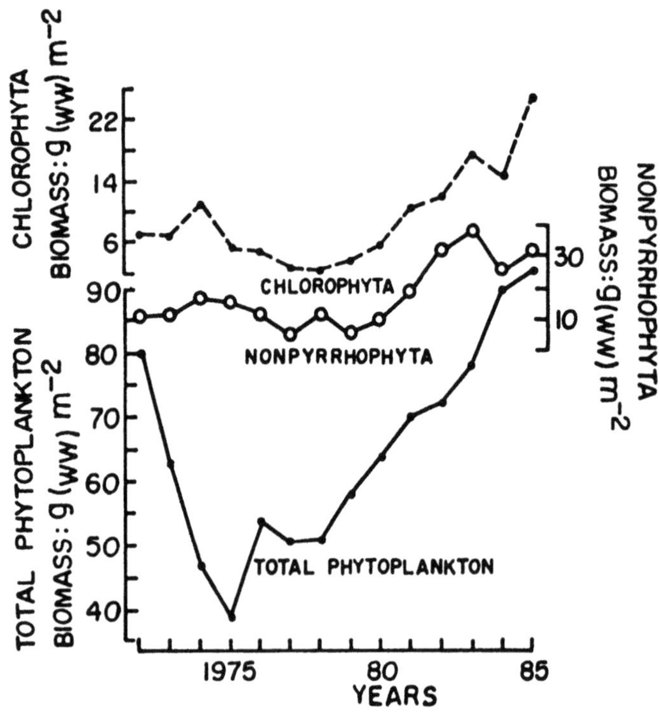


**Figure 35.7** Annual averages of Chlorophyta nonpyrrhophyta, and total phytoplankton biomass in Lake Kinneret during 1972–1985 (Data from Pollingher, 1972–1985).

*soma, Ceriodaphnia,* and large rotifers were not significantly correlated with any phytoplankton variable (see Table 35.2).

The densities of adult copepods, nauplii, *Bosmina,* and *Keratella* were negatively related to Cyanophyta, nonPhyrrhophyta, and total phytoplankton, respectively. Small rotifers showed an inverse relationship with Chlorophyta and total phytoplankton (see Table 35.2).

**Nutrients** Total P increased and the $T_N/T_P$ ratios were lowered significantly (Figure 35.8; also see Table 35.2). Changes of total N over the period of 1969–1985 were not significant, but during 1974–1985, nitrogen declined significantly ($r = -0.83$; $p < 0.0017$) in the lake (see Figures 35.1 and 35.8). The $T_N/T_P$ ratio in the epilimnion (0–16m)was also significantly lowered ($r = -0.72$) when averaged for the summer (June–October) months.

**Phytoplankton-Nutrients Correlations** Phytoplankton variables were insignificantly correlated with total P but significantly and inversely related to the $T_N/T_P$ ratio. NonPyrrhophyta biomass and total phytoplankton were inversely correlated with total N (see Figure 35.1).

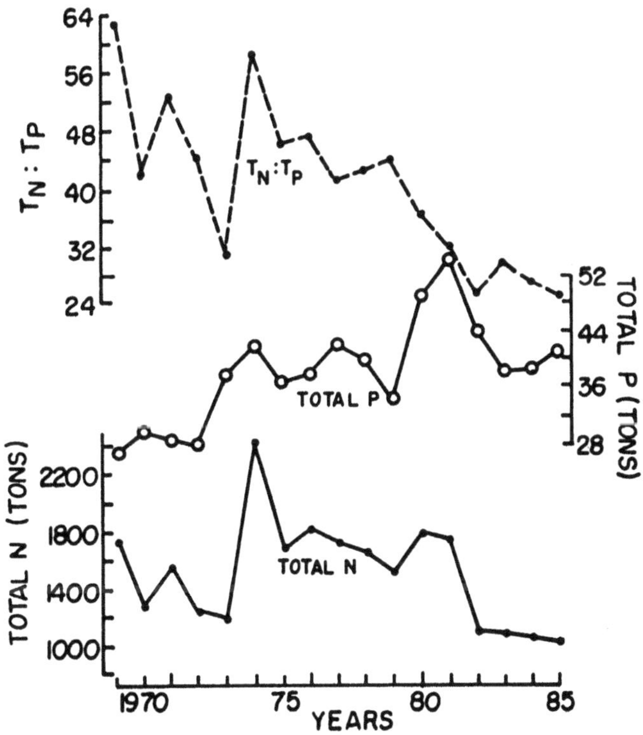


**Figure 35.8** Annual averages of mass content of total N and total P and of their ratio in the epilimnion (0–16 m) of Lake Kinneret during 1969–1985 (Serruya, S., Kinneret Data base—unpublished observations).

## 35.4 Discussion

The deleterious effect on nanoplankton on water quality in Lake Kinneret is one of the main concerns of managers of the drinking water supply. Consequently, understanding the nature of the long-term changes in the lake is important for the future design of lake management (Pollingher, 1986).

The results presented here indicate a decline in density of large-bodied copepods and cladocerans (see Figures 35.4 and 35.6), as well as of small organisms like nauplii (see Figure 35.5) and *Keratella* (see Table 35.2). It is suggested that this zooplankton suppression was caused by an increase of fish predation pressure in the Lake Kinneret ecosystem (Gophen, 1986, 1987, 1988).

Zooplankton reduction (see Figures 35.1 to 35.6) was accompanied by an increase in the biomass of total phytoplankton and nonpyrrhophyta, particularly Chlorophyta (see Figure 35.7). Nevertheless, temporal changes of nutrient contents occurred simultaneously: total P increased during 1969–1985 and total-N declined between 1974 and 1985 (see Figure 35.8). As a result, both the annual and the seasonal (June–October) $T_N/T_P$ ratios were lowered from values of 42–68 during the early 1970's (except 32 in 1973) to $< 29$ in recent years (18 during summer 1985). The impact of $T_N/T_P$ decline on the phytoplankton assemblages was presented by McCarthy (1980), Tilman et al. (1982), Smith (1983), Trimbee and Prepas (1987), and McQueen and Lean (1987). It was reported by Smith (1983) that cyanobacteria blooms occurred only, but not necessarily, when epilimnetic $T_N/T_P$ (wt/wt) ratios were less than 29. The effects of $T_N$, $T_P$, temperature, light availability, and the different N-form sources, solely or in combinations, on the relative biomass of Cyanobacteria were also described (Smith, 1986; McQueen and Lean, 1987; Trimbee and Prepas, 1987).

Until recently, top-down (consumer-controlled) and bottom-up (producer-controlled) forces were intensively but independently studied. However, some recent studies have examined the combined effects of the two-directional ecological forces that act on phytoplankton densities in freshwater ecosystems (McQueen et al., 1986; Benndorf, 1987; Carpenter, 1988).

We suggest that long-term alterations caused by fish zooplanktivory, together with bottom-up forces ($T_N$, $T_P$ and $T_N/T_P$ ratio), in Lake Kinneret during 1972–1985 positively affected (enhanced) densities of Chlorophyta. (Serruya et al., 1980). Nanophytoplankton (mostly chlorophytes) enhancement in Lake Kinneret (see Figure 35.7) was simultanously affected by both top-down and bottom-up forces. Increasing particulate and filter-fish zooplanktivory (predation) suppressed large copepods and cladocerans and also reduced the densities of small nauplii and *Keratella*. As a result, the grazing pressure of zooplankton on nanoplankton was lowered and the algae proliferated. The increase of total P, decrease of total N, and therefore the decline of the $T_N/T_P$ ratio (see Figure 35.8) is probably favorable to nanoplankton, especially chlorophytes, and thus, also contributed to the enhancement of these phytoplankters (Pollingher, 1986).

The inverse relationship between Cyanobacteria and several zooplankton groups (see Table 35.2) probably reflect toxic effects. There was no significant correlation between Pyrrhophyta and zooplankton; this is in agreement with previous observations on the nongrazability of *Peridinium* by zooplankton.

## 35.5 Future Perspectives

The long-term changes in the plankton community structure that were observed in Lake Kinneret are similar to those described in several other large lakes: Lake Victoria (Barel et al., 1985), Lake Gatun (Zaret and Pain, 1973), Lake Michigan (Scavia et al., 1986) and the Canadian lakes (McQueen et al., 1986; McQueen and Lean, 1987). In these ecosystems, the top-down (fish) effect was clearly documented, but bottom-up (nutrients) effect were not always analyzed simultaneously. Management design aimed at improvements of ecosystem structure must consider both types of forces, as well as their practical implications, i.e., treatment costs vs. economic benefits. Therefore, fishery manipulation as a management tool to improve water quality is recommended for the Kinneret.

To prevent any future enhancement of nanoplankton in Lake Kinneret, two parallel operations are recommended:

1. Stop the decline of $T_N/T_P$ ratio in the lake by control of external N and P supply. Benndorf (1987) stated that the efficiency of food-web manipulation aimed at improving water quality by algal biomass reduction is highly dependant on a decrease of P content. We suggest that this is also relevant to the Kinneret management policy.
2. Reduce fish predation pressure on zooplankton by changing the policies of fish introduction and regulation.

---

*Acknowledgements* We thank Dr. W. Geller, H. Pauli, and Y. Prarie for their assistance in preparing this manuscript.

## References

Anderson, G., H. Berggren, G. Cronberg, and C. Crelin. 1978. Effects of Planktivorous and benthivorous fish on organisms and water chemistry in eutrophic lakes. *Hydrobiol.* 59:9–15.

Barel, C.D.N., R. Dorit, P.H. Greenwood, G. Fryer, N. Hughes, P.B.N. Jackson, H. Kawanabe, R.H. Low-McConell, M. Nagoshi, A.J. Ribbink, E. Trewavas, F. Witte, and K. Yamaoka. 1985. Destruction of fisheries in Africa's lakes. *Nature* 315:19–20.

Benndorf, J. 1987. Food web manipulation without nutrient control: A useful strategy in lake restoration? *Schweiz. Z. Hydrol.* 49/2:237–248.

Benndorf, J., H. Kneschke, K. Kossatz, and E. Penz. 1984. Manipulation of the pelagic food web by stocking with predacious fishes. *Int. Revue ges. Hydrobiol.* 69:407–428.

Brooks, J.L. and S.I. Dodson. 1965. Predation, body size, and composition of plankton. *Science* 150:28–35.
Carpenter, S.R. (ed.). 1988. *Complex Interactions in Lake Communities.* NSF Workshop, 21-26/3/1987, Notre-Dame, USA. Springer-Verlag.
Carpenter, S.R. J.F. Kitchell, and J.R. Hodgson. 1985. Cascading trophic interactions and lake productivity. *BioScience* 35:634–639.
Drenner, R.W., G.L. Vinyard, M. Gophen, and S.R. McComas. 1982. Feeding behavior of the cichlid, *Sarotherodon galilaeus*: Selective predation on Lake Kinneret zooplankton. *Hydrobiologia* 87:17–20.
Drenner, R.W., J.R. Mummert, F. DeNoyelles, Jr., and D. Kettle. 1984. Selective particle ingestion by a filter-feeding fish and its impact on phytoplankton community structure. *Limnol. Oceanogr.* 29:941–948.
Drenner, R.W., K.D. Hambright, G.L. Vinyard, M. Gophen, and U. Pollingher. 1987. Experimental study of size selective phytoplankton grazing by a filter-feeding cichlid and the cichlid's effects on plankton community structure. *Limnol. Oceanogr.* 32(5):1138–1144.
Goad, J.A. 1984. A biomanipulation experiment in Green Lake, Seattle, Washington, USA. *Arch. Hydrobiol.* 192:137–154.
Gophen, M. 1986. Fisheries management in Lake Kinneret (Israel). In: *Proceedings of the Ann. Mtg. N. Amer. Lake Management Soc.* Lake Geneva, Wisconsin, USA, p. 327–332.
Gophen, M. 1987. Fisheries management, water quality and economical impacts: a case study of Lake Kinneret. *Proc. World Conf. on Large Lakes.* Mackinac Island, Michigan, USA. Vol. II:5–24.
Gophen, M. 1988. Changes of Copepoda populations in Lake Kinneret during 1969–1985. Proc. Third Internat. Conf. Copepoda, London. G.A. Boxhall and K. Schminke (eds.), *Hydrobiologia* 47:375–379.
Gophen, M. 1969–1985. Zooplankton. In: *Kinneret Limnological Laboratory Annual Reports.* Tiberias, Israel.
Gophen, M. and R. Landau. 1977. Trophic interactions between zooplankton and sardines (*Mirogrex terraesanctae*) populations in Lake Kinneret, Israel. *Oikos* 29(1):166–174.
Gophen, M. and A. Scharf. 1981. Food and feeding habits of *Mirogrex* fingerlings in Lake Kinneret (Israel). *Hydrobiologia* 78:3–9.
Gophen, M. and S.T. Threlkeld. 1988. An experimental study of zooplankton consumption by the Lake Kinneret sardine. *Archiv. Hydrobiol.* 115:91–95.
Gophen, M., R.W. Drenner, and G.L. Vinyard, 1983. Cichlid Stocking and the decline of the Galilee St. Peter's Fish (*Sarotherodon galilaeus*) in Lake Kinneret (Israel). *Can. J. Fish. Aquat. Sci.* 40:983–986.
Hall, D.J., W.E. Cooper, and E.E. Werner. 1970. An experimental approach to the production dynamics and structure of freshwater animal communities. *Limnol. Oceanogr.* 15:839–928.
Hansson, L.A., L. Johansson, and L. Persson. 1987. Effects of fish grazing on nutrient release and succession of primary producers. *Limnol. Oceanogr.* 32(3):72333–729.
Hrbacek, J., M. Dvorakova, V. Korinek, and L. Prochazkova. 1961. Demonstration of the effect of the fish stock on the species composition of zooplankton and the intensity of metabolism of the whole plankton assemblage. *Verh. Int. Ver. Theoret. Angew. Limnol.* 14:192–195.
Hrbacek, J., B. Desortova, and J. Popovsky. 1978. The influence of the fish stock on the phosphorus-chlorophyll ratio. *Verh. Internat. Verein. Limnol.* 20:1624–1628.
Hurlbert, S.H., J. Zedler, and D. Fairbanks. 1972. Ecosystem alteration by mosquito fish (*Gambusia affinis*) predation. *Science* 175:639–641.
Kitchell, J.P., H.F. Henderson, E. Grygierek, J. Hrebacek, S.R. Kerr, M. Pedini, T. Petr. J.

Shapiro, R.A. Stein. J. Stenson, and T. Zaret. 1982. *Management of Lakes by Food-Chain Manipulation.* U.N.F.A.O., Rome.
*Kinneret Limnological Laboratory Annual Reports.* 1969–1985. Israel Oceanographic and Limnological Research, Tiberias, Israel.
Landau, R. 1984. Fish population dynamics in Lake Kinneret. In: *Kinneret Limnological Laboratory Annual Report (1983),* p. 12; p. 75–76.
Landau, R., M. Gophen, and P. Walline. 1988. Larval *Mirogrex terraesanctae* (Cyprinidae) of Lake Kinneret (Israel): growth rate, plankton selectivities, consumption rates and interactions with rotifers. *Hydrobiologia* 169:91–106.
Leventer, H. 1981. Biological control of reservoirs by fish. *Bamidgeh* 33(1):3–23.
McCarthy, J.J. 1980. Nitrogen. In: I. Morris (ed.), *The Physiological Ecology of Phytoplankton.* Univ. of California Press, Berkeley, p. 191–233.
McQueen, D.J., J.R. Post and E.L. Mills. 1986. Trophic relations in freshwater pelagic ecosystems. Can. J. Fish. Aquat. Science 43(8):1571–1581.
McQueen, D.J. and D.R.S. Lean. 1987. Influence of water temperature and nitrogen to phosphorus ratio on the dominance of blue-green algae in Lake St. George, Ontario. *Can. J. Aquat. Sci.* 44:598–604.
Pollingher, U. 1986. Phytoplankton periodicity in a subtropical lake (Lake Kinneret, Israel). *Hydrobiologia* 138:127–138.
Pollingher, U. 1972–1985. Phytoplankton. In: *Kinneret Limnological Laboratory Annual Reports.*
Reinertsen, H. and Y. Olsen. 1984. Effects of fish elimination on the phytoplankton community of a eutrophic lake. *Verh. Internat. Verein. Limnol.* 22:649–657.
Rigler, F.H. 1982. The relation between fisheries management and limnology. *Trans. Am. Fish. Soc.* 111:121–132.
Scavia, D., G.L. Fahnenstiel, M.S. Evans, D. Juden, and J.T. Lehman. 1986. Influence of Salmonid predation and weather on long-term water quality trends in Lake Michigan. *Can. J. Fish. Aquat. Sci.* 43:435–443.
Serruya, C., M. Gophen, and U. Pollingher. 1980. Lake Kinneret: carbon flow patterns and ecosystem management. *Arch. Hydrobiol.* 88(3):265–302.
Shapiro, J., V. Lammara, and M. Lynch. 1975. Biomanipulation: an ecosystem approach to lake restoration. In: P.L. Brezonik and J. Fox (eds.), *Proceedings of a Symposium on Water Quality Management through Biological Control.* Univ. Fl., Gainsville, p. 85–96.
Shapiro, J., Forsberg, V. Lammara, G. Lindmark, M. Lynch, E. Smeltzer, and G. Zoto. 1982. Experiments and experiences in biomanipulation. In: *Interim Report No. 19* Limnological Research Center, Univ. of Minnesota, Minneapolis, p. 1–125.
Shapiro, J. and D.I. Wright. 1984. Lake restoration by biomanipulation: Round Lake, Minnesota, the first two years. *Freshwat. Biol.* 14:371–383.
Smith, V.H. 1983. Low nitrogen to phosphorus ratios favor dominance by blue-green algae in lake phytoplankton. *Science* 221:669–671.
Smith, V.H. 1986. Light nutrient effects on the relative biomass of blue-green algae in lake phytoplankton. *Can. J. Fish. Aquat. Sci.* 43:148–153.
Stenson, J.A.E., T. Bohlin, L. Henrikson, B.I. Nilsson, H.G. Nyman, H.G. Oscarson, and P. Larsson. 1978. Effects of fish removal from a small lake. *Theor. Apl. Limnol. Proc.* 20:794–801.
Stockner, J.G. 1988. The role of autotrophic and heterotrophic picoplankton in structuring the food webs of large oligotrophic British Columbia lakes. *Abstract, Conference on Functional and Structural Properties of Large Lakes, Konstanz,* Sep. 18, 1987.
Tatrai, T. and V. Istvanovics. 1986. The role of fish in the regulation of nutrient cycling in Lake Balaton. *Freshwater Biology* 16:417–424.
Threlkeld, S.T. 1987. Experimental evaluation of trophic-cascade and nutrient-mediated effects of planktivorous fish on plankton community structure. In: W.C. Kerfoot and A. Shi (eds.), *Predation: Direct and Indirect Impacts on Aquatic Communities.* Univ. Press New England, Hanover, p. 161–1173.

Tilman, D., S.S. Kilham, and P. Kilham. 1982. Phytoplankton community ecology: the role of limiting nutrients. *Ann. Rev. Ecol. Syst.* 13:349–372.
Trimbee, A.M. and E.E. Prepas. 1987. Evaluation of total phosphorus as a predictor of the relative biomass of blue-green algae with emphasis on Alberta lakes. *Can. J. Aquat. Sci.* 44:1337–1342.
Vinyard, G.L., R.W. Drenner, M. Gophen, U. Pollingher, D.L. Winkelman, and K.D. Hambright. 1988. An experimental study of the plankton community impacts of two filter-feeding cichlids, the Galilee Saint Peter's fish (*Sarotherodon galilaeus)* and Blue Tilapia (*Tilapia aureus). Can. J. Fish. Aquat. Sci.* 45(4):685–690.
Zaret, T.M. and R.T. Pain. 1973. Species introduction in a tropical lake. *Science* 182:449–455.

# Part 6

# Ecological Structure and Function in Large Lakes

The final chapter of this volume attempts to summarize some of the conlusions that have emerged from recent studies of large lakes and to identify major gaps in our knowledge which should be studied in forthcoming investigations.

It is now well established that large lakes do have specific features that are caused by their size. The most important conclusion probably is that physical forcing of biological events is stronger in large than in small lakes. Only large lakes possess true pelagic zones in which the majority of biogeochemical processes take place. Nevertheless, interactions between the free water zone and the lake bottom play an important part, even in deep large lakes. Because, on the average, a smaller area within the watershed is drained by a given volume of lake water, nutrient loading of large lakes is generally less important than in small lakes; consequently, large lakes tend to be more oligotrophic. This fact may also be related to the observation which suggests that the microbial loop appears to be more important in large than in small lakes. The intense and continuous interaction between hyrodynamic events and biological entities is the prime set of processes which determines ecological structure and function in large lakes.

# 36

# Overview: An Appraisal of Concepts

Colette Serruya

## 36.1 Introduction

The 60 lectures and 33 posters presented at the Constance symposium on the functional and structural properties of large lakes highlighted the fact that research on large lakes is no longer centered, as it was in the 1960s and 1970s, on cultural eutrophication, and the reports presented on reoligotrophication were a delayed result of these past efforts.

New and more integrative approaches to large lakes research became apparent in the symposium discussions. A symptomatic sign was the active presence of a number of physical limnologists at this meeting, which was otherwise dominated by biologists. The contribution of the former was crucial in many respects; firstly, it was important because large lakes can best be distinguished from small ones on the basis of physical structural differences (size and depth) and, as a consequence, of distinct hydromechanical behavior. Second, since physical processes affect all chemical and biological events occurring in lakes, the study of the temporal and spatial scales of physical forcing and the biological rates of adaptation to a given physical environment offer new insights into biological processes in lakes.

The dynamics of sediments was given a special emphasis at the meeting. New methodological approaches, especially the use of radionuclides, were demonstrated. A new interest has emerged in the role of sediments as internal sources of nutrients and as sinks for long-lasting pollutants in large lakes. The Chernobyl nuclear power plant accident has given the scientific community an uncalled-for but very powerful and precise tool for looking into both pathways of material within the food web and the long-term dynamics of sediments. In addition, the radionuclide deposits will act as new reference layers in future paleolimnological studies.

From phenomena mostly affected by mechanical laws, we moved into the domain of organisms. In spite of the abundant literature on the composition of algal populations, it is still difficult to predict species composition and phytoplankton bloom development because of the multitude of parameters involved, which, moreover, vary from lake to lake, and because of the wide range of nutrient and light requirements in phytoplankton species. It is easier to predict the global productivity of lakes; photosynthesis and biosynthesis are goverened by light and nutrient availability in the water, and these parameters, mostly controlled by physical processes, are more easily modeled.

The "microbial loop" dominated the debates on microbial processes and food-web structure. The complex organization of interacting communities replaces the simplistic models of trophic levels, and, in this process, new questions arose, mainly concerning the pathways of carbon transfer to higher trophic levels. It is not yet clear whether there are, in this respect, significant differences between large lakes and small ones. It is, however, obvious that this is a very interesting topic which will provide many new insights into lacustrine ecosystems.

Finally, case studies of interactions between two or more trophic levels within the lacustrine food-web were described.

We note and regret that there were few contributions on benthic populations. In large lakes, such communities can be significantly affected by shore currents and by the movement of the thermocline due to internal waves, especially in lakes with anoxic hypolimnia.

Among the 253 lakes having a surface area of more than 500 $km^2$, 122 lakes are of glacial origin and therefore are relatively young in geological terms, while 85 lakes result from tectonic movements (see Chapter 1). The most ancient lakes of the world such as Lake Baikal and Lake Tanganika fall into the latter category. The majority of the large lakes is located north of the 40th parallel (64%, 122 lakes). Most of the presentations dealt with lakes of the temperate zone and thus were representative of this distribution. However, it would have been interesting to hear more about the large lakes of low latitudes. In these lakes, where the Coriolis force is negligible, low thermal stability due to very small thermal differences between epi- and hypolimnion often modifies the mechanisms of mixing and transport. Moreover, being older than most temperate lakes, the low-latitude lakes have more diverse and more structured food-webs. More comprehensive studies on these lakes would allow us to assess the influence of latitude on physical processes and the effects of nearly geological time scales on the structure and function of biological communities.

## 36.2 What is a Large Lake?

Numerous discussions contributed to a refinement of our understanding of the specific characteristics of large lakes, which were put forward at the beginning of the symposium as working hypotheses (see Chapter 2).

**Pelagial and Littoral Zones** For purely geometrical reasons, large lakes tend to have enhanced lake surface to shore-length ratios. Superimposed on the pure size-dependence of this ratio, obviously, is its dependence on the shape of the lake, that is its deviation from a circular shape. This number, equal to half the radius of a circular lake, is above ca. 10 km in large lakes (see Chapter 3). Only these large lakes have a true pelagial zone distinct from the littoral area.

When large lakes are stratified, wind forcing causes a tilting of the thermocline which propagates itself through the lake with a deflection caused by the Coriolis force, the degree of which depends on geographical latitude. Near the shore, this propagating internal disturbance of the field of motion encounters the littoral slope topography. The ensuing readjustment of water masses and flow velocities in the shore areas causes strong littoral currents (jets). The littoral zone thus can be described as the region where these coastal disturbances take place. Its width scale, known as the internal Rossby radius, $R^*$, varies from 2 to 10 km at mean latitudes (see Chapters 4 and 5). In hydromechanical terms, the existence of a true pelagic zone (at least equal to $2R^*$) requires that the total width of the stratified lake be at least equal to $4R^*$. Therefore, for an average value of $R^*$ of 6 km, the lake surface would be about 450 $km^2$.

**Thermocline Depth** The depth of the thermocline depends on the stability of stratification and on the amount of mechanical energy received per unit of lake surface. Since this input of turbulent kinetic energy is greater with increasing lake surface, thermoclines tend to be deeper in large lakes.

However, as the stirring effect results from adjacent small-scale motions excited from the surface, there is an exponential weakening of the mixing force with depth. Consequently, there is a limit to the deepening of the thermocline; it has been observed in the temperate zone that, everything else being equal, especially the stability conditions, the depth of the thermocline remains constant for horizontal-scale fetch values exceeding 25 km (see Chapter 22). It follows that the surface of a circular lake having a diameter corresponding to the minimal fetch (25 km) which can cause the maximal depth of epilimnion mixing (15–20 m) may be considered as the minimal surface area of a large lake. The surface area of such a lake would be 490 $km^2$. These considerations indicate that the value of 500 $km^2$, chosen as the lower areal limit for large lakes in the study of the world distribution of large lakes (see Chapter 1), has a certain physical meaning.

**Conclusion** In summary, a large lake has a pelagial area unaffected by coastal currents and its thermocline develops at the maximal mixing depth. Although an areal lower limit of 500 $km^2$ was chosen for the definition of a large lake, some lakes of smaller surface may also present properties of large lakes because of their special shape and orientation, while some lakes larger than 500 $km^2$ may not present the above-mentioned properties of large lakes.

## 36.3 Some Specific Properties of Large Lakes

**Wind Forcing** The increasing size of a lake augments the wind fetch and, consequently, the water turbulence. This effect is maximized in very large lakes by the increase of wind velocity near the surface due to variations in the dynamical structure of the planetary boundary layer (see Chapter 5). Large lakes have, therefore, greater turbulence per unit area than small lakes and take up more horizontal momentum from the wind. During the warm season, turbulence is mostly confined to the epilimnion, since the mixed layer is "gliding" on the thermocline with relatively low friction, with the general result of stronger horizontal currents at the surface than in small lakes. In general, and with great variability of details due to local specific conditions and especially to thermal stability, large lakes are characterized by rapid, lake-wide, circular, horizontal water movements both within the epilimnion and below the thermocline during the stratified period. For example, in Lake Constance, these horizontal water movements causes an effective exchange between the Obersee and the Überlingersee (see Chapter 7). In contrast, Lake Geneva, because of local wind effects, seems to have a more locally confined circulation pattern (see Chapter 4).

Internal waves in large lakes are affected by the rotation of the earth. However, the Coriolis effect diminishes towards the equator, and less-complicated longitudinal waves characterize large lakes at very low latitudes. Similarly, the intrusions of river inflows are also deflected by the Coriolis force. Consequently, the "interflow facies" are more developed on the shores located at the right of the inflows in the Northern Hemisphere (see Chapter 9).

**Thermal Forcing** In large lakes, due to their size, we can expect a spatial variability in radiant energy inputs and, as a consequence, a more active thermal circulation than in small lakes. Moreover, in lakes of large size, the great variability of depth and width along the lake axes generates a corresponding variation of overall internal exchanges and surface fluxes of energy which brings about a large scale thermal circulation. These phenomena are particularly frequent in shallow areas where differential heating and cooling cause the formation of thermal bars.

Wind and thermal forcing in general occur simultaneously, thus producing a great variety of combinations of wind and temperature. In addition, under certain conditions of cooling at the surface, the water movements caused by turbulent transport in the upper layers of large lakes can create organized flow patterns, such as Langmuir cells, over large areas.

In summary, for a given value of wind and radiation intensity, large lakes develop stronger currents and more rapid circulation than small ones. Their internal waves are rotational.

**How Do Suspended Cells Respond to External Forcing?** For centuries, philosophers have questioned the reliability of our senses. Our senses are a product of the evolution of the physical world; their message may be incomplete and perhaps biased, but not totally wrong. Similarly, algae and bacteria have evolved in the aquatic environment and are tuned to it, or they would not exist. True as this may be, we need to understand at the mechanistic level what happens to cells submitted to changing turbulence (see Chapter 3).

An algal cell needs light, and for a given value of turbidity and surface irradiance, the intensity it receives depends on its depth in the water column which, in turn, is controlled only by turbulence. In order to get used to an altered light climate, an algal cell needs time to modify its photosynthetic machinery. When turbulence is low and the mixing rate is not greater than the photoadaptation rate of the alga, several "populations" of algae develop in layers of different light climates. For very low mixing rates, the danger exists that algae may sink below the mixed layer which keeps them in suspension. In mixed water columns, the rate of change in light intensity within moving parcels of water exceeds the rate at which algae can adapt to changing light levels. The previously separated "populations" are then mixed with each other and transported passively along a wide light gradient, and the "algal stratification" is destroyed. If the mixing depth increases further, the alga may spend too much time below the euphotic zone to maintain a net production. In large lakes, where the turbulence is higher than in small ones, the survival of algae depends primarily on the transparency of the water.

A planktonic bacterium does not need light and, moreover, due to its small size, remains in suspension over a very wide range of mixing rates. Its removal from the euphotic zone or from the mixing zone has no immediate effect on the bacterium. It is true that bacteria depend on algae as a source of organic matter and have a long-term "interest" in remaining in the vicinity of their food suppliers. However, the autonomy of bacteria *vis à vis* the turbulence-controlled processes may be one of the reasons that they can dominate microbial production in large, turbulent lakes.

In summary, the life and death of an algal cell is greatly affected by light and gravity. The differential pressure of those fields on the cells is governed by water motion that is mainly by external forcing. In contrast, planktonic bacteria have, to a certain extent, freed themselves from the dominance of water motion and external forcing.

**Littoral Zone of Large Lakes** From a physical point of view, the littoral zone of large lakes is a very eventful area. This is the area where any lake-wide motion meets a solid boundary, causing readjustments in velocity and direction (see above) which, under certain conditions, generate rapid coastal currents. In the stratified hypolimnion, particles of water bouncing off the "solid" wall induce an oscillating turbulent boundary layer with the formation of intruding tongues toward the hypolimnion (Spigel and Imberger, 1987). In superficial layers of lakes colder than 4°C, differential early spring heating may cause the occurrence of thermal bars which temporarily isolate the littoral water from the rest of the lake.

Biologists have a very different view of the littoral, for them, it is the area of growth of macrophytes (Wetzel, 1983) or, in the absence of macrophytes, the area extending from the shore to the lower limit of the euphotic zone (see Chapter 8). It is also the area where the benthic littoral fauna is found. Are the biotic activities in the littoral affected by the above-mentioned physical processes occurring there and, if so, how?

In large lakes, the littoral zone represents a small fraction of the lake area and contributes relatively little to the lake's primary productivity. In contrast, its role as spawning ground and shelter to animals at certain stages of their life can be crucial. In this respect, water level fluctuations may favor or limit the success of fish species spawning in shallow areas (see Chapter 8) and even favor pelagic species.

**Particles in Suspension, Sediment Resuspension, and Sediment Focusing** In deep large lakes, turbidity from nonalgal material is less than in small lakes. In general, such lakes are also more oligotrophic (see Chapter 17) and, therefore, have smaller concentrations of algae. However, particles and aggregates of variable composition and physical texture are also found in large lakes. They originate from allochtonous material, bottom deposits, and from the microbial flora and fauna. Direct study is difficult since they are often altered by the sampling techniques applied (see Chapter 12). In contrast, the residence time of particles in lakes has been studied by using various radionuclides, both natural and introduced by man, as tracers. These measurements (see Chapters 10, 11, 13, and 14) indicate that particle settling time is rapid, even in large deep lakes (less than one year).

Free-living bacteria smaller than 1 $\mu$m are probably among the smallest particles in lake water and have the largest specific surface. For this reason, and perhaps also because they possess efficient transport systems, they outcompete algae for inorganic nutrient salts in dilute solutions (see Chapter 27). Before their recent direct observation, Bacon et al. (1985), studying the vertical flux of radionuclides in the ocean, predicted the existence of nonsinking, minute particles adsorbing radionuclides. Together with algae, bacteria form clusters wherever nutrient concentrations are especially high. Under these conditions, the algae in such aggregates can outcompete bacteria for phosphorus (see Chapter 24). Within these organic particle aggregations, anoxic microenvironments may develop, which might explain why the production of reduced compounds can take place in epilimnic, oxygenated waters (see Chapter 26).

Resuspension of bottom sediments has been observed even in deep lakes. In Lake Michigan (see Chapter 10), introduction of phosphorus to the open water by resuspension of bottom sediments is more than 20 times greater than external loading.

Radionuclide studies have allowed us to follow the fate of sediments after they have been deposited on the lake bottom. Results (see Chapter 11) indicate a significant horizontal transport of sediment towards the center of lake basins (sediment focusing). Near the mouths of large rivers, unstable underflows generate turbidites, a facies typical of large lake sediments. Turbidite facies disturb the regular sequences of varves (see Chapter 9), a fact which has to be taken into account in paleolimnological studies (see Chapters 15 and 16).

Although rapid water currents have been reported in the deep layers of large lakes (see Chapter 10), little has been said about the hydromechanical conditions which govern the horizontal transport of sediments.

In conclusion, large lakes are characterized by significant lateral relocation of bottom deposits. Moreover, resuspension is an active process, recirculating nutrients and contaminants. In contrast, in small lakes, resuspension is limited to periods of lake mixing. This, together with the enhanced dilution effect of the inflowing water due to the large water volume of large lakes, account for their generally greater transparency.

**Structure of the Pelagic Community of Large Lakes** In a given climatic zone, large lakes have a much lower watershed surface to lake volume ratio than small lakes. For example, this ratio is 0.005, 0.08, and 0.23 in Lake Tahoe, Lake Geneva, and Lake Constance, respectively, but it is generally higher than one for smaller lakes (e.g., 1.07 for Lake Mendota which has an area of 40 $Km^2$).

This means that, in large lakes, one cubic meter of water drains a much smaller watershed area than in small lakes. Consequently, both the concentration of dissolved nutrients and of particles is smaller in large lakes (see Chapter 17). This feature is accentuated by the fact that nutrients stored in the hypolimnion and particles resuspended from the sediments are recirculated only during the mixed period. Stratification which diminishes recirculation of dissolved and particulate matter also delays the sedimentation of particles.

In large lakes, the thermocline is established at a greater mixing depth than in small lakes (see Section 36.2) because of relatively high turbulence, and planktonic organisms are therefore transported over considerable vertical distances.

There is evidence that, in large lakes, phytoplankton are exposed to lower irridiance and are physiologically and selectively adapted to this light climate (see Chapters 17 and 20). Despite this, we have seen that deep mixing is often unfavorable to algae, which may sink below the turbulent layer or respire more than they can photosynthesize. In contrast, bacteria are less affected by these movements.

Because bacteria outcompete algae for inorganic nutrient salt uptake at low salt concentrations, one can expect that, with the exceptions of the overturn period or exceptional external nutrient loads, bacteria would dominate the turbulent and oligotrophic water of large lakes. Accumulating evidence confirms this view; in Lake Constance, for example, the bacterial biomass can exceed phytoplankton biomass (Simon and Tilzer, 1987), and bacterial production represents, in many large lakes, an important fraction of total production (see Chapters 25 and 28). Conversely, during the overturn period, algal production is dominant and depends on the mixing depth, i.e., on the amount of nutrients reintroduced into the photic zone (see Chapter 6). When the lake is stratified, a large fraction of the particulate organic carbon and of the nitrogen and phosphorus pools of large lakes are kept within very minute, nonsinking bacterial particles. In other words, in such lakes, sedimentation of autochthonous organic matter is small, at least in summer, and internal recycling of nutrient salts dominates. In Lake Constance, 70 to 80% of primary production is recycled within the epilimnion (Stabel and Tilzer, 1981).

Bacteria are grazed by an abundant and diversified population of nanoflagellates and ciliates (see Chapter 27). The components of this microbial loop have been described for several lakes (Nagata, 5.1.5., see Chapters 28 and 29).

The algal autotrophic production is the major source of carbon for bacterial production and therefore should be larger than the amount of carbon required by the microbial loop. Numerous discrepancies exist between the autotrophic input and the microbial output (see Chapter 25). Do we underestimate the algal production? Is the microbial loop a carbon sink? Since the ocean, which is dominated by the microbial loop, produces huge amounts of fish, the answer to the last question is probably negative. It has been argued that, due to the regeneration of organic matter within the microbial loop, secondary production as a whole can very well exceed primary production (Strayer, 1988).

The appreciation of the immense importance of the microbial loop for the pelagic food-web should not tempt us to minimize the role and importance of the classical grazer food chain in large lakes. In the new scheme, the traditional diet of metazoan zooplankton has been enriched by a wide variety of nanociliates and microflagellates, but much methodological research is necessary to quantify the contribution of the microbial loop to the grazing food chain. These new concepts deeply modify our understanding of food-web interactions (see Chapter 30).

## 36.4 Future Challenges

For physicists and sedimentologists at the symposium, it was an easy task to address the problem of size-related properties. In contrast, most biologists related to the problem of large lakes only through the fact that their work was carried out on a lake of large size. Moreover, most biological limnologists proved to be dedicated to the concept of "averages" by selecting "a representative station" for the lake as a whole, of which they claimed that it gave them the "average" value of biomass, productivity, respiration, and grazing at any given date.

Biologists encounter objective difficulties; they deal with entities that are affected by a large array of environmental factors, by other organisms, and by an incredible variety of built-in genetic properties. Moreover, in contrast to the physicists, they can rarely benefit from automatic measurements by using *in situ* equipment, but are drastically constrained by time-consuming measurements. This is, of course, also true in small lakes, but to a lesser extent.

Average values of variables and rates are useful in providing orders of magnitude and allowing quantitative budgets. Averages, however, are only valid if the horizontal and vertical variability is small. As we have seen, large lakes offer a wide spectrum of small-, meso-, and large-scale hydrodynamic features which cannot be represented by one single location for measurements and sample collection. Moreover, both algal cells and particles are transported by currents, but, other than that, there is a vast difference between them. Particles could persist in a perfectly homogenous lake. Living cells, with their genetic diversity could not. Structure, i.e., the heterogeneity of the enviornment, was the condition that created biotic diversity, and it is necessary for their persistence. Why then, do aquatic ecologists feel compelled to study averages in systems whose very structure is characterized by their spatial and temporal heterogeneity?

Despite the challenge presented at the symposium in the keynote lecture on physical limnology, entitled "Mixing and transport in lakes: mechanisms and ecological relevance," few biological studies were designed to analyze horizontal heterogeneities in biological variables and on variations caused by external forcing. We had therefore no opportunity to discuss the strategy of sampling which is most appropriate to the object of our studies. How, for instance, do we propose to understand biological patchiness? And first, why should we? These is more than one answer to this question. First, there is good evidence that patches are the essential macrostructure of algal communities, zooplankton populations, fish larval swarms, and adult fish schools. Patches may play a critical role in the survival of species, their growth rates, and their taxonomic composition. If this is true, modelling a large lake should mean dealing with specific structures which do not exist in smaller water bodies. It remains desirable that we master the model of the beaker and of the aquarium. But by doing only this, we will make no progress in understanding heterogeneous environments. The second reason for studying patchiness is that it it often the only way to account for observed facts: for example, in the eastern Mediterranean Sea, if the measured "average" productivity were indeed representative, it would have been too low to permit the existence of fish in the food-web. Fish were only readmitted when productivity was measured in patches developing around hydromechanical structures.

The patchiness of phytoplankton results from a complex interplay between hydromechanical parameters and the responses of algal cells. "Different hydromechanic processes and different biological responses occur on different spatial and temporal scales," as Legendre and Demers (1984) put it in their call to move from traditional oceanography and limnology to "Dynamic Biological Oceanography and Limnology." The coupling of biological and hydrodynamic processes is such that these authors claim that algae are not controlled either by light or nutrients but by hydrodynamics which make these two essential resources available to algae in a wide range of ever-changing combinations. It also seems, that for certain threshold values of physical parameters, orderly structured patterns stem from chaotic turbulent disorder; what are Langmuir cells but a structured pattern of circulation caused by dissipation of unstructured turbulent energy? Algae respond to these scalar reactions of the physical environment by a wide variety of physiological, community organizational, and taxonomic changes.

**From the incessant and rich dialogue between hydromechanics and biological entities, the structures and functions of large lakes emerge—** structures and functions which have to be specific since, as we have seen, large lakes are characterized by diverse hydromechanical properties.

The title of our symposium, "Large Lakes: Ecological Structure and Function," was not trivial and, to a certain extent, we have not succeeded in addressing its deep meaning; we have, therefore, missed the discussion of ways and means which would lead us from the study of processes to the study of complex systems. We only touched this topic upon the occasion of the analysis of the long data set for Lake Tahoe. The following example illustrates this point:

Let us consider the area of a large lake enriched in nutrients by an inflow influenced by the Coriolis force. The resulting differential growth rate of the algal population of this area creates spatial heterogeneity. A patch is formed, provided the wind velocity remains below the threshold value which prevents growth rates from overcoming horizontal, turbulent diffusion. Then a storm occurs, the patch is dispersed, and the algae are removed from high light intensity. The wind subsides and a new flow field develops. The algae reorganize themselves according to the new hydromechanical setup and according to their new physiological state. If wind-induced mixing of the algal cells has diminished the risk of photoinhibition and at the same time increased the probability of nutrient supply from deep water layers, the whole process ends up with an increase of the division rate of the algal cells. How do we "sample" this succession of events?

Let us now assume that the unstable turbulent flow, becoming sensitive to certain otherwise-unimportant external or internal variations, becomes organized into Langmuir cells or any other structured flow pattern. Algal cells again enter a new situation, where light intensity and quality change constantly, and nutrient gradients are periodically modified. By what formalism can we account for the fact that mechanical energy through disorganized turbulence has finally produced order and structure?

Although limnologists have always claimed that lakes could serve as models for oceans, oceanographers are far ahead of us in this respect. In their search of holistic descriptions of marine ecosystems, oceanographers have proposed that five disciplines come to the rescue of marine ecologists: thermodynamics of dissipative structures, statistical mechanics, ataxonomic aggregations of organisms, flow analysis, and information theory (Ulanowicz and Platt, 1985).

I think that limnologists working on large lakes should move into these new teritories. I am convinced that a close collaboration between biologists and physicists who are determined to work at the system level and willing to use adequate formalisms, is the best prescription for exciting new discoveries. It is likely, for example, that certain discrepancies observed in the carbon budget of the microbial loop may disappear when studied at the system level.

The study of transects and events possibly requires new types of equipment. Industrial firms could help us if we had clear ideas ourselves concerning our short-, medium-, and long-term needs. Moving into these new grounds requires more coordination between limnologists of various backgrounds.

In conclusion, the large number of topics discussed during the large lakes symposium, and the novelty of certain concepts and results presented made this meeting a most stimulating experience. It also made us realize that new ways of cooperation between disciplines are needed to carry out the complex but necessary mutations that will be needed for further progress in this field.

---

*Acknowledgements* I gratefully acknowledge the reviews and comments of M.M. Tilzer, E. Hollan, D. Imboden, D. Planas, and G. Moreau.

## References

Bacon, M., Huh, C., Fleer, P., and Denser, W. 1985. Seasonality in the flux of natural radionuclides and plutonium in the deep Sargasso Sea. *Deep-Sea Research* 32:273–286.

Legendre, L. and Demers, S. 1984. Towards dynamic biological oceanography and limnology. *Canadian Journal of Fisheries and Aquatic Sciences* 41:2–19.

Patalas, K. 1984. Mid-summer mixing depths of lakes of different latitudes. *Verhandlungen der Internationalen vereinigung für Theoretische und Angewandte Limnologie* 22:97–102.

Simon, M. and Tilzer, M.M., 1987. Bacterial responses to seasonal primary production and phytoplankton biomass in Lake Constance. *Journal of Plankton Research* 9:535–552.

Spigel, R.H. and Imberger, J. 1987. Mixing processes relevant to phytoplankton dynamics in lakes. *New Zealand Journal of Marine and Freshwater Research* 21:361–377.

Stabel, H.H. and Tilzer, M.M. 1981.

Nährstoffkreisläufe im Bodensee, Überlinger See und ihre Beziehungen zu biologischen Umsetzungen. *Verhandlungen der Gesellschaft für Ökologie* 9:23–32.

Strayer, D., 1988. On the limits to secondary production. *Limnology and Oceanography* 33:1217–1220.

Ulanowicz, R.E. and Platt, T. (eds.). 1985. *Ecosystem Theory for Biological Oceanography.* Canadian Bulletin of Fisheries and Aquatic Sciences, Vol. 213, 260 pp.

Wetzel, R.G. 1983. *Limnology.* 2nd ed. Sounders College Publ., Philadelphia, PA, 767 pp.

# Index

MIX
Papier aus verantwortungsvollen Quellen
Paper from responsible sources
FSC® C105338

If you have any concerns about our products,
you can contact us on
**ProductSafety@springernature.com**

In case Publisher is established outside the EU,
the EU authorized representative is:
**Springer Nature Customer Service Center GmbH**
**Europaplatz 3, 69115 Heidelberg, Germany**

Printed by Libri Plureos GmbH
in Hamburg, Germany